MW00786850

RAPTORS OF NEW MEXICO

RAPTORS
OF NEW MEXICO

EDITED BY
Jean-Luc E. Cartron

UNIVERSITY OF NEW MEXICO PRESS
ALBUQUERQUE

15 14 13 12 11 10 1 2 3 4 5 6

LIBRARY OF CONGRESS CATALOGING-IN-PUBLICATION DATA

Raptors of New Mexico / edited by Jean-Luc E. Cartron.
p. cm.
Includes bibliographical references and index.
ISBN 978-0-8263-4145-7 (cloth : alk. paper)
1. Birds of prey—New Mexico.
2. Birds of prey—New Mexico—Pictorial works.
3. Birds of prey—New Mexico—Identification.
I. Cartron, Jean-Luc E.
QL696.F3R387 2010
598.909789—dc22
2009053811

DESIGN AND LAYOUT: MELISSA TANDYSH
Composed in 9.9/13 Minion Regular Pro
Display type is ITC Legacy Serif Std
Printed and bound in China by Everbest Printing Company, Ltd.
through Four Colour Imports, Ltd.

With the photography of

Robert Shantz **Roger Hogan** **Geraint Smith**

Steve Baranoff
Doug Brown
Doug Burkett
Jim Burns
John P. DeLong
Lana Hays
Tom Kennedy
Wayne Lynch
Steve Metz
David Nichols
Jerry Oldenettel
David A. Ponton
Bill Schmoker
Lee Zieger

Martha Airth-Kindree
Richard Artrip
Julian D. Avery
Jonathan Batkin
Matt Baumann
Nancy Baczek
Jim Bednarz
Warren Berg
Craig Blakemore
Mark Blakemore
Seamus Breslin
Larry Brock
John V. Brown
Jean-Luc E. Cartron
Luke Cole
Steven Cox
Octavio Cruz
Kirsten Cruz-McDonnell
David Dain
Martha Desmond
Robert W. Dickerman
Adam D'Onofrio
Robert H. Doster
Kent Fiala
Jim Findley
Gordon French
Gary K. Froehlich
Fred Gehlbach
Antonio (Tony) Gennaro
Ralph Giles
Mickey Ginn
Tony Godfrey
David J. Griffin
Jean Hand
Doris Hausleitner
Tim Hayden
Susan Hilton
Patty Hoban
Howard Holley
Stephen Ingraham
Jeff Kaake
David Keller
Ron Kellermueller
Patricia L. Kennedy
Joe Kent
Sally King
Jeremey Knowlton
David J. Krueper
Aaron Lamb
Carl Lundblad
Mike Marcus
Kerstan Micone
Dennis Miller
Narca Moore-Craig
Joshua Nemeth
Patrick O'Brien
Jim Ramakka
Jenny Ramirez
Bryan J. Ramsay
John Rawinski
Adam Read
Wade Reed
Tim Reeves
Janet Ruth
Giancarlo Sadoti
Lawry Sager
Spin Shaffer
Hart Schwarz
Robin Sell
Robin Silver
Robert Sivinski
Calvin Smith
Penny Smith
Tim Smith
Lisa Spray
Peter Stacey
Dale W. Stahlecker
Charlotte Stanford
James N. Stuart
Ronald J. Troy
Mark Walthall
Gordon Warrick
Mark L. Watson
Pat Watt
Richard E. Webster
Jim White
Claudia Williams
Elton M. Williams
Warren Williams
Cole Wolf
Mike Yip
James E. Zabriskie

Species Distribution Maps by

David Dean

STATE OF NEW MEXICO
MINING AND MINERALS DIVISION

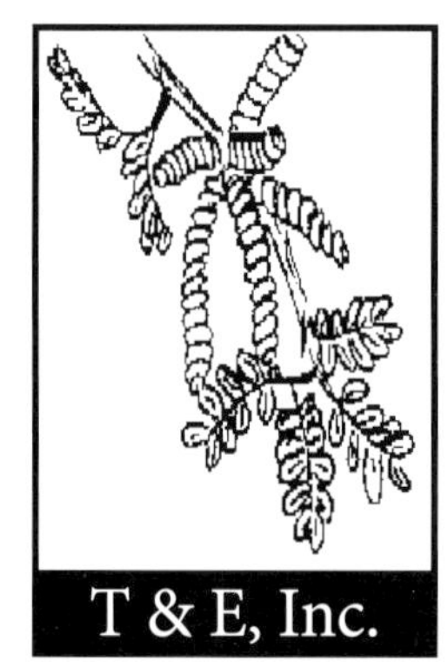
T & E, Inc.

TURNER
FOUNDATION

PNM®
A personal commitment to New Mexico

UNM BIOLOGY

NEW MEXICO
GAME & FISH

POWER
ENGINEERS

Envirological Services

FOREST SERVICE
U S
DEPARTMENT OF AGRICULTURE
Rocky Mountain Research Station

SPONSORS

FRIENDS OF HERON AND
EL VADO LAKES STATE PARKS

EASTERN
NEW MEXICO
UNIVERSITY

New Mexico
Ornithological
Society

Digiscoping
Digest

COLLABORATORS

CONTENTS

FOREWORD

"*T'entends l'oiseau?*" As a toddler in the suburbs of Paris, these were the first words he uttered. "Can you hear the bird?"

From that budding fascination and inquiry an ocean away, through production of *Raptors of New Mexico,* the life of Jean-Luc Cartron has been a world journey. After leaving France, his stops along the way included a one-year post as military physician in the Ivory Coast, residency training and brain tumor research in the United States, and eventually Albuquerque in 1986, where he told me, "I found myself while traveling through the big open spaces of the western U.S. and observing nature."

As it has done with so many others who have stepped into its vastness and inhaled the aridity, the American Southwest aroused him. More important, it informed him of a desire within, a craving to become immersed deeply in examining life on a stark stage of the North American landscape.

Biodiversity and conservation in northern Mexico, nesting seabirds in the Gulf of California, Brown Creepers and Eastern Bluebirds along the Middle Rio Grande, mangroves in Baja, ground arthropods in New Mexico's riparian forests, livestock management in the Southwest, prairie dogs, ecogeographic rules and macroevolutionary paradoxes, a field guide to the plants and animals of the middle Rio Grande bosque, kingbirds in New Mexico and hummingbirds of Sonora . . . all these subjects have fascinated him and have been the topics of his extensive research and publication list. But the raptors have a distinctive hold on him; they arouse a special passion.

During the 1990s he was in the Gulf of California documenting the ecology and conservation needs of Ospreys. Later he studied the impact of power lines on raptors in northern Mexico. One of Jean-Luc's significant field investigations in New Mexico was on Ferruginous Hawks of the Estancia Valley, where as Hawks Aloft's research director he led a study that demonstrated the relationship between the spatial distribution of nests, their productivity, and the distance of nests to prairie dog towns. Various field projects on an assortment of species such as Barn Owls, Swainson's Hawks, and Cactus Ferruginous Pygmy-Owls have rounded out his impressions of Southwestern landscapes and the raptors they harbor.

In my biased view, *Raptors of New Mexico* is the pinnacle of Jean-Luc's accomplishments. Guiding 41 other authors along the slippery path of producing significant text for this book is in itself Herculean! But to do it in a seamless manner that retains individual style and creative expression and at the same time uniformity reflects editorial talent. All 37 regularly occurring raptor species are presented in accounts that are thoughtfully organized and teeming with information.

Two features of this book are extraordinary and merit special mention. The first is the very enlightening chapter on raptor migration in New Mexico. This account relates how the Southwest is linked to raptor movements in the western United States, specifically detailing the important role played by the topography of New Mexico. The maps and graphs are well done and helpful in relating the story of hawk movements.

Migration enthusiasts from other parts of the country largely dismissed raptor concentrations in the Southwest away from familiar landforms that funnel bird movements, such as coastal areas of California and Texas or ridges at Hawk Mountain in Pennsylvania. Steve Hoffman was in the process of identifying Southwestern migration routes in the mid 1980s when he first dragged me up to Aztec Peak in Arizona's Sierra Ancha

Mountains. He had just discovered a significant number of raptors moving along the Sandia and Manzano mountains of New Mexico. Although we never did document the importance of Aztec Peak as a landscape feature facilitating hawk migration, Hoffman did demonstrate that the latitude and relatively mild climatic conditions of the Southwest do not disperse a significant number of migrating raptors from the mountain ridges.

My most memorable events on Aztec were visiting with Ed Abbey near his fire lookout and listening to Steve rave about his New Mexico migration epiphany and plans for establishing an organization to monitor raptor passage in the West . . . the beginnings of HawkWatch International.

The second great aspect of *Raptors of New Mexico* is the spectacular photographs scattered throughout its pages. They are numerous, eye-catching, and informative . . . each is seemingly worth a thousand words. As an attractant, a visual aid, an educational display, a work of art, this collection of photos is unmatched. Collectively they enable the casual reader to "thumb through" the pages, lock on an image, and take away a little bit of knowledge. Or to sit down and enjoy a more engaged trek into the world of raptors.

Surprisingly, only a handful of states have seen production of significant works describing the resident raptors. *Raptors of New Mexico* sets a new standard for state and regional raptor books. Without doubt it is the best one out there.

That Jean-Luc Cartron heard the birds of his homeland and responded by illuminating the enchanting raptors of his current residence is indeed a blessing to us all.

RICHARD L. GLINSKI
Tempe, Arizona

PREFACE

New Mexico . . . the fifth largest U.S. state, with sumptuous vistas of colorful deserts, forested mountains, snowcapped peaks, steep canyon walls, and flat prairies. New Mexico, a largely untamed land that harbors a high diversity and abundance of raptors, from the familiar Red-tailed Hawk, encountered daily in nearly all areas of the state, to rarities such as the Gray Hawk, the White-tailed Kite, or the Mexican Spotted Owl.

I have known since 2000 that I would someday work on a book titled *Raptors of New Mexico*. My own introduction to the state's birds of prey was likely no different from that of numerous other people. It occurred in the fall of 1988, during a brief stay in the state, in the form of a trip to the Bosque del Apache National Wildlife Refuge. Besides numerous Red-tailed Hawks and Northern Harriers, my memories of that trip also include a distant view of an eagle on the ground next to the carcass of a Sandhill Crane. I moved to New Mexico the following fall and became further acquainted with New Mexico's raptors during a visit to HawkWatch International's Manzano migration monitoring site on Capilla Peak—to this day I have never seen so many Sharp-shinned Hawks as on that occasion! In 1992 I started my own raptor research, not in New Mexico, but instead a study of Ospreys in the Gulf of California in Mexico. It was not until 2000 that I began working in New Mexico, first studying Ferruginous Hawks in the Estancia Valley, the Plains of San Agustin, and San Juan County's badlands, as director of research for Hawks Aloft. Today raptor research has lost none of its novelty. All encounters with wild-ranging birds of prey are thrilling moments, even if simply hearing the hiss—reminiscent of an espresso machine—of a Barn Owl nesting in the rafters of an abandoned building.

This book is intended as a dual celebration: of eagles, hawks, and owls and of New Mexico's natural heritage. The information presented in all the chapters draws on countless scientific studies and surveys conducted in the state, particularly on its vast holding of public lands. Until now, much of that information had been published only piecemeal in the technical literature. Perhaps just as much information had not been published, existing instead in the form of reports and entries in the field notes of biologists. Also included in this volume are data from New Mexico wildlife rehabilitators and shelter organizations on injuries and illnesses sustained by raptors. Although those data are generally untapped by conservation biologists, they reveal the many ways in which we humans directly or indirectly harm raptors.

While assembling this volume, an important effort was made to weave together the science of raptor research, the more traditional natural history narratives, and the aesthetics and visual impact of photography. Thus in the midst of research findings are stories from the field, rare observations of skirmishes in the skies, of predation, raptors living in buildings, or clumsy fledglings crashing to the ground. The hundreds of photographs contained in *Raptors of New Mexico* represent more than just visual signposts or entry points for the text. Some also constitute visual records of occurrence in the state or of important or unusual facets of a raptor's life history. Others are used to emphasize important field marks for species identification purposes. Yet others simply capture the regal, almost menacing beauty of birds of prey.

The chapters of *Raptors of New Mexico* are of unequal length, and intentionally so. Very little information other than distributional has been gathered for a number of raptors in New Mexico. The chapters focusing on those species are correspondingly short. For other raptors, much more information is available. In fact, much of what we know about the Northern Goshawk, the Harris's Hawk, the Flammulated Owl, and especially the Mexican Spotted Owl we owe to research conducted by some of the contributing authors in New Mexico. Additional species are monitored closely in the state, especially where they may be affected by some forms of land use such as oil and gas development and mining. This is true for example of the Golden Eagle and the Ferruginous Hawk, whose life habits are now fairly well known in New Mexico. For all of these birds the format of the book allows an in-depth treatment of their distribution, natural history, and population status. Also intentional on my part was letting the authors speak in their own voices and from their own experiences. Contributing authors were given freedom to choose the main thread of their narratives, and, where needed, to insert additional headings or subheadings in the text of their chapters.

Throughout *Raptors of New Mexico*, the vernacular names of bird species are capitalized; the common names of plants, mammals, reptiles, amphibians, and invertebrates are not. Thus the reader will find mention of the Golden Eagle and the Northern Flicker, in contrast to the black-tailed jackrabbit and the prairie rattlesnake. This discrepancy of grammatical usage is admittedly distracting, but in no way does it reflect judgment passed on the relative importance of birds among all life organisms. In an article published in the *Auk* in 1983, the ornithologist Anselm Atkins urged the American Ornithologists' Union to abandon its rule of capitalizing bird species names, a rule enforced by ornithological scientific journals. "Until we do," Atkins wrote humorously, "we ornithologists, with our Important Capitals, continue to look Curiously Provincial."

Raptors are indicators of ecosystem health. Although the outlook for most of New Mexico's raptors seems bright, a few species are declining. Investigating their declines can result in learning moments, a chance to understand and lessen impacts on nature. At the same time, our knowledge of the distribution, taxonomy, and ecology of New Mexico's raptors is a work in progress—and probably always will be. This book is not intended as the final word on any species. Instead, it aims to help educate New Mexicans and others about raptors, inspire the public to spend more time discovering them through their own observations, provide an information resource for management and conservation purposes, and point to some of the questions in need of answers through future research.

ACKNOWLEDGMENTS

In the course of editing this volume, it has been my privilege to work with many very knowledgeable researchers and talented photographers. My gratitude goes first to all the authors who wrote or cowrote the chapters. I am grateful in particular to Dale Stahlecker, whose work in New Mexico has chronicled the rediscovery or return of two raptor species, the Boreal Owl and the Osprey. As the author of multiple chapters, his contribution to *Raptors of New Mexico* has been invaluable. I also would like to extend special thanks to the photographers who contributed their wonderful images to the book. I am immensely grateful to all of them but particularly to Robert Shantz, Roger Hogan, and Geraint Smith, whose many photographs so greatly enhance this book. I am indebted to David Dean, who skillfully prepared the species' distribution maps featured in this volume; Jane Mygatt, who crafted the figure showing the elevational zonation of vegetation types in New Mexico; Jill Root, who expertly copyedited the entire manuscript; Narca Moore-Craig, who contributed some of her wonderful artwork; and Melissa Tandysh, who created the beautiful layout and showed infinite patience for making corrections on the proofs. An important contribution was also made by all those who reviewed chapters of the book: Clint Boal, Carl Bock, Dick Cannings, Carol Finley, Tom Gatz, Fred Gehlbach, Rich Glinski, Chuck Henny, Stuart Houston, Bill Howe, Roy Johnson, Jeff Kelly, David Ligon, Bill Mannan, Carl Marti, Jim Parker, John Parmeter, Chuck Preston, Richard Reynolds, Ron Ryder, Helen Snyder, Karen Steenhof, Jim Watson, and Steve West. Many thanks to Dave Clark and the New Mexico Mines and Minerals Division for providing data on nesting raptors from the McKinley, San Juan, La Plata, and El Segundo mines; Chris Farmer and Keith Bildstein, Hawk Mountain Sanctuary, who kindly gave permission to reproduce a figure originally published in *State of North America's Birds of Prey* (Nuttall Ornithological Club and American Ornithologists' Union); Marianne Wootten and Envirological Services, Inc., for data on Burrowing Owl migration; Shirley Kendall, Dennis Miller, Janelle Harden, and Anne Russell for data on raptor injuries and rehabilitation; Cheryl Bell, Katherine Eagleson, and Sharon Lieber for information on birds kept at the Wildlife Center in Española, including Akaiko, the Eastern Screech-Owl; Diego Villalba of the New Mexico State Land Office for communicating information on cattle tanks and the installation of wildlife escape platforms; Julian Avery, Jonathan Batkin, Doug Burkett, Nancy Cox, Dave Hawksworth, Patty Hoban, John Hubbard, Greg Keller, Ron Kellermueller, Jerry Oldenettel, John Parmeter, Dave Roemer, Lawry Sager, Pat Walsh, Gordon Warrick, Richard Webster, Steve West, Jack Williams, and Sandy Williams for unpublished and published distributional records; Mary Alice Root for providing additional information regarding records contained in the New Mexico Ornithological Society database now accessible online; David Brown for allowing use of his GIS coverages showing the distribution of vegetation communities in the Southwest; and Scott Barnes, Rene Obersole, and Mark Wexler for assistance in tracking out-of-state photographers whose images are included in this volume. Doug Burkett, Eric Greisen, and Steve West all contributed wonderful anecdotal stories on the natural history of some of the raptors in the state. Many thanks also to Hawks Aloft, Inc., Wildlife Center, Inc., and Gila Wildlife Rescue for allowing the use of photographs taken of birds under their care. I am grateful to OPSEC (Operations Security), White Sands Missile Range, and to Dave Mikesic and the Navajo Fish and Wildlife Service for giving permission to publish photographs taken in areas under their administrative authorities. The production of *Raptors of New Mexico* would not have been possible without the financial support of the New Mexico Mining and Minerals Division, the Public Service Company of New Mexico (PNM), T&E, Inc., Power Engineers, Envirological Services, the Turner Foundation, the University of New Mexico (UNM) Department of Biology, the UNM Museum of Southwestern Biology (MSB), The U.S. Forest Service Rocky Mountain Research Station (RMRS), Friends of the Bosque del Apache National Wildlife Refuge, Friends of Heron and El Vado Lakes State Parks (organizers of OspreyFest), the New Mexico Department of Game and Fish, Eastern New Mexico University (ENMU), New Mexico State University (NMSU), and the New Mexico Ornithological Society. Arthurine Pierson and Digiscoping Digest also sponsored *Raptors of New Mexico* project by contributing many of Roger Hogan's raptor photos. Thank you finally to Jim O'Hara and Dave Clark (New Mexico Mining and Minerals Division), John Acklen (PNM), Michael Wilson and Stephen Buchmann (Drylands Institute), Sam Loker (UNM Department of Biology), Jane Mygatt, Tim Lowrey, and Tom Turner (MSB), Beau Turner (Turner Foundation), Tom and Eleanor Wootten (T&E., Inc.), Marianne Wootten (Envirological Services), Deb Finch (RMRS), David Dean (Power Engineers), Leigh Ann Vradenburg (Friends of the Bosque del Apache National Wildlife Refuge), Matt Wunder (New Mexico Department of Game and Fish), Dave Krueper (U.S. Fish and Wildlife Service), Janet Ruth (U.S. Geological Society), Zach Jones (ENMU), Martha Desmond (NMSU), Kathleen Galbraith and Dianna Andrews (Friends of Heron and El Vado Lakes State Parks), and Clark Whitehorn (UNM Press) for their logistic support, encouragements, and generosity.

Introduction

JEAN-LUC E. CARTRON

TODAY, FEW PEOPLE HAVE heard of Johann Karl Wilhelm Illiger (1775–1813) and Nicholas Aylward Vigors (1785–1840), both prominent naturalists of their time. Illiger was a German zoologist and the first curator of the zoological collection at the Berlin Museum of Natural History. He is best known for a revision of Linnaeus' classification in his *Prodromus Systematis Mammalium et Avium* (1811), applying the then-novel concept of family to mammals and birds (Streseman 1975). Vigors was an Irish ornithologist who cofounded the Zoological Society of London in 1826. A friend of John James Audubon, he described several birds new to science—including the Bewick's Wren (*Thryomanes bewickii*) and the Pygmy Nuthatch (*Sitta pygmaea*)—collected during the 1820s in what was then Mexican California by Alexander Collie (Vigors 1839).

Illiger and Vigors may be jointly credited for having coined the word *raptor*. In Illiger's (1811) classification, the birds are divided into seven orders, including one that consists of all hawks, eagles, falcons, vultures, and owls. Before him Linnaeus and Cuvier had grouped these same birds under the order Accipitres, but Illiger named them instead the Raptatores, from the Latin word *raptor*, one who seizes by force. Two decades later, Vigors (1825) proposed his own classification of the birds (fig. I.1) in the 14th volume of the *Transactions of the Linnean Society*. In Vigors' classification, which is based on the long-obsolete Quinary System, birds are divided into five orders, including the Raptores. Vigors generally wished to retain the names of Illiger's orders, but as he stated in a footnote,

> "The term Raptatores of that naturalist [Illiger] I have ventured to alter to that of Raptores, which appears to me more classical." (1825:405)

The name Raptores was adopted in manuals of zoology and annotated checklists of birds during the 19th and early 20th centuries (e.g., Swainson 1836; Owen 1866; Butler 1892; Hoffman 1904; Kermode 1904; Barrows 1912), and it has persisted to this day, even making it into everyday English. Although the Raptores—or Raptors (see Jennings 1828)—represent a now taxonomically defunct order, they continue to be often described and studied together (e.g., Grossman and Hamlet 1964; Weidensaul 1989; Glinski 1998).

The historical origin of the raptors as a grouping is perhaps more than anecdotal in that it helps answer the often-asked question "Should vultures be considered raptors?" After all, vultures are scavengers, though they occasionally can also prey on live animals (e.g., Fowler 1979). The New World vultures may also not be closely related to the rest of the Falconiformes (hawks, eagles, and falcons; see below). Today ornithologists view the denominations "raptors" and "birds of prey" as synonymous, using them interchangeably (e.g., Terres 1980). The modern definition of a raptor or a bird of prey, according to many dictionaries, is that of a predatory or carnivorous bird, one that seizes and carries away. In *Merriam-Webster's Collegiate Dictionary*, however, a bird of prey is one that "feeds . . . on meat taken by hunting or on carrion." Thus there is no clear consensus and yet, if one accepts the premise that treating raptors as one group has no basis other than a historical one, then the answer to the question above is arguably yes, vultures are raptors, and as such they are included in this volume.

In the biological sciences, the discipline that concerns itself with identifying, naming, and classifying living organisms is termed *taxonomy*. In Illiger's and Vigors' time, and dating back to Aristotle during the fourth century BC, taxonomists did not group species on the basis of phylogenetic (evolutionary) descent or common ancestry. After all, these concepts would not even be articulated until the publication in 1859 of Charles Darwin's *On the Origin of Species*. Instead, taxonomists used Aristotle's essentialist method to look for those constant, shared structures that unite classes of organisms. Thus, in retrospect, the grouping of owls and diurnal raptors (vultures, hawks, eagles, and falcons) was based on analogous traits—traits that have arisen through convergent evolution. For example, raptors—whether vultures, hawks, or owls—are characterized by a stout, hooked beak with a cere at its base (hidden under feathers in the owls) and by strong, curved claws or talons. Except in New World vultures, the talons are also very sharp and flexibly articulate with the toes, for seizing and carrying prey and for piercing flesh. Other than the Osprey (*Pandion haliaetus*), diurnal raptors have anisodactyl feet, with the first toe (hallux) pointing back and the other three toes pointing forward. In owls and in the Osprey, the feet are facultatively anisodactyl or zygodactyl: the first toe points backward, the second and third toes point forward, and the fourth toe can swing freely between a backward and a forward position. The similar arrangement of the toes in the Osprey and in owls is thought to be the result of convergence instead of common descent.

Even today, taxonomy—now classifying living organisms based on phylogeny—remains a work in progress. During the lapse of time it required to write and edit *Raptors of New Mexico*, the New World vultures (family Cathartidae) were moved from the order Ciconiiformes back to the order Falconiformes (Banks et al. 2007), where they belonged prior to publication of the seventh edition of the *Check-list of North American Birds* (AOU 1998). Since the 44th supplement to the 7th edition of the *Check-list of North American Birds* (Banks et al. 2003), New World screech-owls have been removed from the genus *Otus* and placed into a separate genus, *Megascops*. The Crested Caracara (*Caracara cheriway*), it was thought, occurred from the United States all the way south to Tierra del Fuego. Based on Dove and Banks (1999), a recently decided taxonomic split resulted in the recognition of two extant species instead of one, all populations of the Crested Caracara south of Amazonia now being known as the Southern Caracara, *Caracara plancus* (AOU 2000). In the 47th supplement to the *Check-list of North American Birds* (Banks et al. 2006), the American Ornithologists' Union merged the genera *Asturina* and *Buteo*. As a result, the Gray Hawk, formerly *Asturina nitida*, was reassigned to the genus *Buteo* and is now recognized as *Buteo nitidus*, a change that required the reordering of all *Raptors of New Mexico*'s chapters, arranged taxonomically! Although subspecies are often considered as arbitrary categories—in some cases the result of inadequate sampling to recognize clinal variation—they are important to identify for conservation purposes, to preserve nature's full potential for speciation. Two of the contributors to *Raptors of New Mexico* (Dickerman and Johnson 2008) proposed a new subspecies of the Great Horned Owl (*Bubo virginianus*) for populations of this species nesting in the southern Rocky Mountains south of the Snake River in Idaho, including the higher elevations of New Mexico.

To date, 44 raptor species have been verified in New Mexico (table I.1). In all likelihood, their current taxonomic designation and arrangement will change. An exciting prospect for the future is the increase in the number of genetic markers now capable of shedding light on the deep branches of what is often called the tree of life. An important possibility is that the New

World vultures will be placed into their own separate order (see Remsen et al. 2009). Recent genetic studies such as those by van Tuinen et al. (2000), Fain and Houde (2004), Ericson et al. (2006), and Hackett et al. (2008) also point to the lack of any close relationship between the Falconidae and the Accipitridae. Thus the Falconiformes as they are known today may be split into three orders (see Remsen et al. 2009). Even in the Strigiformes (the owls), taxonomic changes are likely in the future, though only at low taxonomic levels (genera or species). The Flammulated Owl (*Otus flammeolus*) remains placed in the genus *Otus*, for example, but this placement is only provisional.

With phylogeny at its core, taxonomy has undergone a complete transformation since Johann Illiger's and Nicholas Vigors' time. Modern taxonomists seek lines of descent and the origin and divergence of lineages through evolutionary time. In contrast, Vigors (1825:399) was convinced that at each taxonomic level, lineages were connected by an "uninterrupted chain of affinities," allowing taxonomists to represent them as rings of circles (fig. 1.1). As implicitly or explicitly recognized in the Quinary System conceived by the entomologist Alexander Macleay and adhered to by Vigors, there are five orders within each class of animal, five families within each order, five subfamilies within each family, and so on. The Quinary System, Vigors believed, held true for all subdivisions of animals. Within his Raptores he had identified three families, the Falconidae (itself composed of five subfamilies), the Strigidae, and the Vulturidae. Two families remained to be identified. Ironically, some studies (Takagi and Sasaki 1974; De Boer 1975, 1976; and see above) have lent support for the recognition of not three but four orders of diurnal raptors (Cathartiformes, Sagittariiformes, Accipitriformes, and Falconiformes). And thus, while Vigors' classification has been proven skewed and obsolete, the group of

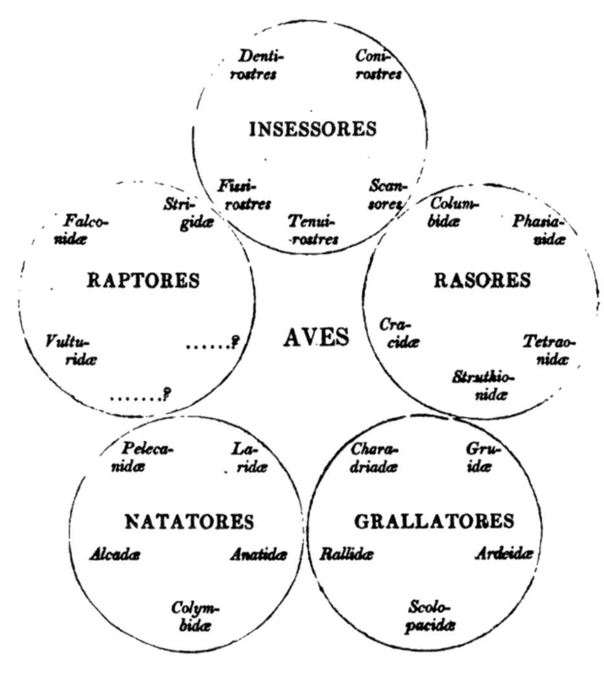

FIGURE 1.1

Nicholas Vigors' Quinary (or Quinarian) Classification of birds (Class: Aves). According to the Quinary System of classification, all animal groups were naturally divisible by five. Thus the class Aves consisted of five orders; each order, including the Raptores, had five families, and each family five subfamilies. According to Jennings (1828), Vigors was confident that the Quinary System would hold true also at the lower taxonomic levels (genera and species). Note that the last two putative families of raptors remain unidentified in Vigors' classification.

birds whose name he and Illiger coined may contain five subdivisions (four orders of diurnal raptors plus the Strigiformes), similar to what Vigors envisioned. Perhaps also, where Vigors' Quinary Classification is concerned, poetry wins over science. In what Jennings (1828:40) calls a "singular coincidence," one of Anna Laetitia Barbauld's (1743–1825) poems, written in 1767, hints at the existence of the same five groups of birds Vigors proposed in 1825.

> Yet who the various nations can declare
> That plough with busy wing the peopled air?
> These cleave the crumbling bark for insect food;
> Those dip the crooked beak in kindred blood;
> Some haunt the rushy moor, the lonely woods;
> Some bathe their silver plumage in the floods;
> Some fly to man, his household gods implore;
> And gather around his hospitable door,
> Wait the known call, and found protection there,
> From all the lesser tyrants of the air . . .

And here, perhaps, we find at last the definition of raptors: birds that "dip the crooked beak in kindred blood."

TABLE I.1. Taxonomy of New Mexico's raptor species. Based on the seventh edition of the *Check-list of North American Birds* (AOU 1998) and supplements.

ORDER	FAMILY	SUBFAMILY	SPECIES
Falconiformes	Cathartidae		Black Vulture (*Coragyps atratus*)
			Turkey Vulture (*Cathartes aura*)
	Accipitridae	Pandioninae	Osprey (*Pandion haliaetus*)
		Accipitrinae	Swallow-tailed Kite (*Elanoides forficatus*)
			White-tailed Kite (*Elanus leucurus*)
			Mississippi Kite (*Ictinia mississippiensis*)
			Bald Eagle (*Haliaeetus leucocephalus*)
			Northern Harrier (*Circus cyaneus*)
			Sharp-shinned Hawk (*Accipiter striatus*)
			Cooper's Hawk (*Accipiter cooperii*)
			Northern Goshawk (*Accipiter gentilis*)
			Common Black-Hawk (*Buteogallus anthracinus*)
			Harris's Hawk (*Parabuteo unicinctus*)
			Red-shouldered Hawk (*Buteo lineatus*)
			Broad-winged Hawk (*Buteo platypterus*)
			Gray Hawk (*Buteo nitidus*)
			Short-tailed Hawk (*Buteo brachyurus*)
			Swainson's Hawk (*Buteo swainsoni*)
			Zone-tailed Hawk (*Buteo albonotatus*)
			Red-tailed Hawk (*Buteo jamaicensis*)
			Ferruginous Hawk (*Buteo regalis*)
			Rough-legged Hawk (*Buteo lagopus*)

TABLE I.1. Taxonomy (*continued*)

ORDER	FAMILY	SUBFAMILY	SPECIES
			Golden Eagle (*Aquila chrysaetos*)
	Falconidae	Caracarinae	Crested Caracara (*Caracara cheriway*)
		Falconinae	American Kestrel (*Falco sparverius*)
			Merlin (*Falco columbarius*)
			Aplomado Falcon (*Falco femoralis*)
			Peregrine Falcon (*Falco peregrinus*)
			Prairie Falcon (*Falco mexicanus*)
Strigiformes	Tytonidae		Barn Owl (*Tyto alba*)
	Strigidae		Flammulated Owl (*Otus flammeolus*)
			Western Screech-Owl (*Megascops kennicottii*)
			Eastern Screech-Owl (*Megascops asio*)
			Whiskered Screech-Owl (*Megascops trichopsis*)
			Great Horned Owl (*Bubo virginianus*)
			Northern Pygmy-Owl (*Glaucidium gnoma*)
			Elf Owl (*Micrathene whitneyi*)
			Burrowing Owl (*Athene cunicularia*)
			Spotted Owl (*Strix occidentalis*)
			Barred Owl (*Strix varia*)
			Long-eared Owl (*Asio otus*)
			Short-eared Owl (*Asio flammeus*)
			Boreal Owl (*Aegolius funereus*)
			Northern Saw-whet Owl (*Aegolius acadicus*)

LITERATURE CITED

[AOU] American Ornithologists' Union. 1998. *Check-list of North American birds.* 7th ed. Washington, DC: American Ornithologists' Union.

———. 2000. Forty-second supplement to the American Ornithologists' Union *Check-list of North American birds. Auk* 117:847–58.

Banks, R. C., R. T. Chesser, C. Cicero, J. L. Dunn, A. W. Kratter, I. J. Lovette, P. C. Rasmussen, J. V. Remsen Jr., J. D. Rising, and D. F. Stotz. 2007. Forty-eighth supplement to the American Ornithologists' Union *Check-list of North American birds. Auk* 124:1109–15.

Banks, R. C., C. Cicero, J. L. Dunn, A. W. Kratter, P. C. Rasmussen, J. V. Remsen Jr., J. D. Rising, and D. F. Stotz. 2003. Forty-fourth supplement to the American Ornithologists' Union *Check-list of North American birds. Auk* 120:923–31.

———. 2006. Forty-seventh supplement to the American Ornithologists' Union *Check-list of North American birds. Auk* 123:926–36.

Barrows, W. B. 1912. *Michigan bird life.* Lansing: Michigan Agricultural College.

Butler, A. W. 1892. *The birds of Indiana.* Indianapolis, IN: Wm. B. Burford. http://www.ulib.iupui.edu/collections/butlerbirds/home.html (accessed 10 April 2009).

De Boer, L. E. M. 1975. Karyological heterogeneity in the Falconiformes (Aves). *Experientia* 31:1138–39.

———. 1976. The somatic chromosome complements of 16 species of Falconiformes (Aves) and the karyological relationships of the order. *Genetica* 46:77–113.

Dickerman, R. W., and A. B. Johnson. 2008. Notes on Great Horned Owls nesting in the Rocky Mountains, with a description of a new subspecies. *Journal of Raptor Research* 42:20–28.

Dove, C. J., and R. C. Banks. 1999. A taxonomic study of Crested Caracaras (Falconidae). *Wilson Bulletin* 111:330–39.

Ericson, P. G. P., C. L. Anderson, T. Britton, A. Elzanowski, U. S. Johansson, M. Kallerrsjo, J. I. Ohlson, T. J. Parsons, D. Zuccon, and G. Mayr. 2006. Diversification of Neoaves: integration of molecular sequence data and fossils. *Biology Letters* 2:543–47.

Fain, M. G., and P. Houde. 2004. Parallel radiations in the primary clades of birds. *Evolution* 58:2558–73.

Fowler, L. E. 1979. Hatching success and nest predation in the green sea turtle, *Chelonia mydas*, at Tortuguero, Costa Rica. *Ecology* 60:946–55.

Glinski, R. L., ed. 1998. *The raptors of Arizona.* Tucson: University of Arizona Press.

Grossman, M. L., and J. Hamlet. 1964. *Birds of prey of the world.* New York: Bonanza Books.

Hackett, S. J., R. T. Kimball, S. Reddy, R. C. K. Bowie, E. L. Braun, M. J. Braun, J. L. Chojnowski, W. A. Cox, K.-L. Han, J. Harshman, C. J. Huddleston, et al. 2008. A phylogenomic study of birds reveals their evolutionary history. *Science* 320:1763–68.

Hoffman, R. 1904. *A guide to the birds of New England and eastern New York.* Boston: Houghton, Mifflin, and Co.

Illiger, J. K. W. 1811. *Prodromus systematis mammalium et avium.* Berlin, Germany: C. Salfeld.

Jennings, J. 1828. *Ornithologia, or the birds.* Part 1. London: Poole and Edwards.

Kermode, F. 1904. *Catalogue of British Columbia birds.* Victoria, BC, Canada: R. Wolfenden.

Owen, R. 1866. *On the anatomy of vertebrates.* Vol. 2, *Birds and mammals.* London: Longmans, Green, and Co.

Remsen, J. V. Jr., C. D. Cadena, A. Jaramillo, M. Nores, J. F. Pacheco, M. B. Robbins, T. S. Schulenberg, F. G. Stiles, D. F. Stotz, and K. J. Zimmer. 2009. *A classification of the bird species of South America.* Version April 2009. Washington, DC: American Ornithologists' Union. http://www.museum.lsu.edu/~Remsen/SACCBaseline.html (accessed 20 April 2009).

Streseman, Erwin. 1975. *Ornithology: from Aristotle to the present.* Cambridge, MA: Harvard University Press.

Swainson, W. 1836. *On the natural history and classification of birds.* Vol. 1. London: Longman, Rees, Orme, Brown, Green, and Longman.

Takagi, N., and M. Sasaki. 1974. A phylogenetic study of bird karyotypes. *Chromosoma* 46:91–120.

Terres, J. K. 1980. *The Audubon Society encyclopedia of North American birds.* New York: Alfred A. Knopf.

van Tuinen, M., C. G. Sibley, and S. B. Hedges. 2000. The early history of modern birds inferred from DNA sequences of nuclear and mitochondrial ribosomal genes. *Molecular Biology and Evolution* 17:451–57.

Vigors, N. A. 1825. Observations on the natural affinities that connect the orders and families of birds. *Transactions of the Linnean Society of London* 14:395–517.

———. 1839. Ornithology. In *The Zoology of Captain Beechey's voyage compiled from the collections and notes made by Captain Beechey, the officers and naturalist of the expedition, during a voyage to the Pacific and Behring's Straits performed in his Majesty's Ship Blossom, under the command of Captain F. W. Beechey, R.N., F.R.S. &c., &c., in the years 1825, 26, 27 and 28*, by J. Richardson, N. A. Vigors, G. T. Lay, E. T. Bennett, R. Owen, J. E. Gray, W. Buckland, and G. B. Sowerby. London.

Weidensaul, S. 1989. *North American birds of prey.* New York: Shooting Star Press.

Raptor Morphology

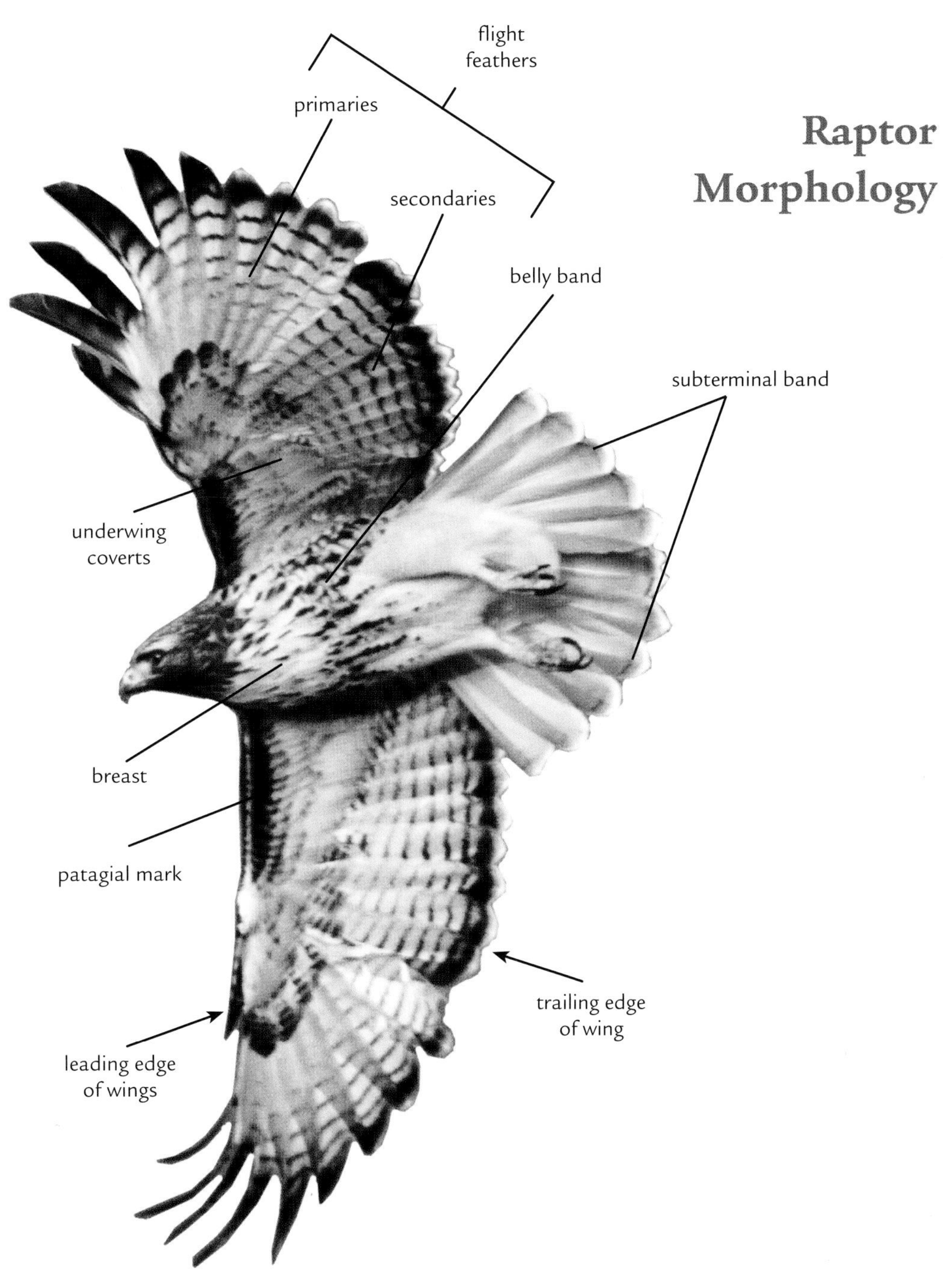

FIGURE RM.1

Red-tailed Hawk

© GERAINT SMITH.

FIGURE RM.2

Peregrine Falcon

© ROBERT SHANTZ.

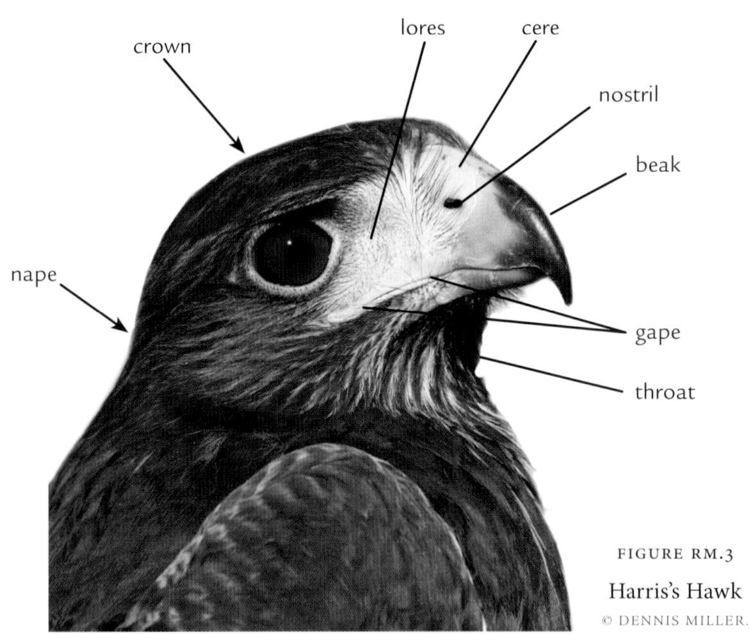

FIGURE RM.3

Harris's Hawk

© DENNIS MILLER.

FIGURE RM.4

Long-eared Owl

© ROBERT SHANTZ.

FIGURE RM.5

Northern Pigmy-owl

© SALLY KING.

MAP M.1

New Mexico's Main Mountain Ranges and Rivers

For clarity, New Mexico's smaller mountain ranges and subranges are not shown on this map. Note that there are two mountain ranges named "San Mateo Mountains," one in Socorro County in the southwestern quadrant of the state, the other only a short distance to the north and encompassing Mount Taylor. Similarly, two mountains are named Guadalupe Mountains. The larger Guadalupe Mountains lie in southeastern New Mexico, the smaller Guadalupe Mountains (not shown on this map) in the extreme southwestern corner of the state.

Maps

General New Mexico Maps

WITH A TOTAL area of 314,917 km^2 (121,590 mi^2), New Mexico is the fifth largest U.S. state. The main rivers flowing through the state consist of the Rio Grande, Pecos, Canadian, San Juan, and Gila. The elevation ranges from 866 m (2,842 ft) at Red Bluff Reservoir in Eddy County to 4,013 m (13,167 ft) on Wheeler Peak in the Sangre de Cristo Mountains. New Mexico is divided into 33 counties and harbors 7 national wildlife refuges and 5 national forests. Much of the human population is concentrated in the Albuquerque-Rio Rancho metropolitan area. As of 2008, all other New Mexico towns harbor populations of less than 100,000.

Species' Distribution Maps

For most species, the reader will find a geographic distribution map based on actual records of occurrences, with considerations also of habitat and elevation. Three colors are used to distinguish breeding (orange), nonbreeding (blue), and year-round occurrence (green). Blue arrows are shown in areas where a species is found only in migration. Where a species is found irregularly and/or very sparsely, the distribution appears in the form of diagonal lines instead of a solid color. Question marks are shown where the status of the species is uncertain, or where the exact distribution contours remain unclear. For one species, the Short-eared Owl, orange was replaced by yellow to indicate that the species occurs during the breeding season, but with no actual breeding yet documented anywhere in New Mexico. For the Mexican Spotted Owl, which is thought to be declining, we added another fill (black X's) to show areas where the bird was present historically but now seems extirpated.

In a few cases, the maps were designed altogether differently. This is true, for example, for the Bald Eagle. Nest location information is still considered "sensitive," for fear of attracting crowds to the few nests existing in the state and disturbing reproduction. On the Bald Eagle distribution map, the breeding distribution is presented only at the county level. The Bald Eagle winters throughout New Mexico. In green on the map are those counties where the species is not only present during winter, but where nesting pairs have been recorded. For three additional species—the White-tailed Kite, Mississippi Kite, and Elf Owl—the map shows individual records of occurrence rather than the actual distribution, which is not well understood. In the Aplomado Falcon chapter, a distribution map is presented, but another map is added to present individual records by decades, to track visually some of the apparent shifts in distribution through time.

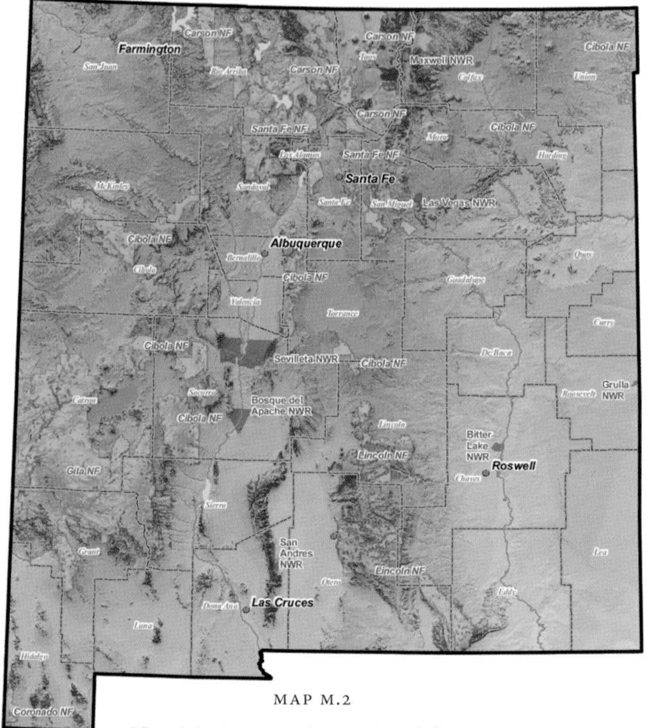

MAP M.2

New Mexico counties, national forests, and national wildlife refuges

MAP M.3

New Mexico's main towns and main roads

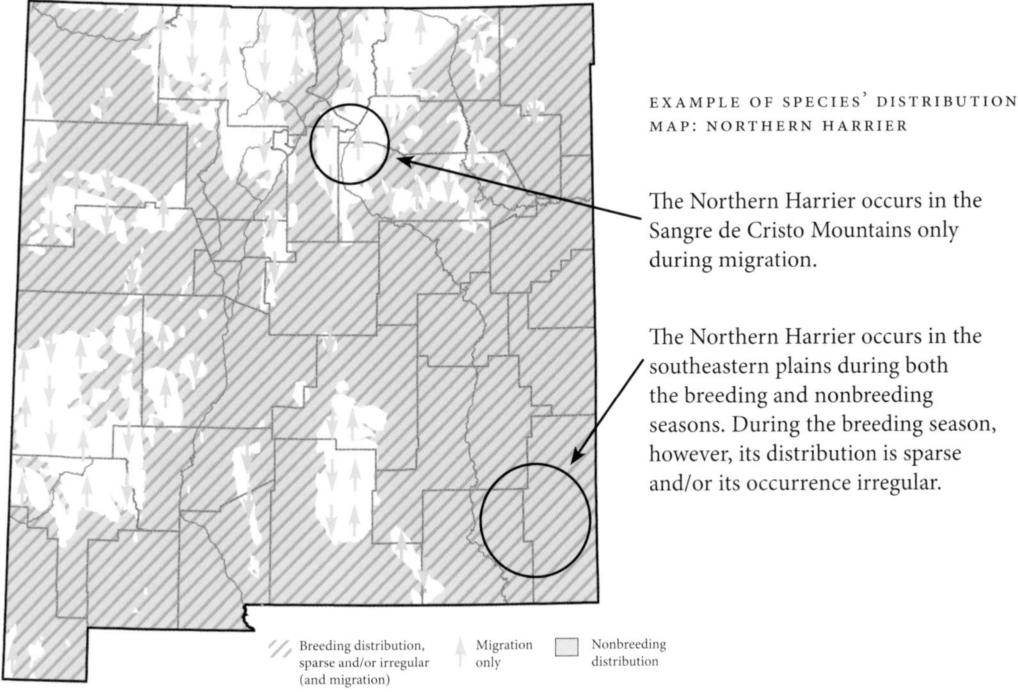

EXAMPLE OF SPECIES' DISTRIBUTION MAP: NORTHERN HARRIER

The Northern Harrier occurs in the Sangre de Cristo Mountains only during migration.

The Northern Harrier occurs in the southeastern plains during both the breeding and nonbreeding seasons. During the breeding season, however, its distribution is sparse and/or its occurrence irregular.

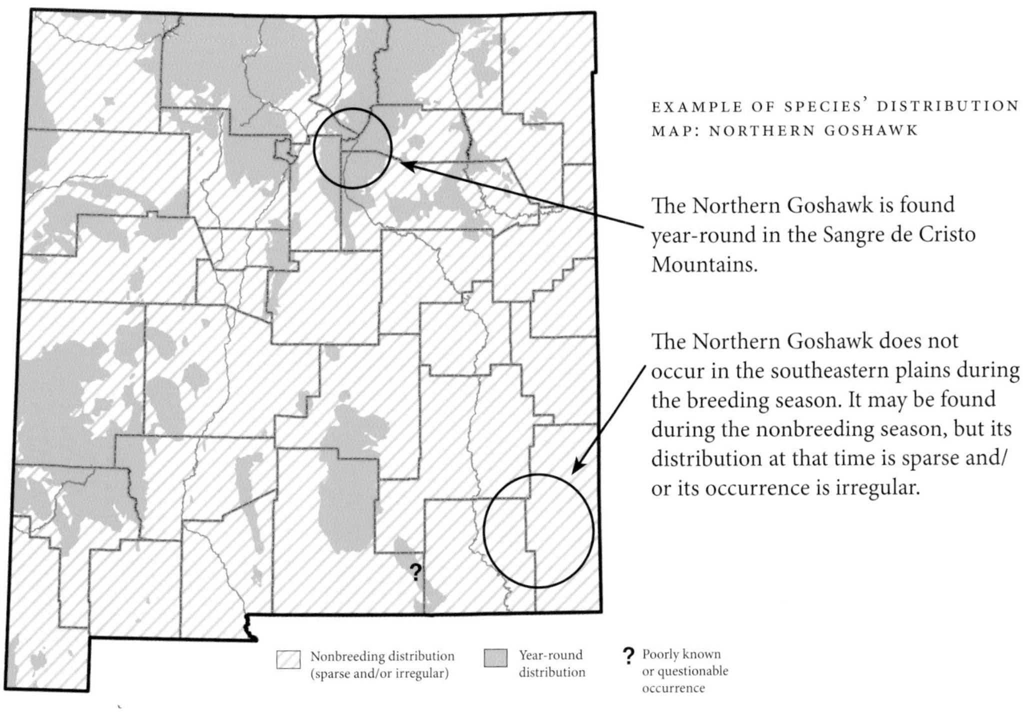

EXAMPLE OF SPECIES' DISTRIBUTION MAP: NORTHERN GOSHAWK

The Northern Goshawk is found year-round in the Sangre de Cristo Mountains.

The Northern Goshawk does not occur in the southeastern plains during the breeding season. It may be found during the nonbreeding season, but its distribution at that time is sparse and/or its occurrence is irregular.

PART I

Introductory Chapters

1

An Introduction to New Mexico's Floristic Zones and Vegetation Communities

TIMOTHY K. LOWREY

NEW MEXICO BOASTS one of the most diverse landscapes in the United States. This landscape diversity is due in large part to an important volcanic history, nonvolcanic mountain building, the presence of three major river systems, and extensive erosion. Geologic and hydrologic processes have produced high mountain peaks, mountain chains, riparian corridors, mesas and arroyos, closed basins with playas, and a number of other topographic features such as badlands. The topographic and geologic diversities interact with the climatic features of temperature, wind, and precipitation to determine plant diversity in New Mexico. In terms of size, New Mexico is the fifth largest state in the union while it has the fourth highest plant diversity in terms of numbers of species.

Floristic Zones and Main Vegetation Communities

Plant diversity in any particular region is often described in terms of vegetation diversity. Vegetation has two components: structure (or physiognomy) and floristic composition. As descriptors of vegetation structure we use such terms as *forest*, *grassland*, or *shrubland*, all of which only give an indication of physical appearance. An area's floristic composition is the actual taxonomic diversity of the different local structural types. For example, all continents except Antarctica have forests but the families, genera, and species of plants they harbor are very different based on exact geographic location.

Thus the description of plant diversity relies on the use of both physiognomic and floristic classification systems. In this chapter the fundamental organization is based on McLaughlin's (1989) floristic classification system for the American Southwest, and secondarily, Brown et al.'s (2007) mapping of vegetation communities in Arizona and New Mexico. We can recognize five floristic zones (districts or subprovinces *sensu* McLaughlin [1989]) and 11 vegetation communities in New Mexico. McLaughlin's zones are the Colorado Plateau, Great Plains (referred to here as Short Grass Prairie, which is the southwestern floristic subdivision of McLaughlin's Great Plains), Chihuahuan Desert, Apachian, and Southern Rocky Mountain–Mogollon. Brown et al.'s vegetation communities consist of (1) alpine tundra; (2) Chihuahuan desertscrub; (3) encinal forest and woodland; (4) Great Basin conifer woodland; (5) Great Basin desertscrub; (6) Great Basin shrub-steppe grassland; (7) montane conifer forest; (8) plains grassland—midgrass and tallgrass;

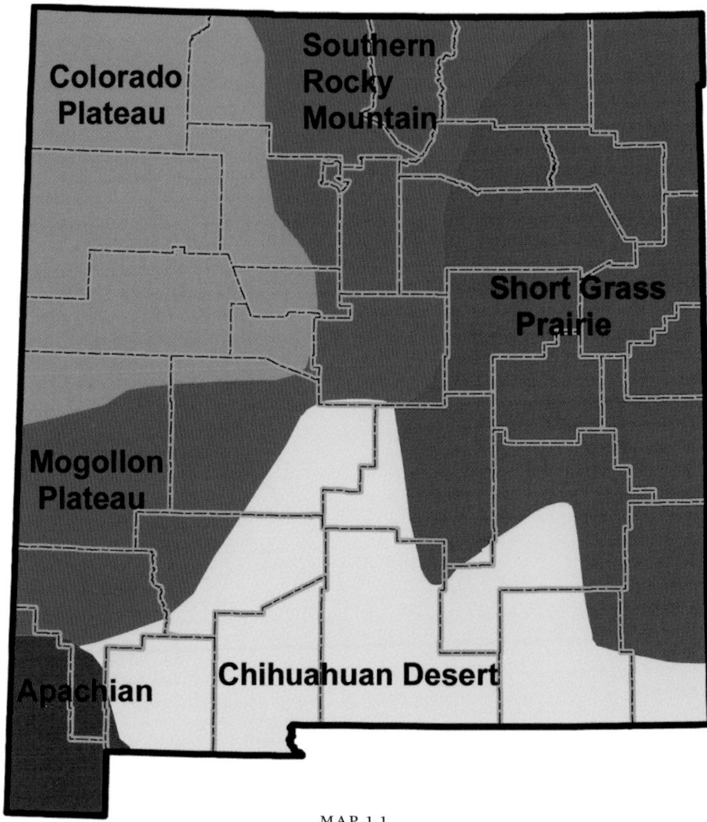

MAP 1.1

New Mexico's five floristic zones. Modified from McLaughlin (1989). The Short Grass Prairie Floristic Zone is the southwestern floristic subdivision of McLaughlin's Great Plains region. The northern boundary of the Chihuahuan Desert floristic zone was redrawn to better follow the contours of the distribution of creosote bush (*Larrea tridentata*) / soaptree yucca (*Yucca elata*) in the Pecos Basin.

(9) plains grassland—shortgrass; (10) semidesert grassland; and (11) subalpine conifer forest.

McLaughlin's 1989 and Brown et al.'s 2007 floristic classification systems are both widely used. The differences between them are largely the result of the spatial scale at which similarities in floristic composition are analyzed. Some of Brown et al.'s vegetation communities are confined to one of McLaughlin's floristic zones. Others, including Great Basin conifer woodland (pinyon-juniper woodland) occur in more than one floristic zone when the primary and secondary determinants of climate and geology (see below) are favorable.

Within New Mexico's floristic zones there may be some or all of the six major structural or physiognomic vegetation types as defined by Dick-Peddie (1993): tundra, forest, woodland, grassland, scrubland, and riparian. The floristic zones—and the vegetation communities—are defined by dominant plant species associations whereas the vegetation structure types are based on growth form and abiotic features of climate (primarily precipitation and temperature), geography (elevation and latitude), and geology (soils, slope, and aspect). In New Mexico the greatest influence on vegetation is precipitation, followed by temperature (Dick-Peddie 1993). Precipitation and temperature represent

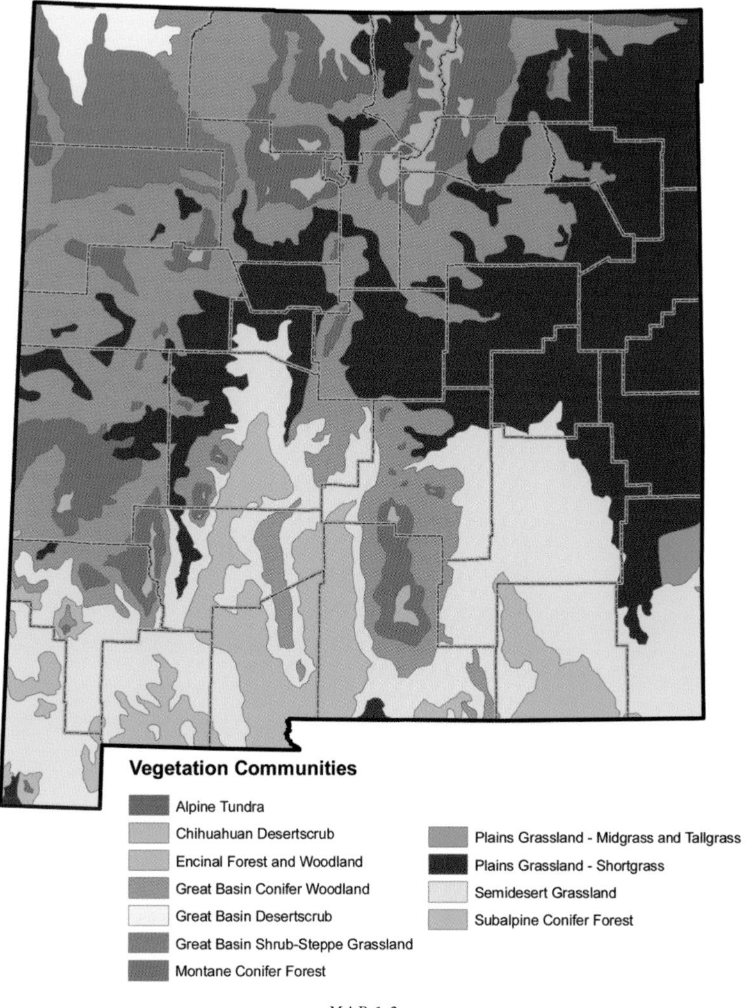

MAP 1.2

New Mexico's 11 vegetation communities. Based on Brown et al. (2007).

the primary determinants of vegetation patterning. Other components of climate, geography, and geology such as wind, soil type, slope, aspect, elevation, latitude, and so forth are considered secondary determinants of vegetation patterns.

Colorado Plateau Floristic Zone

The Colorado Plateau Floristic Zone lies in northwestern New Mexico. It includes all of San Juan, McKinley, and Cibola counties, in addition to large portions of Rio Arriba, Sandoval, and Bernalillo counties, reaching its southern limit in northern Catron County and northern Socorro County. The region receives an average of less than 25 cm (10 in) of precipitation, although the higher elevations may receive an average of 38 cm (15 in). The precipitation falls mostly as snow during the winter and early spring. The Colorado Plateau is characterized by cold winters and warm summers. The soils are largely derived from sedimentary rock and may be highly alkaline or saline. The zone is mainly a shrubland—corresponding to Brown et al.'s (2007) Great Basin desertscrub—although there are also extensive patches of pinyon-juniper woodland (or Great Basin conifer woodland) and Great Basin shrub-steppe grassland. All of Brown et al.'s Great Basin shrub-steppe grassland

PHOTO 1.1

Colorado Plateau Floristic Zone. Pinyon (*Pinus edulis*)-juniper (*Juniperus osteosperma*) woodland with an understory of big sagebrush (*Artemisia tridentata*) and Bailey's yucca (*Yucca baileyi*), Vaca Canyon, San Juan Co. PHOTOGRAPH: © RON KELLERMUELLER/HAWKS ALOFT, INC.

and Great Basin desertscrub are found entirely in the Colorado Plateau Floristic Zone.

The flora in the Colorado Plateau is dominated by shrubby species in the sunflower (Asteraceae) and goosefoot (Chenopodiaceae) families. In particular, dominant species include big sagebrush (*Artemisia tridentata*), four-wing saltbush (*Atriplex canescens*), shadscale (*Atriplex confertifolia*), greasewood (*Sarcobatus vermiculatus*), and winterfat (*Krascheninnikovia lanata*). Saltbush and greasewood are obligate halophytes, meaning they only occur on salty or saline soils.

Short Grass Prairie Floristic Zone

The Short Grass Prairie (SGP) is a phase of the Great Plains grassland but is dominated by grasses of much shorter stature than those in Great Plains proper. This floristic zone occurs in eastern New Mexico from the Colorado border in the northeast south to Otero County. The southeastern corner of New Mexico in Eddy County and portions of Lea and Chaves counties are not part of the SGP but lie instead in the Chihuahuan Desert Floristic Zone and will be treated in that section. The SGP also corresponds largely—though by no means perfectly—to Brown et al.'s (2007) plains grassland-shortgrass and includes an area in northeastern Lea County defined by Brown et al. as plains grassland-midgrass and tallgrass.

There are two interesting subregions within this zone: (1) the Llano Estacado and (2) shinnery sands. The Llano Estacado (from Spanish, meaning "staked plain") is a strikingly flat, treeless, plain. It straddles the Texas–New Mexico border extending west into New Mexico and east into Texas. In New Mexico, the Llano Estacado lies between the Canadian River in the north and the Pecos River Valley in the west, reaching into Otero and Eddy counties along the south boundary. It is separated from the lower elevational portions of the rest of the SGP by the Caprock Escarpment, which is a steep cliff formation reaching heights of 100 meters (330 ft) on the north and west portions of the Llano Estacado.

The shinnery sands are beds or dunes of deep sand largely north of Portales in Roosevelt County and along the Mescalero Escarpment east of Roswell in Chaves County (Allred 2008). The sand beds are dominated by shinnery oak (*Quercus havardii*). Shinnery oak is an

PHOTO 1.2

Short Grass Prairie Floristic Zone north of Hayden, Union Co., with grama grasses (*Bouteloua gracilis* and *Buchloe dactyloides*) and soapweed yucca (*Yucca glauca*).

PHOTOGRAPH: © ROBERT SIVINSKI.

PHOTO 1.3

Short Grass Prairie Floristic Zone. A shinnery oak (*Quercus havardii*) community including large oak specimens (2–5 m tall) with an old raptor nest and sand sage (*Artemisia filifolia*), Eddy Co.

PHOTOGRAPH: © DAVID J. GRIFFIN.

interesting oak species that is a true shrub rather than a tree, with an extensive root system that can extend 16 meters (50 ft) to reach the water table. It forms dense scrub thickets in southeastern New Mexico. Shinnery oak scrub is the main habitat for the Lesser Prairie Chicken (*Tympanuchus pallidicinctus*).

The grasses in the SGP are normally only 7.5–18 cm (3–7 in) tall. The climate is semiarid, with precipitation averaging 25–40.5 cm (10–16 in) per year, falling largely as rain from May to August. The SGP zone has some of the windiest areas in the United States due to the lack of topographic relief and due to the occurrence of downslope winds from the southern Rocky Mountains. The combination of relatively low rainfall and high winds is mainly responsible for the short stature of the vegetation. The SGP grassland is dominated by grama grasses (*Bouteloua gracilis* and *B. hirsuta*) and buffalo grass (*Buchloe dactyloides*). Other common grasses include little bluestem (*Schizachyrium scoparium*), sand dropseed (*Sporobolus cryptandrus*), and purple three-awn (*Aristida purpurea*). Common herbaceous plants also include plains sunflower (*Helianthus petiolaris*), scarlet globemallow (*Sphaeralcea coccinea*), and Lambert's crazyweed (*Oxytropis lambertii*). Besides shinnery oak, sand sagebrush (*Artemisia filifolia*) and winterfat (*Krascheninnikovia lanata*) are common shrubs in the region. The grasses of the SGP are well adapted to grazing since they coevolved with a number of large mammal grazers including the American bison (*Bison bison*) and the pronghorn (*Antilocapra americana*) as well as the smaller prairie dogs (*Cynomys*).

Chihuahuan Desert Floristic Zone

The Chihuahuan Desert Floristic Zone is a diverse desert ecoregion that stretches from just north of Mexico City in Mexico to about 75 km (47 mi) south of Albuquerque, New Mexico, and extends from southeastern Arizona to western Texas. With the exception of New Mexico's "Bootheel," the extreme southern portion of the state is all included in the Chihuahuan Desert Floristic Zone. In New Mexico, the average precipitation in the Chihuahuan Desert is only 18–30.5 cm (7–12 in) per year with most of it falling in summer thunderstorms from June to September. The winters are cool, with nighttime temperatures dropping below freezing at least 100 times per year, but the cool winters are counterbalanced by the very hot summers with many days over 38° C (100° F). The elevation is relatively high even outside the mountains (e.g., Sacramento, Capitan, and Guadalupe mountains), or 910–1,370 m (3,000–4,500 feet). As shown by Brown et al.'s

PHOTO 1.4

Chihuahuan Desert Floristic Zone. A semidesert grassland dominated by needle-and-thread grass (*Heterostipa comata*) with view of the Cornudas Mountains, Otero Co. PHOTOGRAPH: © DAVID J. GRIFFIN.

PHOTO 1.5

(*top right*) Chihuahuan Desert Floristic Zone. Wildflowers (*Lepidium montanum*, *Physaria fendleri*) on Otero Mesa, Otero Co., with soaptree yucca (*Yucca elata*), cholla (*Cylindropuntia imbricata*), beargrass (*Nolina texana*), and littleleaf sumac (*Rhus microphylla*). PHOTOGRAPH: © ROBERT SIVINSKI.

PHOTO 1.6

(*bottom right*) Chihuahuan Desert Floristic Zone. View of the Organ Mountains with sotol (*Dasylirion wheeleri*) in the foreground. Doña Ana Co. PHOTOGRAPH: © JERRY OLDENETTEL.

(2007) vegetation communities, most of the Chihuahuan Desert Floristic Zone consists of semidesert grasslands and shrublands. The grasslands have many of the same species as the short grass prairie but also include big sacaton (*Sporobolus wrightii*), black grama (*Bouteloua eriopoda*), and bush muhly (*Muhlenbergia porteri*). The Chihuahuan Desert has the largest number of cactus species of any of the North American warm deserts. However, the woody plants provide the most characteristic species that most people readily recognize. Dominant shrubs include creosote bush (*Larrea tridentata*), tarbush (*Flourensia cernua*), honey mesquite (*Prosopis glandulosa*), sotol (*Dasylirion wheeleri*), and lechuguilla (*Agave lechuguilla*).

Apachian Floristic Zone

The Apachian Floristic Zone includes portions of southeastern Arizona and the "Bootheel" of southwestern New Mexico. It shows considerable affinity with the flora of the Sierra Madre Occidental in northwestern Mexico (McLaughlin 1989). It is floristically intermediate between the Sonoran and Chihuahuan Deserts, and in Brown et al. (2007) it corresponds to an area covered mainly by semidesert grassland and, to a lesser extent, Chihuahuan desertscrub, with also some plains grassland and encinal forest and woodland. The Apachian zone occupies an elevational range of 1,830–2,440 m (6,000–8,000 ft). It is characterized by hot summers and moderate winters with most of the precipitation in the summer, though it also does receive significant winter rainfall. Although there are vast expanses of grasslands, the vegetation has a large woody component and is considered to include such vegetation structural types as woodland-savanna, scrub, and riparian. Two notable sky island mountain ranges, the Animas and the Peloncillo mountains, occur in this zone. Mountain ranges in the southwestern United States are called "sky islands" because they are like islands in water except the water is replaced by surrounding seas of semidesert grassland or scrub. The sky islands have plant species different from the surrounding grasslands or scrublands. Often they have plants that occur only on the particular mountain range and nowhere else in the world.

Characteristic trees of the Apachian Zone are Arizona cypress (*Cupressus arizonica*), Mexican pinyon pine (*Pinus cembroides*), Chihuahua pine (*Pinus leiophylla* var. *chihuahuana*), and Arizona white oak (*Quercus arizonica*). Interesting riparian trees include Arizona walnut (*Juglans major*), Fremont cottonwood (*Populus fremontii*), and Arizona sycamore (*Platanus wrightii*). Dominant shrubs include longleaf ephedra (*Ephedra trifurca*), ocotillo (*Fouquieria splendens*), and agave (*Agave* spp.).

PHOTO 1.7

Apachian Floristic Zone. A view of the Animas Valley from Deer Creek, Hidalgo Co. Dominant plants are sotol (*Dasylirion wheeleri*) and oak (*Quercus* sp.). PHOTOGRAPH: © NARCA MOORE-CRAIG.

Southern Rocky Mountain–Mogollon Floristic Zone

The Rocky Mountain–Mogollon Floristic Zone includes the Mogollon Rim portion of southwestern New Mexico and the northern part of the central mountain chain from the Sangre de Cristo and Jemez Mountains south through the Sandia and Manzano mountains. The vegetation consists largely of montane and subalpine conifer forest although there are considerable patches of pinyon-juniper woodland at the base of the mountains. Quaking aspen forests, which occur at the same altitude as spruce-fir, are an early succession plant community that is maintained entirely by fire. Moving up the mountainside from pinyon-juniper woodlands, one encounters ponderosa pine forest, mixed-conifer woodland, spruce-fir forest and/or aspen forest, subalpine forest, and finally tundra on the highest mountains (fig. 1.1). This is a typical zonation pattern on the high mountains in New Mexico that results from the combined effects of altitude and exposure. Increasing altitude results in lower temperatures and higher amounts of precipitation. Particular forest tree species have specific requirements of temperature and precipitation. Exposure or aspect (northern versus southern) has a major effect on the vegetation zonation as well. A particular vegetation type on the south side of a mountain will occur at higher elevations than on the north side (Dick-Peddie 1993). This difference is due to the higher amount of solar radiation that impacts southern exposures, resulting in higher temperatures and increased evaporation. Thus, southern exposures are generally hotter and drier than northern exposures at the same altitude.

PHOTO 1.8

Southern Rocky Mountain–Mogollon Floristic Zone. Old-growth ponderosa pine (*Pinus ponderosa*) forest with widely spaced trees and an herbaceous layer of mountain muhly (*Muhlenbergia montana*) and Idaho fescue (*Festuca idahoensis*). Tusas Mountains below Hopewell Lake, Rio Arriba Co. PHOTOGRAPH: © ROBERT SIVINSKI.

PHOTO 1.9

(*top right*) Southern Rocky Mountain–Mogollon Floristic Zone. Fall in a spruce-fir forest (*Picea engelmannii–Abies arizonica*) with quaking aspen (*Populus tremuloides*), Sangre de Cristo Mountains, Santa Fe Co.

PHOTOGRAPH: © ROBERT SIVINSKI.

PHOTO 1.10

(*bottom right*) Southern Rocky Mountain–Mogollon Floristic Zone. Treeline in a subalpine forest community looking toward Serpent Lake and spruce-fir forest, Taos Co. PHOTOGRAPH: © TOM KENNEDY.

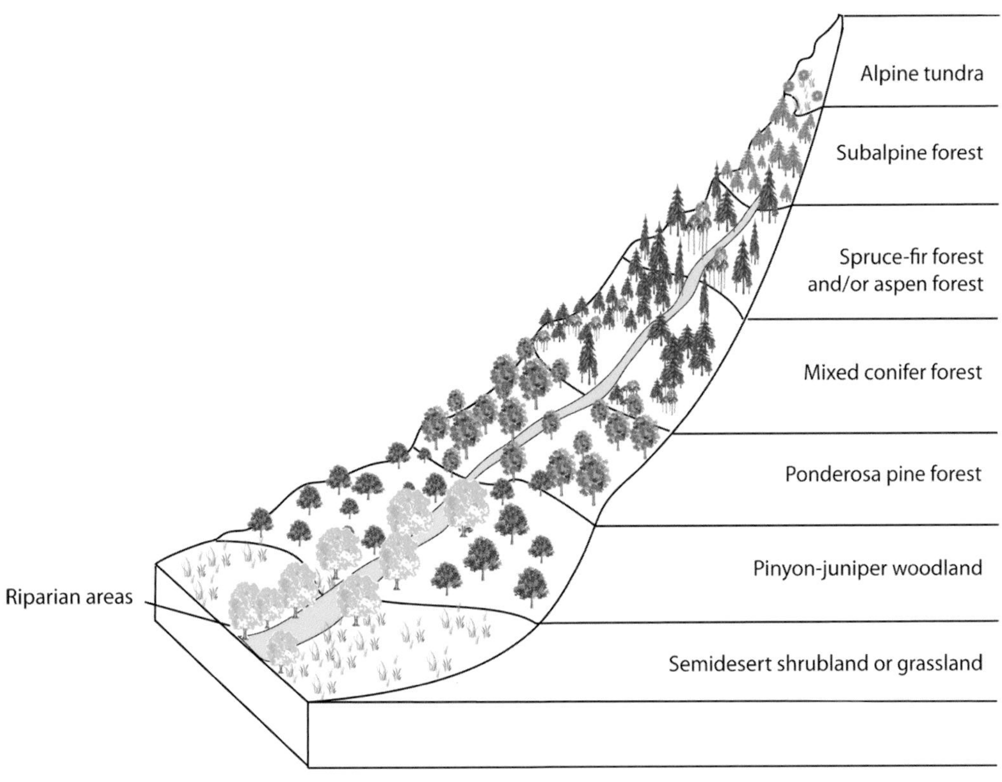

FIGURE 1.1

Elevational zonation of the vegetation in the Southern Rocky Mountain–Mogollon Floristic Zone.

ILLUSTRATION © JANE E. MYGATT.

The dominant tree species in the Rocky Mountain–Mogollon Floristic Zone include ponderosa pine (*Pinus ponderosa*), Rocky Mountain juniper (*Juniperus scopulorum*), Gambel oak (*Quercus gambelii*), Douglas fir (*Pseudotsuga menziesii*), blue spruce (*Picea pungens*), Engelmann's spruce (*Picea engelmannii*), white fir (*Abies concolor*), quaking aspen (*Populus tremuloides*), bristlecone pine (*Pinus aristata*), limber pine (*Pinus flexilis*), and cork-bark fir (*Abies arizonica*). These trees provide important nesting sites for birds as well as sources of food.

Riparian Vegetation Communities

In all floristic zones, the presence of water—especially perennial rivers—allows specialized plant communities to develop. Such vegetation communities are called riparian and act as magnets for animals. Depending on altitude and location in New Mexico, one may encounter montane riparian, floodplain-plains riparian (often referred to as bosque), arroyo riparian, and playa-alkalai sink riparian communities. All riparian communities have distinctive floristic components. Many of the woody species are phreatophytes, meaning that they must have their roots in the water table to survive. Given the dry climate in New Mexico, phreatophytes are typically restricted to riparian zones, where the water tables are shallow. Common trees in riparian areas include cottonwoods (*Populus* spp.), willows (*Salix* spp.), boxelder (*Acer negundo*), water birch (*Betula occidentalis*), and mountain alder (*Alnus tenuifolia*). Shrubs include redosier dogwood (*Cornus sericea*), iodine bush (*Allenrolfea occidentalis*), and seepwillow (*Baccharis glutinosa*). Sedges and grasses dominate the herbaceous taxa. The lowland riparian plant communities have suffered greatly from

PHOTO 1.11

Riparian vegetation in the Rio Grande Valley near Los Lunas, Valencia Co. Dominant plants include Rio Grande cottonwood (*Populus deltoides* subsp. *wislizenii*) and cattail (*Typha* sp.). PHOTOGRAPH: © MIKE MARCUS.

the incursion of invasive species, particularly the two woody plant species Russian olive (*Elaeagnus angustifolia*) and saltcedar (*Tamarix* spp.) (e.g., Cartron et al. 2008). These weeds can dominate riparian areas, leading to major changes in hydrology and soils as well as their associated plants and animals.

Concluding Remarks

Each of the floristic zones hosts a varied fauna. Plants are the primary producers in most habitats in New Mexico due to their ability to carry out photosynthesis. Plants provide basic foods that support food chains and ultimately influence raptors and other carnivorous animals that feed on the herbivores. Plants also provide shelter and nesting sites for a variety of animals. Any major change or disruption of plant communities and their constituent species generally leads to changes in the associated animal communities.

The Endangered Species Act of 1973 has led to the successful recovery of several raptor species whose distribution includes New Mexico. As in the case of the Peregrine Falcon (*Falco peregrinus*; see chapter 25), however, recovery efforts have been successful mostly where the initial anthropogenic effect was contamination of food chains by dangerous pollutants. Today, perhaps the main threat to raptors—and most other animals—consists of habitat loss and degradation. In such cases, the organisms first affected are the plants. With the realization that plants provide the underpinning for most ecosystems on earth, conservation efforts should therefore focus on plant and animal communities jointly.

LITERATURE CITED

Allred, K. A. 2008. *Flora neomexicana I: the vascular plants of New Mexico*. Available at www.lulu.com.

Brown, D. E., P. J. Unmack, and T. C. Brennan. 2007. Digitized map of biotic communities for plotting and comparing distributions of North American animals. *Southwestern Naturalist* 52:610–16.

Cartron, J.-L. E., D. C. Lightfoot, J. E. Mygatt, S. L. Brantley, and T. K. Lowrey. 2008. *A field guide to the plants and animals of the Middle Rio Grande bosque*. Albuquerque: University of New Mexico Press.

Dick-Peddie, W. A. 1993. *New Mexico vegetation: Past, present, and future*. Albuquerque: University of New Mexico Press.

McLaughlin, S. P. 1989. Natural floristic areas of the western United States. *Journal of Biogeography* 16:239–48.

© STEVE BARANOFF.

2

Raptor Migration Dynamics in New Mexico

JEFF P. SMITH

SITUATED TOWARD THE southern end of the Rocky Mountains, New Mexico lies within the heart of one of three major migration corridors for diurnal raptors in western North America: from west to east, Pacific Coast, Intermountain (between the Sierra Nevada–Cascade and Rocky Mountain ranges), and Rocky Mountain (Hoffman et al. 2002). Each spring and autumn, thousands of migrating raptors pass through the state on their way to and from summer ranges farther north and winter ranges farther south. Moreover, because the year-round climate is relatively mild, the state accommodates a complex mix of permanent residents, summer-only or winter-only residents, and transient migrants, similar to most southern-latitude states. For example, Red-tailed Hawks (*Buteo jamaicensis*) can be found in all areas of the state throughout the year. To a large extent, this reflects the presence of year-round residents, but some summer breeders also vacate the state for wintering grounds in Mexico only to be replaced by birds that vacate northern breeding grounds to winter in New Mexico (Preston and Beane 1993). In fact, only 3 of the 24 diurnal raptor species that occur in the state *cannot* be found there on a year-round basis (table 2.1). These consist of the Rough-legged Hawk (*Buteo lagopus*), which occurs as a transient or winter resident; the Broad-winged Hawk (*B. platypterus*), which only passes through the state on migration; and the Mississippi Kite (*Ictinia mississippiensis*), which breeds in the state but winters in South America.

Much of what is known about the migration ecology of diurnal raptors in New Mexico derives from two ongoing, long-term migration studies in the Sandia Mountains just east of Albuquerque (spring study) and in the Manzano Mountains (autumn study) about 55 km (35 mi) farther south (fig. 2.1). Both began in 1985 and involve standardized, annual counts for monitoring long-term population trends (Hoffman and Smith 2003; Smith et al. 2008), as well as extensive trapping, banding, and related research designed to track migration routes, identify source populations, and generate other insights about the ecology of selected species (DeLong and Hoffman 1999, 2004; Hoffman et al. 2002; Smith et al. 2003, 2004; McBride et al. 2004; Lott and Smith 2006; Goodrich and Smith 2008; HawkWatch International 2009a). These projects are part of a network of 12 similar migration studies coordinated by HawkWatch International (HWI) in nine western states from Washington to Texas (HWI 2009b), with other complementary projects coordinated by various organizations in the remaining western states of California, Idaho, and Colorado, as well as in Alberta,

TABLE 2.1. New Mexico's diurnal and nocturnal raptors, with migratory status and seasons of occurrence.

SPECIES	MIGRATORY STATUS[a]	SPRING	SUMMER	AUTUMN	WINTER
Turkey Vulture	Partial migrant	x[b]	x	x	(x)
Osprey	Complete migrant[c]	x	x	x	(x)
White-tailed Kite	Partial migrant	x	x	x	x
Mississippi Kite	Complete migrant	x	x	x	-
Bald Eagle	Partial migrant	x	x	x	x
Northern Harrier	Partial migrant	x	x	x	x
Sharp-shinned Hawk	Partial migrant	x	x	x	x
Cooper's Hawk	Partial migrant	x	x	x	x
Northern Goshawk	Partial migrant	x	x	x	x
Gray Hawk	Partial migrant[d]	x	x	x	(x)
Common Black-Hawk	Partial migrant[d]	x	x	x	(x)
Harris's Hawk	Resident	x	x	x	x
Broad-winged Hawk	Complete migrant	x	–	x	-
Swainson's Hawk	Complete migrant	x	x	x	(x)
Zone-tailed Hawk	Partial migrant[d]	x	x	x	(x)
Red-tailed Hawk	Partial migrant	x	x	x	x
Ferruginous Hawk	Partial migrant	x	x	x	x
Rough-legged Hawk	Complete migrant	x	–	x	x
Golden Eagle	Partial migrant	x	x	x	x
American Kestrel	Partial migrant	x	x	x	x
Merlin	Partial migrant	x	(x)	x	x
Aplomado Falcon	Resident	x	x	x	x
Peregrine Falcon	Partial migrant	x	x	x	x

PHOTO 2.1

Sandia Mountains.

PHOTOGRAPH: © HOWARD HOLLEY.

TABLE 2.1. *(continued)*

SPECIES	MIGRATORY STATUS[a]	SPRING	SUMMER	AUTUMN	WINTER
Prairie Falcon	Partial migrant	x	x	x	x
Barn Owl	Resident/nomadic	x	x	x	x
Flammulated Owl	Partial migrant	x	x	x	(x)[e]
Western Screech-Owl	Resident	x	x	x	x
Whiskered Screech-Owl	Resident	x	x	x	x
Great Horned Owl	Resident	x	x	x	x
Northern Pygmy-Owl	Resident	x	x	x	x
Elf Owl	Partial migrant	x	x	x	-
Burrowing Owl	Partial migrant	x	x	x	x
Mexican Spotted Owl	Partial/altitudinal migrant	x	x	x	x
Long-eared Owl	Partial migrant	x	x	x	x
Short-eared Owl	Partial migrant	x	(x)[f]	x	x
Boreal Owl	Resident	x	x	x	x
Northern Saw-whet Owl	Partial migrant	x	x	x	x

[a] Migratory status of a species as a whole. **Complete migrant**: 90% or more of all individuals migrate to a different geographic area; **partial migrant**: fewer than 90% of all individuals migrate; **resident**: non-migratory; **nomadic**: large-scale movements sporadic depending on climatic and habitat conditions; **altitudinal migrant**: shifts from high to low elevations for the winter. See Bildstein (2006) for further discussion of relevant migration terminology; see also other chapters of this volume for a discussion of migratory status in New Mexico.

[b] x = regularly occurs during season; (x) = rarely occurs during season; – = not known to occur during season.

[c] Notable exceptions are the Osprey populations of the Caribbean, northwestern Mexico, and Belize, which are resident.

[d] Northern birds only.

[e] One Flammulated Owl was found dead in northern New Mexico in January 1996 (Williams 2007). See chapter 28 for a brief discussion of the significance of that winter record.

[f] No confirmed breeding records in New Mexico, but a territorial pair was recently observed and suspected of nesting (see chapter 37).

PHOTO 2.2

Manzano Mountains.

PHOTOGRAPH: © JOHN P. DELONG.

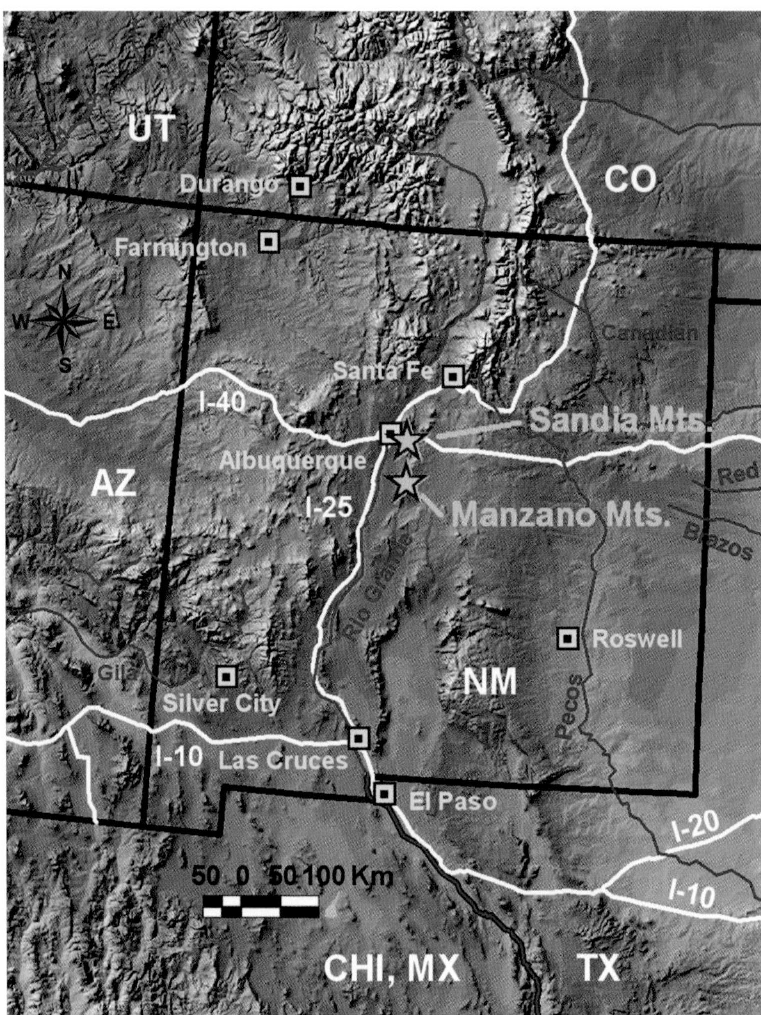

FIGURE 2.1

Location of the Sandia and Manzano mountains raptor migration project sites in New Mexico.

Canada, and Veracruz, Mexico. Other information about New Mexico as a raptor migration corridor and relevant range connections derives from several other banding and satellite-tracking studies that connect various breeding and wintering populations to the state (Steenhof et al. 1984, 2005; Schmutz and Fyfe 1987; Harmata and Stahlecker 1993; Fuller et al. 1998; Martell et al. 2001; Watson and Banasch 2005).

Comparatively little is known about the migrations of nocturnal raptors through New Mexico, or through the Southwest for that matter. Only six of the 13 owl species that occur regularly in New Mexico are known to be truly migratory, though the Mexican Spotted Owl (*Strix occidentalis lucida*) may undertake short-distance, mostly altitudinal "migrations," and the migratory status of Barn Owls (*Tyto alba*) is uncertain but may involve at least extensive dispersal and possible nomadic behavior (table 2.1). Most insight about owl migration in the Southwest derives from limited band-return information and seasonal patterns of occurrence (e.g., Phillips 1942; Best 1969; Martin 1973; Balda et al. 1975). More recently, however, further valuable information concerning the population dynamics and movement ecology of selected species has resulted from additional intensive banding and mark-recapture studies, as well as from application of cutting-edge techniques such as analysis of hydrogen stable-isotope

PHOTO 2.3

(*above*) A HawkWatch International observer at the Sandia Mountains monitoring site in spring 1996. Each year trained observers keep track of numbers of each species flying by the monitoring site. PHOTOGRAPH: © JOHN P. DELONG.

PHOTO 2.4

(*below*) HawkWatch International studies also involve trapping of raptors with use of mist nets and bow nets. PHOTOGRAPH: © JOHN P. DELONG.

PHOTO 2.5

All birds caught are measured, sexed, aged, and banded before being released.

PHOTOGRAPH: © JOHN P. DELONG.

ratios in feathers and radio-tracking (e.g., Gutiérrez et al. 1995; DeLong et al. 2005; Arsenault et al. 1997, 2005; Linkhart and Reynolds 2006, 2007).

In this chapter, I provide a broad overview of what is known about the migration ecology of primarily diurnal raptors that routinely occur in New Mexico. I focus on the state of knowledge concerning the overall spatial and temporal dynamics of movements through the state, as well as relevant migration corridors and connections between New Mexico's raptors and various summer and winter ranges outside the state. My treatment of owl migration is comparatively limited due to the relative dearth of information about owl migration, and focuses most heavily on what HWI and other researchers have learned recently about the movement ecology of Flammulated Owls (*Otus flammeolus*). More detailed information about individual species of both nocturnal and diurnal raptors can be found in the various species-specific chapters that follow in this volume.

Diurnal Raptors

Factors Contributing to Migratory Concentrations

RESPONSES TO SPECIFIC LANDSCAPE FEATURES

Autumn bird migrations typically begin as "broad-front" movements. That is, individuals or small groups depart from scattered breeding territories and begin to move south in broadly dispersed waves of activity. In many cases, as these dispersed individuals move farther and farther south, they begin to converge and aggregate for various reasons. In essence, much as our interstate freeways concentrate long-distance human travelers, this tendency results in development of distinct "raptor migration highways" at certain locations across the continental landscape. Major raptor highways tend to form in relation to specific landscape characteristics that function in one or more of the following ways: (1) serve as navigation aids or "leading lines" that migrants follow to stay on an appropriate course; (2) funnel otherwise broad-front movements along certain "diversion lines" due to barrier effects; (3) facilitate energy-efficient travel by producing air currents that allow migrants to travel long distances with minimum effort; and (4) provide favorable habitat corridors for migrants to follow and remain in proximity to necessary foraging and stopover environments (Kerlinger 1989; Bildstein 2006). In turn, accessible points along these raptor highways comprise ideal opportunities for humans to observe multispecies concentrations on an annual basis and to use standardized counts conducted across networks of such sites as a valuable tool for monitoring regional population trends (Zalles and Bildstein 2000; Hoffman and Smith 2003; Bildstein 2006; Bildstein et al. 2008).

Long north-south mountain ranges are a classic example of leading-line landscape features that migratory raptors often follow. Such ranges serve as effective navigation aids for long-distance migrants, and typically produce strong wind-driven updrafts that migrants can ride for long periods with little effort. Especially in the arid interior West, mountain ranges also often provide the only suitable habitat corridors for forest-dwelling species to follow. Other leading-line features may include coastlines, major river corridors, and other distinct and extensive habitat ecotones that lead in appropriate directions.

Most raptor species are reluctant to cross large expanses of open water due to the absence of favorable updrafts that help keep them aloft (Kerlinger 1989; Bildstein 2006; Goodrich and Smith 2008). Exceptions include species such as the Peregrine Falcon (*Falco peregrinus*) and the Osprey (*Pandion haliaetus*), which routinely rely on powered flight as a primary travel mode and therefore are not deterred by poor lift over water. Most species are also reluctant to cross large expanses of open desert or otherwise inhospitable habitat due to the extreme environmental conditions and attendant lack of food and shelter. At a subtler level and more germane to New Mexico, even extensive treeless prairies and steppe habitats may comprise a barrier to movements of forest- and woodland-dwelling species such as the accipiters (*Accipiter* spp.). Accordingly, due to barrier effects, substantial migration highways can also be found along diversion lines created by the Pacific, Atlantic, and Gulf coasts, the Great Lakes, and the Great Salt Lake and Desert complex in Utah and Nevada, while only a few open-country species routinely move through the Great Plains region (Bildstein 2006). Moreover, although useful updrafts are generally absent out over large water bodies, differential heating of the land and adjacent water often produces favorable air currents along their margins.

In many areas, particularly noteworthy concentrations may occur at certain sites and along specific leading and diversion lines, and thereby comprise great opportunities for intensive long-term monitoring. That does not necessarily mean, however, that all birds moving through such regions follow the primary highway or that the aggregate volume of migration away from the highway is not also substantial. For example, a study of migration within the Appalachian Mountains region of Pennsylvania revealed that although the greatest site-specific concentration occurred at the well-known watch site at Hawk Mountain Sanctuary along the easternmost ridge in the studied complex, the aggregate volume of migration that dispersed along several parallel ridges farther west actually exceeded the volume at Hawk Mountain (Van Fleet 2001). Though not well studied and quantified, it is highly likely that such a scenario also applies across the many parallel ridges of the Great Basin. The best known concentration occurs in the Goshute Mountains of eastern Nevada situated along a major diversion line created by the Great Salt Lake and Desert complex (Hoffman and Smith 2003); however, lesser concentrations are known along ridges such as the Ruby Mountains about 100 km (~60 mi) west of the Goshutes (HWI, unpublished data), in the Egan and Schell Creek ranges near Ely, Nevada (Smith 2005), and in the Spring Mountains in southwestern Nevada (Millsap and Zook 1983). Moreover, although it is clear that species such as Sharp-shinned Hawks (*Accipiter striatus*) and Cooper's Hawks (*A. cooperii*), which typically comprise the most abundant migrants at western watch sites (Hoffman and Smith 2003), are strongly tied to following montane leading lines through regions such as New Mexico and utilizing the wind-driven updrafts present along such ridgelines (Bildstein 2006; Goodrich and Smith 2008), there may well also be significant numbers of scattered individuals that move, at least temporarily, along other kinds of leading lines, such as those provided by extensive, wooded river corridors (e.g., the Rio Grande through central New Mexico).

SPECIES-SPECIFIC VARIATION

The tendency to aggregate on migration varies considerably across species. For example, although territorial during the breeding season, Swainson's Hawks (*Buteo swainsoni*) are highly gregarious at other times of year because group food-finding efforts are particularly effective for locating the unpredictable but locally abundant insect prey (mainly grasshoppers) the species favors outside of the breeding season (England et al. 1997). Similarly, although not particularly gregarious on either their summer or winter ranges, Broad-winged Hawks (*Buteo platypterus*) aggregate in huge numbers during migration to collectively capitalize on unpredictable thermals (i.e., large "bubbles" of hot, rising air) that they rely upon heavily for energy-saving lift that helps them achieve their long migratory journey to Central and South America (Goodrich et al. 1996). Because of their highly gregarious nature, the fact that most if not all individuals of these species vacate North America for the winter, and their common need to utilize wind- or thermal-driven air currents to save energy on their long journeys, species such as these two buteos and other highly gregarious and migratory species such as Turkey Vultures (*Cathartes aura*) and Mississippi Kites are routinely observed in large concentrations at many sites around the continent, especially at southern latitudes where regional and continental convergence is maximized (e.g., see Inzunza et al. 2000; Smith et al. 2001a; Bildstein and Zalles 2001; Bildstein 2006; Goodrich and Smith 2008).

Although not inherently gregarious by nature, most of the other species of diurnal raptors that migrate significant distances also routinely concentrate along common migration routes to take advantage of mutually beneficial landscape features or climatic conditions (Kerlinger 1989). Among the other raptors commonly encountered in large numbers at migration watch sites across the country—including in the Sandia and Manzano mountains of New Mexico—are the Sharp-shinned Hawk, Cooper's Hawk, Red-tailed Hawk, and American Kestrel (*Falco sparverius*). Most individuals of these species typically migrate at least several hundred if not thousands of kilometers each year between summer and winter ranges, and clearly take full advantage of popular "raptor highways." For other species, however, their migratory tendency, distances traveled, and use of common migratory corridors may vary considerably from year to year depending on the latitude at which they breed, climatic conditions,

FIGURE 2.2

Tracking results for three Northern Goshawks (*Accipiter gentilis*; each track color represents a different bird) outfitted with satellite transmitters during autumn migration in the Manzano Mountains, New Mexico, between 2000 and 2002.

and fluctuations in regional prey availability (Bildstein 2006).

For example, northern breeding populations of Golden Eagles (*Aquila chrysaetos*) tend to be migratory whereas southern populations may be entirely sedentary or at most wandering regional residents during the winter (Kochert et al. 2002). Moreover, the distances northern breeders travel south for the winter may be significantly reduced during mild winters when prey availability remains high at northern latitudes. Similarly, when populations of key prey species such as snowshoe hares (*Lepus arcticus*) crash on a cyclical basis every decade or so, Northern Goshawks (*Accipiter gentilis*) that breed at northern latitudes and normally are comparatively sedentary often vacate their breeding ranges en masse and invade southern latitudes in large numbers during fall and winter (Mueller et al. 1977; Squires and Reynolds 1997). In contrast, during any given year, goshawks that breed farther south show widely varying migratory tendencies. Recent satellite tracking of breeding adults in the mountains of Utah indicated that some routinely migrate south each year (albeit relatively short distances of 100–400 km [~60–250 mi]), others undertake only minor altitudinal "migrations" to more favorable lower-elevation winter foraging grounds, and still others remain sedentary on their montane breeding territories (Sonsthagen et al. 2006). Other recent satellite

PHOTO 2.6

Adult male Prairie Falcon (*Falco mexicanus*) captured at the Sandia Mountains monitoring site, spring 1996.

PHOTOGRAPH: © JOHN P. DELONG.

tracking confirmed that most juvenile goshawks captured at migration-monitoring sites across the western United States (including in New Mexico) are dispersing or wandering regional residents that typically remain within 150 km (~90 mi) of their capture locations (e.g., see fig. 2.2; Goodrich and Smith 2008; HWI 2009a).

Exactly which landscape features tend to concentrate migrants also varies from species to species. Some Prairie Falcons (*Falco mexicanus*) and Ferruginous Hawks (*Buteo regalis*) are known to follow looping migration routes designed to progressively track temporal variations in the availability of key prey species such as ground squirrels (*Spermophilus* spp.) and prairie dogs (*Cynomys* spp.). For example, in late summer Prairie Falcons that nest in the Snake River area of central Idaho often travel northeast initially out onto the open prairies of the northwestern Great Plains, then progressively move south along the western margins of the plains, tracking the continued availability of emergent prey (Steenhof et al. 1984, 2005). Eventually they spend the bulk of their winter on the southern prairies wherever prey remain available year-round, and then in the spring routinely cut back up through the southern Rocky Mountains and eastern Great Basin to return to Idaho for the summer. Recent satellite tracking has shown similar movement patterns among Ferruginous Hawks that nest west of the Rockies, while those that nest east of the Rockies tend to remain within the plains region, simply moving back and forth north to south as seasonal climatic and foraging conditions dictate (Schueck et al. 1998; Watson 2003; Watson and Banasch 2005). Because of these complex movement patterns and strategies, as well as the fact that both of these species routinely rely primarily on powered flight and therefore are not strongly tied to seeking out energy-saving lift to facilitate their movements, both species are relatively uncommon at typical mountaintop watch sites like those in the Sandia and Manzano mountains (Hoffman and Smith 2003).

The Swainson's Hawk is another species that is not particularly abundant at typical mountaintop watch sites but often can be found in large numbers moving through open grassland habitats (England et al. 1997). Unlike the Prairie Falcon and the Ferruginous Hawk, this species benefits greatly from energy-saving lift to facilitate its long migration to and from Argentina each year; however, a similar preference for open grassland habitats where favored insect as well as small rodent prey are abundant tends to concentrate the migratory movements of this species along broad north-south-trending prairies and grassland-dominated valleys and foothills. In such areas, instead of relying on wind-driven mountain updrafts for lift, the hawks rely primarily on thermals to help move them along their way with minimal effort (Bildstein 2006).

Another classic species-specific variant concerns Bald Eagles (*Haliaeetus leucocephalus*). Although these eagles can be seen in reasonable numbers migrating along many mountain ridges and taking advantage of the lift they provide, due to their focus on fish and waterfowl prey they are most commonly found moving in close proximity to major river corridors or following paths that specifically connect a series of lakes and reservoirs that remain open during winter and provide food resources (Buehler 2000). Similarly, although large numbers of Golden Eagles routinely concentrate along and follow the Rocky Mountains as they move south out of Alaska and northern Canada (Sherrington 2003; Hoffman and Smith 2003; McIntyre 2004; McIntyre et al. 2006), satellite tracking has shown that other birds drop down from Alaska through the Pacific Northwest and then veer to the southeast across the Great Basin and the southern Rockies to reach similar wintering grounds in eastern New Mexico and west Texas (Goodrich and Smith 2008; HWI 2009a). Like Great Plains Ferruginous Hawks, still other Golden Eagles simply remain within the prairie regions of the western plains, moving only as far south in the winter as needed to find suitable prey.

SEASONAL VARIATION

Most long-term raptor-migration monitoring projects occur in autumn for two primary reasons. First, overall migrant abundance is higher and autumn monitoring affords the opportunity to gauge annual population levels before high winter mortality takes its toll, especially on the year's crop of juvenile birds. Second, migrants often appear to concentrate more along popular flight lines during autumn. This is partly because during autumn, weather and wind factors typically are more conducive to concentrating migrants along mountain ridges, coastlines, and other landscape barriers

TABLE 2.2. Average annual counts of migrating raptors in the Sandia Mountains (spring; 1985–2006) and Manzano Mountains (autumn; 1985–2005) of New Mexico.

SPECIES	SPECIES CODE	SANDIA MTS.	MANZANO MTS.
Turkey Vulture	TV	1,407	396
Osprey	OS	65	30
Northern Harrier	NH	58	58
White-Tailed Kite	WK	<1	–
Mississippi Kite	MK	<1	–
Sharp-shinned Hawk	SS	509	1,489
Cooper's Hawk	CH	768	1,029
Northern Goshawk	NG	12	16
Unknown accipiter	UA	86	245
Common Black-Hawk	CB	<1	–
Broad-winged Hawk	BW	6	7
Swainson's Hawk	SW	55	553
Zone-tailed Hawk	ZT	2	1
Red-tailed Hawk	RT	346	657
Ferruginous Hawk	FH	12	13
Rough-legged Hawk	RL	1	<1
Unidentified buteo	UB	12	24
Golden Eagle	GE	365	118
Bald Eagle	BE	14	3
Unidentified eagle	UE	1	1
American Kestrel	AK	204	562
Merlin	ML	10	25
Aplomado Falcon	AF	<1	–
Prairie Falcon	PR	25	21
Peregrine Falcon	PG	44	49
Unknown falcon	UF	5	9
Unidentified raptor	UU	40	47
TOTAL	–	4,032	5,226

(Bildstein 2006). For example, in many areas west to northwest winds prevail throughout the year and the resulting mountain updrafts are much more conducive to southward than to northward movement. Moreover, the onset of winter storm tracks during autumn is a more dramatic trigger for mass movements to occur and heightens the prevalence of strong northwesterly winds that provide favorable mountain updrafts and tail winds for migrants to exploit. In addition, during spring, not only are experienced adults more prevalent in the returning population due to disproportionately high overwinter mortality of juveniles, they are also driven to return to their established breeding territories as quickly as possible to reclaim them before anyone else moves in. Accordingly, they are likely to be less concerned about waiting for and taking advantage of ideal, energy-saving wind or weather conditions, and may take a variety of shortcuts across the landscape that they would otherwise avoid during autumn (Bildstein 2006).

Such factors as these are why overall migration counts at the Sandia Mountains spring site average 23% less than those at the Manzano Mountains autumn site (table 2.2). In fact, much greater differences are shown for many common raptors (e.g., 65% lower counts in spring for Sharp-shinned Hawks), but this relationship does not apply to all species. For example, Golden Eagle counts in the Sandias average roughly three times higher than in the Manzanos. One possible explanation for this difference is that many Golden Eagles do not end up as far south as southern New Mexico or Texas until after the Manzano autumn monitoring season ends in early November, but those birds do move back north during the standard Sandias monitoring period of late February through early May. Another possibility is that the eagles may follow different routes during spring and autumn. It may be that proportionately more birds move gradually south along the fringes of the Great Plains in autumn away from the Manzano–Sandia flight line, but then may move north more expeditiously in spring by following the main Rocky Mountain corridor where they can take advantage of montane updrafts. Counts of Turkeys Vultures also average roughly 3.5 times higher in the Sandias than in the Manzanos, again perhaps due to variations in the flight lines the species chooses to use during the different seasons (Bildstein 2006).

Range Connections, Migration Corridors, and Concentration Areas in New Mexico

A combination of seemingly ideal montane features led HWI to explore the Sandia and Manzano mountains of central New Mexico for a possible raptor highway in the early 1980s. The Manzano Mountains, in particular, are a relatively isolated, narrow, and well-defined north-south range that creates beneficial updrafts and serves as a distinct flight path for migrating raptors to follow (fig. 2.1). The specific watch site at Capilla Peak atop the crest of the Manzanos lies near the southern end of the range where its concentration effect for southbound migrants is maximized. Besides the wind-driven updrafts that the ridge routinely creates, during calmer periods Capilla Peak itself and two other peaks nearby to the north provide excellent sources of thermal lift arising from heating of the exposed rocky surfaces.

During autumn, the logical place to look for major migratory concentrations is at the southern end of long, leading-line ranges (or other relevant landscape features) where the concentration effect of the feature is maximized. In contrast, the northern ends of such features are the logical places to look for potential spring concentration points. For this reason, the Sandia Mountains monitoring site lies just north of the Manzano range so as to capture the maximum concentration of northbound spring migrants that have moved up along the Manzanos. The site also lies just north of where migrants must leave the Manzano range and cross the east-west-trending expanse of Tijeras Canyon (the I-40 corridor) where the availability of favorable wind-driven and thermal updrafts subsides temporarily. Like Capilla Peak in the Manzanos, the rocky "shields" that comprise the upper crest of the Sandias and lie directly above the count site are a prime target for migrants that need to regain lift after losing altitude across Tijeras Canyon. The shields routinely provide a great source of either wind-driven or thermal updrafts, depending on conditions.

Band-return data collected since 1985 for several species ($n = 106$: 44% Cooper's Hawks, 23% Sharp-shinned Hawks, 14% Red-tailed Hawks, 4% American Kestrels, 4% Peregrine Falcons, 3% Northern Goshawks, and 1% each of Golden Eagles, Merlins [*Falco columbarius*], and Prairie Falcons) and satellite-tracking data for Red-tailed Hawks and Golden Eagles suggest that the Sandia-Manzano flight corridor lies at the apex of a large funnel that collects southbound migrants from the central and eastern Rocky Mountains of Montana and Wyoming (and farther north), filters them through the mountains of northeastern Utah and Colorado, and draws them together into northern New Mexico as the San Juan Mountains converge from the northwest and the Sangre de Cristo Mountains converge from the northeast along the prominent Rio Grande corridor (figs. 2.3–2.6; Hoffman et al. 2002; HWI 2009a).

Some migrants that drop into New Mexico along the Sangre de Cristo range may continue straight south along a line that leads through the Sacramento Mountains. This path requires traversing a broad expanse of

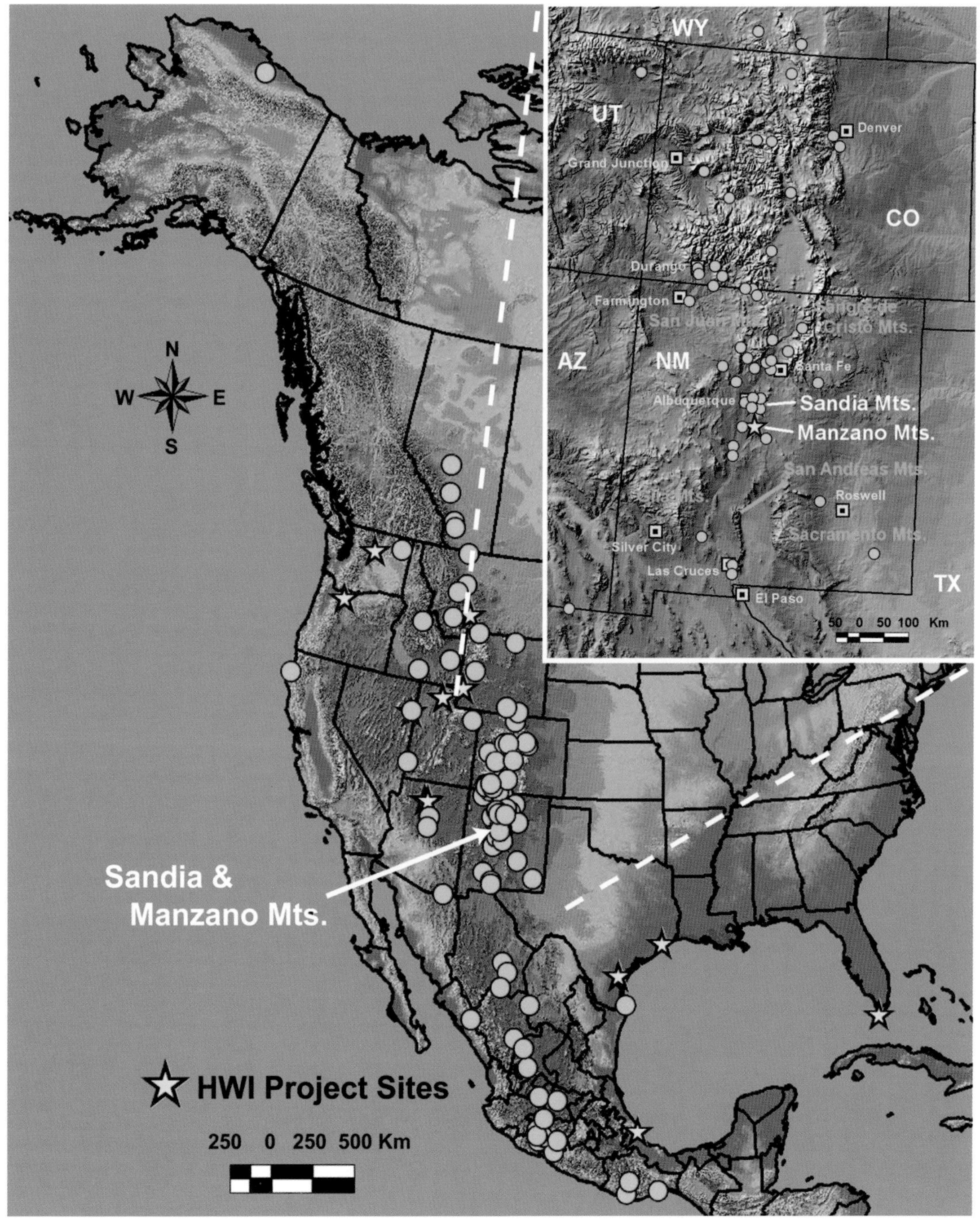

FIGURE 2.3

Distribution of band-return locations for raptors (all species combined) banded during migration in the Sandia and Manzano mountains, New Mexico, between 1985 and 2006.

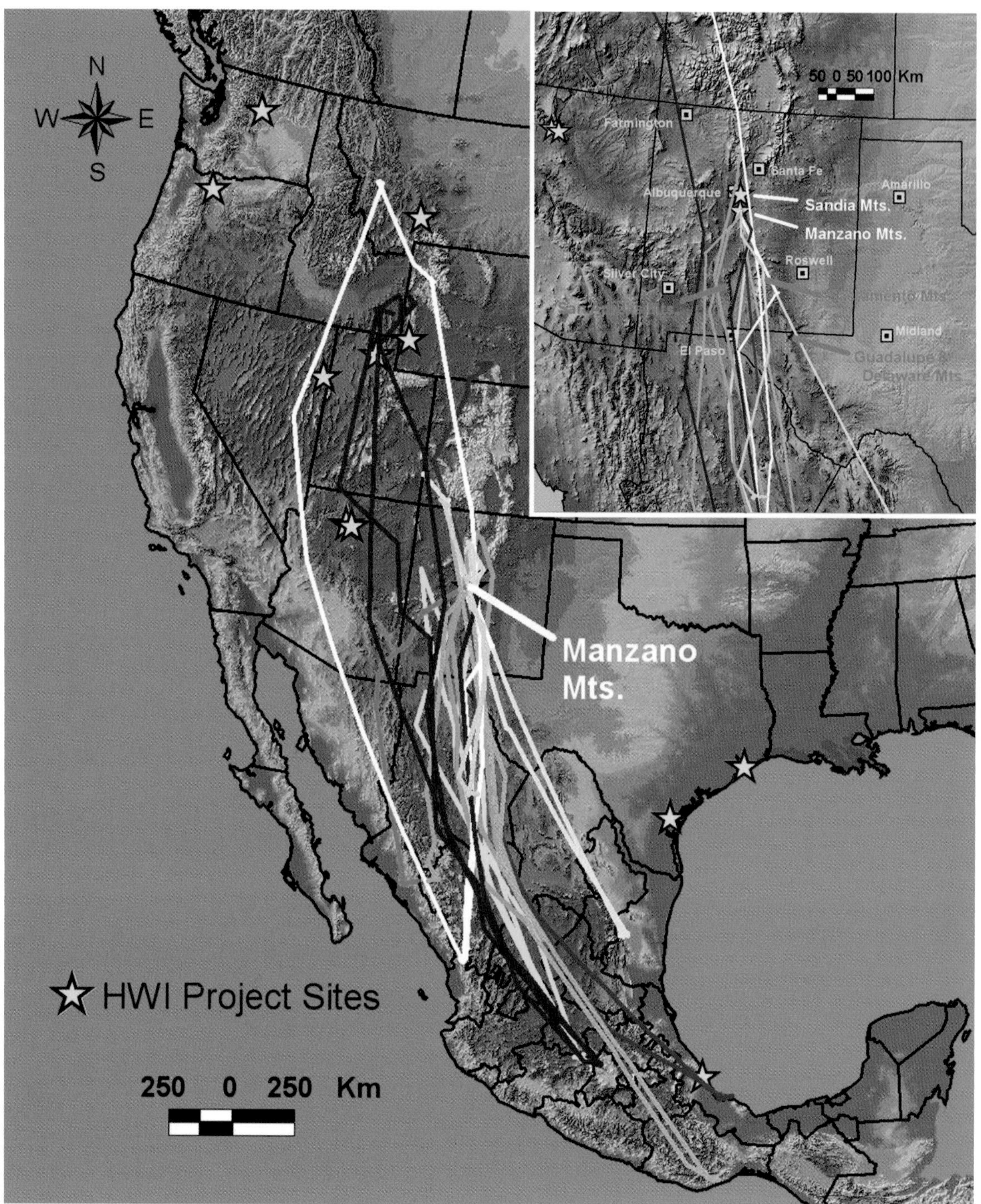

FIGURE 2.4

Tracking results for 11 Red-tailed Hawks (*Buteo jamaicensis*; each track color represents a different bird) outfitted with satellite transmitters during autumn migration in the Manzano Mountains, New Mexico, between 1999 and 2002. Inset displays fall migration tracks only.

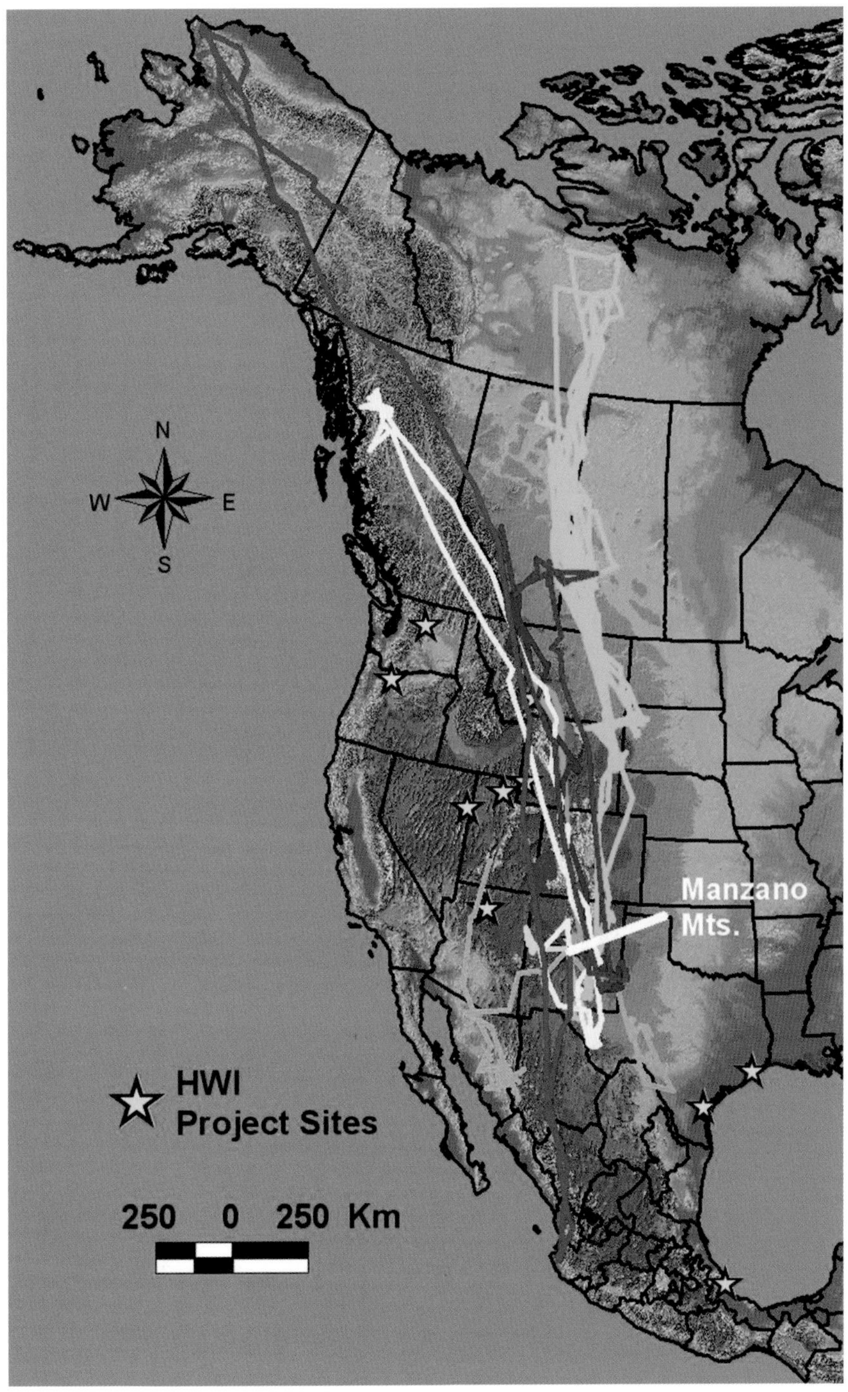

FIGURE 2.5

Tracking results for 11 Golden Eagles (*Aquila chrysaetos*; each track color represents a different bird) outfitted with satellite transmitters during autumn migration in the Manzano Mountains, New Mexico, between 2001 and 2005.

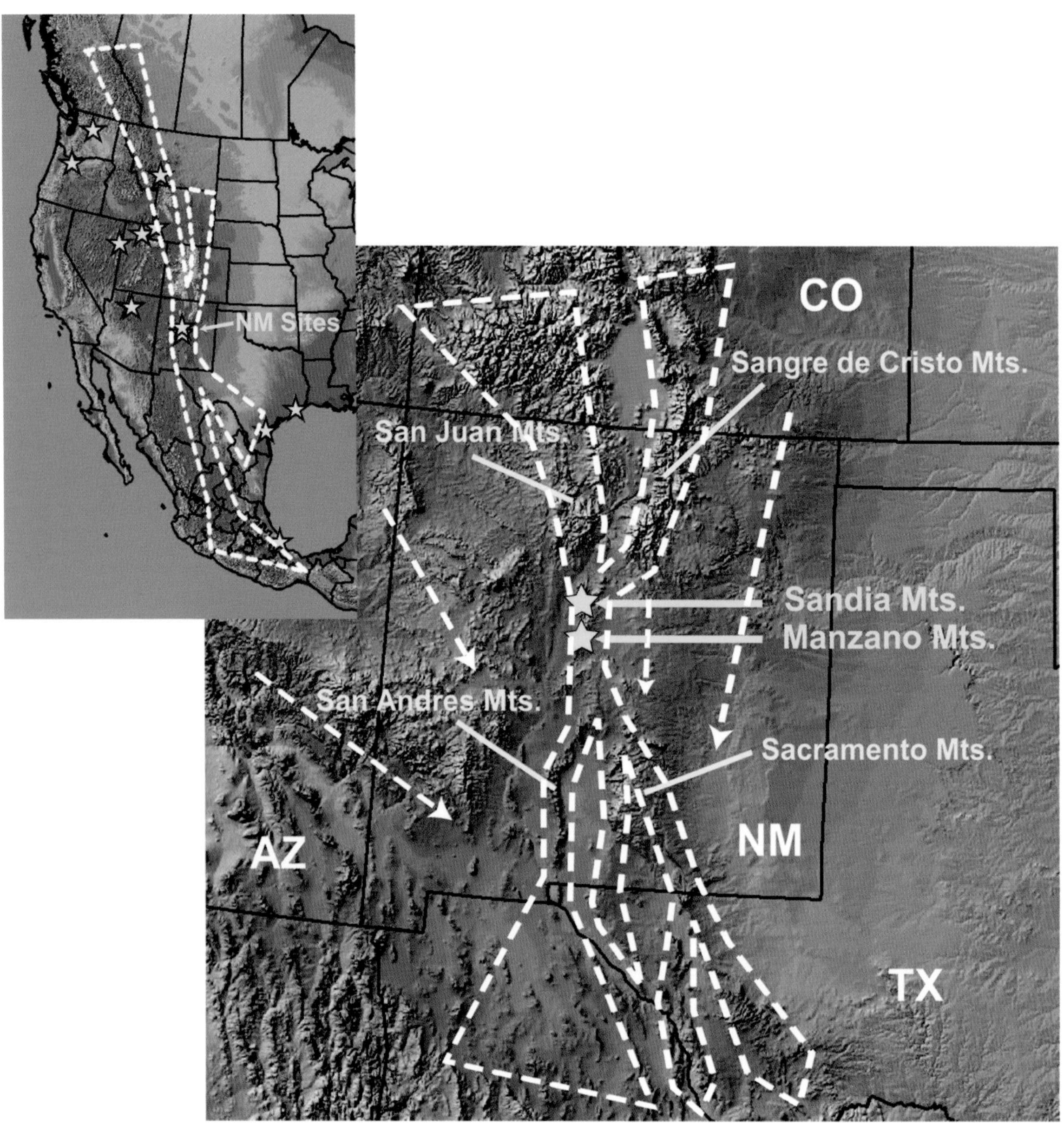

FIGURE 2.6

Flight paths and corridors used by raptors migrating through New Mexico, emphasizing pathways leading through the Sandia and Manzano mountains migration study sites.

TABLE 2.3. Average annual raptor migration counts at 12 full-season, autumn monitoring sites in the western United States and Canada.

SITE	YEARS	ACCIPITERS	BUTEOS	EAGLES	FALCONS	VULTURES	OTHER	TOTAL
Marin Headlands, CA[1]	1995–2004	6,429	9,984	23	914	9,177	1,919	29,256
Goshute Mts., NV	1983–2005	8,214	3,478	271	2,050	339	385	14,738
Boise Ridge, ID[2]	1995–2005	2,219	1,162	59	1,293	992	392	6,117
Lipan Pt., AZ	1991–2005	2,751	1,720	46	1,182	—[3]	224	6,043
Manzano Mts., NM	1983–2005	2,657	1,255	122	661	396	135	5,226
Yaki Pt., AZ	1997–2005	2,918	1,097	31	886	—[3]	111	5,042
Mt. Lorette, ALB[4]	1993–2001	331	137	3,699	31	<1	27	4,225
Commissary Ridge, WY	2002–2005	1,604	1,171	405	368	110	124	3,781
Wellsville Mts., UT	1987–2004	1,515	832	174	857	24	340	3,743
Bonney Butte, OR	1994–2005	1,566	664	146	104	302	123	2,904
Bridger Mts., MT	1992–2005	568	163	1,560	112	1	85	2,488
Chelan Ridge, WA	1998–2005	1,158	412	135	124	31	279	2,139

[1] Data provided by the Golden Gate Raptor Observatory, San Francisco, CA. Due to complicated flight dynamics and customized counting protocols, "counts" at this site represent activity indices rather than precise counts of individuals.

[2] Data provided by the Idaho Bird Observatory, Boise, ID.

[3] Vultures are not tallied at these sites because of difficulties separating migrant and resident birds.

[4] Data provided by the Rocky Mountain Eagle Research Foundation, Calgary, Alberta.

relatively featureless terrain before reaching the primary high-mountain portion of the Sacramento range in southern New Mexico (fig. 2.6). Alternatively, only a relatively short jump across open terrain is required for birds to continue along the prominent Manzano range to the southwest, therefore increasing the chance of a more substantial convergence of migrants along this relatively isolated range. Once migrants reach the southern end of the Manzanos, most probably continue to the southeast along the Sierra Oscura Mountains. Then they are again faced with two options for continuing their journey along prominent montane leading lines. The first path continues south roughly following the Rio Grande corridor and along the San Andres Mountains, eventually feeding through the Franklin Mountains in far western Texas (fig. 2.6). The second diverts more to the southeast and follows down through the Sacramento Mountains and then along the western margins of the Diablo Plateau (Guadalupe and Delaware mountains complex) through west Texas.

Exploratory surveys confirmed at least modest autumn and spring flights through the Franklin Mountains near El Paso, Texas (Kiseda 2005). Band returns from Sandia and Manzano migrants further confirm at least some use of the Rio Grande and San Andres–Franklin Mountains corridor (fig. 2.3). Satellite-tracking data for Red-tailed Hawks also indicate use of this pathway and more generally the Rio Grande corridor, but suggest proportionately greater use of the Sacramento Mountains corridor, which leads more directly along the western flanks of the Sierra Madre Oriental into central Mexico (fig. 2.4; HWI 2009a). Thus, although no specific raptor migration monitoring has occurred along this latter corridor, nor along the San Juan or Sangre de Cristo ranges in northern New Mexico, substantial migration undoubtedly occurs along each of these montane corridors. Nevertheless, the particular combination of leading-line convergences and other favorable characteristics renders the migratory concentrations in the Manzano Mountains noteworthy. Among 12 similar monitoring projects in the western

PHOTO 2.7

Immature Golden Eagle (*Aquila chrysaetos*) fitted with satellite transmitter at the Manzano Mountains monitoring site, fall 2003. PHOTOGRAPH: © JOHN P. DELONG.

PHOTO 2.8

The same eagle being released. Together with the analysis of band returns, satellite-tracking is an important tool for better understanding raptor migration routes. PHOTOGRAPH: © JOHN P. DELONG.

United States and Canada, the Manzanos flight consistently ranks in the top five or six for overall abundance of migrants (table 2.3).

Band return data for Sharp-shinned and Cooper's Hawks, American Kestrels, and Red-tailed Hawks suggest that for these moderate-distance migrants the main Sandia-Manzano flight corridor continues south into Mexico following the passage between the Sierra Madre Oriental to the east and the Sierra Madre Occidental to the west, eventually spilling forth and spreading out across winter ranges in far southern Mexico (fig. 2.3). By contrast, migrants of these species that follow the Intermountain Flyway down through the Great Basin tend to funnel down into Mexico along the western flanks of the Sierra Madre Occidental and winter along the upper and central west coast of Mexico (Hoffman et al. 2002). Band returns and satellite tracking of Red-tailed Hawks and Golden Eagles also indicate that some Manzano migrants veer farther southeast and either end up in western Texas or continue down along the eastern flanks of the Sierra Madre Oriental and into eastern Mexico (figs. 2.4 and

2.5). Other tracking studies of long-distance migrants bound for Central and South America further confirm a major continental convergence of eastern, midwestern, and western flight lines through Veracruz along the southeast coast of Mexico (Fuller et al. 1998; Martell et al. 2001; also see Inzunza et al. 2000; Bildstein and Zalles 2001).

Fair numbers of migrants likely move through western New Mexico as well, but most likely in a relatively dispersed fashion. The landscape of western New Mexico is topographically complex and presents no prominent leading lines for migrants to concentrate along. Therefore, much as proved true upon exploring western Colorado for possible concentration points suited to long-term monitoring (Harrington 1997), whatever migration does occur through western New Mexico is likely to be broadly dispersed across a variety of dynamic pathways. Similarly, the relatively open, prairie landscapes of eastern New Mexico and other broad grassland valleys such as the Estancia Valley just east of the Manzano range are both popular summering and wintering areas and undoubtedly attract significant migratory and transient concentrations of a variety of raptor species. Swainson's Hawks, Prairie Falcons, Ferruginous Hawks, Golden Eagles, and Northern Harriers (*Circus cyaneus*) all frequent such habitats during most times of the year. Based on more than a decade of annual monitoring at Dinosaur Ridge west of Denver, Colorado, we also know that a significant spring migration of a variety of species occurs along the eastern front range of the Rocky Mountains, with this site particularly well known for relatively large counts of Ferruginous Hawks (Rocky Mountain Bird Observatory, pers. comm.). Canadian band-return data and more recent satellite tracking of Ferruginous Hawks outfitted with transmitters on breeding ranges in eastern Alberta, Saskatchewan, and eastern Wyoming also confirm substantial movements of this species along the plains fringing the eastern margins of the Rockies from Canada all the way down into eastern Mexico (Schmutz and Fyfe 1987; Watson and Banasch 2005). Similarly, satellite tracking of Swainson's Hawks outfitted on a variety of breeding ranges in the West demonstrated movement through New Mexico both along the eastern Rockies and fringing plains, and along a track leading down into New Mexico from the Intermountain region and western Rockies, with both tracks and others ultimately converging in southeastern Mexico (Fuller et al. 1998).

Species Representation and Relative Abundance in the Manzano and Sandia Mountains

The four most common diurnal raptor species seen at most western migration sites are the Sharp-shinned Hawk, Cooper's Hawk, Red-tailed Hawk, and American Kestrel (Hoffman and Smith 2003). Though mostly true in the Sandia and Manzano mountains, a few noteworthy variations are apparent (table 2.2). Turkey Vultures average two to three times more abundant than any of these species and Golden Eagles are roughly as abundant as kestrels and red-tails during spring at the Sandias. Additionally, the long-term average count for Swainson's Hawks in the Manzanos is comparable to that of kestrels. At least northern populations of the first four species are highly migratory and typically move hundreds if not thousands of kilometers between summer and winter ranges each year (Preston and Beane 1993; Rosenfield and Bielefeldt 1993; Bildstein and Meyer 2000; Smallwood and Bird 2002). Moreover, all are primarily forest- or woodland-dwelling species as breeders (though in some areas red-tails and kestrels may occupy very open habitats as long as suitable isolated trees or rocky outcrops are available for nesting substrate) and likely rely on montane migration routes, especially in the West, because they provide favorable forested habitat corridors for migrants to follow, and these species also readily exploit, if not rely upon, the energy-saving lift afforded by mountain ridges. The only other species whose counts average more than 100 birds per season at both sites are the Turkey Vulture and Golden Eagle, both proportionately much more abundant in spring at the Sandias than in autumn at the Manzanos. At the Manzanos, the long-term average count for Swainson's Hawks also is more than 500 birds and counts have exceeded 100 birds in 9 of 21 years; however, the true "trademark" for this species is extreme variability. Counts of Swainson's Hawks in the Manzanos have ranged from a low of 3 birds in 1988 to more than 5,000 birds in 45 minutes one evening in 1993, and to more than 7,000 birds during the 2006 season! Such variability reflects the species' highly gregarious nature outside of the breeding season, and the fact that its migrations are not strongly tied to mountain ridgelines but under the right conditions may converge along such pathways (e.g., when strong east winds blow big flights, which otherwise would remain over the eastern prairies, up onto the central mountain ranges of New Mexico).

With many subtle variations, similar patterns of species-specific proportional abundance apply at most other monitoring sites from the Rocky Mountains westward (e.g., see Hoffman and Smith 2003). Migrating Golden Eagles, however, tend to be most abundant along the Rocky Mountains, with average autumn counts ranging from several thousand birds in central Alberta (Sherrington 2003) to 1,500–2,000 birds in southwestern Montana (Hoffman and Smith 2003), 200–300 birds in western Wyoming (Smith and Neal 2006), and finally 75–200 birds in New Mexico (table 2.2). The progressive diminishment of Golden Eagle numbers with decreasing latitude in autumn undoubtedly reflects the fact that migratory eagles moving down out of Alaska and northern Canada along the Rocky Mountains begin fanning out onto open rangelands as soon as they reach southern Canada and especially once they reach Montana and Wyoming (Kochert et al. 2002; McIntyre et al. 2006). Again, however, a Sandias spring count that averages three times higher than the Manzanos autumn count suggests that more eagles typically end up wintering as far south as southern New Mexico and Texas than is suggested by the Manzanos count, most likely because the species' southward movements continue well after snowfall curtails the Manzanos count in early November.

Compared to the aforementioned species, relatively low counts are recorded for most other species that are commonly observed at western monitoring sites like the Sandias and Manzanos (i.e., Northern Harriers; Ospreys; Bald Eagles; Ferruginous, Broad-winged and Rough-legged Hawks; and falcons other than American Kestrels). For species such as the Broad-winged Hawk, low counts reflect simple comparative rarity. This species is one of the most abundant migrants in the East and continuing down through Texas, eastern Mexico, and Central America, but it is rare in the West (Goodrich et al. 1996). Broad-wings appear to be expanding their breeding range westward in Canada, however, and are showing increasing trends at migration sites throughout the West (Smith et al. 2001b; Hoffman and Smith 2003).

Ospreys and Bald Eagles also are inherently less common than many other species, especially in arid regions, because of their reliance on aquatic environments (Buehler 2000; Poole et al. 2002). Moreover, although sightings of migrating Bald Eagles may occur as early as late August, significant movements generally do not begin until mid to late October (see below; see also Smith and Neal 2006 and other site-specific reports available at www.hawkwatch.org) and continue long after most mountaintop migration sites in the West are shut down due to heavy snowfall. For example, the 400–500 Bald Eagles that winter around the Great Salt Lake each year do not begin to amass in earnest until late November (Wilson 1999). The late-season-migrant effect also contributes to low counts of Rough-legged Hawks in the Manzanos (only five migrants recorded in 21 years), as again significant numbers of this species typically do not begin appearing in the lower 48 states until mid to late October and into November (see below; Bechard and Swem 2002). Additionally, although scattered Rough-legged Hawks are seen most years as far south as coastal Texas and the winter range of northern-migrant Bald Eagles may extend into Mexico, at the latitude of the New Mexico sites northern species such as these tend to be inherently uncommon (e.g., only 12 Rough-legged Hawks have been recorded as migrants at the Sandias since 1985; also see Buehler 2000; Bechard and Swem 2002).

Comparatively low counts of the three larger falcons (Merlins and Prairie and Peregrine Falcons) at mountaintop migration sites like the Sandia and Manzano mountains likely reflect a combination of three primary factors: (1) inherently low overall relative abundance across the continent; (2) greater preference for lowland, open-country habitats during migration and winter; and (3) strong powered-flight capability that obviates heavy reliance on energy-saving thermal or mountain updrafts to facilitate long-distance movements (Sodhi et al. 1993; Steenhof 1998; White et al. 2002). In other words, in areas like New Mexico the chance of seeing migrating large falcons is as great or greater out over many open shrubland, prairie, and marshland habitats or along major river corridors.

PHOTO 2.9

A rare Harlan's Red-tailed Hawk (*Buteo jamaicensis harlani*) captured, banded, and released at the Sandia Mountains monitoring site, spring 1996.

PHOTOGRAPH: © JOHN P. DELONG.

Otherwise, many of the most significant migratory and wintering concentrations of Merlins and Peregrine Falcons occur along coastlines or along the shores of large inland water bodies like the Great Lakes, where aggregated shorebirds and waterfowl offer readily available food sources (Bildstein 2006).

Diel Migration Timing

Under typical conditions, most western monitoring sites show similar overall daily activity patterns, with activity rising quickly during mid-morning, remaining high through midday, and then gradually tapering off as evening approaches. This pattern generally tracks availability of wind currents and thermals produced during the warmest parts of the day. In the Manzano Mountains, the overall combined-species diel activity pattern follows a fairly typical unimodal pattern, but is a bit skewed to later afternoon than is true at other western sites (fig. 2.7). Examination of species-specific data revealed two dissimilar patterns. The first group of six species (Northern Harrier, the three accipiters, Ferruginous Hawk, and American Kestrel) showed peak activity from about 1000 to 1300 hours Mountain Standard Time (MST), with activity tapering off gradually thereafter (fig. 2.7). In contrast, the common

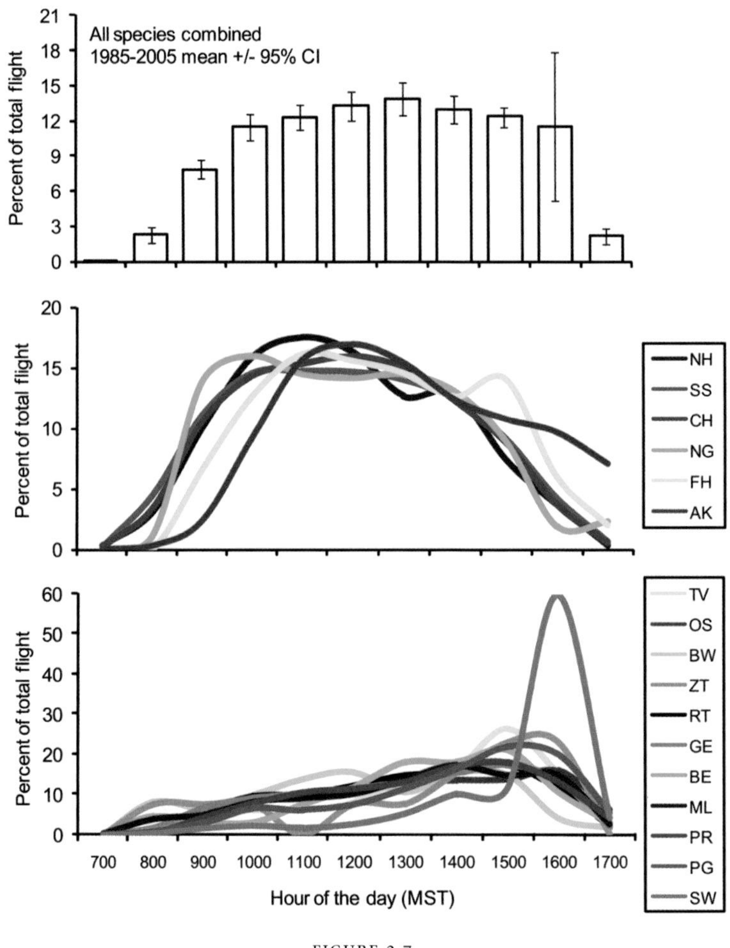

FIGURE 2.7

Aggregate (1990–2006) diel activity patterns for migrating raptors during autumn in the Manzano Mountains, New Mexico. Note that the area under each curve sums to 100%. See Table 2.2 for explanation of species codes.

pattern for all other species was gradually increasing activity through early afternoon, peak activity during the 1400–1700-hour period, and a quick tapering off thereafter.

A different overall activity pattern applies during spring in the Sandia Mountains, showing a bimodal distribution with activity peaks during both mid-morning and mid-afternoon and a slight lull during midday (fig. 2.8). In this case, species could be categorized into three distinct groupings. The first group of eight species showed peak activity during mid- to late morning and included the same six species in the Manzanos morning group plus Broad-winged and Swainson's Hawks (fig. 2.8). At the opposite extreme, Golden and Bald Eagles, Merlins, and Peregrine Falcons showed peak activity during late afternoon, similar to the pattern shown for these species in the Manzanos. The remaining six species showed more variable and/or more broadly distributed activity patterns in the Sandias.

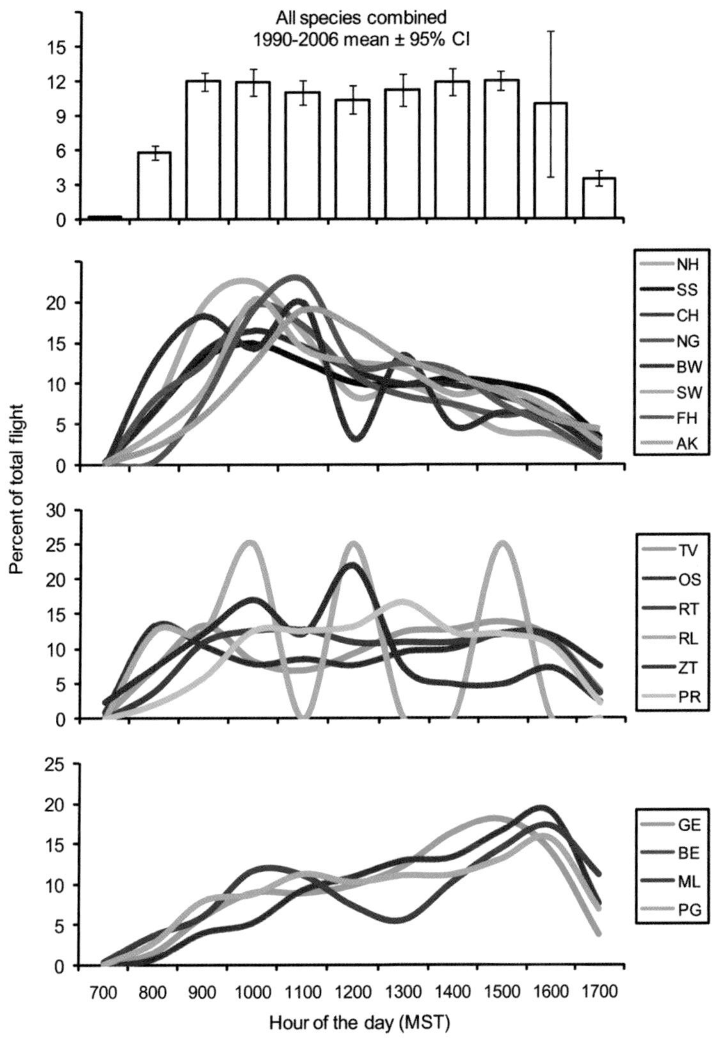

FIGURE 2.8

Aggregate (1990–2007) diel activity patterns for migrating raptors during spring in the Sandia Mountains, New Mexico. Note that the area under each curve sums to 100%. See table 2.2 for explanation of species codes.

Exactly why some species tend to favor late-afternoon movement is uncertain, but at least for the eagles most likely reflects capitalizing on maximum thermal lift. Previous studies indicated that Bald Eagles typically migrate during the warmest parts of the day when thermal lift is maximized (Harmata 1984; Hunt et al. 1992). Similarly, at least adult Golden Eagles are most apt to move through the Bridger Mountains of Montana during the afternoon when thermal lift is generally maximized (Omland and Hoffman 1996). In the Manzano Mountains during October and in the Sandia Mountains during March—peak eagle-migration seasons at the two sites—peak daily temperatures tend to occur from 1300 to 1600 hours, which means that this period of the day is often characterized by maximum thermal production or, alternatively, favorable winds. Explaining why the two falcons tend to fly late is more difficult, since neither species relies heavily on thermal lift to facilitate their largely powered-flight migrations (Sodhi et al. 1993; White et al. 2002). Nevertheless, Clark (1985) indicated that late-afternoon soaring flights are commonly observed among Merlins migrating through Cape May, New Jersey, and White et al. (2002) comment that the Peregrine Falcon is "considered a low-altitude migrant . . . but soars more than generally recognized." So perhaps late-afternoon passage of these two species at mountaintop migration sites also reflects exploitation of late-afternoon thermals or unique wind conditions.

Seasonal Migration Timing

The standardized spring monitoring period in the Sandia Mountains extends from 24 February through 5 May, and the autumn season in the Manzano Mountains runs from 27 August through 5 November (Hoffman and Smith 2003). These periods are designed to capture the majority of activity for all commonly occurring species, constrained by site-access issues pertaining to excess snow cover. Examination of aggregate species-specific passage data for the >20-year periods of record at each site shows obviously truncated distributions for only a few species, primarily Golden and especially Bald Eagles in both seasons (figs. 2.9 and 2.10). Only a few other species show as much as 5–10% of average activity during the first or last five days of the Sandia and Manzano monitoring seasons.

Most species show unimodal or hill-shaped seasonal activity patterns, but the specific timing and extent of each species' passage period varies (figs. 2.9 and 2.10). The 95% bulk-passage periods (i.e., the dates between which the central 95% of aggregate sightings occurred) for some species are relatively narrow, typically reflecting gregarious behavior (e.g., 30 days for Swainson's Hawks and 33 days for Broad-winged Hawks during autumn). Movements of other species may extend over much broader periods (e.g., 60 days for Red-tailed Hawks and 63 days for Northern Harriers in autumn). Species whose peak activity periods tend to occur earlier in autumn generally pass through later in the spring, with most such species being those that migrate in large numbers well down into Central and South America for the winter (e.g., Turkey Vultures, Ospreys, Broad-winged Hawks, and Swainson's Hawks). Conversely, classic late-season autumn species such as Rough-legged Hawks and Golden and Bald Eagles usually are the first to return north in the spring. Most individuals of these species winter north of Mexico, and the species' summer ranges extend to high northern latitudes. With data for all species combined, the central 50% peak-activity periods are 30 March through 14 April in the spring, and 18 September through 3 October in autumn.

Characteristic patterns of age- and sex-specific variation in seasonal passage timing also are evident. Among eight species with easily distinguished age classes and that occur in sufficient numbers to yield useful data, average median passage dates (i.e., date by which 50% of the flight passes by) for adults are 3–16 days later than for immatures of the same species in autumn, and are 3–25 days earlier than for younger birds in the spring (table 2.4, and see for example fig. 2.11). Routine sexing of migrants in flight is possible only for American Kestrels and Northern Harriers, with no simultaneous age distinctions possible for kestrels and sexing restricted to distinguishing among adult harriers. The median passage date for male kestrels is three days later than for females in autumn, and one day later than for females in the spring. Banding data, which include both age and sex distinctions for kestrels, further suggest that in autumn the peak capture period for immature birds is slightly earlier than for adults, and that within age

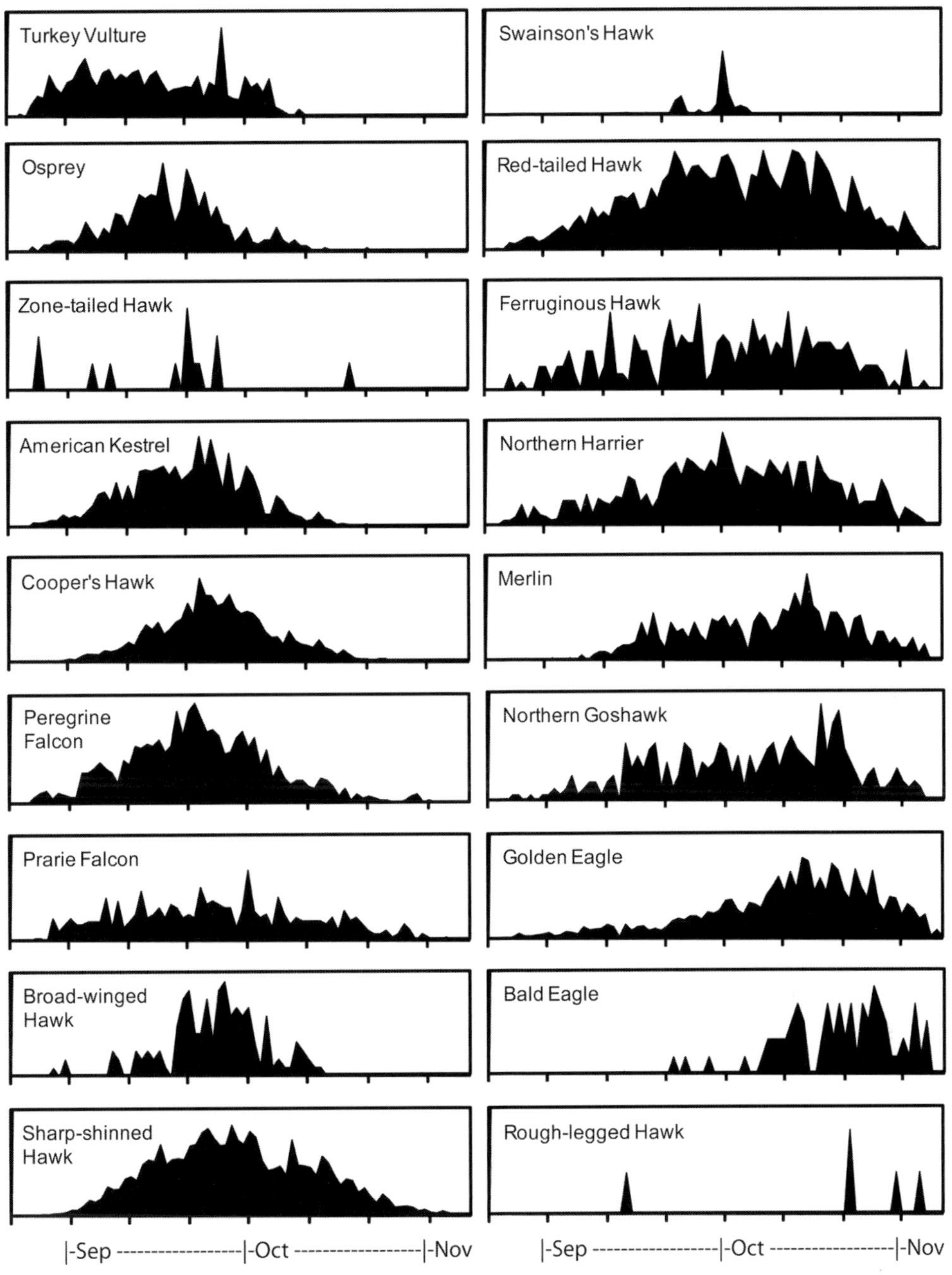

FIGURE 2.9

Aggregate seasonal activity patterns (proportion of total birds counted from 1985 to 2006 by Julian date) for migratory diurnal raptors during autumn migration in the Manzano Mountains, New Mexico.

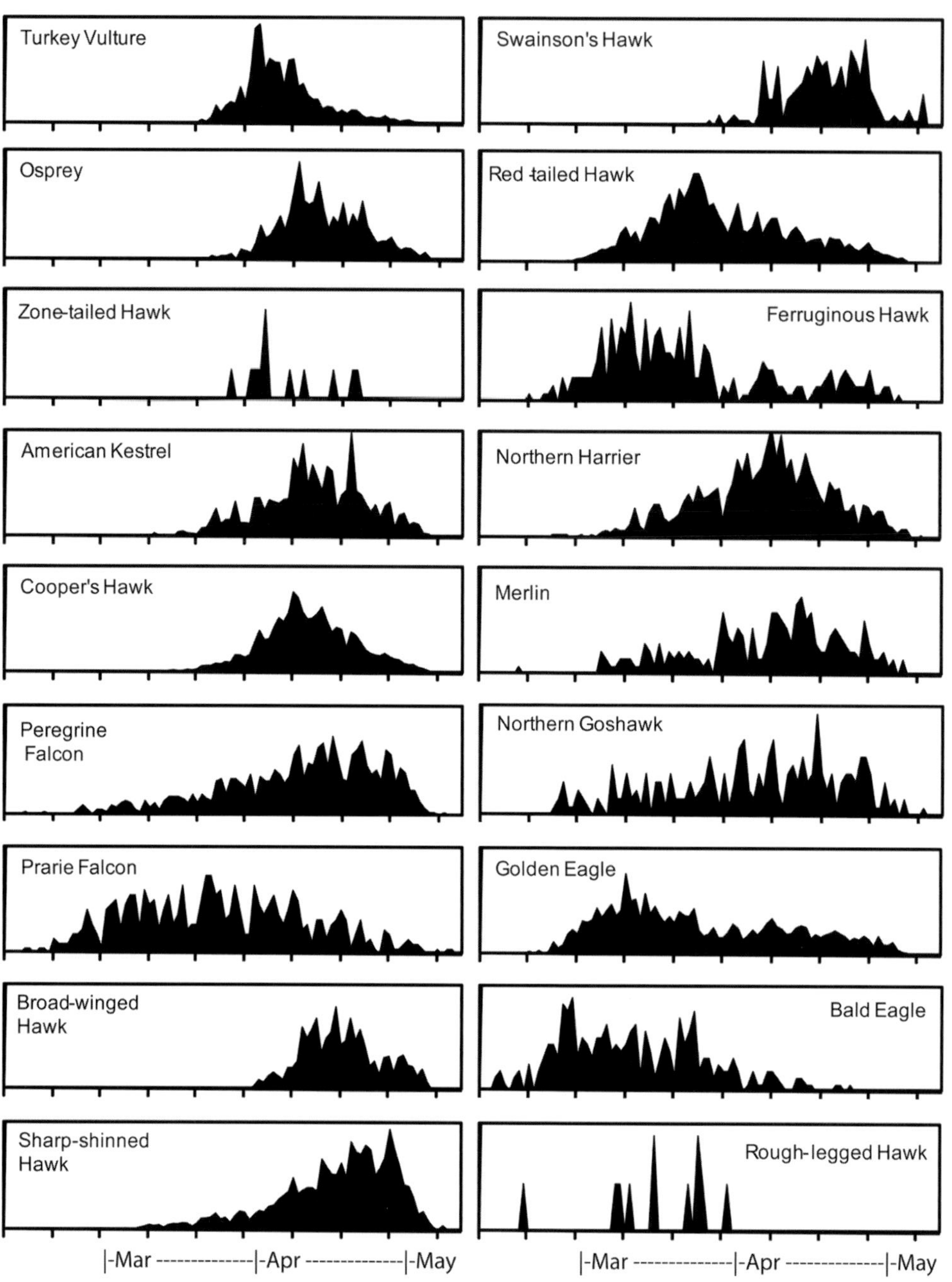

FIGURE 2.10

Aggregate seasonal activity patterns (proportion of total birds counted from 1985 to 2007 by Julian date) for migratory diurnal raptors during spring in the Sandia Mountains, New Mexico.

TABLE 2.4. Age-specific median passage dates for migrating raptors in the Manzano and Sandia mountains, New Mexico.

	MANZANO MTS.						SANDIA MTS.					
	ADULT			IMMATURE			ADULT			IMMATURE		
Northern Harrier	8 Oct	±	4.3[1]	30-Sep	±	2.3	04-Apr	±	2.4	09-Apr	±	7.8
Sharp-shinned Hawk	5-Oct	±	1.5	19-Sep	±	1.5	17-Apr	±	2.4	23-Apr	±	5.0
Cooper's Hawk	28-Sep	±	2.3	21-Sep	±	2.1	09-Apr	±	1.3	20-Apr	±	2.2
Northern Goshawk	5-Oct	±	4.1	2-Oct	±	6.9	28-Mar	±	5.2	20-Apr	±	5.7
Red-tailed Hawk	7-Oct	±	2.3	26-Sep	±	1.9	23-Mar	±	1.3	17-Apr	±	2.4
Ferruginous Hawk	4-Oct	±	8.6	25-Sep	±	6.1		–			–	
Golden Eagle	15-Oct	±	2.4	12-Oct	±	1.7	10-Mar	±	1.6	01-Apr	±	4.8
Bald Eagle		–		16-Oct	±	6.9	07-Mar	±	4.8	11-Mar	±	6.5
Peregrine Falcon	25-Sep	±	2.3	17-Sep	±	3.4	10-Apr	±	3.3	13-Apr	±	4.7

[1] The median passage date is the date by which the first 50% of a season's migrants pass through. Values given are means of annual values (1985–2005) ± SD in days.

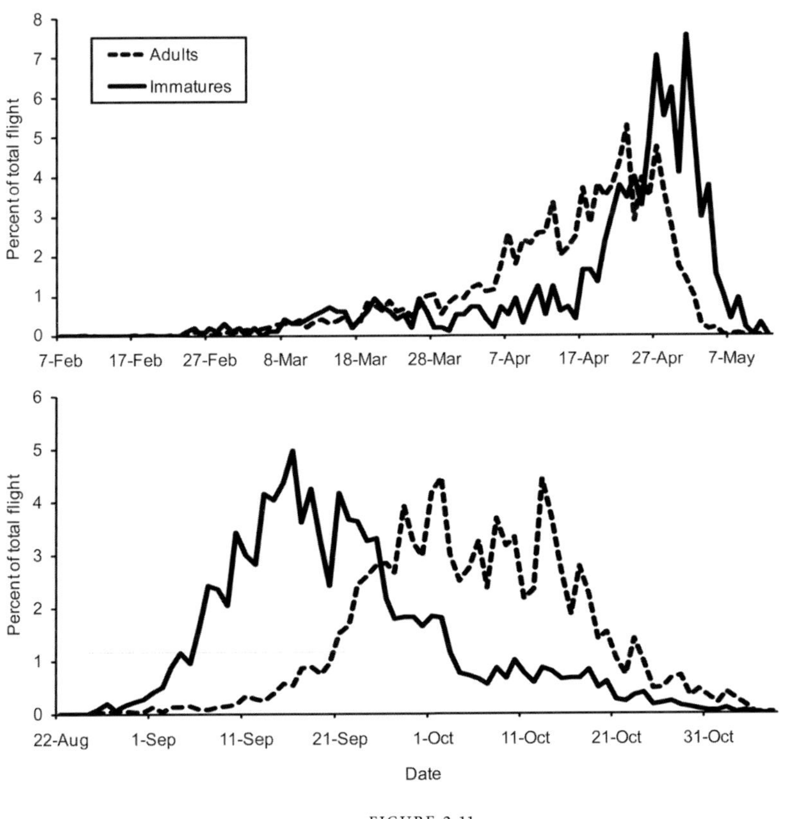

FIGURE 2.11

Aggregate age-specific activity patterns (proportion of all birds counted from 1985 to 2006 within age groups by Julian date) for migrating Sharp-shinned Hawks (*Accipiter striatus*) during spring migration in the Sandia Mountains (upper panel; adults = after-second-year birds, immatures = second-year birds) and during autumn migration in the Manzano Mountains (lower panel; adults = after-hatch-year birds, immatures = hatch-year birds) of New Mexico.

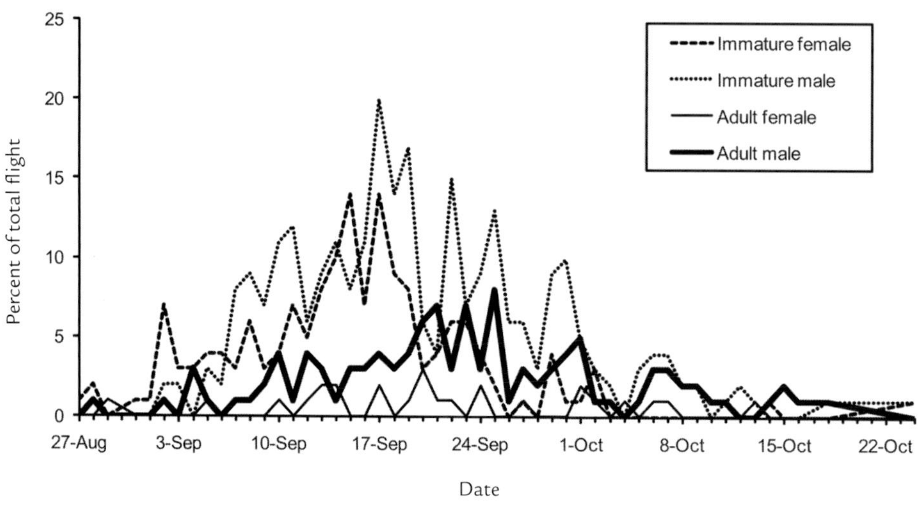

FIGURE 2.12

Aggregate age–sex-specific seasonal activity patterns (total captures from 1990 to 2006 within age-sex groups [adults = after-hatch-year birds, immatures = hatch-year birds] by Julian date) for American Kestrels (*Falco sparverius*) captured during autumn migration in the Manzano Mountains, New Mexico.

groups, the peak capture period for males tends to be slightly later than for females, especially among immature birds (fig. 2.12).

One other novel study conducted using data from the Manzano Mountains migration project sheds light on seasonal migration timing as it relates to representation of different source populations. Combining data derived from analysis of band recovery patterns and hydrogen stable-isotope ratios in migrant feathers, Smith et al. (2003) confirmed a "chain migration" pattern for Sharp-shinned Hawks. That is, the evidence indicated that migrants deriving from northern breeding populations passed through later and wintered farther north than birds deriving from southern breeding ranges. Although such patterns are not well documented for most species (Bildstein 2006), another raptor for which a decent understanding of such dynamics is available is the Peregrine Falcon. Contrasting with the situation for Sharp-shinned Hawks, peregrines often show a "leapfrog migration" pattern, with northern breeders migrating farther, leapfrogging over and wintering farther south than birds originating at more southerly latitudes (White et al. 2002).

Nocturnal Raptors

Data concerning owl migration in the West are comparatively scarce, while possibilities for simultaneous studying and monitoring of multiple species at individual sites are generally limited. Further, only eight species are of interest because of known or possible migratory tendencies. For all three reasons, I present the following information in species-specific fashion rather than trying to integrate understanding in topical fashion as done for diurnal raptors.

Larger Owls

BARN OWL

The migration/dispersal ecology of Barn Owls is poorly known (Marti 1992). Reports by Stewart (1952) and Duffy (1985) suggested that the species is a partial migrant at least in the northern United States. Band returns have also confirmed substantial southward movements among some breeding and nonbreeding adults in Colorado and Texas (Bolen 1978; Millsap and Millsap 1987); however, another long-term study indicated year-round residency for breeding adults in northern Utah (Marti 1988). Moreover, immature Barn Owls are known to disperse over long distances (up to 1,900 km [~1,180 mi]) in just about any direction (Soucy 1980), and such movements may be mistaken for migration (Marti 1992). In addition, possible nomadic behavior is not well documented in the species, but anecdotal evidence from an extensive multispecies raptor nesting study in northwest Utah indicated recent emergence of Barn Owl nesting activity following five years during which no such activity was noted (HWI, unpubl. data). Accordingly, due to a paucity of hard data, whether or not Barn Owls that breed in New Mexico are at all migratory or if any significant true migration through the state occurs is unknown; nevertheless, although limited, the weight of evidence from studies in both North America and Europe suggests that the species typically is nonmigratory but may wander or disperse widely at times (Marti 1992).

MEXICAN SPOTTED OWL

Radio-tracking of Mexican Spotted Owls from breeding areas in the southwestern United States has suggested that most adult birds are year-round residents, but 10–15% of breeding individuals may move up to 50 km (~30 mi) between distinct summer and winter ranges, generally reflecting primarily altitudinal shifts in distribution (Gutiérrez et al. 1995). Most adult movements documented to date through radio-tracking and band recoveries of both Mexican Spotted Owls and closely related California Spotted Owls (*S. o. occidentalis*) have been constrained to within individual mountain ranges; however, one adult female Mexican Spotted Owl, thought to have been a breeder in a New Mexico population, was later recovered dead 187 km (116 mi) away in another New Mexico range (Gutiérrez et al. 1996). In contrast, young Mexican Spotted Owls often move widely during their first and second years, often over great distances and into areas and habitats not normally occupied by adult birds. As a result, adult owls rarely end up breeding in the same mountain ranges where they were fledged (see chapter 35). For example, two females banded as juveniles in the Tularosa Mountains of New Mexico were later documented as probable breeders 22 km (14 mi) away in the Mogollon Mountains of New Mexico, and 56 km (35 mi) away in the San Francisco Mountains of Arizona (Gutiérrez et al. 1996).

LONG-EARED OWL

Specific information about the ecology of Long-eared Owls (*Asio otus*) in New Mexico is scarce in the published literature. In general, Long-eared Owls are known to occur throughout the year across all but the northernmost reaches of the species' breeding range, but substantial annual spring and autumn migrations also are known to occur in eastern North America (Marks et al. 1994). Although band-returns have confirmed substantial movements for a few owls (e.g., Saskatchewan, Canada, to Oaxaca, Mexico), most recoveries of banded Long-eared Owls have occurred in the same state or province where they were originally banded, and several breeders banded in Idaho and Montana were later recaptured at nearby communal winter roosts (Marks et al. 1994). In Europe, Long-eared Owls are known to be nomadic in response to prey fluctuations, but such behavior has not been conclusively documented in North America (Marks et al. 1994). Analyses of passage data from eastern and midwestern North America suggest that Long-eared Owls may utilize similar geographic features (e.g., ridgelines) and respond to weather in similar ways as diurnal migratory raptors (Marks et al. 1994). Several seasons of Flammulated Owl luring and netting during autumn migration in the Manzano Mountains, New Mexico, opportunistically yielded only one to two Long-eared Owls per season (e.g., see DeLong 2003a). Thus, although some New Mexico Long-eared Owls may be migratory, based on knowledge derived from other regions, it seems reasonable to presume that most are probably year-round residents.

SHORT-EARED OWL

Short-eared Owls (*Asio flammeus*) have not been confirmed as breeders as far south as New Mexico (but see chapter 37), but winter throughout the southern states and northern Mexico (Holt and Leasure 1993). Northern populations are generally thought to be highly migratory, but banding data suggest that even mid-latitude breeders often undertake regular north-south migrations (Clark 1975). The species also exhibits substantial nomadic behavior in response to regional prey fluctuations (Holt and Leasure 1993). Accordingly, occurrence of Short-eared Owls during migration seasons and winter among the grasslands of New Mexico is quite likely, though perhaps highly variable due to the influence of periodic nomadism. Otherwise, however, no further specific detail about their movement ecology in the state is currently available.

Smaller Owls

FLAMMULATED OWL

Flammulated Owl migration ecology is poorly understood (McCallum 1994); however, recent advances have begun to yield additional insight (DeLong et al. 2005; DeLong 2006; Stock et al. 2006; Linkhart and Reynolds 2007). Flammulated Owls were once thought to be nonmigratory (Phillips 1942), but the species is now commonly considered a long-distance, north-south migrant (McCallum 1994). HWI began a Flammulated Owl migration study in the Manzano Mountains in 1999. During the first two years of intensive luring and netting (see Reynolds and Linkhart 1984), the project banded 250 Flammulated Owls (DeLong 2000, 2001). No single-site effort has ever resulted in more captures of this species in such a short time, and the project has since revealed much about the morphology and movement ecology of the species (DeLong 2003a, 2003b, 2004, 2006; DeLong et al. 2005). Moreover, this New Mexico study is now complemented by similar HWI research in the Goshute Mountains of northeastern Nevada (Smith 2006), and additional work by the Idaho Bird Observatory atop a ridge near Boise (Stock

PHOTO 2.10

A Flammulated Owl (*Otus flammeolus*) posing on a branch shortly after being banded and released at the Manzano Mountains monitoring site, fall 2001.

PHOTOGRAPH: © JOHN P. DELONG.

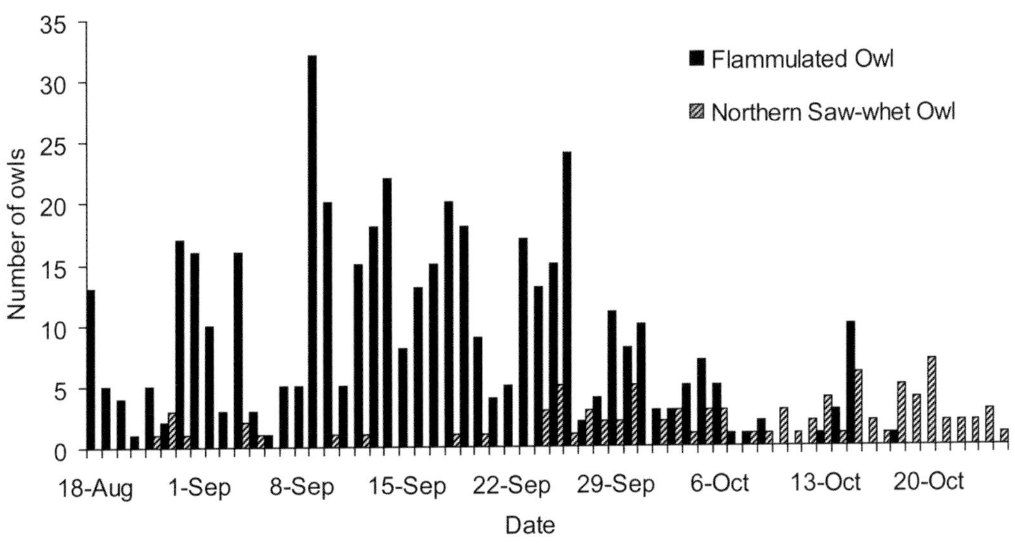

FIGURE 2.13

Aggregate seasonal activity patterns (total captures from 2000 to 2005 by Julian date; all ages included) for Flammulated Owls (*Otus flammeolus*) and Northern Saw-whet Owls (*Aegolius acadicus*) captured during autumn migration in the Manzano Mountains, New Mexico.

et al. 2006). Together these three efforts, along with new population demographic studies by Arsenault et al. (2005) and Linkhart and Reynolds (2006), are quickly increasing our collective knowledge of the demography and migratory characteristics of this little owl.

Across six seasons of netting in the Manzano Mountains, captures of Flammulated Owls occurred from 19 August through 18 October. The aggregate seasonal activity pattern suggests a bell-shaped curve typical of migration passage, with peak activity occurring in mid September and most activity concluded by the first week of October; however, a small disjunct pulse of activity in mid October may be noteworthy (fig. 2.13). Data from several seasons of netting in northeastern Nevada and western Idaho indicate similar bell-shaped seasonal patterns, overall ranges of capture dates, and peak activity periods from mid to late September (Smith 2006; Stock et al. 2006).

Ageing and sexing Flammulated Owls based on morphological characteristics is difficult, but data derived from owls captured in the Manzanos yielded new tools to facilitate both endeavors (DeLong 2003b, 2004). Among 165 owls that could be aged confidently, 89% of those captured at this site by HWI personnel between 2000 and 2005 were immatures (immature : adult ratio of 8.2 : 1). DNA sexing results for 76 owls captured across two seasons yielded a male : female ratio of 1.24, which was not significantly different from 1 : 1 (DeLong 2003b). HWI capture data from Nevada also indicate a high average immature : adult ratio of 3.8, but with a range of annual values from 1.2 to 5.4 (Smith 2006). Sexing of Nevada birds has not yet been possible because significant size variation among birds captured in Nevada and New Mexico precludes application of the discriminant function DeLong (2003b) developed for sexing New Mexico birds (Smith and Neri 2004). Contrary to the New Mexico data, DNA analysis of a sample of 47 owls captured in Idaho indicated a sex ratio that was slightly skewed in favor of females (56% females, 44% males); thus far ageing of Idaho birds has been too limited to yield insight about age-specific details (Stock et al. 2006).

Analysis of New Mexico data revealed significant variation in average capture dates across years (ranging from 9 September in 2002 to 22 September in 2000), but no significant age or sex-specific variation in seasonal timing (3-way ANOVA, model $r^2 = 0.13$; parameter significance evaluated at $P \leq 0.05$; also see DeLong

TABLE 2.5. Measurements (mean ± SD) of Flammulated Owls (*Otus flammeolus*) captured during autumn migration in the Goshute Mountains of northeastern Nevada and the Manzano Mountains of central New Mexico.

MEASURE	GOSHUTE MOUNTAINS (n = 23–91)	MANZANO MOUNTAINS (n = 405–406)
Mass (g)[a]	59.0 ± 6.2	53.2 ± 6.2
Wing chord (mm)[a]	132.4 ± 4.0	129.6 ± 3.5
Standard tail length (mm)[a]	61.5 ± 4.4	61.0 ± 2.7
Tarsus length (mm)[b]	23.1 ± 1.2	22.6 ± 1.2
Hallux length (mm)[c]	6.3 ± 0.4	6.2 ± 0.4
Culmen length (mm)[a]	9.4 ± 0.8	8.7 ± 0.6

[a] t-test, $P < 0.01$.

[b] t-test, $P = 0.05$.

[c] t-test, $P = 0.06$.

[2003a] and DeLong et al. [2005]). Analysis based on a limited, one-season sample of birds sexed by DNA in Idaho also revealed no significant difference in sex-specific timing (Stock et al. 2006). Nevada data also indicate no obvious age-specific timing differences (Smith 2006). However, a nesting study in Colorado suggested that males are more likely to return first to their breeding territories in the spring (Reynolds and Linkhart 1987).

Analysis of hydrogen stable-isotope ratios in feathers collected from 57 hatch-year owls in the Manzano Mountains suggested that the majority of owls captured at the site derive from relatively local origins in northern and central New Mexico and southern Colorado; however, signatures from two birds indicated the possibility of more northerly origins, potentially extending as far north as Washington (DeLong et al. 2005). Preliminary investigations of size differences revealed that owls captured during autumn migration in New Mexico consistently averaged smaller than owls captured in Nevada (table 2.5). These data appear to match the smooth cline of increasing wing length and mass from southeast to northwest across the species' range noted by Marshall (1967). This pattern also appears to provide further support for the notion that most owls encountered in New Mexico derive from relatively nearby in the central Rocky Mountains, rather than comprising a diverse mix of Rocky Mountain birds and long-distance migrants from the northwestern portion of the species' range. In fact, based on migration patterns discerned for many diurnal raptors (Hoffman et al. 2002), one might expect that most owls moving south from the northern Intermountain region would pass down through the Great Basin and then down along the Sierra Madre Occidental through Mexico, rather than cross over into the Rocky Mountains. Recent work by Arsenault et al. (2005) revealed little genetic differentiation among several populations sampled in New Mexico and in the Wasatch Range of northern Utah, however, so apparently significant mixing of populations occurs at least across this range of latitudes within the Rocky Mountains.

NORTHERN SAW-WHET OWL

The migration ecology of Northern Saw-whet Owls (*Aegolius acadicus*) has been studied fairly extensively in eastern North America (Cannings 1993; Brinker et al. 1997), but intensive studies in western North America have arisen only recently (e.g., Stock et al. 2006). The accumulated evidence suggests that Saw-whet Owls can be found during winter throughout their breeding range. Biannual migrations occur across the continent, but some individuals or populations may be seminomadic and periodic large-scale southward irruptions also occur. Relatively short-distance movements appear to be the rule, but nomadic and irruptive movements may cover greater distances (Duffy and Kerlinger 1992; Cannings 1993; Brinker et al. 1997; Marks and Doremus 2000; Stock et al. 2006). Essentially no information is available concerning the specific migratory pathways followed or the winter-summer range connections of western populations.

Readily available data on the migration ecology of Saw-whet Owls in New Mexico are limited to autumn data collected by HWI in the Manzano Mountains (DeLong 2000, 2001, 2003a, 2003b). This banding study was focused on Flammulated Owls and involved primarily luring with broadcast calls of this species; only limited late-season luring included broadcasting Saw-whet Owl calls. As a result, the effort yielded relatively few Saw-whet Owl captures (5–28 per season) and may reflect a biased sample. Nevertheless, scattered captures from late August through September followed by a primary activity pulse in October (fig. 2.13) match findings from other studies in Idaho (Stock et al. 2006) and Nevada (Smith 2006). The primary autumn movement period at Cape May, New Jersey, also begins in mid October (Duffy and Kerlinger 1992).

Data from Cape May, New Jersey, suggested that adults move slightly ahead of immature birds in autumn (Duffy and Kerlinger 1992). Data from Idaho also indicated that immature birds preceded adults during autumn and that males preceded females, but only during a major irruption year; outside that single year, no significant age- or sex-specific variation in timing was evident (Stock et al. 2006). Analysis of the 52 captures recorded between 1999 and 2005 in the Manzano Mountains, with age and sex positively determined, indicated no overall significant differences in passage timing between males and females or between adults and immatures; however, a significant age-sex interaction was indicated (two-way factorial ANOVA with YEAR factor eliminated as nonsignificant; significance equated with $P \leq 0.05$). Bonferroni post hoc comparisons indicated that the average capture date for immature females (27 September) did not differ significantly from that of adult males (6 October), but was marginally earlier ($P = 0.084$) than for immature males (10 October) and significantly earlier ($P = 0.049$) than for adult females (13 October). Analysis of 124 Saw-whet Owl captures in the Goshute Mountains, Nevada, between 2003 and 2005 indicated a similar pattern except that the average capture date for immature females (30 September) was slightly later than for both immature (29 September) and adult (27 September) males, and only the late passage timing of adult females (13 October) differed significantly from that of the other sex-age groups.

ELF OWL

All breeding populations of Elf Owls (*Micrathene whitneyi*) that occur along the U.S.-Mexico border from Arizona to southwestern Texas are considered migratory and generally vacate their breeding ranges to spend the winter in southwestern Mexico (Henry and Gehlbach 1999). Thermoregulation and prey-availability (this owl is almost exclusively insectivorous) issues are believed to drive the species' migratory habits (Ligon 1968; Henry and Gehlbach 1999). As for most migratory owls, virtually nothing is known about the specific pathways these tiny owls follow between their breeding and winter ranges or concerning the specific conditions under which they migrate. Information about the seasonality of the species' migration derives from breeding studies and general observations within U.S. and northern Mexico summer ranges, which typically indicate absence from summer ranges between mid to late October and late February or early March (Henry and Gehlbach 1999). Breeding studies also indicate that the earliest arrivals preferentially occupy lower-elevation habitats and nest earliest, with later arrivals pushed up to higher elevations where less early-season prey is available (Ligon 1968).

BURROWING OWL

One banding study of Burrowing Owls (*Athene cunicularia*) breeding in central New Mexico suggested that most owls from this area disperse during winter, although a small percentage remained resident year-round and winter surveys in the study area revealed no influx of migrants (Martin 1973). The winter destinations of New Mexico breeders are mostly unknown (Arrowood et al. 2001); however, limited winter recoveries of birds banded in south-central New Mexico all came from within the state (Best 1969). Other studies of the species have suggested a leapfrog migration pattern; in other words, northern populations tend to migrate farther and winter farther south than southern populations (Haug et al. 1993). Further new information about the winter ranges of central New Mexico breeders may emerge in the next few years from a new study begun in 2005 that seeks to use radio-tracking to confirm connections to probable wintering grounds across Mexico (Envirological Services 2006).

Migration Studies and Raptor Conservation

The science underpinning use of migration counts as a means of monitoring population trends in migratory diurnal raptors is still evolving (Dunn and Hussell 1995; Hoffman and Smith 2003; Farmer et al. 2007) and the method is not universally applicable, depending on the complexities of species-specific movement ecologies, distributional patterns, and abundances. Nevertheless, given appropriate standardization, rigorous effort, and geographic coverage, the approach clearly represents a viable, efficient, and valuable option for monitoring regional population trends in multiple species (Zalles and Bildstein 2000). Migration counts are now recognized as at least a second-tier priority approach for achieving effective range-wide monitoring and for improving knowledge of population trends for 13 diurnal raptor species recognized in the recent Partners in Flight (PIF) document addressing high-priority needs for range-wide monitoring of North American landbirds (Dunn et al. 2005). The *PIF North American Landbird Conservation Plan* (Rich et al. 2004) further indicates that of the 18 species of migratory diurnal raptors regularly counted at migration sites, only three are well monitored using the traditional Breeding Bird Survey (BBS) method for regional and continental monitoring of avian populations. The PIF plan recognizes this gap in North American bird monitoring and recommends improved migration monitoring to meet this critical conservation goal. Many state wildlife action plans also identify raptors as species for which improved monitoring data are urgently needed.

In the interest of further advancing the science of raptor migration monitoring and developing an efficient system for public reporting of continental status and trends data derived from migration counts, in 2004 HWI joined a formal partnership involving the Hawk Migration Association of North America (HMANA) and Hawk Mountain Sanctuary Association (managers of the world's oldest and most respected long-term raptor migration study site in Pennsylvania) in an effort called the Raptor Population Index (RPI) Project (see http://rpi-project.org/). Key components of this effort include fostering significant improvements in the methods used to derive population trends from migration count data (Farmer et al. 2007), promoting increased standardization and rigor across the continent's substantial network of migration-monitoring projects, and developing a centralized Web-based data-entry, storage, and public-accessible retrieval and display system for organizing North American migration count data (see http://www.hawkcount.org). Most important, the project has produced the first-ever comprehensive report on the continent-wide status and population trends of migratory raptors in North America based on migration counts (Bildstein et al. 2008). Long-term data from the Sandia and Manzano mountains comprise key components of this report, effectively representing the southern Rocky Mountain region. Knowledge of the status and populations trends of raptors is of particular importance for achieving effective ecosystem management and conservation because as top-level, wide-ranging predators known to be sensitive to a variety of human disturbances, raptors can serve as effective biological indicators of overall ecosystem health (Bildstein 2001).

Tracking migration routes and documenting connections between specific summer- and winter-range populations of migratory species provide further conservation benefits. Such information helps ensure accurate, geographically representative interpretations of population trends documented through migration counts. It also provides critical information about the full range of conditions and potential disturbance factors to which different species and populations are exposed throughout the annual cycle, which in turn helps identify probable causes of observed population declines and direct priorities for further research and specific conservation action. Again, information derived from HWI's long-term band-return, satellite-tracking, and stable-isotope research is greatly expanding awareness of the movement ecology and range dynamics of migratory raptors in western North America, with data derived from the two New Mexico projects of key importance for understanding the dynamics and conservation needs of Rocky Mountain populations.

Finally, HWI welcomes public visitation at all of its migration study sites and staffs both of the New Mexico projects with on-site education specialists whose job is to facilitate visitor interactions and provide environmental education at the sites. We also routinely incorporate substantial volunteer participation in most aspects of the associated research and monitoring endeavors. Accordingly, I encourage interested readers to come join us at one of our sites, to experience the marvel of raptor migration and the dynamic beauty of our feathered friends firsthand, and if so motivated to join our team of staff and volunteer researchers in investigating the migration of raptors through New Mexico.

Eyes to the Skies!

Acknowledgments

HWI founder Steve Hoffman, whose tireless efforts effectively guided the projects through their first 15 years of existence, initiated HWI's Sandia and Manzano mountains migration projects. John DeLong also played a key long-term role in both the diurnal trapping and banding programs and, in particular, in development of Flammulated Owl studies at the Manzanos site. The projects also benefited over the years from the participation of hundreds of seasonal and local volunteers. Past and current core financial, in-kind, and logistical sponsors of the projects include: USDA Forest Service–Cibola National Forest and Southwest Region, U.S. Fish and Wildlife Service–Region 2 and Neotropical Migratory Bird Conservation Act grant program, New Mexico Game and Fish–Share with Wildlife Program, National Fish and Wildlife Foundation, Albuquerque Community Foundation, Kerr Foundation, LaSalle Adams Fund, Intel Corporation, Public Service Company of New Mexico, Wild Oats and Whole Foods Markets in Albuquerque, New Belgium Brewing Company, Central New Mexico Audubon Society, and a rich cadre of HWI private donors and members. This manuscript benefited from a thoughtful review by Jeff Kelly.

Arrowood, P. C., C. A. Finley, and B. C. Thompson. 2001. Analyses of Burrowing Owl populations in New Mexico. *Journal of Raptor Research* 35:362–70.

Arsenault, D. P., A. Hodgson, and P. B. Stacey. 1997. Dispersal movements of juvenile Mexican spotted owls (*Strix occidentalis lucida*) in New Mexico. In *Biology and conservation of owls of the northern hemisphere; second international symposium, February 5–9, 1997, Winnipeg, Manitoba, Canada*, ed. J. R. Duncan, D. H. Johnson, and T. H. Nicholls, 47–56. GTR NC-190. St. Paul, MN: USDA Forest Service, North Central Forest Experiment Station.

Arsenault, D. P., P. B. Stacey, and G. A. Hoelzer. 2005. Mark-recapture and DNA fingerprinting data reveal high breeding-site fidelity, low natal philopatry, and low levels of genetic population differentiation in Flammulated Owls (*Otus flammeolus*). *Auk* 122:329–37.

Balda, R. P., B. C. McKnight, and C. D. Johnson. 1975. Flammulated Owl migration in the southwestern United States. *Wilson Bulletin* 87:520–33.

Bechard, M. J., and J. K. Schmutz. 1995. Ferruginous Hawk (*Buteo regalis*). No. 172. In *The birds of North America*, ed. A. Poole and F. Gill. Philadelphia, PA: Academy of Natural Sciences, and Washington, DC: American Ornithologists' Union.

Bechard, M. J., and T. R. Swem. 2002. Rough-legged Hawk (*Buteo lagopus*). No. 641. In *The birds of North America*, ed. A. Poole and F. Gill. Philadelphia, PA: Birds of North America, Inc.

Best, T. R. 1969. Habitat, annual cycle, and food of Burrowing Owls in south central New Mexico. M.Sc. thesis. New Mexico State University, Las Cruces.

Bildstein, K. L. 2001. Why migratory birds of prey make great biological indicators. In *Hawkwatching in the Americas*, ed. K. L. Bildstein and D. Klem Jr., 169–79. North Wales, PA: Hawk Migration Association of North America.

———. 2006. *Migrating raptors of the world: Their ecology and conservation*. Ithaca, NY: Cornell University Press.

Bildstein, K. L., and K. Meyer. 2000. Sharp-shinned Hawk (*Accipiter striatus*). No. 482. In *The birds of North America*, ed. A. Poole and F. Gill. Philadelphia, PA: Birds of North America, Inc.

Bildstein, K. L., and J. Zalles. 2001. Raptor migration along the Mesoamerican land corridor. In *Hawkwatching in the Americas*, ed. K. L. Bildstein and D. Klem Jr., 119–41. North Wales, PA: Hawk Migration Association of North America.

Bildstein, K. L., J. P. Smith, E. Ruelas Inzunza, and R. R. Veit, eds. 2008. *State of North America's birds of prey*. Series in Ornithology no. 3. Cambridge, MA: Nuttall Ornithological Club, and Washington, DC: American Ornithologists' Union.

Bolen, E. C. 1978. Long-distance displacement of two southern Barn Owls. *Bird-Banding* 49:78–79.

Brinker, D. F., K. E. Duffy, D. M. Whalen, B. D. Watts, and K. M. Dodge. 1997. Autumn migration of Northern Saw-whet Owls (*Aegolius acadicus*) in the middle Atlantic and northeastern United States; what observations from 1995 suggest. In *Biology and conservation of owls of the northern hemisphere; second international symposium, February 5–9, 1997, Winnipeg, Manitoba, Canada*, ed. J. R. Duncan, D. H. Johnson, and T. H. Nicholls, 74–89. GTR NC-190. St. Paul, NM: USDA Forest Service, North Central Forest Experiment Station.

Buehler, D. A. 2000. Bald Eagle (*Haliaeetus leucocephalus*). No. 506. In *The birds of North America*, ed. A. Poole and F. Gill. Philadelphia, PA: Birds of North American, Inc.

Cannings, R. J. 1993. Northern Saw-whet Owl (*Aegolius acadicus*). No. 42. In *The birds of North America*, ed. A. Poole and F. Gill. Philadelphia, PA: Academy of Natural Sciences, and Washington, DC: American Ornithologists' Union.

Clark, R. J. 1975. A field study of the Short-eared Owl (*Asio flammeus*) Pontoppidan in North America. *Wildlife Monographs* 47:1–67.

Clark, W. S. 1985. Migration of the Merlin along the coast of New Jersey. *Raptor Research* 19:85–93.

DeLong, J. P. 2000. *Fall 2000 Flammulated Owl banding study in the Manzano Mountains of New Mexico*. Salt Lake City, UT: HawkWatch International.

———. 2001. *Fall 2001 Flammulated Owl banding study in the Manzano Mountains of central New Mexico*. Salt Lake City, UT: HawkWatch International.

———. 2003a. *Flammulated Owl migration project: Manzano Mountains, New Mexico—2003 report*. Salt Lake City, UT: HawkWatch International.

———. 2003b. *Flammulated Owl migration project: Manzano Mountains, New Mexico—2002 report*. Salt Lake City, UT: HawkWatch International.

———. 2004. Age determination and preformative molt in hatch-year Flammulated Owls during the fall. *North American Bird Bander* 29(3): 111–15.

———. 2006. Pre-migratory fattening and mass gain in Flammulated Owls in central New Mexico. *Wilson Journal of Ornithology* 118:187–93.

DeLong, J. P., and S. W. Hoffman. 1999. Differential autumn migration of Sharp-shinned and Cooper's Hawks in western North America. *Condor* 101:674–78.

———. 2004. Fat stores of migrating Sharp-shinned and Cooper's Hawks in New Mexico. *Journal of Raptor Research* 38:163–68.

DeLong, J. P., T. Meehan, and R. Smith. 2005. Investigating fall movements of Flammulated Owls (*Otus flammeolus*) in central New Mexico using stable hydrogen isotopes. *Journal of Raptor Research* 39:19–25.

Duffy, K. E. 1985. Fall migration of Barn Owls at Cape May Point, New Jersey. *Proceedings of hawk migration conference IV, Rochester, New York, March 24–27, 1983*, ed. M. Harwood, 193–205. North Wales, PA: Hawk Migration Association of North America.

Duffy, K. E., and P. Kerlinger. 1992. Autumn owl migration at Cape May Point, New Jersey. *Wilson Bulletin* 104:312–20.

Dunn, E. H., and D. J. T. Hussell. 1995. Using migration counts to monitor landbird populations: review and evaluation of status. *Current Ornithology* 12:43–88.

Dunn, E. H., B. L. Altman, J. Bart, C. J. Beardmore, H. Berlanga, P. J. Blancher, G. S. Butcher, D. W. Demarest, R. Dettmers, W. C. Hunter, et al. 2005. High priority needs for range-wide monitoring of North American landbirds. Partners in Flight Technical Series no. 2. http://www.partnersinflight.org/pubs/ts/02-MonitoringNeeds.pdf (accessed December 2006).

England, A. S., M. J. Bechard, and C. S. Houston. 1997. Swainson's Hawk (*Buteo swainsoni*). No. 265. In *The birds of North America*, ed. A. Poole and F. Gill. Philadelphia, PA: Academy of Natural Sciences, and Washington, DC: American Ornithologists' Union.

Envirological Services. 2006. *Collaborative efforts to determine wintering grounds and habitat quality for Burrowing Owls (*Athene cunicularia*) in México*. Albuquerque, NM: Envirological Services.

Farmer, C. J., D. J. T. Hussell, and D. Mizrahi. 2007. Detecting population trends in migratory birds of prey. *Auk* 124:1047–62.

Fuller, M. R., W. S. Seegar, and L. S. Schueck. 1998. Routes and travel rates of migrating Peregrine Falcons *Falco peregrinus* and Swainson's Hawks *Buteo swainsoni* in the western hemisphere. *Journal of Avian Biology* 29:433–40.

Goodrich, L. J., and J. P. Smith. 2008. Raptor migration in North America. In *State of North America's birds of prey*, ed. K. L. Bildstein, J. P. Smith, E. Ruelas Inzunza, and R. R. Veit, 37–150. Series in Ornithology no. 3. Cambridge, MA: Nuttall Ornithological Club, and Washington, DC: American Ornithologists' Union.

Goodrich, L. J., S. C. Crocoll, and S. E. Senner. 1996. Broad-winged Hawk (*Buteo platypterus*). No. 218. In *The birds of North America*, ed. A. Poole and F. Gill. Philadelphia, PA: Academy of Natural Sciences, and Washington, DC: American Ornithologists' Union.

Gutiérrez, R. J., A. B. Franklin, and W. S. Lahaye. 1995. Spotted owl (*Strix occidentalis*). No. 179. In *The birds of North America*, ed. A. Poole and F. Gill. Philadelphia, PA: Academy of Natural Sciences, and Washington, DC: American Ornithologists' Union.

Gutiérrez, R. J., M. E. Seamans, and M. Z. Peery. 1996. Intermountain movement by Mexican Spotted Owls (*Strix occidentalis lucida*). *Great Basin Naturalist* 56:87–89.

Harmata, A. R. 1984. Bald Eagles of the San Luis Valley, Colorado: their winter ecology and spring migration. Ph.D. Diss. Montana State University, Bozeman.

Harmata, A. R., and D. W. Stahlecker. 1993. Fidelity of migrant Bald Eagles to wintering grounds in southern Colorado and northern New Mexico. *Journal of Field Ornithology* 64:129–34.

Harrington, D. 1997. *Exploratory raptor migration survey in western Colorado—fall 1997*. Salt Lake City, UT: HawkWatch International.

Haug, E. A., B. A. Millsap, and M. S. Martell. 1993. Burrowing Owl (*Speotyto cunicularia*). No. 61. In *The birds of North America*, ed. A. Poole and F. Gill. Philadelphia, PA: Academy of Natural Sciences, and Washington, DC: American Ornithologists' Union.

[HWI] HawkWatch International. 2009a. Satellite-tracking program summaries and results. http://www.hawkwatch.org/satelliteprogram.php. December.

———. 2009b. Multi-site migration monitoring and research project summaries. http://www.hawkwatch.org/migration/migration.php. December.

Henry, S. G., and F. R. Gehlbach. 1999. Elf Owl (*Micrathene whitneyi*). No. 413. In *The birds of North America*, ed. A. Poole and F. Gill. Philadelphia, PA: Academy of Natural Sciences, and Washington, DC: American Ornithologists' Union.

Hoffman, S. W., and J. P. Smith. 2003. Population trends of migratory raptors in western North America, 1977–2001. *Condor* 105:397–419.

Hoffman, S. W., J. P. Smith, and T. D. Meehan. 2002. Breeding grounds, winter ranges, and migratory routes of raptors in the Mountain West. *Journal of Raptor Research* 36:97–110.

Holt, D. W., and S. M. Leasure. 1993. Short-eared Owl (*Asio flammeus*). No. 62. In *The birds of North America*, ed. A. Poole and F. Gill. Philadelphia, PA: Academy of Natural Sciences, and Washington, DC: American Ornithologists' Union.

Hunt, W. G., R. E. Jackman, J. M. Jenkins, C. G. Thelander, and R. N. Lehman. 1992. Northward post-fledging migration of California Bald Eagles. *Journal of Raptor Research* 26:19–23.

Inzunza, E. R., S. W. Hoffman, L. J. Goodrich, and R. Tingay. 2000. Conservation strategies for the world's largest known raptor migration flyway: Veracruz the River of Raptors. In *Raptors at Risk*, ed. R. D. Chancellor and B.-U. Meyburg, 591–96. Berlin, Germany: World Working Group on Birds of Prey and Owls, and British Columbia, Canada: Hancock House Publishers.

Kerlinger, P. 1989. *Flight strategies of migrating hawks*. Chicago: University of Chicago Press.

Kiseda, J. J. 2005. *Seasonal raptor migration observations at the Wyler Aerial Tramway State Park, El Paso, Texas*. El Paso, TX: New Mexico State University Agricultural Experiment Station.

Kochert, M. N., K. Steenhof, C. L. McIntyre, and E. H. Craig. 2002. Golden Eagle (*Aquila chrysaetos*). No. 684. In *The birds of North America*, ed. A. Poole and F. Gill. Philadelphia, PA: Academy of Natural Sciences, and Washington, DC: American Ornithologists' Union.

Ligon, J. D. 1968. *The biology of the Elf Owl,* Micrathene whitneyi. Miscellaneous publ. 136. Ann Arbor: University of Michigan Museum of Zoology.

Linkhart, B. D., and R. T. Reynolds. 2006. Lifetime reproduction of Flammulated Owls in Colorado. *Journal of Raptor Research* 40:29–37.

———. 2007. Return rate, fidelity, and dispersal in a breeding population of Flammulated Owls (*Otus flammeolus*). *Auk* 124:264–75.

Lott, C. A., and J. P. Smith. 2006. A geographic-information-system approach to estimating the origin of migratory raptors in North America using hydrogen stable isotope ratios in feathers. *Auk* 123:822–35.

Marks, J. S., and J. H. Doremus. 2000. Are Northern Saw-whet Owls nomadic? *Journal of Raptor Research* 34:299–304.

Marks, J. S., D. L. Evans, and D. W. Holt. 1994. Long-eared Owl (*Asio otus*). No. 133. In *The birds of North America*, ed. A. Poole and F. Gill. Philadelphia, PA: Academy of Natural Sciences, and Washington, DC: American Ornithologists' Union.

Marshall, J. T. 1967. Parallel variation in North and Middle American Screech-Owls. *Monograph Western Foundation of Vertebrate Zoology* 1:1–72.

Martell, M. S., C. J. Henny, P. E. Nye, and M. J. Solensky. 2001. Fall migration routes, timing, and wintering sites of North American Ospreys as determined by satellite telemetry. *Condor* 103:715–24.

Marti, C. D. 1988. The Common Barn Owl. In *Audubon wildlife report 1988/1989*, ed. W. J. Chandler, 535–50. San Diego, CA: Academic Press.

———. 1992. Barn Owl (*Tyto alba*). No. 1. In *The birds of North America*, ed. A. Poole and F. Gill. Philadelphia, PA: Academy of Natural Sciences, and Washington, DC: American Ornithologists' Union.

Martin, D. J. 1973. Selected aspects of Burrowing Owl ecology and behaviour in central New Mexico. *Condor* 75:446–56.

McBride, T. J., J. P. Smith, H. P. Gross, and M. Hooper. 2004. Blood-lead and ALAD activity levels of Cooper's Hawks (*Accipiter cooperii*) migrating through the southern Rocky Mountains. *Journal of Raptor Research* 38:118–24.

McCallum, D. A. 1994. Flammulated Owl (*Otus flammeolus*). No. 93. In *The birds of North America*, ed. A. Poole and F. Gill. Philadelphia, PA: Academy of Natural Sciences, and Washington, DC: American Ornithologists' Union.

McIntyre, C. L. 2004. Golden Eagles in Denali National Park and Preserve: productivity and survival in relation to landscape characteristics of nesting territories. Ph.D. diss. Oregon State University, Corvallis.

McIntyre, C. L., K. Steenhof, M. N. Kochert, and M. W. Collopy. 2006. Long-term Golden Eagle studies in Denali National Park and Preserve. *Alaska Park Science* 5:42–45.

Millsap, B. A., and P. A. Millsap. 1987. Burrow nesting by Common Barn Owls in north central Colorado. *Condor* 89:668–70.

Millsap, B. A., and J. R. Zook. 1983. Effects of weather on accipiter migration in southern Nevada. *Journal of Raptor Research* 17:43–56.

Mueller, H. C., D. D. Berger, and G. Allez. 1977. The periodic invasions of Goshawks. *Auk* 94:652–63.

Omland, K. S., and S. W. Hoffman. 1996. Seasonal, diel, and spatial dispersion patterns of Golden Eagle autumn migration in southwestern Montana. *Condor* 98:633–36.

Phillips, A. R. 1942. Notes on the migration of Elf and Flammulated Screech Owls. *Wilson Bulletin* 54:132–37.

Poole, A. F., R. O. Bierregaard, and M. S. Martell. 2002. Osprey (*Pandion haliaetus*). No. 683. In *The birds of North America*, ed. A. Poole and F. Gill. Philadelphia, PA: Birds of North America, Inc.

Preston, C. R., and R. D. Beane. 1993. Red-tailed Hawk (*Buteo jamaicensis*). No. 52. In *The birds of North America*, ed. A. Poole and F. Gill. Philadelphia, PA: Academy of Natural Sciences, and Washington, DC: American Ornithologists' Union.

Reynolds, R. T., and B. D. Linkhart. 1984. Method and materials for capturing and monitoring Flammulated Owls. *Great Basin Naturalist* 44:49–51.

———. 1987. The nesting biology of Flammulated Owls in Colorado. In *Biology and conservation of northern forest owls*, ed. R. W. Nero, R. J. Clark, R. J. Knapton, and R. H. Hamre, 239–48. GTR-RM-142. Fort Collins, CO: USDA Forest Service, Rocky Mountain Research Station.

Rich, T. D., C. J. Beardmore, H. Berlanga, P. J. Blancher, M. S. W. Bradstreet, G. S. Butcher, D. W. Demarest, E. H. Dunn, W. C. Hunter, E. E. Iñigo-Elias, et al. 2004. *Partners in Flight North American Landbird Conservation Plan*. Ithaca, NY: Cornell Lab of Ornithology. http://www.partnersinflight.org/cont_plan/ (accessed March 2005).

Rosenfield, R. N., and J. Bielefeldt. 1993. Cooper's Hawk (*Accipiter cooperii*). No. 75. In *The birds of North America*, ed. A. Poole and F. Gill. Philadelphia, PA: Academy of Natural Sciences, and Washington, DC: American Ornithologists' Union.

Schmutz, J. K. and R. W. Fyfe. 1987. Migration and mortality of Alberta Ferruginous Hawks. *Condor* 89:169–74.

Schueck, L., T. Maechtle, M. Fuller, K. Bates, W. S. Seegar, and J. Ward. 1998. Movements of Ferruginous Hawks through western North America. *Journal of the Idaho Academy of Natural Science* 34(1): 11–12.

Sherrington, P. 2003. Trends in a migratory population of Golden Eagles (*Aquila chrysaetos*) in the Canadian Rocky Mountains. *Bird Trends Canada* 9:34–39.

Smallwood, J. A., and D. M. Bird. 2002. American Kestrel (*Falco sparverius*). No. 602. In *The birds of North America*, ed. A. Poole and F. Gill. Philadelphia, PA: Birds of North America, Inc.

Smith, J. P. 2005. *Exploratory raptor migration surveys at sites near Ely, Nevada, with significant windpower generation potential.* Salt Lake City, UT: HawkWatch International, Inc.

———. 2006. *Fall 2005 Flammulated Owl migration study in the Goshute Mountains of northeastern Nevada.* Salt Lake City, UT: HawkWatch International, Inc.

Smith, J. P., and M. C. Neal. 2006. *Fall 2005 raptor migration studies at Commissary Ridge in southwestern Wyoming.* Salt Lake City, UT: HawkWatch International, Inc.

Smith, J. P., and C. Neri. 2004. *Fall 2003 Flammulated Owl migration study in the Goshute Mountains of northeastern Nevada.* Salt Lake City, UT: HawkWatch International, Inc.

Smith, J. P., C. J. Farmer, S. W. Hoffman, K. Z. Woodruff, G. S. Kaltenecker, and P. Sherrington. 2008. Trends in autumn counts of migratory raptors in western North America. In *State of North America's birds of prey*, ed. K. L. Bildstein, J. P. Smith, E. Ruelas Inzunza, and R. R. Veit, 217–52. Series in Ornithology no. 3. Cambridge, MA: Nuttall Ornithological Club, and Washington, DC: American Ornithologists' Union.

Smith, J. P., P. Grindrod, and S. W. Hoffman. 2001b. Migration counts indicate Broad-winged Hawks are increasing in the West: evidence of breeding range expansion? In *Hawkwatching in the Americas*, ed. K. L. Bildstein and D. Klem Jr., 93–106. North Wales, PA: Hawk Migration Association of North America.

Smith, J. P., J. Simon, S. W. Hoffman, and C. Riley. 2001a. New full-season autumn hawkwatches in coastal Texas. In *Hawkwatching in the Americas*, ed. K. L. Bildstein and D. Klem Jr., 67–91. North Wales, PA: Hawk Migration Association of North America.

Smith, R. B., E. C. Greiner, and B. O. Wolf. 2004. Migratory movements of Sharp-shinned Hawks (*Accipiter striatus*) captured in New Mexico in relation to prevalence, intensity, and biogeography of avian hematozoa. *Auk* 121:837–46.

Smith, R. B., T. D. Meehan, and B. O. Wolf. 2003. Assessing migration patterns of Sharp-shinned Hawks *Accipiter striatus* using stable-isotope and band encounter analysis. *Journal of Avian Biology* 34:387–92.

Sodhi, N. S., L. W. Oliphant, P. C. James, and I. G. Warkentin. 1993. Merlin (*Falco columbarius*). No. 44. In *The birds of North America*, ed. A. Poole and F. Gill. Philadelphia, PA: Academy of Natural Sciences, and Washington, DC: American Ornithologists' Union.

Sonsthagen, S. A., R. Rodriguez, and C. M. White. 2006. Satellite telemetry of Northern Goshawks breeding in Utah I: annual movements. In *The Northern Goshawk: a technical assessment of its status, ecology, and management*, ed. M. L. Morrison. *Studies in Avian Biology* 31.

Soucy, L. J. 1980. Three long-distance recoveries of banded New Jersey Barn Owls. *North American Bird Bander* 5:97.

Squires, J. R., and R. T. Reynolds. 1997. Northern Goshawk (*Accipiter gentilis*). No. 298. In *The birds of North America*, ed. A. Poole and F. Gill. Philadelphia, PA: Academy of Natural Sciences, and Washington, DC: American Ornithologists' Union.

Steenhof, K. 1998. Prairie Falcon (*Falco mexicanus*). No. 346. In *The birds of North America*, ed. A. Poole and F. Gill. Philadelphia, PA: Birds of North America, Inc.

Steenhof, K., M. R. Fuller, M. N. Kochert, and K. K. Bates. 2005. Long-range movements and breeding dispersal of Prairie Falcons from southwest Idaho. *Condor* 107:481–96.

Steenhof, K., M. N. Kochert, and M. Q. Moritsch. 1984. Dispersal and migration of southwestern Idaho raptors. *Journal of Field Ornithology* 55:357–68.

Stewart, P. A. 1952. Dispersal, breeding, behavior, and longevity of banded Barn Owls in North America. *Auk* 69:227–45.

Stock, S. L., P. J. Heglund, G. S. Kaltenecker, J. D. Carlisle, and L. Leppert. 2006. Comparative ecology of the Flammulated Owl and Northern Saw-whet Owl during fall migration. *Journal of Raptor Research* 40:120–29.

Van Fleet, K. 2001. Geography of diurnal raptors migrating through the valley and ridge province of central Pennsylvania, 1991–1994. In *Hawkwatching in the Americas*, ed. K. L. Bildstein and D. Klem Jr., 23–41. North Wales, PA: Hawk Migration Association of North America.

Watson, J. W. 2003. *Migration and winter ranges of Ferruginous Hawks from Washington. Final report.* Olympia: Washington Department of Fish and Wildlife.

Watson, J. W, and U. Banasch. 2005. *A Tri-national investigation of Ferruginous Hawk migration. Progress report 1.* Olympia: Washington Department of Fish and Wildlife, and Edmonton, Alberta: Canadian Wildlife Service.

White, C. M., N. J. Clum, T. J. Cade, and W. G. Hunt. 2002. Peregrine Falcon (*Falco peregrinus*). No. 660. In *The birds of North America*, ed. A. Poole and F. Gill. Philadelphia, PA: Birds of North America, Inc.

Williams, S. O. III. 2007. A January specimen of the Flammulated Owl from northern New Mexico. *Wilson Journal of Ornithology* 119:764–66.

Wilson, R. B. 1999. Characteristics of Bald Eagle communal roosts in northern Utah and a survey of eagles occupying roosts. M.Sc. thesis. Utah State University, Logan.

Zalles, J. I., and K. L. Bildstein, eds. 2000. *Raptor watch: a global directory of raptor migration sites.* BirdLife Conservation Series no. 9. Cambridge, UK: BirdLife International, and Kempton, PA: Hawk Mountain Sanctuary Association.

PART II

Species Accounts

3

Turkey Vulture

(*Cathartes aura*)

STEPHEN M. FETTIG AND JEAN-LUC E. CARTRON

TURKEY VULTURES (*Cathartes aura*) are long-winged raptors built for hours of seemingly effortless flight. Their long wings provide the lift needed for flight with relatively little flapping. Dunne et al. (1988:137) make reference to Turkey Vultures being "able to tease lift from thin air while other soaring birds can only sit and watch." Powered flight through flapping seems to be rarely required other than when leaving the ground or roost sites, or when returning to a roost after the winds or rising thermals have ended for the day. Thus Turkey Vulture flight is characterized by gliding and soaring, during which the wings are typically held upward in a shallow but distinct upward "V" often referred to as a "dihedral." This dihedral configuration is accompanied by a rocking or teetering motion from side to side. The dihedral in combination with the teetering motion can provide good long-distance identification clues when viewing Turkey Vultures through binoculars or a spotting scope.

Turkey Vultures are entirely dark brown or nearly black except for the underside of the flight feathers of the wings (remiges) and the underside of the tail. The remiges are silver to gray and contrast greatly with the dark brown underwing coverts. The undertail is also gray. Adults have mostly featherless red heads with ivory-colored bills. Hatch-year birds have gray heads and bills that are mostly or entirely dark. During a Turkey Vulture's second calendar year, the bill becomes increasingly more ivory-colored and the head becomes redder. When perched, Turkey Vultures have a hunched or rounded-back posture. The featherless head gives Turkey Vultures a small-headed appearance.

When viewed at distances where color and feather details cannot be discerned, Turkey Vultures and Zone-tailed Hawks (*Buteo albonotatus*) can be mistaken for one another (Willis 1963; Mueller 1972). Like the Turkey Vulture, the Zone-tailed Hawk shows a side-to-side teetering motion when gliding on dihedral wings, along with gray remiges and dark underwing coverts (chapter 17). Views of the Zone-tailed Hawk's feathered head, yellow cere, or banded tail easily break the illusion of similarity.

Albino Turkey Vultures—with immaculate white plumages—are very rare but they have been documented, including in New Mexico. One was seen and photographed in September 2005 at Heron Lake State Park, Rio Arriba County (S. Lederman, pers. comm.). A partial albino was also observed and photographed in northern Harding County on 25 September 1996 by Jerry Oldenettel and John Parmeter (Williams 1996).

Turkey Vultures lack a syrinx and the airway muscles

PHOTO 3.1

Close-up of an adult Turkey Vulture's head. When seen at close range, Turkey Vultures are unmistakable, with their featherless red heads and ivory-colored bills. Photo taken near Rodeo, Hidalgo Co., 7 April 2007. PHOTOGRAPH: © ROBERT SHANTZ.

PHOTO 3.2

Adult Turkey Vulture, Rio Arriba Co., 25 August 2007. The Turkey Vulture's plumage is mostly dark brown, but the undertail (seen on this perched bird) is silver gray. PHOTOGRAPH: © ROGER HOGAN.

PHOTO 3.3

Turkey Vulture, Frijoles Canyon, northern Sandoval Co., late spring 2006. The back and upperwings are dark brown. The molt sequence for remiges (primary and secondary feathers in the wings) in Turkey Vultures is not well understood. In this photograph several molt limits (old and new feathers side by side) are visible within the remiges, suggesting multiple waves of feather replacement can occur at the same time. PHOTOGRAPH: © SALLY KING.

PHOTO 3.4

Turkey Vulture in flight, Steeple Rock Canyon, Hidalgo Co., 14 May 2008. The silver gray remiges of the underwings contrast with the dark brown underwing coverts. Zone-tailed Hawks (*Buteo albonotatus*) can be mistaken for Turkey Vultures, as they have similarly two-toned underwings and they too can glide with wings held in a strong dihedral. PHOTOGRAPH: © ROBERT SHANTZ.

PHOTO 3.5

Turkey Vulture, Frijoles Canyon, northern Sandoval Co., late spring 2006. Turkey Vultures, like most birds, cannot move their eyes within their sockets. This is because their eyes are cylindrical rather than spherical and thus cannot rotate freely in all directions. This anatomical trait is presumed to be a weight-saving adaptation, since a sphere holds more water than a cylinder of the same long dimension. In this photograph a Turkey Vulture bends its neck in dramatic fashion in order to look up. PHOTOGRAPH: © SALLY KING.

PHOTO 3.6

Juvenile Turkey Vulture, Rio Arriba Co., 7 September 2006. Hatch-year Turkey Vultures have gray heads, black instead of dark brown plumages, and mostly or entirely dark beaks. The bird in the photograph fledged recently, as indicated by the pale feather edges, giving the wing a scaled appearance. PHOTOGRAPH: © ROGER HOGAN.

needed for any well-controlled sound-making abilities (Miskimen 1957). They are typically silent, but can make sounds described as low-pitched, nasal, throaty, and wheezing, whining, hissing, or grunting (Kirk and Mossman 1998). These sounds seem to be most associated with interactions at carcasses and at roosts, or in response to nest disturbance.

Kirk and Mossman (1998) report that three subspecies can be identified in North America with three additional subspecies recognized from Costa Rica southward. Subspecies differences are based mostly on wing and tail measurement, the color of the lesser wing coverts, and variations in overall plumage coloration. *Cathartes aura aura* is the subspecies found in New Mexico and is characterized as being somewhat smaller than the other two North American subspecies.

Distribution

Turkey Vultures can be found from southern Canada south to Tierra del Fuego, Chile (Jaramillo et al. 2003). The species is a partial migrant (chapter 2). It is found year-round throughout most of its range, but northern

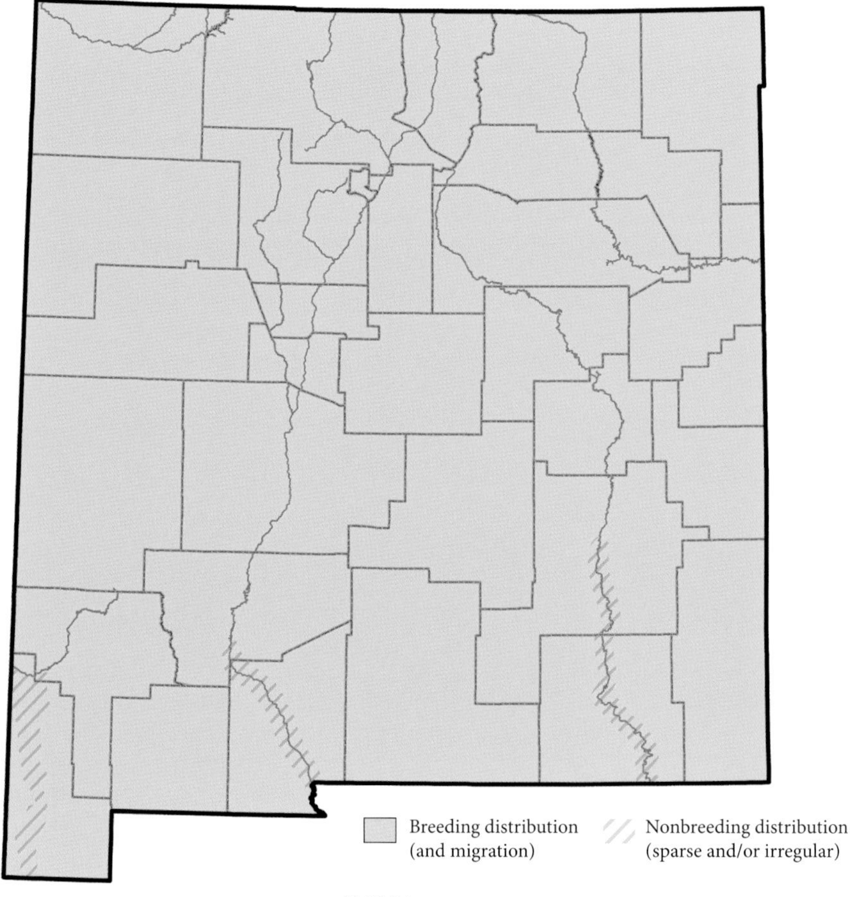

MAP 3.1

Turkey Vulture distribution map

Although Turkey Vultures are found throughout New Mexico during the breeding season, nesting seems strongly or exclusively associated with steep rocky areas, cliffs, or canyons. Note that large numbers of Turkey Vultures likely summer in New Mexico without breeding in most years, and these birds can be found even in large areas of low-elevation, flat grassland and desert-like habitats where breeding is very unlikely.

TABLE 3.1. Observations of Turkey Vultures (*Cathartes aura*) during the months of December and January in New Mexico as reported to the New Mexico Ornithological Society (2009), 1962–2007.

MONTH	DAY	YEAR	COUNTY	LOCALITY	NUMBER OF TURKEY VULTURES REPORTED
January	4	1970	Socorro	Bernardo	2
December	22	1970	Chaves	Bitter Lake NWR	1
January	19	1975	Socorro	Bosque del Apache NWR	1
January	25	1984	Otero	Sacramento River	1
December	21	1995	Eddy	Otis	1
December	26	1995	Sierra	Percha Dam State Park	1
January	1	1998	Sierra	Caballo Lake State Park	1
January	2	1998	Sierra	Las Animas Creek	1
January	4	2001	Sandoval	Corrales	1
December	14	2002	Chaves	Roswell	2
December	14	2002	Doña Ana	Las Cruces	1
December	2	2003	Otero	White Sands Missile Range	1
December	20	2003	Chaves	Roswell	7
January	2	2004	Hidalgo	Animas	2
December	18	2004	Chaves	Roswell	2
January	15-17	2005	Chaves	Roswell	8
January	30	2005	Eddy	Carlsbad	1
December	24-31	2005	Chaves	Roswell	2
January	1-4	2006	Chaves	Roswell	2
January	7	2006	Chaves	Roswell	14
December	16	2006	Chaves	Roswell	2
January	12	2007	Hidalgo	Animas	1

populations—those that breed in Canada and throughout most of the western, central, and northeastern United States—are migratory (Kirk and Mossman 1998). The northern limit of the Turkey Vulture's year-round distribution lies in coastal northern California, the U.S.-Mexico borderlands, and eastern Texas, extending east and north to New York along the Atlantic coast.

In New Mexico, the Turkey Vulture is considered essentially a breeding summer resident and a spring and fall migrant, although winter sightings have also been reported, mainly in the south (see Table 3.1). Hubbard (1978) reported that summering Turkey Vultures can be rare to common statewide at various elevations. Road survey data from 1974 to 1978 for March through October suggested that Turkey Vultures are in fact more common in the southeast (mean of 5.0 per 100 km [~60 mi]) and least common in the northwest (mean of 1.1 per 100 km) (Hubbard 1983).

In winter in the western United States, Turkey Vultures can typically be found only in California both along the coast and in the Central Valley, as well as in

TABLE 3.2. Observations of Turkey Vultures (*Cathartes aura*) during Christmas Bird Counts in Roswell, Chaves Co., 2000–2006 (compiled by Bixler 2008).

DATE	NUMBER OF TURKEY VULTURES REPORTED
16 December 2000	0
15 December 2001	0
14 December 2002	2
20 December 2003	7
18 December 2004	2
17 December 2005	0
16 December 2006	2
15 December 2007	3

southernmost Arizona (Kirk and Mossman 1998). In winter in New Mexico, Turkey Vultures were until recently considered casual in the far south (Parmeter et al. 2002), with even more irregular reports as far north as the San Juan Valley and Maxwell National Wildlife Refuge (Hubbard 1978). Winter sightings are not new to New Mexico, dating at least as far back as the second decade of the 20th century (Bailey 1928). However, the frequency of winter sightings has increased very recently. From 2001 through 2007, observers reported Turkey Vultures a total of 14 times during the months of December and January, or nearly twice the total for the preceding 39 years (table 3.1). In fact, Turkey Vultures are now regular in winter in small numbers in Roswell based on both Christmas Bird Count data from 2000 to 2007 (table 3.2) and multiple observations during January 2008, including a high of 18 birds on 15 January of that year (N. Cox, pers. comm.).

Turkey Vultures are the only New World vultures showing extensive north-south migration as part of their life history (Howell and Webb 1995; Wilbur 1983). Much of the North American population migrates through and winters in Mexico, Central America, and northern South America (Howell and Webb 1995; Kirk and Mossman 1998). The migration through Veracruz, Mexico, can be one of the migration wonders of the world with thousands per day at the peak (Liguori 2005). In New Mexico, mean passage date for spring migrants in the Sandia Mountains is 4 April based on 1985 to 2006 data, while 16 September is the mean passage date for fall migration in the Manzano Mountains (1985 to 2005 data) (Smith and Neal 2007a; see also chapter 2). Travis (1992) reports that Turkey Vultures arrive in Los Alamos County on average on 31 March (median date based on 28 years of data) and depart in the fall on 1 October (median date based on 19 years of data).

Habitat Associations

Turkey Vultures seem equally at home in western deserts, eastern forests, and tropical lowlands (Wilbur 1983). Their main habitat requirements consist of tall perches for roosting and access to open habitats for foraging in search of carrion, whether wild or domestic dead animals (Kirk and Mossman 1998). The species tends to avoid urban and suburban areas and especially avoids areas of row-crop agriculture (Kirk and Mossman 1998).

Breeding Bird Atlas work in Arizona documented Turkey Vultures associated with desertscrub or pinyon (*Pinus edulis*)-juniper (*Juniperus monosperma*) woodlands more than 50% of the time (Corman 2005). Other Arizona habitats used by Turkey Vultures included grasslands and forested mountain ridges (Corman 2005). Similarly, *Atlas* work in Colorado documented pinyon-juniper woodlands, particularly below 2,440 m (8,000 ft), as being most frequently used (Dexter 1998). However, Turkey Vultures were also found summering in all other Colorado habitats except high mountain valleys and peaks, and the croplands and grasslands of

PHOTO 3.7

Turkey Vultures in typical habitat for New Mexico: canyon walls in the Lake Heron area, Rio Arriba Co., 22 July 2006. In New Mexico the species often nests along canyon walls. Cliffs also provide updrafts for soaring. PHOTOGRAPH: © ROGER HOGAN.

northeastern Colorado (Dexter 1998). In New Mexico, Turkey Vultures occur in similar habitats as reported for Arizona and Colorado. Eakle et al. (1996) found Turkey Vultures in southwestern New Mexico to be habitat generalists, using all habitat types encountered along the survey route, including pinyon-juniper woodland, ponderosa pine (*Pinus ponderosa*) forest, plains grassland, Chihuahuan desertscrub, and semidesert grassland. Along the Middle Rio Grande, Turkey Vultures often roost in snags, particularly in burned areas (Cartron et al. 2008).

In New Mexico, nesting habitat has not been described with any degree of detail regarding surrounding vegetation. Elsewhere it is described as forested or partly forested areas with rock outcrops, caves, or fallen trees serving as nest sites. According to Jackson (1983:264), the nest sites of Turkey Vultures are best described as "dark recesses." Of 392 nest records from west of longitude 100° W in North America, 75% were in caves (Jackson 1983). The rest were on cliff ledges, on bare ground, in tree cavities, beside or under logs, beside trees, and in thickets. As described by Ligon (1961), nests in New Mexico are along canyon or mountain slopes and located in caves, under boulders, among slide rocks, or under overhanging ledges. Travis (1992) reported two nests for Los Alamos County, both in caves in tuff cliffs, one approximately 21 m (70 ft) above the floor of Barranca Canyon and the other about 11 m (35 ft) above the floor of a side canyon to Sandia Canyon. Two nests in sandstone cliffs were mentioned by Schmitt (1976) for the San Juan Valley.

Life History

Nesting

The species is a solitary nester (Jackson 1983), in contrast to the Black Vulture (*Coragyps atratus*), which is known to nest in loose groups. Turkey Vultures lay their eggs on bare substrate, soil, or litter without any nest construction. Most clutches consist of two eggs, although there are rare reports—none from New Mexico—of fewer and more than two eggs per nest (Jackson 1983). Incubation lasts 28 to 41 days and the length of the nestling period is approximately 56 to 88 days (Jackson 1983); young begin to perch outside the nest at 40 to 60 days of age (Kirk and Mossman 1998).

Turkey Vultures show very little territoriality or courtship activities away from nests sites. In addition, adults feed young by regurgitation and often leave a

PHOTO 3.8

Young Turkey Vulture, Hidalgo Co., 8 July 2003. Biologist Charlie Painter holds a nestling found in a rocky-canyon wall among several large boulders that formed a small cave approximately 1-1.5 m (3-5 ft) deep. PHOTOGRAPH: AARON LAMB (© CHARLIE PAINTER).

nest during the morning and may not return until evening (Dexter 1998), although the frequency of nest visits can be highly variable ranging from three in three hours to none in twenty-six hours (Davis 1993). These behaviors decrease movement to and from nest sites, making Turkey Vulture nests much more difficult to locate than those of raptors intermittently feeding young throughout the day (Dexter 1998).

During preparation of the *Arizona Breeding Bird Atlas* 16 nests were found in that state (Corman 2005). Nests with young were reported between 17 April and 6 August (Corman 2005). During Colorado *Breeding Bird Atlas* work 12 nests were found (Dexter 1998). Nests with eggs were documented from 16 May to 25 May; nests with young were found from 10 July to 25 July, and observations of fledged young ranged in time from 9 July to 4 August (Dexter 1998). The first nest recorded in New Mexico—by E. F. Pope, about 20 km (12 mi) southwest of Vaughn, Guadalupe County—was a nest that had eggs on 25 May [1919]. Egg dates for four nests found by Jens Jensen (Bailey 1928; Ligon 1961) east of Cerrillos, Santa Fe County, all similarly corresponded to mid to late May (18 May [1930], 19 May [1929], 26 May [1933], and 27 May [1928]). A nest with eggs was also reported by S. Williams from Twin Buttes Canyon in Chaves County on 3 June 1997 (Williams 1997). However, yet another nest in Walnut Canyon near Carlsbad Caverns had downy young in late April (Ligon 1961), indicating that some pairs may lay their eggs as early as March in the state. More typical is one downy young found at a nest in Chaves County on 2 June 1993 by S. Williams (Snider 1993a). Two nests found west of Antelope Wells, Hidalgo County—one along Deer Creek and the other at McKinney Flats—had "large" young on 10 and 12 July 2003, respectively (Williams 2003, *fide* C. Painter; photo 3.8).

As seen by the first author, birds with extensive amounts of down on their bodies are able to fly and join communal roosts. In northern Sandoval County, for example, young of the year are seen at a historic communal roost as early as mid to late August (S. Fettig, pers. obs.).

Jackson (1983), based on a review of information available at the time, suggested that Turkey Vultures apparently do not re-nest. Kirk and Mossman (1998), based on 15 years of work with marked birds in Wisconsin, point out that Turkey Vultures may lay a second clutch in an alternate nest site if the first nesting attempt fails.

Communal Roosting

Turkey Vultures are social birds, often gathering at communal roosts in the evening or flying together in large numbers, particularly during migration. Flock size increases in the tropics. The largest concentrations of Turkey Vultures observed in New Mexico include 500 birds at Carlsbad Caverns National Park on 23 September 1993 (Snider 1993b, *fide* S. West), 250 birds at Bitter Lake National Wildlife Refuge, Chaves County, on 1 April 1974 (Hubbard 1974a, *fide* D. Boggs), 175 at a roost at Rattlesnake Springs, Eddy County, on 16 April 1995 (Snider 1995, *fide* S. West), 165 at Holloman Lake, Otero County, on 3 May 1995 (Snider 1995, *fide* G. Ewing), and 150 birds again at Bitter Lake National Wildlife Refuge on 16 August 1974 (Hubbard 1974b, *fide* D. Boggs).

Frijoles Canyon in northern Sandoval County has had its Turkey Vulture communal roost since approximately 1939. The roost was monitored during the 1980s by Jim Travis (pers. comm.), who at the time found Turkey Vultures to roost in the cottonwoods along the

PHOTO 3.9

(*left*) Turkey Vulture roost site near Chama, Rio Arriba Co., in 1997. PHOTOGRAPH: © JOHN P. DELONG.

PHOTO 3.10

(*top right*) Turkey Vulture roost in the Heron Lake area, Rio Arriba Co., 10 April 2006. PHOTOGRAPH: © ROGER HOGAN.

PHOTO 3.11

(*bottom right*) Turkey Vulture roost in the Heron Lake area, Rio Arriba Co., 24 August 2007. PHOTOGRAPH: © ROGER HOGAN.

local creek near the visitor center. Since about 2001 some (typically 10–15) of the vultures have started to use a twin-topped conifer in the same area, while spring migrants in 2009 roosted also on the canyon walls. The first author has monitored the Frijoles Canyon roost since 1995, 10 times a year in recent years. The largest number of vultures counted was 95, on 11 June 2000. The historical mean number of vultures at the roost in Frijoles Canyon is around 80 (Travis 1992).

Diet and Foraging

Unlike most birds, Turkey Vultures have an excellent sense of smell that is put to good use in finding carrion while foraging opportunistically in flight. If food is concealed from view, Turkey Vultures rely on their sense of smell and approach from the downwind side. Turkey Vultures eat food almost entirely on the ground where it is found or drag food a few meters at most.

Turkey Vultures prefer relatively fresh carrion within a day or two of death. Carrion that is very old with large volumes of insect larvae may be inedible for these birds. Carrion typically consists of dead mammals and much less commonly of dead birds, reptiles, and fish (Kirk and Mossman 1998). The species takes live prey occasionally in unnatural situations, such as in captivity, although there are also reports of Turkey Vultures attacking young birds or green sea turtle hatchlings (Fowler 1979; Kirk and Mossman 1998).

PHOTO 3.12

Turkey Vultures on a cow carcass in Wagon Mound, Mora Co., July 2009. Turkey Vultures are scavengers with a keen sense of smell that helps them locate carrion. PHOTOGRAPH: © JEAN-LUC CARTRON.

Predation and Interspecific Interactions

According to Kirk and Mossman (1998), Turkey Vultures have few documented predators. Tyler (1937, cited by Kirk and Mossman 1998) reports a Bald Eagle (*Haliaeetus leucocephalus*) killing a Turkey Vulture. Based on work in southern Pennsylvania and northern Maryland, Jackson (1983) suggests that predation on adults may be highest at roost sites, with fox (*Vulpes* sp. or *Urocyon* sp.), opossum (*Didelphis virginiana*), and possibly raccoon (*Procyon lotor*) being the main predators. Predation on eggs, nestlings, and adults at nests has also been documented, though rarely, involving opossums, foxes, and domestic dogs (Howes 1926; Jackson 1983).

PHOTO 3.13

Turkey Vulture and Black-billed Magpie (*Pica hudsonia*) in a dead tree on 25 July 2005 near Paseo del Pueblo Norte in Taos, Taos Co. The Turkey Vulture in the photo is being mobbed by three magpies, only one of which is visible here. PHOTOGRAPH: © GERAINT SMITH.

PHOTO 3.14

Turkey Vulture and one of two Black-billed Magpies (*Pica hudsonia*) mobbing it in a field in Arroyo Hondo, Taos Co., 28 August 2005. Turkey Vultures are mobbed by small birds less frequently than other raptors. Intriguingly, however, mobbing of Turkey Vultures by Black-billed Magpies appears to be common in northern New Mexico.

PHOTOGRAPH: © GERAINT SMITH.

Turkey Vultures are less territorial than other raptors, and they do not seem to defend territories against other birds of other species in particular (see Kirk and Mossman 1998). Instead, most nonpredatory interspecific interactions involving Turkey Vultures have been observed at carcasses. Turkey Vultures are displaced at fresh carcasses by Red-tailed Hawks (*Buteo jamaicensis*) but dominant over American Crows (*Corvus brachyrhynchos*) (Prior and Weatherhead 1991). Turkey Vultures are also the victims of kleptoparasitic attacks by eagles. Both Bald Eagles and Golden Eagles (*Aquila chrysaetos*) have been observed chasing after Turkey Vultures, causing them to regurgitate food that is then consumed by the eagles (Oberholser 1906; Coleman and Fraser 1986; Evens 1991).

Status and Management

Rich et al. (2004) provided the first estimate of the total Turkey Vulture population, 4.5 million, 29% of which are in the United States and Canada during the summer. Rich et al. (2004) suggest the estimate is in the correct order of magnitude and give it a very high rating on repeatability because it is based on the standard Breeding Bird Survey (BBS) methods.

Butcher and Niven (2007) combined Breeding Bird Survey (BBS) and Christmas Bird Count (CBC) data to look for continent-wide trends on 550 species. For Turkey Vultures they found a 1.74% annual increase during the breeding season based on BBS data and a 4.8% annual winter increase based on CBC data. Overall Butcher and Niven (2007) suggest that Turkey Vultures may have experienced a 2.43% average population increase per year over the last 40 years (1965–2005). This amounts to an estimated 161% population increase over 40 years.

Wilbur (1983) indicates that historically Turkey Vulture populations have expanded and contracted. Cameron (1907) points to the occurrence of thousands of Turkey Vultures in eastern Montana during the slaughter of bison (*Bison bison*) in 1883. Through most of the subsequent 25 years, however, Turkey Vultures were rare in eastern Montana (Cameron 1907). In the Northeast, Turkey Vultures were not reported regularly in Vermont until about 1960 and now are a regular part of the summer avifauna of the state (Martin 1985). This range expansion appears to have been due to an increase in reliable food supply from road-killed animals and increased mortality of deer from starvation due to overpopulation (Sutton 1928; Bagg and Parker 1951; Meade 1988). A general increase in Turkey Vultures has also been reported for Michigan (Eastman 1991) since the 1890s and for New York (Meade 1998) since approximately the 1920s.

Hoffman and Smith (2003) examined migration counts from six sites in the western United States including the Sandia (spring 1985–2001) and Manzano (fall 1985–2001) mountains in New Mexico and made comparisons with BBS and CBC data from 1977 to 2000 (BBS) or 2001 (CBC). For Turkey Vultures they found a strong increasing trend at four of the six sites through the mid 1990s and a strong increasing trend

PHOTO 3.15

Turkey Vulture along State Road 92, Hidalgo Co., 15 October 2006. Turkey Vultures are known to incur mortality from collisions with both automobiles and aircraft, but these human-related mortality causes have not been studied in New Mexico.

PHOTOGRAPH: © ROBERT SHANTZ.

from BBS and CBC data for the same period. After 1998, CBC data show no distinct trend while migration counts for the six sites begin to decline. Hoffman and Smith (2003) suggest that the onset of severe drought across much of the interior western United States may be a major factor in the recent declines of Turkey Vultures as measured by migration counts.

A closer look at migration counts in the Sandia and Manzano mountains suggests that the Turkey Vulture population that migrates through central New Mexico has declined from the relatively high counts in the mid and late 1990s. Fall counts of Turkey Vultures during 1985–2006 in the Manzano Mountains peaked with 1,116 birds in 1998 but averaged only 317 during 2003–6, or below the annual mean of 396 birds during the 22-year period of records (Smith and Neal 2007a); spring counts of Turkey Vultures during 1985–2007 in the Sandia Mountains peaked with 3,245 birds in 1998 but averaged only 963 during 2004–7, or below the 23-year annual mean of 1,407 birds (Smith and Neal 2007b).

In eastern North America (east of 100° W longitude) there has been an apparent change in nest-site selection over time. Jackson (1983) found the use of tree cavities—relative to other nest sites—decreased from 53.0% prior to 1920 to 22.3% after 1920 while thicket use increased from 11.0% to 38.3%. Jackson (1983) attributes the change to the fact that most tree-cavity nests are in trees 150–200 years old while commercially valuable trees are now harvested before they are 100 years old. Similar changes in nest site selection have not been detected in western North America (Jackson 1983), probably because of the abundance and use of cave and rocky-habitat nest sites.

Coleman and Fraser (1989) suggest that human disturbance and canid predation may be important causes of nesting failure. Kirk and Mossman (1998) report that Turkey Vultures feeding on road-killed animals can commonly be killed or injured by motor vehicles along roadways. Collisions with aircraft are also of high management concern, especially for the U.S. Air Force (DeFusco 1993, as cited by Kirk and Mossman 1998). None of these potential sources of vulture mortality or nest failure has been studied in New Mexico.

The importance of communal roosting stands out as needing more research and understanding (Kirk and Mossman 1998). Dunn et al. (2005) suggest that expanding BBS routes in Mexico would be valuable for monitoring Turkey Vultures since more than two-thirds of the species' range is south of the current BBS coverage in the United States. Any management actions should consider nest-site and communal roost-site protection, minimizing deaths by collisions both on highways and in the air, and expanding citizen monitoring efforts. Whether flying gracefully through our skies or sunbathing with their wings spread out in the early morning at communal roosts, Turkey Vultures are an integral part of nearly all New Mexico landscapes. They are also scavengers, serving an important ecological function. As such they deserve our respect and protection.

Bagg, A. M., and H. M. Parker. 1951. The Turkey Vulture in New England and Eastern Canada up to 1950. *Auk* 68:315–33.

Bailey, F. M. 1928. *Birds of New Mexico*. Santa Fe: New Mexico Department of Game and Fish.

Bixler, S., comp. 2008. Christmas Bird Count data for Roswell, New Mexico. http://www.audubon.org/bird/cbc/ (accessed March 13, 2008).

Butcher, G. S., and D. K. Niven. 2007. *Combining date from the Christmas Bird Count and the Breeding Bird Survey to determine the continental status and trends of North American birds*. Ivyland, PA: Audubon Science Office, National Audubon Society.

Cameron, E. S. 1907. The birds of Custer and Dawson counties, Montana. *Auk* 24:241–70.

Cartron, J.-L. E., D. C. Lightfoot, J. E. Mygatt, S. L. Brantley, and T. K. Lowrey. 2008. *A field guide to the plants and animals of the Middle Rio Grande bosque*. Albuquerque: University of New Mexico Press.

Coleman, J. S., and J. D. Fraser. 1986. Predation on Black and Turkey Vultures. *Wilson Bulletin* 98:600–601.

———. 1989. Habitat use and home ranges of Black and Turkey Vultures. *Journal of Wildlife Management* 53:782–92.

Corman, T. E. 2005. Turkey Vulture. In *Arizona breeding bird atlas*, ed. T. E. Corman and C. Wise-Gervais, 118–19. Albuquerque: University of New Mexico Press.

Davis, D. 1993. Breeding behavior of Turkey Vultures. In *Vulture biology and management*, ed. S. R. Wilbur and J. A. Jackson, 271–86. Berkeley: University of California Press.

DeFusco, R. P. 1993. Modeling bird hazards to aircraft: A GIS application study. *Photogrammetric Engineering and Remote Sensing* 59:1481–87.

Dexter C. 1998. Turkey Vulture. In *Colorado breeding bird atlas*, ed. H. E. Kingery, 66–67. Denver: Colorado Wildlife Heritage Foundation.

Dunn, E. H., B. L. Altman, J. Bart, C. J. Beardmore, H. Berlanga, P. J. Blancher, G. S. Butcher, D. W. Demarest, R. Dettmers, W. C. Hunter, et al. 2005. High priority needs for range-wide monitoring of North American landbirds. Partners in Flight technical series no. 2. http://www.partnersinflight.org/pubs/ts/02-MonitoringNeeds.pdf.

Dunne P., D. Sibley, and C. Sutton. 1988. *Hawks in flight*. Boston: Houghton Mifflin.

Eakle, W. L., E. L. Smith, S. W. Hoffman, D. W. Stahlecker, and R. B. Duncan. 1996. Results of a raptor survey in southwestern New Mexico. *Journal of Raptor Research* 30:183–88.

Eastman, J. 1991. Turkey Vulture. In *The atlas of breeding birds of Michigan*, ed. R. Brewer, G. A. McPeek, and R. J. Adams Jr., 156–57. East Lansing: Michigan State University Press.

Evens, J. G. 1991. Golden Eagle attacks Turkey Vulture. *Northwestern Naturalist* 72:27.

Fowler, L. E. 1979. Hatching success and nest predation in the green sea turtle, *Chelonia mydas*, at Tortuguero, Costa Rica. *Ecology* 60:946–55.

Hoffman, S. W., and J. P. Smith. 2003. Population trends of migratory raptors in western North America, 1977–2001. *Condor* 105:397–419.

Howell, S. N. G., and S. Webb. 1995. *A guide to the birds of Mexico and northern Central America*. New York: Oxford University Press.

Howes, P. G. 1926. A Turkey Vulture nest in the state of New York. *Bird Lore* 28:175–80.

Hubbard, J. P., ed. 1974a. *New Mexico Ornithological Society Field Notes* 13(2): 1 June–30 November.

———, ed. 1974b. *New Mexico Ornithological Society Field Notes* 13(1): 1 December 1973–31 May.

———. 1978. *Revised check-list of the birds of New Mexico*. Publ. no. 6. Albuquerque: New Mexico Ornithological Society.

———. 1983. Roadside raptor counts as an indicator of the status of the Turkey Vulture in New Mexico. In *Vulture biology and management*, ed. S. R. Wilbur and J. A. Jackson, 375–84. Berkeley: University of California Press.

Jackson, J. A. 1983. Nesting phenology, nest site selection, and reproductive success of Black and Turkey Vultures. In *Vulture biology and management*, ed. S. R. Wilbur and J. A. Jackson, 245–70. Berkeley: University of California Press.

Jaramillo, A., P. Burke, and D. Beadle. 2003. *Birds of Chile*. Princeton, NJ: Princeton University Press.

Kirk, D. A., and M. J. Mossman. 1998. Turkey Vulture (*Cathartes aura*). In *The birds of North America*, ed. A. Poole and F. Gill. Philadelphia, PA: Birds of North America, Inc.

Ligon, J. S. 1961. *New Mexico birds and where to find them*. Albuquerque: University of New Mexico Press.

Liguori, J. 2005. *Hawks from every angle*. Princeton, NJ: Princeton University Press.

Martin, N. 1985. Turkey Vulture. In *The atlas of breeding birds of Vermont*, ed. S. B. Laughlin and D. P. Kibbe, 70–71. Hanover, NH: University Press of New England.

Meade, G. M. 1988. Turkey Vulture. In *The atlas of breeding birds in New York State*, ed. R. F. Andrle and J. R. Carroll, 96–97. Ithaca, NY: Cornell University Press.

Miskimen, M. 1957. Absence of syrinx in the Turkey Vulture (*Cathartes aura*). *Auk* 74:104–5.

Mueller, H. 1972. Zone-tailed Hawk and Turkey Vulture: Mimicry or aerodynamics? *Condor* 74:221–22.

[NMOS] New Mexico Ornithological Society. 2009. *NMOS Field Notes* database. http://nhnm.unm.edu/partners/NMOS/ (accessed 2 May 2009).

Oberholser, H. C. 1906. *The North American eagles and their economic relations*. Bulletin 27. Washington, DC: U.S. Biological Survey.

Parmeter, J., B. Neville, and D. Emkalns. 2002. *New Mexico bird finding guide*. Albuquerque: New Mexico Ornithological Society.

Prior, K. A., and P. J. Weatherhead. 1991. Competition at the carcass: Opportunities for social foraging by turkey vultures in southern Ontario. *Canadian Journal of Zoology* 69:1550–56.

Rich, T. D., C. J. Beardmore, H. Berlanga, P. J. Blancher, M. S. W. Bradstreet, G. S. Butcher, D. W. Demarest, E. H. Dunn, W. C. Hunter, E. E. Iñigo-Elias, et al. 2004. *Partners in Flight North American landbird conservation plan*. Ithaca, NY: Cornell Laboratory of Ornithology.

Schmitt, C. G. 1976. *Summer birds of the San Juan Valley, New Mexico*. Publ. no. 4. Albuquerque: New Mexico Ornithological Society.

Smith, J. P., and M. C. Neal. 2007a. *Fall 2006 raptor migration studies in the Manzano Mountains of central New Mexico*. Salt Lake City, UT: HawkWatch International, Inc.

———. 2007b. *Spring 2007 raptor migration study in the Sandia Mountains of central New Mexico*. Salt Lake City, UT: HawkWatch International, Inc.

Snider, P. R., ed. 1993a. *New Mexico Ornithological Society Field Notes* 32(3): 1 June–31 July.

———, ed. 1993b. *New Mexico Ornithological Society Field Notes* 32(4): 1 August–30 November.

———, ed. 1995. *New Mexico Ornithological Society Field Notes* 34(2): 1 March–31 May.

Sutton, G. M. 1928. Extension of the breeding range of the Turkey Vulture in Pennsylvania. *Auk* 45:501–3.

Travis, James R., ed. 1992. *Atlas of the breeding birds of Los Alamos County, New Mexico*. Pajarito Ornithological Survey, LA-12206, Los Alamos, NM: Los Alamos National Laboratory.

Tyler, W. M. 1937. *Cathartes aura septentrionalis*, Wied. Turkey Vulture. In *Life histories of North American birds of prey*. Vol. 1, ed. A. C. Bent, 12–28. Bulletin 167. Washington, DC: U.S. National Museum.

Wilbur, S. R. 1983. The status of Vultures in the Western hemisphere. In *Vulture Biology and Management*, ed. S. R. Wilbur and J. A. Jackson, 113–23. Berkeley: University of California Press.

Williams, S. O. III, ed. 1996. *New Mexico Ornithological Society Field Notes* 35(4): Fall 1996.

———, ed. 1997. *New Mexico Ornithological Society Field Notes* 36(3): Summer 1997.

———, ed. 2003. *New Mexico Ornithological Society Field Notes* 42(3): Summer 2003.

Willis, E. O. 1963. Is the Zone-tailed Hawk a mimic of the Turkey Vulture? *Condor* 65:323–17.

4

Osprey
(*Pandion haliaetus*)

DALE W. STAHLECKER

THE OSPREY (*Pandion haliaetus*) is so unique among diurnal birds of prey that it is taxonomically placed in its own monotypic subfamily (Pandioninae), having diverged from the rest of the Accipitridae (most hawks and eagles) 24–30 million years ago (Poole et al. 2002). Strikingly marked with a black "mask" over piercing, bright yellow eyes, an Osprey with raised hackles appears comical and majestic at the same time. Ospreys are larger than hawks but smaller than the true eagles. Sibley (2000) gives averages for the species as 58.5 cm (23 in) in length, a wingspan of 160 cm (63 in), and a weight of 1,600 g (3.5 lb). North American Ospreys range in weight from 1,400 to 2,000 g (3–4.5 lb.), with females averaging 25% heavier (Poole et al. 2002).

PHOTO 4.1

Osprey pair, Heron Lake, Rio Arriba Co., 5 April 2007. The female (right) is larger and typically shows conspicuous dark speckling on the chest, forming a necklace. In males, the necklace tends to be less pronounced or absent. PHOTOGRAPH: © ROGER HOGAN.

PHOTO 4.2

(*top left*) Adult Osprey, Heron Lake, Rio Arriba Co., 25 March 2008. Note the dark chocolate upperparts and the white head with the dark eye-line. Wingtips extend beyond tail tip. PHOTOGRAPH: © ROGER HOGAN.

PHOTO 4.3

(*top right*) Adult male Osprey, Heron Lake, Rio Arriba Co., 2006. Both males and females have white underparts, but males often lack the necklace shown by females. PHOTOGRAPH: © ROGER HOGAN.

PHOTO 4.4

(*left*) Adult Osprey in flight, Rio Arriba Co., 25 June 2007. Note the dark "elbow" or carpal patch. PHOTOGRAPH: © ROGER HOGAN.

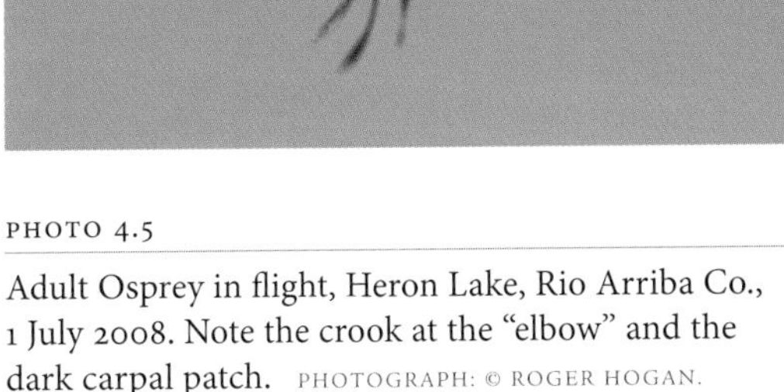

PHOTO 4.5

Adult Osprey in flight, Heron Lake, Rio Arriba Co., 1 July 2008. Note the crook at the "elbow" and the dark carpal patch. PHOTOGRAPH: © ROGER HOGAN.

PHOTO 4.6

Osprey fledgling near its nest, Heron Lake, Rio Arriba Co., 24 August 2007. With their conspicuous light-buff feather edges, fledglings are easily recognized from adults. PHOTOGRAPH: © ROGER HOGAN.

However, females only average 6% larger for four other body measurements (Prevost 1983; Poole 1989).

Four subspecies are recognized, but only one, *P. h. carolinensis*, occurs in North America. An adult *P. h. carolinensis* is dark chocolate-brown above and mostly white below. The head is also mostly white, with a dark line through the eye, which has a dark pupil surrounded by the yellow iris. Females, and some males, have dark speckling on the chest forming a "necklace." Feet and cere are pale blue-gray. Immatures are like adults, except that their irises are orange-red for about a year, and their back and wing covert feathers have light-buff edges, giving them a dappled appearance. In nestlings the eye-line feathers also have light edges, making them appear almost blond. As the edging wears away, immatures become difficult to distinguish from adults.

In flight, Ospreys of all ages are uniquely shaped and marked. Even at a distance, a distinctive crook in the wing at the "elbow," actually the carpal joint, is diagnostic. They are also uniformly dark above and pale below with dark "elbow" patches. The distinctive white head with the dark eye-line is usually easily seen. While Ospreys give a variety of calls (Poole et al. 2002), their sharp whistled guard call is distinctive, and often given when any intruder comes too close to a nest or sometimes even to a hunting perch in migration or on the wintering grounds (Poole et al. 2002).

Distribution

Ospreys occur on every continent but Antarctica. Except for Australia, their breeding range is restricted to the Northern Hemisphere (Poole et al. 2002). Most migrants winter north of the equator, but some migrate far into the Southern Hemisphere. Strays occasionally appear in the Hawaiian Islands, the world's most remote archipelago (Pratt et al. 1987). Most Ospreys breeding at lower latitudes are resident (Poole 1989; Poole et al. 2002).

In North America Ospreys are particularly abundant along the Gulf of Mexico, the southern Atlantic Coast, and the northern Atlantic Coast from Chesapeake Bay (>3,000 pairs; Watts et al. 2004) north through New England and well into eastern Canada (Poole et al. 2002). Their inland breeding range covers much of the boreal forests of Canada into central Alaska, with concentrations in the Great Lakes states and provinces. In western North America, nesting populations range along the Pacific Coast from Alaska south to San Francisco Bay, California, and inland to the Rocky Mountains of Idaho, Montana, and northwestern Wyoming. More isolated populations are scattered through Utah, Colorado, eastern Arizona, and north-central New Mexico, mostly in mountainous regions with rivers and natural and man-made lakes. Northwestern Mexico has resident breeding populations both along the Pacific Coast of Baja California and along the Gulf of California (Henny and Anderson 2004).

More northern breeding Ospreys winter from Texas and southern Oregon southward into South America (Poole et al. 2002). Band recoveries (Poole and Agler 1987) and satellite telemetry (Martell et al. 2001) have shown that Ospreys from each geographical region in North America occupy relatively distinct wintering areas. Northeastern Osprey populations winter in the Caribbean and northern South America. Western Osprey populations winter primarily in southern Mexico and Central America. Finally, the winter distribution of Ospreys from central North America overlaps that of the other two sets of populations.

While New Mexico is commonly thought of as a desert state, rivers and lakes are found throughout. Ospreys reportedly nested along the West Fork of the Gila River in 1916 and 1920 (Bailey 1928; Ligon 1961). An egg set housed at the University of New Mexico Museum of Southwestern Biology (egg set MSB E-68) was collected on 23 April 1916 by J. S. Ligon, the first documented case of nesting by the Osprey in New Mexico. The locality given for the egg set, "0.5 mi above T6 Ranch" in what was then Socorro County, likely corresponds to the West Fork of the Gila, corroborating Bailey (1928) and Ligon (1961). Catron County,

PHOTO 4.7

Osprey on nest at Elephant Butte Reservoir in 2001.

PHOTOGRAPH: © GORDON FRENCH.

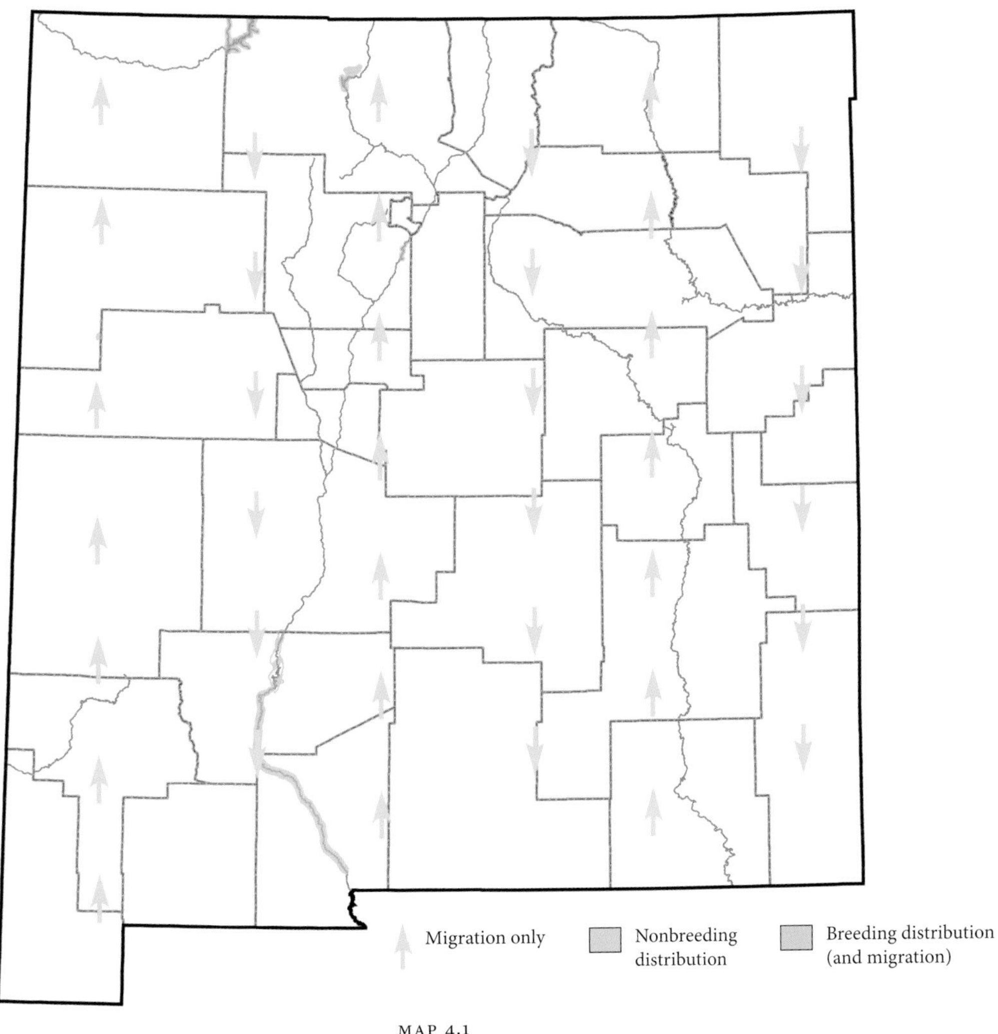

MAP 4.1

Osprey distribution map

which includes all of the West Fork of the Gila, was split off from Socorro County only in 1921.

Between 1920 and 1990, however, there were few reports of nesting from New Mexico, and none of them was substantiated. The first tantalizing note suggesting additional nesting was provided by Ligon (1961:77): Osprey "nests have been broken up near the Seven Springs State Fish Hatchery northwest of Jemez Springs, where the birds preyed on fish in the several large brood ponds," but with no details as to when and how often. Ospreys reportedly nested on a rock outcrop near Elephant Butte Dam in 1980 (*fide* C. Sandell, letter to the New Mexico Department of Game and Fish, 1992), nested in a tree in the upper reaches of the lake in 1989 (Goodman 1989, *fide* R. A. Fischer), built a nest in a side canyon below the dam in 1992 (C. Sandell, op. cit.), and nested again in an upper lake tree in 2001. While some of these nesting reports were secondhand accounts and again lacking in documentation, the 2001 nest was photographed by Gordon French (photo 4.7), a retired geoscientist and wildlife photographer. In fact, two large young were also photographed, but fledging success was never determined as dropping lake levels made it impossible to reach

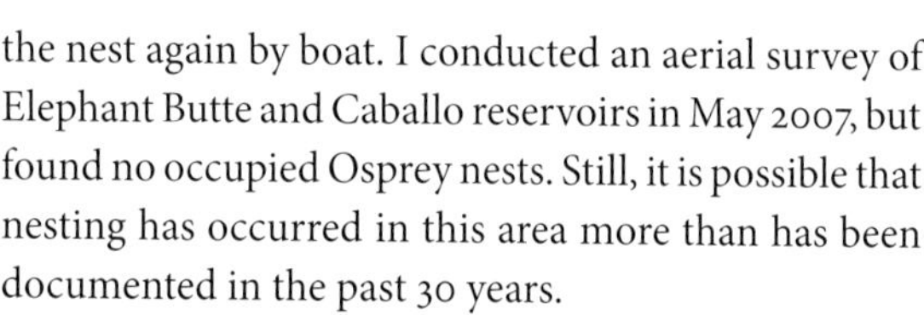

PHOTO 4.8

(*above*) Osprey pair hovering over a nesting platform along Highway 64 near Tierra Amarilla in the Chama Valley, Rio Arriba Co., 2 June 2006. PHOTOGRAPH: © GERAINT SMITH.

PHOTO 4.9

(*right*) Moonset over an Osprey in nest at Heron Lake, Rio Arriba Co., 2006. PHOTOGRAPH: © ROGER HOGAN.

the nest again by boat. I conducted an aerial survey of Elephant Butte and Caballo reservoirs in May 2007, but found no occupied Osprey nests. Still, it is possible that nesting has occurred in this area more than has been documented in the past 30 years.

Ospreys were first found to be nesting consistently in north-central New Mexico in the 1990s (Stahlecker 2002). By 2008, nesting had been documented in five different counties, including Sandoval starting in 1997, McKinley in 2001, Sierra in 2001, and San Juan in 2008. But only in Rio Arriba County has more than one nest been found. There population growth was slow through the 1990s, but gained momentum late in the decade. By the 2008 breeding season, 18 pairs were present in Rio Arriba County (see below), 9 at Heron Lake, 5 at El Vado Lake, and 4 in the Chama Valley in the vicinity of Tierra Amarilla, with also a nesting attempt on the nearby Jicarilla Nation in 2007.

A nesting pair first appeared at Cochiti Lake, Sandoval County, in 1997. A pair had returned every year until 2005, when only a male was present through the spring. No resident Ospreys were recorded at Cochiti Lake in 2006 and 2007 but a pair reoccupied the nest site in 2008. In western New Mexico, nesting Ospreys were first documented at Ramah Lake, McKinley County, in 2002. The newest pioneering pair appeared below Navajo Dam, San Juan County, in 2008 near the Quality Trout Waters of the San Juan River.

Ospreys, especially males, are likely to return to their natal area to breed (Poole et al. 2002). In southern New England all males ($n = 33$) and 80% of females ($n = 39$) first bred within 50 km (~30 mi) of their natal nests (Poole 1989). In Michigan, the mean distance from the natal nest to the first breeding nest was 14.5 km (9 mi) ($n = 37$) for males and 38 km (24 mi) ($n = 31$) for females (Postupalsky 1989). Both populations were increasing, with many artificial nest sites available. Even in an established population in Finland, where the availability of nest sites was presumably more limited, the median and maximum distances from natal to breeding nests were 47 and 145 km (29 and 90 mi) for males and 135 and 303 km (84 and 188 mi) for females (Saurola 1995). Therefore it is most probable that the Ospreys that pioneered into north-central New Mexico were from nearby southwestern Colorado, where breeding by Ospreys has been documented since at least 1961 (Bailey and Niedrach 1965). With 18 pairs, the northern Rio Arriba County population now equals its presumed southwestern Colorado source population, where 14–15 pairs were confirmed in 2004 after two years of surveys (Stahlecker and Brady, pers. obs.).

The first nesting pair at Cochiti Lake likely originated from either Heron/El Vado or southwestern Colorado. However, it is more probable that the Ramah Lake Osprey pair pioneered into New Mexico from the central Arizona population, which is much closer. The Arizona population was reported as eight pairs in the mid 1980s (Vahle et al. 1988), and as more than 50 pairs by the late 1990s (Dodd and Vahle 1998). The intermittently found nests at Elephant Butte Reservoir also were closer to the Arizona population (~200 km [~125 mi]) than Heron/El Vado (~370 km [~230 mi]), suggesting the possibility of other undocumented pairs at scattered lakes and, perhaps again, on the Gila River in the intervening region. Finally, the latest pioneers at Navajo Dam represent an expansion of the "Four Corners" Osprey population, which now has more than 30 pairs (see Stahlecker 2002).

Ospreys are also regular in New Mexico during migration. Between 1985 and 2001 on average 26 (+/- 12) Ospreys were counted over the Manzano Mountains during autumn migration and 62 (+/- 30) were counted over the Sandia Mountains in spring migration (Hoffman and Smith 2003). One adult with a satellite transmitter that had nested in eastern Oregon passed through New Mexico during fall migration (Martell et al. 2001).

PHOTO 4.10

(*above*) Osprey pair at Cochiti Lake, Sandoval Co., April 2004. A pair first appeared at Cochiti Lake in 1997 and returned to that location every year through 2004. In 2004, the pair did not breed, and no female returned the following year. No Osprey was present in 2006 or 2007. An apparently new pair occupied the site in 2008 but likely did not lay eggs. PHOTOGRAPH: © MARK L. WATSON.

PHOTO 4.11

(*left*) Osprey near Sumner Lake, De Baca Co., during the fall migration season, 28 September 2008. PHOTOGRAPH: © ROBERT SHANTZ.

Migrating Ospreys are attracted to New Mexico lakes and rivers during their passage through the state, particularly in areas where open water is scarce. On 11 April 2001, I counted 14 Ospreys at Quemado Lake in west-central New Mexico, though none currently nests there. Quemado Lake is only 53 ha (130 ac) in size but it is more than 64 km (40 mi) from any other large body of water. It supports a thriving, largely introduced fish population, including significant numbers of goldfish (*Cyprinus auratus*), illegally released there in the past. While March migrants are likely returning to New Mexico and other interior breeding locations, April migrants are probably returning to Canada and Alaska, where breeding begins later.

A few Ospreys winter each year in New Mexico. Most sightings reported by birdwatchers are from Elephant Butte Reservoir and its vicinity. However, Ospreys are irregularly reported in winter at other lakes statewide and as far north as Navajo Reservoir on the Colorado border.

Habitat Associations

As a near-obligate fish predator, Ospreys occur mainly at lakes or along rivers. Currently breeding Ospreys are found mostly at mid-elevation (>2,130 m [>7,000 ft]) lakes, where near-shore ponderosa pines (*Pinus ponderosa*), pinyon pines (*P. ebulis*), and power poles provide nest sites and perches. The one documented nest site at Elephant Butte was much lower, 1,370 m (4,500 ft), and in a snag surrounded by water (G. French, pers. comm.). This is an unusual nest site for New Mexico, but not atypical for the species rangewide, as it often nests in flood-killed trees in reservoirs (Poole et al. 2002). Of course, migrants can be observed along mountain ridges, utilizing thermals and winds to conserve energy during long flights across xeric landscapes (see chapter 2).

PHOTO 4.12

Heron Lake with the sun setting over the Brazos Cliffs, Rio Arriba Co., May 2005.

PHOTOGRAPH: © DALE W. STAHLECKER.

Life History

Ospreys must be at least three years old to breed (Poole et al. 2002). Early banding studies showed that most young from North America remained on their wintering grounds for 18 months, including a full spring and summer following the year they were fledged. This "teenage" period in the life history of the species is least known, though satellite telemetry has the potential to soon provide insight. Young Ospreys migrate back to North America in the spring before turning two years old, but may wander widely. After one more migration, they are likely to return to their natal area when three years old (Henny and Van Velzen 1972) and will attempt to breed if they can find a mate and a nest site.

Nesting

Adult Ospreys begin returning to New Mexico in mid March, but some nest sites are not reoccupied until mid April. The nest is important to the pair bond, and even pairs that do not lay eggs will work on one or more nests through the summer. Nests are built with a collection of sticks and lined with bark, grasses, flotsam, or other trash. Nests on man-made platforms are relatively short and flat, perhaps because winter winds blow away materials not interlaced with supports. However, nests in snags are usually noticeably larger. All reoccupied nests are repaired each year before eggs are laid. Favored nest sites are usually repeatedly used until lost; the power pole used by the first New Mexico pair at El Vado Lake was reused for 16 consecutive years.

New Mexico's core population in northern Rio

Arriba County is largely dependent upon artificial nest platforms, and accelerated growth of this population (see below) coincided with the initiation of active nest site management in 1997. Nest platforms have been placed above crossbars on power poles selected by Osprey pairs, or have been placed in snags or dead-top ponderosa pines to attract the birds away from power poles. Since 2000, about 75% of northern New Mexico Osprey nests have been on man-made platforms on poles, and the remaining 25% split equally between natural snags and platforms placed on topped trees or snags. Between 1990 and 2005, 85 of 127 (67%) nesting attempts statewide have been on platforms on poles, 17 (14%) on platforms in trees, and 25 (19%) on

PHOTO 4.13

(*top left*) Nest platform on a topped tree, El Vado Lake, Rio Arriba Co., 13 June 2007. PHOTOGRAPH: © ROGER HOGAN.

PHOTO 4.14

(*bottom left*) Nest platform on a pole, El Vado Lake, Rio Arriba Co., 10 July 2007. PHOTOGRAPH: © ROGER HOGAN.

PHOTO 4.15

(*above*) Osprey nesting platform on a pole at the El Vado Lake dam, Rio Arriba Co., 20 April 2007. A very heavy wind blew the nest apart in the summer of 2008. The pole and the platform were left intact. PHOTOGRAPH: © ROGER HOGAN.

totally natural sites (trees or snags). But while overall in North America Osprey pairs using man-made nesting substrates fledge significantly more young than those at natural nest sites (Poole et al. 2002), in New Mexico the fledging rates are almost equal at natural (1.25/attempt) and completely man-made (1.31/attempt) sites, and much lower on platforms in trees (0.70/attempt).

No nests in New Mexico have been visited during incubation. Clutches range in size from one to four eggs, with three the mode in most populations (Poole et al. 2002). In a migratory population in Michigan, average clutch size was 2.92 (n = 537), with 76% of clutches containing three eggs (Postupalsky 1989). The Osprey egg set housed at the University of New Mexico Museum of Southwestern Biology consists of two eggs (see Ligon 1961).

Bent (1937:361) describes Osprey eggs as "the handsomest of all hawk's eggs"; base color varies from creamy white to cinnamon, with reddish markings especially on the large end of the eggs (Poole et al. 2002). In a well-studied population, older females nested as much as two weeks earlier than younger females (Poole 1989). While no females in New Mexico were individually marked, females at nest sites occupied at least three years were often incubating by mid April, while at newer nest sites initiation of incubation was delayed as much as a month. Incubation requires 36–40 days (Steeger et al. 1992); therefore the first hatching in New Mexico generally occurs during the week before Memorial Day. Young fledge when 50–55 days old (Poole 1989); in New Mexico fledging usually occurs from mid July to early August.

PHOTO 4.16

(*above*) Sticks brought to the nest can sometimes be large and cumbersome! Heron Lake, Rio Arriba Co., 20 June 2007. PHOTOGRAPH: © ROGER HOGAN.

PHOTO 4.17

(*top right*) Pair during copulation at nest at Heron Lake, Rio Arriba Co., 22 March 2008. PHOTOGRAPH: © ROGER HOGAN.

PHOTO 4.18

(*right*) Adult with three-quarter-grown nestlings, Heron Lake State Park, Los Ojos, Rio Arriba Co., 3 July 2007. PHOTOGRAPH: © ROGER HOGAN.

PHOTOS 4.19a and b

Aerial duel between a paired male, holder of the nest territory, and an intruder, Rio Arriba Co., 28 March 2009. The female of the pair watched the fight from a tree approximately 300 m (~1,000 ft) away.

PHOTOGRAPHS: © ROGER HOGAN.

PHOTO 4.20

(*top left*) Male delivering fish to the nest, Heron Lake, Rio Arriba Co., summer 2006. PHOTOGRAPH: © ROGER HOGAN.

PHOTO 4.21

(*middle left*) Nestling in nest, Heron Lake, Rio Arriba Co., 25 June 2008. PHOTOGRAPH: © ROGER HOGAN.

PHOTO 4.22

(*bottom left*) Adult bringing fish to a nest near Tierra Amarilla in the Chama Valley, Rio Arriba Co., 1 August 2005. PHOTOGRAPH: © GERAINT SMITH.

PHOTO 4.23

(*top right*) Osprey with fish, Heron Lake, Rio Arriba Co., 19 June 2007. PHOTOGRAPH: © ROGER HOGAN.

PHOTO 4.24

Three very large nestlings ready to fledge from what has become a rather messy nest (with sticks dangling over the edge of the nest platform). Near Tierra Amarilla in the Chama Valley, Rio Arriba Co., 15 August 2005.

PHOTOGRAPH: © GERAINT SMITH.

Diet and Foraging

Ospreys will hunt from a perch if it is close to a prime foraging area, but more often they hunt from the wing, flapping, hovering, or gliding approximately 10–30 m (~30–100 ft) above the surface of the water (Poole et al. 2002). It is not uncommon, at Heron and El Vado lakes, to scan across the water and see two or three males in flight. Meanwhile, the females wait on the nest, with eggs or young, patiently or impatiently depending upon the time elapsed since their last meals. Ospreys normally plunge into the water to capture fish, then must push off the water with their wings, shake off wet plumage, align the prey aerodynamically, and carry it to a perch or to the nest.

Range-wide, live fish have represented more than 99% of all prey recorded (Poole et al. 2002), though fish species taken vary greatly geographically. More than 80 prey species have been recorded in North America. However, in many cases only two to three abundant species make up the bulk of prey taken (Poole et al. 2002). In northern New Mexico, it is likely that introduced trout and salmon (*Oncorhynchus* sp.) are used heavily, though nongame fish are likely often taken as well (photos 4.25–4.27). Several Osprey nests are close

PHOTO 4.25

(*top*) Osprey with catfish at Heron Lake, Rio Arriba Co., 25 June 2008. PHOTOGRAPH: © ROGER HOGAN.

PHOTO 4.26

(*bottom left*) Osprey with rainbow trout (*Oncorhynchus mykiss*), El Vado Lake, Rio Arriba Co., 30 March 2007. PHOTOGRAPH: © ROGER HOGAN.

PHOTO 4.27

(*bottom right*) Osprey with carp, Heron Lake, Rio Arriba Co., 8 May 2007. PHOTOGRAPH: © ROGER HOGAN.

to a state fish hatchery, which the males reportedly regularly raid.

On 27 June 2006 an almost complete fish die-off occurred at Ramah Lake at a time when water level was very low. Dry conditions were responsible for underwater plant decomposition, which itself caused extremely low oxygen levels and, in turn, the fish die-off. Despite this drastic drop in prey numbers, the resident Osprey pair successfully fed itself and two young through July and remained at the lake through at least mid August. While the number of fish drastically dropped, at least initially those that were still alive probably remained near the surface, weakened and thus more vulnerable (Stahlecker et al., unpubl. data).

Predation and Interspecific Interactions

Outside of New Mexico, documented predators of eggs or young have included Bald Eagles (*Haliaeetus leucocephalus*), Great Horned Owls (*Bubo virginianus*), and raccoons (*Procyon lotor*; Poole et al. 2002). However, there are no documented nestling or adult losses in New Mexico. Half-grown nestlings disappeared or died in one nest on a platform at Heron Lake in 2005; predation could not be ruled out (but see below). Otherwise and overwhelmingly, most nestlings appeared to have fledged in all New Mexico nests monitored between 1993 and 2008, based on reproductive outcome at nests where evidence of hatching existed.

Where Bald Eagles co-occur with Ospreys, they are well known for stealing freshly caught fish from Ospreys (Poole et al. 2002). Although the two species now have overlapping breeding distributions in northern New Mexico, no piracy by Bald Eagles has yet been observed in the state . At Heron Lake on 4 April 2005, a male Osprey carrying a fish toward its nest was repeatedly dived upon by a Common Raven (*Corvus corax*). Both birds were quite vocal. When last seen, the Osprey appeared to have dropped the fish.

Status and Management

New Mexico Ospreys, like their conspecifics elsewhere throughout the world, are often tolerant of human activity near their nest sites. However, since 1990 occupied sites less than 50 m (~160 ft) (hereafter "exposed sites") from highways, campgrounds, or other busy areas (table 4.1) have had a noticeably lower rate of nesting success and fledged on average only half the number of young compared to those isolated from potential human disturbance by at least 100 m (~330 ft) ("buffered sites"). Surprisingly, "remote" sites (nesting sites at least 200 m [~660 ft] from any human activities) have had a success rate and a fledging rate more comparable to those of exposed sites. However, more than half of the nests have been at buffered sites, many of which have had long (>3 years) occupancy histories. Thus buffered sites were likely occupied by more experienced adults, who on average fledge more young than less experienced pairs (Poole et al. 2002).

Which factor has had the most pronounced effect on reproductive outcome in New Mexico in recent years? A logistic regression (with Akaike's Information Criterion) was used to examine the relative contribution

TABLE 4.1. Effect of proximity to human activity on Osprey (*Pandion haliaetus*) nesting success and productivity in New Mexico, 1990–2005.[a]

	DESCRIPTION	NO. NESTING ATTEMPTS	NO. (%) SUCCESSFUL ATTEMPTS	PRODUCTIVITY
Exposed	<50 m (~160 ft) from regular human activity	35	14 (40)	0.80
Buffered	>100 m (~330 ft) from most human activity	71	49 (69)	1.51
Remote	>200 m (~660 ft) from most human activity	20	10 (50)	0.90
Total		126	73 (58)	1.21

[a] Productivity is the number of young fledged per nesting attempt. Successful nesting attempts are those that led to fledging of at least one young. The Elephant Butte Reservoir nest found occupied in 2000 is excluded here, as reproductive outcome could not be determined (G. French, pers. comm.).

of three variables toward determining reproductive outcome. Besides proximity to human activity and the number of years since first occupancy (a surrogate variable for age and experience of the nesting pair), reservoir level was also included as one of the three variables, given that El Vado and Heron lakes in particular have had low water levels since 2000 (and as recently as 2005). The best model identified by the logistic regression was the one that combined territory age and proximity to human activity, with no interactions between those two variables (likelihood ratio test Chi-square = 17.5, df = 3, P = 0.001). Reservoir water levels were not a significant factor in Osprey nesting success. As elsewhere in the species' range (Poole et al. 2002), the odds of a New Mexico territory being successful in fledging young increased markedly after the first year of occupancy. Of 15 sites monitored from the time of first occupancy, only 2 (13%) were successful the first year, while 4 (27%) were first successful the second year and 5 (33%) first succeeded in their third year. Still, 4 (27%) had not yet been successful through 2006, even after as many as five years of occupancy. Meanwhile, the odds of success at buffered sites were 1.7 greater than at exposed sites. This suggests that future platforms should be placed more than 100 m (~330 ft) from potential human activities and current platforms within 50 m (~160 ft) of human activities should be moved farther from them if possible.

Productivity in the northern Rio Arriba County subpopulation (fig. 4.1) has averaged 1.2 young/occupied territory (1990–2008). Recovering Osprey populations in New England (Poole 1989) and Michigan (Postupalsky 1989) stabilized at 0.8–0.9 young/occupancy. In Chesapeake Bay, where Ospreys start breeding (on average) at 5.7 years of age, a rate of 1.15 young/occupancy appears necessary for stability, and it was calculated that if an Osprey began breeding at 6.7 years of age instead, a rate of 1.3 young/occupancy would be required (Poole 1989). Northern Rio Arriba County Ospreys have likely been entering the breeding population at a younger age, though this should change as the habitat approaches saturation. Saturation could come earlier under drought conditions than under an average rainfall regime.

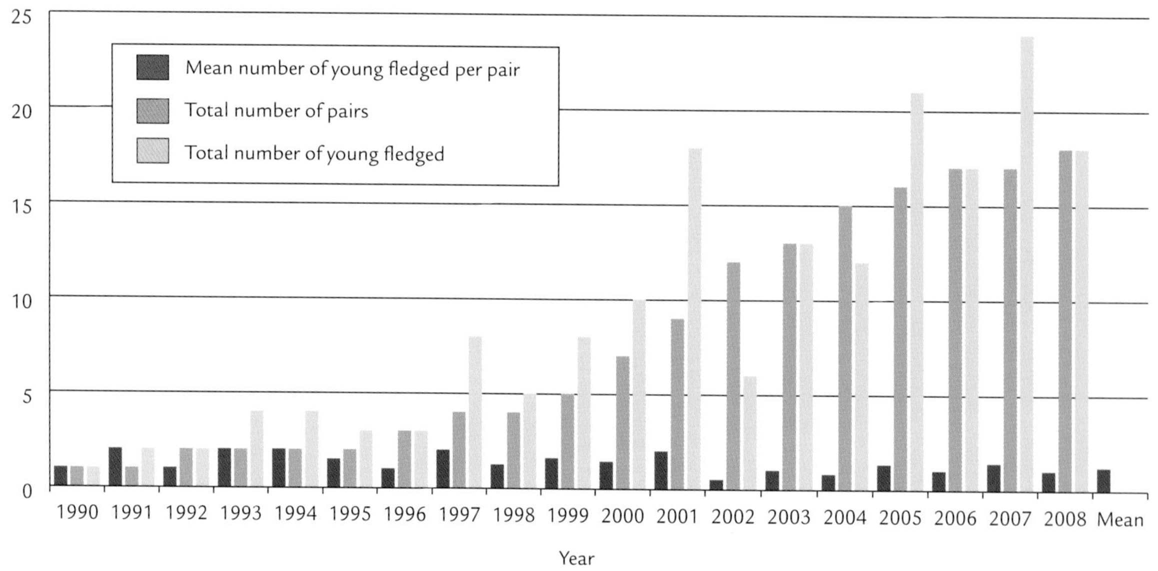

FIGURE 4.1

Number of Osprey pairs, young fledged, and young fledged/occupied nest in Rio Arriba County, New Mexico, 1990–2008.

PHOTO 4.28

Adult Osprey, Heron Lake spillway, Rio Arriba Co., 27 March 2008. In the future, New Mexico's Osprey breeding population may continue to grow, but West Nile Virus and global climate change represent potential threats.

Occupancy of the outlier territory at Cochiti Lake remains tenuous due to low average productivity (0.5 young/year). The nest site was also characterized by great proximity (<50 m [<160 ft]) to human activity (table 4.1), a possible contributing factor. No nesting or resident Ospreys were noted in 2006 or 2007, but a pair reoccupied the site in 2008. The outpost at Elephant Butte Reservoir has also "blinked in and out," but the pioneering pair at Navajo Dam seemed long overdue given nearby Colorado and Heron/El Vado populations. Finally, the Ramah Lake pair fledged two young/year in 2002–7. Thus Ramah Lake may soon serve as the source for further expansion in western New Mexico. At the same time, the limited number of lakes and the existing long-term drought may limit population growth in that part of the state.

North American Osprey populations are growing throughout most of the continent (Poole et al. 2002). In New Mexico, numbers of Ospreys observed migrating over the Sandia and Manzano mountains increased significantly between 1985 and 2001 (Hoffman and Smith 2003). New Mexico's Osprey nesting population is also growing. Since colonization in 1990, the northern Rio Arriba County population has grown to 18 pairs. The short-term future of nesting "populations" in the other four New Mexico counties besides Rio Arriba County is still uncertain, but in the long term Ospreys may well be found nesting at lakes throughout the state. The speed at which this occurs will depend on how often a male becomes a pioneer at a new lake. It will also depend on his success in attracting a mate and in producing young that will return to the new natal area when reaching breeding age.

Continued cooperation with electric utility companies in areas where Ospreys now occur, or where they next appear, is essential. Ospreys in northern New Mexico would not have been nearly so successful without the cooperation of Northern Rio Arriba Electric Cooperative (NORA Co-op). That utility company has provided safe sites for nesting on poles along functioning power lines or on individual poles set aside where the possibility of electrical outages and Osprey electrocution is eliminated.

The mysterious disappearance of half-grown young from three Heron Lake nest sites in 2008 is suggestive of West Nile Virus, since similar disappearances also occurred in Peregrine Falcon (*Falco peregrinus*) and Golden Eagle (*Aquila chrysaetos*) nests in the same general area. Another unknown is climate change. New Mexico has been in a drought cycle since the late 1990s. If climate change permanently limits rainfall, the availability of surface water and the prey it supports may in the future have profound effects on Osprey numbers, productivity, and distribution in New Mexico.

> After this chapter was completed, in 2009, Ospreys built a new nest at McGaffey Lake, McKinley County, 25 km (16 mi) north of Ramah Lake (D. Mikesic, in litt., with photos). McGaffey Lake is only 4–5 ha (8.5–12.5 acres), and there are no other lakes in the vicinity, but it is regularly stocked by the New Mexico Department of Game and Fish.

LITERATURE CITED

Bailey, A. M., and R. J. Niedrach. 1965. *Birds of Colorado*. 2 vol. Denver: Museum of Natural History.

Bailey, F. M. 1928. *Birds of New Mexico*. Santa Fe: New Mexico Department of Game and Fish.

Bent, A. C. 1937. *Life histories of North American birds of prey, Part 1*. U. S. National Museum Bulletin 167. Washington, DC: Smithsonian Institution.

Dodd, N. L., and J. R. Vahle. 1998. Osprey. In *The raptors of Arizona*, ed. R. L. Glinski, 37–41. Tucson: University of Arizona Press.

Goodman, R. A., ed. 1989. *New Mexico Ornithological Society Field Notes* 28(3).

Henny, C. J., and C. J. Anderson. 2004. Status of nesting Ospreys in coastal Baja, Sonora, and Sinoloa, Mexico, 1977 and 1992–1993. *Bulletin of the Southern Academy of Sciences* 103:95–114.

Henny, C. J., and W. T. Van Velzen. 1972. Migration patterns and wintering localities of American Osprey. *Journal of Wildlife Management* 36:1133–41.

Hoffman, S. W., and J. P. Smith. 2003. Population trends of migratory raptors in western North America, 1977–2001. *Condor* 105:397–419.

Ligon, J. S. 1961. *New Mexico birds and where to find them*. Albuquerque: University of New Mexico Press.

Martell, M. S., C. L. Henny, P. E. Nye, and M. J. Solensky. 2001. Fall migration routes, timing, and wintering sites of North American Ospreys, as determined by satellite telemetry. *Condor* 103:715–24.

Poole, A. F. 1989. *Ospreys: A natural and unnatural history*. Cambridge, UK: Cambridge University Press.

Poole, A. F., and B. Agler. 1987. Recoveries of Ospreys banded in the United States, 1918–84. *Journal of Wildlife Management* 51:148–55.

Poole, A. F., R. O. Bierregaard, and Mark S. Martell. 2002. Osprey (*Pandion haliaetus*). No. 683. In *The birds of North America*, ed. A. Poole and F. Gill. Philadelphia, PA: Birds of North America, Inc.

Postupalsky, S. 1989. Osprey. In *Lifetime reproduction in birds*, ed. I. Newton, 297–313. London: Academic Press.

Pratt, H. D., P. L. Bruner, and D. G. Berrett. 1987. *The birds of Hawaii and the tropical Pacific*. Princeton, NJ: Princeton University Press.

Prevost, Y. 1983. Osprey distribution and subspecies taxonomy. In *Biology and management of Bald Eagles and Ospreys*, ed. D. M. Bird, 157–74. Ste. Anne de Bellevue, Quebec: Harpell Press.

Saurola, P. 1995. Finnish Ospreys (*Pandion haliaetus*) in 1971–1994. *Vogelweldt* 116:199–204.

Sibley, D. A. 2000. *The Sibley guide to the birds*. New York: Chanticleer Press.

Stahlecker, D. W. 2002. Look out, here they come! Growth of the Four Corners Osprey population and its implications to New Mexico. *New Mexico Ornithological Society Bulletin* 30:35 (abstract only).

Steeger, C., H. Esselink, and R. C. Ydenberg. 1992. Comparative feeding ecology and reproductive performance of Ospreys in different habitats of southeastern British Columbia. *Canadian Journal of Zoology* 70:470–75.

Vahle, J. R., N. L. Dodd, and S. Nagiller. 1988. Osprey. In *Proceedings of the southwest raptor management symposium and workshop*, ed. R. L. Glinski et al., 37–47. Washington, DC: National Wildlife Federation.

Watts, B. D., M. A. Byrd, and M. U. Watts. 2004. Status and distribution of breeding Ospreys in the Chesapeake Bay: 1995–96. *Journal of Raptor Research* 38:47–54.

5

White-tailed Kite

(*Elanus leucurus*)

JANET M. RUTH AND DAVID J. KRUEPER

THE WHITE-TAILED KITE (*Elanus leucurus*) is a small to medium-sized neotropical raptor that reaches its northernmost distribution along the Pacific coast of the United States and is also found in the Southwest borderland states eastward to southern Texas, and in Florida. This falcon-shaped raptor's major field marks are a long white tail, black marginal coverts on a gray back, a white ventral area, and long, relatively thin, moderately pointed wings with darker tips. The White-tailed Kite can appear gull-like in flight, and is often seen hovering, or "kiting," while hunting. When soaring or gliding, it holds its wings in a high dihedral; in powered flight it shows moderately fast, buoyant wing beats. In contrast to many other raptor species, there is considerable overlap in the sizes of the sexes, although females average slightly larger. Size and body mass appear to decrease with latitude, but this recorded pattern is based on measurements of a limited number of individuals (Dunk 1995). Adult total length: 32–41 cm (13–16 in); wingspan: 94–102 cm (37–40 in); mass: 294–350 g (10–12 oz) (Dunk 1995; Wheeler 2003).

Adult plumage is acquired at one year of age with a complete prebasic molt. The crown and nape are pale gray and the rest of the head is white; the eyes are large, the iris is red; the inner lores are black; the smallish bill is black with yellow cere and gape; and the rest of the back from the nape to the uppertail coverts is gray, darker in females than in males. The dorsal wing exhibits a large black "shoulder patch" that includes the median coverts and lesser coverts; the primaries and secondaries are medium gray. The ventral wing sports a black rectangular spot in the carpal region; the underwing coverts are white; and the primaries are dark gray grading to a paler gray in the inner primaries and secondaries. The tail is moderately long; the central two tail feathers are pale gray and the rest of the tail feathers are white. The long tail extends just beyond the primaries when perched. The underbody of the bird from throat to vent is white, and the legs and feet are yellow.

The juvenal plumage is acquired within 40 days after hatching and is only retained for a few months. During this time, the juvenile exhibits a white crown with a tawny wash and narrow brown streaks, and the iris of the eye is brown. The back and scapular feathers are a medium brownish-gray with white tips, which give them a scalloped look. The breast sports a tawny necklace, with the tawny streaks extending onto the upper belly and along the flanks. The white tail has a narrow, dusky terminal band.

The subadult plumage is achieved with the first prebasic, or preformative, molt when the juvenile is 3.5

PHOTO 5.1

Adult White-tailed Kite in flight, San Francisco Bay area, California, February 2008. The underparts are mostly white. The outer primaries are dark gray, appearing almost black, grading to a paler gray in the inner primaries and secondaries. Note also the black rectangular spot in the carpal region of the underwings.

PHOTOGRAPH: © STEVE BARANOFF.

PHOTO 5.2

Adult White-tailed Kite, dorsal view, near Gilchrist, Texas, 16 March 2007. The back from the nape to the uppertail coverts is gray, as are the primaries and secondaries. The median and lesser coverts form a contrasting black shoulder patch. The head is almost entirely white. The eyes are large, with red irises. The tail is white.

PHOTOGRAPH: © STEVE METZ.

PHOTO 5.3

White-tailed Kite in subadult plumage, Rodeo, Hidalgo Co., 2 January 2006. Note the rufous wash on the breast, the tawny feathers and dark streaks remaining on the crown, the remaining scalloping on the back and scapular feathers, and the brownish-orange eyes.

PHOTOGRAPH: © JERRY OLDENETTEL.

to 4 months old (Palmer 1988); the molt period can extend from summer through fall. During this period, the subadult continues to show a rufous wash on the breast while also exhibiting a gradual loss of crown streaking, eventual loss of the white tips on the back and scapular feathers due to feather wear, a gradual fading of the terminal tail band, and a gradual change in eye color from brown to orangish. Juvenal remiges are retained in subadult plumage.

At one time, the White-tailed Kite was considered to be conspecific with the Black-winged Kite (*Elanus caerulus*) of Eurasia and the Black-shouldered Kite (*E. axillaris*) of Australia, but these forms have now been split out as distinct species (Clark and Banks 1992; AOU 1998).

Distribution

White-tailed Kites are considered to be resident throughout much of their breeding range, although individuals disperse during the nonbreeding season, resulting in some winter range expansion. Additionally, there is some evidence of nomadism during times of low prey abundance (Stendell 1972).

White-tailed Kites are a widely distributed neotropical species, reaching the northern edge of their breeding range along the northern Pacific coast of the United States, in the Southwest borderland region eastward to southern Texas, and in Florida. They range south through Mexico, Central America, and much of South America except at high elevations, reaching the northern two-thirds of Argentina and Chile (Ferguson-Lees and Christie 2001).

This species' distribution has exhibited dramatic changes within the past 100 years, not only in North America but also in Central America, where the White-tailed Kite was historically unknown (Eisenmann 1971). From a wider range in the southern United States prior to 1900, it had contracted to small populations in western California, the coast of extreme southern Texas, and the Gulf Coast of Mexico by about 1935 (Eisenmann 1971). A substantial population increase and range expansion were observed in California by the 1950s, and in Texas, Mexico, and Central America by the 1960s (Eisenmann 1971; Larson 1980; Pruett-Jones et al. 1980). It has been suggested that birds dispersing from the expanding California populations were the source for the population increases and range expansions in Mexico, Central America, and Oregon in the second half of the 20th century (Eisenmann 1971; Pruett-Jones et al. 1980). Eisenmann (1971) suggests a number of reasons for these population increases and range expansions: (1) more suitable habitat for White-tailed Kites created by agricultural development, particularly the spread of irrigated agriculture, and by clearing of forests and woodlands for both highways and agriculture in Central America; (2) reductions in human persecution of raptors; (3) kite tolerance of human disturbance; and (4) the life history of the species itself, which can promote rapid population growth and/or colonization of new areas through reduced intraspecific territoriality, high reproductive potential, and nomadism in response to local and seasonal fluctuations in prey abundance. Pruett-Jones et al. (1980) expand on the tie to prey abundance and argue that the population increases were in response to favorable densities of microtine prey populations. Consequently, they also suggest that declines seen after 1975 were a result of reduced prey populations caused by drought; they found a statistically significant positive association between mean precipitation and kite numbers. Interestingly, Pruett-Jones et al. (1980) noted similar patterns of population declines in the early 1900s followed by some recovery for three kite species: White-tailed Kite, Mississippi Kite (*Ictinia mississippiensis*), and Snail Kite (*Rostrhamus sociabilis*).

The White-tailed Kite's current distribution in North America is larger than at any time in its known history and there is reason to expect continuing expansion (Dunk 1995). Breeding distributions in the United States include a stronghold in California throughout the entire coastal region and in the Central Valley, as well as more recent expansions into the arid portions of southern California, western Oregon, and southwest Washington. The species is a fairly common to locally common resident in South Texas (Wheeler 2003), and there is a small breeding population in south Florida (Dunk 1995). To set the context for describing the range of White-tailed Kite populations in New Mexico, we provide additional details about recent distributions in the regions surrounding the state, all of which are potential sources for birds moving into New Mexico.

Eisenmann (1971) summarizes historical occurrences for Mexico that include the states of Baja California, Nuevo León, Tamaulipas, Tabasco, and Quintana Roo, with breeding in Veracruz and Campeche. He reports an expansion of White-tailed Kite observations by the late 1960s to the Gulf Coast, including eastern San Luis Potosí, the humid lowlands of the southeastern Pacific slope of Oaxaca and Chiapas, and again in Baja California. Additional analyses of Christmas Bird Count (CBC) data in the 1970s and a synthesis of reported occurrences (Pruett-Jones et al. 1980) recorded White-tailed Kites in five of the 10 CBC count sites throughout Mexico and noted particular records for Tamaulipas, the coastal plains of Veracruz, and Chiapas. Russell and Monson (1998) report the recent range extension of White-tailed Kites into Sonora, with the first documented record in 1979. Through the 1980s and early 1990s these birds were recorded in all parts of Sonora except the far northeast, although no nests had been found by the time Russell and Monson (1998) were completing *The Birds of Sonora*. The White-tailed Kite has been recorded more recently in northern Sonora along the Río Concepcion, Río Busani, and Río San Pedro (Flesch 2008). Howell and Webb (1995) also considered it to be increasing in Mexico, describing it as a

PHOTO 5.4

White-tailed Kite, San Francisco Bay area, California, November 2004. White-tailed Kites reach the northern edge of their breeding distribution along the northern Pacific coast of the United States.

PHOTOGRAPH: © STEVE BARANOFF.

PHOTO 5.5

White-tailed Kite, San Pedro Riparian National Conservation Area, Arizona, 13 April 2006. The White-tailed Kite is considered an irregular and local resident in Arizona, where it was first reported in 1972 and where breeding was first documented in 1983. PHOTOGRAPH: © TONY GODFREY/ ARTFULLBIRDS.COM.

common to fairly common resident to 1,500 m (4,900 ft) in Baja California Norte and from Sonora to Tamaulipas and south. They also noted particular records from the border states of Chihuahua and Coahuila.

White-tailed Kites are irregular and localized residents throughout southern Arizona (Corman and Wise-Gervais 2005). The species was first reported in 1972 in Cochise County and first documented in Pima County in August 1978 (Ellis and Monson 1979). First documented breeding for Arizona was recorded in Pinal County in 1983 (Gatz et al. 1985). The White-tailed Kite is now considered a local and irregular resident and breeder in southeastern Arizona (Gatz 1998). Breeding has been confirmed in Cochise, Graham, Santa Cruz, Pima, Pinal, Maricopa, and La Paz counties (Wheeler 2003; Corman and Wise-Gervais 2005) and was considered probable in Yuma County (Corman and Wise-Gervais 2005). Although Rosenberg et al. (1991) suggest that the most likely source of White-tailed Kites appearing in the Lower Colorado River Valley in the early 1980s was southern California, they acknowledge the possibility of a Mexican source. The fact that kites were reported in southeastern Arizona prior to appearing in the western part of the state lends support to the argument for a Mexican origin (T. A. Gatz, pers. comm.).

Oberholser (1974) reports historical records for White-tailed Kites in southern Texas around the beginning of the 20th century as far up the Rio Grande as Starr County and inland to Lee County. However, by the time Oberholser (1974) was completing *The Bird Life of Texas*, the species' distribution had contracted southward, and that author describes the White-tailed Kite as a locally fairly common to scarce bird only in extreme South Texas (Cameron and Hidalgo counties). Oberholser (1974) also reports nonbreeding records from throughout much of the southeastern one-third of Texas. He describes a modest increase in White-tailed Kite numbers in Texas during the 1960s as compared to substantial population increases in Mississippi Kites in Texas and in White-tailed Kites in California. He suggests that the modest, limited increase in the resident White-tailed Kite (compared to the migratory Mississippi Kite) in Texas is due to the adverse effects of the cold winter weather that at times sweeps through Texas. He also suggests that the more substantial increases in White-tailed Kites in California were due to the species' positive response to the more intensive agriculture and thus higher rodent populations in California.

The White-tailed Kite's primary current range in Texas (where the species is now described as an uncommon to common resident) is in the southern and southeastern parts of the state—coastal prairies, southern post oak savannah, and eastern South Texas brush country (Lockwood and Freeman 2004). It is only a rare, local resident along the Rio Grande to Valverde County (Lockwood and Freeman 2004). The White-tailed Kite is also a rare or casual visitor as far west as the Trans-Pecos (Peterson and Zimmer 1998; Lockwood and Freeman 2004), the part of Texas closest to New Mexico. Brush (2005) reports that White-tailed Kite populations in South Texas have increased in the last 30–35 years.

Unless noted otherwise, all of the following data for New Mexico are from the *New Mexico Ornithological Society* (NMOS) *Field Notes*, the *NMOS Field Notes* database (NMOS 2007), and *NMOS Field Notes* files (the latter referenced as the name of the observer, *in litt.*). The White-tailed Kite was not reported from the state by either Bailey (1928) or Ligon (1961), and Hubbard (1978) described the species as only occasional in the south and casual at Las Vegas, San Miguel County (1961), with an unverified description in April 1973 at Bitter Lake National Wildlife Refuge, Chaves County. The first and second verified records for the state, documented with photos, were at Bitter Lake National Wildlife Refuge (July 1975, April 1983).

A number of interesting spatial and temporal patterns emerge when examining all of the NMOS records. The White-tailed Kite has now been reported in almost all counties in the southern half of New Mexico (Catron, Chaves, Doña Ana, Eddy, Grant, Hidalgo, Lea, Luna, Otero, Sierra, and Socorro counties); the exceptions are Curry, De Baca, Lincoln, and Roosevelt counties. It has also been reported in Bernalillo County in the central part of the state (photos 5.7a–c). White-tailed Kite records have been verified with photos (date of first photographic record in parentheses) in Chaves (July 1975), Doña Ana (July 1985), Socorro (October 1992), Luna (March 1993), Grant (March 1997), Hidalgo (June 2001), Eddy (December 2006), and Bernalillo (April 2008) counties. It has been verified through the collection of a specimen (July–August 2006) in Catron County (Division of Birds, Museum of Southwestern Biology, University of New Mexico). By far the largest numbers of reports are from Hidalgo (>50 records from 1982 to the present) and Luna (>25 records from 1974 to the present) counties in the southwestern corner of the state. This spatial pattern suggests that the sources of birds dispersing into New Mexico are most likely northwestern Mexico and/or southeastern Arizona.

Most records were of one or two birds; a smaller number were of three or four individuals. Most interesting were a roost of 11 birds reported only as in the "southwest quadrant of the State" (October 2000) and the seven pairs monitored in southwestern New Mexico (May–June 2001), which successfully fledged 10 young.

Following the first verified record in 1975, White-tailed Kite sightings have been sporadic, showing year-to-year and month-to-month variation. In the 1980s, for example, there were from zero to six reports per year, indicating a first small pulse into the state; from

PHOTO 5.6

White-tailed Kite, Hidalgo Co., July 2003. The White-tailed Kite has been reported from most of New Mexico's southern counties. Most records have been from Hidalgo and Luna Counties.

PHOTOGRAPH: © BILL SCHMOKER.

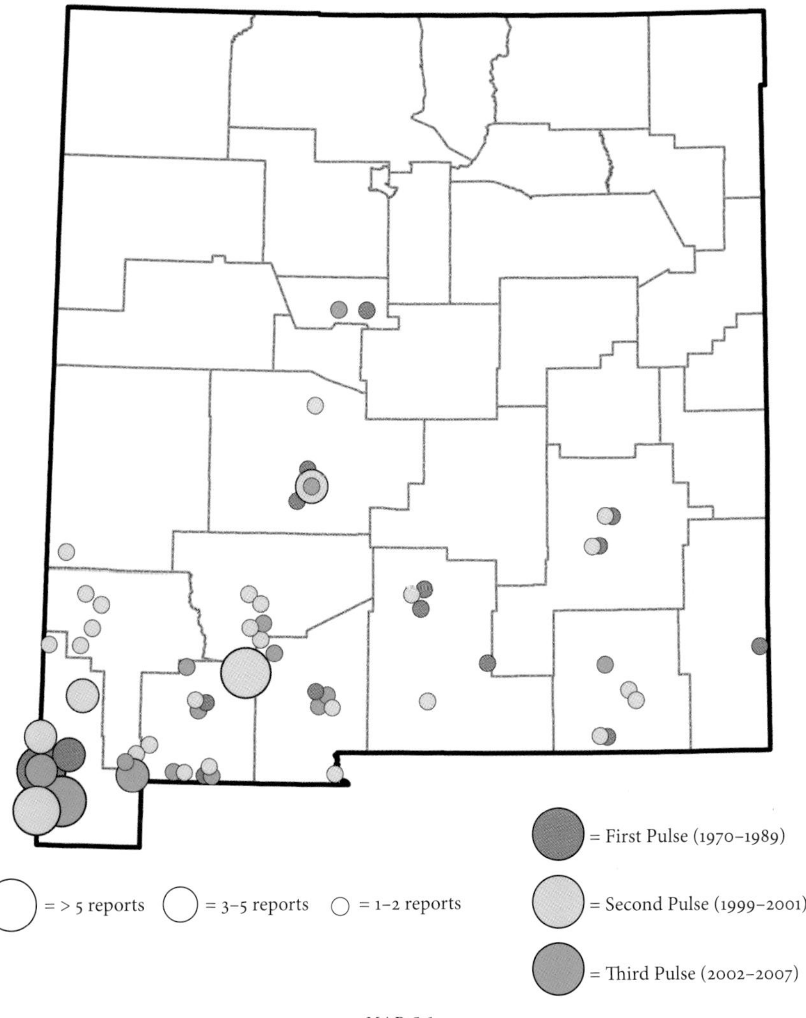

MAP 5.1

White-tailed Kite distribution map

the 1990s to the present, there continued to be a few years with no records at all, and other years with two to 15 reports per year; in all cases the records are scattered throughout all months of the year. It was not until 2000, with the observed second pulse of birds, that nest building was first reported for New Mexico, followed in 2001 by successful fledging of young. There appears to be a third pulse of observations recorded in 2006 and 2007. These pulse patterns are all consistent with a population that is in the process of expanding its range into and within the state; the continued sporadic occurrence of years when there are no observations is also consistent with a species at the periphery of its range that shows nomadic capabilities in response to precipitation-driven changes in rangeland and prey conditions. White-tailed Kites may vacate the periphery of their newly expanded range in New Mexico during harsh years, while there may be an influx of birds into the state in years with good conditions.

The monthly pattern of observations pooled across years—multiple records in all months—is consistent with the status of the White-tailed Kite as a resident

PHOTOS 5.7a, b, and c

White-tailed Kite at Montessa Park in the Albuquerque metropolitan area, Bernalillo Co., 5 April 2008. Northernmost verified record of the species in New Mexico.

PHOTOGRAPHS: © GARY K. FROEHLICH.

species in most of its range. Not surprising for a species that is still expanding its range into the state, there has not yet been a single year with records in all 12 months. The relatively large number of fall and winter records is also consistent with a supplementary influx of individuals into the state during postbreeding and wintering dispersal.

Habitat Associations

White-tailed Kites typically inhabit low-elevation open and semiopen grasslands, both wet and dry, both natural and human-altered, as well as pasturelands, meadows, idle fields, grassy coastal dunes, marshes, oak-woodlands, savannahs, and agricultural areas with ample unfarmed edge habitat. In arid regions, they are often associated with irrigated croplands such as alfalfa and grass hay fields (Dunk 1995; Wheeler 2003). In fact, 45% of all White-tailed Kite records from Arizona were from irrigated agricultural areas or their edges (Corman and Wise-Gervais 2005). Natural open habitats in Arizona include grasslands and wetlands with occasional trees or shrubs for perching and nesting (Gatz 1998; Corman and Wise-Gervais 2005). Most grassland observations were from semiarid grasslands with mesquite (*Prosopis* sp.), paloverde *(Parkinsonia* sp.), and soaptree yucca *(Yucca elata*), as well as cottonwood and oak stringers adjacent to grasslands (Corman and Wise-Gervais 2005).

In New Mexico, White-tailed Kites have been observed foraging in open desert grasslands with yucca and shrub components and have been seen frequently perching in cottonwood riparian stringers adjacent to these grasslands (JMR and DJK, pers. obs.; R. Meyer, *in litt.*). They have also been reported in tobosa (*Hilaria mutica*) swales with surrounding creosote bush (*Larrea divaricata*) (S.O. Williams III, *in litt.*), in a scrubby, weedy draw at a riparian edge with prickly pear (*Opuntia* spp.), creosotebush, and cottonwood (M. Ristow photo, *in litt.*), in riparian bosque habitat, and in semiurbanized habitat surrounded by a golf course and off-road vehicle recreational sand scrub habitat.

PHOTO 5.8

Plains grassland habitat in the Animas Valley, Hidalgo Co., just south of the Diamond A Ranch headquarters. White-tailed Kites have been observed foraging in the area where this photo was taken, as well as perching on trees in nearby cottonwood riparian stringers.

PHOTOGRAPH: © NARCA MOORE-CRAIG.

Nesting habitat in New Mexico has consisted of (1) open grassland with scattered yuccas; (2) a tobosa grassland draw surrounded by creosote bush, littleleaf sumac (*Rhus microphylla*), yucca, and mesquite; (3) a sacaton (*Sacaton* spp.) draw (all R. Meyer, *in litt.*), (4) open Chihuahuan Desert rangeland (C. Britt, photo *in litt.*); and (5) areas near farmland or irrigated farmland (S. O. Williams III, *in litt.*). In Arizona, nesting has also been described in a wash on abandoned farmland and in oak savannah habitats (Gatz et al. 1985).

Specific plant associations do not appear to be as important as vegetative structure and prey abundance; lightly grazed or ungrazed fields are more suitable due to higher prey populations (Bammann 1975; Dunk 1995). Topography of breeding habitat can vary from flat to moderately hilly. Elevation varies regionally: up to 600 m (2,000 ft) in California; up to 1,200 m (3,900 ft) in Arizona grasslands; and up to 1,030 m (3,400 ft) in Sonora, Mexico (Wheeler 2003). White-tailed Kite wintering habitat is similar to breeding habitat, although proximity to nest trees is obviously no longer important.

Life History

The natural history of White-tailed Kites has not been well studied within New Mexico; however, recent monitoring of a number of kite pairs in southwestern New Mexico (R. Meyer, *in litt.*) will hopefully result in additional valuable information specific to the state. Most research on the natural history of the species in the United States has been conducted on California populations. The natural history of White-tailed Kites in New Mexico is likely similar to that in Arizona, as population expansion patterns and habitat have been very similar in the two states (Gatz et al. 1985).

Nesting

White-tailed Kites nest in a wide variety of shrubs and trees ranging in height from less than 3 m (9 ft) tall (e.g., *Atriplex* and *Baccharis*) (Stendell 1972) to more than 50 m (~160 ft) tall (Dunk 1995). The first nesting records for White-tailed Kite in Arizona document kites nesting in Fremont cottonwood (*Populus*

PHOTO 5.9

Pair during courtship: male transferring a rodent to the female in midair, San Pedro Riparian National Conservation Area, Arizona, 13 April 2006.

PHOTOGRAPH: © TONY GODFREY/ ARTFULLBIRDS.COM.

fremontii) and Emory oak (*Quercus emoryi*) (Gatz et al. 1985). More recently in Arizona, the species has also been found nesting in pecan trees (*Carya pecan*), mesquite, paloverde, cholla (*Opuntia* sp.) and netleaf hackberry (*Celtis reticulata*) (Corman and Wise-Gervais 2005). In New Mexico, kites have placed their nests in yuccas (*Yucca* spp.) (R. Meyer and C. Britt photo, *in litt.*).

The breeding season of White-tailed Kites is quite long. Courtship behavior has been observed from December through September and nest building from January through August (Dunk 1995; Wheeler 2003). In California, most egg-laying occurs in February and March (Dunk 1995), but this is likely somewhat different in New Mexico (see below). Clutch size ranges from three to six (typically four), with an incubation period of 30–32 days (Dunk 1995). The female is primarily involved in nest construction and incubation; incubation takes about 30 days (Gatz 1998). The young fledge when 35–40 days old (Gatz 1988), and White-tailed Kites can raise two or more broods in a single year (Hawbecker 1940; Dixon et al. 1957).

In Arizona (Corman and Wise-Gervais 2005), the following nest-related dates were recorded: 30 March for the earliest nest with recently hatched nestlings, 7 July for the latest occupied nest (but see below), and 6 August or earlier for the observation of fledglings. The first documented nest in Arizona was an example of late nesting. Nest building was observed on 5 August, and apparent incubation of the eggs on 12 August; unfortunately the nest was destroyed in mid August. Fledglings have been observed in Arizona in June, July, and August; observations of juvenal-plumaged birds have been noted in mid June and from September to November (Gatz et al. 1985).

All of the following data are again from *NMOS Field Notes*, the *NMOS Field Notes* database (NMOS 2007), and *NMOS Field Notes* files (the latter referenced as name of observer, *in litt.*). Breeding has been reported and verified with a photo of a nest with three young in May 2007 for Hidalgo County. Two birds, one of which was perched on a nest structure in April 2000, and a nest with four eggs in April 2007, were reported for Luna County, but these records were not verified with photos. In addition, there was a report of White-tailed Kites building a nest in Luna County (August 2000—verified with photo) (R. Meyer, *in litt.*); the nest later contained an abandoned egg. Seven pairs of White-tailed Kites were monitored in southwestern New Mexico (May–June 2001) (R. Meyer, *in litt.*), mostly in Luna County although some may have been in Grant or Hidalgo counties (S. O. Williams III, *in litt.*); of the seven pairs, three successfully fledged a total of ten young, one failed (twice), and the status of three pairs was unknown. One pair laid a first clutch of eggs and failed, re-nested, and laid a second clutch that also failed. A manuscript providing additional valuable data from these nest observations is currently in preparation. The following breeding-related dates were extracted from these records: 1 August for nest-building; 19 April, 28 April, and 9 June for nests with eggs (9 June was a second nest); 28 April and 13 May for nests with young; and 16 May, 25 May, 2 June, and 23 June for fledgings. Clutch size appeared to range from two to four based on observed egg, nestling, and fledgling numbers. However, the actual numbers of eggs and nestlings may have been underestimated, as some of the nests that were monitored could have failed before full clutches were laid or not all eggs may have hatched or all young successfully fledged.

Diet and Foraging

Although White-tailed Kites can have a varied diet, they feed predominantly on small mammals. In each of 12 different studies, small mammals comprised >95% of all documented prey items (Dunk 1995). Prey mass typically falls in the range of 20–70 g (0.7–2.5 oz) and, based on the same set of studies as above, the predominant prey taxa are voles (*Microtus*), house mice (*Mus musculus*), and harvest mice (*Reithrodontomys*), as well as some other mammals and a few birds and insects. Foraging observations and pellet analyses suggest that in Arizona, South Texas, and perhaps Florida, White-tailed Kites rely on cotton rats (*Sigmodon*) as their main prey (Gatz 1998). Gatz (1998) noted that cotton rat populations, like kites, probably benefit from the expansion of irrigated agriculture in Arizona. It has been suggested that White-tailed Kites are adapted to wander in search of food as rodent populations fluctuate (Eisenmann 1971) and that kite population fluctuations may actually result in part from fluctuating prey (vole) populations (Larson 1980).

White-tailed Kites are aerial hunters, foraging at 5–25 m (15–80 ft) above the ground. They can hover on shallow-beating wings for over one minute (longer in strong winds) (Dunk 1995), and often lower their legs while hovering. Having spotted prey, they will dive headfirst with wings fully extended above their body (Wheeler 2003), and if successful will fly to a preferred perch to consume the prey.

Status and Management

The White-tailed Kite has no special conservation status, although like all other New Mexico raptors it is protected under the Migratory Bird Treaty Act. The population of White-tailed Kites in the United States (about 20% of the global population) is estimated at 53,000 birds (Rich et al. 2004). North American Breeding Bird Survey (BBS) data for the United States from 1966 to 2006 (Sauer et al. 2007) show no significant population increases or decreases for White-tailed Kites. For the second half of that time period (1986–2006), BBS data for the United States even suggest a possible decline, though not significant. The data for California, where most of the BBS data were collected (Sauer et al. 2007), show similar patterns both for 1966–2006 and for 1986–2006 (i.e., no population trend overall but a possible decline during the last two decades). In contrast, for 1966–2006, Texas shows an almost significant population increase (Sauer et al. 2007). Another tool for monitoring the status of species over time, Christmas Bird Count (CBC) data are consistent with the patterns of expanding White-tailed Kite distributions in the United States. Data from winter 1900–1901 to winter 2006–7 reveal a spike in the number of observations of White-tailed Kite in California from 1940 to 1960 (up to 79 birds on up to 11 counts), with a smaller spike in Texas from 1950 to 1960 (up to 9 birds on up to 2 counts). Beginning in 1990, CBC data capture the more recent and ongoing large increase in White-tailed Kite numbers in California (1,000–2,000 birds on 58–84 counts); in Arizona (up to 19 birds on up to 7 counts); and in Texas (200–500 birds on 20–31 counts), with coinciding observations in New Mexico in 1992–93 (1 bird), 2000–2001 (1 bird), 2005–6 (3 birds), and 2006–7 (1 bird).

Little is known about threats to the White-tailed Kite in New Mexico. The species is a relatively new addition to the state's avifauna, with breeding documented only during the last decade, and little research has been conducted on what is still a very small population in New Mexico. Dunk (1995) notes that degradation of habitat, especially loss of nest trees and foraging habitat with high prey populations, poses a significant threat. Corman and Wise-Gervais (2005) suggest that populations in the arid Southwest will continue to fluctuate in response to annual precipitation and local prey availability. Paradoxically, observations that White-tailed Kites appear to have benefited from increases in irrigated agriculture in California, Arizona, and Texas raise concerns for the species' future throughout the Southwest. Increasing and conflicting demands on water resources due to drought, expanding human populations, and the specter of global climate change make it likely that irrigated agriculture will have to be scaled back. The resulting loss of irrigated fields as a relatively new habitat source for White-tailed Kites, along with the simultaneous loss and degradation of natural grassland habitat and prey populations due to the same factors, would not bode well for kites in this region on the periphery of the species' range. Any efforts to improve management of desert grasslands and to protect them from conversion to other uses by federal and state natural resource agencies, conservation organizations, and private landowners would benefit White-tailed Kites and the prey populations on which they rely. The White-tailed Kite is one of several raptor species expanding its range into New Mexico. As such, it requires more careful attention from land managers and additional research and monitoring to better understand its life history and ecology.

LITERATURE CITED

[AOU] American Ornithologists' Union. 1998. *Check-list of North American birds.* 7th ed. Washington, DC: American Ornithologists' Union.

Bailey, F. M. 1928. *Birds of New Mexico.* Santa Fe: New Mexico Department of Game and Fish.

Bammann, A. R. 1975. Ecology of predation and social interactions of wintering White-tailed Kites. M.S. thesis. Humboldt State University, Arcata, CA.

Brush, T. 2005. *Nesting Birds of a tropical frontier: The Lower Rio Grande Valley of Texas.* College Station: Texas A&M University Press.

Clark, W. S., and R. C. Banks. 1992. The taxonomic status of the White-tailed Kite. *Wilson Bulletin* 104:571–79.

Corman, T. E., and C. Wise-Gervais. 2005. *Arizona breeding bird atlas.* Albuquerque: University of New Mexico Press.

Dixon, J. B., R. E. Dixon, and J. E. Dixon. 1957. Natural history of the White-tailed Kite in San Diego County, California. *Condor* 59:156–65.

Dunk, J. R. 1995. White-tailed Kite (*Elanus leucurus*). No. 178. In *The birds of North America*, ed. A. Poole and F. Gill. Philadelphia, PA: Academy of Natural Sciences, and Washington, DC: American Ornithologists' Union.

Eisenmann, E. 1971. Range expansion and population increase in North and Middle America of the White-tailed Kite (*Elanus leucurus*). *American Birds* 25:529–36.

Ellis, D. H., and G. Monson. 1979. White-tailed Kite records for Arizona. *Western Birds* 10:165.

Ferguson-Lees, J., and D. A. Christie. 2001. *Raptors of the world.* Boston: Houghton Mifflin Co.

Flesch, A. D. 2008. Distribution and status of breeding landbirds in northern Sonora, Mexico. *Studies in Avian Biology*, no. 37:28–45.

Gatz, T. A. 1998. White-tailed Kite. In *The raptors of Arizona*, ed. R. L. Glinski, 42–45. Tucson: University of Arizona Press.

Gatz, T. A., M. D. Jakle, R. L. Glinski, and G. Monson. 1985. First nesting records and current status of the Black-shouldered Kite in Arizona. *Western Birds* 16:57–61.

Hawbecker, A. C. 1940. The nesting of the White-tailed Kite in southern Santa Cruz County, California. *Condor* 42:106–11.

Howell, S.N.G., and S. Webb. 1995. *A guide to the birds of Mexico and northern Central America.* Oxford, UK: Oxford University Press.

Hubbard, J. P. 1978. *Revised check-list of the birds of New Mexico.* Publ. no. 6. Albuquerque: New Mexico Ornithological Society.

Larson, D. 1980. Increase in the White-tailed Kite populations of California and Texas: 1944–1978. *American Birds* 34:689–90.

Ligon, J. S. 1961. *New Mexico birds and where to find them.* Albuquerque: University of New Mexico Press.

Lockwood, M. W., and B. Freeman. 2004. *The Texas Ornithological Society handbook of Texas birds.* College Station: Texas A&M University Press.

[NMOS] New Mexico Ornithological Society. 2007. *NMOS Field Notes* database. http://nhnm.unm.edu/partners/NMOS/.

Oberholser, H. C. 1974. *The bird life of Texas.* Austin: University of Texas Press.

Palmer, R. S. 1988. *Handbook of North American birds.* Vol. 4. New Haven: Yale University Press.

Peterson, J., and B. R. Zimmer. 1998. *Birds of the Trans-Pecos.* Austin: University of Texas Press.

Pruett-Jones, S. G., M. A. Pruett-Jones, and R. L. Knight. 1980. The White-tailed Kite in North and Middle America: current status and recent population changes. *American Birds* 34:682–88.

Rich, T. D., C. J. Beardmore, H. Berlanga, P. J. Blancher, M.S.W. Bradstreet, G. S. Butcher, D. W. Demarest, E. H. Dunn, W. C. Hunter, E. E. Iñigo-Elias, et al. 2004. *Partners in Flight North American landbird conservation plan.* Ithaca, NY: Cornell Laboratory of Ornithology.

Rosenberg, K. V., R. D. Ohmart, W. C. Hunter, and B. W. Anderson. 1991. *Birds of the Lower Colorado River Valley.* Tucson: University of Arizona Press.

Russell, S. M., and G. Monson. 1998. *The birds of Sonora.* Tucson: University of Arizona Press.

Sauer, J. R., J. E. Hines, and J. Fallon. 2007. *The North American breeding bird survey, results and analysis 1966–2006.* Version 10.13.2007. Laurel, MD: U.S. Geological Survey, Patuxent Wildlife Research Center.

Stendell, R. C. 1972. The occurrence, food habits, and nesting strategy of White-tailed Kite in relation to a fluctuating vole population. Ph.D. diss. University of California, Berkeley.

Wheeler, B. K. 2003. *Raptors of western North America.* Princeton, NJ: Princeton University Press.

6

Mississippi Kite

(*Ictinia mississippiensis*)

ANTONIO (TONY) GENNARO

ONE DOES NOT have to be a professional ornithologist or skilled birdwatcher to recognize the Mississippi Kite (*Ictinia mississippiensis*) as it displays its dynamic and graceful flight of flapping, soaring, and diving against a background of New Mexican sky. The falcon-like bird has a body length of 0.33 m (1 ft) or more and pointed wings with a span of about 1 m (3 ft). Adults are mostly light to medium gray below, with a uniformly black—or nearly so—tail and a narrow but noticeable whitish band along the trailing edge of the inner wing. The primaries, dorsally black (often with some rufous), also appear darker from below. The back and flight coverts are slate gray, contrasting with pale gray to silvery white secondaries. The head and nape are lighter, whitish in males and the same shade of gray as for the breast in females. The eyes are red, surrounded by a circle of black feathers forming a noticeable spot toward the front. The feet are mostly yellow to orange. Females have some whitish or pale gray areas along shafts and inner webs of rectrices (Wheeler and Clark 1995; Parker 1999). Yearlings (subadults) are similar to adults, except for the three white bars on the underside of their tails, as well as variable amounts of retained juvenal body contours and wiry lining feathers. Juveniles have a typically near-white to pale buff head, neck, and venter heavily streaked with brown and rufous (see Parker 1999).

PHOTO 6.1

Adult Mississippi Kite west of Eunice, Lea Co., June 2006. The adult is mostly medium to light gray. The back and flight coverts are slate gray, contrasting with pale gray to silvery white secondaries. The primaries (not visible here) are black. The eyes are red, with a circle of black feathers around them, most prominent toward the front. PHOTOGRAPH: © MARK L. WATSON.

The Mississippi Kite is usually silent, and only two calls are recognized (Parker 1999). The vocalization most often heard is a high-pitched, whistled "*phee-phew*," given to indicate presence or as an alarm call.

The Latin binomial of the Mississippi Kite is *Ictinia mississippiensis. Ictinia* is a Greek derivative meaning "a kite." The specific epithet, *mississippiensis*, refers to the state where Alexander Wilson collected the type specimen (AOU 1998). Wilson assigned a Latin binomial to the bird and published the new name in his *American Ornithology* in 1811. According to Bolen and Flores (1993), Wilson's kite may not have been the first to be viewed in hand by a naturalist. In 1806, Peter Curtis, assigned to the Foreman and Curtis expedition by President Thomas Jefferson, collected a Mississippi Kite in Louisiana just north of Shreveport. For various reasons, however, Curtis did not publish a Latin binomial for the bird, and therefore the species did not make its official debut to the world of ornithology (Bolen and Flores 1993). Consequently, we shall continue to say Mississippi Kite for a species that officially could have been called the Louisiana Kite.

Distribution

The Mississippi Kite is a resident only of the Americas. It breeds in North America and winters in South America as far south as Argentina and Paraguay (Blake 1949; Eisenmann 1963; Hayes et al. 1990; Parker 1999). The North American breeding range is in the United States, and restricted almost entirely to the Southeast, southern

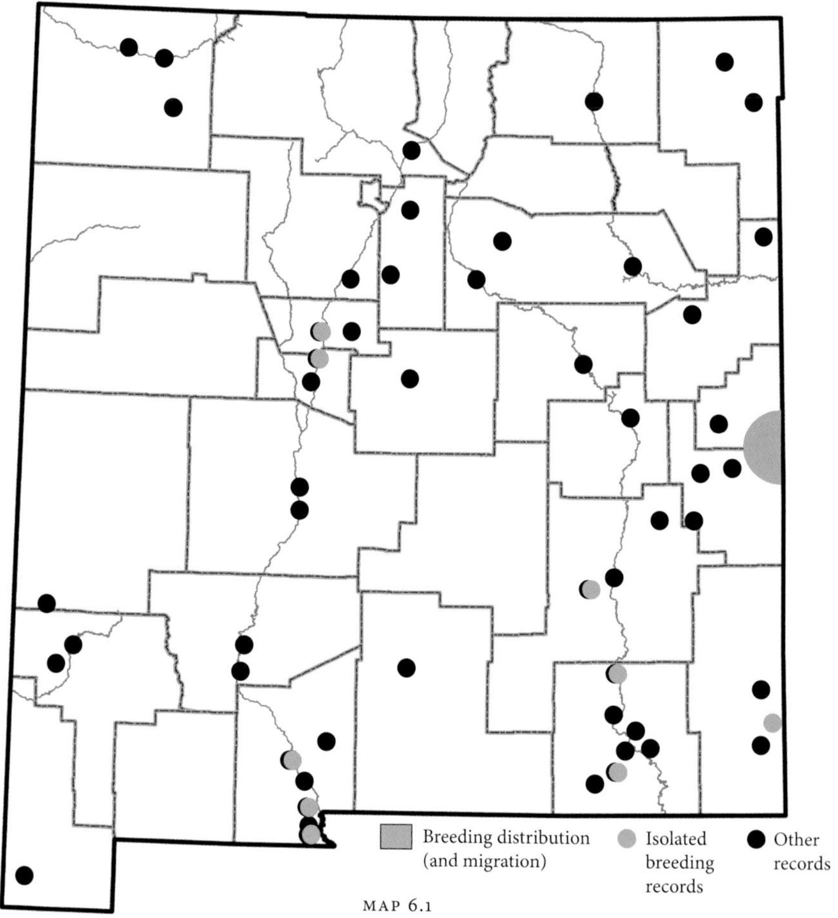

MAP 6.1

Mississippi Kite distribution map

Great Plains, and Southwest, with a few scattered, isolated populations farther north (Parker 1999).

Mississippi Kites have been observed in numerous locations in New Mexico, but recorded breeding sites are only in the Rio Grande Valley from Corrales southward, in the Pecos River Valley from Roswell southward, and on the eastern plains from Clovis southward to Hobbs (Ligon 1961; Travis 1962; McKnight and Snider 1966; Hubbard 1978, 1979; Hubbard and Eley 1985; map 6.1). Bailey (1928) mentions three specimens of this species taken in New Mexico by Samuel Washington Woodhouse in 1851, but there is no trace of those specimens and the records are at best doubtful. For nearly a century thereafter, no Mississippi Kite was reported from New Mexico. The earliest accepted records for the state are all observations dating back to the mid and late 1950s, suggesting that the Mississippi Kite is a newcomer to New Mexico (Ligon 1961). Following the first record, one Mississippi Kite from the upper Animas Valley about 6 km (~4 mi) east of Cloverdale on 13 May 1955, early records include several kites over a cottonwood grove a few miles south of Carlsbad in September 1955, one 5 km (3 mi) west of Clayton on 25 May 1957, two feeding above an irrigated field near Canutillo on 9 July 1957, one at Bosque del Apache on 21 May 1958, and one about 11 km (7 mi) northwest of Roswell on 26 June 1958 (Ligon 1961).

Jay M. Sheppard's 1957 sighting of a family of Mississippi Kites circling over riparian woodland just north and east of Black River Village in the Pecos River Valley, Eddy County, was the first evidence of breeding kites in New Mexico. Sheppard observed two juveniles, and possibly a third, being fed by two adults (Sheppard 2006). The second sighting of breeding Mississippi Kites was during the summer of 1960 in the Rio Grande Valley near Anthony, Doña Ana County (Zimmerman 1962). The first recorded breeding on the eastern plains was in Portales, Roosevelt County, in 1970 (Hubbard 1971).

PHOTO 6.2

(*top*) Mississippi Kite nesting and foraging habitat in Clovis, Curry Co. Nests are placed in trees along fairways of the golf course. PHOTOGRAPH: © TONY GENNARO.

PHOTO 6.3

(*bottom*) Mississippi Kite nesting and foraging habitat at Corrales, Sandoval Co. in the Middle Rio Grande Valley. Nesting has been recorded on the edge of the Corrales Bosque Preserve (wooded area "below" the river) where it borders the Corrales Riverside Drain. Foraging is over fields, woodlots, yards, and ditches. PHOTOGRAPH: © JIM FINDLEY.

Habitat Associations

Despite their limited and often local distribution, Mississippi Kites nest in a variety of habitats, with notable differences between southeastern and Great Plains or southwestern populations. In the Southeastern

United States, Mississippi Kites nest in large tracts of mature lowland and floodplain hardwoods, but farther west and including in New Mexico the species is associated with human-dominated habitats. They are primarily urban nesters on the eastern New Mexico plains and rural or semiurban nesters in the Rio Grande and Pecos River valleys.

Urban kites on the eastern plains of New Mexico forage and nest in association with open fields, residential neighborhoods, city parks, cemeteries, and golf course fairways; planted Siberian elms (*Ulmus pumila*) are the predominant nest trees (Gennaro 1988a). In the Rio Grande and Pecos River valleys, nest trees have not been documented, but probably consist mainly of cottonwoods (*Populus* spp.). One active nest in Corrales in the Middle Rio Grande Valley was on the levee side of the local cottonwood riparian forest; foraging was observed over fields, woodlots, yards, and ditches (J. Findley, pers. comm.).

(6.4)

(6.5)

(6.6)

Life History

Nesting

The Mississippi Kite is known to often nest in loose colonies (Parker 1999). This is true in particular on the eastern plains of New Mexico, where my research on the species included thorough nest surveys from six different communities: Cannon Air Force Base and Clovis in Curry County; Portales, Roosevelt County; Roswell, Chaves County; Artesia, Eddy County; and Hobbs in Lea County. A total of 110 nests were counted in 1992, but nesting chronology reported here is from 54 nests monitored at an urban site including part of Hillcrest Park and the municipal golf course in Clovis (Gennaro 1988a; photos 6.4–6.9).

Mississippi Kites began to arrive in early May, and all were present by mid May (Gennaro 1988a). Adults appeared to be paired upon arrival, as noted elsewhere in many earlier studies (Parker 1999). Copulations were frequently noted in Clovis during the first two weeks of May, and most pairs were incubating eggs by the end of that month. During both the incubation and the nestling periods, Mississippi Kites showed aggressive nest-defense behavior, at times dive-bombing people on the golf course (see photo 6.10).

(6.7)

(6.8)

(6.9)

PHOTOS 6.4–6.9

Mississippi Kites at their nests, Clovis, Curry Co. (6.4) nest building (PHOTOGRAPH: JOE KENT; © TONY GENNARO); (6.5) male delivering a prey item to the female (PHOTOGRAPH: MARTHA AIRTH-KINDREE; © TONY GENNARO); (6.6) both adults and downy young (barely visible) in the nest (PHOTOGRAPH: MARTHA AIRTH-KINDREE; © TONY GENNARO); (6.7) two downy young with adult holding arthropod prey item (PHOTOGRAPH: JEAN HAND; © TONY GENNARO); (6.8) adult feeding a nestling (PHOTOGRAPH: MARTHA AIRTH-KINDREE; © TONY GENNARO); (6.9) adult feeding an old nestling an arthropod (PHOTOGRAPH: RICHARD ARTRIP; © TONY GENNARO).

According to Parker (1999), Mississippi Kites generally lay two eggs in a clutch. One-egg clutches are uncommon and often the result of loss of one egg in a two-egg clutch; three-egg clutches are very rare (Parker 1999). Although the information was not published in Gennaro (1988a), clutch size was recorded for 24 nests in 1983. Nine (37.5%) of the nests contained one egg, while the other 15 (62.5%) had two eggs.

Hatchlings were first observed in mid June, with most present by early July. Incubation required approximately 32 days, an interval similar to that reported by many others (see Parker 1999). Nestlings remained in the nest for 15–18 days after hatching. For the next 18–22 days they spent time both in the nest and on branches adjacent to the nest. Young kites fledged in 33–40 days, a period of time similar to that found in other studies (Parker 1999). The kite population decreased considerably by the end of August, and all had departed from Hillcrest Park and the Clovis Municipal Golf Course by the first week in October.

Diet and Foraging

Mississippi Kites are well known for their midair captures of medium to large insects, a feat accomplished during soaring and hawking (see Parker 1999). Hawking involves a short and direct flight from a perch to capture an insect, before returning to the perch where the prey is then eaten. Insects may also be consumed in flight as they often are after capture during soaring, or they may be presented to a mate, nestlings, or young fledglings (photos 6.5–6.9). Kites may also capture bats and swallows in midair, eating them on a perch, while other prey may be captured on the ground or on vegetation (Parker 1999).

Prey capture data show that insects are numerically dominant in the Mississippi Kite's diet (Parker 1999). Glinski and Ohmart (1983) reported that cicadas were the main prey delivered by Mississippi Kites to three nests in Arizona. During my studies of breeding kites in Clovis, New Mexico, during the 1980s, I noted that their prey were primarily cicadas, grasshoppers, and occasionally dragonflies. Airth-Kindree (1988) reported that cicadas and grasshoppers made up the majority of prey brought to a kite nest she investigated at Colonial Park in Clovis, with small toads also frequently captured. J. Findley (pers. comm.) observed a kite carrying a small bird in Corrales.

Predation and Interspecific Interactions

Interspecific interactions have not been documented in New Mexico. Parker (1999) reports that large raptors such as Great Horned Owls (*Bubo virginianus*) and mammals such as raccoons (*Procyon lotor*) are the greatest threat to kites in rural areas. Parker (1999) suggests possible competition with Swainson's Hawks (*Buteo swainsoni*) for reoccupancy of old nests, but in eastern New Mexico I never noted the two species nesting in close proximity to each other.

Status and Management

Mississippi Kites nesting in urban areas are known to have higher reproductive success than those nesting in rural areas, in part due to a lower incidence of predation and weather-related nest failures (Parker 1996, 1999). Parker's (1996) finding was based in part on data from Clovis, New Mexico, where Gennaro (1988a) reported an average of 1.2 fledglings/nesting attempt. Similar productivity levels were reported by Shaw (1985) from San Angelo, Texas (1.1 fledglings/nesting attempt) and by Parker (1996) primarily from Meade, Kansas (1.0–1.3 fledglings/nesting attempt). In contrast, observed productivity in rural areas was only 0.63 on the Great Plains (Parker 1974), 0.60 in Arizona (Glinski and Ohmart 1983), and 0.61 in southern Illinois (Evans 1981). From the studies above it appears that breeding pairs are twice as productive in urban settings compared to rural areas (Parker 1996). Calculations by Parker (1974) indicated that a success rate of 0.5 fledgling per nesting attempt is high enough for a population of Mississippi Kites to remain stable. With a mean number of 1.2 fledglings produced per nest, the population of Mississippi Kites nesting in Clovis would therefore far exceed minimum replacement level and could be expected to grow rapidly or serve as a source population for colonization elsewhere. In 1992, as mentioned above, I counted 110 pairs nesting at Cannon Air Force Base and Clovis in Curry County; Portales, Roosevelt County; Roswell, Chaves County; Artesia, Eddy County; and Hobbs in Lea County. No survey of the Mississippi Kite nesting population in eastern New Mexico has been conducted since then. However, counts of Mississippi Kites in flight during more recent visits to a few of the areas above (i.e., Cannon Air Force Base, Clovis, Portales) all suggested stable population numbers.

PHOTO 6.10

Warning sign at Cannon Air Force Base, Curry Co. The success of Mississippi Kites in urban areas, particularly golf courses and city parks, has led to management issues. Mississippi Kites are aggressive near their nests and do not hesitate to dive at people, occasionally striking them with their talons. Warning signs and education have proven the best tools to mitigate the problem and ensure continued coexistence of people and kites. PHOTOGRAPH: © PENNY SMITH.

PHOTO 6.11

"Mexico," an adult Mississippi Kite at least 11–12 years old and one of Hawks Aloft's educational birds. He was sent to the Wildlife Center in fall 1999 from Portales, Roosevelt Co. He had a broken wing due to an unknown cause and could not be released back in the wild. He was transferred to Hawks Aloft in fall 2000. Mexico must be kept indoors all winter. All Mississippi Kites leave New Mexico for warmer climates in the fall, and captive kites have a history of poor tolerance to low temperatures.

PHOTOGRAPH: © DOUG BROWN.

In New Mexico as elsewhere, the success of urban kite populations generates an unusual but important management issue. When they have eggs or nestlings, Mississippi Kites dive at people who venture too close to their nests. Some kites hit people with their talons, and in some cases blood has been drawn. Even near misses can startle individuals under attack. Gennaro (1988b) studied kite nest-defense behavior on the municipal golf course in Clovis. Members of his research team subjected themselves to 903 dives and were struck 24 (3%) times. Three (27%) of the 11 breeding pairs studied were nonaggressive, while three others (27%) were moderately aggressive, and five (46%) were highly aggressive, the latter typically attacking from behind and making contact with the head, toppling hats but not causing injuries.

Aggressive nest-defense behavior warrants management. Individuals in kite nesting areas should be informed that Mississippi Kites are protected by state and federal regulations such as the Migratory Bird Treaty Act, and that management must be conducted only by state and/or federal agencies or their designees. On golf courses and in city parks, cemeteries, and recreational areas where kites are known to nest, signs should be posted to indicate the presence of diving kites. Education is also an important tool in kite management. The benefits kites provide to humans can and should be emphasized to the public with help from the media. Kites should also be presented by skilled individuals to students in schools and to personnel in civic clubs. Information presented under a positive light fosters curiosity and tolerance by people toward nesting Mississippi Kites.

Intensive management, however, may be necessary wherever aggressive kites intimidate people. Eggs and/or nestlings should be removed from nests in trees located near trails and pathways, and placed in foster parent kite nests located in remote areas. Parker (1980) removed eggs and nestlings from kite nests in Ashland, Kansas, and transferred them to kite foster parents. Nesting kites chosen for egg and nestling removal departed from breeding sites. My attempt to manage aggression on the Clovis golf course using Parker's (1980) method was met with some initial success, as dives at people ceased at sites where eggs had been removed (Gennaro 1988b). But each pair re-nested and continued to display aggressive behavior. As a result, I used fake Mississippi Kites constructed from balsa to discourage re-nesting in areas frequented by humans (Gennaro 1988b). My studies from 1982 through 1985 showed that kites build only one nest per tree, and that kites frequently reuse old nests. Therefore before kites arrived in spring, I placed fake kites in old nests to prevent re-nesting and placed both man-made nests and fake kites in trees not previously used for nesting. That method successfully manipulated nest placement of breeding kites. Consequently, Mississippi Kite aggression and human-kite conflict were reduced.

In summary, Mississippi Kites have colonized several areas of New Mexico. Urban nesting populations in particular are healthy, exhibiting high reproductive success. Meanwhile, aggressive nest-defense behavior represents a potential source of conflict with people, but this problem can be managed successfully. For those reasons, the long-term outlook for the species in New Mexico seems reasonably bright. Mississippi Kites can be expected to flap, soar, and dive across the New Mexican sky for years to come.

LITERATURE CITED

Airth-Kindree, M.M.A. 1988. Nesting developmental behavior of a Mississippi Kite (*Ictinia mississippiensis)* from an urban population at Colonial Park, Clovis, New Mexico. M.S. thesis. Eastern New Mexico University, Portales.

[AOU] American Ornithologists' Union. 1998. *Check-list of North American birds*. 7th ed. Washington, DC: American Ornithologists' Union.

Bailey, F. M. 1928. *Birds of New Mexico*. Santa Fe: New Mexico Department of Game and Fish.

Blake, E. R. 1949. *Ictinia mississippiensis* collected in Paraguay. *Auk* 66:82.

Bolen, E. G., and D. L. Flores. 1993. *The Mississippi Kite*. Austin: University of Texas Press.

Eisenmann, E. 1963. Mississippi Kite in Argentina: with comments on migration and plumages in the genus *Ictinia*. *Auk* 80:74–77.

Evans, S. A. 1981. Ecology and behavior of the Mississippi Kite (*Ictinia mississippiensis*) in southern Illinois. M.S. thesis. Southern Illinois University, Carbondale.

Gennaro, A. L. 1988a. Breeding biology of an urban population of Mississippi Kites in New Mexico. In *Proceedings of the southwest raptor management symposium and workshop*, ed. R. L. Glinski et al., 188–90. Washington, DC: National Wildlife Federation.

———. 1988b. Extent and control of aggressive behavior toward humans by Mississippi Kites. In *Proceedings of the Southwest Raptor Management Symposium and Workshop*, ed. R. L. Glinski et al., 249–52. Washington, DC: National Wildlife Federation.

Glinski, R. L., and R. D. Ohmart. 1983. Breeding ecology of the Mississippi Kite in Arizona. *Condor* 85:200–207.

Hayes, F. E., S. M. Goodman, J. A. Fox, T. G. Tamayo, and N. E. Lopez. 1990. North American bird migrants in Paraguay. *Condor* 92:947–60.

Hubbard, J. P., ed. 1971. *New Mexico Ornithological Society Field Notes* 10(2): 1 June–30 November.

———. 1978. *Revised checklist of the birds of New Mexico*. Publ. no. 6. Albuquerque: New Mexico Ornithological Society.

———, ed. 1979. *New Mexico Ornithological Society Field Notes* 18(2): 1 June–30 November.

Hubbard, J. P., and J. W. Eley 1985. *Handbook of species endangered in New Mexico*. Santa Fe: New Mexico Department of Game and Fish.

Ligon, J. S. 1961. *New Mexico birds and where to find them*. Albuquerque: University of New Mexico Press.

McKnight, B. C., and P. R. Snider. 1966. *New Mexico Ornithological Society Field Notes* 5(2): 1 June–30 November.

Parker, J. W. 1974. The breeding biology of the Mississippi Kite in the Great Plains. Ph.D. diss. University of Kansas, Lawrence

———. 1980. Kites of the prairies. *Bird Watcher's Digest*, July/August, 86–94.

———. 1996. Urban ecology of the Mississippi Kite. In *Raptors in human landscapes*, ed. D. Bird, D. Varland, and J. Negro, 45–52. New York: Academic Press.

———. 1999. Mississippi Kite (*Ictinia mississippiensis*). No. 402. In *The birds of North America*, ed. A. Poole and F. Gill. Philadelphia, PA: Academy of Natural Sciences, and Washington, DC: American Ornithologists' Union.

Shaw, D. M. 1985. The breeding biology of urban-nesting Mississippi Kites (*Ictinia mississippiensis*) in west central Texas. M.S. thesis. Angelo State University, San Angelo.

Sheppard, J. M. 2006. First nesting of the Mississippi Kite (*Ictinia mississippiensis*) and early status in New Mexico. *New Mexico Ornithological Society Bulletin* 34:16–18.

Travis, J. R. 1962. *New Mexico Ornithological Society Field Notes*, Report no. 1, June through September. Albuquerque: NMOS.

Wheeler, B. K., and W. S. Clark. 1995. *A photographic guide to North American raptors*. London: Academic Press.

Zimmerman, D. A., ed. 1962. Nesting season, June 1 to August 15, Southwest Region. *Audubon Field Notes* 16:497.

7

Bald Eagle

(*Haliaeetus leucocephalus*)

DALE W. STAHLECKER AND HIRA A. WALKER

THE BALD EAGLE (*Haliaeetus leucocephalus*) is arguably New Mexico's most strikingly marked raptor. The adult's white head and tail, golden bill and feet, and almost black body make for an impressive sight, whether it is seen close or far, alone or with several other eagles. Full adult plumage is attained by the age of 5.5 years (Buehler 2000), and juvenile and immature plumages are less dramatic. Juveniles depart the nest almost completely chocolate brown, nearly black, and with a black beak. With succeeding molts, immatures initially show significant white in the underwings and belly, then less as they molt toward adulthood. At 3.5 years, the head is mostly white, but with a dark eye-line. At this age, Bald Eagles also have considerable white in their tails and mostly yellow bills with some remaining black streaks. As is true of all of the world's "fish eagles," which are adapted to catch their prey near the water's surface, the lower legs of Bald Eagles are bare at all ages.

PHOTO 7.1

Adult Bald Eagle, Reserve, Catron Co., 14 February 2007. With its completely white head and white tail, the adult Bald Eagle is unmistakable. PHOTOGRAPH: © RALPH GILES.

PHOTO 7.2

(*top left*) Adult Bald Eagle (educational bird). The head is white, with a bright yellow beak and pale yellow eyes. Other than the head and the tail, the rest of the plumage is dark brown. PHOTOGRAPH: © GERAINT SMITH.

PHOTO 7.3

(*bottom left*) Juvenile Bald Eagle, Rio Arriba Co., 19 November 2008. The bird in the photograph is less than one year old. Juveniles are dark brown when they leave the nest. The dark brown fades on the belly, giving it a tawny color. The undertail can be almost completely dark brown or, as on this bird, buffy white with dark feather edges and a dark terminal band (obscured by shadow). The head is brown; the eyes have dark brown irises. PHOTOGRAPH: © ROGER HOGAN.

PHOTO 7.4

(*bottom right*) Immature Bald Eagle in flight, Rio Arriba Co., 24 March 2008. This immature is 1–2 years of age and called a "White Belly I" (basic I plumage). The head is still mostly dark; the eyes have light brown irises. On the breast is a conspicuous dark brown bib, contrasting with a mostly white belly with many brown spots and streaks. The axillaries are white. The rectrices on the undertail are white with dark edges and a brown terminal band. The new primaries and secondaries are shorter, but some of the secondaries have not molted yet, resulting in the wings having a conspicuously uneven trailing edge. In the next plumage (basic II), the molt of secondaries will be completed. PHOTOGRAPH: © ROGER HOGAN.

PHOTO 7.5

(*top*) Immature Bald Eagle, Rio Grande Gorge, Taos Co., 14 February 2008. This bird is 2–3 years of age and called a "White Belly II" (basic II plumage). The contrast between the dark brown of the breast forming a bib and the white belly is still distinct. However, note the mostly white throat, the mottled white and brown crown, and the dark eye-line. PHOTOGRAPH: © GERAINT SMITH.

PHOTO 7.6

(*bottom*) Subadult Bald Eagle, Rio Arriba Co., 24 December 2006. Subadults (3.5–4.5 years) are very similar to adults, but the mostly white head shows some dark flecking, including a faint dark eye-line; the tip of the tail is also mottled, dark brown. PHOTOGRAPH: © ROGER HOGAN.

Bald Eagles are found only in North America and those that breed at northerly latitudes (roughly north of 40°N) are noticeably larger than those that are resident at southerly latitudes to such extent that they are classified as two separate subspecies (*H.1. alascanus* and *H.1. leucocephalus*, respectively) (Buehler 2000). However, distinguishing between the two subspecies solely on the basis of size is problematic. The increase in size with increasing latitude is gradual and, as with most birds of prey, females from a particular latitude are approximately 25% larger than their male counterparts. In addition, the two subspecies can co-occur when the northern *alascanus* move southward for the winter and when the southern *leucocephalus* move northward during postbreeding season dispersal. In New Mexico, a Bald Eagle found in winter could be either subspecies and, thus, could be anywhere from 71 to 96 cm (28–38 in) long with a wingspan of 168–244 cm (5.5–8 ft), and could weigh anywhere from 3 kg (6.5 lbs) for the smallest southern males to 6.5 kg (14 lbs) for the largest northern females (Palmer et al. 1988).

The dark body with white head and tail makes adult Bald Eagles hard to confuse with any other large bird of prey, but subadult Bald Eagles with an eye-line could be confused with Ospreys (*Pandion haliaetus*) and dark juveniles could be confused with adult Golden Eagles (*Aquila chrysaetos*). Both Bald Eagles and Ospreys regularly occur near water in New Mexico, but most Ospreys have migrated south out of New Mexico by winter when

PHOTO 7.7

Adult Bald Eagle in flight over the Chama River, Rio Arriba Co., 20 January 2009. Note the completely white head and tail, dark body, and golden bill and feet.

PHOTOGRAPH: © ROGER HOGAN.

Bald Eagles are most abundant. In addition, Ospreys are smaller, with much smaller gray bills and legs (Sibley 2000). The Golden Eagle, which is a year-round resident of mostly open country throughout New Mexico (chapter 21), is closest in size to the Bald Eagle, whether perched or in flight. However, Golden Eagles soar with wings held in a dihedral shape or slight V-shape, whereas Bald Eagles have noticeably broader wings that are held almost flat. On close examination, Golden Eagles also have feathered legs, smaller heads, and slightly smaller beaks (Buehler 2000; Sibley 2002). Furthermore, immature Bald Eagles after their first molt (> one year old) usually have extensive white on the underwings and body, unlike any Golden Eagle plumages.

Bald Eagles also can be identified by their unique vocalizations. Adults call during exchanges at nests and there are often multiple vocalizations over food or roost perches when adults and immatures gather in large groups in winter. Their most common call is a rolling, high-pitched (for their size) "chatter" call, consisting of three to four introductory notes, followed by six to nine more rapid and descending notes (Buehler 2000).

Distribution

The eight species within the genus *Haliaeetus*, for the most part, have disjunct distributions with limited overlap in range. Although the widespread White-tailed Eagle (*H. albicilla*) of Eurasia has nested in the far western Aleutian Island of Attu in Alaska (Tobish and Balch 1987), the Bald Eagle is the only member of its genus that regularly occurs in the Western Hemisphere. Bald Eagles are confined almost totally to Canada and the United States, with only limited breeding and wintering populations in northern Mexico (Henny et al. 1993). Bald Eagle population estimates range from 70,000 to 100,000 individuals, with 75–90% breeding in Canada and Alaska (Buehler 2000). The population in the continental United States has increased from fewer than 500 to more than 4,000 pairs from the 1960s to 2005 (U.S. Fish and Wildlife Service 2007).

Bald Eagles that breed in interior Canada and Alaska move south to winter in warmer climes. Only small numbers of these eagles migrate through New Mexico to winter at more southerly latitudes; so few were counted in the Manzano Mountains during autumn migration and in the Sandia Mountains during spring migration that statistical analyses could not be completed (Hoffman and Smith 2003). Instead, large numbers of Bald Eagles move into New Mexico after peak migration has concluded to winter throughout the state (chapter 2). The number of wintering Bald Eagles in the state appears to be gradually increasing. An annual average of 231 Bald Eagles were counted in New Mexico during annual January aerial counts along rivers and at reservoirs between 1979 and 1986 (Hubbard et al. 1988). From 1987 to 1991, the annual average count had increased to 370 and, from 1992 to 1996, when the last aerial counts by New Mexico Department of Fish and Game were completed, the average was 421. The highest count was 512 in January 1990 (NMOS 2007). During approximately the same period (1960–2006), Christmas Bird Count data showed much year-to-year variation but overall an increase in the number of Bald Eagles counted (fig. 7.1).

The majority of Bald Eagles wintering in New Mexico are adults. Between 1979 and 1986, adults represented 56% of wintering Bald Eagles counted from aircraft (Hubbard et al. 1988). By river drainage, they averaged >61% of eagles counted on the San Juan River, Rio Chama, and Upper Rio Grande (north of Albuquerque), and >54% on the Upper and Lower Canadian River. In contrast, immature eagles averaged >71% of eagles counted on the Lower Rio Grande and >62% of eagles counted on the Upper Pecos River. At Cochiti Reservoir (Sandoval County) between 1979 and 1985, immature eagles were most abundant when more waterfowl were available as prey, and both adult

PHOTO 7.8

Bald Eagle in Rio Arriba Co., 13 December 2007. The vast majority of New Mexico's Bald Eagles are winter residents and leave the state to breed farther north.

PHOTOGRAPH: © ROGER HOGAN.

PHOTO 7.9

Two adult Bald Eagles perched along the upper Rio Grande, Taos Co., 19 December 2006. In New Mexico, Bald Eagles are found primarily at lakes or reservoirs and along rivers (e.g., Rio Grande, Rio Chama, San Juan River, Canadian River, and Upper Pecos River).

PHOTOGRAPH: © GERAINT SMITH.

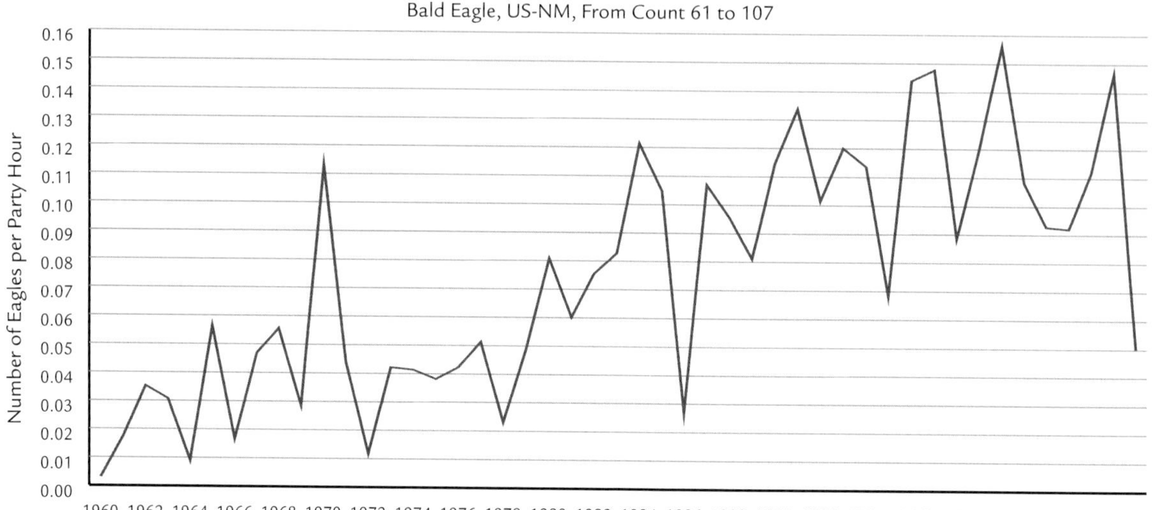

FIGURE 7.1

Number of Bald Eagles per party hour reported on New Mexico Christmas Bird Counts, 1960–2006 (National Audubon Society 2008).

and immature eagles were more numerous during colder winters (Johnson 1988).

Techniques used to track Bald Eagle movements, such as fitting eagles with radio transmitters or with plastic color-markers on their wings, have provided evidence that migrant Bald Eagles wintering in New Mexico show year-to-year fidelity to wintering grounds. Bald Eagles marked in the 1980s were often found in successive years, as much as six years later, in the same wintering areas in southern Colorado and northern New Mexico where they had been first captured (Harmata and Stahlecker 1993). Radio transmitters also have helped to link New Mexico's wintering Bald Eagles with their probable breeding grounds. An adult male that was radio-tagged at Abiquiu Reservoir, Rio Arriba County, New Mexico, in 1988 was tracked migrating northward in spring through the San Luis Valley just north of the New Mexico/Colorado border (Stahlecker and Smith 1993). It is likely that it was returning to central Canada to breed, similar to three eagles that were trapped in the San Luis Valley and tracked to breeding territories 1,900 km (1,200 mi) farther to the north in central Canada (Harmata 2002).

Although the Bald Eagle is most abundant and widespread in New Mexico in winter, the species also breeds in the state. The earliest breeding records are suspect. Bailey (1928:180) reported that "east, west and north, in the three states bordering New Mexico, the Bald Eagle is known to nest, and it has been reported to Mr. Ligon as nesting on the Frisco and east Gila Rivers. It is fairly common in western Socorro [Catron] County, he [Ligon] says." Ligon (1961:75), for his part, only said, "the author has but two nesting records for the Bald Eagle in the state. One nest was situated on a pinnacle in the gorge of the Middle Fork, and the other in a cliff on the East Fork of the Gila River." Together these statements suggest that Ligon never observed these nests, even though he collected a set of Osprey eggs from a treetop nest on the West Fork of the Gila on 24 April 1916 (Ligon 1961). Because both Bald Eagles (in Arizona; Hunt 1998) and Ospreys (in Mexico; Henny and Anderson 2004) have been found to regularly nest in cliffs, it is uncertain whether nests of Osprey were mistaken for nests of Bald Eagles by an unreliable observer and, thus, these reports of nesting remain questionable.

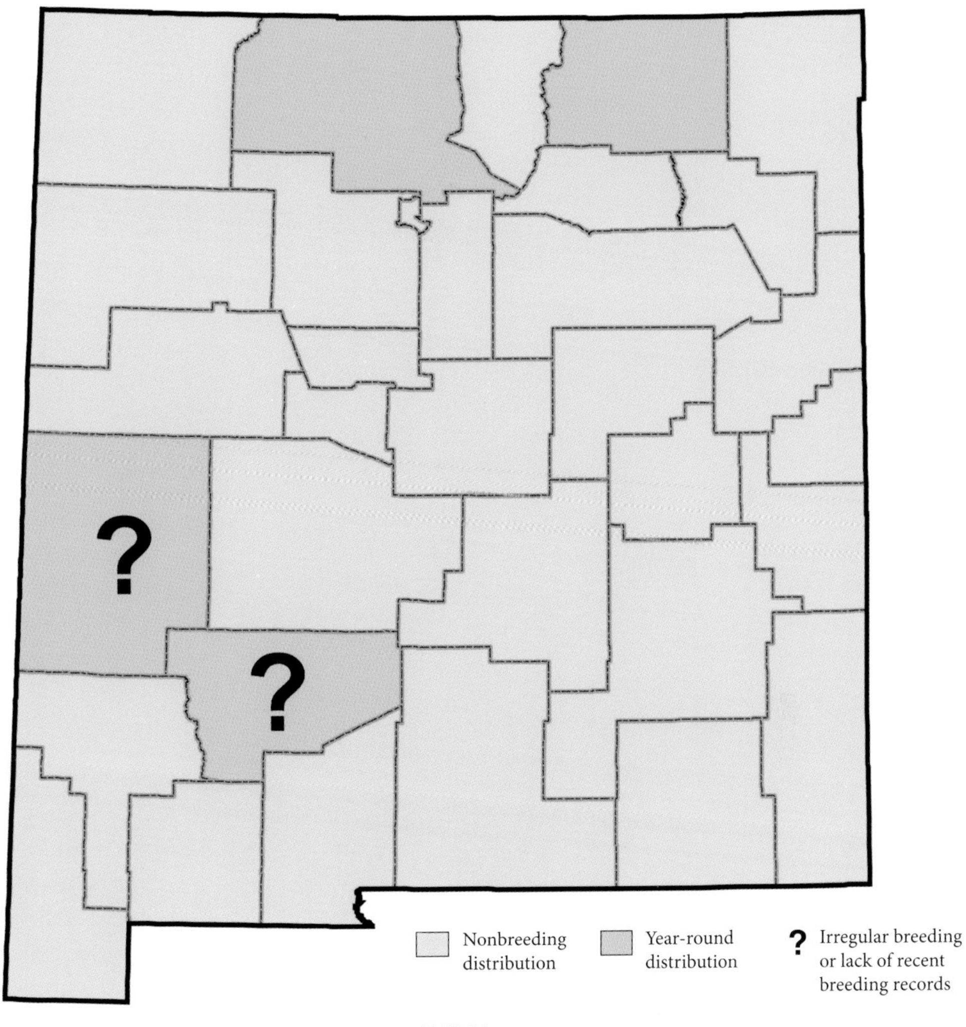

MAP 7.1

Bald Eagle distribution map

Bald Eagles winter throughout New Mexico. However, the number of Bald Eagle nesting territories in the state is very small. Because the exact location of these territories is considered sensitive information, the "breeding distribution" is given here as the entire counties harboring the known territories. Whether southwestern New Mexico still harbors active territories is uncertain.

The first confirmed New Mexico Bald Eagle nest was found in a tree in central Catron County in 1979 (distant from the Gila drainage), but that site has apparently not been occupied since (Williams 2000). By 2008, nesting had been confirmed in 4 of New Mexico's 33 counties: Catron (two territories), Colfax (three territories), Sierra (one territory), and Rio Arriba (two territories) (New Mexico Department of Game and Fish, unpubl. data; table 7.1). The second Catron County territory was occupied in 2005 and 2009. Long-term occupancy of nest sites of Bald Eagles began in Colfax County in 1987 and in Sierra County in 1988, but by 2009 the first Colfax County territory had been unoccupied for three years and the only Sierra County territory had not been occupied for more than 10 years. The two additional eagle breeding territories in Colfax

TABLE 7.1. History of occupancy and mean annual productivity (mean number of young fledged per year) for the eight known Bald Eagle (*Haliaeetus leucocephalus*) territories in New Mexico, 1979–2009.

TERRITORY	YEAR DISCOVERED	YEARS OCCUPIED	MEAN ANNUAL PRODUCTIVITY
Catron #1	1979	1979–1980	0
Catron #2	2005	2005, 2009	0
Colfax #1	1987	1987–2005	1.6[1]
Colfax #2	1998	1998–2009	1.4
Colfax #3	1998	1998–2009	1.0[2]
Rio Arriba #1	2006	2006–2009	0.75
Rio Arriba #2	2006	2006–2009	0.75
Sierra #1	1988	1988–1997	1.4

[1] No data for 1989, number of nestlings fledged unknown in 1990.

[2] Not monitored in 2006.

County were discovered in 1998. Nesting of Bald Eagles in Rio Arriba County was first documented in May 2006, when an adult female was found incubating in an old Osprey nest in a snag. The nest was abandoned by the end of the month. However, in 2007, one young fledged from a more typical Bald Eagle nest in a live Douglas fir (*Pseudotsuga menziesii*) 8 km (5 mi) away. In 2008, an attempt in that nest was unsuccessful, but another nest, only 3 km (2 mi) from the 2006 nest, fledged two young, confirming that two pairs were present in 2008 and possibly all three years. The presence of nesting Bald Eagles in Rio Arriba County did not come as a surprise. Southwestern Colorado, which has had breeding Bald Eagles since at least 1974 (Winternitz 1998), supported a breeding population of at least 10 pairs in 2003 and 2004 (DWS, unpubl. data). Also not surprising are reports—though unsubstantiated—of adult Bald Eagles and possible nests on nearby Navajo Reservoir and the Animas River in San Juan County, New Mexico. Note that in this chapter we do not provide the exact location of any historic or current Bald Eagle nests in New Mexico. Because the breeding population is so small, we do not want excessive human visitation of nest areas to result in nest failure or abandonment (see also Status and Management in this chapter).

Habitat Associations

Bald Eagles are found near water throughout most of their range during the breeding season. When northern waters freeze, the summer residents head south. When they arrive in New Mexico for the winter, most still are associated with water, foraging along the state's rivers and reservoirs (Hubbard et al. 1988). The radio-tagged adult male at Abiquiu Reservoir had a late winter home range of only 16 km^2 (6 mi^2), consistently staying close to the reservoir (Stahlecker and Smith 1993).

In the 1980s, in southern Colorado farmlands, where most water was frozen all winter, rabbits and domestic carrion were very important Bald Eagle food items

PHOTO 7.10

Typical habitat in New Mexico includes reservoirs. In this photograph, Bald Eagles on partly frozen Eagle Nest Lake, Colfax Co., 15 December 2006. PHOTOGRAPH: © ELTON M. WILLIAMS.

PHOTO 7.11

Typical habitat in Rio Arriba Co.: a lakeshore with some live trees and snags. PHOTOGRAPH: © ROGER HOGAN.

PHOTO 7.12

Typical habitat in Rio Arriba Co.: a lakeshore with dense trees and snags. PHOTOGRAPH: © ROGER HOGAN.

PHOTO 7.13

Bald Eagle in large ponderosa pine (*Pinus ponderosa*) in Taos Co. Large trees or snags serve as vantage points that are important components of the Bald Eagle's habitat in New Mexico. PHOTOGRAPH: © GERAINT SMITH.

PHOTO 7.14

Bosque del Apache National Wildlife Refuge, 23 November 2005. At the Bosque del Apache, Bald Eagles are often observed perched on emergent snags in ponds or flying over fields and wet meadows. PHOTOGRAPH: © STEVE BARANOFF.

PHOTO 7.15

Bald Eagle on the ground in a prairie dog (*Cynomys*) town, Maxwell National Wildlife Refuge, 9 March 2005. PHOTOGRAPH: © PATTY HOBAN.

(Harmata 2002) and eagles ranged away from water. How much this also occurs in New Mexico's agricultural areas is largely unknown. On the Jicarilla Apache Nation in north-central New Mexico, big-game hunting seasons are nearly continuous from October to mid January, attracting wintering Bald Eagles that feed on remains from field-dressed kills and on wounded animals that die later. While as many as 152 Bald Eagles were counted in January 1997 during a week of big-game counts by helicopter (T. Watts, Jicarilla Department of Game and Fish, unpubl. data), the amount of recounting is unknown. Still, it is likely that some years 30–50 Bald Eagles spend much of the winter in these uplands and far from open water.

The habitat within the eight Bald Eagle nesting territories occupied between 1979 and 2009 in New Mexico is quite diverse. Elevation ranges from 1,400 m (4,600 ft) in Sierra County to 2,650 m (8,700 ft) in Colfax County. Five of the nests have been between a few meters (yards) and 800 m (half a mile) from a major river or lake, but one nest, while in a riparian corridor with a small perennial stream, was 11.5 km (7 mi) from a reservoir. Surrounding vegetation types varied from Chihuahuan desertscrub in Sierra County and short grass prairie and farm fields in Colfax County to mixed conifer forest edges in Colfax and Rio Arriba counties. Black-tailed prairie dog (*Cynomys ludovicianus*) towns are commonly found near territories, especially in Colfax County (Williams 2000).

PHOTO 7.16

An adult female Bald Eagle and her two large nestlings, Rio Arriba Co., 15 June 2009. The young fledged a week later. While New Mexico's pairs of Bald Eagles have generally been successful in fledgling young, the nesting population has remained small (2-4 pairs) over the past two decades. PHOTOGRAPH: © DALE W. STAHLECKER.

Life History

Nesting

Because the lives of Bald Eagles generally revolve around water, their nests are usually within 2.5 km (1.5 mi) of a lake or river (Buehler 2000). Most nests are placed below the crown of the largest live tree on the slope or shore in the area. Coniferous trees are preferred, but deciduous trees are readily used if they are the only trees available. In areas where trees are limited, as along Arizona desert rivers (Hunt 1998) or on Aleutian island shores (Sherrod et al. 1976), cliffs are used. In New Mexico, cottonwoods (*Populus* spp.) have been used at the low-elevation territories in Colfax and Sierra counties, while ponderosa pines (*Pinus ponderosa*) and Douglas firs have been used at the higher-elevation territories in Catron, Colfax, and Rio Arriba counties.

Bald Eagle clutches are normally one to three eggs, rarely four, and are laid over three to six days. Eggs are large and oval-shaped, and average 7–7.6 cm (~3 in) long and 5.3–5.6 cm (~2 in) at greatest width (Stalmaster 1987), but, as with body size, their size increases gradually from south to north (Buehler 2000). The eggs are dull white and usually unmarked, though rarely they can show brown spots (Stalmaster 1987). New Mexico nests have not been entered during incubation, so clutch size has not been documented. Nevertheless, Colfax County pairs sometimes have fledged three young (Williams 2000) and, thus, it can be deduced that at least three eggs and, perhaps, even four are sometimes laid. Incubation for Bald Eagles is generally 35 days. Young fledge between 8 and 14 weeks after hatching (Buehler 2000), but most fledge when 11–13 weeks (75–90 days) old (e.g., Hunt et al. 1992).

Fledged juveniles can be dependent upon their parents for 4–10 weeks (Buehler 2000).

Eagles that migrate out of the frozen north spend almost half the year away from their nesting territories. However, New Mexico's few breeding pairs likely remain on and defend their territories year-round. Colfax and Rio Arriba county pairs are usually incubating eggs in late February or early March, are feeding young by early April, and have large, all dark young by mid June that fledge before the end of the month (New Mexico Department of Game and Fish, unpubl. data). Dispersal of juveniles reared in New Mexico within and, possibly, away from New Mexico is unknown. However, fledglings from southern nesting populations in Florida (Broley 1947), California (Hunt et al. 1992), Texas (Mabie et al. 1994), and Arizona (Hunt et al. 2009) migrate rapidly north after independence, likely seeking concentrations of prey (i.e., spawning fish) that are easier to capture for these inexperienced hunters. Most California and Arizona fledglings returned to their natal regions the same autumn, and survivors were found in the local breeding populations when they reached maturity, though one eagle fledged from Arizona ultimately bred in southern California, about 500 km (300 mi) west of its natal region (Hunt et al. 2009).

Bald Eagles breeding in New Mexico have a strong tendency to reuse the same nest each year unless the nest tree falls. From 1987 to 2009, four New Mexico territories, three in Colfax County and one in Sierra County, were occupied ten or more years (table 7.1). The other four New Mexico territories have only one- to four-year histories. New Mexico territories have been successful in fledging between 0.9 and 1.6 young/year, within the range reported elsewhere for stable populations (Buehler 2000). However, nesting success has varied greatly among territories. Neither of the two Catron County territories fledged young in their short periods of known occupancy. Both Rio Arriba County territories were successful twice between 2006 and 2009, their average reproductive output approaching rangewide averages (table 7.1), the two pairs also likely part of a larger breeding population in southwestern Colorado. Of the eight known breeding territories in New Mexico, one long-occupied territory (1987–2005) in Colfax County successfully fledged the most young.

Diet and Foraging

Bald Eagles evolved to capture fish and prefer to take fish whenever available (Buehler 2000). Despite the fact that fish are more easily digested than other vertebrates and are often underrepresented in prey remains and in regurgitated pellets (Mersman et al. 1992), in a summarization of 20 different reports on food habits from across the species' range, the diet of nesting Bald Eagles consisted of 56% fish, 28% birds, 14% mammals, and 2% other (Stalmaster 1987).

PHOTOS 7.17a and b

(a) Adult Bald Eagle on the carcass of a Snow Goose (*Chen caerulescens*) at the Bosque del Apache National Wildlife Refuge, 15 December 2008; (b) taking off with the carcass in its talons. PHOTOGRAPHS: © JAMES N. STUART.

PHOTO 7.18

Young Bald Eagles sparring over the carcass of a goose, Bosque del Apache National Wildlife Refuge, 12 February 2004.

PHOTOS 7.21a and b

(*above*) Bald Eagle pair on a winter nest in the Chama Valley, Rio Arriba Co., 6 March 2007. A "winter nest" is built and used in courtship/pair maintenance, but the pair does not stay to breed locally and instead migrates north in spring. PHOTOGRAPHS: © ROGER HOGAN.

PHOTO 7.19

(*top left*) Bald Eagle communal roost, Rio Arriba Co., 8 November 2008. PHOTOGRAPH: © ROGER HOGAN.

PHOTO 7.20

(*bottom left*) Bald Eagle communal roost, Rio Arriba Co., 10 January 2007. PHOTOGRAPH: © ROGER HOGAN.

Unlike their northern and eastern cousins, resident Bald Eagles in the southwestern United States apparently prey more on small mammals and birds than fish (Hunt 1998). In Colorado and nearby Wyoming, 53% of 237 prey items were mammals, 25% were fish (admittedly underrepresented), and 22% were birds (Kralovec et al. 1992). Similarly, in New Mexico, although all three Colfax County nest sites were within a few km (mi) of lakes or rivers and prey remains at the nests included fish, all three nest sites were also near black-tailed prairie dog colonies and prey remains at nests had a higher percentage of mammals, particularly prairie dogs, than reported elsewhere (Williams 2000).

In winter, Bald Eagles are opportunists that will take advantage of prey or carrion abundances. Thus, in some parts of their range, large numbers of eagles can concentrate, particularly during salmon spawning runs or below dams where stunned or dead fish float on the surface (Buehler 2000). In New Mexico, Johnson (1988) found that regurgitated pellets of wintering Bald Eagles that foraged principally at or near Cochiti Reservoir, Sandoval County, were comprised mostly of birds, less of mammals, and least of the more easily digested fish.

Intraspecific Interactions

Bald Eagles, along with other members of the genus *Haliaeetus*, are among the very few of the world's birds of prey that sometimes congregate in large numbers. In wintering areas with abundant prey or carrion, tens, hundreds, or even thousands of eagles can occur within a relatively small area (Buehler 2000). These foraging groups gather at dusk in large, usually super-canopy trees to roost. Roost trees are generally protected from prevailing winds (e.g., Stalmaster and Gessaman 1984). Bald Eagle aggregations in New Mexico generally consist of fewer than 10 individuals, occasionally approaching 50 at southern reservoirs (Hubbard et al. 1988). While communal roosts in New Mexico have been documented (Johnson 1988, DWS, unpubl. data) and are likely as common as in other states, publication of roost locations again has been discouraged to limit human visitation, which could cause eagles to abandon a roost.

Pairs have been documented on the wintering grounds building nests, copulating, but then still migrating north in the spring (Harmata 2002; Buehler 2000; DWS, unpubl. data). An unusual dimension of these winter pairings was illustrated along the Rio Chama below Abiquiu on 12–13 February 1987. Late in the afternoon, the pair was perched together, but the male flew downriver; he was soon joined by a second adult female. The first female became agitated, calling and then flying closer and closer to the other female. During the most intense 17 minutes of the interaction, the paired female dove on the other female five times, twice nearly hitting her, and landed within 1.5 m (5 ft) of her twice, once involving a 15-sec stare-down. Eventually the male rejoined his mate, and they flew back to their usual roost tree. The following morning, the paired female returned to chase the other female downriver. While male raptors will protect their mates from other males near nests, this was apparently the first known case of a female raptor "claiming" a mate on the wintering grounds (Stahlecker 2007).

Predation and Interspecific Interactions

Most documented interactions with other birds and mammals have been in relation to food. In New Mexico, perched Bald Eagles that are feeding on fish are often pestered by Black-billed Magpies (*Pica hudsonia*) hoping to steal a morsel. Common Ravens (*Corvus corax*) also harass Bald Eagles with prey or at carcasses.

PHOTO 7.22

Young Bald Eagle being mobbed by American Crows (*Corvus brachyrhynchos*), Rio Arriba Co., 19 November 2008. PHOTOGRAPH: © ROGER HOGAN.

PHOTOS 7.23a and b

Adult Bald Eagles along the west bank of the upper Rio Grande in Taos Co., December 2006. In New Mexico, the species does not appear to be limited by foraging habitat, and yet the number of nests remains very small compared to those in neighboring states. PHOTOGRAPHS: © GERAINT SMITH.

PHOTO 7.24

Adult Bald Eagle, Rio Arriba Co., 10 January 2007. The Bald Eagle was removed from the Endangered Species List on 28 June 2007. PHOTOGRAPH: © ROGER HOGAN.

There was one case of a Golden Eagle being dominant over several Bald Eagles at a carcass in New Mexico (chapter 21), but no clear dominance was reported in Utah (Sabine and Gardener 1987). On 27 March 1987, in Rio Arriba County, a Bald Eagle and a Northern Harrier (*Circus cyaneus*) both pursued an adult female Peregrine Falcon (*Falco peregrinus*) that was carrying a teal; the eagle was the first to reach the teal when the falcon dropped it (DWS).

In Arizona, Great Horned Owls (*Bubo virginianus*) killed a Bald Eagle juvenile that was roosting on the nest cliff four weeks after fledging (Hunt et al. 2009). In general, however, Bald Eagles have few natural predators. Directly (shooting, poisoning) or indirectly (collisions, electrocutions), humans or their activities have caused the majority of *documented* eagle deaths in North America (Franson et al. 1985).

Status and Management

A little over forty years ago, the Bald Eagle was at risk of extirpation from the 48 contiguous states largely due to habitat destruction and degradation, illegal shooting, and dichloro-diphenyl-trichloroethane (DDT) contamination of its prey base. Since then, the species has made a dramatic recovery. In response to conservation actions such as federal banning of DDT and federal listing of the species as Endangered or Threatened in all states within its range but Alaska, and concurrent reduction in threats, Bald Eagle population sizes and reproduction significantly increased to levels that prompted its removal from the Endangered Species List on 28 June 2007 (U.S. Fish and Wildlife Service 2007).

The species is still protected from "take," which includes disturbance as well as killing, by the Federal Bald and Golden Eagle Protection Act. Delisting set into motion a need to allow Take Permits on a case-by-case review (chapter 21). In addition, the species remains listed on several state endangered species lists. In New Mexico, the Bald Eagle remains listed as Threatened, after first being placed on the list protected by the state's Wildlife Conservation Act in 1975 (NMDGF 2008). The state's known breeding population of Bald Eagles is still quite small, with never more than four active nests in a year, despite two decades of documented nesting in the state. In contrast to the Bald Eagle, the Osprey has seen its breeding population grow to 20 pairs since its almost simultaneous discovery as a New Mexico breeding species (chapter 4). Unlike the breeding population, New Mexico's wintering population of Bald Eagles grew steadily through the 1980s and 1990s. It is unknown whether this trend has continued since then, as systematic aerial counts have not been conducted for more than a decade, but Christmas Bird Count data and opportunistic observations indicate that New Mexico remains an important wintering area for migrant Bald Eagles (National Audubon Society 2008).

Factors limiting Bald Eagle productivity and population growth in New Mexico are unknown. The species does not appear to be habitat-limited (unoccupied habitat appears suitable, particularly in northern New Mexico), illegal shooting is not prevalent, and DDT has been federally banned since 1972. Although starvation is likely the greatest, but least documented, cause of eagle deaths across the species' range (chapter 21), lead poisoning due to scavenging, especially in winter, on game killed with lead bullets could also decrease survival. Harmata and Restani (1995) found lead in the blood of 97% of migrant Bald Eagles captured in the spring in Montana. Each autumn in Arizona, lead levels in the blood of reintroduced California Condors (*Gymnogyps californianus*) increase, sometimes to lethal levels, when carcasses and entrails of deer and elk from big-game hunting are readily available (Hunt et al. 2006). The Arizona Game and Fish Department is providing coupons for the purchase of nonlead ammunition in the northern Arizona management units where the condors are most active (Sullivan et al. 2007). It is recommended that New Mexico, along with other western states, aggressively work to decrease and eventually eliminate lead ammunition not only to protect scavenging birds, but also the humans that consume the big game they kill. In addition, research and monitoring efforts in New Mexico should focus on gathering information needed to develop best management practices for increasing the state's breeding population of Bald Eagles.

It is clear that New Mexico continues to be an important wintering area for Canadian-nesting Bald Eagles. Ironically, to expand on our earlier quotation of Bailey (1928:180), to the "east, west and north, in the three states bordering New Mexico, the Bald Eagle is known to nest" in much higher numbers, with 30–50 occupied territories/state (Winternitz 1998; Hunt 1998; Mabie et al. 1994), or tenfold the number currently known in New Mexico. It would seem a reasonable goal for state and federal land management agencies, that within a decade, Bald Eagle breeding pairs in New Mexico should number in the teens instead of the single digits. Public interest and involvement can help to make this goal attainable.

LITERATURE CITED

Bailey, F. M. 1928. *Birds of New Mexico*. Santa Fe: New Mexico Department of Game and Fish.

Broley, C. L. 1947. Migration and nesting of Florida Bald Eagles. *Wilson Bulletin* 59:3–20.

Buehler, D. A. 2000. Bald Eagle (*Haliaeetus leucocephalus*). No. 506. In *The birds of North America*, ed. A. Poole and F. Gill. Philadelphia, PA: Birds of North America, Inc.

Franson, J. C., L. Sileo, and N. J. Thomas. 1995. Cause of eagle deaths. In *Our living resources*, ed. B. T. LaRoe, G. S. Farris, C. E. Puckett, P. D. Dorna, and M. J. Mac, 68. Washington, DC: U.S. Department of the Interior.

Harmata, A. R. 2002. Vernal migration of Bald Eagles from a southern Colorado wintering area. *Journal of Raptor Research* 316:256–64.

Harmata, A. R., and M. Restani. 1995. Environmental contaminants and cholinesterase in blood of vernal migrant Bald and Golden Eagles in Montana. *Intermountain Journal of Sciences* 1:1–15.

Harmata, A. R., and D. W. Stahlecker. 1993. Fidelity of migrant Bald Eagles to wintering grounds in southern Colorado and northern New Mexico. *Journal of Field Ornithology* 64:129–34.

Henny, C. J., and D. W. Anderson. 2004. Status of nesting Osprey in coastal Baja California, Sonora and Sinaloa, Mexico, 1977 and 1992–93. *Bulletin of the Southern California Academy of Science* 103:95–114.

Henny, C. J., B. Conant, and D. W. Anderson. 1993. Recent distribution and status of nesting Bald Eagles in Baja California, Mexico. *Journal of Raptor Research* 27:203–9.

Hoffman, S. W., and J. P. Smith. 2003. Population trends of migratory raptors in western North America, 1977–2001. *Condor* 105:397–419.

Hubbard, J. P., W. H. Baltosser, and C. G. Schmitt. 1988. Midwinter aerial surveys of Bald Eagles in New Mexico. In *Proceedings of the southwest raptor management symposium and workshop*, ed. R. L. Glinski et al., 289–94. Tech. series no. 11. Washington, DC: National Wildlife Federation.

Hunt, W. G. 1998. Bald Eagle. In *The raptors of Arizona*, ed. R. L. Glinski, 50–54. Phoenix: Arizona Game and Fish Department.

Hunt, W. G., W. Burnham, C. N. Parish, K. Burnham, B. Mutch, and G. L. Oaks. 2006. Bullet fragments in deer remains: implications for lead exposure in avian species. *Wildlife Society Bulletin* 34:169–71.

Hunt, W. G., D. E. Driscoll, R. I. Mesta, J. H. Barclay, and R. E. Jackman. 2009. Migration and survival of juvenile Bald Eagles from Arizona. *Journal of Raptor Research* 43:121–26.

Hunt, W. G., R. E. Jackman, J. M. Jenkins, C. G. Thelander, and R. K. Lehman. 1992. Northward post-fledging migration of California Bald Eagles. *Journal of Raptor Research* 26:19–23.

Johnson, T. H. 1988. Effects of reservoir habitat changes on wintering Bald Eagles at Cochiti Reservoir, northern New Mexico. In *Proceedings of the southwest raptor management symposium and workshop*, ed. R. L. Glinski, B. A. Giron Pendleton, M. B. Moss, M. N. LeFranc, B. A. Millsap, and S. W. Hoffman, 182–87. Washington, DC: National Wildlife Federation.

Kralovec, M. L., R. L. Knight, G. R. Craig, and R. G. McLean. 1992. Nesting productivity, food habits, and nest sites of Bald Eagles in Colorado and southeastern Wyoming. *Southwestern Naturalist* 37:356–61.

Ligon, J. S. 1961. *New Mexico birds and where to find them.* Albuquerque: University of New Mexico Press.

Mabie, D. W., M. T. Merendino, and D. H. Reid. 1994. Dispersal of Bald Eagles fledged in Texas. *Journal of Raptor Research* 28:213–19.

Mersman, T. J., D. A. Buehler, J. D. Fraser, and J. D. K. Seegar. 1992. Techniques used in Bald Eagle food habit studies. *Journal of Wildlife Management* 56:73–78.

National Audubon Society. 2008. The Christmas Bird Count historical results. http://www.audubon.org/bird/cbc (accessed 29 April 2008).

[NMDGF] New Mexico Department of Game and Fish. 2008. *Threatened and endangered species of New Mexico. 2008 Biennial Review and Recommendations.* Santa Fe: New Mexico Department of Game and Fish.

[NMOS] New Mexico Ornithological Society. 2007. *NMOS Field Notes* database. http://nhnm.unm.edu/partners/NMOS (accessed 22 January 2007).

Palmer, R. S., J. S. Gerrard, and M. V. Stalmaster. 1988. Bald Eagle. In *Handbook of North American birds.* Vol. 4, ed. R. S. Palmer, 187–237. New Haven: Yale University Press.

Sabine, N., and K. Gardener. 1987. Agonistic encounters between Bald Eagles and other raptors wintering in west central Utah. *Journal of Raptor Research* 21:118–20.

Sherrod, S. K., C. M. White, and F. S. L. Williamson. 1976. Biology of the Bald Eagle on Amchitka Island, Alaska. *Living Bird* 15:145–82.

Sibley, D. A. 2000. *The Sibley guide to the birds.* New York: Chanticleer Press.

Stahlecker, D. W. 2007. Was a migrant Bald Eagle defending her mate on her New Mexico wintering grounds? *Journal of Raptor Research* 41:80–81.

Stahlecker, D. W., and T. G. Smith. 1993. A comparison of home range estimates for a bald eagle wintering in New Mexico. *Journal of Raptor Research* 27:42–45.

Stalmaster, M. V. 1987. *The Bald Eagle.* New York: Universe Books.

Stalmaster, M. V., and J. A. Gessaman. 1984. Ecological energetics and foraging behavior of overwintering Bald Eagles. *Ecological Monographs* 54:407–28.

Sullivan, K., R. Sieg, and C. Parish. 2007. Arizona's efforts to reduce lead exposure in California Condors. *Series in Ornithology* 2:109–21.

Tobish, T. G. Jr., and L. G. Balch. 1987. First North American nesting and occurrence of *Haliaeetus leucocephalus* on Attu Island, Alaska. *Condor* 89(2): 433–34.

[USFWS] U.S. Fish and Wildlife Service. 2007. Endangered and threatened wildlife and plants; removing the Bald Eagle in the lower 48 states from the list of endangered and threatened wildlife. *Federal Register* 72:37346–72.

———. 2008. *Draft environmental assessment: proposal to permit take under the Bald and Golden Eagle Act.* Arlington, VA: Branch of Policy and Permits, USFWS.

Williams, S. O. III. 2000. History and current status of Bald Eagles nesting in New Mexico. *New Mexico Ornithological Society Bulletin* 28:43–44 (abstract only).

Winternitz, B. L. 1998. Bald Eagle. In *Colorado breeding bird atlas*, ed. H. Kingery, 108–9. Denver: Colorado Bird Atlas Partnership.

8

Northern Harrier

(*Circus cyaneus*)

ROBERT H. DOSTER

FORMERLY KNOWN IN North America as the Marsh Hawk, the Northern Harrier (*Circus cyaneus*) is a medium-sized, slim-bodied raptor with long wings and a long tail that typically occurs in grasslands, open fields, and wetlands throughout its extensive range in North America and Eurasia (Ferguson-Lees and Christie 2001). The Northern Harrier is unique among North American raptors in both form and flight. It has a well-defined auricular disk or facial ruff formed of stiff feathers, used for locating prey acoustically (MacWhirter and Bildstein 1996) and giving the bird an owl-like appearance. Additionally unique is the Northern Harrier's flight, slow, low over the ground, with distinctive buoyant, tilting glides and wings held in a shallow V-configuration (MacWhirter and Bildstein 1996).

Adults are sexually dimorphic in both plumage and size. Males are about 12.5% smaller and nearly 50% lighter than females (MacWhirter and Bildstein 1996). Males measure 41–46 cm (16–18 in) with a wingspan of 97–109 cm (38–43 in); females are 41–51 cm (16–20 in) in length and have a wingspan in the range of 109–122 cm (43–48 in) (Wheeler 2003). Bildstein (1988) describes mean Northern Harrier body mass from breeding birds in Wisconsin as follows: adult male 336 g (12 oz) (308–387 g (11–14 oz), $n = 57$); subadult male 346 g (12 oz) (337–363 g (12–13 oz), $n = 6$); adult female 513 g

PHOTO 8.1

Adult male Northern Harrier, northwest of Lordsburg, Hidalgo Co., 7 March 2004. The adult male plumage is gray and white; the irises are lemon yellow. Note the facial disk, used for locating prey by sound.

PHOTOGRAPH: © ROBERT SHANTZ.

PHOTO 8.2

(*top left*) Adult male in flight, Farmington Bay Wildlife Management Area, Utah, 6 February 2008. The underparts are mainly white. The head and neck are gray, giving adult males a hooded appearance. Wing tips are black. PHOTOGRAPH: © DOUG BROWN.

PHOTO 8.3

(*bottom left*) Adult female in flight, Bosque del Apache National Wildlife Refuge, Socorro Co., 16 November 2008. The brown head and neck give a hooded appearance; the buffy underparts are streaked with dark brown; and the underwings show prominent white bands through the secondaries. Another distinctive field mark of all Northern Harriers is the white rump, visible in this photo. PHOTOGRAPH: © DOUG BROWN.

PHOTO 8.4

(*above*) Close-up of an adult female, Sandia Mountains (HawkWatch International migration study site), spring 1995. Northern Harriers of all ages, males and females, have a distinct owl-like facial disk. PHOTOGRAPH: JOHN P. DELONG.

(18 oz) (432–621 g (15–22 oz), $n = 93$); subadult female 500 g (18 oz) (435–654 g (15–23 oz), $n = 24$).

The general coloration of adult males is more reminiscent of a gull than a raptor. Plumage is medium gray on the head, upper breast, front of the neck, back and scapulars, and upperwing surface while the rest of the body (lower breast, belly, flanks, leg feathers, and undertail coverts) is white with a variable amount of rufous flecking. Wing tips are black on both surfaces with the under surface of the wings white to light gray and the trailing edge of the inner secondaries black-tipped. The tail is dark gray above and light gray to whitish below with dark gray barring that is darkest on

PHOTOS 8.5a and b

(*top left and top right*) Immature male, Manzano Mountains (HawkWatch International migration study site), fall 1991. Immature males and females have dark brown heads, necks, and backs similar to adult females. Note the white rump (uppertail coverts) and the facial disk, seen in all ages of both males and females. Immatures have brown irises instead of the yellow irises of adults. PHOTOGRAPHS: © JOHN P. DELONG.

the outer rectrices. The uppertail coverts form a distinctive white rump patch.

In contrast to adult males, adult females are predominantly brown with a brown crown and auriculars streaked with tan. Together with the dark head, the brown, tan streaked neck and hindneck give the appearance of a hood. The facial ruff is outlined with darkly streaked white feathers that highlight the facial disk. The supraorbital ridge and cheek below the eye are also pale, and thus highlighted, in most females. The underparts are streaked with dark blackish-brown and show tawny edging to many feathers. The uppertail coverts and rump feathers are white. The feathers

PHOTO 8.6

(*above*) Immature female, Manzano Mountains (HawkWatch International migration study site), fall 2002. Immature males and females are similar and show bright orange underparts. PHOTOGRAPH: © JOHN P. DELONG.

of the breast, sides, and flanks often have dark shaft streaks. Dark brown upperwing coverts match the back and scapulars. Underwing feather webs are whitish; the primaries have a dark brown tip and four to six dark brown bars; and the secondaries are whitish with three dark brown bars. Underwing linings are similar to those on the breast and sides. The central tail feathers are brown with gray-brown barring and tawny tips. Other rectrices are tinged buff with each pair progressively darker toward the outside and have four to five dark bars present.

The plumage of juvenile Northern Harriers of both sexes resembles the adult female's but with a bright rufous-orange coloration to the belly, flanks, and undertail coverts. The head, neck, and hindneck have dark rufous-tawny streaks that form a hood that is more pronounced than that on the adult female. The underwing lining is similarly rufous-orange and the secondaries are strongly barred, darker in color than the primaries.

In all sexes and ages the Northern Harrier has a greenish-yellow or yellow cere and, as already mentioned, a well-defined auricular disk. Both whitish, a short supercilium and a streak below the eye form pale spectacles around the eyes. The tarsi are yellow in both sexes and all ages.

Distribution

According to MacWhirter and Bildstein (1996), the North American subspecies (*C. c. hudsonius*) of the Northern Harrier breeds widely across the continent from northern Alaska and Canada south to the northern part of the Baja California Peninsula in Mexico, and, eastward, southern Nevada, southern Utah, northern New Mexico, northern Texas, southern Kansas, central Iowa, central Wisconsin, southern Michigan, northern Ohio, southern Pennsylvania, southeastern Virginia, and northeastern North Carolina. South of this main breeding range the species is described as a rare and erratic breeder (but see below). In winter the species primarily occurs from southern Canada south through the United States, Mexico, Central America, and the Caribbean islands (MacWhirter and Bildstein 1996). Outside North America another subspecies of the Northern Harrier, *C. c. cyaneus*, occurs throughout Europe and Asia. It breeds in Eurasia from Portugal to Lapland, east to northern China, Russia, Siberia, and the Kamchatka Peninsula. In winter it ranges throughout northern Africa and tropical Asia (del Hoyo et al. 1995).

PHOTO 8.7

(*top*) Northern Harrier at the Bosque del Apache National Wildlife Refuge, 22 November 2006. The flight of a harrier is distinctive, being slow and close to the ground, with the wings held in a pronounced dihedral or V-configuration. PHOTOGRAPH: © STEVE BARANOFF.

PHOTO 8.8

(*bottom*) Bosque del Apache National Wildlife Refuge, 11 November 2007. The Bosque del Apache is one of the prime locations for observing harriers in winter in New Mexico. The species is described as abundant at the refuge at that time of year. PHOTOGRAPH: © GERAINT SMITH.

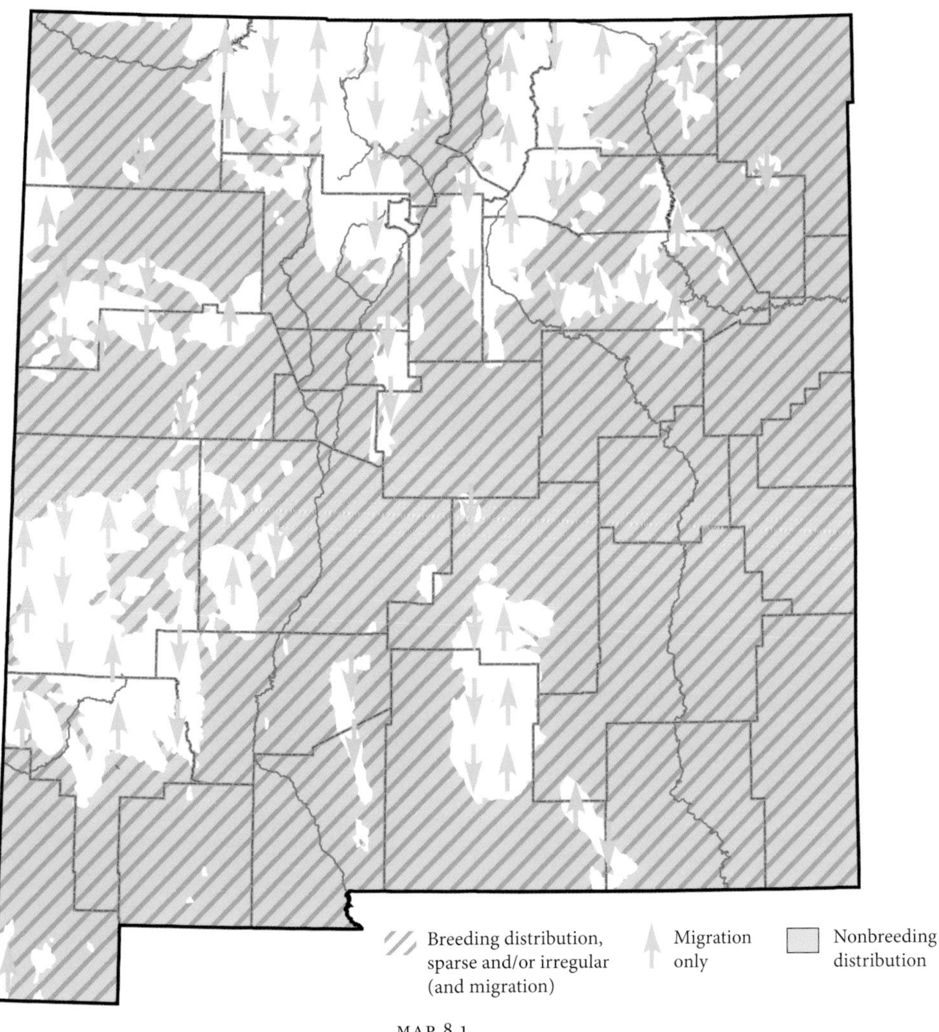

MAP 8.1

Northern Harrier distribution map

The Northern Harrier is found in New Mexico chiefly during migration and in winter, at which times the species may be described as common (Bailey 1928). In winter, the Northern Harrier occurs across the lower elevations of the state, while during migration its distribution even spreads to the highest mountains. The species is also a sparse breeder in New Mexico, but with no clear indication that nesting is less regular in the southern half of the state. Breeding has been reported in particular from the San Juan Valley in the northwest and from Maxwell National Wildlife Refuge and Las Vegas National Wildlife Refuge in the northeast, but also farther south, at Artesia, the Bosque del Apache National Wildlife Refuge, Estancia, Portales, Rattlesnake Springs, the Roswell area, and San Simon Cienega in Hidalgo County (Ligon 1961; Hubbard 1978; Rosenberg et al. 1980). In fact, outside of the San Juan Valley, Maxwell National Wildlife Refuge, and Las Vegas National Wildlife Refuge, reports of Northern Harriers during the breeding season have occurred nearly statewide in New Mexico with the majority of these reports being from the lowland areas

PHOTO 8.9

Female Northern Harrier on a tumbleweed near water at the Bosque del Apache National Wildlife Refuge, 10 February 2007. PHOTOGRAPH: © TOM KENNEDY.

of the Pecos River Valley and in deserts of the southern tier of counties (NMOS 2007). It is true that Northern Harrier sightings during the breeding season have not always been accompanied by confirmation of nesting but nests can be quite difficult to find. The somewhat irregular breeding of Northern Harriers throughout New Mexico may be related, in part, to cyclical prey availability (Hamerstrom 1979). In adjacent Arizona, Northern Harriers are even scarcer as breeders, with only six nesting attempts documented from 1872 to 2000 (Mikesic and Duncan 2000).

Habitat Associations and Life History

During breeding, Northern Harriers typically occupy wet, open areas including marshy meadows, wet pastures, old fields, and freshwater and brackish marshes (MacWhirter and Bildstein 1996). In New Mexico in particular, they nest in marshlands composed of rushes as well as in dense grasslands at middle and lower elevations (Ligon 1961; Hubbard 1978). Wintering Northern Harriers are associated with a broader variety of open, lower-elevation habitats dominated by herbaceous cover, including deserts, dry plains, upland and lowland grasslands, old fields, open-habitat floodplains, and freshwater marshes (Temeles 1986; Bildstein 1987; Collopy and Bildstein 1987). In New Mexico in particular, wintering occurs in most open, lowland habitats from the Chihuahuan Desert basins and grasslands of the south to the Colorado Plateau and southwestern tablelands in the north. Migrating Northern Harriers are known to forage in most open habitat types in New Mexico, such as large openings in lowland riparian forests to montane meadows above 3,650 m (11,975 ft) (Bailey 1928). While making long-distance migratory movements across the state, which can be up to 160 km (100 mi) (Root 1988) in a single day, harriers will soar at high elevations to cross New Mexico's tall mountain ranges.

On the breeding grounds Northern Harrier males advertise their territories and court females by performing a sky-dancing display. In this aerial display males engage in a sequence of (up to 74) deep, U-shaped undulations at a height of 10–300 m (30–980 ft) across an area up to 1 km (0.6 mi) wide (MacWhirter and Bildstein 1996). These displays usually end with the male disappearing into a potential nest location. A female

PHOTO 8.10

Maxwell National Wildlife Refuge, Colfax Co., 7 September 2004. Breeding has been reported at the refuge. PHOTOGRAPH: © PATTY HOBAN.

attracted by this display typically follows the male and displaces him from the site, suggesting that the display plays a role in nest-site selection (Simmons 1991). Males will sky-dance over suitable nesting areas while on migration, so this behavior may not always reflect ownership of a breeding territory (Hamerstrom 1969).

Across North America, Northern Harriers begin nesting in mid May or early June in northern regions, but mid March at southern latitudes (Wheeler 2003), including presumably in New Mexico, where Ligon (1961) found three nests with eggs as early as mid April. Nests are built on the ground, most frequently within patches of dense, often tall, vegetation in undisturbed areas (MacWhirter and Bildstein 1996). In New Mexico, Northern Harriers breed mainly in marshes with rank rushes (Ligon 1961), but not exclusively so. Schmitt (1976:7) reported harriers nesting in a "large clump of Russian thistle" (*Salsola tragus*) at Kirtland in the San Juan Valley of northwestern New Mexico. Mikesic and Duncan (2000) documented a harrier nest—with four eggs—found on the ground on 9 June 2000, also in San Juan County, in a fallow field with vegetation again consisting exclusively of Russian thistle. In Eddy County in 2005, Mark Walthall and I observed a nest with six eggs near Lakewood that was placed on the ground in a large field of tall kochia (*Kochia scoparia*) and constructed of kochia stalks.

Harrier nests are built by both sexes and are composed of reeds, rushes, grasses, forbs, or readily available plant material. Eggs are laid at two- to three-day intervals with clutches ranging from four to six eggs and the female doing all of the incubation (Baicich and Harrison 1997). Incubation is typically 30 to 32 days in length (Breckenridge 1935). Nestlings begin to move about at two weeks of age at which time they disperse into surrounding vegetation, creating narrow paths between the nest and their resting and feeding areas. In four to five weeks post-hatching, juveniles begin to make brief flights near the nest (MacWhirter 1994). Northern Harriers, and other harriers of the genus *Circus*, are unique among raptors in that they regularly, though not always, have a polygynous breeding system (one male mating with several females) (Simmons 2000). Thus far, however, polygynous behavior has not been observed at breeding locations in New Mexico.

PHOTO 8.11

(*top*) Example of nesting habitat in San Juan Co. A pair nested in 2004 in the marsh/wetland area, which is surrounded by badlands. PHOTOGRAPH: © RON KELLERMUELLER/HAWKS ALOFT, INC.

PHOTO 8.12

(*bottom*) Old McMillan lake bed, just north of Brantley Lake, Eddy Co. A pair nested at that location in 2005. Note the tall, weedy vegetation consisting of kochia (*Kochia scoparia*). PHOTOGRAPH: © MARK WALTHALL.

Across North America, the primary diet of Northern Harriers is composed of small and medium-sized mammals, mostly rodents. Harriers will also take birds (mainly passerines and small water birds), reptiles, and frogs (Collopy and Bildstein 1987). At northern latitudes of the harrier's range, *Microtus* voles are almost the exclusive component of diet. A harrier nearly always hunts on the wing, coursing low (<5 m) (<16 ft)

PHOTO 8.13

Northern Harrier nest found by the author and Mark Walthall in April 2005 in the old McMillan lake bed. A tall stand of kochia (*Kochia scoparia*) surrounds the nest, which contains six eggs. PHOTOGRAPH: © MARK WALTHALL.

PHOTO 8.14

Two old nestlings in a nest in San Juan Co., 25 June 2004. The nest fledged a total of four young. Two other young not in the photo had already fledged and were perched nearby. PHOTOGRAPH: © RON KELLERMUELLER/HAWKS ALOFT, INC.

over the ground with a buoyant, gliding flight, flapping intermittently (MacWhirter and Bildstein 1996) and with its face looking down at the ground below. The owl-like facial ruff and structures facilitate prey detection by sound, even in the absence of visual cues (Rice 1982). Occasionally Northern Harriers will practice prey-robbery of other, smaller raptors but seldom of conspecifics. This behavior is more prevalent during periods of severe weather and where raptor densities are high (MacWhirter and Bildstein 1996). Harriers will also take wounded or ill prey (Bildstein 1987) and, in New Mexico, the species has even been observed feeding on carrion (photos 8.16 and 8.17).

During the nonbreeding season, Northern Harriers form communal roosts on the ground. Roosts are typically located in dense vegetation (grass and other herbaceous plants), and within these roosts individual harriers will occupy patches of open ground. These roosts are often traditional and have been known to be occupied for over a decade (MacWhirter and Bildstein 1996). The average number of harriers in a roost is 20, though this number can vary widely from 2 to 85 individuals (MacWhirter and Bildstein 1996). At several locations in the lower Mississippi River alluvial valley of the southeastern United States, I have observed Northern Harriers sharing ground roosts with Short-eared Owls (*Asio flammeus*), often exchanging places during the crepuscular period. That interaction has yet to be observed in New Mexico, but the two species share similar wintering habitats in the state (see chapter 37).

Northern Harriers are a partial, but often long-distance (>1,500 km (930 mi), especially in northern portions of range) migrant. Most individuals migrate alone (Beske 1982). This species occasionally soars in migration but typically conducts active flapping flight. Migratory movements are associated with low atmospheric pressure and the approach of a cold front that induces rising air and southerly winds. Harriers typically fly in front of an atmospheric low in spring and behind this weather pattern in the fall (Haugh and Cade 1966; MacWhirter and Bildstein 1996). In New Mexico, where few birds of this species are presumably resident year-round, an influx of fall and spring migrants is detected every year (see chapter 2). Most fall migrants make their appearance in the northern part of the state in early August; spring migration in

PHOTO 8.15

Immature hunting in the manner typical of a Northern Harrier (on the wing, low over the ground), Bosque del Apache National Wildlife Refuge, 11 November 2005. PHOTOGRAPH: © STEVE BARANOFF.

PHOTO 8.16

Adult male eating a black-tailed jackrabbit (*Lepus californicus*) roadkill northwest of Lordsburg, Hidalgo Co., 7 March 2004. PHOTOGRAPH: © ROBERT SHANTZ.

PHOTO 8.17

Northern Harrier feeding on the fresh carcass of a Snow Goose (*Chen caerulescens*) or Ross's Goose (*Chen rossii*), Bosque del Apache National Wildlife Refuge, December 2001. PHOTOGRAPH: © DAVID NICHOLS.

PHOTO 8.18

An unusual observation of a Cooper's Hawk (*Accipiter cooperii*) and a Northern Harrier hunting together in Socorro Co. On 4 February 2009, Doug Burkett watched a female Northern Harrier as she flushed a group of Scaled Quails (*Callipepla squamata*) and began to maneuver trying to catch one of the birds. Suddenly "a gray bolt" zipped through Doug Burkett's field of view in pursuit of the same quail. They all went down in the tall grass and very soon the gray bolt (an adult Cooper's Hawk) came back up empty-handed. It landed on a nearby yucca (*Yucca* sp.). The Northern Harrier then circled around and landed within about 3 m (10 ft) of the Cooper's Hawk (photo). When Doug Burkett approached the two birds with his camera, the Cooper's Hawk flushed first. The Northern Harrier took off immediately after, following closely behind the Cooper's Hawk. PHOTOGRAPH: © DOUG BURKETT.

New Mexico occurs largely before the end of March (Ligon 1961). Wintering birds can linger and be seen as late as mid May, as has been observed nearby in Arizona (H. Snyder, pers. comm.).

Status and Management

The global population of Northern Harriers has recently been estimated at 1.3 million with 35% of that population occurring in the United States and Canada (Rich et al. 2004). Breeding Bird Survey (BBS) data show a downward population trend of -1.7% per year during the interval 1966–2005, which is statistically significant ($P < 0.05$) (Sauer et al. 2005). However, this survey-wide BBS trend estimate may not take into account natural population cycles in parts of the species' range (Sauer et al. 2005). Nonetheless, the U.S. Fish and Wildlife Service (2002) has shown concern over the declines detected by the BBS, coupled with the vulnerability of the harrier's habitat, and has designated the species as a Bird of Conservation Concern across several Bird Conservation Regions in North America.

According to Ligon (1961), Northern Harriers are

PHOTO 8.19

Northern Harrier release by Gila Wildlife Rescue, 30 July 2007. The bird in the photo had sustained a lacerating, infected wound under its wing caused by a barbed wire fence.

PHOTOGRAPH: © DENNIS MILLER.

PHOTO 8.20

Adult female Northern Harrier, north of Lordsburg, Hidalgo Co., 7 January 2005. Range-wide, the species is threatened by the loss of wetlands, native prairies, and pastures, and by the widespread use of pesticides.

PHOTOGRAPH: © ROBERT SHANTZ.

the most abundant hawk wintering in New Mexico, but they are limited as a breeding species in the state because of the scarcity of marshes supporting rank rushes. Destruction of wetlands poses a threat to breeding and wintering populations of Northern Harriers throughout the species' range. However, given the scarcity of wetlands in arid New Mexico, this habitat threat is of particularly grave concern in the state. At the scale of the species' range, conversion of native prairies for farming has also contributed to local population declines and continues to be a major threat (Toland 1985). Overgrazing of pastures and the advent of larger crop fields with fewer fencerows, combined with the widespread use of insecticides and rodenticides, have reduced prey availability and thus the amount of appropriate habitat for the species (Hamerstrom 1986). Thus the long-term outlook for the Northern Harrier seems uncertain, but in New Mexico as elsewhere the species can be viewed as an indicator of ecological impacts associated with any further, widespread alteration of grasslands and wetlands.

LITERATURE CITED

Baicich, P. J., and C. J. O. Harrison. 1997. *A guide to the nests, eggs, and nestlings of North American birds*. San Diego, CA: Academic Press.

Bailey, F. M. 1928. *Birds of New Mexico*. Santa Fe: New Mexico Department of Game and Fish.

Beske, A. E. 1982. Local and migratory movements of radio-tagged juvenile harriers. *Raptor Research* 16:39–53.

Bildstein, K. L. 1987. Behavioral ecology of Red-tailed Hawks (*Buteo jamaicensis*), Rough-legged hawks (*Buteo lagopus*), Northern Harriers (*Circus cyaneus*), and American Kestrels (*Falco sparverius*) in south central Ohio. *Ohio Biological Survey Biological Notes*, no. 18.

———. 1988. Northern Harrier *Circus cyaneus*. In *Handbook of North American birds*. Vol. 4, *Diurnal raptors (part 1)*, ed. R. S. Palmer, 251–303. New Haven: Yale University Press.

Breckenridge, W. J. 1935. An ecological study of some Minnesota Marsh Hawks. *Condor* 37:268–76.

Collopy, M. W., and K. L. Bildstein. 1987. Foraging behavior of Northern Harriers wintering in southeastern salt and freshwater marshes. *Auk* 104:11–16.

del Hoyo, J., A. Elliot, and J. Sargatal, eds. 1995. *Handbook of the birds of the world*. Vol. 2, *New world vultures to Guineafowl*. Barcelona, Spain: Lynx Edicions.

Ferguson-Lees, J., and D. A. Christie. 2001. *Raptors of the world*. Boston: Houghton Mifflin Co.

Hamerstrom, F. 1969. A harrier population study. In *Peregrine Falcon populations, their biology and decline*, ed. J. J. Hickey, 367–83. Madison: University of Wisconsin Press.

———. 1979. Effect of prey on predator: voles and harriers. *Auk* 96:370–74.

———. 1986. *Harrier, hawk of the marsh: the hawk that is ruled by a mouse*. Washington, DC: Smithsonian Institution Press.

Haugh, J. R., and T. J. Cade. 1966. The spring hawk migration around the southeastern shore of Lake Ontario. *Wilson Bulletin* 78:88–110.

Hubbard, J. P. 1978. *Revised check-list of the birds of New Mexico*. Publ. no. 6. Albuquerque: New Mexico Ornithological Society.

Ligon, J. S. 1961. *New Mexico birds and where to find them*. Albuquerque: University of New Mexico Press.

MacWhirter, R. B. 1994. Offspring sex ratios, mortality, and relative provisioning of daughters and sons in the Northern Harrier (*Circus cyaneus*). Ph.D. diss. Ohio State University, Columbus.

MacWhirter, R. B., and K. L. Bildstein. 1996. Northern Harrier (*Circus cyaneus*). No. 210. In *The birds of North America*, ed. A. Poole and F. Gill. Philadelphia, PA: Academy of Natural Sciences, and Washington, DC: American Ornithologists' Union.

Mikesic, D. G., and R. B. Duncan. 2000. Historical review of Arizona's nesting Northern Harriers, including the most recent confirmed nesting in 1998. *Western Birds* 31:243–48.

[NMOS] New Mexico Ornithological Society. 2007. *NMOS Field Notes* database. http://nhnm.unm.edu/ partners/ NMOS (accessed 19 August 2007).

Rice, W. R. 1982. Acoustical location of prey by the Marsh Hawk: adaptation to concealed prey. *Auk* 99:403–13.

Rich, T. D., C. J. Beardmore, H. Berlanga, P. J. Blancher, M. S. W. Bradstreet, G. S. Butcher, D. W. Demarest, E. H. Dunn, W. C. Hunter, E. E. Inigo-Elias, et al. 2004. *Partners in Flight North American landbird conservation plan*. Ithaca, NY: Cornell Laboratory of Ornithology.

Root, T. 1988. *Atlas of wintering North American birds: and analysis of Christmas Bird Count data*. Chicago: University of Chicago Press.

Rosenberg, K. V., J. P. Hubbard, and G. H. Rosenberg. 1980. The spring migration: southwest region. *American Birds* 34:803–6.

Sauer, J. R., J. E. Hines, and J. Fallon. 2005. *The North American breeding bird survey, results and analysis 1966–2005*. Version 6.2.2006. Laurel, MD: U.S. Geological Survey, Patuxent Wildlife Research Center. http://www.mbr-pwrc.usgs.gov/ bbs/bbs.html (accessed 20 August 2007).

Schmitt, C. G. 1976. *Summer birds of the San Juan Valley, New Mexico*. Publ. no. 4. Albuquerque: New Mexico Ornithological Society.

Simmons, R. 1991. Comparisons and functions of sky-dancing displays of *Circus* harriers: untangling the Marsh Harrier complex. *Ostrich* 62:45–51.

———. 2000. *Harriers of the world: their behaviour and ecology*. Oxford, UK: Oxford University Press.

Temeles, E. J. 1986. Reversed sexual size dimorphism: effects on resource defense and foraging behaviors of nonbreeding Northern Harriers. *Auk* 103:70–78.

Toland, B. 1985. Nest site selection, productivity, and food habits of Northern Harriers in southwest Missouri. *Natural Areas Journal* 5:22–27.

U.S. Fish and Wildlife Service. 2002. Birds of conservation concern 2002. Arlington, VA: Division of Migratory Bird Management. http://migratorybirds.fws.gov/ reports/BCC 2002.pdf (accessed 29 March 2008).

Wheeler, B. K. 2003. *Raptors of western North America*. Princeton, NJ: Princeton University Press.

9

Sharp-shinned Hawk

(*Accipiter striatus*)

JOHN P. DELONG

I ONCE HEARD the Sharp-shinned Hawk (*Accipiter striatus*) referred to as "100 grams of trouble." From a small bird's perspective, this phrase says it all. Sharp-shinned Hawks are small but capable predators of small birds such as sparrows, warblers, and thrushes. Like all accipiters, Sharp-shinned Hawks have short, rounded wings and long tails, and they are nearly identical in plumage to the Cooper's Hawk (*Accipiter cooperii*). Sharp-shinned Hawk males and females are more separated in size (a phenomenon known as reversed sexual size dimorphism—reversed because males are larger than females in many other species of bird) than any other North American raptor (Storer 1966). In the interior West of North America, females weigh on average 170 g (6 oz), whereas males weigh in at a little more than half of a female, at 98 g (3 oz) (Pearlstine and Thompson 2004). Similarly, length and wingspan of Sharp-shinned Hawks are 24–27 cm (9–11 in) and 53–56 cm (21–22 in) for males and 29–34 cm (11–13 in) and 58–65 cm (23–26 in) for females (Bildstein and Meyer 2000). Sharp-shinned Hawks are slightly smaller in the western than in the eastern parts of their range (Smith et al. 1990; Pearlstine and Thompson 2004).

Sharp-shinned Hawks are handsome birds, with striking deep-red eyes in the adult and long slender yellow legs for which they are named. Above, adult

PHOTO 9.1

Adult female (left) and adult male (right) Sharp-shinned Hawks in hand. Sharp-shinned Hawk males weigh only a little more than half the weight of females. The two birds in this photo were caught during the 1996 fall migration at HawkWatch International's Manzano Mountains study site. PHOTOGRAPH: © JOHN P. DELONG.

PHOTO 9.2

(*top left*) Adult male Sharp-shinned Hawk, HawkWatch International's Manzano Mountains study site, fall 1991. The upperparts of adult males are blue-gray whereas in adult females they are gray-brown. PHOTOGRAPH: © JOHN P. DELONG.

PHOTO 9.3

(*top right*) Immature male Sharp-shinned Hawk trapped during the 1995 fall migration at HawkWatch International's Manzano Mountains study site. The upperparts of immature birds are brownish above, but with rufous tips to the covert feathers that can give their backs a slightly reddish cast. On the underparts are vertical brown to rufous-brown streaks extending from the throat through most of the belly. Note the full crop. PHOTOGRAPH: © JOHN P. DELONG.

PHOTO 9.4

(*left*) Immature Sharp-shinned Hawk in flight, Carlisle Canyon, Hidalgo Co., 28 January 2007. In flight, Sharp-shinned Hawks have S-shaped leading and trailing wing edges, whereas Cooper's Hawks (*Accipiter cooperii*) have straighter wing edges. PHOTOGRAPH: © ROBERT SHANTZ.

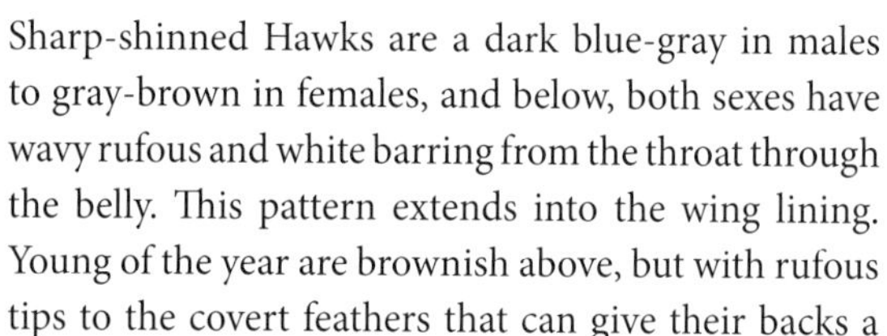

Sharp-shinned Hawks are a dark blue-gray in males to gray-brown in females, and below, both sexes have wavy rufous and white barring from the throat through the belly. This pattern extends into the wing lining. Young of the year are brownish above, but with rufous tips to the covert feathers that can give their backs a slightly reddish cast. Underneath, they have vertical brown to rufous-brown streaks extending from the throat through most of the belly. Undertail coverts in adults and young are white. Flight and tail feathers are barred with brown and white-gray.

Given their similarity to Cooper's Hawks, some

differences are worth pointing out. Tails tend to be more squared-off in Sharp-shinned Hawks and more rounded in Cooper's Hawks. Belly markings extend farther toward the undertail coverts in Sharp-shinned Hawks than in Cooper's Hawks. Sharp-shinned Hawk heads are small and rounded, and have eyes that appear centered on the face, whereas Cooper's Hawk heads have a large, blocky appearance with eyes that appear to be set forward. In flight, Sharp-shinned Hawks have S-shaped wings (leading and trailing edges of wing curve like an "S") and flap with quick, almost uncountable beats, interspersed with bouts of gliding. Cooper's Hawks have straighter wing edges and flap more slowly, as though their flight is labored. It can be difficult to distinguish between Sharp-shinned Hawks and Cooper's Hawks, even for experienced observers, so multiple traits should be used to make a positive identification of one or the other.

Distribution

Sharp-shinned Hawks have a large geographic range, occurring during the breeding season from Alaska and Canada south to northern Argentina, including most areas of North America where coniferous and

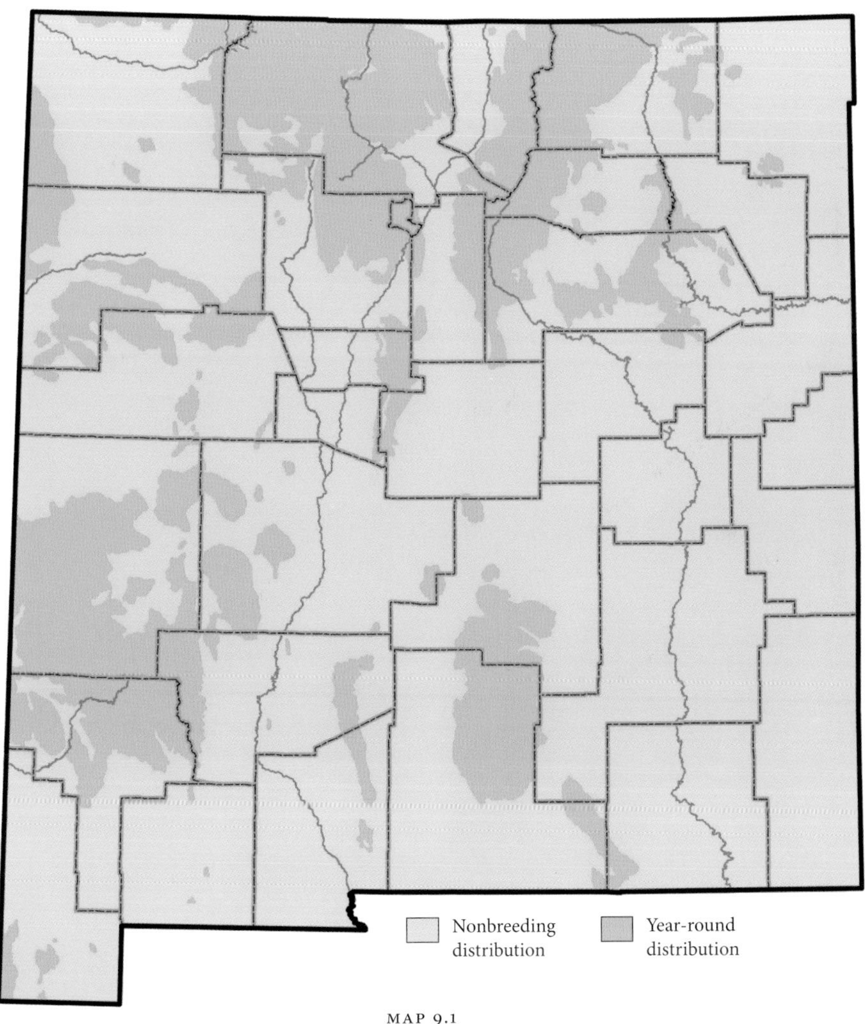

MAP 9.1

Sharp-shinned Hawk distribution map

PHOTO 9.5

(*top left*) Immature female Sharp-shinned Hawk in a yard in Las Cruces, Doña Ana Co., 24 November 2008. Sharp-shinned Hawks are more widely distributed and conspicuous during the nonbreeding season, at which time they can be observed anywhere in New Mexico where tree cover and prey are available. PHOTOGRAPH: © DAVID J. GRIFFIN.

PHOTO 9.6

(*top right*) Sharp-shinned Hawk in Santa Fe Co., 28 December 2008. PHOTOGRAPH: © WARREN BERG.

PHOTO 9.7

(*left*) Sharp-shinned Hawk in winter weather, Datil, Catron Co., 9 March 2009. PHOTOGRAPH: © LISA SPRAY.

mixed deciduous-coniferous forests are present (Bildstein and Meyer 2000; Wheeler 2003). They have been recorded in most mountain ranges of New Mexico during the breeding season, including the Animas, Jemez, Manzano, Mogollon, Pinos Altos, Sacramento, Sandia, Sangre de Cristo, San Mateo (Mt. Taylor), San Mateo (Socorro County), and Zuni mountains (Bailey 1928; Ligon 1961; Parmeter et al. 2002; NMOS 2007). In summer, they appear to be generally uncommon and hard to detect, though more common in the northern than the southern part of the state, and are much more widespread in the state in fall and winter (Bailey 1928; Ligon 1961; Parmeter et al. 2002). In winter, Sharp-shinned Hawks are somewhat more conspicuous, and they can be found through the state wherever suitable tree cover and prey are available (Bailey 1928; Ligon 1961; Parmeter et al. 2002; NMOS 2007).

The Sharp-shinned Hawk is a partial migrant: most but not all birds migrate southward in fall (see chapter 2 for more details regarding raptor migration in New Mexico). It is likely that breeding areas in New Mexico are occupied in winter both by year-round residents and by overwintering birds that have replaced breeding birds. Sharp-shinned Hawks may begin dispersing out of breeding or natal habitats in August, and full-blown southward migration generally begins in September. Peak passage of birds in New Mexico is in late September and early October (DeLong and Hoffman 1999; Bildstein and Meyer 2000; chapter 2). On average, young birds migrate through the state approximately 15 days before adults, and females precede males by approximately 4 days (DeLong and Hoffman 1999). The differences in timing may result from differences in departure or from differences in speed of travel—this is an open question. Although Sharp-shinned Hawks may be found virtually anywhere in the state during migration, they concentrate in montane areas because they use wind shear along mountain ridges to stay aloft with minimal energy expenditure. Northbound spring migrants pass through the state mostly in March and April, peaking in early April, again mostly in montane areas (Smith and Neal 2007b). Concentrations of migrating Sharp-shinned Hawks are well known from the Manzano Mountains in fall and the Sandia Mountains in spring (chapter 2), and, based on my personal experience, also occur in the Animas Mountains in spring.

PHOTO 9.8

8 October 2007. Sharp-shinned Hawk caught during the 2007 fall migration at HawkWatch International's Manzano Mountains study site. The Sharp-shinned Hawk is the raptor most often recorded during fall migration in the Manzano Mountains. PHOTOGRAPH: © TOM KENNEDY.

Sharp-shinned Hawks migrating through the Manzano and Sandia mountains breed from New Mexico north through Alberta, Canada, and winter from New Mexico south through southern Mexico (Hoffman et al. 2002). A unique melding of band-recovery and stable-hydrogen isotope approaches was used to show that Sharp-shinned Hawks display "chain migration." Hawks that migrate through New Mexico early in the fall migration season breed at lower latitudes and winter farther south than later migrants who breed at higher latitudes and winter farther north (Smith et al. 2003).

Habitat Associations

Sharp-shinned Hawk are forest dwellers, and they are generally absent from grasslands and shrublands during the breeding season. Nesting habitat includes primarily various conifer forests, but deciduous forests

PHOTO 9.9

Sharp-shinned Hawk, Bosque del Apache National Wildlife Refuge, March 1994. In New Mexico, winter habitat includes floodplain riparian woodlands. PHOTOGRAPH: © GARY K. FROEHLICH.

may also be used (Bildstein and Meyer 2000). In New Mexico, Sharp-shinned Hawks typically nest in dense mixed conifer montane forests (e.g., Ligon 1961), but they also may nest in small patches of spruce-fir forest surrounded by aspen stands, as documented in neighboring Colorado (Joy et al. 1994). Unlike Cooper's Hawks (see chapter 10), Sharp-shinned Hawks generally do not nest in lowland riparian forests such as cottonwood gallery forest. In comparison with their larger relatives and neighbors, Northern Goshawks (*Accipiter gentilis*) and Cooper's Hawks, Sharp-shinned Hawks use smaller, more closely spaced stands of trees for nesting (Siders and Kennedy 1996). They may place their nests in trees either low in drainage bottoms or high up the slopes, unlike Northern Goshawks and Cooper's Hawks, which tend to nest in trees placed in the bottom of drainages (pers. obs.).

In winter, Sharp-shinned Hawks may occupy the same montane coniferous forests used in summer, but they extend their habitat use to include pinyon-juniper woodlands, deciduous woodlots, urban backyards and parks, riparian deciduous forests, and occasionally even shrublands and grasslands (Bildstein and Meyer 2000).

PHOTO 9.10

Sharp-shinned Hawk in a pine tree in Santa Fe Co., 28 December 2008. In New Mexico, Sharp-shinned Hawks inhabit montane forests during the breeding season. During the nonbreeding season, they use a larger number of vegetation types ranging from low to higher elevations. PHOTOGRAPH: © WARREN BERG.

Natural History

As with most predators, the ecology and behavior of Sharp-shinned Hawks closely reflect their choice of food. The breeding season is timed so that nestlings are large and fledging when the prey base—particularly small nestlings of other bird species—is most abundant and vulnerable (Bildstein and Meyer 2000). Migrating Sharp-shinned Hawks seem to follow migrating songbirds along the route of their migration and winter in areas that harbor abundant songbirds. In New Mexico, this usually means areas with abundant sparrows, warblers, and thrushes.

Nesting

Range-wide, breeding begins in April, and young typically fledge in July (Bildstein and Meyer 2000). Nests are flat, made of sticks, and usually placed near the tree trunk near thick branch cover (Bildstein and Meyer 2000). Sharp-shinned Hawks generally do not reuse nests or use nests of other species, but instead build new structures each year in annually reoccupied territories (Bildstein and Meyer 2000). In some cases, however, the new nest is in the same tree used in previous years (Ligon 1961). Nest trees in New Mexico are typically Douglas firs (*Pseudotsuga menziesii*) (Ligon 1961).

Sharp-shinned Hawks have one brood per season, but in Puerto Rico, where the breeding season may be more extended than in New Mexico, the species may re-nest if the first nesting attempt fails (Bildstein and Meyer 2000). Three Sangre de Cristo nests reported by Ligon (1961) had three, four, and five eggs, respectively. Other than those three nests, clutch size is undocumented in New Mexico. Elsewhere, females lay four to five eggs, rarely three to eight, and fledge an average of 2.7 young per nest (Bildstein and Meyer 2000). The incubation period lasts approximately 30–32 days (see Bildstein and Meyer 2000). Based on studies in New Brunswick (Meyer 1987), Oregon (Reynolds and Wight 1978), and Utah (Platt 1976), the length of the nestling period is approximately 21–27 days, with males fledging several days ahead of females. The female incubates the eggs and broods the young nestlings while the male hunts, delivering food at the nest (Bildstein and Meyer 2000). As the nestlings become older, the female spends less time at the nest and begins to hunt as well (Snyder and Snyder 1998; Bildstein and Meyer 2000).

Sharp-shinned Hawks are variably aggressive in their defense of nests. However, it has been my experience that some adults call shrilly and may even dive on intruders, coming within striking range. Although not quite as intimidating as a Northern Goshawk, the speed and agility of a Sharp-shinned Hawk defending its nest in a dense forest is astounding.

Diet and Foraging

Sharp-shinned Hawks specialize in small birds (Bildstein and Meyer 2000). A wide variety of bird species is taken, but Sharp-shinned Hawks are opportunistic enough to also take small mammals and even insects on occasion (Bailey 1928; Bildstein and Meyer 2000). A sample of 159 Sharp-shinned Hawks shot in the late 1800s was examined for stomach contents. Of 107 stomachs that had food in them, 103 (96%) contained songbirds (Fisher 1893, reported in Duncan 1980). Later, in the early 1900s, 110 stomachs were analyzed for food contents (Duncan 1980). Of 86 stomachs containing food remnants, 73 (85%) contained parts of birds, and five (6%) contained remnants of mammals. In southwestern Colorado, Sharp-shinned Hawks took bird species in proportion to their abundance, but took more mammals than expected from their abundance in the area (9% of all prey items observed; Joy et al. 1994). In

PHOTO 9.11

Immature Sharp-shinned Hawk at Luis Lopez, Socorro Co., 13 January 2008. Sharp-shinned Hawks are often noted in winter near residences, where they may hunt small passerine birds. PHOTOGRAPH: © JERRY OLDENETTEL.

PHOTO 9.12

Adult male eating an Inca Dove (*Columbina inca*) in a yard in Las Cruces, Doña Ana Co., 19 December 2008.
PHOTOGRAPH: © DAVID J. GRIFFIN.

the breeding season, adults may feed their young with a high proportion of nestling birds taken from open-cup nests of songbirds (Joy et al. 1994; Bildstein and Meyer 2000). The small size of a Sharp-shinned Hawk, especially that of the male, constrains the species to preying upon small birds and mammals. However, females may take prey as large as quail and Northern Flickers (*Colaptes auratus*) (Storer 1966; Duncan 1980). It may say something about the vigor of the Sharp-shinned Hawk predation ethic that females regularly attempt to subdue large pigeons at HawkWatch International's banding sites, birds that are more than twice their size! Bailey (1928) herself referred to the Sharp-shinned Hawk as one of the hawks killing the largest number of game birds, small poultry, and insectivorous birds.

Sharp-shinned Hawks hunt from perches or while in flight (Bildstein and Meyer 2000). They use cover to mask their approach to prey, relying on surprise and quick capture (Bailey 1928; Bildstein and Meyer 2000). Although Sharp-shinned Hawks may chase prey into dense vegetation, they also may give up the chase if the element of surprise is lost. They forage in the upper canopy, especially during the breeding season, or anywhere that prey may be found. For example, I once observed a Sharp-shinned Hawk at the University of New Mexico campus chase a House Sparrow (*Passer domesticus*) out of some trees and then pin the bird on the ground out in the open in front of a bus stop along Central Avenue. The Sharp-shinned Hawk's small size, long toes and legs, swift flight, and tremendous maneuverability allows it to snag prey from the air, from branches, or from the ground, and to slice through dense vegetation in the process. Sharp-shinned Hawks truly are "100 grams of trouble!"

Predation and Interspecific Interactions

Range-wide, Sharp-shinned Hawks share their forest habitats with Cooper's Hawks, Northern Goshawks, Broad-winged Hawks (*Buteo platypterus*), and a variety of forest owls. Although potentially in competition with other forest hawks, their small size gives them access to prey species and types that are more difficult for the other, larger hawks (Reynolds 1972; Reynolds and Meslow 1984). On the other hand, their small size puts them in the prey size range of larger raptors. At a Northern Goshawk nest on Mt. Taylor, I found the remains of a Sharp-shinned Hawk below the nest, along with similar-sized prey such as jays and woodpeckers. As documented outside New Mexico, migrating Sharp-shinned Hawks can also be attacked and killed by Peregrine Falcons (*Falco peregrinus*) (Bildstein and Meyer 2000), while in New Mexico I have observed them being attacked by Cooper's Hawks during migration.

Status and Management

Sharp-shinned Hawk populations are apparently stable. Counts at New Mexico HawkWatch International sites, which monitor populations that breed from

PHOTO 9.13

Adult Sharp-shinned Hawk on a fence in a yard in Las Cruces, Doña Ana Co., 10 January 2009. During the nonbreeding season, Sharp-shinned Hawks can collide with window panes as they chase after small birds in residential areas.

New Mexico north into Canada and Alaska, do not suggest any long-term directional trend for the species (Hoffman and Smith 2003; Smith and Neal 2007a, 2007b). Nonetheless, there are some anthropogenic threats to the species. Most notably, some Sharp-shinned Hawks collide with cars and windows, especially when chasing prey. Birds brought to a Colorado wildlife rehabilitation center from the surrounding area were generally diagnosed with some type of trauma, often the result of a collision (Wendell et al. 2002). Recovery of banded birds also can give some indication of what causes mortality in Sharp-shinned Hawks. Sources of mortality for Sharp-shinned Hawks banded by HawkWatch International at four sites in the western United States, including both the Sandia and Manzano sites in New Mexico, consisted mainly of collision with human structures (33 birds), followed by shooting (25), unspecified injury or sickness (2), poisoning (1), and predation by cat or dog (1) (Hoffman et al. 2002). An additional 40 birds were found dead but the cause of death was unknown. Notably, however, no Sharp-shinned Hawks banded by HawkWatch International and later recovered dead were electrocuted. Generally, small hawks are less prone to being electrocuted because their wings are unlikely to simultaneously connect both hot wires and grounds on power poles. Secretive and swift, Sharp-shinned Hawks should be a difficult species to persecute directly. However, shooting still is a conservation concern for many raptor species, including the Sharp-shinned Hawk (Hoffman et al. 2002), and shooting was historically a major problem in the eastern United States (Bailey 1928).

Sharp-shinned Hawks wintering in Mexico and central America may be exposed to some pesticides not currently used in the United States, some of them highly toxic for wildlife. In the 1970s, when chlorinated hydrocarbons were still used in the United States, high levels of DDE and associated significant eggshell thinning were reported in Sharp-shinned Hawks from Arizona–New Mexico and Oregon (Snyder et al. 1973). Lead levels for birds in or migrating through New Mexico are not known, but Cooper's Hawks using the same migratory flyways appear to be exposed to higher levels of lead in their winter ranges than in their breeding ranges (McBride et al. 2004).

LITERATURE CITED

Bailey, F. M. 1928. *Birds of New Mexico*. Santa Fe: New Mexico Department of Game and Fish.

Bildstein, K. L., and K. Meyer. 2000. Sharp-shinned Hawk (*Accipiter striatus*). No. 482. In *The birds of North America*, ed. A. Poole and F. Gill. Philadelphia, PA: Birds of North America, Inc.

DeLong, J. P., and S. W. Hoffman. 1999. Differential autumn migration of Sharp-shinned and Cooper's hawks in western North America. *Condor* 101:674–78.

Duncan, S. 1980. An analysis of the stomach contents of some Sharp-shinned Hawks (*Accipiter striatus*). *Journal of Field Ornithology* 51:178.

Hoffman, S. W., and J. P. Smith. 2003. Population trends of migratory raptors in western North America, 1977–2001. *Condor* 105:397–419.

Hoffman, S. W., J. P. Smith, and T. D. Meehan. 2002. Breeding grounds, winter ranges, and migratory routes of raptors in the mountain west. *Journal of Raptor Research* 36:97–110.

Joy, S. M., R. T. Reynolds, R. L. Knight, and R. W. Hoffman. 1994. Feeding ecology of Sharp-shinned Hawks nesting in deciduous and coniferous forests in Colorado. *Condor* 96:455–67.

Ligon, J. S. 1961. *New Mexico birds and where to find them*. Albuquerque: University of New Mexico Press.

McBride, T. J., J. P. Smith, H. P. Gross, and M. J. Hooper. 2004. Blood lead and ALAD activity levels of Cooper's Hawks (*Accipiter cooperii*) migrating through the southern Rocky Mountains. *Journal of Raptor Research* 38:118–24.

Meyer, K. D. 1987. Sexual size dimorphism and the behavioral ecology of breeding and wintering Sharp-shinned Hawks. Ph.D. diss. University of North Carolina, Chapel Hill.

[NMOS] New Mexico Ornithological Society. 2007. *NMOS Field Notes* database. http://nhnm.unm.edu/partners/NMOS/ (accessed 16 November 2007).

Parmeter, J., B. Neville, and D. Emkalns. 2002. *New Mexico bird finding guide*. 3rd ed. Albuquerque: New Mexico Ornithological Society.

Pearlstine, E. V., and D. B. Thompson. 2004. Geographic variation in morphology of four species of migratory raptors. *Journal of Raptor Research* 38:334–42.

Platt, J. B. 1976. Sharp-shinned Hawk nesting and nest site selection in Utah. *Condor* 78:102–3.

Reynolds, R. T. 1972. Sexual dimorphism in accipiter hawks: a new hypothesis. *Condor* 74:191–97.

Reynolds, R. T., and E. C. Meslow. 1984. Partitioning of food and niche characteristics of coexisting Accipiter during breeding. *Auk* 101:761–79.

Reynolds, R. T., and H. M. Wight. 1978. Distribution, density, and productivity of accipiter hawks breeding in Oregon. *Wilson Bulletin* 90: 182–96.

Siders, M. S., and P. L. Kennedy. 1996. Forest structural characteristics of accipiter nesting habitat: is there an allometric relationship? *Condor* 98:123–32.

Smith, J. P., and M. C. Neal. 2007a. Fall 2006 raptor migration studies in the Manzano Mountains of central New Mexico. Unpublished report. Salt Lake City, UT: HawkWatch International, Inc.

———. 2007b. Spring 2007 raptor migration studies in the Sandia Mountains of central New Mexico. Unpublished report. Salt Lake City, UT: HawkWatch International, Inc.

Smith, J. P., S. W. Hoffman, and J. A. Gessaman. 1990. Regional size differences among fall-migrant accipiters in North America. *Journal of Field Ornithology* 61:192–200.

Smith, R. B., T. D. Meehan, and B. O. Wolf. 2003. Assessing migration patterns of Sharp-shinned Hawks *Accipiter striatus* using stable-isotope and band encounter analysis. *Journal of Avian Biology* 34:387–92.

Snyder, N. F. R., and H. A. Snyder. 1998. Northern Goshawk. In *The raptors of Arizona*, ed. R. L. Glinski, 58–62. Tucson: University of Arizona Press.

Snyder, N. F. R., H. A. Snyder, J. L. Lincer, and R. T. Reynolds. 1973. Organochlorines, heavy metals, and the biology of North American accipiters. *BioScience* 23:300–305.

Storer, R. W. 1966. Sexual dimorphism and food habits in three North American accipiters. *Auk* 83:423–36.

Wendell, M. D., J. M. Sleeman, and G. Kratz. 2002. Retrospective study of morbidity and mortality of raptors admitted to Colorado State University Veterinary Teaching Hospital during 1995 to 1998. *Journal of Wildlife Diseases* 38:101–6.

Wheeler, B. K. 2003. *Raptors of western North America*. Princeton, NJ: Princeton University Press.

10

Cooper's Hawk

(*Accipiter cooperii*)

JEAN-LUC E. CARTRON, PATRICIA L. KENNEDY, ROB YAKSICH, AND SCOTT H. STOLESON

THE COOPER'S HAWK (*Accipiter cooperii*) is intermediate in size between the Northern Goshawk (*Accipiter gentilis*) and the Sharp-shinned Hawk (*A. striatus*), northern North America's other two accipiters. The two sexes are almost alike in plumage, but as in both of the other species, the female is noticeably larger. According to Wheeler and Clark (1995), a female Cooper's Hawk has a mean body length of 45 cm (18 in) (range: 42–47 cm [16–19 in]), a mean wingspan of 84 cm (33 in) (range: 79–87 cm [31–34 in]), and a mean body weight of 528 g (19 oz) (range: 479–678 g [17–24 oz]). This is in contrast to the male's mean length of 39 cm (15 in) (range: 37–41 cm [14–16 in]), wingspan of 73 cm (29 in) (range: 70–77 cm [28–30 in]), and body weight of 341 g (12 oz) (range: 302–402 g [10–14 oz]). The size difference is typically sufficient to tell male and female Cooper's Hawks apart. However, because the female Sharp-shinned Hawk approaches in size the male Cooper's Hawk, distinguishing between the two can be quite difficult in the field. Identification of juvenile accipiters to species is particularly daunting, and juvenile Cooper's Hawks can be confused with the juveniles of both other species.

No subspecies of the Cooper's Hawk are recognized despite some size differences between eastern and western populations (Whaley and White 1994). In adults, the underparts are barred crosswise with rufous. Upperparts are slightly different in the male and female. In the male, the back is bluish gray. The crown is darker than the back, almost black, with a pale line of contrast at the nape. In the female, upperparts are gray with a brownish hue. In both sexes, the cheeks are rufous and the iris is orange, turning to red in older birds. Juveniles are brown above, white below with brown longitudinal streaking on the chest becoming sparser on the belly. The undertail coverts are pure white, unlike the distinctly streaked coverts shown by young goshawks. The eyes are pale gray at fledging, soon changing to yellow. Some juveniles show the white superciliary (eyebrow) that is more characteristic of the Northern Goshawk, adult or juvenile. While perched, both adult and juvenile Cooper's Hawks can give the appearance of having squarish heads, with the eyes placed farther forward than those of Sharp-shinned Hawks. The wings are short and rounded and the tail is relatively long, as is typical of accipiters. The tail shows four or more black transverse bars and a white terminal band that is often noticeably wide. The tip of the tail tends to be rounded rather than square, with shorter outer feathers. The Cooper's Hawk has slower wing beats than the Sharp-shinned Hawk. When gliding, it also projects its head in front of its wrists, unlike the Sharp-shinned Hawk.

PHOTO 10.1

Adult Sharp-shinned Hawk (*Accipiter striatus*) female (left) and adult Cooper's Hawk male (right), trapped and banded at HawkWatch International's Manzano Mountains migration monitoring site, fall 1995. Note the similarity in size between the two birds. PHOTOGRAPH: © JOHN P. DELONG.

PHOTOS 10.2a and b

Hatch-year Cooper's Hawk netted, banded, and released at the HawkWatch Manzano migration study site, 6 October 2007. Immatures have white underparts with brown longitudinal streaking most prominent on the chest. PHOTOGRAPHS: © TOM KENNEDY.

PHOTO 10.3

(*left*) Immature Cooper's Hawk. Photographed at the Rio Grande Nature Center, Albuquerque, Bernalillo Co. PHOTOGRAPH: © DOUG BROWN.

PHOTO 10.4

(*above*) Adult Cooper's Hawk (head shot). Bird netted, banded, and released at the HawkWatch Manzano migration study site, 6 October 2007. PHOTOGRAPH: © TOM KENNEDY.

PHOTO 10.5

Adult male Cooper's Hawk. Note the blue-gray back, darker cap, and pale line of contrast at the nape. Bird netted, banded, and released at the HawkWatch Manzano migration study site, 6 October 2007. PHOTOGRAPH: © TOM KENNEDY.

PHOTO 10.6

(*above*) Adult female Cooper's Hawk. Note the brown-tinged upperparts. Photographed at the Bosque del Apache National Wildlife Refuge in November 2006.

PHOTOGRAPH: © LANA HAYS.

PHOTO 10.7

(*top right*) Adult Cooper's Hawk photographed at White Rock, Los Alamos Co., May 2007. The underparts of adults are barred crosswise with rufous. The long tail has conspicuous brown bars in both adults and immatures.

PHOTOGRAPH: © SALLY KING.

PHOTO 10.8

(*bottom right*) Adult Cooper's Hawk with fanned tail feathers, Santa Fe Co., 1 December 2007.

PHOTOGRAPH: © WARREN BERG.

Cooper's Hawks are silent throughout most of the year, but they become vocal during the breeding season, uttering a wide variety of calls. Most often heard is the alarm call, a noisy, repetitive staccato "*cak-cak-cak*."

Distribution

The Cooper's Hawk has a wide distribution that extends from southern Canada south to Central America, and from the Pacific east to the Atlantic (Rosenfield and Bielefeldt 1993). It is a partially migratory species, with breeding and nonbreeding distributions that overlap extensively. According to Palmer (1988), northernmost populations vacate their breeding range during winter (but see Rosenfield and Bielefeldt 1993). Southward, some Cooper's Hawks migrate, but others are year-round residents (Rosenfield and Bielefeldt 1993; chapter 2). Breeding is not known to occur south of northern Mexico.

The Cooper's Hawk occurs year-round throughout most of New Mexico, ranging from low to middle or even high elevations primarily in well-wooded areas (Hubbard 1972, 1978). Breeding populations are distributed east to the northeastern quadrant of the state but only to the Pecos Valley farther south (Hubbard 1978).

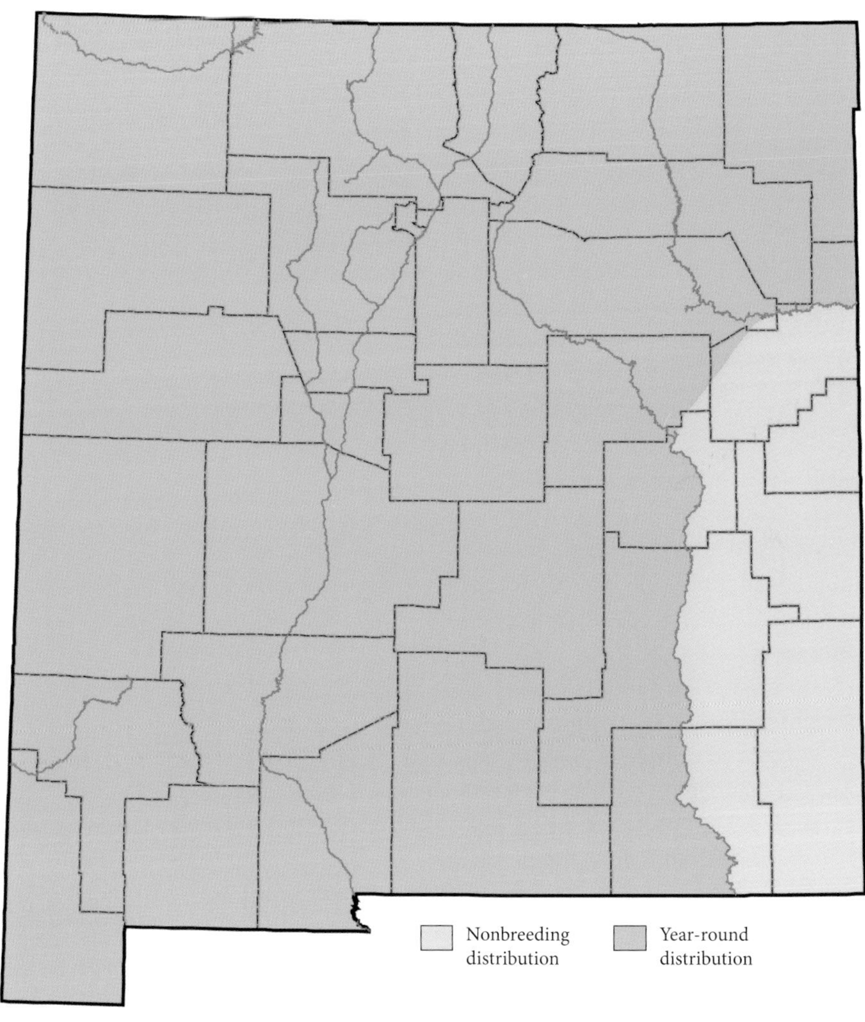

MAP 10.1

Cooper's Hawk distribution map

During migration and in winter, the species occurs statewide (Hubbard 1978), although likely not ranging to the higher elevations they use during breeding.

Large numbers of migrating Cooper's Hawks fly over the Sandia and Manzano Mountains each year (chapter 2). Most of those migrants appear to nest in the southern Rocky Mountains of northern New Mexico and Colorado. Their winter distribution extends to Mexico east of the Sierra Madre Occidental and along that country's southwestern Pacific coast (Hoffman et al. 2002).

Habitat Associations

In New Mexico, the Cooper's Hawk is mainly a bird of floodplain riparian woodlands, wooded mountain foothills and canyons, aspen and coniferous montane forests, and urban areas with trees (Bailey 1928; Ligon 1961; Tatschl 1967; Hubbard 1971a; Schmitt 1976; Stacey 1984; Goodman 1990; Kennedy 1991; Siders and Kennedy 1996). At the lower elevations of the state, we have found the species to be strongly associated with riparian woodlands, especially during the nesting season. Along the Middle Rio Grande (from Cochiti Dam to Elephant Butte Reservoir), the Cooper's Hawk nests in mature cottonwood (*Populus deltoides*) stands with a tall canopy. The understory can be sparse or, conversely, very dense, and is composed of native but also exotic vegetation, mainly saltcedar (*Tamarix chinensis*) and Russian olive (*Elaeagnus angustifolia*) (Cartron et al., 2008). Here it is easily the most common diurnal raptor. In 2007 in the Albuquerque area, one of us (JLEC) counted a total of 11 occupied territories along a 24-km (15-mi) stretch of the river, or an average of one territory every 2.18 km (1.38 mi). An even greater density of nesting pairs was documented in Corrales, Sandoval County. At that location in 2003, 13 nesting pairs were recorded along an 11-km (7-mi) stretch of the west bank of the river (Williams 2003), the organization Hawks Aloft (2008) reporting local densities of 2.85 occupied nests per 40 ha (99 ac) of bosque. To the south, in Bosque Farms in 2002, two occupied nests on the same side of the river were separated by only 300 m (~980 ft) (JLEC et al., unpubl. data). The Cooper's Hawk is also a common species in cottonwood (*Populus fremontii*) riparian woodland along the Gila River in Grant County, where four occupied nests in 2002 were spaced approximately every 3 km (2 mi) on alternating sides of the river (SHS, unpubl. data). At three of the nests, the vegetation consisted of dense stands of fairly young (<50 cm [<20 in] dbh) cottonwoods with little understory (SHS, unpubl. data). At Rattlesnake Springs, in yet another part of the state (Eddy Co.), Cooper's Hawks nest in riparian woodlots dominated by cottonwood or netleaf hackberry (*Celtis reticulata*) and characterized by a tall canopy and dense understory (D. Roemer, pers. comm.).

PHOTO 10.9

(*top*) Cooper's Hawk's nesting habitat along the Middle Rio Grande in central New Mexico. PHOTOGRAPH: © JEAN-LUC CARTRON.

PHOTO 10.10

(*bottom*) Golf course with its Cooper's Hawk nesting pair, Albuquerque, Bernalillo Co., April 2008. PHOTOGRAPH: © JEAN-LUC CARTRON.

PHOTO 10.11

(*top left*) Cooper's Hawk in wooded habitat, Madrid, Santa Fe Co., 2008. PHOTOGRAPH: © LAWRY SAGER.

PHOTO 10.12

(*bottom left*) Cooper's Hawk in low-density residential area in Santa Fe Co., 23 December 2007. PHOTOGRAPH: © WARREN BERG.

PHOTO 10.13

(*above*) Cooper's Hawks on power line in low-density residential area in Santa Fe Co., 18 July 2008. PHOTOGRAPH: © WARREN BERG.

Besides riparian floodplain woodlands, Cooper's Hawks also nest in upland habitats ranging in elevation from pinyon-juniper woodland to mixed coniferous forest (Tatschl 1967; Kennedy 1988; Siders and Kennedy 1996). In the Jemez Mountains and on the adjacent Pajarito Plateau (hereafter Jemez Mountains), Cooper's Hawks nest chiefly at elevations of 2,300–2,380 m (~7,550–7,810 ft) in ponderosa pine (*Pinus Ponderosa*) and mixed conifer stands (Siders and Kennedy 1996). As determined by Siders and Kennedy (1996), overstory trees in nest areas are spaced 4.2–7.0 m (13.8–23.0 ft) apart, for an overall tree density of 750–1,650 per hectare (~1,850–4,080 per acre).

In Arizona, Cooper's Hawks often nest in urban areas with trees (Boal and Mannan 1998). The same is true in New Mexico, where the Cooper's Hawk has increasingly become an urban nesting species. Harden et al. (2006) analyzed Wildlife Rescue Inc. intake data for the Albuquerque region for 15 years and found numbers of Cooper's Hawks increased more than six-fold from 1990 through 2004! Cooper's Hawks build their nests in parks, at golf courses, and on school campuses of Albuquerque, often along arroyos and typically surrounded by open habitat with no understory (JLEC, unpubl. data).

Outside the nesting season, Cooper's Hawks generally occupy the same habitats as during breeding. However, they also more readily occur in less wooded vegetation types such as semidesert and plains grasslands, and Chihuahuan desertscrub (Eakle et al. 1996). In fact, migrating individuals, most of which follow montane forested corridors (see chapter 2), may occasionally be found in areas completely devoid of trees (e.g., Hubbard 1971b). Even during the nesting season, it is possible to observe Cooper's Hawks in nonforested habitats, such as in the Gila River Valley, perched on a pole over mesquite thickets or grasslands or cruising through these habitats (SHS, pers. obs.). Overall, however, we find that Cooper's Hawks in nonforested areas gravitate toward clusters of trees or large shrubs, whether along arroyos or in small planted groves.

Life History

Migration

The proportion of Cooper's Hawks residing year-round in New Mexico instead of migrating south for the winter is unknown. From 1985 through 2006, an annual average of 768 Cooper's Hawks were observed passing over the Sandia Mountains in spring, while the annual average count in fall over the Manzano Mountains was 1,029 individuals (see chapter 2). Some of those migrants were shown to nest in northern New Mexico; others nested farther north (Hoffman et al. 2002).

The median spring passage date for Cooper's Hawks over the Sandias is 9 April for adults and 20 April for immatures (chapter 2). However, spring migration may be as late as June even in the extreme southern part of the state (Hubbard 1978). In fall over the Manzanos, adults tend to migrate later compared to immatures. The median passage date for adults is 28 September; the median passage date for immatures is 21 September (chapter 2). These age-specific patterns of migration are typical for most raptor species at most migration stations (e.g., Ueta et al. 2000; Mueller et al. 2003).

Nesting

Cooper's Hawks typically begin breeding at the age of two years or older (Rosenfield and Bielefeldt 1993). However, up to 22% of breeding pairs consist of an adult male and a year-old immature female, with pairing of adult females and year-old immature males also reported but not nearly as common (Rosenfield and Wilde 1982). We have observed breeding pairs with immature females in New Mexico, both along the Middle Rio Grande and in the Jemez Mountains. Pairs with immature males have yet to be confirmed in the state.

Cooper's Hawks nest in a wide variety of trees in New Mexico: ponderosa pine, Douglas fir (*Pseudotsuga menziesii*), white fir (*Abies concolor*), and pinyon pine (*Pinus edulis*) in montane forests (Kennedy 1988; PLK, unpubl. data); Fremont cottonwood (*Populus fremontii*), Rio Grande cottonwood (*P. deltoides* ssp. *wislizenii*), Arizona sycamore (*Platanus wrightii*), and, more rarely, Siberian elm (*Ulmus pumila*) along riparian corridors (JLEC and PLK, unpubl. data); and cottonwood and elm trees in urban areas (JLEC, unpubl. data). At Rattlesnake Springs in southeastern New Mexico, Cooper's Hawk nests have been found in netleaf hackberry (D. Roemer, pers. comm.). Bailey (1928) also mentions oaks (*Quercus*), boxelders (*Acer negundo*), and walnuts as additional Cooper's Hawk nest trees in the state.

Nests are typically placed in the crotch of a tree, but Cooper's Hawks may use mistletoe as a supporting platform and, presumably, for additional nest concealment (Kennedy 1988). Nest trees measured by Siders and Kennedy (1996) in the Jemez Mountains were 17–25 m (~60–80 ft) tall with 58–74% canopy closure at the nest site. In that same area, nest trees were more likely to be found on east-facing slopes and nests were placed predominantly on the east and north sides of the nest trees (Kennedy 1988).

In New Mexico, Cooper's Hawks typically lay their eggs in May (Ligon 1961). Most clutches consist of four and occasionally five eggs (Ligon 1961; appendix 10-1); clutches of three eggs probably do occur but have not been recorded in the state. Broods of five young have been observed, and one pair in the South Valley of Albuquerque and one pair outside of Los Alamos are known to have produced five fledglings (Williams 2003; PLK, unpubl. data). Fledging occurs

(a)

(c)

(b)

(d)

PHOTOS 10.14a, b, c, and d

Copper's Hawk nest in riparian woodland (nest in cottonwood [*Populus* sp.], approximately 10 m [30 ft] high) in Animas Park along the Animas River in Farmington, San Juan Co. Note that the parent female is a yearling (immature plumage). (a) Parent female and 11-day-old chick (age estimates based on first date parent seen feeding hatchling; that date taken as the day of hatching), 21 June 2007. (b) Parent female feeding 19-day-old nestling, 29 June 2007. (c) Parent female (in immature plumage) and 20-day-old nestling (second nestling in left background), 30 June 2007. Parent has small piece of food. (d) Two 27-day-old nearly grown nestlings, 7 July 2007. PHOTOGRAPHS: © TIM REEVES.

PHOTO 10.15

Thirty-six-day-old fledgling about 75 m (250 ft) from nest (same nest as in photos 10.14a–d, 16 July 2007.

PHOTOGRAPH: © TIM REEVES.

from early July in southern parts of the state to mid or late July in the Jemez Mountains (Kennedy 1988).

Throughout nesting, the male provides most of the food for the pair and, after hatching, also for the young. Females may also begin hunting toward the end of the nestling period and while the fledglings are still dependent on their parents. Interestingly, more than half of all female Cooper's Hawks may desert during the fledgling dependency period (Kelly and Kennedy 1993). In such cases, the adult male assumes the role of sole food provider for the fledglings; the female does not re-nest.

Diet and Foraging

When hunting, Cooper's Hawks typically scan their surroundings from a well-concealed perch. They are not true sit-and-wait predators. Instead they fly to new perches repeatedly throughout the day (saltatory foraging; see also chapter 11). In north-central New Mexico, Kennedy (1991) found that hunting Cooper's Hawks change perches every 8 to 13 minutes. When the right opportunity arises, Cooper's Hawks leave their perch and dash through or around vegetation in a swift and low flight, pulling their legs forward to strike and grasp their prey. Typically, prey is killed by the hawk's driving its talons into the animal with strong squeezing actions of its feet (Meng 1951).

Throughout most of their distribution, Cooper's Hawks primarily eat birds and mammals (Rosenfield and Bielefeldt 1993). Although not typical, reptiles may be numerically dominant in the species' diet in some areas, while amphibians, fish, and insects have been recorded as prey only rarely. In the Jemez Mountains, Kennedy (1991) identified 44 taxa (32 birds, 9 mammals, and 3 reptiles) in the diet of Cooper's Hawk nesting pairs (table 10.1). Northern Flickers (*Colaptes auratus*), Steller's Jays (*Cyanocitta stelleri*), American Robins (*Turdus migratorius*), Mourning Doves (*Zenaida macroura*), chipmunks (*Tamias* sp.), and cottontails (*Sylvilagus* sp.) were the prey most often recorded. Most of the mammalian prey that could be aged were juveniles, whereas most avian prey appeared to be adults. Mammalian prey too heavy to carry were dismembered, presumably at the point of capture, before being delivered to the nest (Kennedy 1991).

Diet information tends to be cursory or anecdotal elsewhere in New Mexico. Ligon (1961) mentions quails, doves, and flickers as apparent favored prey of Cooper's Hawks in the state. Near Rodeo, in Hidalgo County, a Cooper's Hawk was observed killing a Scaled Quail (*Callipepla squamata*) in November 1966 (McKnight and Snider 1966; table 10.1). In the riparian cottonwood forest along the Middle Rio Grande, we have observed Cooper's Hawks hunting a variety of birds including Mourning Doves and House Finches (*Carpodacus mexicanus*), following them in close pursuit or snatching them from a branch. Along the Gila River, the diet of Cooper's Hawks included Black-headed Grosbeaks (*Pheucticus melanocephalus*), Mourning Doves, and

PHOTO 10.16

Cooper's Hawk with prey, Cimarron Canyon State Park, Colfax Co. September 2007.

PHOTOGRAPH: © ELTON M. WILLIAMS.

PHOTOS 10.17a and b

(*left and bottom*) Cooper's Hawks on water dishes near Santa Fe, Santa Fe Co., 12 June 2008 and 1 November 2008. Cooper's Hawks are notorious for hunting at bird feeders and water dishes. Occasionally they learn to drive small birds toward glass windows. They only have to pick up their prey once these have collided with the window and lie inert on the ground.

PHOTOGRAPHS: © WARREN BERG.

TABLE 11.1 Documented prey of Cooper's Hawks (*Accipiter cooperii*) in New Mexico. Data from Ligon 1961; McKnight and Snider 1966; Hubbard 1981; Kennedy 1991; and RY pers. obs.

CLASS: ORDER	SPECIES	LOCATION
Aves	Wood Duck (*Aix sponsa*)	Middle Rio Grande Bosque
	Domestic fowl (*Gallus* sp.)	Jemez Mountains/Pajarito Plateau
	Scaled Quail (*Callipepla squamata*)	Near Rodeo, Grant Co.
	American Kestrel (*Falco sparverius*)	Jemez Mountains/Pajarito Plateau; Sandia Mountains
	Mourning Dove (*Zenaida macroura*)	Jemez Mountains/Pajarito Plateau; Gila River Valley
	Band-tailed Pigeon (*Columba fasciata*)	Jemez Mountains/Pajarito Plateau
	Western Screech Owl (*Megascops kennicottii*)	San Mateo Mountains
	Burrowing Owl (*Athene cunicularia*)	Johnson Mesa, Colfax Co.
	Northern Flicker (*Colaptes auratus*)	Jemez Mountains/Pajarito Plateau
	Red-naped Sapsucker (*Sphyrapicus nuchalis*)	Jemez Mountains/Pajarito Plateau
	Williamson's Sapsucker (*Sphyrapicus thyroideus*)	Jemez Mountains/Pajarito Plateau
	Hairy Woodpecker (*Picoides villosus*)	Jemez Mountains/Pajarito Plateau
	Cordilleran Flycatcher (*Empidonax occidentalis*)	Jemez Mountains/Pajarito Plateau
	Ash-throated Flycatcher (*Myiarchus cinerascens*)	Jemez Mountains/Pajarito Plateau
	Violet-green Swallow (*Tachycineta thalassina*)	Jemez Mountains/Pajarito Plateau
	Plumbeous vireo (*Vireo plumbeus*)	Gila River Valley
	Steller's Jay (*Cyanocitta stelleri*)	Jemez Mountains/Pajarito Plateau
	Scrub Jay (*Aphelocoma coerulescens*)	Jemez Mountains/Pajarito Plateau
	Clark's Nutcracker (*Nicifraga columbiana*)	Jemez Mountains/Pajarito Plateau
	Black-billed Magpie (*Pica pica*)	Jemez Mountains/Pajarito Plateau
	Pygmy Nuthatch (*Sitta pygmaea*)	Jemez Mountains/Pajarito Plateau
	Western Bluebird (*Sialia mexicanus*)	Jemez Mountains/Pajarito Plateau
	Mountain Bluebird (*Sialia currucoides*)	Jemez Mountains/Pajarito Plateau
	European Starling (*Sturnus vulgaris*)	Jemez Mountains/Pajarito Plateau
	Yellow-rumped Warbler (*Dendroica coronata*)	Jemez Mountains/Pajarito Plateau
	Spotted Towhee (*Pipilo maculofasciatus*)	Jemez Mountains/Pajarito Plateau
	Dark-eyed Junco (*Junco hyemalis*)	Jemez Mountains/Pajarito Plateau
	Chipping Sparrow (*Spizella passerina*)	Jemez Mountains/Pajarito Plateau
	Brown-headed Cowbird (*Molothrus ater*)	Jemez Mountains/Pajarito Plateau
	Western Tanager (*Piranga ludoviciana*)	Jemez Mountains/Pajarito Plateau
	Black-headed grosbeak (*Pheuctitus melanocephalus*)	Jemez Mountains/Pajarito Plateau
	Pine Siskin (*Carduelis pinus*)	Gila River Valley
	Lesser Goldfinch (*Carduelis psaltria*)	Jemez Mountains/Pajarito Plateau
	Red Crossbill (*Loxia curvirostra*)	Jemez Mountains/Pajarito Plateau
	House Finch (*Carpodacus mexicanus*)	Jemez Mountains/Pajarito Plateau; Middle Rio Grande Valley
	Evening Grosbeak (*Coccothraustes vespertinus*)	Jemez Mountains/Pajarito Plateau; Raton

TABLE 11.1 *(continued)*

CLASS: ORDER	SPECIES	LOCATION
Mammalia	Cottontail (*Sylvilagus* sp.)	Jemez Mountains/Pajarito Plateau
	Abert's squirrel (*Sciurus aberti*)	Jemez Mountains/Pajarito Plateau
	Red squirrel (*Tamiasciurus hudsonicus*)	Jemez Mountains/Pajarito Plateau
	Chipmunk (*Tamias* sp.)	Jemez Mountains/Pajarito Plateau
	Golden-mantled ground squirrel (*Spermophilus lateralis*)	Jemez Mountains/Pajarito Plateau
	Rock squirrel (*Spermophilus variegatus*)	Jemez Mountains/Pajarito Plateau
	Woodrat (*Neotoma* sp.)	Jemez Mountains/Pajarito Plateau
	Deer mouse (*Peromyscus maniculatus*)	Jemez Mountains/Pajarito Plateau
Reptilia	Eastern fence lizard (*Sceloporus undulatus*)	Jemez Mountains/Pajarito Plateau
	Western terrestrial garter snake (*Thamnophis elegans*)	Jemez Mountains/Pajarito Plateau
	Little striped whiptail (*Cnemidophorus inornatus*)	Jemez Mountains/Pajarito Plateau

Plumbeous Vireos (*Vireo plumbeus*). In urban and semiurban environments, Cooper's Hawks often hunt small birds near feeders. Based on prey remains, doves appeared to be an important portion of a nesting pair's diet in a residential neighborhood of Albuquerque (JLEC, unpubl. data).

Birds larger than a dove or a quail and including small raptors are by no means immune from predation by Cooper's Hawks. McKnight and Niles (1964) mention a Cooper's Hawk diving at an avocet in August 1964 in the Roswell area. Kennedy (1991) and H. Schwarz (pers. comm.) found Cooper's Hawks to consume American Kestrels (*Falco sparverius*) in the Jemez and Sandia mountains (at the latter location, based on prey remains found under one nest), respectively. Ligon (1961) mentions one case of predation on a [Western] Screech Owl (*Megascops kennicottii*) he observed in the San Mateo Mountains. Hubbard (1981) describes a Cooper's Hawk killing a Burrowing Owl (*Athene cunicularia*) on Johnson Mesa, Colfax County. At the Rio Grande Nature Center in Albuquerque, one of us (RY) once witnessed a female Cooper's Hawk dive at a Wood Duck (*Aix sponsa*) hen that had just taken flight. The two tumbled to the ground, and quite a struggle ensued. But the Cooper's Hawk might have been wary of human presence and did not continue the fight for long. She released her grip on the Wood Duck and flew off, while the duck struggled into nearby thickets. Later that day, while walking in the same vicinity, RY flushed a female Cooper's Hawk off a dead female Wood Duck on the ground. The hawk had consumed quite a bit of breast meat, indicating that she had been feeding for a while. How soon she returned to finish off the duck after the initial attack is unknown.

Kennedy (unpubl. data) radio-tracked 24 adult Cooper's Hawks (12 females and 12 males) during the 1984, 1986, and 1988 breeding seasons in the Jemez Mountains. Home range size averaged larger for females ($\bar{X}$ = 2,803 ha [6,926 ac]) than for males ($\bar{X}$ = 1,206 ha [2,980 ac]). Home range size also varied enormously, more so among females (87 to 8,620 ha [215 to 2,130 ac]) than among males (169 to 3,032 ha [418 to 7,492 ac]). This variation was likely due to the very different parental care strategies selected by the females during the breeding season. In the Jemez Mountains, some females stayed home and defended the nest while the male hunted. Other females hunted to augment male prey deliveries after the young were old enough to thermoregulate without brooding. Finally, some females defended the nest and/or hunted until the fledging period and then deserted (Kelly and Kennedy 1993). These different strategies resulted in widely different movement patterns among females and thus influenced variation in home range size. As primary food providers, males selected parental care strategies that did not vary as much as those of the females. Variation in the size of their home range was probably a result of: (1) a male's experience with its home range; (2) food requirements (which vary seasonally and as a function of brood size); and (3) the local availability of prey.

Predation and Interspecific Interactions

The Cooper's Hawk has its own natural enemies including several other raptor species (Rosenfield and Bielefeldt 1993). In New Mexico, predation by the Northern Goshawk has been documented in the Jemez Mountains. Here the local nesting and foraging habitats of the two accipiter species overlap, as do their respective diets, and remains of Cooper's Hawks have been found at goshawk nests (Kennedy 1991). Predation by Great Horned Owls (*Bubo virginianus*) and Red-tailed Hawks (*Buteo jamaicensis*) has been documented outside of New Mexico (Peyton 1945; Rosenfield 1988). Occasional predation by Great Horned Owls on Cooper's Hawks likely occurs in the riparian forest along the Middle Rio Grande, where the two species often have adjacent nesting territories (JLEC, unpubl. data). All of the mesocarnivores common in the state are also likely predators of Cooper's Hawks. Mammals known to prey on accipiters outside of New Mexico include pine martens (*Martes americana*) and raccoons (*Procyon lotor*) (Squires and Kennedy 2006).

Cooper's Hawks can be aggressive near their nests and attack other raptors (they also dive-bomb people who venture too close to their nests). Along the Gila River, we have observed brief attacks on Red-tailed Hawks, and once on an inadvertently flushed Barn Owl (*Tyto alba*). As is typical among raptors, Cooper's Hawks are often mobbed or chased by other birds (Rosenfield and Bielefeldt 1993). In New Mexico, we have observed Cooper's Hawks being harassed by birds such as Black-chinned Hummingbirds (*Archilochus alexandri*), American Crows (*Corvus brachyrhynchos*), Brewer's Blackbirds (*Euphagus cyanocephalus*), and Western Kingbirds (*Tyrannus verticalis*). Along the Gila River, Cooper's Hawks were occasionally mobbed or chased by flocks of Mexican Jays (*Aphelocoma ultramarina*) (SHS, unpubl. data), and in the Jemez Mountains they were regularly mobbed by flocks of Pinyon Jays (*Gymnorhinus cyanocephalus*) (PLK, unpubl. data). Although Pinyon Jays are common birds in this area, their absence from the observed diet of local Cooper's Hawks (table 10.1) may be a testament to the effectiveness of their mobbing behavior as a predator detection mechanism (Marzluff and Balda 1992).

Status and Management

There have been essentially no long-term changes in observed Cooper's Hawk numbers in New Mexico (Sauer et al. 2008; Smith et al. 2008; J. Smith, pers. comm.). Through 2005, migrating Cooper's Hawks were showing a significant long-term increase in the Manzano Mountains (Smith et al. 2008). However, that trend did not continue after 2005, and in the fall of 2009 HawkWatch International even recorded the second lowest Cooper's Hawk count to date (J. Smith, pers. comm.).

Three decades ago, Hubbard (1978:14) described the species as "rare to fairly common" during both breeding and nonbreeding. As seen in this chapter, however, densities of nesting pairs can be quite high in riparian floodplain woodlands, where observed nesting productivity is within the range of what has been observed for Cooper's Hawks elsewhere in their range (Rosenfield and Bielefeldt 1993). A total of 11 pairs in Corrales along the Middle Rio Grande produced 26 young in 2003, or a mean of 2.36 fledglings per nest (Williams 2003). From 2004 through 2007, Hawks Aloft (2008) monitored 232 Cooper's Hawk nests in the Middle Rio Grande bosque, with an estimated 189 (81%) of these nests having successfully fledged young, for a total aggregate estimated productivity of 2.2 young fledged per occupied nest and 2.7 young per successful nest. Productivity estimates from the Jemez Mountains are similar to those from the Middle Rio Grande bosque. Out of 29 nesting attempts monitored from 1984 to 1988 in the Jemez Mountains, 86.2% were successful (fledged at least one young); 2.36 young were fledged per successful nest and 2.03 young per nesting attempt (PLK, unpubl. data).

PHOTO 10.18

Cooper's Hawk being treated at the Gila Wildlife Rescue Center in Silver City, Grant Co., after a collision with a window.

PHOTOGRAPH: © DENNIS MILLER.

Once negatively affected by shooting, trapping, and organochlorine contaminants, the Cooper's Hawk is no longer a species of concern in the United States (Rosenfield and Bielefeldt 1993), nor is there currently any threat to its persistence in New Mexico in particular. In urban or semiurban areas, Cooper's Hawks occasionally collide with windowpanes while chasing after birds, resulting in some mortality (S. Kendall, pers. comm.). Acquisition logs from Wildlife Rescue, Inc., also show trichomoniasis—from the ingestion of infected pigeons or doves—to be a somewhat frequent cause of mortality in the Albuquerque area (J. Harden, pers. comm.). Riparian loss and degradation represent additional threats but only at a local level. McBride et al. (2004) investigated lead blood concentrations in Cooper's Hawks captured during migration at HawkWatch International sites in the Sandia and Manzano mountains. They did detect higher lead blood concentrations in spring than in fall migrants, indicating higher environmental exposure to lead in the winter range of the species. However, blood concentrations of lead did not reach toxicity levels.

Naturalists of the early and mid 20th century often showed a strong negative bias against the Cooper's Hawk and, to a lesser degree, the Sharp-shinned Hawk. For example, Bent (1937:112) referred to both species as "blood-thirsty villains," the Cooper's Hawk the worst of the two! At the core of the Cooper's Hawk's infamous reputation was its propensity to attack poultry as well as quail, always a favorite among game bird hunters (Bent

1937; Ligon 1961). Today, however, the Cooper's Hawk is getting recognition for its important position in the food webs of many ecosystems. Birdwatchers have also developed an appreciation for the Cooper's Hawk as a skilled and often spectacular predator. Residents in urban areas typically do not view Cooper's Hawks as a threat or nuisance—unlike coyotes and bobcats—and are often elated to discover these hawks nesting near their homes, a piece of untamed nature brought into their neighborhoods.

LITERATURE CITED

Bailey, F. M. 1928. *Birds of New Mexico*. Santa Fe: New Mexico Department of Game and Fish.

Bent, A. C. 1937. *Life histories of North American birds of prey*. Part 1. U.S. National Museum Bulletin 167. Washington, DC: Smithsonian Institution.

Boal, C. W., and R. W. Mannan. 1998. Nest-site selection by Cooper's hawks in an urban environment. *Journal of Wildlife Management* 62:864–71.

Cartron, J.-L. E., D. C. Lightfoot, J. E. Mygatt, S. L. Brantley, and T. K. Lowrey. 2008. *A field guide to the plants and animals of the Middle Rio Grande Bosque*. Albuquerque: University of New Mexico Press.

Eakle, W. L., E. L. Smith, S. W. Hoffman, D. W. Stahlecker, and R. B. Duncan. 1996. Results of a raptor survey in southwestern New Mexico. *Journal of Raptor Research* 30:183–88.

Goodman, R. A., ed. 1990. *New Mexico Ornithological Society Field Notes* 29(3): 1 June–31 July.

Harden, J., R. W. Dickerman, and E. P. Elliston. 2006. Collection, value and use of wildlife rehabilitation data. *Journal of Wildlife Rehabilitation* 28:10–28.

Hawks Aloft, Inc. 2008. Raptor monitoring in the Middle Rio Grande bosque of central New Mexico. Unpublished report submitted to the Middle Rio Grande Bosque Initiative, U.S. Fish and Wildlife Service, and U.S. Bureau of Reclamation, Albuquerque, NM.

Hoffman, S. W., J. P. Smith, and T. D. Meehan. 2002. Breeding grounds, winter ranges, and migratory routes of raptors in the Mountain West. *Journal of Raptor Research* 36:97–110.

Hubbard, J. P. 1971a. The summer birds of the Gila Valley, New Mexico. *Occasional Papers of the Delaware Museum of Natural History, Nemouria* 2:1–35.

———, ed. 1971b. *New Mexico Ornithological Society Field Notes* 10(2): 1 June–30 November.

———, ed. 1972. *New Mexico Ornithological Society Field Notes* 11(2): 1 June–30 November.

———. 1978. *Revised check-list of the birds of New Mexico*. Publ. no. 6. Albuquerque: New Mexico Ornithological Society.

———. 1981. Cooper Hawk attacks Burrowing Owl. *New Mexico Ornithological Society Bulletin* 9:25.

Kelly, E. J., and P. L. Kennedy. 1993. A dynamic state variable model of mate desertion in Cooper's hawks. *Ecology* 72:351–66.

Kennedy, P. L. 1988. Habitat characteristics of Cooper's Hawks and Northern Goshawks nesting in New Mexico. In *Proceedings of the southwest raptor management symposium and workshop*, ed. R. L. Glinski et al., 218–27. Washington, DC: National Wildlife Federation.

———. 1991. Reproductive strategies of Northern Goshawks and Cooper's Hawks in north-central New Mexico. Ph.D. diss. Utah State University, Logan.

Ligon, J. S. 1961. *New Mexico birds and where to find them*. Albuquerque: University of New Mexico Press.

Marzluff, J. M., and R. P. Balda. 1992. *The Pinyon Jay: behavioral ecology of a colonial and cooperative Corvid*. London, UK: T and AD Poyser.

McBride, T. J., J. P. Smith, H. P. Gross, and M. J. Hooper. 2004. Blood-lead and ALAD activity levels of Cooper's Hawks (*Accipiter cooperii*) migrating through the southern Rocky Mountains. *Journal of Raptor Research* 38:118–24.

McKnight, B. C., and D. M. Niles, eds. 1964. *New Mexico Ornithological Society Field Notes* 6: 1 June–30 November.

McKnight, B. C., and P. R. Snider, eds. 1966. *New Mexico Ornithological Society Field Notes* 5(2): 1 June–30 November.

Meng, H. K. 1951. Cooper's Hawk, *Accipiter cooperii* (Bonaparte). Ph.D. diss. Cornell University, Ithaca, New York.

Mueller, H. C., D. D. Berger, and N. S. Mueller. 2003. Age and sex differences in the timing of spring migration of hawks and falcons. *Wilson Bulletin* 115:321–24.

Palmer, R. S., ed. 1988. *Handbook of North American birds.* Vol. 4, part 1, Diurnal Raptors. New Haven: Yale University Press.

Peyton, S. B. 1945. Western Red-tailed Hawk catches Cooper's Hawk. *Condor* 47:167.

Rosenfield, R. N. 1988. Cooper's Hawk. In *Handbook of North American birds.* Vol. 4, part 1, Diurnal Raptors, ed. R. S. Palmer, 328–30, 331–32, 349–52. New Haven: Yale University Press.

Rosenfield, R. N., and J. Bielefeldt. 1993. No. 75. In *The birds of North America*, ed. A. Poole and F. Gill. Philadelphia, PA: Birds of North America, Inc.

Rosenfield, R. N., and J. Wilde. 1982. Male Cooper's Hawk breeds in juvenal plumage. *Wilson Bulletin* 94:213.

Sauer, J. R., J. E. Hines, and J. Fallon. 2008. *The North American Breeding Bird Survey, Results and Analysis 1966–2007.* Version 5.15.2008. Laurel, MD: USGS Patuxent Wildlife Research Center.

Schmitt, C. G. 1976. *Summer birds of the San Juan Valley, New Mexico.* Publ. no. 4. Albuquerque: New Mexico Ornithological Society.

Siders, M. S., and P. L. Kennedy. 1996. Forest structural characteristics of Accipiter nesting habitat: is there an allometric relationship? *Condor* 98:123–32.

Smith, J. P., C. J. Farmer, S. W. Hoffman, G. S. Kaltenecker, K. Z. Woodruff, and P. F. Sherrington. 2008. Trends in autumn counts of migratory raptors in western North America. In *State of North Americas Birds of Prey,* eds. K. L. Bildstein, J. P. Smith, E. Ruelas Inzunza, and R. R. Veit, 217–51. Series in Ornithology 3. Cambridge, MA: Nuttall Ornithological Club, and Washington, DC: American Ornithologists' Union.

Squires, J. R., and P. L. Kennedy. 2006. Northern goshawk ecology: an assessment of current knowledge and information needs for conservation and management. *Studies in Avian Biology* 31:8–62.

Stacey, P. B. 1984. The birds of Water Canyon, Magdalena Mountains, New Mexico. *New Mexico Ornithological Society Bulletin* 11:47–60.

Tatschl, J. L. 1967. Breeding birds of the Sandia Mountains and their ecological distributions. *Condor* 69:479–90.

Ueta, M., F. Sato, H. Nakagawa, and N. Mita. 2000. Migration routes and differences of migration schedule between adult and young Steller's Sea Eagles, *Haliaeetus pelagicus. Ibis* 142:35–39.

Whaley, W. H., and C. M. White. 1994. Trends in geographic variation of Cooper's Hawk and Northern Goshawk in North America: a multivariate analysis. *Proceedings of the Western Foundation of Vertebrate Zoology* 5:1–209.

Wheeler, B. K., and W. S. Clark. 1995. *A photographic guide to North American raptors.* London: Academic Press.

Williams, S. O. III. 2003. *New Mexico Ornithological Society Field Notes* 42(3): Summer 2003.

11

Northern Goshawk

(*Accipiter gentilis*)

PATRICIA L. KENNEDY AND JEAN-LUC E. CARTRON

A TOP PREDATOR of montane forests, the Northern Goshawk (*Accipiter gentilis*) is the embodiment of fierce tenacity, power, and stealth. It is the largest North American accipiter, with a body length of 46–51 cm (18–20 in) in males and 53–62 cm (21–24 in) in females, and a mean body weight of 816 g (1.8 lbs) in males and 1,059 g (2.3 lbs) in females (Wheeler and Clark 1995). Up close, the adults are readily identified from the two other North American accipiters, the Sharp-shinned Hawk (*A. striatus*; chapter 9) and Cooper's Hawk (*A. cooperii*; chapter 10), by their large size; white superciliaries (eyebrows) that contrast with the blackish crown and blackish cheek patch; a pale chest and pale belly both finely barred with gray; and white undertail coverts that often appear fluffed out. In flight, an adult Northern Goshawk is again distinctive by its size, but also shows characteristics shared with other accipiters, including the bluish gray back and upperwing coverts, the rather long banded tail, barred flight feathers, yellow legs and feet, and rapid wing beats alternating with short glides. Compared to Cooper's Hawks, however, Northern Goshawks have broader wings and, relative to the length of their body, a shorter tail. The Northern Goshawk tail has a narrower terminal band than the Cooper's Hawk tail and the undertail coverts are not spotted or

PHOTO 11.1

Adult Northern Goshawk trapped in fall 2002 at HawkWatch International's Manzano Mountains study site. The head is distinctive, with white superciliaries (eyebrows) that contrast with the blackish crown and blackish cheek patch. PHOTOGRAPH: © JOHN P. DELONG.

PHOTO 11.2

Adult female Northern Goshawk trapped and banded during the 1991 fall migration, HawkWatch International's Manzano Mountains study site. Both male and female adults have a pale chest and belly finely barred with gray, barred flight feathers, a long banded tail, and white undertail coverts. Note also the blackish crown and cheek patch and contrasting white superciliary.

PHOTOGRAPH: © JOHN P. DELONG.

streaked. Reversed size dimorphism (i.e., larger body size for females than males) is not as pronounced in the Northern Goshawk as it is in the other accipiters. Besides being larger, however, the female also tends to have coarser gray barring on the underparts, and the upperparts are slightly more brown.

Similar to other accipiters, the juvenal plumage, which is retained through the first winter, is markedly different from the adult plumage. The upperparts are mottled brown, and the chest and belly are pale with heavy brown streaking. The whitish supercilium is wide but not always distinct. The undertail coverts are streaked or spotted instead of pure white. The barring on the tail creates a wavy pattern, and the dark bands have narrow white edges. Also, the eyes of juveniles are yellow (in contrast to adults, which have yellow-orange to deep red eyes). Juvenile Northern Goshawks and Cooper's Hawks can be difficult to distinguish in the field. However, the juvenile Northern Goshawk is larger, with longer wings that appear more tapered while gliding. Besides the wavy, dark tail bands with narrow white edges, another good field characteristic is the tawny diagonal bar on the upperwings (see also Palmer 1988; Johnsgard 1990; Wheeler and Clark 1995; and reviews in Squires and Reynolds 1997).

Taxonomy experts disagree over the number of Northern Goshawk subspecies present in North America. The two subspecies recognized by the American Ornithologists' Union (AOU 1983) are *A. g. atricapillus* and *A. g. laingi*.

A third subspecies, *A. g. apache*, was proposed for populations inhabiting the sky islands of southeast Arizona into northwest Mexico (Van Rossem 1938; Friedmann 1950; Snyder and Snyder 1991; Hubbard 1992; Whaley and White 1994). It is described as darker in plumage, almost black dorsally, larger in size, and long winged (Van Rossem 1938). Because the subspecies *A. g. apache* had been proposed based on the examination of only very few birds, it is not recognized by the American Ornithologists' Union as legitimate. The U.S. Fish and Wildlife Service (1998) also considers the validity of this subspecies to be unresolved. In recent genetic studies, however, Bayard de Volo (2008) used molted feathers to examine the genetic relatedness of 21 populations across the range of the species. Her results indicate that the sky islands of southeast Arizona and northwest Mexico (see chapter 1) harbor a set of populations geographically isolated from more northern populations. Together with the morphological differences already mentioned, these results support the validity of the *apache* subspecies. The only feathers analyzed from New Mexico were contributed by PLK from the nest sites she studied in the Jemez Mountains and adjacent Pajarito Plateau (hereafter referred to as the Jemez Mountains). Bayard de Volo's analysis of these feathers suggests the Jemez Mountain population is genetically distinct from the Sky Island and Rocky Mountain Northern Goshawks. However, only 12 feathers were analyzed from the Jemez Mountains and feathers were not analyzed from any other populations in New Mexico; thus it is not clear how to interpret these results, however intriguing they may be. As noted by Bayard de Volo (2008), subspecific designations are difficult to make and, given their implications in terms of policies, should be based on multiple lines of evidence. Her genetic results are based on a single mitochondrial region and she recommends additional genetic analyses be conducted before subspecific designations are changed.

PHOTO 11.3

Pat Kennedy holding a nesting pair of Northern Goshawks trapped at their nest site in the Jemez Mountains, June 1988. Both the female (left) and the male (right) have bluish gray upperparts (more brown in females, grayer in males) and a rather long banded tail. This photo illustrates the reversed sexual size dimorphism typical of accipiters and many other raptors: the male is smaller than the female. The birds in this photo were fitted with radio transmitters, banded, and released at the nest site. PHOTOGRAPH: © PATRICIA L. KENNEDY.

Northern Goshawks are usually silent except during certain periods of the breeding season. During courtship they are very vocal particularly around dawn and their vocalizations at this time are best described as "chattering" or "wails" (Penteriani 2001; Dewey et al. 2003). If the nest area is disturbed by a potential predator or conspecific intruder, the adults will also utter a loud cackling "*kye kye kye*" that is described as their alarm call (Palmer 1988; Kennedy and Stahlecker 1993). During the fledgling-dependency period juveniles are very vocal, either begging or uttering the alarm call at a nest site intruder. Juvenile begging and alarm calls resemble a high-pitched version of the female's wail and alarm call, respectively. Listening for the species' distinct vocalizations and playbacks of these

PHOTO 11.4

Immature Northern Goshawk trapped at HawkWatch International's Manzano Mountains study site, fall 1993. The upperparts are mottled brown. The barring on the tail creates a wavy pattern. The whitish supercilium is wide. It is conspicuous on this bird, but on other immature birds it may not be distinct. The eyes of juveniles are yellow, not orange or red as in the adults. PHOTOGRAPH: © JOHN P. DELONG.

vocalizations are commonly used to conduct surveys of breeding goshawks (Kennedy and Stahlecker 1993; Dewey et al. 2003).

Distribution

The Northern Goshawk is a resident or partial resident of boreal and montane forests throughout much of the Holarctic, reaching as far north as the Arctic Circle both in North America and Eurasia (del Hoyo et al. 1994; Squires and Reynolds 1997; Ferguson-Lees and Christie 2001; Kenward 2006). In Eurasia the southern limit of the species' distribution lies around the Mediterranean Basin and in Asia Minor, Iran, the Himalayas, eastern China, and Japan (del Hoyo 1994; Kenward 2006). In North America, Northern Goshawks breed from boreal forests of north-central Alaska to Newfoundland and south to western and southwestern montane forests in the United States, and locally in the mountains of northwestern and western Mexico. In central to eastern North America, Northern

Goshawks breed in the western Great Lakes region and eastward to Pennsylvania, central New York, northwestern Connecticut, and locally south in montane habitats at least to West Virginia and possibly eastern Tennessee and western North Carolina (Squires and Kennedy 2006).

Although many Northern Goshawks are year-round residents throughout the species' extensive range, some seasonal movements are observed, especially in the North during prey population crashes and long periods of inclement winter weather. As a result the geographic range of the Northern Goshawk expands during the winter, with the species observed outside the breeding range in places like Texas and the northern part of the Gulf States (Squires and Kennedy 2006).

The distribution of the Northern Goshawk in New Mexico is not well known. The species occurs year-round in at least some of the state's forested mountain ranges (Ligon 1961; Taschl 1967; Hubbard 1978). Breeding has been documented in the Burro, Gallinas, Jemez, Peloncillo, Pinos Altos, Sacramento, San Juan, Sangre de Cristo, and Zuni mountains (Ligon 1961; Johnson and Harris 1967; Shuster 1977; Hubbard 1978; Kennedy 1988, 1991, unpubl. data; NMOS 2008). Based on breeding season records, breeding also likely occurs in the Mogollon, Sandia, San Francisco, and

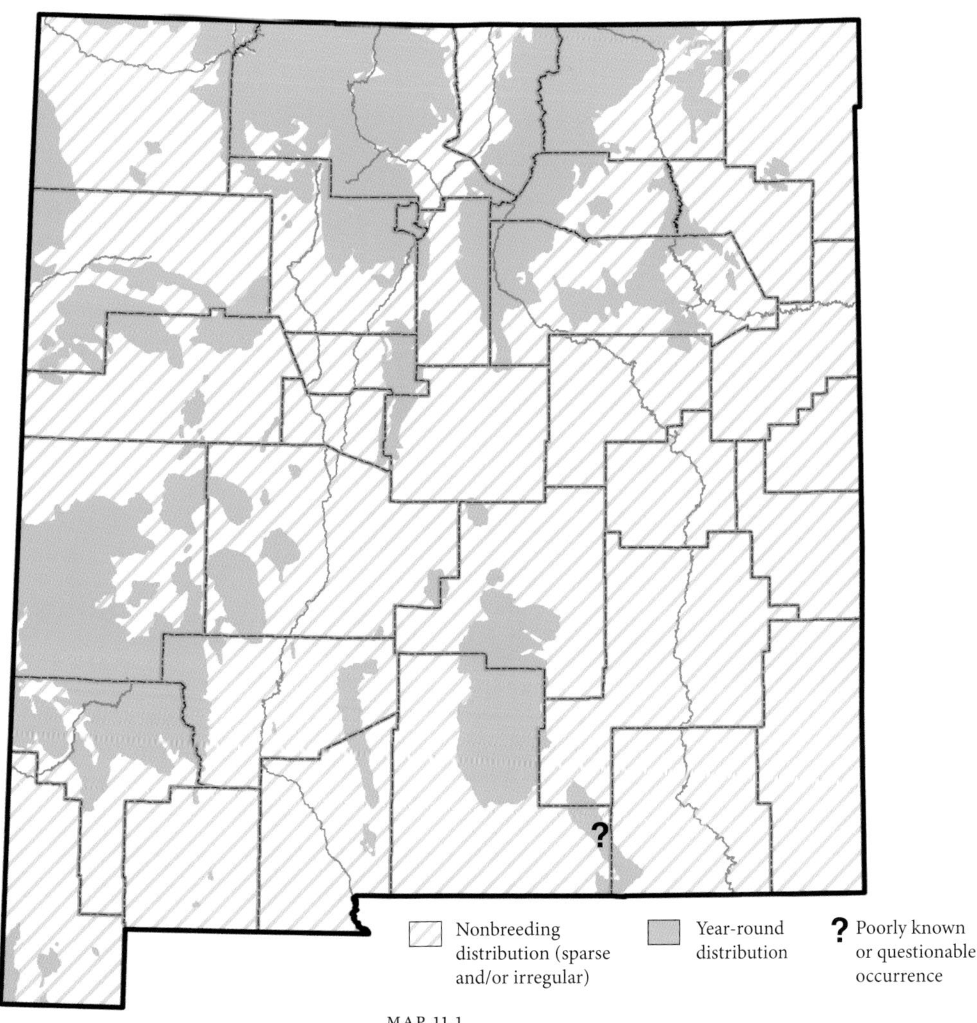

MAP 11.1

Northern Goshawk distribution map

San Mateo mountains (Hubbard 1978; NMOS 2008), although no nests have been found yet in those mountain ranges. There are records of occurrence from additional mountain ranges during the nonbreeding season (NMOS 2008), likely indicating in part an influx of spring and fall transients and winter residents in nonbreeding areas, and/or an incomplete knowledge of the distribution of resident birds. There is at least one fall record for the species from the Guadalupe Mountains of southeastern New Mexico, but no evidence of breeding or wintering at that location (NMOS 2008). Outside of the breeding season, there have been sightings in low-elevation areas, but very few are from the eastern plains (Hubbard 1978; NMOS 2008). During fall and spring, small numbers of Northern Goshawks are also observed flying over the Manzano and Sandia mountains, their migration farther north likely taking them over ridges of the Sangre de Cristo and San Juan mountains (see Migration section below and chapter 2 for more details).

PHOTO 11.5

Adult Northern Goshawk on a snag at Sugarite Canyon State Park, Colfax Co., 2 April 2007. Perhaps a spring migrant, especially as the species is not known to nest at that location. However, breeding pairs can easily escape detection unless broadcast surveys are conducted.

PHOTOGRAPH: © JEAN-LUC CARTRON.

Habitat Associations

The Northern Goshawk is a forest-dwelling raptor. It occurs in most forest types found across its geographic range, including North America's boreal, eastern deciduous, and montane coniferous forests (reviewed in Squires and Reynolds 1997). Within heavily forested areas, Northern Goshawks are generally considered habitat generalists but tend to nest in mature to old-growth stands with high canopy closure (Reynolds et al. 1982; Speiser and Bosakowski 1987; Crocker-Bedford and Chaney 1988; Hayward and Escaño 1989; Siders and Kennedy 1994; Squires and Ruggiero 1996). In Arizona, Crocker-Bedford and Chaney (1988) reported that Northern Goshawks nested 5.8 times more often in stands with at least 80% canopy closure than expected by chance alone.

In New Mexico as elsewhere in the Southwest, Northern Goshawks breed in most montane and subalpine forest cover types (Kennedy 1988; Reynolds et al. 1994). The most extensive nesting habitat evaluation for the state was conducted by Siders and Kennedy (1994, 1996) in the Jemez Mountains. Here goshawks nest predominantly in ponderosa pine (*Pinus ponderosa*), aspen (*Populus tremuloides*)/mixed conifer, and mixed conifer stands at 2,460–2,530 m (8,070–8,300 ft) in elevation. Nest sites have 60–71% canopy closure with an average tree density of 800–1,400 trees per ha (~1,980–3,460 per ac). These nest site characteristics are within the range of what has been reported in other areas (Kennedy 1988; Siders and Kennedy 1994, 1996). Because of the wide variety of nesting habitats used by the Northern Goshawk and the Cooper's Hawk, it is difficult to differentiate their nest sites strictly based on forest characteristics (see also chapter 10). In the Jemez Mountains the only characteristics that were significantly different between the nest sites of the two species were stand basal area and mature tree density. Northern Goshawk nest sites had more basal area and higher densities of mature trees than did Cooper's Hawk sites (Siders and Kennedy 1996).

Although Northern Goshawks appear to select relatively closed-canopy forests for nesting, exceptionally they will nest in more open forests. Northern Goshawks nest in tall willow communities along major drainages in arctic tundra, riparian cottonwood (*Populus* spp.) stands, and small stands of aspen in shrub-steppe habitat (Squires and Kennedy 2006). This pattern has been documented in New Mexico by Kennedy (1988), who found a nest in an open stand of Chihuahua pine (*Pinus leiophylla*)/Arizona oak (*Quercus arizonica*) in the Peloncillo Mountains. Shuster (1977) also found a nest in a mixed stand of ponderosa pines, pinyons (*Pinus edulis*), and junipers (*Juniperus*) with a very open canopy in the Sacramento Mountains.

Northern Goshawk winter habitat preferences are unclear due to a paucity of studies on this topic. Winter habitat studies have been conducted primarily in Europe (Squires and Kennedy 2006), with one published study conducted in North America (Arizona: Drennan and Beier 2003). These studies suggest that winter habitat used by the Northern Goshawk is more variable than breeding habitat and that the species' winter habitat use is likely influenced by its local migratory status (see below for more details on migration). In areas where Northern Goshawks are residents, breeding pairs can remain on their breeding season home ranges during the nonbreeding season (Boal et al. 2003). However, migratory populations may overwinter in habitats very different from their breeding season home ranges such as low-elevation shrub-steppe (Squires and Kennedy 2006).

PHOTO 11.6

(*top*) Pat Kennedy holding an adult male Northern Goshawk, trapped at its nest site in the Jemez Mountains, June 1988. Note the habitat in the background. The Northern Goshawk is a forest-dweller. In the Jemez Mountains, it nests mainly in ponderosa pine (*Pinus ponderosa*), aspen (*Populus tremuloides*)/mixed conifer, and mixed conifer stands at 2,460–2,530 m (8,070–8,300 ft). The bird was fitted with a radio transmitter, banded, and released at the nest site.

PHOTOGRAPH: © PATRICIA L. KENNEDY.

PHOTO 11.7

(*bottom*) Adult female Northern Goshawk, trapped at its nest site in the Jemez Mountains, June 1988.

PHOTOGRAPH: © PATRICIA L. KENNEDY.

Life History

Most of the life history information for the Northern Goshawk in New Mexico is based on Kennedy's (1991) five-year study of nesting pairs in the Jemez Mountains. In 1992 and 1993, one of us (PLK) and Johanna Ward conducted additional research to study the effect of food supply on the size, survival, and postfledging dispersal of young. They supplemented some broods—and later the fledglings—with food, fitted young with miniature radio transmitters, and then compared size, survival, and dispersal of young raised in supplemented nests versus nonsupplemented nests (Ward and Kennedy 1994, 1996; Ward et al. 1997; Kennedy and Ward 2003). Although very limited, some natural history information also exists from the Pinos Altos and Peloncillo mountains (Shuster 1977; Kennedy 1988).

Migration

Northern Goshawks are partial migrants (Squires and Reynolds 1997), meaning that some individuals maintain year-round occupancy of nest territories (Boal et al. 2003) while other individuals in the population undergo seasonal movements to wintering areas. The degree to which populations are partially migratory may relate to food availability on breeding areas during winter. In the Yukon, goshawks were year-round residents during periods of high snowshoe hare (*Lepus americanus*) abundance, but winter sightings sharply declined when hare densities were low (Doyle and Smith 1994). Approximately every 10 years, large numbers of goshawks are observed migrating to southern wintering areas, apparently in response to low prey abundance at northern latitudes (Mueller and Berger 1968; Mueller et al. 1977; Doyle and Smith 1994); these incursions usually last at least two years (Squires and Reynolds 1997). Irruptive movements of goshawks are composed primarily of adults (Sutton 1931; Mueller et al. 1977); juvenile proportions are variable, probably dependent on reproductive success during the previous nesting season. Thus birds observed in New Mexico during the winter are likely a combination of territory holders and birds from northern latitudes.

Fall migration generally commences after young disperse from natal areas and occurs between mid September and mid December. Spring migration is more difficult to observe because birds are hard to follow on the wintering grounds. Thus spring migration is poorly understood (Squires and Reynolds 1997). The most detailed study of Northern Goshawk migratory movements was conducted by Sonsthagen (2002). She used satellite telemetry to monitor the fall migration of 34 female goshawks breeding throughout the state of Utah. The goshawks moved throughout Utah and the 34 female goshawks exhibited a variety of movement patterns. However, her data support previously reported patterns based on band returns (Reynolds et al. 1994; Hoffman et al. 2002) and radio telemetry (Squires and Ruggerio 1995; Stephens 2001) that goshawk migrations involve short-distance movements (<500 km [<310 mi]). Of the 34 birds fitted with satellite transmitters, 19 wintered near their breeding area and 15 were migrants. The migrants moved 49–613 km [30–381 mi] to wintering areas and only two birds moved >500 km (>310 mi). In New Mexico, passage of Northern Goshawks in fall over the Manzanos is from late August through early November, mainly from around mid September through late October; in spring over the Sandias, migrating Northern Goshawks have been recorded from late February through early May, mainly in March and especially April (chapter 2). Habitat used by Northern Goshawks during migration has never been documented.

Nesting

Northern Goshawks nest in trees, typically coniferous trees in western North America (Squires and Reynolds 1997). The nest tree is usually among the tallest in the surrounding forest stand (Reynolds et al. 1982; Hargis et al. 1994; Squires and Ruggiero 1996). However, the nest itself is rarely above the lower half of the canopy, though nest height is positively correlated with nest tree height (e.g., Reynolds et al. 1982; Speiser and Bosakowski 1987). Nest trees in New Mexico include ponderosa pine, aspen, and Chihuahuan pine (*Pinus leiophylla*) (Ligon 1961; Kennedy 1988). In the Jemez Mountains goshawks use nest trees 25–31 m (82–102 ft) tall and 43.3–56.7 cm (17–22 in) in diameter with 58–74% canopy closure at the nest tree (Siders and Kennedy 1996). The nest described by Shuster (1977) in the Sacramento Mountains was approximately 30 m (100 ft) above ground in a ponderosa pine.

Typical goshawk breeding areas contain several alternative nests that are used over several years (Squires and Kennedy 2006). The reason for using alternative nests is unknown, but may reduce exposure to disease

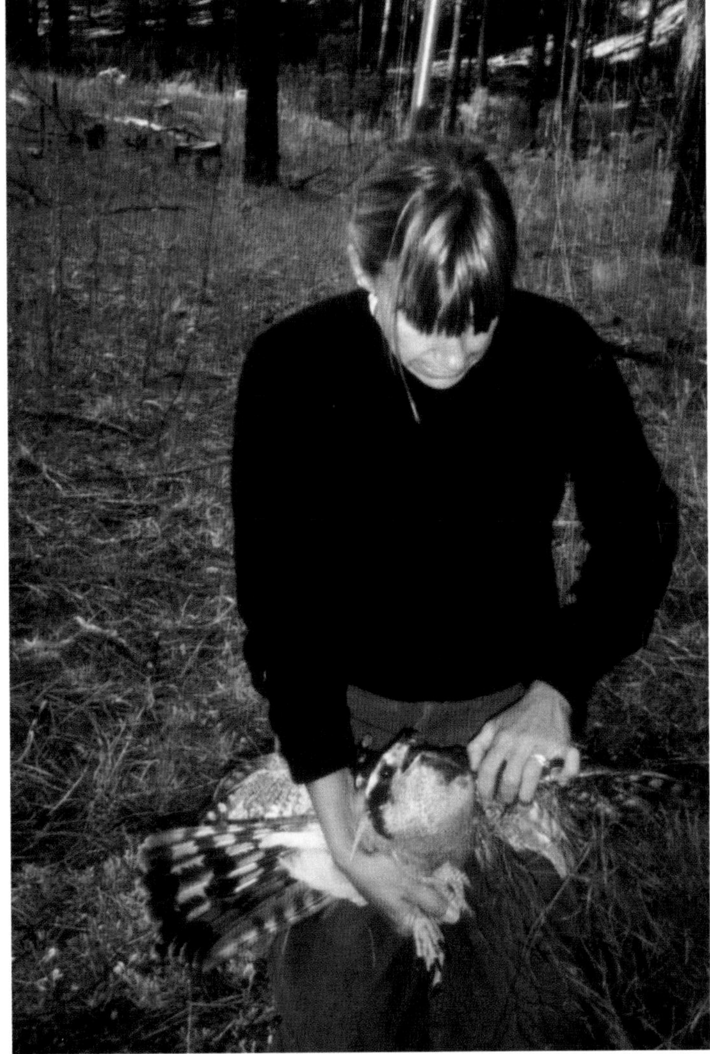

PHOTO 11.8

Pat Kennedy taking an adult Northern Goshawk out of a Dho-gaza net, Jemez Mountains, June 1988. Dho-gaza nets with stuffed Great Horned Owls (*Bubo virginianus*) are commonly used by researchers to trap adult accipiters at their nests (a U.S. Fish and Wildlife special permit is required for using these nets). Nesting Northern Goshawks see the owl as a threat to the nestlings, mob it, and get caught in the nets surrounding the owl. The bird in the photograph was fitted with a radio transmitter, banded, and released at the trap site. PHOTOGRAPH: © PATRICIA L. KENNEDY.

PHOTO 11.9

Likely migrating immature Northern Goshawk in the Manzano Mountains, fall 2002. PHOTOGRAPH: © JOHN P. DELONG.

PHOTO 11.10

(*top left*) During courtship a pair of Northern Goshawks will either construct a new nest or remodel an old nest. This particular nest had been used for several years by the same pair and the female is adding fresh green boughs of Douglas fir (*Pseudotsuga menziesii*) to line the nest cup. The nest tree is an aspen (*Populus tremuloides*) in the Jemez Mountains and the body of the nest is made out of aspen twigs. April 1989. PHOTOGRAPH: © DAVID A. PONTON AND PATRICIA L. KENNEDY.

PHOTO 11.11

(*bottom left*) After placing the Douglas fir (*Pseudotsuga menziesii*) bough, the female spends some time moving aspen (*Populus tremuloides*) twigs around, as pictured here. Jemez Mountains, April 1989. PHOTOGRAPH: © DAVID A. PONTON AND PATRICIA L. KENNEDY.

PHOTO 11.12

(*top right*) This same adult female photographed during courtship (photo 11.10) is standing up on the nest after having laid her first egg for the season. The egg is in the green boughs lining the nest cup in the lower right corner. This female laid two eggs but only one egg was viable. Jemez Mountains, April 1989. PHOTOGRAPH: © DAVID A. PONTON AND PATRICIA L. KENNEDY.

PHOTO 11.13

(*bottom right*) Two young nestlings in a nest in the Jemez Mountains, July 1993. PHOTOGRAPH: © JOHN P. DELONG.

PHOTOS 11.14a and b

(*top and middle*) A female nestling is waiting patiently to be fed. The adult female is piecing the prey, which is analogous to a parent's cutting up a toddler's food at the dinner table. Jemez Mountains, June 1989. PHOTOGRAPHS: © DAVID A. PONTON AND PATRICIA L. KENNEDY.

PHOTO 11.15

(*right*) The female is passing the last morsel of food to the nestling. The nestling is within three to five days of fledging and still has quite a few white downy feathers that will be replaced by juvenile feathers during the next 30–60 days (the fledgling-dependency period). Jemez Mountains, June 1989. PHOTOGRAPH: © DAVID A. PONTON AND PATRICIA L. KENNEDY.

and parasites. Although goshawks may use the same nest in consecutive years, nest areas may include from one to eight alternative nests that are usually located within 0.4 km (0.25 mi) of each other. Alternative nests can be clumped in one- to three-nest stands or widely distributed throughout the bird's home range (Squires and Kennedy 2006).

The adult female builds a new nest or repairs an old one, assisted by the male only occasionally (Zirrer 1947; Schnell 1958; Lee 1981). Clutch size has not been well documented in New Mexico. It varies geographically but typically ranges from two to four; clutches of one or five eggs are rare but have been reported (see Squires and Reynolds 1997). The eggs are laid at an interval of two to three days (Beebe 1974; Cramp and Simmons 1980). Incubation may begin with the first or second egg and lasts 28–38 days (Squires and Reynolds 1997). The eggs are incubated primarily by the female, while the male provisions her with food (Zirrer 1947; Lee 1981). After the eggs have hatched, the female broods the young while the male continues to hunt. Some but not all females begin hunting as the young approach fledging (Palmer 1988). The number of young raised to fledging is highly variable, but the productivity of successful pairs in North American populations typically ranges from 1.4 to 2.7 (Squires and Reynolds 1997), although a mean productivity of 3.9 fledglings was documented in Yukon based on eight nests (Doyle and Smith 1994). Nests are reported to fail only rarely after the early nestling period (Squires and Reynolds 1997). In Arizona (n = 98 nests), 85% of nests successfully fledged young, 3% did not appear to ever have eggs, 6% had eggs but failed during incubation, and 6% failed during the nestling period (Reynolds et al. 1994). Siblicide and cannibalism have been documented in New Mexico nests especially during periods of low food availability (Estes et al. 1999).

In the Jemez Mountains, nestlings fledged at the age of approximately 40 days (Kennedy and Ward 2003). For about 40 more days, fledglings remained dependent on their parents for food while acquiring sustained flight capabilities and learning to hunt. All juveniles became independent by the age of 15 weeks, or sometime during a period extending from 4 August through 9 September (Kennedy and Ward 2003). After the dependency period, juveniles from nests that had been supplemented with food tended to remain within 10 km (6.2 mi) of their natal nests, in contrast to juveniles from nonsupplemented nests, which tended to disperse not only farther from the nest but also earlier (Kennedy and Ward 2003). The authors concluded that juvenile Northern Goshawks make decisions to disperse far or not based on local food availability.

Diet and Foraging

In the Jemez Mountains, Kennedy (1991) reported 21 avian prey species and 8 mammalian prey species in the diet of goshawk breeding pairs (table 11.1). Cottontail (*Sylvilagus* sp.), Northern Flicker (*Colaptes auratus*), Abert's squirrel (*Sciurus aberti*), Steller's Jay (*Cyanocitta stelleri*), and red squirrel (*Tamiasciurus hudsonicus*) accounted for 65% of all documented prey. Kennedy (1991) revealed predation on four other raptor species: Cooper's Hawk, American Kestrel (*Falco sparverius*), Prairie Falcon (*F. mexicanus*), and Northern Pygmy-Owl (*Glaucidium gnoma*). Also notable in the diet of Northern Goshawks, based on size, were Common Ravens (*Corvus corax*) and one cat (*Felis* sp.).

Other diet information from New Mexico is qualitative, if not anecdotal, but provides some interesting information on the breadth of the diet of this raptor. According to Ligon (1961), Northern Goshawks in northern New Mexico prey on Blue Grouse (*Dendragapus obscurus*). Additionally, he mentions two Northern Goshawks catching a large domestic chicken at a private residence in Walker Canyon in the Sacramento Mountains on 15 August 1951. Predation on domestic chickens by Northern Goshawks was also recorded along upper Eagle Creek near Alto in Lincoln County (Ligon 1961). Jim Ramakka (pers. comm.) observed a Northern Goshawk kill a cottontail outside the front window of his house near Aztec, San Juan County, in the winter of 2005–6. The goshawk dragged the partially eaten rabbit under a sagebrush (*Artemisia* sp.) and came back to feed on it again the next day.

Northern Goshawks hunt from a perch but frequently change perches in the absence of prey detection, a foraging strategy known as "saltatory" (Kennedy

TABLE 11.1. Documented prey of Northern Goshawks (*Accipiter gentilis*) in the Jemez Mountains and on the Pajarito Plateau in New Mexico. All data from Kennedy (1991).

CLASS: ORDER	SPECIES
Aves	Cooper's Hawk (*Accipiter cooperii*)
	American Kestrel (*Falco sparverius*)
	Prairie Falcon (*Falco mexicanus*)
	Rock Dove (*Columba livia*)
	Northern Pygmy-Owl (*Glaucidium gnoma*)
	Belted Kingfisher (*Ceryle alcyon*)
	Northern Flicker (*Colaptes auratus*)
	Red-naped Sapsucker (*Sphyrapicus nuchalis*)
	Williamson's Sapsucker (*Sphyrapicus thyroideus*)
	Hairy Woodpecker (*Picoides villosus*)
	Steller's Jay (*Cyanocitta stelleri*)
	Western Scrub-Jay (*Aphelocoma californica*)
	Common Raven (*Corvus corax*)
	Clark's Nutcracker (*Nucifraga columbiana*)
	Western Bluebird (*Sialia mexicanus*)
	Mountain Bluebird (*Sialia currucoides*)
	Hermit Thrush (*Catharus guttatus*)
	American Robin (*Turdus migratorius*)
	European Starling (*Sturnus vulgaris*)
	Spotted Towhee (*Pipilo maculatofasciatus*)
	Evening Grosbeak (*Coccothraustes vespertinus*)
Mammalia	Cottontail (*Sylvilagus* sp.)
	Abert's squirrel (*Sciurus aberti*)
	Red squirrel (*Tamiasciurus hudsonicus*)
	Chipmunk (*Tamias* sp.)
	Golden-mantled ground squirrel (*Spermophilus lateralis*)
	Rock squirrel (*Spermophilus variegatus*)
	Unidentified microtine rodent
	Cat (*Felis* sp.)

1991; see also chapter 10). In the Jemez Mountains, the median time spent at a perch by hunting male and female Northern Goshawks was 3.5 minutes. Between perches the mean flight bout duration was 2.3 minutes for males but only 1.1 minutes for females, and 75% of all flights between perches were less than 20 seconds in females (Kennedy 1991). Similar to Cooper's Hawks, Northern Goshawks will also try to flush ground-dwelling prey such as quail and cottontails from shrubs and brush piles by entering the vegetation from the ground and then capturing them as they exit their refugia (PLK, unpubl. data).

Predation and Interspecific Interactions

The extent to which interspecific competition for habitat as well as prey from potential competitors, such as the Red-tailed Hawk (*Buteo jamaicensis*) and the Great Horned Owl (*Bubo virginianus*), affects Northern Goshawk habitat use is not well understood. In addition, these potential competitors also function as potential predators, making the effect of their presence on Northern Goshawks difficult to interpret. Northern Goshawks may be excluded from nest areas by other raptors, although it is not uncommon for the species to nest close to other raptors. As noted earlier, Cooper's Hawks and Northern Goshawks show a preference for similar nesting habitat and they have been recorded nesting in the same stands in New Mexico (PLK, unpubl. data; chapter 10). But because Cooper's Hawks are smaller than Northern Goshawks and begin nesting later, they are unlikely to be effective nest site competitors (Squires and Kennedy 2006).

In addition to nest site competition, several species of hawks and owls, and numerous mammalian predators, can potentially compete with goshawks for prey. The Red-tailed Hawk and the Great Horned Owl prey on many of the same species as Northern Goshawks, although neither has the same degree of dietary overlap with this species as does the Cooper's Hawk, which also forages in the same habitat and takes many of the same prey species (Squires and Kennedy 2006; chapter 10). Because both the Red-tailed Hawk and the Great Horned Owl are more abundant in open habitats such as meadows, edge, forest openings, and woodlands, the extent to which they coexist and compete for food with goshawks probably varies as a function of the openness of forest types and extent of natural and anthropogenic fragmentation of forests (Squires and Kennedy 2006).

Although Northern Goshawks are formidable predators, they are occasionally killed by other predators. The literature describing predation on Northern Goshawks mostly consists of anecdotal observations, with little information regarding population responses. For example, we know that Great Horned Owls kill adults and nestlings throughout the species' range (Squires and Kennedy 2006). Shuster (1977) reported likely predation by a Great Horned Owl in the Sacramento Mountains, having found the remains of a Northern Goshawk on the ground under a tree amidst whitewash and large owl pellets. Other avian predators in North America include Golden Eagles (*Aquila chrysaetos*) and Bald Eagles (*Haliaeetus leucocephalus*), both of which have been documented to kill Northern Goshawks on their wintering areas (Squires and Ruggerio 1995). Mammalian predators include pine martens (*Martes americana*), fishers (*Martes pennanti*), wolverine (*Gulo gulo*), and raccoons (*Procyon lotor*) (Squires and Kennedy 2006).

One-half of nestling mortalities ($n = 12$) in New Mexico were attributed to predation (Ward and Kennedy 1996). Several studies, including one in New Mexico, have indicated that predation on goshawk nestlings may increase during periods of low goshawk food availability because female goshawks may be required to spend more time away from the nest foraging instead of protecting young (Zachel 1985; Rohner and Doyle 1992; Ward and Kennedy 1996; Dewey and Kennedy 2001).

Status and Management

During the 1990s, the Northern Goshawk was at the center of a bitter controversy that generated extensive litigation, including in the Southwest (see Squires and Kennedy 2006 for a detailed review). In the early 1990s, environmental groups petitioned the U.S. Fish and Wildlife Service to list the Northern Goshawk under the Endangered Species Act while also threatening the U.S. Forest Service Region 3—including all national forests in New Mexico—with legal action unless all timber harvesting was suspended in Northern Goshawk territories. At issue was the perception that (1) Northern Goshawk populations were declining; (2) the species was highly dependent on old-growth forest; and (3) timber harvesting practices in old-growth forest were not sustainable, negatively affecting the goshawk and contributing to its decline. A recent extensive review of the Northern Goshawk literature by Squires and Kennedy (2006) addressed these perceptions.

Are Northern Goshawk populations declining? Squires and Kennedy (2006) concluded that even with the wealth of information collected on the Northern Goshawk in the past 20+ years, there is still no evidence that North American goshawk populations are declining. However, because of the nature of the scientific information they also could not rule out the possibility that declines have occurred but that they have not been measured.

Are Northern Goshawks old-growth specialists? Squires and Kennedy (2006) concluded that Northern Goshawks have a strong preference for mature and old-growth forests, but this preference is dependent on nest density, spatial scale (the structure and composition of the landscape, both locally and regionally), and season. For example, preference for mature and old-growth forest seems strongest within a 100-ha (~250-ac) area surrounding the nest stand. As nest density increases, low-quality habitats are more likely to be occupied, and thus nesting habitat diversity used by a population may increase. As spatial scale increases from the nest site to the landscape in which home ranges are embedded, habitat diversity increases; thus Northern Goshawks are more of a habitat generalist at these larger spatial scales than at the scale of the nest site. Finally, the limited data on nonbreeding habitat use patterns suggests Northern Goshawks are more of a habitat generalist during the nonbreeding season than during the breeding season.

Does timber harvesting negatively impact Northern Goshawks? Squires and Kennedy (2006) concluded that forest management (i.e., cutting, thinning, controlled burning) is the primary human source of potential impacts on Northern Goshawk populations. Forest management may improve or degrade habitat depending on implementation, especially as management practices affect the density of large trees and canopy closure. Forest management practices that reduce the size of nest stands may decrease territory occupancy rates. However, as Squires and Kennedy (2006) note, few studies have directly assessed the impacts of timber management on Northern Goshawk populations, although limited data suggest Northern Goshawks can tolerate some timber harvest near nest stands. The effects of forest management on prey populations and potential predators and competitors vary by species, and specific effects are poorly documented.

Acknowledgments

The authors wish to thank Richard Reynolds for his very insightful comments on a draft of the chapter, as well as Colette Coiner for further comments on the chapter and her help checking the literature cited section.

[AOU] American Ornithologists' Union. 1983. *Check-list of North American birds.* 6th ed. Washington, DC: American Ornithologists' Union.

Bayard de Volo, S. 2008. Genetic studies of Northern Goshawks (*Accipiter gentilis*): genetic tagging and individual identification from feathers, and determining phylogeography, gene flow and population history for Goshawks in North America. Ph.D. thesis. Colorado State University.

Beebe, F. L. 1974. Goshawk. In *Field studies of the Falconiformes of British Columbia: vultures, eagles, hawks, and falcons.* British Columbia Provincial Museum Occasional Papers, series no. 17:54–63.

Boal, C. W., D. E. Andersen, and P. L. Kennedy. 2003. Home range and residency status of northern goshawks breeding in Minnesota. *Condor* 105:811–16.

Cramp, S., and K. E. L. Simmons, eds. 1980. *Handbook of the birds of Europe, the Middle East, and North Africa: the birds of the Western Palearctic.* Vol. 2, *Hawks to Bustards.* Oxford, UK: Oxford University Press.

Crocker-Bedford, D. C., and B. Chaney. 1988. Characteristics of Goshawk nesting stands. In *Proceedings of the southwest raptor management symposium and workshop, 21–24 May 1986, University of Arizona, Tucson*, ed. R. L. Glinski, B. G. Pendleton, M. B. Moss, M. N. LeFranc Jr., B. A. Millsap, and S. W. Hoffman, 210–17. Scientific and Technical Series no. 11. Washington, DC: National Wildlife Federation.

del Hoyo, J., A. Elliott, and J. Sargatal, eds. 1994. *Handbook of the birds of the world.* Vol. 2. Barcelona, Spain: Lynx Edicions.

Dewey, S. R., and P. L. Kennedy. 2001. Effects of supplemental food on parental care strategies and juvenile survival of Northern Goshawks. *Auk* 118:353–65.

Dewey, S. R., P. L. Kennedy, and R. M. Stephens. 2003. Are dawn vocalization surveys effective for monitoring goshawk nest-area occupancy? *Journal of Wildlife Management* 67:390–97.

Doyle, F. I., and J. M. N. Smith. 1994. Population responses of Northern Goshawks to the 10-year cycle in numbers of snowshoe hares. *Studies in Avian Biology* 16:122–29.

Drennan, J. E., and P. Beier. 2003. Forest structure and prey abundance in winter habitat of Northern Goshawks. *Journal of Wildlife Management* 67:177–85.

Estes, W. A., S. R. Dewey, and P. L. Kennedy. 1999. Siblicide at Northern Goshawk nests: does food play a role? *Wilson Bulletin* 111:432–36.

Ferguson-Lees, J., and D. A. Christie. 2001. *Raptors of the world.* New York: Houghton Mifflin.

Friedmann, H. 1950. The birds of North and Middle America, part XI. *United States National Museum Bulletin* 50:150–63.

Hargis, C. D., C. McCarthy, and R. D. Perloff. 1994. Home ranges and habitats of Northern Goshawks in eastern California. *Studies in Avian Biology* 16:66–74.

Hayward, G. D., and R. E. Escaño. 1989. Goshawk nest-site characteristics in western Montana and northern Idaho. *Condor* 91:476–79.

Hoffman, S. W., J. P. Smith, and T. D. Meehan. 2002. Breeding grounds, winter ranges, and migratory routes of raptors in the Mountain West. *Journal of Raptor Research* 36:97–110.

Hubbard, J. P. 1978. *Revised check-list of the birds of New Mexico.* Publ. no. 6. Albuquerque: New Mexico Ornithological Society.

———. 1992. A taxonomic assessment of the Northern Goshawk in southwestern North America. Unpublished report. Santa Fe: New Mexico Department of Game and Fish.

Johnsgard, P. A. 1990. Goshawk. In *Hawks, eagles, and falcons of North America: biology and natural history.* 176–82. Washington, DC: Smithsonian Institution Press.

Johnson, R. R., and B. K. Harris. 1967. Unusual nesting of a Goshawk in southern New Mexico. *Condor* 69:209–10.

Kennedy, P. L. 1988. Habitat characteristics of Cooper's Hawks and Northern Goshawks nesting in New Mexico. In *Proceedings of the Southwest raptor management symposium and workshop, 21–24 May 1986, University of Arizona, Tucson*, ed. R. L. Glinski, B. G. Pendleton, M. B. Moss, M. N. LeFranc Jr., B. A. Millsap, and S. W. Hoffman, 218–27. Scientific and Technical Series no. 11. Washington, DC: National Wildlife Federation.

———. 1991. Reproductive strategies of Northern Goshawks and Cooper's Hawks in north-central New Mexico. Ph.D. diss. Utah State University, Logan.

Kennedy, P. L., and D. W. Stahlecker. 1993. Responsiveness of northern goshawks to taped broadcasts of 3 conspecific calls. *Journal of Wildlife Management* 57:249–57.

Kennedy, P. L., and J. M. Ward. 2003. Effects of experimental food supplementation on movements of juvenile Northern Goshawks (*Accipiter gentilis atricapillus*). *Oecologia* 134:284–91.

Kenward, R. 2006. *The Goshawk.* London: T & A. D. Poyser.

Lee, J. A. 1981. Comparative breeding behavior of the Goshawk and Cooper's Hawk. Thesis. Brigham Young University, Provo, Utah.

Ligon, J. S. 1961. *New Mexico birds and where to find them.* Albuquerque: University of New Mexico Press.

Mueller, H. C., and D. D. Burger. 1968. Sex ratios and measurements of migrant Goshawks. *Auk* 85:431–36.

Mueller, H. C., D. D. Berger, and G. Allez. 1977. The periodic invasions of goshawks. *Auk* 94:652–63.

[NMOS] New Mexico Ornithological Society. 2008. *NMOS Field Notes* database. http://nhnm.unm.edu/partners/NMOS/ (accessed 5 December 2008).

Palmer, R. S., ed. 1988. *Handbook of North American birds*. Vol. 4, part 1, *Diurnal raptors*. New Haven: Yale University Press.

Penteriani, V. 2001. The annual and diel cycles of Goshawk vocalizations at nest sites. *Journal of Raptor Research* 35:24–30.

Reynolds, R. T., E. C. Meslow, and H. M. Wight. 1982. Nesting habitat of coexisting Accipiters in Oregon. *Journal of Wildlife Management* 46:124–38.

Reynolds, R. T., S. M. Joy, and D. G. Leslie. 1994. Nest productivity, fidelity, and spacing of Northern Goshawks in northern Arizona. *Studies in Avian Biology* 16:106–13.

Rohner, C., and F. I. Doyle. 1992. Food stressed Great Horned Owl kills adult Goshawk: exceptional observation or community process? *Journal of Raptor Research* 26:261–63.

Schnell, J. H. 1958. Nesting behavior and food habits of Goshawks in the Sierra Nevada of California. *Condor* 60:377–403.

Shuster, W. C. 1977. Northern Goshawk nesting in southern New Mexico. *Western Birds* 8:29–32.

Siders, M. S., and P. L. Kennedy. 1994. Nesting habitat of *Accipiter* hawks: is body size a consistent predictor of nest habitat characteristics? *Studies in Avian Biology* 16:92–96.

———. 1996. Forest structural characteristics of accipiter nesting habitat: is there an allometric relationship? *Condor* 98:123–32.

Snyder, N.F.R., and H. A. Snyder. 1991. *Birds of prey: natural history and conservation of North American raptors*. 1st ed. Stillwater, MN: Voyageur Press.

Sonsthagen, S. A. 2002. Year-round habitat, movement, and gene flow of Northern Goshawks breeding in Utah. Thesis. Brigham Young University, Provo, Utah.

Speiser, R., and T. Bosakowski. 1987. Nest site selection by Northern Goshawks in northern New Jersey and southeast New York. *Condor* 89: 387–394.

Squires, J., and P. L. Kennedy. 2006. Northern Goshawk ecology: an assessment of current knowledge and information needs for conservation and management. *Studies in Avian Biology* 31:8–62.

Squires, J. R., and R. T. Reynolds. 1997. Northern Goshawk (*Accipiter gentilis*). No. 298. In *The birds of North America*, ed. A. Poole and F. Gill. Philadelphia, PA: Academy of Natural Sciences.

Squires, J. R., and L. F. Ruggiero. 1995. Winter movements of adult Northern Goshawks that nested in south central Wyoming. *Journal of Raptor Research* 29:5–9.

———. 1996. Nest site preference of Northern Goshawks in south central Wyoming. *Journal of Wildlife Management* 60:170–77.

Stephens, R. M. 2001. Migration, habitat use, and diet of Northern Goshawks that winter in the Uinta Mountains, Utah. Thesis. University of Wyoming, Laramie.

Sutton, G. M. 1931. The status of the Goshawk in Pennsylvania. *Wilson Bulletin* 43:108–13.

Tatschl, J. L. 1967. Breeding birds of the Sandia Mountains and their ecological distributions. *Condor* 69:479–90.

U.S. Fish and Wildlife Service. 1998. Status review of the Northern Goshawk in the forested West. Portland, OR: Office of Technical Support, Forest Resources. http://pacific.fws.gov/news/pdf/gh_sr.pdf (accessed 24 March 2009).

van Rossem, A. J. 1938. A Mexican race of the Goshawk (*Accipiter gentilis*; Linnaeus). *Proceedings of the Biological Society of Washington* 51:99–100.

Ward, J. M., and P. L. Kennedy. 1994. Approaches to investigating food limitation hypotheses in raptor populations: an example using the Northern Goshawk. *Studies in Avian Biology* 16:114–18.

———. 1996. Effects of supplemental food on growth and survival of juvenile Northern Goshawks. *Auk* 113:200–208.

Ward, J. M., B. Booth, and P. L. Kennedy. 1997. A motorized food box for use in supplemental feeding experiments. *Journal of Field Ornithology* 68:69–74.

Whaley, A. S., and C. M. White. 1994. Trends in geographic variation of Cooper's Hawk and Northern Goshawk: a multivariate analysis. *Western Foundation of Vertebrate Zoology* 5:161–209.

Wheeler, B. K., and W. S. Clark. 1995. *A photographic guide to North American raptors*. London: Academic Press.

Zachel, C. R. 1985. Food habits, hunting activity, and post-fledging behavior of Northern Goshawks (*Accipiter gentilis*) in interior Alaska. Thesis. University of Alaska, Fairbanks.

Zirrer, F. 1947. The Goshawk. *Passenger Pigeon* 9:79–94.

© ROBERT SHANTZ.

12

Common Black-Hawk

(*Buteogallus anthracinus*)

GIANCARLO SADOTI

AN UNCOMMON DENIZEN of some of the more idyllic streamside forests of New Mexico, the Common Black-Hawk (*Buteogallus anthracinus*; hereafter black hawk) is a neotropical raptor with many distinctive traits, whether morphological, behavioral, or dietary. It is a large hawk with a body length of 51–56 cm (20–22 in), a wingspan of 102–127 cm (40–50 in), and a weight of 635–1,315 g (1.4–2.9 lbs) (Wheeler and Clark 1995). The genus name, *Buteogallus*, translates to "chicken-hawk" and likely refers to this group's long-legged appearance. The black hawk's adult plumage is black, as indicated by the species name, *anthracinus*, with also a slate gray tinge, fading to brown. The wings are notably broader than in most buteonine hawks and the undersides appear gray or mottled brown (Schnell 1994). The most distinguishing field mark of this species is the conspicuous white median band and a less conspicuous white terminal band on the tail feathers. Other noteworthy features include the yellow to yellow-orange skin on the cere, gape, and legs. Juvenile birds (postfledging to one to two years) have a streaked white and brown plumage with five to seven dark bands alternating with light bands on the tail feathers. In the second to third year, the adult black plumage replaces the streaked juvenal plumage, with the tail feathers usually molting last and replaced by the distinct two-band pattern.

PHOTO 12.1

Common Black-Hawk soaring near Alamos in Sonora, Mexico, March 2007. Best field marks for the adult include the overall coal black plumage, the very broad wings, and the median white tail band. The tail also shows a white terminal band, but it is less conspicuous.

PHOTOGRAPH: © MARK L. WATSON.

PHOTO 12.2

Immature Common Black-Hawk, Doña Ana Co., 27 April 2007. Note the brown and white streaked plumage and the banded tail. PHOTOGRAPH: © JAMES E. ZABRISKIE.

The vocalization most frequently heard is the "alarm" call, which is divided into 8–14 short, sharp notes that build in pitch and volume, peak at the fourth to fifth note, and then diminish in volume, pitch, and duration while accelerating in rate. While sounding whistled due to high-frequency overtones (Schnell 1994), the lowest tone (approximately 3 kHz) is near that of a Red-tailed Hawk's (*Buteo jamaicensis*) common call (Preston and Beane 1993). When birds are agitated, the alarm call may be repeated several times and sometimes without clear breaks. In the vicinity of an active nest, this call is often heard long before the bird is seen.

The taxonomic status of the black hawk in parts of its range is currently in flux. Wiley and Garrido (2005) recently concluded that the Cuban Black-Hawk (previously *B. a. gundlachii*) is a full species (*B. gundlachii*). The status of the Utila black hawk (*B. utilensis*) is still unclear. The black hawk occurring in coastal areas from southern Mexico to northern Peru is now considered the Mangrove Black-Hawk (*B. subtilis*) (Schnell 1994). In contrast to coastal and Caribbean areas, the taxonomic status of the migratory population of black hawks in New Mexico has been firmly established for some time. Along with all other inland populations of this species, the black hawk in New Mexico is treated as the subspecies *B. a. anthracinus*.

Distribution

Broad-scale

The accepted continental form (*B. anthracinus*) is found from northern South America north to the southwestern United States. The species is largely nonmigratory,

PHOTO 12.3

Common Black-Hawk along the Rio Mayo near Alamosa, Sonora, Mexico, March 2006. PHOTOGRAPH: © MARK L. WATSON.

but toward the northern end of its range—from Mexico's Sierra Madre Occidental east to west Texas and north through New Mexico and Arizona to extreme southwest Utah—it is a migratory summer resident. Throughout this largely arid region, the black hawk has a patchy distribution closely linked to the presence of wooded riparian areas along perennial watercourses. Where migratory breeders spend their winters is almost entirely unknown, with only one band recovered in western Durango, Mexico, of a first-year bird from southeastern Arizona. Information on migratory routes is similarly limited (Schnell 1994).

New Mexico

CORE AREAS

New Mexico's breeding black hawk population is centered on four drainages (in descending order of abundance): the upper Gila and San Francisco watershed, the Mimbres River, the Rio Hondo, and several tributaries of the Middle Rio Grande. Additional peripheral breeding areas are discussed below.

The upper Gila River (Catron, Grant, and Hidalgo counties; from the Gila Wilderness to the Arizona border) and San Francisco River (Catron County; from the vicinity of Reserve to the Arizona border), along

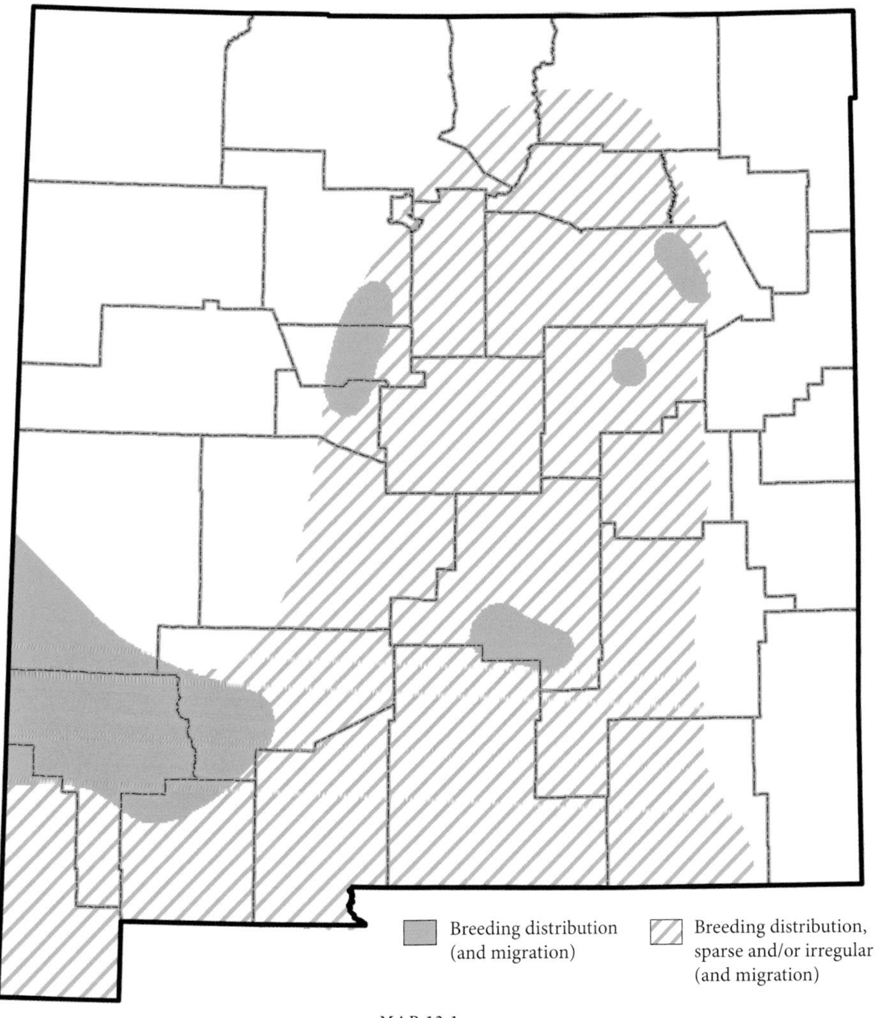

MAP 12.1

Common Black-Hawk distribution map

PHOTO 12.4

(left) Common Black-Hawk at Bear Canyon Lake in the Mimbres Valley, Grant Co., 30 July 2007.
PHOTOGRAPH: © JERRY OLDENETTEL.

PHOTO 12.5

(above) Common Black-Hawk over water, State Road 92 bridge near Virden, Hidalgo Co., 3 April 2005.
PHOTOGRAPH: © ROBERT SHANTZ.

with tributaries, are characterized by alternating steep-walled canyons and broad valleys. Where breeding habitat is available, black hawks are distributed along most of the Gila River–San Francisco watershed, from narrow montane canyons at the headwaters of the two rivers, agricultural floodplain valleys along middle reaches, and arid canyons as the two rivers exit the state before merging in Arizona.

The Mimbres River (Grant County) drains the eastern Gila Wilderness and the western Black Range before disappearing into the Chihuahuan Desert near Deming. Nesting has been documented from San Juan to north of Mimbres along this drainage, with possible nesting into northern Luna County (Williams 2004). A 1994–95 inventory of black hawk breeding territories in southwestern New Mexico (Skaggs 1996) estimated 65–80 breeding territories in the Gila, San Francisco, and Mimbres watersheds combined.

The Rio Hondo watershed (Lincoln County), where black hawks were first documented breeding in 1996 (Williams 1996), drains the Sierra Blanca and the

PHOTO 12.6

Rio Hondo Valley in Lincoln County. A small breeding population of Common Black-Hawks is known from along the Rio Bonito, Rio Ruidoso, and Rio Hondo in the Rio Hondo Valley. Land use consists primarily of farming (pastures) and, increasingly, residential development. Common Black-Hawk territory density is low compared to other areas with known breeding populations.
PHOTOGRAPH: © RONALD J. TROY.

Capitan Mountains through two major tributaries, the Rio Bonito and the Rio Ruidoso. Black hawks are found in agricultural floodplain valleys between Hondo and Glencoe (to the west), and Capitan (to the northwest). An estimated 8–10 pairs were breeding in this area in 2002 and 2003 (Troy and Stahlecker 2008).

The final area of regular breeding is centered on several Rio Grande tributaries in Sierra County draining the eastern slope of the Black Range (Williams 1994b). Located north of the towns of Kingston and Hillsboro, these drainages are characterized by narrow wooded canyons. Black hawks have also been regularly observed during the spring and summer months in the vicinity of Percha Dam (e.g., Hubbard 1988; Williams 2002), though breeding has not been confirmed.

PERIPHERAL BREEDING DISTRIBUTION

Though breeding is restricted to a limited range of riparian habitats, records of confirmed and probable breeding outside traditional breeding areas over the last 40 years illustrate the opportunistic nesting habits of the black hawk. Areas of successful nesting have included several noncontiguous sections of the Middle Rio Grande from northern Valencia County (Williams 2003) through Bernalillo County (where nesting was first documented at Alameda in 1971–72 [Hundertmark 1974], a first north of the Gila River watershed, and later in Albuquerque in 1989–90 [Williams and Hubbard 1990a]). Other peripheral locations of known nesting include at least two locales on the Canadian River in San Miguel County (Williams 2003, 2004), with also probable nesting on Santa Rosa Lake in Guadalupe County (Williams 2004). Individual black hawks have been observed in the breeding season near apparently suitable breeding habitat on the Upper Pecos River (San Miguel County; Hubbard 1985; Williams 2004) and in several locales along the Rio Grande in Socorro County (Hubbard 1987; Williams 2004), including a likely fledgling at Bosque del Apache National Wildlife Refuge (Williams and Hubbard 1990b).

WINTER RECORDS

Black hawks, though neotropical migrants, are occasionally observed in New Mexico during winter, but whether those birds are overwintering breeders is unclear. Single birds were observed near breeding areas in the Cliff–Gila Valley (Grant County) in December 1994 (Williams 1993) and on the San Francisco River in January 1991 (Williams and Hubbard 1991) and February 1992 (Williams 1992), while another was seen near Vado, Doña Ana County, in December 1993 (Williams 1994a). Even more unusual for New Mexico were two adults observed together by Doug Burkett (*in litt.*) and others at White Sands Missile Range, Doña Ana County, on 31 January 2009.

Habitat Associations

Black hawks are closely associated with mature riparian woodlands along perennial streams where they nest in sturdy forks of mature trees. Southwestern New Mexico nest locations documented by Skaggs (1996) ranged in elevation between 1,150 and 2,000 m (~3,770 to 6,560 ft). At lower elevations mature riparian corridors, often called gallery forests, are typically dominated by cottonwoods (*Populus* spp.), willows (*Salix* spp.), Arizona sycamore (*Platanus wrightii*),

PHOTO 12.7

Common Black-Hawk in typical nesting habitat consisting of mature riparian woodland. Lower box of the Gila River, Hidalgo Co., late May 2007.
PHOTOGRAPH: © JOSHUA NEMETH.

PHOTO 12.8

(*top left*) Common Black-Hawk habitat with high density of nesting pairs: Gila River Valley near the southern end of the Gila Bird Area. This roughly 50-ha (124-ac) patch of riparian woodland was known by the author to have two Common Black-Hawk nesting pairs. PHOTOGRAPH: © GIANCARLO SADOTI.

PHOTO 12.9

(*bottom left*) Common Black-Hawk habitat with lower density of nesting pairs: Gila River Valley south of Riverside and north of the Gila Bird Area. Note that along this stretch of the river cottonwoods (*Populus fremontii*) form scattered groves rather than a continuous forest. A Black-Hawk pair nested in the cottonwood cluster in the lower right corner in 2000–2001. Also in 2000–2001 another Common Black-Hawk pair nested in the far left side about ⅓ from the top of the photo (the farthest discernible riparian stand). PHOTOGRAPH: © GIANCARLO SADOTI.

PHOTO 12.10

(*above*) Rio Hondo Common Black-Hawk nesting habitat consists of streamside riparian vegetation with cottonwoods (*Populus* sp.), exotic poplars (*P. alba*), black walnuts (*Juglans rupestris*), boxelders (*Acer negundo*), and the exotic Siberian elm (*Ulmus pumila*). PHOTOGRAPH: © RONALD J. TROY.

boxelder maple (*Acer negundo*), and other deciduous trees. Higher-elevation riparian forests, representing a smaller proportion of the nesting habitat, are typically less deciduous-dominated but include Arizona sycamore, alder (*Alnus* spp.), narrowleaf cottonwood (*Populus angustifolia*), and boxelder, together with ponderosa pine (*Pinus ponderosa*) and/or Douglas fir (*Pseudotsuga menziesii*).

Migratory habitat and routes are poorly understood throughout the northern range of the black hawk, and almost completely unknown in New Mexico. Most of the birds observed during periods of likely migration are found in areas of current or past breeding, suggesting that most, if not all, of these individuals were early arrivals or late departures. Migration routes are believed to generally follow riparian corridors, though migrants have been observed over forested mountain ridgelines in southeastern and central Arizona (Snyder and Snyder 1991; GS), with also a single migrating bird observed in spring 2005 at the Sandia Mountains site operated by HawkWatch International (Williams 2005), the first in 21 years of monitoring. The geography of drainages in Arizona, New Mexico, and Texas dictates that migrating birds would have to cross considerable distances containing little to no riparian vegetation.

Life History

Nesting

Black hawks typically arrive on New Mexico nesting grounds sometime during the second to fourth week of March. Repeat territory occupation by individuals or pairs has been verified for one site in Arizona (Schnell 1998), and likely occurs also in New Mexico (GS; R. A. Fisher, unpubl. data). Courtship and territory establishment begin shortly after arrival on nesting grounds and feature extensive aerial displays. In Arizona, nest selection appears to begin as early as five days following arrival of the birds, with nest construction or rebuilding occurring through early April (Schnell 1998). Several nests are typically maintained within each territory, with different nests occupied in

PHOTO 12.11

(*top left*) Nest in a sycamore (*Platanus wrightii*), Gila River Valley, Grant Co., 15 July 2001. Although Common Black-Hawk nests in the Gila River Valley are typically in cottonwood (*Populus fremontii*) trees, two were in sycamores in 2000. The nest in the photo was successful in 2000. It was used again early during the 2001 breeding season before being abandoned for another, nearby nest. PHOTOGRAPH: © GIANCARLO SADOTI.

PHOTO 12.12

(*bottom left*) Common Black-Hawk nest in a cottonwood (*Populus fremontii*), Gila River Valley, Grant Co., 30 July 2001. The attending pair was successful in 2000 and 2001. PHOTOGRAPH: © GIANCARLO SADOTI.

PHOTO 12.13

(*above*) In the Rio Hondo Valley all occupied nests documented by Ron Troy and Dale Stahlecker during their 2002–3 study were in large cottonwood (*Populus* sp.) trees. PHOTOGRAPH: © RONALD J. TROY.

PHOTO 12.14

(*top*) Nest failure, Gila River Valley, Grant Co., early August 2001. The nest in the photo was used in 2000 and 2001, and the attending pair was successful in 2000 but not in 2001. The photograph was taken on the day after a particularly windy afternoon rainstorm. A large dead branch was lying on the nest. It presumably killed an adult (whose tail feathers are visible) and two nestlings when it fell. The nestlings, visible the week before, were two to three weeks from fledging age. No activity was observed on the day the photo was taken, nor on a subsequent visit two to three days later. PHOTOGRAPH: © GIANCARLO SADOTI.

PHOTO 12.15

(*bottom*) Juvenile Common Black-Hawk in the nest, Gila River, Grant Co., 25 July 2003. PHOTOGRAPH: © LUKE COLE.

alternate years. In Arizona, eggs are typically laid in mid April, with hatching near the end of May. Young are provisioned with food for six to seven weeks while still in the nest, and another six to eight weeks after leaving the nest. Most birds leave their nesting areas by mid October, migrating to unknown winter locations (Schnell 1994).

Schnell (1998) suspects that nest locations are selected in the tallest tree near preferred foraging areas, though birds at territories in the Cliff–Gila Valley—whether returning or new occupants—remained faithful to or reused established nests despite substantial changes to terrestrial and aquatic habitats following shifts of the active river channel (GS). Results of comparisons between vegetation structure at nest trees and non-nest trees within territories in the Cliff–Gila Valley suggest black hawks select nest trees with smaller trunk diameters and broader canopies and surrounded by sparser and shorter understory trees than non-nest trees within territories (Sadoti 2008).

Diet and Foraging

Black hawks are typically "sit-and-wait" hunters, perching on a low branch or rock before swooping or dropping to snatch prey with their talons. Birds occasionally search actively for prey, either by hopping from perches or by wading in shallow waters (Schnell 1994; GS). Black hawks are easily flushed while hunting, typically flying out of sight of observers (Snyder and Snyder 1991; GS).

Though only anecdotal observations of black hawks with prey have been made in New Mexico, diets of birds elsewhere in the southwestern United States (usually ascertained by observing deliveries to nestlings) indicate that a broad range of species are taken. While the overall diet of black hawks is weighted toward aquatic vertebrates and macroinvertebrates, prey appear to be opportunistically selected (Schnell 1994). The majority of food biomass delivered to two nests in west-central Arizona was fish (Millsap 1981), with fish and amphibians making up 34% and 27%, respectively, of prey items delivered to nests in Aravaipa Canyon, also in Arizona (Schnell 1998). Birds returning to nests with prey in the Cliff–Gila Valley were observed carrying garter snake (*Thamnophus cyrtopsis*), bullfrog (*Rana catesbiena*), crayfish (nonnative *Orconectes* or *Procambarus* spp.),

PHOTO 12.16

Common Black-Hawk with crayfish along State Road 92 near the Arizona state line, Hidalgo Co., 31 May 2006.

PHOTOGRAPH: © ROBERT SHANTZ.

and one fledgling passerine. Crayfish exoskeletons were occasionally encountered on rocks or logs adjacent to watercourses in black hawk nesting territories in the Cliff–Gila Valley, suggesting these remains were from birds foraging on these crustaceans. Additionally, castings found on large branches below two nests in the Cliff–Gila Valley, though only visually inspected (GS, unpubl. data), appeared to have proportionally high biomass in crayfish exoskeletons. Little information is available on potential impacts of prey status on black hawks. However, Millsap (1981) found no nesting black hawks in areas where fish were abundant but amphibian and reptile prey were scarce, while in contrast Schnell (1994) observed a positive relationship between annual fledging rates and fish abundance in Aravaipa Canyon, Arizona.

An anecdotal observation of an adult black hawk following a riding mower in Glenwood, Catron County, in 1996 (Snider 1996) suggests that this bird was using a foraging strategy commonly seen in the closely related Savannah Hawk (*Buteogallus meridionalis*) in the llanos of Venezuela (S. Stoleson, pers. comm.), and also reminiscent of American Kestrels (*Falco sparverius*) and Red-tailed Hawks among tilling tractors in Michigan (Caldwell 1986). Additionally, black hawks in El Salvador are reported to follow grass fires, probably searching for grasshoppers (Dickey and van Rossem 1938). An interesting observation was made by Boal and Mannan (1996) of an abandoned black hawk nestling raised in captivity who, when presented with a large pan of water, perched on the rim and appeared to search for food. This behavior suggests some innate associating of prey with water.

Predation and Interspecific Interactions

From observations in Arizona (Schnell 1994), black hawks are generally intolerant of other large birds in the vicinity of nests, including Turkey Vulture (*Cathartes aura*), Red-tailed Hawk, and Common Raven (*Corvus corax*). Very few incidents of interspecific contact are documented, though a Peregrine Falcon (*Falco peregrinus*) was observed striking a black hawk in midair in Arizona (T. Myers, pers. comm., in Boal and Mannan 1996). Few inter-raptorial observations have been documented in New Mexico, except for the harassment of an immature black hawk by two Mississippi Kites (*Ictinia mississippiensis*) in Escondida (Socorro County) in 1986 (Hubbard 1987) and an apparent defense of a nesting area by an adult Swainson's Hawk (*Buteo swainsoni*) against a second-year black hawk in the Cliff–Gila Valley in 2000 (GS). Undoubtedly most interactions go unobserved, and it is likely that the presence of black hawks in the immediate vicinity of other raptor nests, and vice versa, often provokes aerial skirmishes. Few encounters have been observed in the Cliff–Gila Valley, though displacement of a black hawk pair by Red-tailed Hawks during initial nest establishment was reported by R. Shook (pers. comm.). The displaced pair relocated 108 m (354 ft) away in the same contiguous patch of forest and both nests fledged successfully. Elsewhere in the

Cliff–Gila Valley, a pair of Swainson's Hawks nested 402 m (1,319 ft) from a pair of black hawks, occupying a nest used by black hawks during the previous breeding season (GS). In one case of particularly close interspecific neighboring in this same area, a black hawk was observed nesting 39 m (128 ft) from the nest of a Zone-tailed Hawk (*Buteo albonotatus*). One possible explanation for the close proximity of these nests and others observed in the area may be the partitioning of prey among raptor species, with the black hawk being the only bird to favor aquatic prey.

Very few observations of predation upon adult or nestling individuals have been reported in the United States. An adult Golden Eagle (*Aquila chrysaetos*) is reported to have captured and carried away an adult black hawk from a nest area, presumably in Arizona (S. Morgan, pers. comm., in Schnell 1994). Occupation by Great Horned Owls (*Bubo virginianus*) of nests previously occupied by black hawks has been documented in Texas (Webster 1976) and New Mexico (Hundertmark 1974), though the presence of owls in black hawk nests cannot be directly attributed to depredation. Arboreal mammals such as raccoons (*Procyon lotor*), coatis (*Nasua narica*), and ringtails (*Bassariscus astutus*) may also pose a risk to black hawk nestlings and eggs (Boal and Mannan 1996).

Status and Management

The historical status of the black hawk in New Mexico is nearly unknown. However, despite being a victim of at least a century of human encroachment and habitat alteration along the state's mid-elevation waterways, New Mexico's black hawk population appears to be currently stable—and even increasing in some areas—while also expanding its overall distribution. In particular, the number of "peripheral" breeders seems to have increased through the latter half of the 20th century.

No definite black hawk record exists for the state prior to 1918 (Bailey 1928; Phillips 1968). The species was reportedly observed along the Gila River in 1879, but no specimen was secured at the time (Mearns 1886). Bailey (1928:176) notes a report by J. S. Ligon of a specimen collected in October 1915 "20 miles east of Silver City." However, the whereabouts of that specimen is not known. Finally, a report (also in Bailey 1928:176) of a bird collected by Mearns from the San Luis Mountains in the summer of 1892 is attributed to New Mexico, but the location is unclear and may be from Sonora, Mexico (J. Hubbard, *in litt.*). There is disagreement over the suggestion that the scarcity of early records reflects low population levels and a small distribution during the mid and late 19th century (Phillips 1968; Hubbard 1971; Palmer 1988; J. Hubbard, *in litt.*). In retrospect, the black hawk could easily have been overlooked by early naturalists, due to its association with remote riparian canyons, its secretive nature away from nesting territories, and the common confusion of this species with the Zone-tailed Hawk through the late 19th and early 20th centuries (J. Hubbard, *in litt.*) and even through the 1950s (e.g., Bohl and Traylor 1958). All early New Mexico records lacking specimen evidence should be generally considered suspect.

Both Schnell (1998) and Murphy (1978), citing the disappearance of Texas's Lower Rio Grande black hawk population by the 1940s (Oberholser 1974) and the loss in quality and extent of breeding habitat elsewhere in the United States, believed that black hawks were likely more abundant before widespread alterations to riparian areas in the Southwest. This conclusion was also reached by the New Mexico Department of Game and Fish (NMDGF 1988).

While broader-scale population changes are difficult to assess, local changes are more reliably estimated for particular areas. The accessibility and overall ornithological significance of the Cliff–Gila Valley has attracted the interest of many, including the late Ralph A. Fisher, who dedicated countless hours to observing black hawks in his "backyard" between the mid-1960s and his death in 1999. Complete surveys of the valley's riparian forests were first conducted in the 1970s by R. Glinski (1976–77, *in litt.*) and Fisher (1979, *in litt.*), followed by surveys in the 1980s by Egbert (1981) and Montgomery et al. (1985), the mid-1990s (Skaggs 1996), and early 2000s by GS (2000–2001) and S. Stoleson (2002, *in litt.*). Results of these surveys illustrate a clear and steady increase in the number of breeding pairs from an estimated 11 in 1976–77 to an estimated 27 in 2002. Surveys before the 1970s were incomplete, but searches for black hawks in the 1960s found very few pairs, and observed between-season territory abandonment was generally attributed to shooting (Zimmerman 1962, 1970). The cause of this increase from the 1960s through the early 2000s is not immediately clear but could be due, in part, to a combination of decreased human persecution; maturation of new riparian forest patches (potentially providing additional nest substrates); an increase in successional riparian scrub cover following decreased grazing pressures from livestock (potentially providing additional cover and perching substrates); the conversion of agricultural fields to pasture; and geohydrological changes to the river channel (both potentially improving prey abundance and/or availability) (J. Hubbard, *in litt.*). The Hondo Valley population may also have experienced a black hawk population increase during a similar time period, though the increase in known nesting territories in this area is undoubtedly due in part to the survey efforts of R. and Y. Troy and D. Stahlecker (NMDGF 2002; Troy and Stahlecker 2008).

The New Mexico black hawk population in 2005 was likely over 100 breeding pairs. This number is based on the statewide estimate of 70–90 pairs by Skaggs (1996) and adjusted for known increases in the Cliff–Gila Valley population and likely increases in the Hondo Valley. Range-wide surveys by Richard Glinski and Helen Snyder in 1975–77 were incomplete in New Mexico (H. Snyder, pers. comm.), but nearly 200 pairs were estimated for Arizona (Schnell et al. 1988). With 10–20 pairs breeding in west Texas (Boal and Mannan 1996), and only a negligible number in southwestern Utah (Schnell 1994), New Mexico accounts for roughly one-third of the U.S. population.

Skaggs (1996) monitored 25 nesting attempts in 1994–95 in southwestern New Mexico. Mean brood size at fledging was 1.00 young per nest and 84% of nesting attempts were successful. New Mexico nest records from 1965 to 1993 ($n = 38$) compiled by Skaggs (primarily from southwestern New Mexico) reveal productivity levels of 1.24 young per successful nest. In the Cliff–Gila Valley in 2000–2001, 66% of 38 nesting attempts were successful, though brood size at fledging was not accurately determined (Sadoti 2008). These numbers are similar to the apparently stable Aravaipa Canyon, Arizona, population where, based on 168 nesting attempts, a fledging rate of 0.98 birds per nest and an overall 78% success rate were observed between 1976 and 1994 (Schnell 1994). In contrast, a fledging rate of only 0.54 birds per nest (46% nesting success) was observed from a total of 13 nesting attempts in the Hondo Valley in 2002–3 (Troy and Stahlecker 2008).

Little public concern for the welfare of New Mexico's black hawks is documented before the mid 20th century. However, by the mid 1960s, largely in response to habitat reduction in the preceding half-century, Hubbard (1965:474) predicted that "at the present rate it is only a matter of time until the black hawk is extirpated from the United States." Observations of two birds found shot in the Cliff–Gila Valley (of three territories in 1968–69), the losses of riparian forests there and elsewhere (Zimmerman 1970), and the apparently small statewide population bolstered concern for the persistence of the species and prompted its state listing as an Endangered Species (Group 2, Threatened) in 1975 (NMDFG 1988).

There is now general consensus that the greatest ongoing threat to the persistence of the black hawk in the United States is the alteration and loss of riparian habitat for nesting and foraging (e.g., Porter and White 1977; Snyder and Snyder 1991; Schnell 1994). These effects are discussed in detail in Cartron et al. (2000). Beyond direct conversion of streamside forests to agriculture, mining, and commercial, residential, recreational, or other uses, local nesting habitat may be lost as nesting trees age and die in the absence of forest regeneration nearby. Though cultural views on raptors are more tolerant now than fifty years ago, black hawks still face direct threats from shooting in New Mexico, as evidenced by the wanton killing of an incubating female in 1997 (D. Zimmerman, pers. comm.). The accelerated loss of habitat in the presumed winter range of northern Mexico during the last quarter-century represents another cause of concern (D. Zimmerman, pers. comm.). Ultimately the persistence of this species likely depends on the coordination of monitoring and conservation efforts with researchers and managers south of the U.S. border.

The overall outlook for the black hawk in New Mexico and other Southwest states is cautiously optimistic (Snyder and Snyder 1991; Schnell 1994). Recommendations for a database or other monitoring system to measure population status and trends were made by both LeFranc and Glinski (1988) and the New Mexico Department of Game and Fish (NMDGF 1996), but no such system has been developed to date, nor has any baseline statewide survey of breeding territories been conducted. Natural Heritage New Mexico at the University of New Mexico currently tracks this species in a modern database, but it is limited to voluntarily contributed records. Ultimately, the New Mexico Department of Game and Fish is responsible for effective monitoring of the status of black hawks in the state. Additionally, a long-term and wide-range banding and monitoring program for southwestern breeders and fledglings is necessary to better assess demographic rates (including survivorship and breeding area fidelity) and to answer basic questions about the species' natural history (LeFranc and Glinski 1988; Schnell 1994, 1998). This program should include monitoring of breeding areas south of the U.S. border into the state of Sonora, where similar habitats and conservation issues have been noted (Rodríguez-Estrella and Brown 1990). Finally, studies of breeding-season movements using very-high-frequency (VHF) telemetry and migration studies using satellite telemetry are needed to provide basic information on resource use or home ranges and to determine where New Mexico black hawks spend their winters.

Acknowledgments

S. Stoleson provided assistance and suggestions during field studies, shared unpublished data, and made helpful comments during the preparation of this manuscript. C. Frazier, R. McCollough, T. Neville, and B. H. Smith contributed Natural Heritage New Mexico data. B. Howe and M. A. Root provided NMOS records. C. Boal, P. Boucher, J. Hubbard, R. Shook, S. Williams, S. Schwartz, and D. and M. Zimmerman were also helpful through suggestions, sharing of unpublished reports, and other information. J. DeLong, K. Johnson, and J.-L. Cartron encouraged my early interest in the species. R. Glinski and H. Snyder provided comments on an earlier draft and contributed additional information. D. and T. Ogilvie, The Nature Conservancy, and several private landowners graciously granted access to areas for field studies.

LITERATURE CITED

Bailey, F. M. 1928. *Birds of New Mexico*. Santa Fe: New Mexico Department of Game and Fish.

Boal, C. W., and R. W. Mannan. 1996. *Conservation assessment for the common black-hawk (*Buteogallus anthracinus*)*. Final report to USDA Forest Service, Tonto National Forest. Tucson: University of Arizona.

Bohl, W. H., and E. Traylor. 1958. Correction in identification of the zone-tailed hawk as a Mexican black hawk. *Condor* 60(2): 139.

Caldwell, L. D. 1986. Predatory bird behavior and tillage operations. *Condor* 88:93–94.

Cartron, J.-L., S. H. Stoleson, P. L. L. Stoleson, and D. Shaw. 2000. Riparian areas. In *Ecological and socioeconomic aspects of livestock management in the Southwest*, ed. R. Jemison and C. Raish, 281–327. Amsterdam: Elsevier Press.

Dickey, D. R., and A. J. van Rossem. 1938. *The birds of El Salvador*. Publications of the Field Museum of Natural History. *Zoological Series* 23.

Egbert, J. 1981. Field inventories in New Mexico of selected Gila Valley birds. Unpublished report to the New Mexico Department of Game and Fish, Santa Fe.

Hubbard, J. P. 1965. Bad days for the black hawk. *Audubon Field Notes* 19(4): 474.

———. 1971. The summer birds of the Gila Valley, New Mexico. *Occasional Papers of the Delaware Museum of Natural History, Nemouria*, no. 2.

———. 1985. Southwest region: New Mexico. *American Birds* 39(3): 336–38.

———. 1987. Southwest region: New Mexico. *American Birds* 41(1): 128–30.

———. 1988. Southwest region: New Mexico. *American Birds* 42(1): 113–19.

Hundertmark, C. 1974. Breeding range extensions of some birds in New Mexico. *Wilson Bulletin* 86(3): 298–300.

LeFranc, M. N. Jr., and R. L. Glinski. 1988. Southwest raptor management issues and recommendations, In *Proceedings of the southwest raptor management symposium and workshop*, ed. R. L. Glinski et al., 375–92. Washington, DC: National Wildlife Federation.

Mearns, E. A. 1886. Some birds of Arizona. *Auk* 3:60–73.

Millsap, B. A. 1981. Distributional status of falconiformes in west central Arizona—with notes on ecology, reproductive success and management. Tech. Note no. 355. Washington, DC: U.S. Department of the Interior, Bureau of Land Management.

Montgomery, J. M., G. S. Mills, S. Sutherland, and R. B. Spicer. 1985. *Wildlife and fishery studies, Upper Gila water supply project*. Part 1, *Terrestrial wildlife*. Boulder City, NV: U.S. Bureau of Reclamation.

Murphy, J. 1978. Management considerations for some western hawks. *Transactions of the North American Wildlife and Natural Resource Conference* 43:241–51.

[NMDGF] New Mexico Department of Game and Fish. 1988. *Handbook of species endangered in New Mexico*. F-185, 1–2. Santa Fe.

———. 1996. *Threatened and endangered species of New Mexico—1996 biennial review and recommendations*. Santa Fe.

———. 2002. "Uncommon" black-hawk perseveres in southeast New Mexico. *Share With Wildlife*, Fall 2002, 1, 13–15.

Oberholser, H. C. 1974. *The bird life of Texas*. Austin: University of Texas Press.

Palmer, R. S., ed. 1988. *Handbook of North American birds*. Vol. 4. New Haven: Yale University Press.

Phillips, A. R. 1968. The instability of the distribution of land birds in the Southwest. In *Collected papers in honor of Lyndon Land Hargrave*, by A. H. Schroeder, ed. R. L. Glinski et al., 375–92. Vol. 1, *Papers of the Archeological Society of New Mexico*. Santa Fe: Museum of New Mexico Press.

Porter, R. D., and C. M. White. 1977. Status of some rare and lesser known hawks in western United States. In *Proceedings of the world conference on birds of prey, Vienna*, ed. R. D. Chancellor, 39–57. Cambridge, UK: International Council for Bird Preservation.

Preston, C. R., and R. D. Beane. 1993. Red-tailed Hawk (*Buteo jamaicensis*). No. 52. In *The birds of North America*, ed. A. Poole and F. Gill. Philadelphia, PA: Academy of Natural Sciences, and Washington, DC: American Ornithologists' Union.

Rodríguez-Estrella, R., and B. T. Brown. 1990. Density and habitat use of raptors along the Rio Bavispe and Rio Yaqui, Sonora, Mexico. *Journal of Raptor Research* 24:47–55.

Sadoti, G. 2008. Nest-site selection by Common Black-Hawks in southwestern New Mexico. *Journal of Field Ornithology* 79:11-19.

Schnell, J. H. 1994. Common Black-Hawk (*Buteogallus anthracinus*). No. 122. In *The birds of North America*, ed. A. Poole and F. Gill. Philadelphia, PA: Academy of Natural Sciences, and Washington, DC: American Ornithologists' Union.

———. 1998. Common Black-Hawk. In *The raptors of Arizona*, ed. R. L. Glinski, 73–76. Tucson: University of Arizona Press.

Schnell, J. H., R. L. Glinski, and H. Snyder. 1988. Common Black-Hawk. In *Proceedings of the southwest raptor management symposium and workshop*, ed. R. L. Glinski et al., 65–70. Washington, DC: National Wildlife Federation.

Skaggs, R. 1996. The common black-hawk (*Buteogallus anthracinus*) in New Mexico: 1994–95 inventories. Unpublished report to the New Mexico Department of Game and Fish, Santa Fe.

Snider, P. R., ed. 1996. *New Mexico Ornithological Society Field Notes* 35(4): 1 August–30 November.

Snyder, N., and H. Snyder. 1991. *Birds of prey, natural history and conservation of North American raptors*. Stillwater, MN: Voyager Press.

Troy, R. J., and D. W. Stahlecker. 2008. Status of a disjunct population of Common Black-Hawks in southeastern New Mexico: 2002–2003. *New Mexico Ornithological Society Bulletin* 36:14–22.

Webster, F. S. 1976. South Texas region. *American Birds* 23:975–78.

Wheeler, B. K., and W. S. Clark. 1995. *A photographic guide to North American raptors*. New York: Academic Press.

Wiley, J. W., and O. H. Garrido. 2005. Taxonomic status and biology of the Cuban Black-Hawk, *Buteogallus anthracinus gundlachii* (Aves : Accipitridae). *Journal of Raptor Research* 39(4): 351–64.

Williams, S. O. III. 1992. Southwest region: New Mexico. *American Birds* 46(2): 298–301.

———. 1993. Southwest region: New Mexico. *Field Notes* 47(2): 178–81.

———. 1994a. Southwest region: New Mexico. *Field Notes* 48(2): 236–38.

———. 1994b. Southwest region: New Mexico. *Field Notes* 48(5): 973–76.

———. 1996. Southwest region: New Mexico. *Field Notes* 50(5): 980–83.

———. 2002. New Mexico. *North American Birds* 56(3): 340–43.

———. 2003. New Mexico. *North American Birds* 57(3): 383–85.

———. 2004. New Mexico. *North American Birds* 58(3): 410–14.

———. 2005. New Mexico. *North American Birds* 59(3): 472–76.

Williams, S. O. III., and J. P. Hubbard. 1990a. Southwest region: New Mexico. *American Birds* 44(3): 476–79.

———. 1990b. Southwest region: New Mexico. *American Birds* 44(5): 1167–69.

———. 1991. Southwest region: New Mexico. *American Birds* 45(2): 301–3.

Zimmerman, D. A. 1962. Southwest region. *Audubon Field Notes* 16(4): 436–39.

———. 1970. Birds and bird habitat on national forest lands in the Gila River Valley, southwestern New Mexico. Unpublished report to U.S. Forest Service.

13

Harris's Hawk

(*Parabuteo unicinctus*)

JAMES C. BEDNARZ

THE HARRIS'S HAWK (*Parabuteo unicinctus*) is, perhaps, the most compelling and enigmatic bird of prey native to the Land of Enchantment. Not only does this hawk exhibit a striking rich "chocolate" and rufous coloration, but it often occurs in family social groups that engage in complex cooperative hunting behaviors to capture lagomorphs (jackrabbits and cottontails) in the desert lands of southern New Mexico. This hawk is boldly marked, tricolored (dark chocolate brown, rufous, and contrasting white), medium-large in size, and characterized by long legs and somewhat paddle-shaped wings in flight. At a distance, Harris's Hawks will appear very dark and show some resemblance to a dark-morph or rufous-morph Red-tailed Hawk (*Buteo jamaicensis*; see chapter 18). However, if you stop to observe one of these dark hawks, you typically will see a second or third chocolate-colored hawk perched on nearby power poles or trees, often making short flights from one perch to another—a definite indication of Harris's Hawks. These groups represent family units usually consisting of a breeding male, a breeding female, older offspring that may be up to three years of age, and the most recent brood of juveniles.

The plumage coloration of adults is bold, monotypical, and diagnostic. In good light, the overall body color is chocolate brown to almost sooty black. The upperwing coverts, wing linings, and flanks are rusty to a rich chestnut red. Contrasting with these dark colors are the "snowy" white upper and lower tail, most visible in flight. The tail is dark brown to almost black, but has a white base and relatively broad terminal white tip. Also, the bill of this hawk is relatively large and appears to project outward more than in typical buteo hawks because of its naked lores. The Harris's Hawk also has relatively long bright yellow legs and a long tail. The brightly colored yellow legs, ceres, lores, and orbital region all markedly contrast with the dark plumage of this hawk.

Juveniles resemble adults from a distance, except that their underparts are streaked with cream or buffy coloration. The amount of whitish coloration is variable, and the rufous patches on the wing coverts and legs are reduced and dull colored. The underwing is distinctive in flight, showing finely barred gray secondaries and whitish primaries with dark tips. The tail is crossed with many fine dusky bars, and the tip is white. As primaries are mostly replaced in the prebasic molt, typically occurring in the spring and summer following the year of hatching, the distinctive whitish primaries of immature hawks may be visible in some individuals that are as old as 16 months of age (Bednarz 1995).

PHOTO 13.1

Adult Harris's Hawk (educational bird). The plumage is chocolate brown to sooty brown. Contrasting with the head's dark plumage are the bright-yellow cere and gape. The beak is bluish with a black tip. PHOTOGRAPH: © JOHN V. BROWN.

PHOTO 13.2

Adult Harris's Hawk (educational plumage) with lighter plumage. PHOTOGRAPH: © DENNIS MILLER.

Harris's Hawks are distinctively sexually size dimorphic with females averaging 47% heavier than males. Female mass in New Mexico ranged from 775 to 1,278 grams (27 to 45 oz) (n = 54) and male mass from 515 to 800 grams (18 to 28 oz) (n = 57) (Bednarz 1995). If mixed-gender hawks are perched side by side, their gender can often be reliably distinguished by size in the field at a distance. Adult females and males average about 10% and 6.5% heavier in mass than immature Harris's Hawks of the same gender, respectively. This age difference in size also holds true for wing length, but not for tail length, which averages slightly greater in immatures compared to adults (Bednarz 1995).

No other hawk has similar coloration to that of the Harris's Hawk. Dark-morph buteos are most similar (e.g., dark- or rufous-morph Red-tailed Hawks), but generally lack the chestnut thighs and wing patches, and the white on both tail coverts and tail. The Snail Kite (*Rostrhamus sociablilis*) may appear similar at a distance, but also lacks the chestnut patches, has a much thinner and finely hooked bill, and does not occur in New Mexico.

Typical flight is somewhat accipiter-like, and is characterized by flap-flap-glide, flap-flap-glide, usually about 10–100 m (~30–330 ft) above the ground. Wings normally exhibit a slightly cupped appearance, and wing tips are pointed down when gliding (Bednarz 1995; Clark and Wheeler 2001). When not hunting, flight may appear sluggish (Palmer 1988); however, the Harris's Hawk is agile, can accelerate rapidly and "hug" the landscape contour, and maneuver around obstacles (McElroy 1977; Bednarz 1995). Males are especially agile, and under the right conditions they may fly backward and hover briefly. Females are more directional in flight and give the impression of speed and power (McElroy 1977). Flight speeds have been measured from 4.13 m/sec (13.5 ft/sec) when climbing (Pennycuick et al. 1989) to 16.2 m/sec (53 ft/sec) in a 7° glide (Tucker and Heine 1990). Also, the Harris's Hawk may soar with flattened wings at high altitudes beyond the range of the naked eye, typically in the middle of

PHOTO 13.3

(*left*) Adult Harris's Hawk, Los Medaños Area, southeastern New Mexico, early June 1986. Contrasting with the overall dark plumage, the upper wing coverts, wing linings, and flanks are rusty to a rich chestnut red. Note also the long, yellow legs. PHOTOGRAPH: © DAVID A. PONTON.

PHOTO 13.4

(*above*) Harris's Hawk in flight, south of Monument, Lea Co., 26 December 2008. Important field marks include the paddle-shaped wings and the tail's white base and terminal band. PHOTOGRAPH: © MARK L. WATSON.

PHOTO 13.5

Free-flying, hatch-year male Harris's Hawk captured in September 1982 as part of Bednarz's research program in the Los Medaños area in southeastern New Mexico. The underparts of juveniles are streaked with cream or buff coloration. The underwings show finely barred gray secondaries and whitish primaries with dark tips. The tail is characterized by many fine dusky bars and its tip is white. PHOTOGRAPH: © JIM BEDNARZ.

the day, and will engage in dramatic dives from these heights (McElroy 1977; Bednarz 1995).

Two subspecies of Harris's Hawk have been described in the United States, *P. u. superior* in California (extirpated) and Arizona, and *P. u. unicintus* in Texas and much of Central America (van Rossem 1942). Bednarz (1988a), in an attempt to identify the subspecies occurring in New Mexico and to more clearly define the described subspecies, examined 91 Harris's Hawk specimens from the Southwest. The eight Harris's Hawk specimens examined from New Mexico showed intermediate characteristics between those of putative *superior* and *uncinctus* specimens, and regression analysis suggested latitudinal clinal variation among all the specimens measured. Bednarz (1988a) questioned the validity of the subspecific designation of *P. u. superior* and suggested that a clinal pattern more accurately reflected morphological variation in this species than designation of two distinct subspecies.

Distribution

The Harris's Hawk is patchily distributed from the southwestern United States south through Mexico and Central America, reaching into South America as far as Argentina and Chile in isolated areas of appropriate desert scrub and savanna vegetation. Within the southwestern United States, isolated populations are currently only found within the southern portions of three states, Arizona, New Mexico, and Texas (Bednarz 1995). Historically, a population of Harris's Hawks nested along the Colorado River in both Arizona and California; however, this population is now extirpated. Currently the Harris's Hawk is described as a casual winter visitor in the Salton Sea region of California (Patten et al. 2003).

In New Mexico, mostly isolated breeding populations of Harris's Hawks are very spottily distributed in the southern tier of counties (Eddy, Lea, Luna, Hidalgo, and Otero counties). The species' breeding distribution commonly expands and retracts through time depending upon abundance of desert prey populations (Jaksic et al. 1992). This phenomenon is exacerbated along the northern periphery of the species' range, which again lies in southern New Mexico. Therefore a pattern of isolated colonization by nesting pairs, followed by local extinctions, appears to be common in New Mexico.

New Mexico's core breeding area of this species is found in southeastern Eddy County and southern Lea County, east of the Guadalupe Mountains and south of Hobbs and Lake McMillan (see map). In a personal communication to me, Steve West suggested that the Harris's Hawk might be a relatively recent invader of New Mexico. However, J. Stokley Ligon, probably the first ornithologist to visit the southeastern part of the state (in 1919), reported the species as a common breeder in the area surrounding Carlsbad, Eddy County (Bailey 1928; Ligon 1961). Also, based on the descriptions of "large and plentiful mesquites" in the diaries of the Pope expedition, which passed through the region in 1854 (Byrne 1854), Bednarz et al. (1988) concluded that the vegetation was suitable for breeding Harris's Hawks more than 150 years ago in southeastern New Mexico.

Perhaps the densest concentration of breeding Harris's Hawks found in the state is within the Los Medaños area in east-central Eddy County and west-central Lea County between Loving and Jal. It is also the area where several associates and I conducted research over a 10-year period on the behavior and ecology of Harris's Hawks (e.g., Bednarz 1987a, 1987b; Bednarz 1988b, 1988c; Bednarz and Ligon 1988; Bednarz et al. 1990; Bednarz and Hayden 1991; Gerstell and Bednarz 1999). Indeed, much of the ecology and natural history of the Harris's Hawk described in this chapter is based on data collected in the Los Medaños area. The majority of this area is characterized by a mesquite-oak shrubland with rolling sandy soils. Major cover types include low, vegetated shinnery oak and mesquite dunes; yucca-grasslands; creosote bush (*Larrea tridentata*) flatlands; and sparsely vegetated active dunes. Although the latter is the "signature" cover-landscape type of this region, it only occupies a small proportion of the Los Medaños area. Harris's Hawks primarily use and nest in the mesquite-oak communities, particularly in areas with large mesquites. Climatically, this area is subject to extreme hot temperatures in the summer. Winters are cold, with temperatures regularly dropping to -10°C at night.

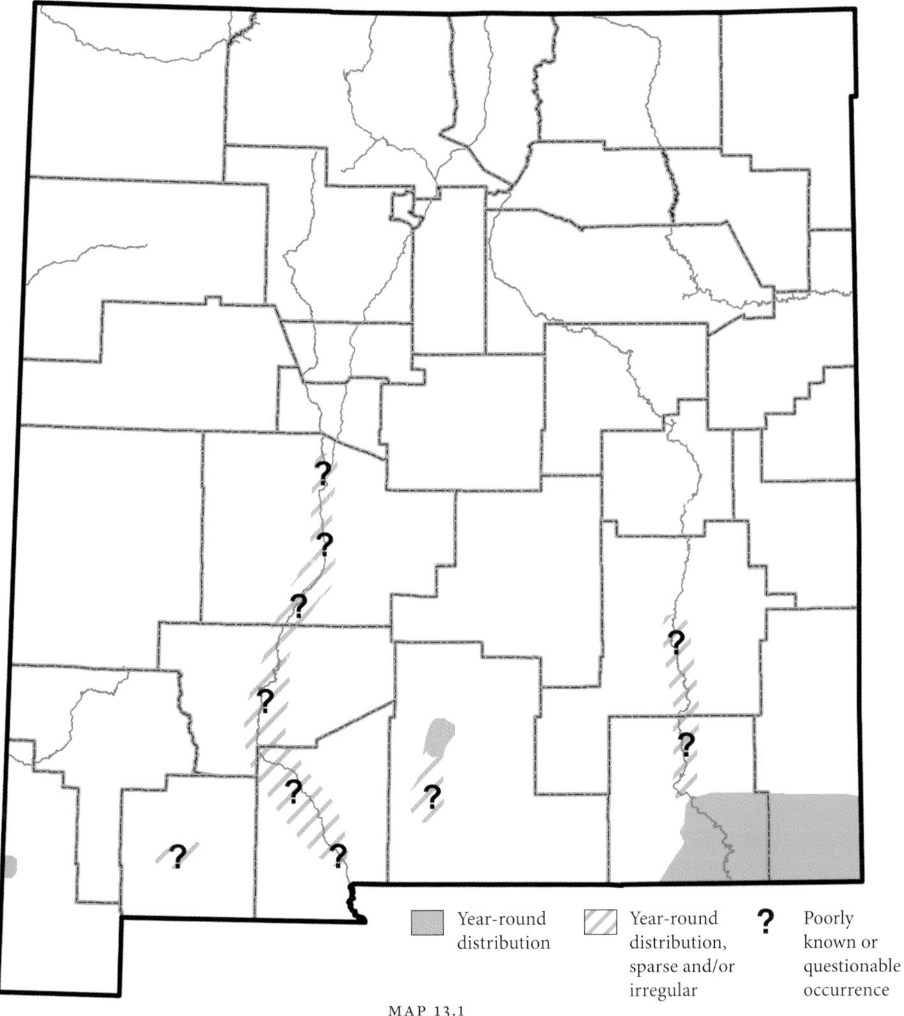

MAP 13.1

Harris's Hawk distribution map

Over the past three decades, several isolated nests or small satellite populations of breeding Harris's Hawks have been documented in southwestern and south-central New Mexico. The first satellite population was documented in Hidalgo County in the vicinity of Cotton City, San Simon Cienega, and Rodeo. It is in this area in 1980 that L. Siebert documented a nest that successfully fledged three young, with additional reports of at least five separate breeding groups in the early 1980s (Bednarz et al. 1988). Ralph Fisher, who regularly monitored the Hidalgo County population, and Dale Zimmerman (pers. comm.) both suggested that this population originated after 1976. Bednarz et al. (1988) proposed that the Harris's Hawk population along the San Pedro River in Arizona may have represented the source population for this invasion into southwestern New Mexico.

Subsequent to documentation of the Hidalgo County population, isolated nests were reported in Otero County near Alamogordo (Hubbard 1980a), Luna County near Deming (Williams 1999), and Sierra County near Truth or Consequences (Williams 2000a). In Otero County, Harris's Hawks now have been seen regularly—though not yet frequently—since

PHOTO 13.6

(*top*) Harris's Hawk near Loving, Eddy Co., 20 December 2007. The core of the Harris's Hawk distribution in New Mexico lies in the southeastern corner of the state in Eddy and Lea Counties.

PHOTOGRAPH: © JERRY OLDENETTEL.

PHOTO 13.7

(*bottom*) Jim Bednarz using a mirror pole to check a stick nest built in one of 10 artificial nesting platforms constructed by the U.S. Department of Energy (DOE) (Waste Isolation Pilot Plant) and collaborators in the Los Medaños area in southeastern New Mexico (photo probably taken in 1987). Platforms were designed by Bednarz and Tim Hayden, materials for platforms were provided by the DOE, and the actual construction of platforms was the work of volunteers. All 10 platforms were used by Chihuahuan Ravens (*Corvus cryptoleucus*), Harris's Hawks, and Great Horned Owls (*Bubo virginianus*) within three years of construction.

PHOTOGRAPH: © TIM HAYDEN.

1980. Sightings of isolated nests have been reported from the Tularosa Basin from Tularosa to about 8 km (5 mi) south of Alamogordo (e.g., Williams 2001). Also, birds have been seen as far south as Orogrande (Hubbard 1980b). Thus, Harris's Hawks perhaps nest intermittently in the Tularosa Valley from Tularosa to Orogrande (indicated by green diagonal lines; see map 13.1).

Harris's Hawks have also been sighted regularly—though infrequently—between Deming and Gage along I-10 since 1988, and two nests were located in that area in 1999 (Williams 1999). Finally, a Harris's Hawk nest with one young was reported in November 1999 near Truth or Consequences (Williams 2000a). Birds were seen infrequently, but again regularly, along the Rio Grande Valley from the upper end of Elephant Butte Lake State Park south to Percha Dam State Park between 1989 and 2000 (e.g., Snider 1994, 1995; Williams 2000b). Whether nesting still occurs in that area is unknown. Surprisingly, no Harris's Hawk nests have been documented along the Rio Grande corridor in Doña Ana County, in spite of almost annual reports of up to three hawks from 1994 through 1999 (e.g., Snider 1994, 1995). Thus the distribution map shows the Rio Grande Valley from Elephant Butte Lake south

PHOTO 13.8

Two Harris's Hawks on a power pole near Cotton City, Hidalgo Co., 7 December 2002. The small breeding population found in southwestern New Mexico is probably of recent origin (post-1976). PHOTOGRAPH: © ROBERT SHANTZ.

PHOTO 13.9

Unusual recent record from Doña Ana Co., 12 November 2008. PHOTOGRAPH: © DOUG BURKETT.

past Mesilla as an area of likely, though only occasional, nesting by Harris's Hawks (green diagonal lines; see map 13.1).

Although nonmigratory (Bednarz 1995), the Harris's Hawk has a reputation for being nomadic (e.g., Earl 1918; Bunker 1919; Snyder 1919; Parmalee and Stephens 1964; Griffin 1976; Bednarz et al. 1988). The majority of extralimital records are from what usually corresponds to the nonbreeding period (October–March) and thus likely represent dispersing birds. Some extralimital sightings could be of escapees from falconers, but such birds generally have jesses on their tarsi. Thus, especially during the winter period, raptor enthusiasts should be aware that Harris's Hawks could show up in almost any location in New Mexico in desert, open, or semiopen environments. Extralimital records have been relatively common up through the Rio Grande Valley, especially at the Bosque del Apache National Wildlife Refuge (including the bird beautifully photographed by David Nichols on 27 December 2001 [full-page photo at start of the chapter]; see also map 13.1). Extreme records in the state (not shown on the distribution map) include observations of single birds near Bernalillo (18 August 1981) and Corrales (27 August 1981) in Sandoval County, east of the Manzano Mountains in Torrance

County (9 October 1978), east of Tolar (10 December 2001), and near Portales (fall 1982, no exact date given) in Roosevelt County (Witzeman et al. 1979; Hubbard et al. 1982; Hubbard 1983; Williams 2002).

Habitat Associations

Harris's Hawks are associated with a variety of desertscrub, savanna, grassland, and desert wetland vegetation types (Bednarz 1995). Scattered trees and other vertical structures (e.g., power poles, windmills, oilfield extraction facilities) provide important perch sites used by foraging hawks and hunting groups, as well as structures for supporting nests (Bednarz and Ligon 1988). Grasslands and other open environments without suitable perches cannot support nesting Harris's Hawks.

In southeastern New Mexico, Harris's Hawks are found in areas with large (>4 m [>13 ft] in height) mesquite (*Prosopis* spp.) and soapberry trees (*Sapindus drummondii*). Bednarz and Ligon (1988) found little quantitative difference in the characteristics of habitat used by pairs and those of habitat used by larger groups in New Mexico. However, occupied habitats had greater densities of larger-diameter mesquite trees, more sagebrush (*Artemisia* spp.), and less litter than random sites not occupied by hawks. Also, data from New Mexico indicated that Harris's Hawks occupy habitats with greater densities of large mesquite, less grass cover, and more exposed ground than do Swainson's Hawks (*Buteo swainsoni*) (Bednarz 1988b). More generally, Bednarz et al. (1988) described the core habitat used by Harris's Hawks in New Mexico as mesquite-oak shrubland with rolling sandy soils often characterized by the presence of shinnery oak (*Quercus havardii*).

Throughout southern New Mexico, Harris's Hawks are also sporadically found nesting in riparian corridors adjacent to desert or mostly open environments. Their nests may be found in narrow but well-developed riparian belts of large mesquite, hackberry (*Celtis* spp.) trees, and occasionally cottonwoods (*Populus* spp.) along drainages or waterways (Bednarz et al. 1988). For example, Harris's Hawks nest in these types of habitats along the Pecos River and the Rio Grande, as well as their tributaries.

PHOTO 13.10

Harris's Hawk nesting habitat northeast of Carlsbad, Eddy Co., 21 May 2008. The vegetation is dominated by shinnery oak (*Quercus havardii*) and mesquites (*Prosopis* sp.). Note the Harris's Hawk nest in a mesquite tree on the left. PHOTOGRAPH: © DAVID J. GRIFFIN.

Life History

Nesting and Social Biology

Breeding pairs of Harris's Hawks are probably most often formed in the spring when levels of reproductive hormones are elevated in response to lengthening periods of daylight (Mays et al. 1991). In New Mexico, courtship and nest-building activities typically begin in March (Bednarz 1995). At this time, Harris's Hawk pairs and groups construct or refurbish one or more alternative nests (Whaley 1986; Bednarz 1995). The onset of egg-laying depends on available prey resources. In relatively high-prey years, first clutches are most often laid in March (Bednarz 1987a, 1995). However, in relatively low-prey years, egg-laying is more asynchronous and most clutches are produced in April or May (Bednarz 1987a, unpubl. data). In relatively "average" years, most Harris's Hawk young hatch in late April and fledge in early June (Bednarz 1987a; Bednarz et al. 1988).

In New Mexico, and other parts of the species' range, Harris's Hawk pairs lay second and even third clutches after previous breeding attempts (Mader 1977; Brannon 1980; Whaley 1986; Bednarz 1987a). In fact, Harris's Hawks have been found nesting year-round; in New Mexico in particular, occupied nests have been reported during every month of the year except January (Pache 1974; Bednarz 1987a, 1995; Bednarz et al. 1988). Bednarz (1987a) documented that 55% of 20 spring/early-summer breeding attempts were followed by second efforts in 1981 (a relatively high-prey year). He estimated that as much as 70% of a sample of 10 breeding groups may have bred in the autumn of 1980 (also a high-prey year). However, when available prey is in short supply, which occurred in 1988–90 in southeastern New Mexico, no nesting occurred in the autumn (JCB, unpubl. data). In the high-prey year of 1981, 8 (50%) of 16 groups that successfully fledged young in the spring nested a second time and 6 (37.5%) of these successfully produced a second brood (Bednarz 1987a). One New Mexico Harris's Hawk group that fledged a single young in May 1981 tried to re-nest three additional times in succession, all of which failed (the last attempt failed in late November) (Bednarz 1987a).

Nests are located in almost any relatively tall, sturdy structure in or adjacent to open desert. Bednarz et al. (1988) reported Harris's Hawk nests in eight different species of tree in New Mexico, but mesquite was most commonly used (54.6% of 55 nests). Overall, my research associates and I recorded nests supported by 199 different trees in New Mexico between 1980 and 1991. The majority of nests (51.3%) were placed in mesquite (*Prosopis* spp.), and secondarily, soapberry trees (*Sapindus drummondii*) were used (25.6%) (JCB, unpubl. data). Other trees that supported Harris's Hawk nests included hackberry (*Celtis reticula*), littleleaf sumac (*Rhus microphylla*), juniper (*Juniperus monosperma*), saltcedar (*Tamarix* sp.), crown of thorns (*Koeberlinia spinosa*), Spanish bayonet (*Yucca aloifolia*), and mulberry (*Morus* sp.) (JCB, unpubl. data). Pache (1974) found four nests northeast of Carlsbad in 1973 that were located in a cottonwood (*Populus* sp.), a mulberry tree, and two mesquite trees. Bailey (1928) stated that nests in New Mexico were placed in cactus, Spanish bayonet, chaparral (possibly mulberry?), mesquite, and hackberry. Ligon (1961), for his part, reported that Harris's Hawk nests were usually located in hackberry, native chinaberry trees (probably soapberry trees), and sometimes cottonwoods in New Mexico.

The mean height of nests in small trees in New Mexico was 2.73 m (8.9 ft) ($n = 55$; Bednarz et al. 1988). However, nests can also be found in very tall trees or other structures; Whaley (1986) reported a nest in Arizona at a height of 21.3 m (69.8 ft) in a transmission tower. Harris's Hawks commonly place nests in man-made structures in New Mexico and elsewhere, including electrical transmission towers, power poles, windmill platforms, and artificial nesting platforms, and my associates and I even found a nest in a weather antenna (Bednarz et al. 1988; Bednarz and Hayden 1988; Mader 1988).

Harris's Hawk groups typically use the same general area year after year and display subtle agonistic behavior toward neighboring conspecific groups (Mader 1975a; Brannon 1980; Whaley 1986). Dawson and Mannan (1991a) simulated an intruder at active nest sites by releasing a captive Harris's Hawk within 10 m (33 ft) of occupied nests in Arizona. Experimental resident groups chased intruders at least 500 m (1,640 ft) from the nest in 13 of 14 trials. The possibility that the expulsion of simulated intruders may have represented defense of nestlings against potential predators

PHOTOS 13.11a and b

Harris's Hawk nest east of Buckeye, Lea Co. in 2008. The nest contained three eggs on 27 May (a, *above*) and two downy nestlings on 2 July (b). PHOTOGRAPHS: © JEAN-LUC CARTRON.

was considered and rejected by Dawson and Mannan (1991a). The most convincing support of the existence of territorial behavior in New Mexico is that nests are almost always spaced at least 500 m (1,640 ft) apart (Whaley 1986; Bednarz 1987b, 1995). Territories seem to be mostly maintained by mutual avoidance rather than by conspicuous defense behavior (Bednarz 1995).

Harris's Hawks lay between one and five eggs (Bailey 1928; Ligon 1961; Bednarz 1995). In New Mexico, annual mean clutch sizes varied from 1.84 (n = 13) in 1988 to 3.04 (n = 26) in 1982 over a 10-year period (JCB, unpubl. data). J. Renwald reported a mean clutch size of 3.34 for 36 Harris's Hawk nests in southeastern New Mexico between 1977 and 1979 (Bednarz et al. 1988). The color of freshly laid eggs is very pale bluish, but this rapidly fades to white as incubation proceeds (Bednarz 1995). Occasionally the eggs are marked with a few spots of pale brownish or lavender color (Bent 1937; Mader 1988).

The incubation period of Harris's Hawks has been variously reported as ranging from 31 to 36 days (Mader 1975a; Bednarz 1986). However, the average mode incubation period falls between 34 and 35 days (Nice 1970; Mader 1975a; Bednarz 1986, 1995). Hatching of young occurs asynchronously (Bednarz and Hayden 1991), probably over a two- to four-day period (Bednarz 1995).

Harris's Hawk hatchlings can raise their heads weakly and force their eyes open. Typically the breeding female is present at the nest either brooding or shading young constantly after hatching. At this stage, nestlings are vulnerable to exposure and will die if brooding females are disturbed from nests for more than a few minutes (Bednarz 1995). Direct parental care at the nest drops off dramatically after the nestlings are about one week old and declines further as the brood-rearing period progresses (Mader 1979). The primary growth period for Harris's Hawk nestlings occurs from 8 to 30 days of age; during this period males gain an average of 18.84 g/day (0.66 oz/day) and females 31.39 g/day (1.11 oz/day) (Bednarz and Hayden 1991). Prey is primarily captured by the breeding male, but also by auxiliary group members, and given to the breeding female, who commonly transports the prey to the nest and tears it into small pieces before feeding the chicks directly (Mader 1979; Bednarz 1995). The mean prey-delivery rate in Arizona was 4.9 items/day (range = 2–18 items/day) when young were 1–18 days old (Dawson and Mannan 1991b). Nonbreeding auxiliary hawks occasionally attempt to feed nestlings (Brannon 1980; Bednarz 1987b; Dawson and Mannan 1991b). Nestlings begin to eat prey unaided between 18 and 20 days of age (Mader 1975a; Bednarz 1987b). However, adult Harris's Hawks will occasionally feed nestlings as old as 40 days of age. Young Harris's Hawks begin to climb branches of the nest tree at about 40 days of age (Mader 1988). The lighter males fledge at significantly earlier ages (mean = 44.8 days; n = 55 young) than the heavier females (mean = 47.9 days; n = 35; Bednarz and Hayden 1991).

PHOTOS 13.12a, b, c, and d

Typical scenes of life at a nest in a saltcedar tree (*Tamarix* sp.) in the Los Medaños area of southeastern New Mexico (probably in 1987). (a, *top left*) The adult female lands on the nest tree carrying a prey item; the single nestling in view is about 25 days of age. (b, c, d) The two nestlings are approximately 28 days of age.

PHOTO 13.13

(*top left*) Adult male Harris's Hawk feeding brood of two young (approximately 28 days old) in a nest in a mesquite (*Prosopis* sp.) tree in June 1986 in the Los Medaños area of southeastern New Mexico. PHOTOGRAPH: © DAVID A. PONTON.

PHOTO 13.14

(*bottom left*) Harris's Hawk nestling (approximately 30 days old) in a nest supported by a crown of thorns (*Koeberlinia spinosa*) tree in the Los Medaños area in southeastern New Mexico, 22 June 1988. PHOTOGRAPH: © JIM BEDNARZ.

PHOTO 13.15

(*above*) Jim Bednarz banding and processing a female Harris's Hawk nestling (approximately 35 days of age) on 17 June 1986 in the Los Medaños area of southeastern New Mexico. PHOTOGRAPH: © TIM HAYDEN.

PHOTO 13.16

Banded Harris's Hawk nestling (band no. 877-78905) about 35 days old on 4 July 1981 in the Los Medaños area of southeastern New Mexico. The nestling is marked with a unique combination of color vinyl riveted bands (right leg = silver/yellow, left leg = yellow).

PHOTOGRAPH:

© JIM BEDNARZ

PHOTO 13.17

Two Harris's Hawk young near fledging age (approximately 45 days old) in a brood of three at a nest supported by a soapberry (*Sapindus drummondii*) tree in the Los Medaños area of southeastern New Mexico, June 1988.

PHOTOGRAPH:

© JIM BEDNARZ.

Cooperative Breeding

Perhaps one of the most fascinating aspects of Harris's Hawk breeding biology is that this species typically lives in breeding groups of two to seven hawks (Bednarz 1995). In New Mexico, Bednarz (1987b) documented that 50.8% of 61 breeding groups that he monitored consisted of between three and five members; mean group size was 2.7 hawks (*n* = 61). On average, groups are larger in Arizona (mean = 3.8 hawks; Dawson and Mannan 1989), but groups with more than two members were less frequent in Texas (5–13%; Griffin 1976; Brannon 1980).

Two types of auxiliaries, also called "helpers," have been recognized: (1) offspring that stay on their natal territory and associate with their parents for up to three years (called auxiliaries by Bednarz [1987b] and gamma helpers by Dawson and Mannan [1991b]); and (2) unrelated birds, mostly adults, that actively assist breeding hawks (called beta helpers by Dawson and Mannan [1991b]). In New Mexico, all of the marked helpers observed in nesting groups (*n* = 105 auxiliaries) were offspring of the breeders (Bednarz 1995). However, Bednarz (1987b) genetically documented the case of a nest in New Mexico attended by three adults, a breeding female and two males that were not father and son and thus were likely not related. If true, the nonbreeding male at that nest would have been a beta helper, the only one found so far at a nest in New Mexico. Overall, however, data collected by Bednarz (1987b, unpubl. data) over a 10-year period suggested that extra birds tending Harris's Hawk nests in New Mexico were almost all offspring that deferred dispersal from their natal territory for up to three years after fledging. In contrast, unrelated helpers are much more common in Arizona (38% of 108 helpers at 64 nests; Dawson and Mannan 1991b).

All helpers, whether beta or gamma helpers, participate in cooperative hunts and defend nestlings from potential predators (Mader 1975a; Bednarz 1987b; Dawson and Mannan 1991b). However, unrelated male helpers, which again are rare in New Mexico and relatively common in Arizona, attempt polyandrous mating with the breeding female and participate in incubation, brooding, shading, and direct feeding of young, though less frequently than the breeding pair (Mader 1975a, 1975b; Dawson and Mannan 1991b; Bednarz 1995). Helping did not affect clutch size, number of young produced/nest, or number of offspring fledged/year in New Mexico (Bednarz 1987b). Groups with helpers did rear slightly larger nestlings and initiated second nests more frequently than did pairs of Harris's Hawks in New Mexico (Bednarz 1987b). Breeders and other group members also benefited from increased foraging success due to cooperative hunting (see below).

Boom-and-Bust Population Dynamics

The Harris's Hawk is uniquely adapted to the highly variable environment of the Chihuahuan Desert, as well as to other desert environments. Numerous features of the natural history of this hawk allow it to respond numerically nearly instantaneously to changes in its available food base. In New Mexico, the key prey resources that Harris's Hawks depend upon and respond to are lagomorph populations, especially desert cottontails (*Sylvilagus audubonii*) (Bednarz 1987a, 1988b, 1988c; Gerstell and Bednarz 1999). Rabbit populations vary wildly through time and may quickly go from boom to bust populations. For example, mean counts of cottontails along a 16-km (10-mi) census route in my Los Medaños study area varied from a high of 16.8/count in 1981 to 1.6/count in 1987 (Gerstell and Bednarz 1999). Harris's Hawks respond to any declines in available prey in a number of ways—declines in clutch sizes, lower success rate of nests, reduction in the number of fledglings produced per nest, cessation of all fall breeding activity, and suspension of nesting altogether (JCB, unpubl. data). These downward adjustments in reproductive performance have substantial consequences on local populations. For example, the number of Harris's Hawk nests found in two systematically searched plots totaling 171-km^2 (66-mi^2) in the Los Medaños area declined from 35 nests in 1985 to 5 nests in 1989 and 1990 (fig. 13.1). Likewise, Jaksic et al. (1991, 1992) documented that Harris's Hawks disappeared from a site in Chile in apparent response to a dramatic decline in prey.

Conversely, groups of Harris's Hawks have a tremendous capacity to respond rapidly to favorable prey conditions. Specifically, groups begin laying eggs in February and possibly earlier and produce larger

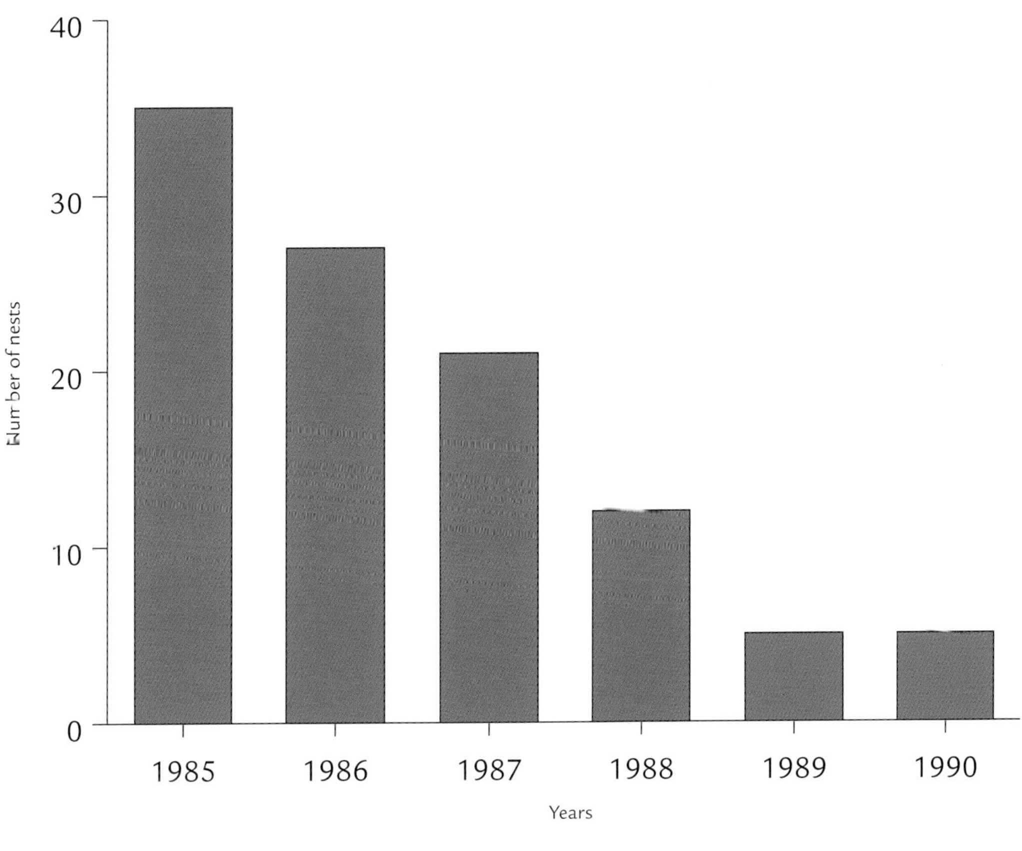

FIGURE 13.1.

Number of Harris's Hawk active nests (nests with viable eggs or young) found in two systematically searched plots totaling 171 km^2 (66 mi^2) between 1985 and 1990 in the Los Medaños area of southeastern New Mexico.

clutches (typically four eggs), while nest success improves and more young fledge per nest. However, perhaps most influential of all are the adaptations that result in double brooding and fall nesting, and allow groups to produce two or even possibly three broods in an annual period (Brannon 1980; Bednarz 1987a, 1995). Thus Harris's Hawk populations demonstrate a rather flexible boom-and-bust strategy, breeding as much as possible (up to three broods) in good (boom) years and deferring nesting in bad (bust) years (Bednarz 1995).

Although this is not fully understood, I suggest that such boom-and-bust periods associated with widely fluctuating prey availability probably also trigger emigration out of areas where prey populations are in decline. I witnessed the disappearance of several groups of Harris's Hawks (some consisting of as many as four hawks) during the lean prey years in the Los Medaños area in the late 1980s. I suspect it is very unlikely that all members of such groups perished in a short period of time. Rather, I hypothesize that at some point as prey populations diminish in one area, groups of hawks emigrate and search for a new area with suitable numbers of prey to support nesting (see fig. 13.1). It is unknown whether Harris's Hawk groups move as wholes across the landscape searching for a new breeding area, or whether the groups disband and each individual goes its own way. I believe that it is during lean periods that Harris's Hawks disperse and colonize new habitats (with better populations of prey), resulting sporadically in new nesting records in areas where the species had not previously been reported (see Distribution).

Diet and Foraging

In the southwestern United States, species commonly recorded in the diet of Harris's Hawks include the desert cottontail, the black-tailed jackrabbit (*Lepus californicus*), ground squirrels (*Ammospermophilus* and *Spermophilus* spp.), woodrats (*Neotoma* spp.), kangaroo rats (*Dipodomys* spp.), and pocket gophers (*Geomys* and *Thomomys* spp.), in addition to Gambel's Quail (*Callipepla gambelii*), Scaled Quail (*C. squamata*), Northern Bobwhite (*Colinus virginianus*), Cactus Wren (*Campylorhynchus brunneicapillus*), and Northern Mockingbird (*Mimus polyglottos*) among birds, and desert spiny lizards (*Sceloporus magister*) and skinks (*Eumeces* spp.) (Mader 1975a; Brannon 1980; Whaley 1986; Bednarz 1988b). Seasonal analysis of prey remains in New Mexico shows that cottontails and jackrabbits are the most important prey species consumed by Harris's Hawks throughout the year (spring biomass = 91.4%, summer = 89.6%, autumn = 91.6%; Bednarz 1987a). The frequency of insects and lizards taken increases in the summer, but these prey items contribute relatively little biomass to the diet (Bednarz 1988b).

At New Mexico nests, desert cottontails made up 61.4% of all vertebrate prey remains, compared to quail 5.6%, other birds 3.7%, woodrats 5.4%, ground squirrels 4.4%, kangaroo rats 2.8%, and other rodents 0.7% (Bednarz 1988b; table 13.1). In terms of estimated biomass, desert cottontails made up 87.1% of the diet, jackrabbits and unidentified lagomorphs 4%, woodrats 3%, quail 2.5%, other birds 1.3%, ground squirrels 1.3%, kangaroo rats 0.4%, and other rodents 0.2% (Bednarz 1988b). Pache (1974) identified six types of invertebrates in the pellets collected at four New Mexico Harris's Hawk nests: beetles (Coleoptera), ants (Formicidae), bees (Hymenoptera), grasshoppers (Orthoptera), woodticks (Ixodidae), and Centipedes (*Scolopendra*). However, invertebrates contribute almost nothing to the diet in terms of biomass (Bednarz 1988b; Gerstell and Bednarz 1999). Data based on observations of prey deliveries to nests have fewer biases and typically provide a more accurate representation of the diet of raptors at least during the breeding period. Observations that Tim Hayden and I made over a six-year period clearly indicate that lagomorphs (63.4% of the diet by biomass) and total mammals (94.3% of the biomass) are the most important prey items taken by Harris's Hawks during the nesting period in the Los Medaños area of New Mexico (table 13.1).

Harris's Hawks exhibit two primary modes of hunting: (1) sit-and-wait foraging and (2) short-flight-perch hunting. Sit-and-wait foraging is most often employed by lone hawks, though also occasionally by groups (Dawson 1988; Bednarz 1995). Short-flight-perch hunting involves hawks making short flights (~60–300 m [~200–980 ft]) between perches, then active scanning from a perch for <5 min, and again moving to a new perch (Bednarz 1988c; Dawson 1988). Harris's Hawks will attack prey both from perches (most often) and while in flight (JCB, pers. obs.). The short-flight-perch movement strategy is commonly employed during cooperative hunts (Bednarz 1988c).

Cooperative Hunting

Harris's Hawks engage in the most sophisticated cooperative hunting tactics thus far documented in birds (Ellis et al. 1993). Perhaps the first report indicating cooperative hunting by this species was by Major Allan Brooks as reported in Bent (1937:145): "Very often a pair of these hawks combine to secure their quarry, and I have seen a snowy heron shared amicably after it had fallen a victim to one of these raptores." Mader (1975a) provided the first quantitative data on cooperative hunting by one trio in Arizona, whereby he observed team hunting (>1 hawk) in 48% of 61 capture attempts. However, the success was relatively low (16%) and did not correlate with the size of the hunting group (Mader 1975a).

Tim Hayden and I initiated a relatively intensive study of Harris's Hawk behavior and ranging patterns in the Los Medaños area of New Mexico in 1985 and found that cooperative hunting was much more frequent in the nonbreeding season (October–March) than in the primary nesting period (Bednarz 1995). Typically, cooperative hunting bouts are initiated by

PHOTO 13.18

Harris's Hawk nestling about seven days old in a nest with fresh desert cottontail (*Sylvilagus audubonii*) prey in the Los Medaños area of southeastern New Mexico, 19 May 1983. The nest is about 5.5 m (18 ft) above the ground and supported by a hackberry (*Celtis reticulata*) tree. PHOTOGRAPH: © JIM BEDNARZ.

PHOTO 13.19

Two downy Harris's Hawk nestlings (approximately 12 days old) and one addled egg with two fresh spotted ground squirrel (*Spermophilus spilosoma*) prey in a nest supported by a mesquite (*Prosopis* sp.) tree in the Los Medaños area of southeastern New Mexico, 5 July 1981. PHOTOGRAPH: © JIM BEDNARZ.

PHOTO 13.20

Adult male Harris's Hawk delivering whiptail lizard (*Aspidoscelis* sp.) to a brood of two young (approximately 28 days old) in a nest in a mesquite (*Prosopis* sp.) tree in June 1986 in the Los Medaños area of southeastern New Mexico. PHOTOGRAPH: © DAVID A. PONTON.

TABLE 13.1. Prey brought to Harris's Hawk (*Parabuteo unicinctus*) nests during 461 hours of observations from blinds in the Los Medaños area of New Mexico, 1985–1990.

PREY	NUMBER	PERCENT FREQUENCY	PERCENT BIOMASS[a]
Desert cottontail (*Sylvilagus audubonii*)	19	16.8	41.4
Black-tailed jackrabbit (*Lepus californicus*)	2	1.8	7.8
Unidentified lagomorph	6	5.3	14.2
Total Lagomorphs	27	23.9	63.4
Unidentified large mammal[b]	7	6.2	10.6
Woodrat (*Neotoma* spp.)	15	13.3	12.7
Ground squirrel (*Spermophilus* spp.)	3	2.6	1.4
Kangaroo rat (*Dipodomys* spp.)	4	3.6	0.8
White-footed mouse (*Peromyscus* spp.)	1	0.9	0.1
Unidentified small mammal	11	9.7	1.5
Unidentified mammals	3	2.6	3.8
Total mammals	71	62.8	94.3
Whiptail lizard (*Aspidoscelis* spp.)	5	4.4	0.3
Unidentified lizard	28	24.8	1.9
Horned Lizard (*Phrynosoma* spp.)	1	0.9	0.1
Gopher snake (*Pituophis melanoleucus*)	1	0.9	1.9
Total reptiles	35	31.0	4.2
Scaled Quail (*Callipepla squamata*)	2	1.8	1.4
Unidentified bird	1	0.9	0.1
Total birds	3	2.7	1.5
Unidentified	4	3.5	–

[a] Biomass expansion factors used follow Bednarz (1986).

[b] Lagomorph or woodrat.

an "assembly ceremony," in which members of an established social group assemble and perch together, sometimes all on one branch (Bednarz 1988c; Dawson 1988). During our work, T. Hayden and I observed three distinct cooperative tactics: (1) *surprise pounce*: several hawks typically coming from different directions and converging on prey away from cover; (2) *flush-and-ambush*: one or more hawks penetrating the cover while other hawks alertly watch from nearby perches and attack when the prey is flushed from cover; and (3) *relay attack*: a relatively long chase of jackrabbits in which the lead "chase" position is sequentially alternated among hunting party members (Bednarz 1988c). Importantly, and despite Mader's (1975a) earlier report, groups of Harris's Hawks seem to be more successful in capturing prey than lone hawks. In Arizona, Dawson (1988) reported that one hawk had a 20% success rate (n = 65 attempts), two hawks 32.1% (n = 53), three hawks 40.4% (n = 2), four hawks 37.9% (n = 29), and five hawks 50% (n = 8). In New Mexico, I documented that capture rates of rabbits (cottontails and jackrabbits) significantly increased with hunting party size: two to three hawks had no lagomorph kills/50 hr, four hawks made 1.7 kills/50 hr, five hawks 3.1 kills/50 hr, and six hawks 3.9 kills/50 hr (Bednarz 1988c). Based on average hawk mass, average daily temperature in winter, and the amount of time spent flying, I calculated the daily energy requirement for an individual Harris's Hawk in New Mexico to be 147.8 kcal/day (Bednarz 1988c). Thus, given the average kill rate observed (0.63 lagomorph kills/day), I estimated that cooperatively taken lagomorphs accounted for 88.8% of the combined maintenance needs of the average-sized Harris's Hawk social group (4.8 hawks, 709.4 kcal/day). Harris's Hawks kill other prey during winter, particularly quail and medium-sized birds (JCB, pers. obs.), and these prey undoubtedly meet the remaining dietary needs of the hawk in winter. Moreover, the individual benefits, in terms of average energy intake available, from just the rabbit take, for groups of five hawks (148.1 kcal/day) and six hawks (148.6 kcal/day), slightly exceeded the needed daily energy budget to subsist (147.8 kcal/day; Bednarz 1988c). As the individual benefits of cooperative hunting seemed to maximize at a group size of five hawks, which also approximated the mean hunting party size (4.8 hawks) and the most common size of hunting party (five hawks) in New Mexico, I have suggested that the social nature of Harris's Hawks may be related to the adaptive advantages of cooperative hunting (Bednarz 1988c).

Predation and Interspecific Interactions

Primary predators of Harris's Hawk eggs and nestlings include coyotes (*Canis latrans*), Great Horned Owls (*Bubo virginianus*), Common Ravens (*Corvus corax*), Chihuahuan Ravens (*C. cryptoleucus*), and humans (Whaley 1986; Dawson and Mannan 1991b; Bednarz 1995). Coyotes are able to snatch nestlings from low nests (1–2 m [3.3–6.6 ft] in height) or cause young about to fledge to jump from their nest and run them down (JCB, pers. obs.). Any mammal or snake capable of climbing to nests probably occasionally kills and eats eggs and small young. Predation of flying immature and adult hawks is probably rare. However, I documented at least three cases in which Great Horned Owls captured and consumed adult male Harris's Hawks in New Mexico (pers. obs.).

Because of the direct threat of predation they represent, Great Horned Owls are mobbed vigorously by Harris's Hawks (Dawson and Mannan 1991b; Bednarz 1995). Additionally, Harris's Hawk groups defend lagomorph carcasses from other predators and scavengers for up to two days after the kill was made. I witnessed groups of Harris's Hawks successfully drive away Northern Harriers (*Circus cyaneus*), Red-tailed Hawks, and Ferruginous Hawks (*Buteo regalis*) from their rabbit kills (pers. obs.; also see Bent 1937). Harris's Hawks share their breeding habitat throughout their range in southern New Mexico with Swainson's Hawks (see chapter 16). Relatively intense interactions ensue between these two species. They almost always involve the Swainson's Hawk as the aggressor, often stooping or diving on the Harris's Hawk and chasing it away (Gerstell and Bednarz 1999). Simultaneous telemetry of one nesting Swainson's Hawk and three neighboring Harris's Hawk groups suggested that the territories of these two species mostly do not overlap, but nests have been found less than 175 m (~570 ft) apart (Gerstell and Bednarz 1999; JCB, unpubl data). In one case, a Swainson's Hawk pair caused a Harris's Hawk to abandon a newly constructed nest (it is unknown whether the nest had eggs; JCB, pers. obs.). Finally, interspecific interactions also involve nonraptors. Harris's Hawks are vigorously mobbed by passerines including Scissor-tailed Flycatchers (*Tyrannus forficatus*), Western Kingbirds (*T. verticalis*), Ash-throated Flycatchers (*Myiarchus cinerascens*), Northern Mockingbirds, and probably others (Bednarz 1995).

Status and Management

Quantitative assessments of Harris's Hawk numbers in New Mexico are lacking. The overall population trend for this species in the state is not known, largely because there are no data available on historical distribution and abundance (Bednarz et al. 1988). As provided by Bailey (1928), the earliest descriptions of Harris's Hawks in New Mexico seem to suggest that these birds were fairly common south and east of Carlsbad as they are today.

Between 1981 and 1983, Bednarz et al. (1988) reported a nesting density of Harris's Hawks of 2.4 nests/10 km² (/3.86 mi²) in the Los Medaños area of New Mexico. In the same area, after a prey and hawk population decline, Hayden and Bednarz (1991) found that nest densities varied from 0.29 nest/10 km² to 2.0 nests/10 km² (see fig. 13.1). Based on road counts conducted from November 1985 through December 1987 in the Los Medaños area and adjusted for detectability (Buckland et al. 2001), Bednarz et al. (1990) estimated that winter densities varied from 1.12 Harris's Hawks/10 km² to 2.60 Harris's Hawks/10 km². Christmas Bird Count data from New Mexico do not show any consistent trends, except that Harris's Hawks have been present, apparently in good numbers, since the counts were initiated in the southeastern part of the state.

Declines of Harris's Hawk populations have occurred in California—where the species is currently extirpated—and in Texas due to habitat modification and widespread mesquite control programs (Bednarz et al. 1988). In Arizona, Harris's Hawk populations initially declined due to loss of paloverde-saguaro habitats to urban development between Tucson and Phoenix (Bednarz et al. 1988). Yet, since 1975 the urban population of Harris's Hawks in Tucson likely has increased from about 10 to 62 breeding groups (Dwyer and Mannan 2007). Thus, the acclimation of the species to urban environments in Arizona has somewhat compensated for the decline in this species caused by habitat modification. In recent years, however, the urban population of Harris's Hawks in Tucson also suffered high rates of mortality by electrocution (Dwyer 2006; Dwyer and Mannan 2007). Thus far, Harris's Hawks have not become urbanized anywhere in New Mexico. The largest urban area occurring within the primary range of the Harris's Hawk in New Mexico is Carlsbad, and the species is mostly observed on the outskirts of town in habitats that are more natural or at least have had only limited rural development.

The population of Harris's Hawks in southeastern New Mexico could be adversely affected by the cumulative elimination of rangelands caused by widespread development of oil and gas resources as well as by development of the Waste Isolation Pilot Plant (WIPP, a permanent repository for nuclear waste) (Bednarz 1995). Hayden and Bednarz (1991) examined the Harris's Hawk population in the vicinity of the WIPP facility and a nearby reference area and found a number of subtle differences. Specifically, Harris's Hawks used larger home ranges in the WIPP area and showed a slight reduction in reproductive success when compared to the reference area (Hayden and Bednarz 1991). Overall, Hayden and Bednarz (1991) suggested that the combined stresses of the presence of human intruders, modification of natural desert vegetation by oil and gas and other development, and occasional shooting and trapping by humans, along with natural stresses (especially low prey populations) could cause a long-term and permanent decline in the Harris's Hawk population in New Mexico.

Human activity near Harris's Hawk nests causes nest attendants to flee from nests, and most adult hawks in New Mexico will not return to their nests to feed and care for their young as long as humans remain present (Bednarz et al. 1988; Bednarz 1995). Hayden and Bednarz (1991) reported that the average distance that Harris's Hawks would flee from their nests in southeastern New Mexico in response to an approaching pedestrian was 216 m (709 ft) (SD = 149, n = 74 approaches) during the incubation period and 303 m (994 ft) (SD =154, n = 110 approaches) during the

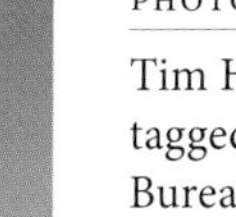

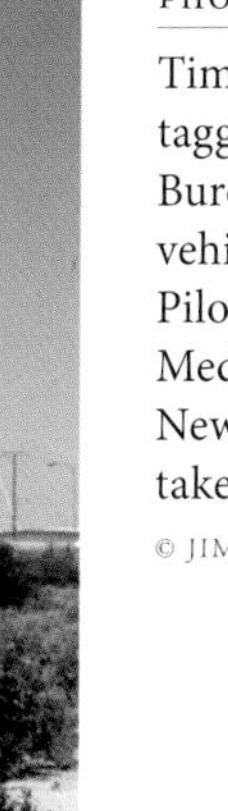

PHOTO 13.21

Tim Hayden tracking a radio-tagged Harris's Hawk from a Bureau of Land Management vehicle near the Waste Isolation Pilot Plant (WIPP) in the Los Medaños area of southeastern New Mexico (photo probably taken in 1987). PHOTOGRAPH: © JIM BEDNARZ.

PHOTO 13.22

Four Harris's Hawks on a power line and power pole east of Loco Hills, Eddy Co., 27 December 2009. Large numbers of Harris's Hawks were electrocuted on power poles in the Tucson area in Arizona before power poles were retrofitted to address the problem. Even in New Mexico, the author has found electrocuted Harris's Hawks. PHOTOGRAPH: © MARK L. WATSON.

brood-rearing period. Based on these data, Hayden and Bednarz (1991) suggested a buffer zone with a radius of 400 m (~1,310 ft) be maintained around nests. Within this buffer zone human activities should be restricted to minimize disruption of nesting attempts.

Human-caused Harris's Hawk mortality documented in New Mexico and elsewhere includes shooting and trapping, drowning in livestock water tanks, and electrocution (Whaley 1986; Bednarz et al. 1988; Bednarz 1995; Dwyer 2006). Of 44 Harris's Hawks radio-tagged in southeastern New Mexico, 9 mortality events were documented, of which 4 were caused by shooting and 1 by an illegal trap (Hayden and Bednarz 1991; Bednarz 1995). I also found three electrocuted Harris's Hawks in southeastern New Mexico (JCB, pers. obs.). Mortality by electrocution is a particularly severe problem in urban areas of Arizona (Dwyer 2006). For example, Dwyer and Mannan (2007) documented that a minimum of 1.4 Harris's Hawks were electrocuted per monitored nest prior to retrofitting of

PHOTO 13.23

Harris's Hawk, Wildlife Rescue, Inc., Albuquerque, 26 April 2008. Originally a falconer's bird, this Harris's Hawk had a hunting accident and as a result suffered the loss of metacarpals. It could not be released and serves now as one of Wildlife Rescue's educational birds.

PHOTOGRAPH: © JAMES N. STUART.

selected power poles to minimize the risk of electric shock injuries. When the removal of carcasses by scavengers was factored in, the estimated rate of electrocuted Harris's Hawks increased to 1.89 hawks/nest. The technical solution for reducing or eliminating the risk of raptor electrocution exists, though always requiring financial commitment: existing power poles can be retrofitted to either insulate or separate energized conductors >60 cm (24 in) (Lehman 2001; APLIC 2006; Dwyer and Mannan 2007). In Tucson, after selected dangerous power poles were retrofitted using "raptor-safe" designs, Dwyer and Mannan (2007) documented an 83% reduction in numbers of electrocuted Harris's Hawks, or only 0.2 electrocutions/nest.

Between 1979 and 1989, nearly 200 Harris's hawks were released in an effort to restore an extirpated population along the lower Colorado River in California and southwestern Arizona (Walton et al. 1988). At least seven pairs attempted to nest in 1988 and 1989, five of which hatched young. Occasional nests were also found in this region throughout the 1990s (S. Henry, pers. comm.). The Colorado River population has since dwindled, probably because of lack of suitable habitat and because proposed restoration of habitat was not implemented. Currently Harris's Hawks are considered casual winter visitors in southeastern California (Patten et al. 2003) and are again functionally extirpated as a breeding species in California.

A number of management recommendations for Harris's Hawk populations have been proposed by

Bednarz et al. (1988), LeFranc and Glinski (1988), Hayden and Bednarz (1991), Bednarz (1995), Dwyer and Mannan (2007), and others. Most of these recommendations have not been implemented. Primary management proposals include: (1) classifying the Harris's Hawk as a U.S. Fish and Wildlife Service Species of Special Concern; (2) establishing raptor-management areas where high densities of breeding Harris's Hawks exist and detrimental human activities can be restricted; (3) systematically monitoring population trends; (4) controlling off-road vehicle use and hunting during peak nesting periods; (5) retrofitting problem power poles and transformers; (6) using captive-bred birds instead of removing wild birds for falconry; (7) developing drown-proof water sources for livestock; (8) restoring degraded riparian areas and rangelands; and (9) establishing public education programs to inform the public of the fascinating ecology of the Harris's Hawk and its contributions to healthy ecosystem function, and to promote conservation of this species and the desert ecosystems upon which it depends.

[APLIC] Avian Power Line Interaction Committee. 2006. *Suggested practices for avian protection on power lines: the state of the art in 2006*. Washington, DC: Edison Electric Institute, APLIC, and Sacramento: California Energy Commission.

Bailey, F. M. 1928. *Birds of New Mexico*. Santa Fe: New Mexico Department of Game and Fish.

Bednarz, J. C. 1986. The behavioral ecology of the cooperatively breeding Harris's Hawk in southeastern New Mexico. Ph.D. diss. University of New Mexico, Albuquerque.

———. 1987a. Successive nesting and autumnal breeding in the Harris' Hawk. *Auk* 104:85–96.

———. 1987b. Pair and group reproductive success, polyandry, and cooperative breeding in Harris' Hawks. *Auk* 104:393–404.

———. 1988a. Harris' Hawk subspecies: is *superior* larger or different than *harrisi*? In *Proceedings of the southwest raptor management symposium and workshop*, ed. R. L. Glinski, B. G. Pendleton, M. B. Moss, M. N. LeFranc Jr., B. A. Millsap, and S. W. Hoffman, 294–300. Washington, DC: National Wildlife Federation.

———. 1988b. A comparative study of the breeding ecology of Harris' and Swainson's Hawks in southeastern New Mexico. *Condor* 90:311–23.

———. 1988c. Cooperative hunting in Harris' Hawks (*Parabuteo unicinctus*). *Science* 239:525–27.

———. 1995. Harris' Hawk (*Parabuteo unicinctus*). No. 146. In *The birds of North America*, ed. A. Poole and F. Gill. Philadelphia, PA: Academy of Natural Sciences, and Washington, DC: American Ornithologists' Union.

Bednarz, J. C., and T. J. Hayden. 1988. The Los Medaños cooperative raptor research and management program, final report 1985–1987. Unpublished report, Department of Biology, University of New Mexico, Albuquerque.

———. 1991. Skewed brood sex ratio and sex-biased hatching sequence in Harris' Hawks. *American Naturalist* 137:116–32.

Bednarz, J. C., and J. D. Ligon. 1988. A study of the ecological bases of cooperative breeding in the Harris' Hawk. *Ecology* 69:1176–87.

Bednarz, J. C., J. W. Dawson, and W. H. Whaley. 1988. Raptor status reports: Harris' Hawk. In *Proceedings of the southwest raptor management symposium and workshop*, ed. R. L. Glinski, B. G. Pendleton, M. B. Moss, M. N. LeFranc Jr., B. A. Millsap, and S. W. Hoffman, 71–82. Washington, DC: National Wildlife Federation.

Bednarz, J. C., T. J. Hayden, and T. Fischer. 1990. The raptor and raven community of the Los Medaños area in southeastern New Mexico: a unique and significant resource. In *Proceedings of the ecosystem management: rare species and significant habitats symposium*, ed. R. Mitchell, C. Shevaik, and D. Leopold, 92–101. Bulletin no. 471. Albany: New York State Museum.

Bent, A. C. 1937. *Life histories of North American birds of prey*. Part 1, U.S. National Museum Bulletin 167. Washington, DC: Smithsonian Institution.

Brannon, J. D. 1980. The reproductive ecology of a Texas Harris' Hawk (*Parabuteo unicinctus harrisi*) population. M.S. thesis. University of Texas, Austin.

Buckland, S. T., D. R. Anderson, K. P. Burnham, J. L. Laake, D.L. Borchers, and L. Thomas. 2001. *Introduction to distance sampling: estimating abundance of biological populations*. Oxford, UK: Oxford University Press.

Bunker, C. D. 1919. Harris' hawk (*Parabuteo unicinctus harrisi*) in Kansas. *Auk* 36:285.

Byrne, J. H. 1854. Diary of the expedition. In *Report of exploration of a route for the Pacific railroad, near the thirty-second parallel of north latitude, from the Red River to the Rio Grande*, by J. Pope, 51–92. Washington, DC: U.S. War Department, Corps of Topographical Engineers.

Clark, W. S., and B. K. Wheeler. 2001. *A field guide to hawks of North America*. 2nd ed. Boston: Houghton Mifflin Co.

Dawson. J. W. 1988. The cooperative breeding system of the Harris' Hawk in Arizona. M.S. thesis. University of Arizona, Tucson.

Dawson, J. W., and R. W. Mannan. 1989. A comparison of two methods of estimating breeding group size in Harris's Hawks. *Auk* 106:480–83.

———. 1991a. The role of territoriality in the social organization of Harris's Hawks. *Auk* 108:661–72.

———. 1991b. Dominance hierarchies and helper contributions in Harris's Hawks. *Auk* 108:649–60.

Dwyer, J. F. 2006. Electric shock injuries in a Harris's Hawk population. *Journal of Raptor Research* 40:193–99.

Dwyer, J. F., and R. W. Mannan. 2007. Preventing electrocutions in an urban environment. *Journal of Raptor Research* 41:259–67.

Earl, J. M. 1918. Harris's Hawks in Ohio. *Wilson Bulletin* 39:15–16.

Ellis, D. H., J. C. Bednarz, D. G. Smith, and S. P. Flemming. 1993. Social foraging classes in raptorial birds. *Bioscience* 43:14–20.

Gerstell, A. T., and J. C. Bednarz. 1999. Competition and patterns of resources use by two sympatric raptors. *Condor* 101:557–65.

Griffin, C. R. 1976. A preliminary comparison of Texas and Arizona Harris' Hawks (*Parabuteo unicinctus*) populations. *Raptor Research* 10:50–54.

Hayden, T. J., and J. C. Bednarz. 1991. The Los Medaños cooperative raptor research and management program, final report 1988–1990. Unpublished report, Department of Biology, University of New Mexico, Albuquerque.

Hubbard, J. P., ed. 1980a. *New Mexico Ornithological Society Field Notes* 19(1): 1 December 1979 to 31 May.

———, ed. 1980b. *New Mexico Ornithological Field Notes* 19(2): 1 June 1980 to 30 November.

———, ed. 1983. New Mexico. *American Birds* 37:211–13.

Hubbard, J. P., K. V. Rosenberg, and G. H. Rosenberg, eds. 1982. Southwest region. *American Birds* 36:204–7.

Jaksic, F. M., J. E. Jimenez, S. A. Castro, and P. Feinsinger. 1992. Numerical and functional response of predators to a long-term decline in mammalian prey at a semi-arid Neotropical site. *Oecologia* 89:90–101.

Jaksic, F. M., J. E. Jimenez, and P. Feinsinger. 1991. Dynamics of guild structure among predators: competition or opportunity? In *Proceedings of the XX International Ornithological Congress*, ed. B. D. Bell et al., 1480–88). Wellington, New Zealand: Hutchenson, Bowman, and Stewart.

LeFranc, M. N. Jr., and R. L. Glinski. 1988. Southwest raptor management issues and recommendations. In *Proceedings of the southwest raptor management symposium and workshop*, ed. R. L. Glinski, B. G. Pendleton, M. B. Moss, M. N. LeFranc Jr., B. A. Millsap, and S. W. Hoffman, 375–92. Washington, DC: National Wildlife Federation.

Lehman, R. N. 2001. Raptor electrocution on power lines: current issues and outlook. *Wildlife Society Bulletin* 29:804–13.

Ligon, J. S. 1961. *New Mexico birds and where to find them.* Albuquerque: University of New Mexico Press.

Mader, W. J. 1975a. Biology of the Harris' Hawk in southern Arizona. *Living Bird* 14:59–85.

———. 1975b. Extra adults at Harris' Hawk nests. *Condor* 77:482–85.

———. 1977. Harris's Hawks lay three clutches of eggs in one year. *Auk* 94:370–71.

———. 1979. Breeding behavior of a polyandrous trio of Harris' Hawks in southern Arizona. *Auk* 96:776–88.

———. 1988. Bay-winged Hawk. In *Handbook of North American birds.* Vol. 4, part 1, *Diurnal raptors*, ed. R. S. Palmer, 394–401. New Haven: Yale University Press.

Mays, N. A., C. M. Vleck, and J. W. Dawson. 1991. Plasma luteinizing hormone, steroid hormones, behavioral role, and nest stage in cooperatively breeding Harris' Hawks (*Parabuteo unicinctus*). *Auk* 108:619–37.

McElroy, H. 1977. *Desert hawking II.* Yuma, AZ: Privately printed.

Nice, M. M. 1970. In review no. 48. *Bird Banding* 41:57.

Pache, P. H. 1974. Notes on prey and reproductive biology of Harris' Hawk in southeastern New Mexico. *Wilson Bulletin* 86:72–74.

Palmer, R. S. 1988. Bay-winged Hawk. In *Handbook of North American birds.* Vol. 4, part 1, *Diurnal raptors*, ed. R. S. Palmer, 390–94. New Haven: Yale University Press.

Parmalee, P. F., and H. A. Stephens. 1964. Status of the Harris's Hawk in Kansas. *Condor* 66:443–45.

Patten, M. A., G. Mccaskie, and P. Unitt. 2003. *Birds of the Salton Sea.* Berkeley: University of California Press.

Pennycuick, C. J., M. R. Fuller, and L. McAllister. 1989. Climbing performance of Harris' Hawks (*Parabuteo unicinctus*) with added load: implications for muscle mechanics and for radiotracking. *Journal of Experimental Biology* 142:17–29.

Snider, P. R., ed. 1994. *New Mexico Ornithological Society Field Notes* 33(4): 1 August to 30 November.

———. 1995. *New Mexico Ornithological Society Field Notes* 34(1): 1 December to 28 February.

Snyder, L. 1919. Harris's Hawk in Kansas. *Auk* 36:567.

Tucker, V. A., and C. Heine. 1990. Aerodynamics of gliding flight in a Harris' Hawk, *Parabuteo unicinctus. Journal of Experimental Biology* 149:469–89.

van Rossem, A. J. 1942. Notes on some Mexican and Californian birds, with description of six undescribed races. *Transactions of San Diego Society of Natural History* 9:377–84.

Walton, B., J. Linthicum, and G. Steward. 1988. Release and re-establishment techniques developed for Harris's Hawks—Colorado River 1979–1986. In *Proceedings of the southwest raptor management symposium and workshop*, ed. R. L. Glinski, B. G. Pendleton, M. B. Moss, M. N. LeFranc Jr., B. A. Millsap, and S. W. Hoffman, 318–20. Washington, DC: National Wildlife Federation.

Whaley, W. H. 1986. Population ecology of the Harris' Hawk in Arizona. M.S. thesis. University of Arizona, Tucson.

Williams, S. O. III, ed. 1999. New Mexico region. *North American Birds* 53:312–14.

———, ed. 2000a. New Mexico. *North American Birds* 54 (1): 86–89.

———. 2000b. New Mexico. *North American Birds* 54(3): 312–15.

———, ed. 2001. New Mexico. *North American Birds* 55:336–38.

———, ed. 2002. New Mexico. *North American Birds* 56:207–9.

Witzeman, J., J. P. Hubbard, and K. Kaufman, eds. 1979. The autumn migration: southwest region. *American Birds* 33:202–4.

14

Broad-winged Hawk

(*Buteo platypterus*)

JEAN-LUC E. CARTRON

A FAMILIAR SIGHT in spring and summer in large forested areas of eastern North America, the Broad-winged Hawk (*Buteo platypterus*) occurs in New Mexico, but only during migration and only in small numbers. As a result, inexperienced observers may overlook or misidentify migrating Broad-winged Hawks simply because they do not expect to see the species in our state. And yet, the Broad-winged Hawk is distinctive enough. It is smaller than all the other buteos found in northern North America, measuring only 34–42 cm (13–17 in) in length and weighing 308–483 g (11–17 oz) (Wheeler and Clark 1995), while at the same time lacking the long tail and rounded wings of accipiters.

In flight, the adult light morph is readily identified by the conspicuous, broad black band on the trailing edge of the otherwise pale underwings, the dark tips of the primaries, and especially the tail, showing broad black bands alternating with white bands, one broad, the other one or two narrow. The broad white band between the two equally broad black bands remains visible on the undertail of the perched adult light-morph. The underparts are mostly pale, with rufous-brown barring across the belly becoming denser across the breast, even forming a bib on some birds. The back and wings are dark brown. The immature light-morph lacks most of the distinguishing characteristics of the adult. The border along the trailing edge of the underwings is narrow and dusky rather than black, the alternating dark and pale bands on the tail are narrower and less contrasting, and the rufous-brown crossbars on the belly and breast are missing, replaced by variable amounts of dark streaking. Helpful for identifying juveniles are a pale supercilium, often distinct, and a dark malar stripe contrasting with the white throat. Dark-morph Broad-winged Hawks are rare but have been observed in New Mexico (Ligon 1961; Parmeter 2007). The adult's body and wing coverts are dark brown, with contrasting pale flight feathers. As in the adult light-morph, the trailing edge of the underwing shows the wide black band and the tail the alternating black and white bands. Immature dark-morphs are similar to adult dark-morphs but without the broad bands on the tail and without the broad black band along the trailing edge of the underwings. In all Broad-winged Hawks, the wings show pointed tips especially while birds are gliding. Neither the immature nor the adult Broad-winged Hawk shows the white crescent across the primaries of the otherwise often similar Red-shouldered Hawk (*Buteo lineatus*), casual in New Mexico (chapter 40). Some juvenile light-morph Broad-winged Hawks may have faint abdominal bands but can be distinguished from juvenile Red-tailed

PHOTO 14.1

Adult light-morph Broad-winged Hawk, Creek Co., Oklahoma, 10 June 2005. Note the broad black and white bands on the undertail and the dense, rufous-brown barring across the breast, almost forming a bib.

PHOTOGRAPH: © STEVE METZ.

PHOTO 14.2

(*right*) Immature light-morph Broad-winged Hawk taking off, Chico Basin Ranch, El Paso Co., Colorado, 11 May 2003. Immatures are more difficult to identify than adults. The tail bands are narrower and less contrasting; the trailing edge of the underwings is dusky rather than black. PHOTOGRAPH: © BILL SCHMOKER.

PHOTO 14.3

(*below*) Immature Broad-winged Hawk rescued by Rio Rancho Animal Control in early October 2002. Helpful field marks for identifying immatures include a pale, often distinct supercilium (eyebrow) and a dark malar stripe contrasting with the white throat. PHOTOGRAPH: © JOHN P. DELONG.

Hawks (*Buteo jamaicensis*) by the absence of a dark patagial mark (see chapter 18).

The main call of the Broad-winged Hawk is a "*peeeurr*," a high-pitched whistle uttered at all times of year (Burns 1911; Bent 1937; Matray 1974; Palmer 1988).

Six subspecies of Broad-winged Hawks are recognized. The subspecies breeding in mainland North America and migrating through New Mexico is the nominate form, *Buteo platypterus platypterus* (Goodrich et al. 1996).

Distribution

The Broad-winged Hawk breeds on mainland North America and in the Caribbean (Goodrich et al. 1996). On mainland North America, the breeding range extends from southern Canada south through much of the eastern United States to the Gulf Coast and northern Florida. Westward, the edge of the main breeding distribution is in central Alberta, northeastern North Dakota, Minnesota, Iowa, Missouri, eastern Oklahoma, and northeastern Texas, with small disjunct populations in northeastern Kansas and in southwestern and central Iowa (Goodrich et al. 1996).

The Broad-winged Hawk is highly migratory. All birds of the nominate form depart from their breeding grounds in the fall, mainly toward Mexico and Central and South America, but also southern Florida, the Florida Keys, and probably some of the Caribbean islands (Goodrich et al. 1996). In Mexico, the wintering range is mainly from Colima eastward along the Pacific slope and from northern Chiapas eastward and southward along the Gulf of Mexico and Caribbean slope (Ferguson-Lees and Christie 2001). The southern edge of the wintering distribution is in northern and eastern Peru, Bolivia, and southern Brazil (AOU 1983). Most of the birds wintering in southern Florida

PHOTO 14.4

Adult Broad-winged Hawk on its wintering grounds in Costa Rica, 28 February 2007. PHOTOGRAPH: © DOUG BROWN.

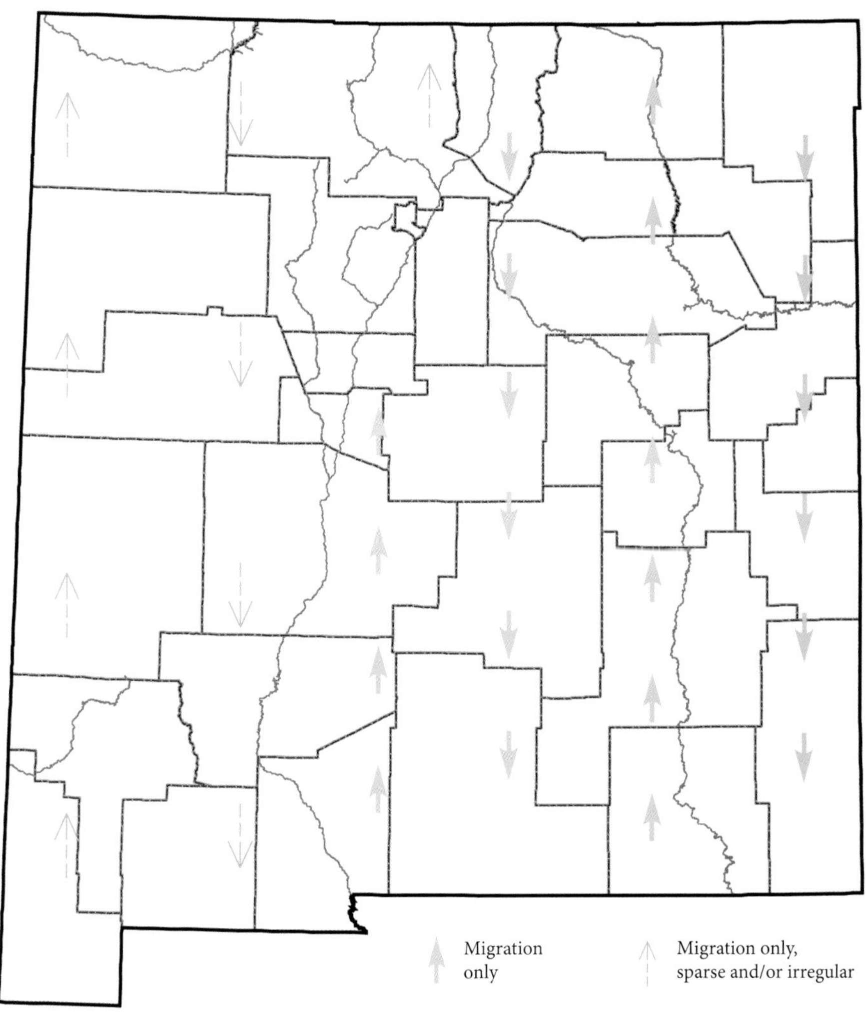

MAP 14.1

Broad-winged Hawk distribution map

are immatures, with occasional winter records from the lower Mississippi Delta and coastal Texas also involving immatures (Goodrich et al. 1996).

During migration in New Mexico, the species has been reported primarily from the eastern part of the state west to the Rio Grande Valley (Ligon 1961; Parmeter et al. 2002; Parmeter 2007; NMOS 2008). Records of occurrence from west of the Rio Grande Valley are mainly from the southwestern corner of the state and include sightings of single birds near Gila, Grant County, on 12 May 1973 (Monson 1973, *fide* D. Zimmerman); on the Zuni Reservation, McKinley County, on 3 May 1981 (Rosenberg et al. 1981, *fide* J. Trochet); in the Jemez Mountains, Los Alamos County, on 5 June 1991 (Williams and Hubbard 1991, *fide* J. Black and B. Black, tentative identification only); the Florida Mountains, Luna County, on 7 September 1998 (Williams 1999, *fide* L. Malone); Mangas Springs, Grant County, on 4 May 2000 (Williams 2000, *fide* J. Paton); Katfish Kove, Luna County, on 20 April 2003

PHOTO 14.5

Immature Broad-winged Hawk, Rodeo, Hidalgo Co., 15 October 2006. Broad-winged Hawks are less common west of the Rio Grande Valley.
PHOTOGRAPH: © RICHARD E. WEBSTER.

(Williams 2003, *fide* M. Scott and J. Zabriskie); and Rodeo, Hidalgo County, on 15 October 2006 (Williams 2007, *fide* R. Webster; photo 14.5).

During migration, the Broad-winged Hawk uses both ridge updraft and thermals (Goodrich et al. 1996). For this reason it does not just follow mountain ranges but instead can also be observed over flat topography. Small numbers of Broad-winged Hawks are observed every year in spring and fall over the Sandia and Manzano mountains (chapter 2). However, it is at migration stopovers of the eastern plains (e.g., Rattlesnake Springs in Eddy Co. and Boone's Draw in Roosevelt Co.) that the Broad-winged Hawk is most likely encountered in New Mexico (see Parmeter et al. 2002).

The occurrence of dark-morph birds in New Mexico—two of six Broad-winged Hawks recorded at the North Roosevelt Trap in Roosevelt County through 2006 were dark-morphs (Parmeter 2007)—strongly suggests that New Mexico lies in the path of migrants from the northwestern portion of the species' breeding distribution, where most dark-morphs are found (Wheeler and Clark 1995; Goodrich et al. 1996). Migrating dark-morph birds have been recorded mainly along the eastern Great Plains but also throughout the West (Wheeler and Clark 1995).

Habitat Associations and Life History

On its breeding and wintering grounds the Broad-winged Hawk is a bird of forested areas (Goodrich et al. 1996). However, the eastern plains of New Mexico harbor only a few isolated wooded areas typically surrounded by vast grasslands. It is largely in these small woodlands, often referred to as migrant traps because they attract many migrating forest-dwelling birds, that Broad-winged Hawks find stopover habitat in New Mexico. One migrant trap where Broad-winged Hawks are sighted regularly in migration is Boone's Draw, which is located about 23 km (14 mi) southwest of Portales in Roosevelt County. Designated as an Important Bird Area, it is a 61-ha (151-ac), privately owned woodland dominated by cottonwoods (*Populus fremontii*), saltcedar (*Tamarix* sp.), and seep-willow (*Baccharis* sp.). Broad-winged Hawks can also be observed in another migrant trap of the eastern plains, the North Roosevelt (or Melrose) Trap. It is located on State Trust land about 16 km (10 mi) west of Melrose in Roosevelt County. Here again the trees are mainly cottonwoods, with also some dense silver poplars (*Populus alba*) (Parmeter 2007). Small towns with city parks and tree-lined streets (e.g., Tatum and Jal, Lea Co.) can also serve as stopover habitat in the eastern plains, with Broad-winged Hawks reported from such locations (Parmeter et al. 2002; J. Parmeter, pers. comm.).

Perhaps the best location for observing Broad-winged Hawks in New Mexico is Rattlesnake Springs in Eddy County (Parmeter et al. 2002). Much like Boone's Draw and the North Roosevelt Trap farther north, Rattlesnake Springs attracts many spring and fall migrants. It is an oasis with permanent water and large cottonwood trees surrounded by vast Chihuahuan Desert plains. Farther west, Broad-winged Hawks also use stopover habitat in the Rio Grande Valley, both the Middle Rio Grande bosque and, downstream from Elephant Butte Reservoir, Percha Dam State Park's extensive stands of cottonwoods (NMOS 2008; J. Parmeter, pers. comm.).

PHOTO 14.6

(*top left*) Example of Broad-winged Hawk stopover habitat in New Mexico: the North Roosevelt (or Melrose) Trap. 28 April 2007.
PHOTOGRAPH: © JERRY OLDENETTEL.

PHOTO 14.7

(*bottom left*) The North Roosevelt (or Melrose) Trap is one of eastern New Mexico's most popular destinations for birding during the migration season. Surrounded by vast expanses of grassland, that location attracts many migrating species that are forest-dwellers, including the Broad-winged Hawk.
PHOTOGRAPH: © JAMES N. STUART.

PHOTO 14.8

(*above*) Broad-winged Hawk at the North Roosevelt (or Melrose) Trap, Roosevelt Co., 5 October 2007.
PHOTOGRAPH: © JERRY OLDENETTEL.

PHOTOS 14.9a and b

Rattlesnake Springs, Eddy Co. Perhaps the best location for observing Broad-winged Hawks in New Mexico, especially 20–30 April.

PHOTOGRAPHS: © DAVID J. GRIFFIN.

PHOTO 14.10

Percha Dam State Park, Sierra Co. Broad-winged Hawk stopover habitat includes cottonwood (*Populus* sp.) stands such as those found along the Rio Grande, both in the Middle Rio Grande bosque and at Percha Dam State Park. Downstream from Elephant Butte Reservoir, most of the cottonwood groves once found along the Rio Grande have disappeared. One exception is Percha Dam, where migrating Broad-winged Hawks have been recorded. PHOTOGRAPH: © JERRY OLDENETTEL.

PHOTO 14.11

Migrating Broad-winged Hawk "kettle," Santa Anna National Wildlife Refuge, Texas, March 2005. Kettles can consist of thousands of birds migrating together.

PHOTOGRAPH: © BILL SCHMOKER.

PHOTO 14.12

Bosque Redondo Lake, De Baca Co. (winter view). Bosque Redondo consists of a string of oxbow lakes with reeds, cottonwoods (*Populus fremontii*), and saltcedars (*Tamarix* sp.) in the Pecos River floodplain. Most New Mexico records of Broad-winged Hawks are single birds, but a "kettle" of five Broad-winged Hawks was observed at Bosque Redondo by John Parmeter, 5 October 2002. PHOTOGRAPH: © JERRY OLDENETTEL.

PHOTO 14.13

Adult Broad-winged Hawk during migration at National Audubon Society's Sabal Palm Sanctuary in Brownsville, Texas, 13 March 2004. A migrating flock had dropped down for the night and to feed. The Broad-winged Hawks hunted from the trees, swooping down into the tall grass, where presumably they found rodents and grasshoppers, both important in the species' diet. PHOTOGRAPH: © LEE ZIEGER.

Migration of Broad-winged Hawks in both spring and fall tends to be highly synchronized, compressed into a shorter period of time compared to that for other raptors (Goodrich et al. 1996). Over the Manzano Mountains in the fall, Broad-winged Hawks may be recorded from late August to mid October, but mostly from mid September to early October (chapter 2), and at the North Roosevelt Trap, all six fall records given by Parmeter (2007) fall within a 15-day period, from 22 September to 7 October. In spring, Broad-winged Hawks occur over the Sandias from early April to early May, but again mainly over a short time period of about two weeks (chapter 2). Highly synchronized migration presumably reflects the Broad-winged Hawk's need to save energy by finding thermals for soaring during the long migratory journey, at a time when thermals are most reliably found. Migrating birds ride up thermals, then glide downward before entering the next thermal. Highly synchronized Broad-winged Hawk migration often results in the formation of flocks of several thousand birds, named "kettles" (e.g., Goodrich et al. 1996). Such large flocks have never been observed in the skies of New Mexico. Most New Mexico records are of single birds. However, on 5 October 2002 John Parmeter (pers. comm.) observed a kettle of five soaring at Bosque Redondo in De Baca County.

The Broad-winged Hawk's diet often consists mainly of small mammals and amphibians, in addition to birds (mostly nestlings and fledglings during the breeding season), reptiles, and some invertebrates including orthopterans and other insects (Goodrich et al. 1996). No information exists for New Mexico other than Ligon's (1961) mention of finding grasshoppers in the stomach of a bird he collected about 65 km (40 mi) south of Portales. The Broad-winged

PHOTO 14.14

Immature Broad-winged Hawk recovered by Rio Rancho Animal Control and given to wildlife rehabilitator Shirley Kendall on 5 October 2002. The bird was injured, having apparently hit a window. It was banded by Steve Cox (band number 2206-39734) and was successfully released on 28 October 2002 at Capilla Peak in the Manzano Mountains.

PHOTOGRAPH: © JOHN P. DELONG.

Hawk is typically a perch-and-wait predator waiting in a tree, scanning the forest floor, and swooping down on its prey.

Status and Management

Like many other raptors, the Broad-winged Hawk was once the victim of widespread persecution in the United States, and Burns (1911) in particular mentions a record of dubious distinction set by one man in New Jersey after he shot 298 birds in one day! Although shooting still occurs in Mexico and father south, a more serious threat may now consist of deforestation along migration pathways and on wintering grounds (see Goodrich et al. 1996). Possible impacts of harmful chemicals on wintering grounds should also be investigated.

At present, there is no clear population trend at the scale of the entire species' distribution (Goodrich et al. 1996). However, numbers recorded at migration study sites throughout the West have been increasing over time, suggesting that the species may be expanding its range westward in Canada (Smith et al. 2001; Hoffman and Smith 2003; chapter 2).

During migration, the concentration of Broad-winged Hawks in funnel areas makes it possible to estimate the total population size of the Broad-winged Hawk's nominate subspecies. Broad-winged Hawk numbers in mainland North America may not be too far from a count of up to 1.7 million birds recorded in one fall season in eastern Mexico (Goodrich et al. 1996). In comparison, annual averages of only 6 and 7 birds have been counted over the Sandias in spring from 1985 to 2006 and over the Manzanos in fall from 1985 to 2005, respectively (chapter 2). In New Mexico, there is no special threat to the Broad-winged Hawk—nor would any such threat be meaningful at the scale of the species' distribution—although the species in our state will benefit from the preservation of migrant traps in the eastern plains and riparian corridors and stopover habitat elsewhere.

LITERATURE CITED

[AOU] American Ornithologists' Union. 1983. *Check-list of North American birds.* 6th ed. Washington, DC: American Ornithologists' Union.

Bent, A. C. 1937. *Life histories of North American birds of prey.* Part 1. *U.S. National Museum Bulletin* 167. Washington, DC: Smithsonian Institution.

Burns, F. L. 1911. A monograph of the Broad-winged Hawk (*Buteo platypterus*). *Wilson Bulletin* 23 (3/4): 1–320.

Ferguson-Lees, J., and D. A. Christie. 2001. *Raptors of the world.* Boston: Houghton Mifflin Co.

Goodrich, L. J., S. C. Crocoll, and S. E. Senner. 1996. Broad-winged Hawk (*Buteo platypterus*). No. 218. In *The birds of North America*, ed. A. Poole and F. Gill. Philadelphia, PA: Academy of Natural Sciences, and Washington, DC: American Ornithologists' Union.

Hoffman, S. W., and J. P. Smith. 2003. Population trends of migratory raptors in western North America, 1977–2001. *Condor* 105:397–419.

Ligon, J. S. 1961. *New Mexico birds and where to find them.* Albuquerque: University of New Mexico Press.

Matray, P. F. 1974. Broad-winged Hawk nesting and ecology. *Auk* 91:307–24.

Monson, G., ed. 1973. Southwest region. *American Birds* 27:803–6.

[NMOS] New Mexico Ornithological Society. 2008. *NMOS Field Notes* database. http://nhnm.unm.edu/partners/NMOS/ (accessed 9 December 2008).

Palmer, R. S. 1988. *Handbook of North American birds.* Vol. 5. New Haven: Yale University Press.

Parmeter, J. E. 2007. Annotated checklist of the birds of the Melrose Migrant Trap, Roosevelt County, New Mexico. *New Mexico Ornithological Society Bulletin* 35:1–40.

Parmeter, J. E., B. Neville, and D. Emkalns. 2002. *New Mexico bird finding guide.* 3rd ed. Albuquerque: New Mexico Ornithological Society.

Rosenberg, K. V., J. P. Hubbard, and S. B. Terrill, eds. 1981. Southwest region. *American Birds* 35:849–52.

Smith, J. P., P. Grindrod, and S. W. Hoffman. 2001. Migration counts indicate Broad-winged Hawks are increasing in the West: evidence of breeding range expansion? In *Hawkwatching in the Americas*, ed. K. L. Bildstein and D. Klem Jr., 93–106. North Wales, PA: Hawk Migration Association of North America.

Wheeler, B. K., and W. S. Clark. 1995. *A photographic guide to North American raptors.* New York: Academic Press.

Williams, S. O. III, ed. 1999. New Mexico region. *North American Birds* 53:86–88.

———, ed. 2000. New Mexico. *North American Birds* 54:312–14.

———, ed. 2003. New Mexico. *North American Birds* 57:383–85.

———, ed. 2007. New Mexico. *North American Birds* 61:115–18.

Williams, S. O. III, and J. P. Hubbard, eds. 1991. Southwest region: New Mexico. *American Birds* 45:1146–49.

15

Gray Hawk

(*Buteo nitidus*)

DAVID J. KRUEPER

THE GRAY HAWK (*Buteo nitidus*) (AOU 2006) is a medium-sized neotropical buteo that reaches its northernmost distribution in the southwestern United States. It is intermediate in shape between buteos and accipiters, with a longer tail, shorter and rounder wings, and longer legs than other buteo species. The flight is similar to that of an accipiter, with rapid wing beats followed by a short glide, a characteristic that gave the Gray Hawk its older name, the Mexican Goshawk. As with most raptor species, the female is distinctly larger than the male, with the average Arizona adult female weighing 628 g (22 oz) and reaching 44 cm (17 in) in length, while the average Arizona adult male weighs 429 g (15 oz) and reaches about 41 cm (16 in) in length (Oberholser 1974; Bibles et al. 2002). The wingspan ranges from 81 to 94 cm (32 to 37 in) for the species (Howell and Webb 1995).

The adult plumage is typically reached via a complete definitive prebasic molt in April through August beginning in the second calendar year, although some birds have been observed to undergo a partial body molt earlier in the year (Bent 1937; Dickey and van Rossem 1938). The sexes are similar in coloration with the female reported to be slightly darker (Bibles et al. 2002). The head and upperparts are a medium gray coloration with the underparts being finely barred with gray and white throughout. The eyes are light brown and the cere, legs, and feet are yellow. The undertail coverts are conspicuously white. The upper side of the tail is grayish-black overall with two bold white bands while the underside is dark gray with three white bands. The underwings are pale white with gray

PHOTO 15.1

Adult Gray Hawk, near Gomez Farias, Tamaulipas, Mexico, 21 February 2008. Below, adults are finely barred gray and white throughout. The head and upperparts are medium gray. The cere, legs, and feet are yellow.

PHOTOGRAPH: © LEE ZIEGER.

PHOTO 15.2

Immature on power pole near El Fuerte, Sinaloa, Mexico, December 2008. Immature birds show a dark eye-stripe and distinct malar stripe. The underparts are white with strong chocolate vertical streaking. The cere, legs, and feet are yellow.

PHOTOGRAPH: © MARK L. WATSON.

barring throughout with the outermost five primaries on the outer rounded profile of the wing having black tips (Howell and Webb 1995; Sibley 2000; National Geographic Society 2006).

The immature plumage is somewhat variable, but in general the upperparts are dark blackish-brown with lighter buff mottling, while the underparts are buff-brown or light brown with heavy blackish-brown streaking extending from the throat and neck down to the belly. The long tail is grayish-brown with five to nine dark-brown to black tail bands. The undertail and uppertail coverts are whitish with some brown mottling, with the leg feathers brownish-buff with blackish barring. The underwings are also buff-brown with dark barring on the flight feathers. Conspicuous facial patterns in the immature plumage that can be important for identification include the blackish eye-line below a white superciliary and a blackish malar stripe on white throat. Diagnostic in both adult and juvenal plumages is a conspicuous white "U" on the upper rump at the base of the tail feathers.

Distribution

Gray Hawks are a very widely distributed neotropical species, breeding from the extreme southwestern United States and southern Texas south throughout Mexico, Central America, northern Argentina, Paraguay, and southeastern Brazil (AOU 1998; Bibles et al. 2002). The winter range is the same as the breeding range, except for the northernmost populations within Arizona, northern Sonora, New Mexico, and west Texas, which are migratory (Brown and Amadon 1968; Bibles et al. 2002). Most individuals from these northern populations withdraw to the south several hundred kilometers or farther in the nonbreeding season, but wintering individuals have also been noted in southern Arizona and in the Lower Rio Grande Valley of Texas (Oberholser 1974; Monson and Phillips 1981; Lockwood and Freeman 2004; Brush 2005). It appears that this species has expanded its breeding range to the north, east, and west within Arizona during the past half-century (Bibles et al. 2002; Corman 2005), and within the core of its breeding range in Arizona it has increased in abundance since 1986 (Glinski and Millsap 1987; Krueper 1999).

Within New Mexico, Gray Hawks are considered to be rare and very local in distribution. The species was first reported within the state on 23 April 1876 when two nests were located near Fort Bayard, Grant County, by the collector Frank Stephens (Hubbard 1970, 1974). However, these first records were questioned based upon the early nesting date, nesting habitat, materials used in the nest construction, and coloration and size of the eggs themselves, which were collected and catalogued at the American Museum of Natural History. Based on the above, Hubbard (1974) believed that the records pertained to nesting Cooper's Hawks (*Accipiter cooperii*). Zimmerman (1976), however, urged caution when considering the two records, and suggested not to dismiss them because they did not conform to what

PHOTO 15.3

Gray Hawk, Playa de Oro, Jalisco, Mexico, February 2007. Gray Hawks are distributed along the Sierra Madre Occidental from southeastern Arizona and southwestern New Mexico in the west, and along the Sierra Madre Oriental from southern Texas in the east, south to northern Argentina, Paraguay, and southeastern Brazil. PHOTOGRAPH: © PAT WATT.

PHOTO 15.4

Gray Hawk in flight, Arizona, 7 July 2007. In Arizona, Gray Hawks are primarily summer residents to the southeastern corner of the state, from the Altar Valley to the west to the San Bernardino Valley in the east, and north to the vicinity of the Gila River. PHOTOGRAPH: © TOM KENNEDY.

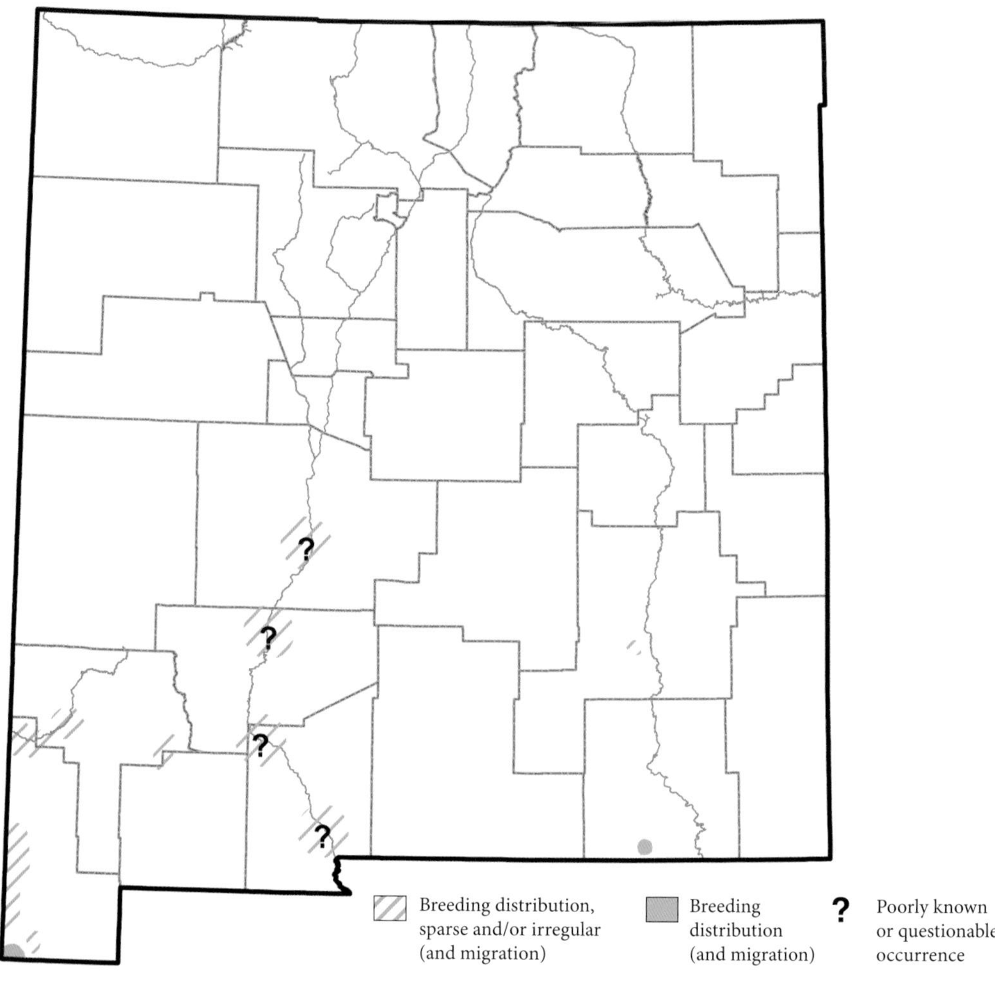

MAP 15.1

Gray Hawk distribution map

is considered "normal" Gray Hawk natural history. An additional early record of an adult with one fledgling was also reported from near Cliff, Grant County, in July 1953, but it was not verified (Hubbard 1978).

The species has since been documented again from within Grant County, but also from extreme northern Luna County, and, on multiple occasions, Hidalgo County, all from the southwestern corner of the state (Hubbard 1978; R. Glinski *in litt.*; map 15.1). Gray Hawks have also been reported from Doña Ana, Chaves, Eddy, Sierra, and Socorro counties (see below). The earliest substantiated records are from San Simon Cienega on 10 April 1961 and the Mimbres Valley on 16 May 1973 (Zimmerman 1976). The first and only record from Luna County dates back to 6 August 1975, when Rich Glinski (*in litt.*) saw and heard at least one adult along the Mimbres River at Walsh Ranch, about 13 km (8 mi) downstream from Dwyer along Highway 61. After a period of time with few substantiated records from the state, reports increased in the early 1990s and have since become annual from the southwestern corner of New Mexico, while also becoming increasingly regular farther east (Williams and Krueper 2008). Within the past decade, the species has been most frequently recorded from Guadalupe Canyon, Hidalgo County, with reports of one or more Gray Hawks every year

PHOTO 15.5

(*left*) Gray Hawk nest, Rattlesnake Springs, Eddy Co., 8 June 2007. Prior to 2007, there were no nesting records for the Gray Hawk from Rattlesnake Springs. In fact, nesting had been reported in New Mexico from only one location: Guadalupe Canyon, Hidalgo Co., near the Arizona state line. The 2007 Gray Hawk nest at Rattlesnake Springs failed. PHOTOGRAPH: © ROBERT H. DOSTER.

PHOTOS 15.6a and b

(*below left and right*) Gray Hawk fledgling, Rattlesnake Springs, Eddy Co., 27 August 2008. Two adults had been recorded at Rattlesnake Springs earlier in the season. Although no nest was found, the photos of the fledgling indicate that nesting did occur in 2008 at Rattlesnake Springs, and that it was successful. PHOTOGRAPHS: © ROBERT H. DOSTER.

since 2004 (Williams 2004, 2005, 2006, 2007, 2008). Another area of the state where Gray Hawks have been regularly recorded consists of the Animas Valley, also in Hidalgo County. In August–September 2007, the species was even sighted at three separate locations within the valley (Williams 2007). Outside Hidalgo County, Gray Hawks have been reported infrequently from the Gila River Valley near Red Rock and Cliff within Grant County since at least 2004 by various observers (e.g., Williams 2006). Since 2006, the Gray Hawk has staged a spectacular eastward extension of its range in New Mexico. The species has been documented in particular at Rattlesnake Springs in Eddy County in 2006, 2007, and 2008 (Williams 2006, 2007, 2008; see below). Additional noteworthy reports from very recent years include possible Gray Hawks at Mesilla Park, Doña Ana County, on 13 April 2006 (*fide* J. Nemeth); Las Animas Creek, Sierra County, on 22 April 2006 (*fide* R. Ketchum); and near San Marcial, Socorro County, 13 May 2006 (*fide* J. Oldenettel) (Williams 2006); and an adult at Mesilla Park, Doña Ana County, 18 May 2008 (Williams 2008, *fide* Joshua Nemeth).

With the 1876 nesting record from Bayard cast in doubt, and the 1953 report of a fledgling not verified, the first confirmed breeding in New Mexico is recent and from Guadalupe Canyon, which is located in the southern Peloncillo Mountains and straddles the Arizona–New Mexico state line. On 25 July 2004, C. Lundblad (*in litt.*) found a full-grown juvenile on a nest in the New Mexico portion of the canyon, about 3.5 km (2 mi) east of the state line. An adult was also perched toward the top of the same tree, and the juvenile appeared to have already fledged. Breeding in the Arizona portion of Guadalupe Canyon had been known to occur since at least 2000 (Williams 2000), while in the New Mexico portion of the canyon it was first suspected in 2002 following observations of one juvenile (Williams 2002; Williams and Krueper 2008).

Since 2004, breeding in the New Mexico portion of Guadalupe Canyon has been documented annually, often by multiple observers. On several occasions in April and May 2005, a pair was documented at a nest 12 m (40 ft) or higher in a cottonwood (*Populus fremontii*) toward the upper end of the canyon. On 9 July 2005, Narca Moore-Craig (pers. comm.) observed an adult delivering a snake at the nest, while on 9 August 2005 Jonathan Batkin (pers. comm.) and others observed two Gray Hawk fledglings perched in a tree in the vicinity of the nest site. An adult flew to them, bringing a small, unidentified mammal, which it then fed to the fledglings. On 8 July 2005 an immature was also seen in a different area of the New Mexico side of the canyon (J. Oldenettel, pers. comm.), suggesting a second nest that year. In August of the next year (2006), in the vicinity of the 2005 nest site, the author observed an adult flush from a cottonwood with a nest, with one immature within 0.5 km (0.3 mi) of the nest. In 2007, two active nests were reported from Guadalupe Canyon in New Mexico (Williams 2007, *fide* Narca Moore-Craig et al.). One pair nested at the upper end of the canyon near the 2005 nest site (N. Moore-Craig, pers. comm.), but the nest tree was a different cottonwood and the nest was placed about 10 m (35 ft) or higher above the ground. A second nest was located about 1.6 km (1 mi) from the first. In 2008, the New Mexico portion of Guadalupe Canyon harbored three occupied nesting territories, with an additional two or three territories within the Arizona portion of the canyon, each spaced again approximately 1.6 km (1 mi) apart (N. Moore-Craig, pers. comm.). Thus Gray Hawks are now breeding annually in Guadalupe Canyon, with a seemingly growing number of nesting pairs present. It remains to be seen what the carrying capacity of the canyon is, but at some point soon the saturation point may be reached.

Nesting was also recently documented at Rattlesnake Springs, Eddy County. An occupied nest was photographed on 6 and 8 June 2007 by Rob Doster (photo 15.5). By 11 July, however, the nest had been abandoned, but the pair remained in the area (Williams 2007). In 2008, a pair was observed at Rattlesnake Springs by multiple observers, but no nest could be located. However, on 27 August 2008, Rob Doster found and photographed a fledgling at Rattlesnake Springs, indicating successful nesting at that location (photos 15.6a and b).

The Gila River drainage near Cliff in Grant County represents the northernmost record of the Gray Hawk in New Mexico. A report of a Gray Hawk in 1989 from farther north, in the Socorro area, is an unconfirmed record. Extreme records for New Mexico are 5 April (2008) in Guadalupe Canyon and 6 October (2004), when an immature was observed near Rodeo, Hidalgo County (Williams 2008).

Habitat Associations and Life History

Gray Hawks typically inhabit arid deciduous and semideciduous woodlands throughout their range, primarily below 1,400 m (4,620 ft) (Bibles et al. 2002). Within Arizona and New Mexico, Gray Hawks are typically found in montane riparian-associated canyons that contain one or more of the following vegetative associations: Fremont cottonwood, netleaf hackberry (*Celtis reticulata*), Goodding's willow (*Salix gooddingii*), velvet ash (*Fraxinus velutina*), Arizona sycamore (*Platanus wrightii*), and/or Arizona walnut (*Juglans major*) (see Dick-Peddie 1993). Velvet mesquite (*Prosopis* sp.) is almost always found in the adjacent uplands surrounding the riparian area (Glinski 1998; Bibles et al. 2002).

The natural history of the Gray Hawk has not been

PHOTOS 15.7a and b

Examples of Gray Hawk habitat in New Mexico: Guadalupe Canyon, located in the southern Peloncillo Mountains. It straddles the New Mexico–Arizona state line, with the upper portion of the canyon in Hidalgo Co., New Mexico. In recent years, Gray Hawks have been recorded nesting in Guadalupe Canyon in both Arizona and New Mexico.

PHOTOGRAPHS: © DAVID J. KRUEPER.

studied in New Mexico. However, it is almost certainly similar to the species' natural history reported from southeastern Arizona, where Gray Hawk habitat is similar. The birds typically arrive in mid to late March and return to the previous year's nesting territories. Courtship and nest building begin almost immediately, with both sexes constructing the nest, usually placed in a Fremont cottonwood about two-thirds of the way up and near the trunk (Bibles et al. 2002). The nest observed in 2004 in the New Mexico portion of Guadalupe Canyon was on a main branch in the mid-canopy of a large Fremont cottonwood (C. Lundblad, *in litt.*).

Two to three eggs are laid typically in early May, with an incubation period of approximately 33 days (Glinski 1998). The young remain in the nest for about six weeks and then fledge within the first two weeks of July (Bibles et al. 2002). The above timeline was determined in Arizona, but as mentioned above it almost certainly applies to New Mexico as well. This is confirmed by C. Lundblad's observation at the 2004 nest of what appeared to be an already fledged young in late July.

In Arizona, the diet consists primarily of whiptail (*Aspidoscelis* spp.) and spiny lizards (*Sceloporus* spp.), with small birds such as Gambel's Quail (*Callipepla gambelii*), doves (*Zenaida* spp.), kingbirds (*Tyrannus* spp.), and towhees (*Pipilo* spp.), and small mammals such as cottontail rabbit (*Sylvilagus* spp.), woodrat (*Neotoma* spp.), and deer mice (*Peromyscus* spp.) being occasionally taken (Bibles et al. 2002). The species is a "perch-and-pounce" predator, typically perching below the canopy or on low branches in or adjacent to mesquite woodlands, waiting until it has spotted a prey item, and then with great maneuverability and speed pursuing and capturing its target prey.

Gray Hawks typically remain in the natal area until late September or early October (Bibles et al. 2002) and then depart southward to spend the nonbreeding season in Mexico. Exceptionally, birds might remain into November and December, and a few were recorded overwintering near Amado along the Santa Cruz River south of Tucson in 2004–5 (Arizona Bird Records Committee files). Glinski (1998) found that of seven birds originally banded as nestlings in Arizona, all midwinter band recoveries came from individuals overwintering in northern Sinaloa, Mexico.

Status and Management

The total population within the United States is estimated at under 100 pairs (Taylor 2006; Williams and Krueper 2008), with about 80 nesting pairs in Arizona alone (Glinski 1998). The species is not considered to be globally threatened and is generally widespread and relatively numerous within the Latin American portion of its range (del Hoyo et al. 1994). At least within Arizona, nesting sites are reoccupied annually and productivity seems to be sustaining the current population (Corman 2005). While the species has shown noticeable increases in abundance and distribution within Arizona in the past two decades, habitat degradation, conversion to agriculture, and illegal shooting, primarily within the tropics, provide continuing threats for the species (Glinski 1998; Taylor 2006).

Within New Mexico, the species appears to be increasing in abundance and distribution (Williams and Krueper 2008). However, as with many other avian species at the periphery of their distribution, any range expansion and population increase could prove temporary. Compared to Arizona, Gray Hawk habitat is also more limited in New Mexico. Thus, while the outlook for the Gray Hawk in New Mexico currently seems positive, any increase in distribution and abundance is likely to be less dramatic than in Arizona.

Acknowledgments

I would like to thank Jonathan Batkin, Brent Bibles, Bill Howe, Jerry Oldenettel, John Parmeter, Christopher Rustay, Sandy Williams, and the many reporting observers in the field for providing valuable information that proved essential to the chapter. Narca Moore-Craig contributed her meticulously detailed field notes for the nesting information within Guadalupe Canyon. Critiques and helpful suggestions from Jean-Luc Cartron, Rich Glinski, and Janet Ruth greatly improved the accuracy and quality of information in this chapter. Finally, I would like to thank Rob Doster, Tom Kennedy, Mark Watson, Pat Watt, and Lee Zieger for providing their many fine photographs that round out the text and visually enhance the chapter.

LITERATURE CITED

[AOU] American Ornithologists' Union. 1998. *Check-list of North American birds*. 7th ed. Washington, DC: American Ornithologists' Union.

———. 2006. Forty-seventh supplement to the American Ornithologists' Union check-list of North American birds. *Auk* 123:926–36.

Bent, A. C. 1937. *Life histories of North American birds of prey*. Part 1. U.S. National Museum Bulletin 167, 264–69. Washington, DC: Smithsonian Institution.

Bibles, B. D., R. L. Glinski, and R. R. Johnson. 2002. Gray Hawk (*Asturina nitida*). No. 652. In *The birds of North America*, ed. A. Poole and F. Gill. Philadelphia, PA: Birds of North America, Inc.

Brown, L., and D. Amadon. 1968. *Eagles, hawks and falcons of the world*. Vol. 2. New York: McGraw-Hill Book Co.

Brush, T. 2005. *Nesting birds of a tropical frontier: the Lower Rio Grande Valley of Texas*. College Station: Texas A&M University Press.

Corman, T. E. 2005. Gray Hawk. In *Arizona breeding bird atlas*, ed. T. E. Corman, and C. Wise-Gervais, 136–37. Albuquerque: University of New Mexico Press.

del Hoyo, J., A. Elliott, and J. Sargatal, eds. 1994. *Handbook of the birds of the world*. Vol. 2, *New World vultures to guineafowl*. Barcelona, Spain: Lynx Edicions.

Dick-Peddie, W. A. 1993. *New Mexico vegetation: past, present, and future*. Albuquerque: University of New Mexico Press.

Dickey, D. R., and A. J. van Rossem. 1938. *Birds of El Salvador*. Zoological Series no. 23, 1–609. Chicago: Field Museum of Natural History.

Glinski, R. L. 1998. Gray Hawk *Buteo nitidus*. In *The raptors of Arizona*, ed. R. L. Glinski, 82–85. Tucson: University of Arizona Press.

Glinski, R. L., and B. A. Millsap. 1987. Status of the Sonora Gray Hawk *Buteo nitidus maximus* (van Rossem 1930). Arizona Game and Fish Department report (Contract no. 14-16-0002-82-216) to U.S. Fish and Wildlife Service, Albuquerque.

Howell, S. N. G., and S. Webb. 1995. *A guide to the birds of Mexico and northern Central America*. New York: Oxford University Press.

Hubbard, J. 1970. *Check-list of the birds of New Mexico*. Publ. no. 3. Albuquerque: New Mexico Ornithological Society.

———. 1974. The status of the Gray Hawk in New Mexico. *Auk* 91:163–66.

———. 1978. *Check-list of the birds of New Mexico*. Publ. no. 6. Albuquerque: New Mexico Ornithological Society.

Krueper, D. J. 1999. *Annotated checklist to the birds of the upper San Pedro River Valley, Arizona*. Safford, AZ: Bureau of Land Management.

Lockwood, M. W., and B. Freeman. 2004. *The TOS handbook of Texas birds*. College Station: Texas A&M University Press.

Monson, G., and A. R. Philips. 1981. *Annotated checklist of the birds of Arizona*. 2nd ed. Tucson: University of Arizona Press.

National Geographic Society. 2006. *National Geographic field guide to the birds of North America*, ed. J. L. Dunn and J. Alderfer. 5th ed. Washington, DC: National Geographic Society.

Oberholser, H. C. 1974. *The bird life of Texas*. Austin: University of Texas Press.

Sibley, D. A. 2000. *The Sibley guide to the birds*. New York: Chanticleer Press.

Taylor, C. 2006. Gray Hawk. In *National Geographic complete birds of North America*, ed. J. Alderfer, 140. Washington, DC: National Geographic Society.

Williams, S. O. III, ed. 2000. *New Mexico Ornithological Society Field Notes* 39(3): Summer 2000.

———, ed. 2002. New Mexico. *North American Birds* 56:467–70.

———, ed. 2004. New Mexico. *North American Birds* 58:410–13.

———, ed. 2005. New Mexico. *North American Birds* 59:472–75.

———, ed. 2006. New Mexico. *North American Birds* 60:412–16.

———, ed. 2007. New Mexico. *North American Birds* 61:487–91.

———, ed. 2008. New Mexico. *North American Birds* 62:452–56.

Williams, S. O. III, and D. J. Krueper. 2008. The changing status of the Gray Hawk in New Mexico and adjacent areas. *Western Birds* 39:202–8.

Zimmerman, D. A. 1976. On the status of *Buteo nitidus* in New Mexico. *Auk* 93:650–55.

16

Swainson's Hawk

(*Buteo swainsoni*)

JAMES C. BEDNARZ, JEAN-LUC E. CARTRON, AND JOE C. TRUETT

THE SWAINSON'S HAWK (*Buteo swainsoni*) is arguably one of the most striking and handsome buteos found anywhere in the world. Often seen soaring high above grassland or other open habitats, adult Swainson's Hawks are fairly conspicuous with wings raised above horizontal in a dihedral position, and with their anterior underwing linings conspicuously cream-colored followed by noticeably dark flight feathers. The only other relatively common raptor in New Mexico that reveals this characteristic two-tone underwing pattern is the Osprey (*Pandion haliaetus*), easily distinguished from Swainson's Hawk by a number of other characteristics (see chapter 4). The overhead flight silhouette of the Swainson's Hawk is also characteristic, showing more narrow and pointed wings than the other common buteos found in New Mexico. The generally inconspicuous tail of this species can also be a useful field mark. Underneath it is whitish with multiple narrow dark bands, but with a wide, dark subterminal band (in adults only).

When perched, the adult Swainson's Hawk is one of the most boldly colored hawks of North America. The contrast of the striking, usually chestnut, chest band or "bib" with the very dark dorsal plumage coloration, contrasting again with a small but conspicuous white face patch (single "headlight") above the bill gives this hawk a dramatic appearance. A useful field characteristic that helps identify all morphs of this species is its relatively long wings, which in adults slightly exceed or at least reach to the tip of the tail when perched. Only Short-tailed Hawks (*B. brachyurus*), White-tailed Hawks (*B. albicaudatus*), and Zone-tailed Hawks (*B. albonotatus*) share this trait among buteos in North America. The White-tailed Hawk is unverified in New Mexico while the Short-tailed Hawk is only accidental (chapter 40). The Zone-tailed Hawk has a conspicuously different plumage, especially in the underwing pattern, and generally occurs in higher-elevation habitats (see chapter 17). Immature Swainson's Hawks are characterized by two-tone underwings and long pointed wings (as in adults), cream or buffy and lightly to heavily streaked underparts, narrow bands on the tail, dark patches on the sides of the breast (shaped like parentheses), and pale eyebrows and cheeks with contrasting dark eyeline and malar streak.

Dark and intermediate morphs of Swainson's Hawk occur especially in California and some West Coast populations (Wheeler and Clark 1995; England et al. 1997). We have only observed a scattered few of these darker versions in New Mexico, much less than 1 percent. Although more difficult to identify than

PHOTO 16.1

Adult light-morph Swainson's Hawk in flight, photographed along State Road 9 in Hidalgo Co., 3 May 2008. Note the contrast between the dark chest band (bib) and the light belly; the shape of the wings, narrow and pointed for a Buteo species; and the two-tone underwings (cream-colored wing coverts but dark flight feathers). PHOTOGRAPH: © ROBERT SHANTZ.

PHOTO 16.2

Adult light-morph Swainson's Hawk (educational bird). Note the white throat and the small but conspicuous white face patch above and alongside the bill (single "headlight") contrasting with the dark plumage on the head and the back. Dark-morph birds have a dark throat but still show the white face patch. PHOTOGRAPH: © JEAN-LUC CARTRON.

PHOTO 16.3

Adult light-morph Swainson's Hawk on a utility pole along State Road 92 south of Virden, Hidalgo Co., 27 April 2003. The conspicuous chestnut bib contrasts with the dark brown plumage on the wings, back, and head and with the white of the throat and of the forehead and outer lores (single "headlight"). PHOTOGRAPH: © ROBERT SHANTZ.

PHOTO 16.4

Adult light-morph Swainson's Hawk (educational bird). The light belly, which contrasts with the dark chest band (bib), can show light to heavy barring. Note also that the bib can be dark brown instead of chestnut, as on this bird. PHOTOGRAPH: © JEAN-LUC CARTRON.

PHOTO 16.5

Adult light-morph Swainson's Hawk along U.S. 70 north of Lordsburg, Hidalgo Co., 29 September 2003. The dorsal plumage of adults is uniformly dark brown. Note also the long, pointed wings. PHOTOGRAPH: © ROBERT SHANTZ.

PHOTOS 16.6a, b, and c

(*top left*) Light-morph immature near Animas, Hidalgo Co., 2 August 2008. Note the pale eyebrow (superciliary) and cheek contrasting with the dark eye-line and mustache (malar streak), and the pale feather edgings on the upperwing.
PHOTOGRAPH: © ROBERT SHANTZ.

(*top right*) Light morph immature in flight, north of Antelope Wells, Hidalgo Co., 6 September 2007; note the buffy underparts and the dark streaking on the chest.
PHOTOGRAPH: © JERRY OLDENETTEL.

(*left*) Light-morph immature near Summit Station, Hidalgo Co., 13 September 2007.
PHOTOGRAPH: © ROBERT SHANTZ.

the common light morph, these darker birds always show the single white "headlight" above the bill when perched, and at least slightly lighter leading-edge wing linings compared to darker trailing flight feathers, as well as typical Swainson's Hawk flight features when in the air (i.e., the relatively narrow, pointed wing shape and the dihedral wing position when soaring).

Measurements provided by Wheeler and Clark (1995) are 43–55 cm (17–22 in) for length, 120–137 cm (47–54 in) for wingspan, and 595–1,240 g (1.3–2.7 lb) for weight. Females tend to be larger than males.

Distribution

The Swainson's Hawk breeds regularly in North America from Alberta, Saskatchewan, and Manitoba (with possible small, disjunct breeding populations occurring even farther north, to Alaska, Yukon, and Northwest Territories) south through much of the western United States into northern mainland Mexico (England et al. 1997). In the fall, nearly all of the Swainson's Hawk population departs on a long migratory journey to South America, primarily Argentina. Small numbers winter in Central America or even remain in North America,

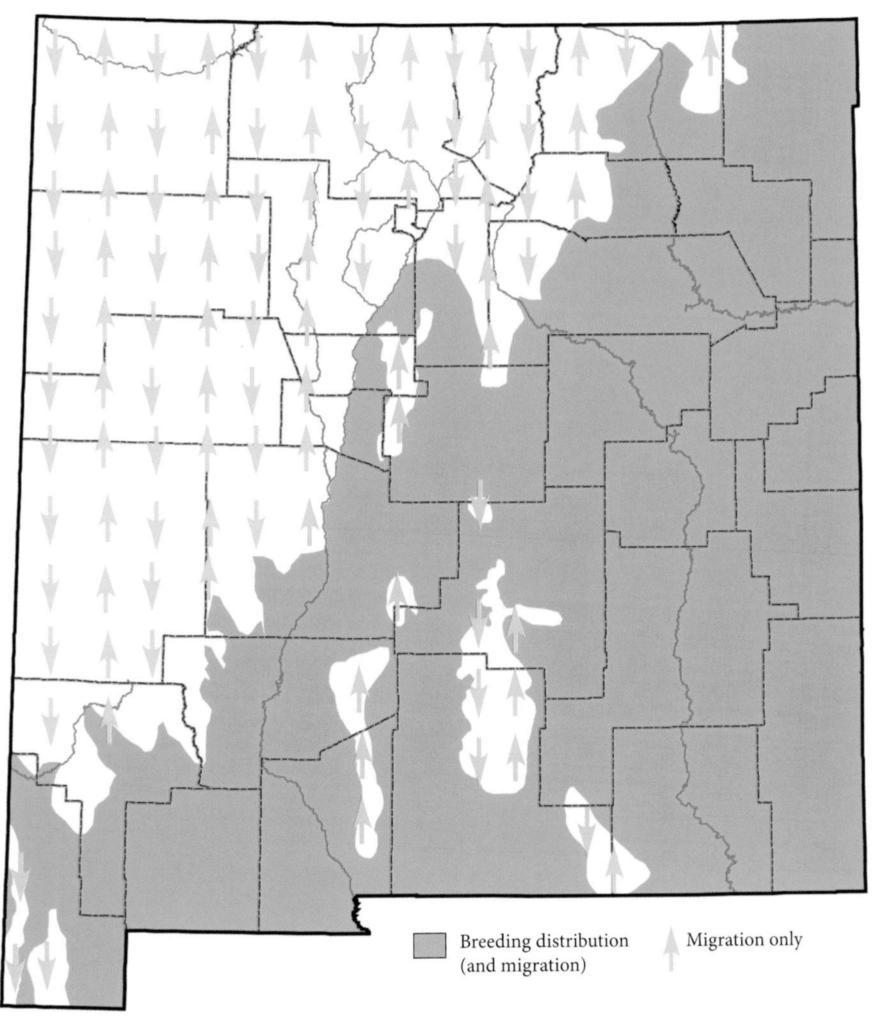

MAP 16.1

Swainson's Hawk distribution map

along the Pacific coast of Mexico north to Nayarit, and in the United States, mainly in Florida and California (England et al. 1997).

As a summer resident, the Swainson's Hawk occurs throughout much of New Mexico, typically ranging in elevation up to about 1,830 m (6,000 ft) (Bailey 1928; Ligon 1961; Hubbard 1978). However, roadside raptor surveys conducted by the New Mexico Department of Game and Fish during the 1970s and 1980s show the species to be much more common in the eastern half and southern third of the state (see Bednarz 1988b). Nesting populations are known especially from the southeastern corner of the state (Bednarz 1988a; Bednarz and Hoffman 1988; Gerstell and Bednarz 1999; Cartron et al. 2009), the Middle Rio Grande Valley (Cartron et al. 2004; JLEC, unpubl. data), the Jornada del Muerto valley (Pilz 1983), the Estancia Valley (Cartron et al. 2002; Cartron 2005), and the Kiowa National Grasslands (Schwarz 2005; Giovanni et al. 2007). In contrast, the species is at best a scarce breeder in west-central, northwestern, and north-central New Mexico. One of us (JLEC) did not find the species in the Plains of San Agustin, which supports a breeding population of another grassland raptor, the Ferruginous Hawk (*Buteo regalis*), but lies above the elevational range of the Swainson's Hawk (see Cartron et al. 2002).

In migration, the Swainson's Hawk ranges throughout the state, and at elevations up to 2,740 m (9,000 ft) or more (Bailey 1928; Ligon 1961). According to Hoffman and Smith (2003), large numbers of Swainson's Hawks migrate in the fall over the Manzano Mountains, but these numbers tend to be highly variable from year to year. Flocks of migrating Swainson's Hawks also occur in the western half of the state. For example, the *New Mexico Ornithological Society Field Notes* (McKnight and Niles 1964) mention a group of 79 individuals on 20 September 1964 at Beaverhead, Catron County. Flocks including hundreds of birds in September have been observed elsewhere in the state (Bailey 1928; Ligon 1961).

By and large, Swainson's Hawks leave New Mexico in the fall. However, along with other southwestern states and Louisiana, New Mexico has regular wintering records of the species (e.g., Bailey 1928; McKnight and Niles 1965; Bednarz 1988b; see England et al. 1997).

PHOTO 16.7

Swainson's Hawk perched on a yucca (*Yucca elata*) in yucca grassland, Otero Mesa, Otero Co., 16 July 2008. PHOTOGRAPH: © DAVID J. GRIFFIN.

PHOTO 16.8

Swainson's Hawk nesting habitat near Wagon Mound, Mora Co. PHOTOGRAPH: © JEAN-LUC CARTRON.

Habitat Associations

In New Mexico, the Swainson's Hawk is primarily associated with low- and mid-elevation open or mostly open country with scattered trees, such as semidesert grassland, Chihuahuan desertscrub, shinnery oak (*Quercus havardii*) associations, and agricultural areas (Bednarz 1988a, 1988b; Eakle et al. 1996; Cartron et al. 2009). It is in those habitat types that most pairs nest. However, small numbers of Swainson's Hawks in New Mexico breed instead in strips of riparian woodlands, typically along fields and pastures (Cartron et al. 2004) and, within urban areas, also along golf courses (JLEC, pers. obs.).

An area supporting a healthy concentration of breeding Swainson's Hawks is Los Medaños in southern Eddy County and southern Lea County between Loving and Jal. Indeed much of the breeding natural history research (e.g., Bednarz 1988a; Bednarz and Hoffman 1988; Gerstell and Bednarz 1999) described in this chapter is based on data collected in Los Medaños. Climatically, this area is subject to extreme hot temperatures in the summer and to cold winters, with temperatures regularly dropping to -10°C at night. The majority of Los Medaños is characterized by a mesquite (*Prosopis*) -oak shrubland with rolling sandy soils. Major cover types include low, vegetated shinnery oak and mesquite dunes; yucca (*Yucca*) -grasslands; creosote bush (*Larrea tridentata*) flatlands; and sparsely vegetated active dunes. Although this latter cover type is characteristic, it only occupies a small proportion of the land surface area. Swainson's Hawks use and nest in all of Los Medaños's vegetation communities.

Life History

Nesting

In New Mexico, early dates recorded for the Swainson's Hawk's return from migration correspond to the latter part of March (chapter 2). Swainson's Hawks migrating through New Mexico are mainly observed throughout April and in early May (Hoffman and Smith 2003; chapter 2), but most pairs breeding in the state seem to be back on their nesting territories by mid April (Pilz 1983; JLEC, unpubl. data). On the Jornada Experimental Range (southern Jornada del Muerto valley), birds were paired within two weeks following their return on their territories, copulation was first observed during the third or fourth week of April, and all pairs were with complete clutches by 11 May in 1974 and 1975 (Pilz 1983). Bednarz (1988b) reported the peak of egg-laying in southeastern New Mexico during the first week of May.

Swainson's Hawks build their nests in a variety of tall and short trees and in yuccas. Ligon (1961) listed tree yuccas (*Yucca* spp.), Spanish bayonet (*Yucca aloifolia*), cottonwood (*Populus* spp.), hackberry (*Celtis* spp.), soapberry (*Sapindus drummondii*), and mesquite as nesting substrates used in New Mexico. On the Jornada Experimental Range, Swainson's Hawks nested almost exclusively in yuccas; only one nest was discovered in a mesquite tree (Pilz 1983). In Los Medaños in southeastern New Mexico, Bednarz (1988a) found most Swainson's Hawk nests in mesquites (84% of nest trees; see also Bednarz 1988b) and secondarily in soapberry. Nest trees recorded by Cartron et al. (2009) in 2008 in northern and central Lea County (also in southeastern New Mexico) consisted mainly of elms (*Ulmus*), with one nest discovered in a shinnery oak and another in a locust (*Robinia* sp.). In the Estancia Valley, Swainson's Hawk nests were typically observed in elm trees, less often in cottonwoods (Cartron 2005; JLEC, unpubl. data). Along the Middle Rio Grande and Middle Gila River, nesting was in tall cottonwoods (Cartron et al. 2004). In Albuquerque, one Swainson's Hawk nest in the late 1990s was in a poplar (*Populus* sp.), and another one in 2006 was in a locust, both along golf courses (JLEC, pers. obs.). Nests have also been observed in cholla (*Cylindropuntia*) and saltcedar (*Tamarix*) in New Mexico (see photos).

Ligon (1961) reported a Swainson's Hawk nest built on a sand dune on 14 May 1934 about 64 km (40 mi) east of Roswell. Although rare, ground or cliff nests are indeed mentioned in old historical accounts of the species (e.g., Bent 1937). However, relatively contemporary research and reviews of nesting data (e.g., Bednarz 1988b; England et al. 1997) all indicate that Swainson's Hawks no longer nest on the ground, a change in nesting habits. Interestingly, when Bednarz (1988a, unpubl. data) worked in southeastern New Mexico from 1980 through 1990, none of the more than 250 Swainson's Hawk nests he observed was on a utility pole, and within its entire distribution the species has been reported to use utility poles and other man-made structures only occasionally (Dunkle 1977; James 1992; Brubaker et al. 2003). This is all in contrast to recent findings by Cartron et al. (2009) in central and northern Lea County and western Chavez County. Of 20 observed nests, 11 were on power poles. Thus it may be that Swainson's Hawks in New Mexico are again changing in their nesting habits, progressively getting accustomed to using man-made structures.

Swainson's Hawks are known to return to the same

PHOTO 16.9

(*top left*) Swainson's Hawk nest in a dead tree near Springer Lake, Colfax Co., 10 June 2008. In New Mexico, pairs often nest in small trees, the nest being placed as low as 1.7 to 2 m (3.3 to 6.5 ft) above the ground. Nests in snags are not rare. One nest JCB found was in a dead mesquite (*Prosopis* sp.) 1.7 m (3.3 ft) from the ground. PHOTOGRAPH: © JEAN-LUC CARTRON.

PHOTO 16.10

(*bottom left*) Swainson's Hawk nest on utility pole in Lea Co., 26 May 2008. PHOTOGRAPH: © JEAN-LUC CARTRON.

PHOTO 16.11

(*top right*) Nest in a cholla cactus (*Cylindropuntia* sp.), Otero Mesa, Otero Co., 15 May 2004. PHOTOGRAPH: © DAVID J. GRIFFIN.

PHOTO 16.12

(*bottom right*) Saltcedar (*Tamarix*) nest with nestling, Chaves Co., 28 June 2006. PHOTOGRAPH: © RON KELLERMUELLER/HAWKS ALOFT, INC.

PHOTO 16.13

(*top left*) Swainson's Hawk nest with clutch of three eggs near Maxwell, Colfax Co., 22 May 2007. PHOTOGRAPH: © JEAN-LUC CARTRON.

PHOTO 16.14

(*bottom left*) Nest with two eggs in cholla cactus (*Cylindropuntia* sp.), Otero Mesa, Otero Co., 15 May 2004. PHOTOGRAPH: © DAVID J. GRIFFIN.

PHOTOS 16.15a and b

(*above*) Swainson's Hawk nest with single nestling in elm (*Ulmus* sp.) tree west of Lovington, Lea Co., 2 July 2008. Also in the nest and not seen is an unhatched egg. PHOTOGRAPHS: © JEAN-LUC CARTRON.

PHOTO 16.16

(*top left*) Two nestlings in a cholla (*Cylindropuntia* sp.) nest, Torrance Co., July 2006. PHOTOGRAPH: © RON KELLERMUELLER/HAWKS ALOFT, INC.

PHOTO 16.17

(*middle left*) A brood of three Swainson's Hawk chicks (approximately six to nine days old; youngest to oldest) temporally removed from their nest and placed in Jim Bednarz's cowboy hat to hold birds prior to measuring and marking, June 1983. Data were used in Bednarz's studies of raptors of the Los Medaños area in southeastern New Mexico. PHOTOGRAPH: © JIM BEDNARZ.

PHOTO 16.18

(*bottom left*) An adult Swainson's Hawk arriving with prey at nest in mesquite (*Prosopis* sp.) tree in the Los Medaños area of southeastern New Mexico (probably in 1987). The nest contained two nestlings (one nestling is partially obscured) about 18 days of age. PHOTOGRAPH: © DAVID A. PONTON.

PHOTO 16.19

(*above*) Near-fledgling Swainson's Hawk chick (approximately 42 days old) removed temporarily from its nest for measuring and marking in the Los Medaños area of southeastern New Mexico, July 1981. PHOTOGRAPH: © JIM BEDNARZ.

nesting territories year after year (Fitzner 1980). Within those nesting territories, however, they appear to often shift annually between alternate nests or, more often, to build new nests. Of 18 occupied nests found in 1975 on the Jornada Experimental Range, 14 (78%) were new. Of the other four, only two had been used the year before; the other two were old nests that presumably had been occupied before 1974 (Pilz 1983).

Swainson's Hawks typically lay two or three, and occasionally four, eggs (Ligon 1961; Pilz 1983). The eggs are creamy or bluish white, with variable amounts of spotting at the large end (Ligon 1961). In southeastern

New Mexico, mean clutch size was 2.17 (n = 196 nests over a 10-yr period) (Bednarz 1988b, unpubl. data), with yearly means ranging from a low of 1.9 in 1988 and 1989 to 2.71 in 1981. On the Jornada Experimental Range mean clutch size was 2.6 in 1974 (n = 8) and 3 in 1975 (n = 6) (Pilz 1983).

Hatching in southeastern New Mexico occurred from 20 May to 3 July (Bednarz 1988b). Bednarz and Hoffman (1988) reported a mean age at fledging of 40.4 days (n = 5 nests; range 39–42 days).

Diet and Foraging

Several studies have focused on the diet of Swainson's Hawks in New Mexico (table 16.1) and, other than during migration, they all show the species to be a prey generalist (see Bednarz 1988a). On the Jornada Experimental Range, for example, nesting pairs brought back to their nests at least 21 prey species. Most frequently taken were the western whiptail lizard (*Aspidoscelis tigris*), Texas horned lizard (*Phrynosoma cornutum*), and spotted ground squirrel (*Spermophilus spilosoma*), with fewer numbers of other mammals and lizards, and snakes, toads, birds, and insects (Pilz 1983). Lagomorphs, both *Lepus* and *Sylvilagus*, provided the greatest proportion of the diet biomass both on the Jornada Experimental Range and in the Los Medaños area (Pilz 1983; Gerstell and Bednarz 1999). The proportion of lagomorphs (many of them juveniles) among prey items in southeastern New Mexico appeared particularly high when compared to other parts of the species' distribution (Bednarz and Hoffman 1988).

Pilz (1983) reported both the glossy snake (*Arizona elegans*) and coachwhip (*Masticophis flagellum*) as prey on the Jornada Experimental Range, and in Los Medaños Bednarz (1988a) also observed snake remains at nests. These findings bring to mind Ligon's (1961) secondhand account of a Swainson's Hawk attempting to take a rattlesnake (*Crotalus* sp.) southwest of Carlsbad, with both animals killing each other in the struggle that ensued. A case of interspecific raptor predation was documented in the South Valley of Albuquerque along the Rio Grande, where Cartron et al. (2004) found an American Kestrel (*Falco sparverius*) among prey remains at a Swainson's Hawk nest.

PHOTO 16.20

Swainson's Hawk eating its prey on a yucca (*Yucca* sp.) near Animas, Hidalgo Co., 9 September 2007.

PHOTOGRAPH: © ROBERT SHANTZ.

PHOTO 16.21

Swainson's Hawk with mammalian prey along State Road 9 west of Animas, Hidalgo Co., 29 April 2007.

PHOTOGRAPH: © ROBERT SHANTZ.

TABLE 16.1 Species documented as Swainson's Hawk (*Buteo swainsoni*) prey in New Mexico.

CLASS	SPECIES	LOCATION				
		NESTING				MIGRATION
		Armendaris Ranch, Jornada del Muerto Valley	Jornada Experimental Range, Jornada del Muerto Valley	Middle Rio Grande Bosque	Los Medaños area	Jornada del Muerto Valley
INSECTA	Orthoptera spp.			X	X	X
	Coleoptera spp.			X	X	
AMPHIBIA	Woodhouse's toad (*Bufo woodhousii*)			X		
REPTILIA	Longnose leopard lizard (*Gambelia wislizenii*)		X			
	Texas horned lizard (*Phrynosoma cornutum*)		X		X	
	Roundtail horned lizard (*Phrynosoma modestum*)		X			
	Desert spiny lizard (*Sceloporus magister*)		X			
	Chihuahuan spotted lizard (*Aspidoscelis exsanguis*)			X		
	Western whiptail (*Aspidoscelis tigris*)		X			
	Great Plains skink (*Eumeces obsoletus*)			X	X	
	Glossy snake (*Arizona elegans*)		X			
	Coachwhip (*Masticophis flagellum*)		X			
AVES	American kestrel (*Falco sparverius*)			X		
	Scaled quail (*Callipepla squamata*)		X		X	
	Northern bobwhite (*Colinus virginianus*)				X	
	Greater roadrunner (*Geococcyx californianus*)				X	
	Loggerhead shrike (*Lanius ludovicianus*)		X			
	Scott's oriole (*Icterus parisorum*)		X			
MAMMALIA	Brazilian free-tailed bat (*Tadarida brasiliensis*)	X				
	Desert cottontail (*Sylvilagus audubonii*)		X		X	
	Black-tailed jackrabbit (*Lepus californicus*)		X		X	
	Banner-tailed kangaroo rat (*Dipodomys spectabilis*)		X			
	Ord's kangaroo rat (*Dipodomys ordii*)		X		X[a]	
	Merriam's kangaroo rat (*Dipodomys merriami*)		X		X[a]	
	Spotted ground squirrel (*Spermophilus spilosoma*)		X			
	Pocket gopher sp. (*Geomys*)				X	
	Woodrat sp. (*Neotoma*)		X		X	
	Hispid cottonrat (*Sigmodon hispidus*)		X		X	
	Harvest mouse sp. (*Reithrodontomys*)				X	
	White-footed mouse sp. (*Peromyscus*)				X	
	Pocket mouse sp. (*Perognathus*)				X	
	House mouse (*Mus musculus*)			X		

[a] *Dipodomys* skulls found at Swainson's Hawk nests were not identified to species. The two common species found in Los Medaños were the Ord's kangaroo rat (*D. ordii*) and, to a lesser extent, the Merriam's kangaroo rat (*D. merriami*). Both species were almost certainly preyed upon by Swainson's Hawks.

References for Jornada Experimental Range (Doña Ana County), Middle Rio Grande Valley (Bernalillo County), and Los Medaños (Eddy and Lea counties) during nesting: Pilz 1983; Bednarz 1986, 1988, unpubl. data; Gerstell and Bednarz 1999; Cartron et al. 2004; and S. Altenbach pers. comm. Reference for Jornada del Muerto Valley (Sierra and Doña Ana Counties) during migration: Ligon 1961.

PHOTO 16.22

Single downy Swainson's Hawk nestling (approximately 10 days old) with recently delivered Great Plains skink (*Eumeces obsoletus*) in nest supported by a mesquite (*Prosopis* sp.) tree in the Los Medaños area of southeastern New Mexico, June 1986.

PHOTOGRAPH: © JIM BEDNARZ.

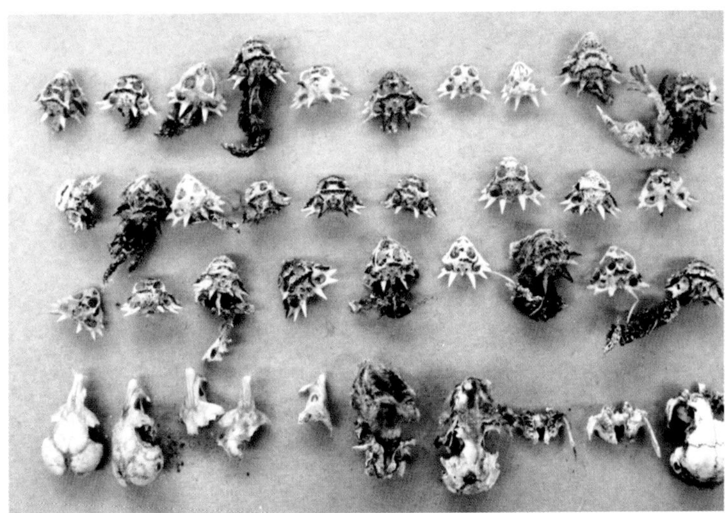

PHOTO 16.23

Prey remains collected at one Swainson's Hawk nest site in August 1983 by Bednarz in the Los Medaños area of southeastern New Mexico. This prey sample included remains from 28 individual Texas horned lizards (*Phrynosoma cornutum*), 5 kangaroo rats (*Dipodomys* spp.), and 5 juvenile desert cottontails (*Sylvilagus audubonii*).

PHOTOGRAPH: © JIM BEDNARZ.

In a remote part of the Armendaris Ranch occurs one of the true spectacles of the natural world. From spring to fall each year at the Jornada Caves in the Jornada del Muerto, Swainson's Hawks assemble daily to catch bats (JCT, pers. obs.). As the swarms of bats leave the caves, they fill the air with soft rattling from the collective flapping of their wings, climbing higher and higher in long streamers to head away for nighttime foraging. They emerge like clockwork, the timing changing little from day to day though it varies over the season with changes in timing of sunsets and probably with food abundance and distribution. The hawks apparently have their own internal clocks; the first arrivals each day coincide remarkably with the first appearance of the bats (JCT, pers. obs.). As the bats exit the caves—actually lava tubes—near sundown, the hawks swoop in, snatching them from the air and consuming them on the wing.

Eight species of bat use the caves, the great majority of which are the migratory Brazilian free-tailed bats (*Tadarida brasiliensis*) that roost there from April to October. Swainson's Hawks prey on the Brazilian

PHOTO 16.24

Swainson's Hawk hunting bats at the Armendaris Ranch in 2004. PHOTOGRAPH: © ROBIN SILVER.

free-tailed bat in particular and perhaps on most or all of the other seven bat species using these caves (S. Altenbach, pers. comm.). Both the free-tailed bats and the Swainson's Hawks generally peak in numbers in late summer. At this time bats may reach 5–8 million strong (S. Altenbach, pers. comm.), and hawks daily preying on bats can range up to about 80 (JCT pers. obs.). But on average over summer, the number of bats approximates 200,000–500,000 and that of hawks 20–40. Similar bat-catching behavior by Peregrine Falcons (*Falco peregrinus*) and other falcons has been observed at dusk in other parts of the world (e.g., Pierson and Donahue 1983).

As shown by their predation on bats, Swainson's Hawks evidently can use a wide variety of hunting techniques, each one adapted to the ecology of a prey species or capitalizing on local conditions. For example, Ligon (1961) describes Swainson's Hawks following tractors and catching mice (Muridae) and kangaroo rats (*Dipodomys* spp.) as the plowing of the soil left them suddenly exposed.

During migration, several observers have reported high densities of Swainson's Hawks feasting on large wingless grasshoppers (Orthoptera) (Ligon 1961). One such observation was in the Jornada del Muerto valley, where the grasshopper species most likely would have been the plains lubber (*Brachystola magna*) (D. Lightfoot, pers. comm.). According to Ligon (1961), concentrations of Swainson's Hawks feeding on grasshoppers in New Mexico could once reach flocks of 1,000 individuals or more. Even today, fairly large gatherings of migrants still occur (Bednarz 1988b). Two hundred Swainson's Hawks feeding on grasshoppers were recorded at La Plata, San Juan County on 27 September 2003 (Williams 2003a).

Predation and Interspecific Interactions

The Swainson's Hawk has its own natural enemies, including the Golden Eagle (*Aquila chrysaetos*) and the Great Horned Owl (*Bubo virginianus*), both of which have been found to kill nestlings or fledglings outside New Mexico (Dunkle 1977; Fitzner 1978; Woodbridge 1991). In New Mexico, loss of nestlings to predation by a Great Horned Owl was suspected in particular at one nest in De Baca County in 2006 (R. Kellermueller, pers. comm.; photo 16.25).

In southeastern New Mexico, Swainson's Hawks share their habitat with Harris's Hawks (*Parabuteo unicinctus*). Telemetry on one male Swainson's Hawk and several Harris's Hawks suggested that the territories

PHOTO 16.25

Nest failure due possibly to predation by a Great Horned Owl (*Bubo virginianus*). In the nest—in a juniper (*Juniperus* sp.) tree—are the remains of a dead, fully feathered nestling. Another nestling was found dead on the ground near the nest. Southern De Baca Co., 26 July 2006. PHOTOGRAPH: © RON KELLERMUELLER/HAWKS ALOFT, INC.

mostly do not overlap but can abut, and nests of the two species have been found less than 175 m (~570 ft) apart (Gerstell and Bednarz 1999; JCB, unpubl. data). Interactions between the two species nearly always involve the Swainson's Hawk as the aggressor, often stooping or diving on the Harris's Hawk and chasing it away.

In the Estancia Valley, Western Kingbird (*Tyrannus verticalis*) nests are often found within close proximity of occupied Swainson's Hawk nests (Cartron 2005). In such cases, the resident Swainson's Hawk pair is chased and harassed by the smaller birds, and yet it is the Swainson's Hawk pair that evidently initiated nesting earlier. Swainson's Hawk nests have also been found in close association with nesting Scissor-tailed Flycatchers (*Tyrannus forficatus*) east of Carlsbad, Eddy County (JCB, unpubl. data). JCB believes that the prey remains accumulating during the brood-rearing stage of the hawk nesting period attract relatively large numbers of insects that provide a source of food for the breeding flycatchers. Swainson's Hawks nesting in the Rio Grande bosque also have to contend with passerines (e.g., red-winged blackbirds [*Agelaius phoeniceus*]) chasing after them (JLEC, pers. obs.).

Status and Management

Assuming equal density of nesting hawks, Clark (1988) multiplied an estimated proportion of the total Swainson's Hawk breeding range found within each U.S. state and Canadian province by the number of migrating birds counted at the Isthmus of Panama (Smith 1980, 1985) to develop crude estimates of the number of Swainson's Hawks found within states and provinces. With this technique, Clark estimated that approximately between 1,519 and 2,009 pairs of Swainson's Hawks were breeding in New Mexico in the 1980s. Bednarz (1988b) suggested that this was a substantial underestimate of the true population in the state because nesting densities documented by Pilz (1983; 95/1,000 km^2) and Bednarz and Hoffman (1988; 142/1,000 km^2) were considerably greater than those derived by Clark (12–16 pairs/1,000 km^2). Furthermore, New Mexico Game and Fish Department road count data from the 1970s and 1980s indicated that densities of nesting Swainson's Hawks were similar over much of the southern (including Pilz's study area) and eastern (including Bednarz and Hoffman's study area) portions of the state (as delineated in that road count study). Conservatively, Bednarz (1988b) concluded that at least 3,000 pairs of Swainson's Hawks nested in New Mexico in the mid 1980s.

During the late 1990s, the Swainson's Hawk received increased attention due to enormous mortality on Argentina wintering grounds as a result of pesticide exposure (Woodbridge et al. 1995; Goldstein et al. 1997). Whether the status of the Swainson's Hawk was ever affected in New Mexico is not known, but at present the population of this species appears stable or perhaps increasing. Hoffman and Smith (2003) found increasing numbers of Swainson's Hawks migrating through New Mexico during the 1990s. Currently we believe that a minimum estimate of 3,000 pairs is still valid, and that 10,000 or more nesting pairs found within the state during most years is not an unreasonable estimate.

Pilz (1983) reported for the Jornada Experimental Range an average of 1.6 young fledged per nest (1.3 in 1974 and 2.2 in 1975). Bednarz and Hoffman (1988) estimated that 1.3–1.9 young fledged per nest/year during a four-year period in Los Medaños of southeastern New Mexico. Tim Hayden and Bednarz continued to monitor the reproductive success of this population of Swainson's Hawks through 1990. They found that annual production varied from 0.65 fledglings/nest to 1.91 fledglings/nest based on observations at 238 nests (JCB and Hayden, unpubl. data). The grand mean was 0.99 fledgling produced per New Mexico Swainson's Hawk nest over the nine study-years.

Few cases of human-related Swainson's Hawk mortality have been reported in New Mexico. Ginter and Desmond (2004) found one dead individual under a tall communication tower in the southern part of the state. The presumed cause of mortality was collision with the tower, and the timing of the discovery coincided with fall migration (Ginter and Desmond 2004). Road traffic is another probable cause of mortality, as one of us (JLEC) found a dead juvenile in 2005 along NM State Route 26, also during fall migration.

JCB banded 313 Swainson's hawks in New Mexico

PHOTO 16.26

Adult light-morph Swainson's Hawk near Carlsbad, 28 April 2007.

PHOTOGRAPH: © DOUG BROWN.

between 1981 and 2002. Four of these hawks, all banded as nestlings, were later recovered; all had been shot. Three of these hawks were found between 200 and 280 km (125 and 175 mi) away from where they were banded as nestlings; two were recovered in New Mexico and one in Texas. The fourth hawk was found shot in Ibague, Colombia, 21 months after it had fledged and approximately 4,020 km (2,500 mi) from its natal area in New Mexico. This bird must have completed one successful round trip to South America and was likely on its return from its second migration when it was shot.

Recently, West Nile Virus was mentioned as the suspected cause of nestling mortality in several nests near Lingo in the eastern part of the state (Williams 2003b). A hail or rain storm could also have caused nestling mortality in those nests. Severe thunderstorms, which can be responsible for complete nest failures (see England et al. 1997; C. Boal, pers. comm.), are common in late spring and summer on New Mexico's eastern plains. An active Swainson's Hawk nest was destroyed by heavy rain in June 1966 near Loving in Eddy County (McKnight and Snider 1966).

The Swainson's Hawk is the "signature" raptor of the lowland grassland and desert habitats of New Mexico. Whether driving through the Chihuahuan Desert or the eastern plains and open valleys in the summer, an alert observer cannot miss the abundance of soaring Swainson's Hawks in their characteristic flight dihedral. Early or late in the day, these hawks are often perched on roadside power poles displaying their handsome plumage coloration. All indications are that populations of this hawk in the state are relatively healthy and that the New Mexico nesting grounds contribute significantly to the global population. With continued prudent management and protection of the open deserts and grasslands, the Swainson's Hawk should persist as the common and symbolic aerial predator of the open rangelands of New Mexico for many centuries to come.

LITERATURE CITED

Bailey, F. M. 1928. *Birds of New Mexico*. Santa Fe: New Mexico Department of Game and Fish.

Bednarz, J. C. 1986. The behavioral ecology of the cooperatively breeding Harris's Hawk in southeastern New Mexico. Ph.D. diss. University of New Mexico, Albuquerque.

———. 1988a. A comparative study of the breeding ecology of Harris' and Swainson's Hawks in southeastern New Mexico. *Condor* 90:311–23.

———. 1988b. Swainson's Hawk. In *Proceedings of the southwest raptor management symposium and workshop*, ed. R. L. Glinski, B. G. Pendleton, M. B. Moss, M. N. LeFranc Jr., B. A. Millsap, and S. W. Hoffman, 87–96. Washington, DC: National Wildlife Federation.

Bednarz, J. C., and S. W. Hoffman. 1988. The status of breeding Swainson's Hawks in southeastern New Mexico. In *Proceedings of the southwest raptor management symposium and workshop*, ed. R. L. Glinski, B. G. Pendleton, M. B. Moss, M. N. LeFranc Jr., B. A. Millsap, and S. W. Hoffman, 253–59. Washington, DC: National Wildlife Federation.

Bent, A. C. 1937. *Life histories of North American birds of prey*. Part 1. U.S. National Museum Bulletin 167. Washington, DC: Smithsonian Institution.

Brubaker, D. L., K. L. Brubaker, and B. C. Thompson. 2003. Raptor and Chihuahuan raven nesting on decommissioned telephone-line poles in the northern Chihuahuan Desert. *Journal of Raptor Research* 37:135–46.

Cartron, J.-L. E. 2005. Nesting associations of the Western Kingbird (*Tyrannus verticalis*) with two *Buteo* species in the Estancia Valley of central New Mexico. *New Mexico Ornithological Bulletin* 33:53–55.

Cartron, J.-L. E., R. R. Cook, G. L. Garber, and K. K. Madden. 2002. Nesting and productivity of ferruginous hawks in two areas of central and western New Mexico, 1999–2000. *Southwestern Naturalist* 47:482–85.

Cartron, J.-L. E., D. A. Dean, S. H. Stoleson, and P. J. Polechla Jr. 2004. Nesting of Swainson's Hawks (*Buteo swainsoni*) in riparian woodlands of New Mexico. *New Mexico Ornithological Society Bulletin* 32:91–94.

Cartron, J.-L. E., L. A. Sager Jr., and Hira A. Walker. 2009. Notes on some breeding raptors of central and northern Lea County, New Mexico. *New Mexico Ornithological Society Bulletin* 37:7–14.

Clark, W. S. 1988. *Estimates of breeding population density of Swainson's hawks based on migration counts*. Proceedings of the Ferruginous and Swainson's Hawk status workshop. Washington, DC: U.S. Department of the Interior, Fish and Wildlife Service.

Dunkle, S. W. 1977. Swainson's Hawks on the Laramie Plains, Wyoming. *Auk* 94:65–71.

Eakle, W. L., E. L. Smith, S. W. Hoffman, D. W. Stahlecker, and R. B. Duncan. 1996. Results of a raptor survey in southwestern New Mexico. *Journal of Raptor Research* 30:183–88.

England, A. S., M. J. Bechard, and C. S. Houston. 1997. Swainson's Hawk (*Buteo swainsoni*). No. 265. In *The birds of North America*, ed. A. Poole and F. Gill. Philadelphia, PA: Academy of Natural Sciences, and Washington, DC: American Ornithologists' Union.

Fitzner, R. E. 1978. Behavioral ecology of the Swainson's Hawk (*Buteo swainsoni*) in southeastern Washington. Ph.D. diss. Washington State University, Pullman.

———. 1980. Behavioral ecology of the Swainson's Hawk (*Buteo swainsoni*) in Washington. PLN-2754. Richland, WA: Pacific Northwest Laboratory.

Gerstell, A. T., and J. C. Bednarz. 1999. Competition and patterns of resources use by two sympatric raptors. *Condor* 101:557–65.

Ginter, D. L., and M. J. Desmond. 2004. Avian mortality during fall 2001 migration at communication towers along the Rio Grande corridor. *Southwestern Naturalist* 49:414–17.

Giovanni, M. D., C. W. Boal, and H. A. Whitlaw. 2007. Prey use and provisioning rates of breeding Ferruginous and Swainson's Hawks on the southern Great Plains, USA. *Wilson Journal of Ornithology* 119:558–69.

Goldstein, M. I., B. Woodbridge, M. E. Zaccagnini, and S. B. Canavelli. 1997. An assessment of mortality of Swainson's Hawks on wintering grounds in Argentina. *Journal of Raptor Research* 30:106–7.

Hoffman, S. W., and J. P. Smith. 2003. Population trends of migratory raptors in western North America. *Condor* 105:397–419.

Hubbard, J. P. 1978. *Revised check-list of the birds of New Mexico*. Publ. no. 6. Albuquerque: New Mexico Ornithological Society.

James, P. C. 1992. Urban-nesting Swainson's Hawks in Saskatchewan. *Condor* 94:773–74.

Ligon, J. S. 1961. *New Mexico birds*. Albuquerque: University of New Mexico Press.

McKnight, B. C., and D. M. Niles, eds. 1964. *New Mexico Ornithological Society Field Notes, Seasonal Report* 6: 1 June–30 November, 1964.

———, eds. 1965. *New Mexico Ornithological Society Field Notes, Seasonal Report* 7: 1 December 1964–31 May 1965.

McKnight, B. C., and P. R. Snider, eds. 1966. *New Mexico Ornithological Society Field Notes* 5(2): 1 June–30 November 1966.

Pierson, J. E., and P. Donahue. 1983. Peregrine falcon feeding on bats in Suriname South America. *American Birds* 37:257–59.

Pilz, W. R. 1983. Nesting ecology and diet of Swainson's Hawk in the Chihuahuan Desert, south-central New Mexico. Master's thesis. New Mexico State University, Las Cruces.

Schwarz, H. R. 2005. Ferruginous Hawk platform monitoring summary for 2005. Unpublished report, Cibola National Forest, Albuquerque, NM.

Smith, N. G. 1980. Hawk and vulture migrations in the neotropics. In *Migrant birds in the neotropics: ecology, behavior, distribution, and conservation*, ed. A. Keast and E. S. Morton, 51–65. Washington, DC: Smithsonian Institution Press.

———. 1985. Counting migrating raptors. In *Proceedings of hawk migration conference IV*, ed. M. Harwood, 239–42. Rochester, NY: Hawk Migration Association of North America.

Wheeler, B. K., and W. S. Clark. 1995. *A photographic guide to North American raptors*. San Diego, CA: Academic Press.

Williams, S. O. III, ed. 2003a. *New Mexico Ornithological Society Field Notes* 42(4).

———, ed. 2003b. *New Mexico Ornithological Society Field Notes* 42(3).

Woodbridge, B. 1991. Habitat selection by nesting Swainson's Hawks: a hierarchical approach. M.S. thesis. Oregon State University, Corvallis.

Woodbridge, B., K. K. Finley, and T. S. Seager. 1995. An investigation of the Swainson's Hawk in Argentina. *Journal of Raptor Research* 29:202–4.

17

Zone-tailed Hawk
(*Buteo albonotatus*)

SCOTT H. STOLESON AND GIANCARLO SADOTI

THE ZONE-TAILED HAWK (*Buteo albonotatus*) might well be dubbed "the Great Pretender," because it so closely resembles the ubiquitous Turkey Vulture (*Cathartes aura*) in appearance and behavior as to be frequently mistaken for it. In the border regions where it lives, it may be confused as well with another "Mexican" raptor, the Common Black-Hawk (*Buteogallus anthracinus*). In fact, a nesting pair of hawks in Harding County was initially reported as the northernmost breeding record for Common Black-Hawk, but subsequently identified correctly as Zone-tailed Hawk (Bohl and Traylor 1958).

Zone-tails are medium-large buteos, averaging 51 cm (20 in) in length, 129 cm (51 in) in wingspread, and 830 g (1.8 lbs) in weight, with females tending to be larger than males (Clark and Wheeler 1987). Compared to Red-tailed Hawks (*Buteo jamaicensis*), Zone-tails are slim and lanky: they are almost equal in length, have a slightly greater wingspan, yet weigh 23% less on average (see Clark and Wheeler 1987). Their wings are long and narrow for a buteo species, shaped much like those of the larger Turkey Vulture instead. The superficially similar Common Black-Hawk has much broader wings and a shorter tail in flight.

Like the Turkey Vulture and the Common Black-Hawk, the Zone-tailed Hawk is predominantly black

PHOTO 17.1

The Zone-tailed Hawk is predominantly blackish in color. The cere and legs are yellow. The lores are gray, whereas they are yellow in the similar-looking Common Black-Hawk (*Buteogallus anthracinus*). Steeple Rock Canyon, Hidalgo Co., 2 May 2003.

PHOTOGRAPH: © ROBERT SHANTZ.

PHOTO 17.2

(top left) Zone-tailed Hawk in flight, Bitter Creek, Grant Co., 23 June 2007. In flight, the Zone-tailed Hawk can be difficult to distinguish from a Turkey Vulture (*Cathartes aura*). The two species have similar silhouettes. Their underwings are two-toned, with dark coverts contrasting with paler flight feathers. Both the Turkey Vulture and the Zone-tailed Hawk also fly in a strong dihedral, rocking from side to side. However, the Turkey Vulture lacks the fully feathered head, the yellow cere, the tail boldly barred in black and white, and the heavy barring on the flight feathers, all found in the Zone-tailed Hawk. PHOTOGRAPH: © ROBERT SHANTZ.

PHOTO 17.3

(bottom left) Zone-tailed Hawk in Steeple Rock Canyon, Hidalgo Co., 30 June 2003. Note the heavy barring on the silvery flight feathers and the bands on the topside of the tail, gray instead of white as on the underside. PHOTOGRAPH: © ROBERT SHANTZ.

PHOTO 17.4

(above) Zone-tailed Hawk in Steeple Rock Canyon, Hidalgo Co., 7 June 2003. The Zone-tailed Hawk has long, narrow wings rather than the broad wings of the Common Black-Hawk (*Buteogallus anthracinus*). The tail of a female Zone-tailed Hawk (photo) has one wide and two narrow white bands. The tail of a male has only one wide and one narrow white band. PHOTOGRAPH: © ROBERT SHANTZ.

or blackish in color. Adult birds have the tail boldly barred in black and white, the females having three white bands, the males just two (Clark and Wheeler 1987). The undersides of the flight feathers are subtly barred in silvery-gray and dusky, and contrast strongly with the black wing coverts, adding to the vulture-like effect. Unlike vultures, however, the head is fully feathered in black, and the cere and legs are bright yellow. When perched, a Zone-tailed Hawk can be distinguished from a Common Black-Hawk by its gray lores (yellow in the Black-Hawk). It also lacks the unusually long legs of the other species (see chapter 12).

In contrast to most other hawks, the plumage of immature Zone-tails differs little from that of adults: some white spotting on the breast and narrower bars in shades of gray in the tail are the main differences. Also unlike other buteos, this species has no light phase. No subspecies are recognized (Johnson et al. 2000).

Distribution

The Zone-tailed Hawk is primarily a raptor of the American tropical and subtropical regions; only the northernmost 5% of its breeding range occurs in the United States (Snyder and Glinski 1988). It breeds locally from the southwestern United States south through Mexico, Central America, and northern South America to Peru west of the Andes and, east of the Andes, to eastern Bolivia, Paraguay, and southeastern Brazil (AOU 1998). Populations in the northernmost part of the range (including all U.S. birds) are essentially migratory; elsewhere they are resident. Birds regularly winter as far north as northern Mexico (Baja California, Sonora) (AOU 1998). Occasional individuals have even been reported in winter from lower-elevation areas of southern California, Arizona, New Mexico, and the Lower Rio Grande region of Texas (Snyder and Glinski 1988), but where the bulk of the migrant population spends the winter months remains unknown.

Within the United States, breeding Zone-tailed Hawks are found locally from western Arizona east across New Mexico into southwestern Texas, and sporadically in southern California (Matteson and Riley 1981; Snyder and Glinski 1988; AOU 1998; Johnson et al. 2000). Additionally, sightings of Zone-tailed Hawks have recently increased in southwestern Utah and in southern Nevada, with nesting documented in 2005 in the latter area (Fridell 2005). Snyder and

PHOTO 17.5

Most of New Mexico's Zone-tailed Hawk breeding population appears to be concentrated in Hidalgo and Grant counties in the southwest. Here, a Zone-tailed Hawk in Steeple Rock Canyon north of Virden, Hidalgo Co., 24 May 2003. Steeple Rock drains westward into the Gila River.

PHOTOGRAPH: © ROBERT SHANTZ.

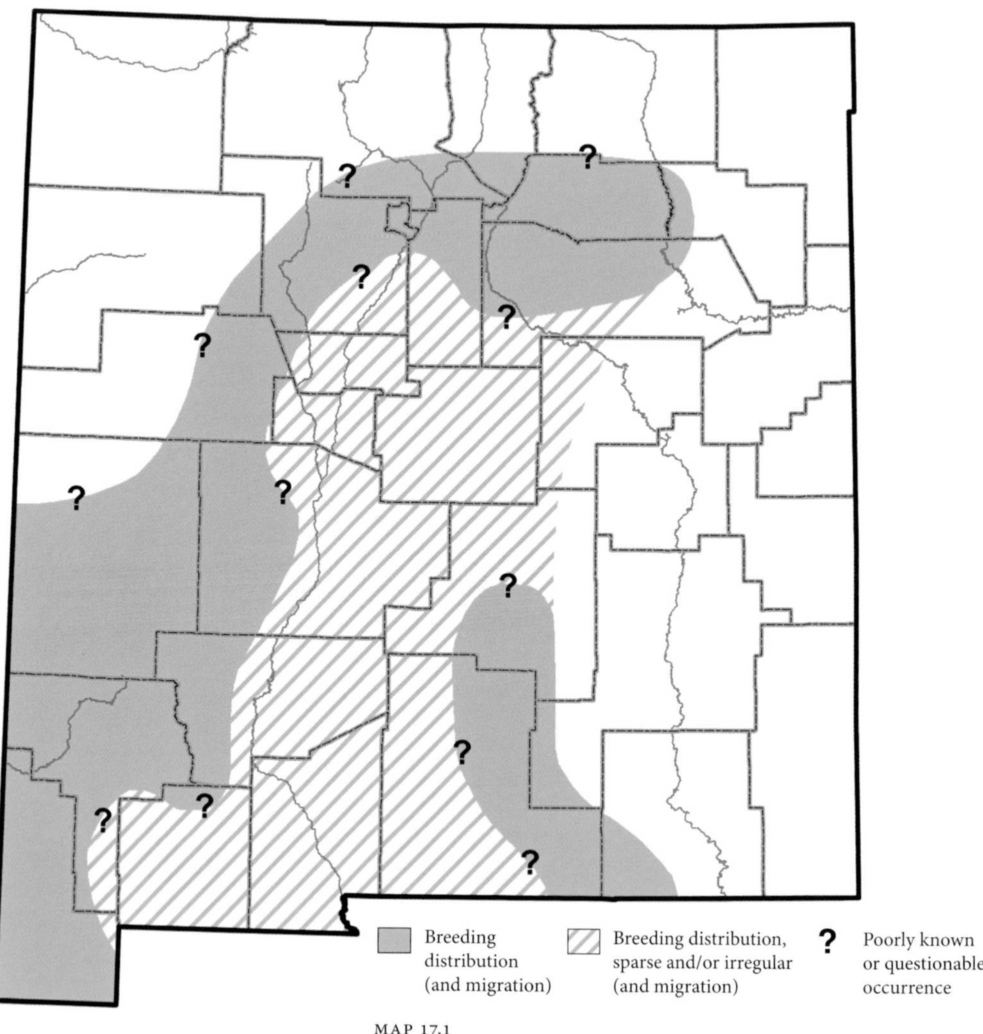

MAP 17.1

Zone-tailed Hawk distribution map

Glinski (1988) estimated a total of just 100 pairs breeding within the United States, most of them occurring in Arizona.

In New Mexico, Zone-tailed Hawks breed in small numbers across much of the southwestern part of the state north very locally as far as the Jemez Mountains, and along the Canadian River, in Mills Canyon in Harding County (Bohl and Traylor 1958; Ligon 1961; Hubbard 1978; Travis 1992; Williams 1992; NMOS 2008). Zone-tail breeding pairs are also reported frequently from the Guadalupe Mountains (especially Turkey Canyon) and the Sacramento and Capitan mountains in the southeastern quadrant of the state (Ligon 1961; NMOS 2008). While nowhere can the species be considered common, most reports of breeding birds—probably indicating where the highest densities are found—come from rugged, wooded, or semiwooded areas of Grant and Hidalgo counties. Zone-tailed Hawk nests have been reported from multiple locations within those two counties, including along the Gila River from well up in the Gila National Forest downstream to the Arizona state line and various canyons in the Peloncillo and Animas mountains (NMOS 2008).

At the northern limit of the species' distribution in New Mexico, a small breeding population has been documented in the Jemez Mountains, including at Bandelier National Monument, Los Alamos County (Southwest Parks and Monuments Association 1986; Travis 1992; Kennedy et al. 1995). Breeding has also been reported fairly regularly at Mills Canyon along the Canadian River since 1956 (e.g., Bohl and Traylor 1958; Ligon 1961; Williams 2005; H. Schwarz, pers. comm.). In the large swath of territory from Grant County north and east to Jemez Springs, Sandoval County (see Hubbard and Baltosser 1978, *fide* B. Heinrich), or between the Jemez Mountains and Mills Canyon, breeding has apparently been documented only in the San Francisco Valley of Catron County (Snider 1992, *fide* S. Williams). However, Hart Schwarz also found a territorial Zone-tailed Hawk in Bear Trap Canyon in the San Mateo Mountains of Socorro County on 24 May 2000 (Williams 2000). Breeding season sightings have been reported from other locations including in Doña Ana, Bernalillo, and San Miguel counties and north to Rio Arriba County (NMOS 2008), and these observations probably represent at least occasional breeding. Although not yet reported, Zone-tailed Hawks may breed in seemingly suitable canyon country in Cibola, McKinley, Valencia, and perhaps even Mora or Taos counties. However, they are absent as breeders from the eastern plains.

Zone-tailed Hawks have been reported as vagrants in much of the state (NMOS 2008), including San Juan County in the northwest (Snider 1995). Most vagrants seem to occur in spring and fall, and so are likely to be migrants that have wandered off course. Migrants appear regularly albeit in very small numbers at HawkWatch International's monitoring sites in the Sandia and Manzano mountains; the maximum seasonal total in recent years was 10 in the spring of 2000; one to three in a season is the norm (Smith 2004; chapter 2). Unlike for many other raptors, most reports of Zone-tails in migration are of single birds, although migration in small groups may be more common than reported because of problems distinguishing them from vultures.

Habitat Associations

Zone-tailed Hawks hunt over a wide range of open and forested habitats (Johnson et al. 2000). For nesting, however, the species shows a strong association with water (Murphy 1978; Johnson et al. 1987) and either gallery forests or rugged topography with some

PHOTO 17.6

Zone-tailed Hawk in flight in Steeple Rock Canyon, Hidalgo Co. Note the rugged topography in the background. 17 May 2003. PHOTOGRAPH: © ROBERT SHANTZ.

PHOTOS 17.7a and b

(*top*) Typical Zone-tailed Hawk habitat in New Mexico, with rugged topography and wooded riparian vegetation in the canyon bottom: Gila Lower Box near Box Canyon confluence, Hidalgo Co., 14 August 2007. PHOTOGRAPH: © ROBERT SHANTZ.

(*bottom*) Zone-tailed Hawk in Gila Lower Box, Hidalgo Co., 19 April 2008. PHOTOGRAPH: © ROBERT SHANTZ.

forest component (Porter and White 1977; Millsap 1981; Johnson et al. 2000). Typical for the species is to find it nesting in large trees or on cliffs situated in riparian woodlands or forested canyons up to 2,200 m (7,220 ft) in elevation (Johnson et al. 2000). Breeding locations have been described across the species' northern range, though studies of breeding habitat are limited to one each in Arizona (Millsap 1981), New Mexico (Kennedy et al. 1995), and Texas (Matteson and Riley 1981). Nests in the Jemez Mountains of north-central New Mexico were situated in montane forests within or near steep-walled canyons, many of them with extensive cliffs (Kennedy et al. 1995). In lower-elevation areas of New Mexico's Gila River drainage, nests are typically located in groves of mature riparian trees, usually cottonwoods (*Populus fremontii*) and again typically in the vicinity of steep slopes or large rock outcrops (Hubbard 1971; Johnson et al. 1973; GS and SHS, pers. obs.). Similarly,

PHOTOS 17.8a and b

(*top*) Steeple Rock, Hidalgo Co. PHOTOGRAPH: © ROBERT SHANTZ.

(*bottom*) Zone-tailed Hawk at Steeple Rock Canyon, 10 May 2003. PHOTOGRAPH: © ROBERT SHANTZ.

all Arizona nests located by Millsap (1981) were in areas of varied topographic relief or immediately adjacent to cliffs or steep talus, while west Texas nests were found in montane forests or rugged canyons (Matteson and Riley 1981). Cliff nesting has been reported in Texas (Matteson and Riley 1981) and Arizona (Bailey 1928).

Twelve nests found in north-central New Mexico (Kennedy et al. 1995) were in stands of ponderosa pine (*Pinus ponderosa*) and oak (*Quercus* spp.) or ponderosa pine, Douglas fir (*Pseudotsuga menziesii*), and oak. Other known nesting areas in New Mexico include montane sites in the Peloncillo, Animas, and Mogollon mountains (Williams 1992, 2003), and wooded canyons on the Canadian River (Bohl and Traylor 1958; Williams 1995) and in the Guadalupe Mountains (Williams 2003). Additional areas of nesting described by Ligon (1961) appear to be primarily montane canyons. Zone-tailed Hawks will occasionally nest in broad, low-gradient floodplains as well, such as along the Gila River in Grant County. Zone-tails are not averse to nesting close to man, and adapt readily to regular low-level activities such as ranching or farming, common in such floodplain areas (Snyder and Glinski 1988). In summary, while the diverse diet of the Zone-tailed Hawk allows exploitation of prey in nonforested areas, preference for protected nest locations in tall trees or rugged topography near waterways presumably limits nesting distribution to a relatively few, dispersed areas in New Mexico and the greater Southwest.

Habitat associations of migrating Zone-tailed Hawks are poorly understood (Johnson et al. 2000). Most observations of birds migrating in the Southwest have been made from montane forest locations such as HawkWatch International's sites in New Mexico's Sandia Mountains (in spring) and Manzano Mountains (in fall) (chapter 2). Several riparian canyon sites in the foothills of mountain ranges have multiple observations during spring or fall migration (NMOS 2008). Occasionally, however, individual migrants are observed in areas of little topographic relief such as the North Roosevelt (or Melrose) Migrant Trap, a small, isolated area of mature cottonwoods in Roosevelt County (Williams 1999).

Life History

Displays

Zone-tail pairs regularly engage in a variety of aerial displays, especially during courtship. These include dives, rolls, loops, and whirling, in which two birds lock talons and spin toward the ground. This whirling behavior was first reported in Zone-tailed Hawks by Hubbard (1974) based on observations made in New Mexico's Gila Valley. Clark (1984) suggested that it represented agonistic behavior between territorial breeding birds and intruders, rather than courtship behavior. However, Kennedy et al. (1995) document frequent displays between paired birds throughout the breeding season. Most displays are accompanied by loud, high-pitched vocalizations.

Human or other intruders in the vicinity of a nest will be met by one or both adults screaming their plaintive nasal cry while soaring overhead. Close approach to an active nest is likely to provoke rather more aggressive and seemingly fearless responses: birds will dive at and even strike at intruders (Johnson et al. 2000). Wilson et al. (1993) reported a Zone-tailed Hawk family diving at a mountain lion (*Felis concolor*) in south Texas.

Nesting

In the Southwest in general, many Zone-tailed Hawks arrive on their breeding territories in late March to early April (e.g., Kennedy et al. 1995). Pairs show strong fidelity to nest sites, and courtship and copulation often take place near old nests (Snyder and Glinski 1988). Nests are constructed of twigs and bark. Like numerous other hawks, Zone-tails decorate their nests with fresh green plant material, presumably to help reduce ectoparasite loads (Wimberger 1984).

While no studies of nest-site selection have been conducted in North America, information is available at the local scale of nests. Arizona nests (Millsap 1981) were located in mature ponderosa pine, Arizona alder (*Alnus obligifolia*), cottonwood, and Arizona sycamore (*Platanus wrightii*) ranging from approximately 10 to 24 m (33 to 79 ft) tall and 51 to 98 cm (20 to 39 in) in diameter. Nests in north-central New Mexico were situated in ponderosa pines averaging 24 m tall (79 ft) and 60 cm (24 in) in diameter (Kennedy et al. 1995). Along the Gila River and in tributary canyons, nest trees included cottonwoods and Arizona sycamores

PHOTO 17.9

Zone-tailed Hawk pair in Steeple Rock Canyon, Hidalgo Co., 30 June 2003. PHOTOGRAPH: © ROBERT SHANTZ.

PHOTO 17.10

Zone-tailed Hawk in a nest in a cottonwood (*Populus fremontii*), Steeple Rock Canyon, Hidalgo Co., 17 May 2003. PHOTOGRAPH: © ROBERT SHANTZ.

PHOTO 17.11

Zone-tailed Hawk in a nest in an Arizona sycamore (*Platanus wrightii*), Bitter Creek, Grant Co., 23 June 2007. PHOTOGRAPH: © ROBERT SHANTZ.

PHOTO 17.12

Zone-tailed Hawk nest with adult feeding nestling, Bitter Creek, Grant Co., 23 June 2007. PHOTOGRAPH: © ROBERT SHANTZ.

(see photos). In both north-central New Mexico (Kennedy et al. 1995) and the Cliff-Gila Valley (SHS and GS, unpubl. data), nests were consistently located within the top 25% of the supporting substrate, with many nests within the uppermost 10%. Nesting studies throughout the southwestern United States also consistently report the prevalence of cliffs or steep slopes in close proximity to nest trees. In Arizona, nests were situated in areas with steeper slopes than nests of Cooper's Hawks (*Accipiter cooperii*), Red-tailed Hawks, and Common Black-Hawks (Millsap 1981).

Egg-laying dates in the United States generally range from late March to mid May (Palmer 1988). First egg dates for the Jemez nests were estimated to be in the first week of May (Kennedy et al. 1995) and are likely to be similar elsewhere in New Mexico. One to three eggs (typically two) are laid and incubated primarily by the female for 28–35 days (Snyder and Glinski 1988; Johnson et al. 2000). The incubating female is fed by her mate. Once eggs hatch, the female continues to brood the young until they reach about 28 days old. Young fledge after 42 to 56 days in the nest, typically in July or August (Kennedy et al. 1995; Johnson et al. 2000). They continue to receive food from the parents until family units break up in September (Kennedy et al. 1995).

Male Zone-tailed Hawks do most of the hunting and provide most of the food through the nesting cycle. During incubation, the male delivers two to three prey items per day to the sitting female, taking her place on the eggs while she eats (Johnson et al. 2000). Once the eggs hatch, males typically deliver prey to the female in an aerial exchange away from the nest; she then returns to the nest to dismember the prey and feed it to the chicks (Kennedy et al. 1995; Johnson et al. 2000). Rates of prey delivery reported from New Mexico nests ranged from 4.8 to 8 prey items per day, depending on brood size and stage (Kennedy et al. 1995). As in many raptors, food shortage may induce siblicidal aggression among nestlings (Johnson et al. 2000).

Diet and Foraging

Zone-tailed Hawks are primarily aerial hunters, scanning for prey while coursing 15–150 m (50–500 ft) over relatively open country such as fields, pastures, desert grasslands, or even open rocky areas in montane forests (Zimmerman 1970; Snyder and Glinski 1988). In Arizona, Millsap (1981) observed that while riparian habitats provided primary nesting, roosting, and resting cover, most foraging occurred in adjacent uplands, up to 26 km (16 mi) from nests. Once a potential victim is spotted, the hawk drops into a steep stoop with partially closed wings. Occasionally Zone-tailed Hawks hunt from perches (Johnson et al. 2000).

The diet of Zone-tailed Hawks includes a variety of small to medium-sized mammals, birds, reptiles, and amphibians, augmented by an occasional invertebrate (Johnson et al. 2000). The relative importance of these various taxa in the diet varies among regions and habitats; typically Zone-tailed Hawks show a rather heavier use of birds and reptiles than do most sympatric *Buteo* hawks (Hiraldo et al. 1991; Johnson

PHOTO 17.13

Zone-tailed Hawk eating a ground squirrel, Steeple Rock Canyon north of Virden, Hidalgo Co., September 1992. The diet of the Zone-tailed Hawk has not been studied in southwestern New Mexico although hispid cotton rats (*Sigmodon hispidus*) may be important prey in parts of the Gila River Valley. In the Jemez Mountains, Zone-tailed Hawks prey on Red Crossbills (*Loxia curvirostra*), Northern Flickers (*Colaptes auratus*), and other birds, in addition to lizards and mammals such as the rock squirrel (*Spermophilus variegatus*). PHOTOGRAPH: © ROBERT SHANTZ.

PHOTO 17.14

The Zone-tailed Hawk's resemblance to the Turkey Vulture (*Cathartes aura*), both physically and behaviorally, may help the Zone-tailed Hawk during hunting. Used to the common and ubiquitous vultures, small animals may not recognize the Zone-tailed Hawk as the predator that it is until it is too late. Zone-tailed Hawks may also mingle with Turkey Vultures while hunting, further increasing prey deception. Zone-tailed Hawk in Steeple Rock Canyon, Hidalgo Co., 7 June 2003. PHOTOGRAPH: © ROBERT SHANTZ.

et al. 2000). Prey taken by Zone-tails range in size from mice and lizards (e.g., *Sceloporus* spp.) to ground squirrels (*Ammospermophilus* spp., *Spermophilus* spp.), cottontails (*Sylvilagus* spp.), and quail (*Callipepla* spp., *Cyrtonyx montezumae*) (Cottam 1947; Zimmerman 1976; Snyder and Snyder 1991; Johnson et al. 2000).

To date, the only published diet information from New Mexico is derived from Kennedy et al.'s (1995) study in the Jemez Mountains. Based on prey remains and castings at nest and roost sites as well as on prey deliveries at nests, the authors of that study documented extensive use of birds, mammals, and reptiles, although the relative importance of each of these taxonomic groups varied greatly with sampling method. Prey taxa heavily utilized by the Jemez Mountains Zone-tailed Hawk population included *Crotaphytus* lizards and *Tamias* chipmunks. Among the birds preyed upon by Zone-tailed Hawks in the Jemez Mountains were the Common Nighthawk (*Chordeiles minor*), Northern Flicker (*Colaptes auratus*), Steller's Jay (*Cyanocitta stelleri*), Western Scrub-Jay (*Aphelocoma californica*), American Robin (*Turdus migratorius*), and Red Crossbill (*Loxia curvirostra*). The rock squirrel (*Spermophilus variegatus*) and Abert's squirrel (*Sciurus aberti*) were also identified as prey of the Zone-tailed Hawk in the Jemez Mountains (Kennedy et al. 1995).

In Durango, Mexico, Zone-tails feed extensively on cotton rats (Hiraldo et al. 1991), and the same may be true in at least some parts of the Gila River Valley. We frequently watched a Zone-tail pair that hunted over riparian pastures in the Cliff-Gila Valley carrying hispid cotton rats (*Sigmodon hispidus*) toward their nest; we never observed them with any other recognizable prey species. An early report by Anthony (1892) also noted that Zone-tails were numerous in April around prairie dog colonies near what is now Hatchita, suggesting they may have preyed on prairie dogs, although these rodents have not been reported as a prey species elsewhere (Johnson et al. 2000).

Zone-tailed Hawks not only resemble Turkey Vultures physically, but also in their flight style, exhibiting a strong dihedral and occasional rocking. Often they soar among a group of Turkey Vultures while foraging, perhaps as camouflage. Willis (1963) postulated that the Zone-tail's strong physical resemblance to the Turkey Vulture may be a form of aggressive mimicry, which allows the hawk to closely approach potential prey that are habituated to the presence of the ubiquitous vultures (but see Mueller 1972). Snyder and Snyder (1991) report the capture success rate of Zone-tails in Arizona was significantly greater when soaring with vultures (30% successful) than when flying alone (7% successful), based on a sample of 55 observations. It is noteworthy that once a Zone-tail flying among vultures has spotted potential prey (as indicated by its locking its gaze on one spot on the ground), it often continues soaring past until well beyond the intended victim, often beyond some cover, at which point it stoops back at an acute angle in a surprise attack (Snyder and Glinski 1988; SHS).

Predation and Interspecific Interactions

In New Mexico, we and others (see Johnson et al. 2000) have found Zone-tailed Hawks sharing their nesting habitats with a variety of other raptors, including Cooper's Hawk, Northern Goshawk (*Accipiter gentilis*), Red-

PHOTO 17.15

Zone-tailed Hawk being harassed by a Western Kingbird (*Tyrannus verticalis*), Bitter Creek, Grant Co., 23 June 2007.

PHOTOGRAPH: © ROBERT SHANTZ.

tailed Hawk, Swainson's Hawk (*Buteo swainsoni*), Common Black-Hawk, American Kestrel (*Falco sparverius*), Peregrine Falcon (*F. peregrinus*), Barn Owl (*Tyto alba*), Western Screech-Owl (*Megascops kennicottii*), Flammulated Owl (*Otus flammeolus*), and Great Horned Owl (*Bubo virginianus*). And yet, other than one Zone-tailed Hawk being chased by an American Kestrel in Grant County (Zimmerman 1976), we are not aware of any report of agonistic interactions with other raptors in the state. Elsewhere, Zone-tailed Hawks have been reported to aggressively defend nest sites from other raptors, ranging in size from Kestrels and Cooper's Hawks to Red-tailed Hawks and Golden Eagles (*Aquila chrysaetos*) (Johnson et al. 2000). A transient Zone-tail attacked an adult Bald Eagle (*Haliaeetus leucocephalus*) carrying a fish on the Verde River in Arizona, but was outmaneuvered by the eagle (Johnson et al. 2000). Miller (1952) reported evidence that nearly fledged young from an Arizona nest were taken by a Great Horned Owl. Apparently no other predation on Zone-tails has ever been documented.

Schnell (1994) noted that Common Black-Hawks occasionally nest in unoccupied Zone-tailed Hawk nests. We found the two species breeding sympatrically in the riparian forest along the Gila River in Grant County (chapter 12), with nests of Zone-tails and Black-Hawks as close as 39 m (128 ft) from each other (GS, unpubl. data). Yet in six years of intensive field work, we never observed any interactions between the two species. While Common Black-Hawks foraged along the river, Zone-tailed Hawks foraged away from the river and in the surrounding uplands. Thus the prey bases of the two species likely showed little overlap, with no severe competition for prey (Zimmerman 1970).

Among the Zone-tailed Hawk nests we monitored in the Gila River Valley, one was unusual, being situated in the midst of a large (25+ pairs) rookery of Great Blue Herons (*Ardea herodias*) in a mature cottonwood grove, apparently in a state of mutual tolerance between the two species. Matteson and others in Texas (see Johnson et al. 2000) also suggest a possible commensal relationship between Zone-tailed Hawks and both the Black-chinned Hummingbird (*Archilochus alexandri*) and the Broad-tailed Hummingbird (*Selasphorus platycercus*). These two hummingbirds are found to regularly visit Zone-tailed Hawk nests to forage for insects around prey remains.

Status and Management

Zone-tailed Hawks occur at notably low densities. Kennedy et al. (1995) reported an apparent density of 1.51 pairs per 100 km² (39 mi²) in ponderosa pine forests of northern New Mexico, compared with estimates of 2.6 pairs per 100 km² in the Davis Mountains of Texas (Johnson et al. 2000) and 0.2 pairs per 100 km² in west-central Arizona (Millsap 1981). The minimum distance between active nests in northern New Mexico was 3.6 km (2.2 mi). In the Gila riparian corridor of Grant County, we found active nests within 2.5 km (1.5 mi) of each other (GS and SHS, unpubl. data). Smaller inter-nest distances have been reported elsewhere: 1.1 km (0.7 mi) in west Texas, and as little as 300 m (980 ft) in a western Arizona riparian area (Johnson et al. 2000). These small inter-nest distances probably result from a limited area of appropriate nesting habitat being concentrated in linear riparian systems coupled with presumably high prey availability in the vicinity.

Snyder (1998) suggested that perhaps a dozen pairs may nest in New Mexico. However, even without conducting comprehensive searches we knew of at least five distinct pairs in the portion of the Gila River Basin from the Burro Mountains east to the upper end of the Cliff-Gila Valley and south to Mangas Springs—an area of <50 km (~30 mi) of river valley representing just a small fraction of the primary breeding range within the southwestern corner of the state. Robert Shantz's photographs, illustrating this chapter, show that additional nesting occurs regularly in side canyons farther downstream, near the Arizona state line (see photos). In 1992, at least three active nests were reported from the Peloncillo Mountains and one active nest from the Animas Mountains (Williams 1992). Also in 1992, Kennedy et al. (1995) found five occupied nests (in eight territories) in the Jemez Mountains and on the adjacent Pajarito Plateau. With more breeding pairs from other locations, such as the Capitan, Sacramento, and Guadalupe mountains, there may be significantly more than a dozen breeding pairs in New Mexico. However, until comprehensive surveys are conducted throughout the state, the true number of Zone-tailed Hawks within New Mexico will remain a matter of speculation. Unfortunately, the Zone-tail's very low population density, cryptic appearance, large home range, and affinity for remote, often rugged areas make it an exceptionally difficult species to census accurately.

The maximum potential reproductive rate of the Zone-tailed Hawk is rather limited by small brood sizes (even for a large raptor) and by having a single brood per season. Birds at the northern limit of their range in north-central New Mexico were further constrained by rather poor nest success: only a third of active territories successfully produced any young, for a meager productivity rate of 0.45 fledglings per occupied territory (Kennedy et al. 1995). In contrast, four of five nests found further south in Hidalgo County in 1992 were successful, although the actual productivity rate is unknown (Williams 1992). Matteson and Riley (1981) reported 0.78 and 1.14 fledglings per pair in two consecutive years in a Texas study. A sample of 22 nests in central Arizona averaged 1.85 fledglings per nest (Millsap 1981). Average productivity in New Mexico and whether it is enough to sustain the current population in the state remain unknown.

The Zone-tailed Hawk is listed in Texas as a Threatened Species (Johnson et al. 2000), and considered a Sensitive Species within U.S. Forest Service Region 3. Otherwise it has no special status. It has a Heritage ranking of S3 in New Mexico (rare or uncommon within the state); in Arizona it ranks S4 (fairly common). Zone-tails are considered "apparently secure" globally (G4). The Zone-tail occurs at such low densities, usually in remote areas, that it is relatively unaffected by human activities. Further, it appears to tolerate low levels of human disturbance around nest sites. For those reasons, Snyder and Glinski (1988) suggest that of the three border specialist raptors (Zone-tailed Hawk, Common Black-Hawk, and Gray Hawk [*Buteo nitidus*]), the Zone-tail may be the least vulnerable to extirpation within the United States. The riparian woodlands favored by these three hawks constitute one of the most endangered habitats in North America (Cartron et al. 2000); however, many of the montane areas where the Zone-tail also

PHOTO 17.16

Zone-tailed Hawk released by Gila Wildlife Rescue, 25 November 2007. This bird was found emaciated near Cliff, Grant Co., and initially had to receive stomach tube feedings until it became stronger. Preparation for the release involved exercise in a flight cage.

PHOTOGRAPH: © DENNIS MILLER.

PHOTO 17.17

Zone-tailed Hawk in Steeple Rock Canyon, Hidalgo Co., 10 May 2003. The New Mexico Zone-tailed Hawk population is small, though population numbers are difficult to estimate.

PHOTOGRAPH: © ROBERT SHANTZ.

occurs are within National Forest or National Park Service lands, and thus protected from development and intensive human activity. The primary threat within the United States is probably illegal shooting; we know little about threats on the species' wintering grounds.

The Zone-tailed Hawk population in New Mexico may consist of substantially more than 12 pairs, but it is nonetheless small. Given also that observed productivity seems low in at least parts of the distribution in New Mexico, the species must be considered at risk in the state. Ensuring that this unique raptor remains a part of New Mexico's avifauna may require proactive management, including identification of the most productive territories for special protection. First and foremost, however, a thorough census of existing and recent Zone-tailed Hawk territories is needed in New Mexico to provide a baseline for monitoring the species into the future.

Acknowledgements

Suggestions and comments by J.-L Cartron and R. R. Johnson greatly improved the first draft of this chapter. The authors' field work was conducted under the auspices of the U.S. Forest Service's Rocky Mountain Research Station, and funded by the Gila National Forest, Phelps-Dodge Corporation, and the National Fish and Wildlife Foundation.

[AOU] American Ornithologists' Union. 1998. *Check-list of North American birds*. 7th ed. Washington, DC: American Ornithologists' Union

Anthony, A. W. 1892. Birds of southwestern New Mexico. *Auk* 9:357–69.

Bailey, F. M. 1928. *Birds of New Mexico*. Santa Fe: New Mexico Department of Game and Fish.

Bohl, W. H., and E. Traylor. 1958. A correction in identification of the Zone-tailed Hawk as a Mexican Black Hawk. *Condor* 60:139.

Cartron, J.-L. E., S. H. Stoleson, P. L. L. Stoleson, and D. W. Shaw. 2000. Riparian areas. In *Livestock management in the American Southwest: ecology, society, and economics*, ed. R. Jemison and C. Raish, 281–327. Amsterdam: Elsevier Science.

Clark, W. S. 1984. Agonistic "whirling" by Zone-tailed Hawks. *Condor* 86:488.

Clark, W., and B. Wheeler. 1987. *Peterson field guides, hawks*. Boston: Houghton Mifflin Co.

Cottam, C. 1947. Zone-tailed Hawk feeds on rock squirrel. *Condor* 49:210.

Fridell, R., ed. 2005. Great Basin. *North American Birds* 59:632–33.

Hiraldo, F., M. Delibes, J. Bustamante, R. Estrella. 1991. Overlap in the diet of diurnal raptors breeding at the Michilia Biosphere Reserve, Durango, Mexico. *Journal of Raptor Research* 25:25–29.

Hubbard, J. P. 1971. The summer birds of the Gila Valley, New Mexico. *Occasional Papers of the Delaware Museum of Natural History, Nemouria*, no. 2.

———. 1974. Flight displays in two American species of *Buteo*. *Condor* 76:214–15.

———. 1978. *Revised check-list of the birds of New Mexico*. Publ. no. 6. Albuquerque: New Mexico Ornithological Society.

Hubbard, J. P., and W. H. Baltosser, eds. 1978. *New Mexico Ornithological Society Field Notes* 17(2).

Johnson, R. R., S. W. Carothers, and D. B. Wertheimer. 1973. The importance of the lower Gila River, New Mexico, as a refuge for threatened wildlife. Unpublished report, U.S. Fish and Wildlife Service, Albuquerque, NM.

Johnson, R. R., R. L. Glinski, and S. W. Matteson. 2000. Zone-tailed Hawk (*Buteo albonotatus*). No. 529. In *The birds of North America*, ed. A. Poole and F. Gill. Philadelphia, PA: Birds of North America, Inc.

Johnson, R. R., L. T. Haight, and J. M. Simpson. 1987. Endangered habitats versus endangered species: a management challenge. *Western Birds* 18:89–96.

Kennedy, P. L., D. E. Crowe, T. F. Dean. 1995. Breeding biology of the Zone-tailed Hawk at the limit of its distribution. *Journal of Raptor Research* 29:110–16.

Ligon, J. S. 1961. *New Mexico birds and where to find them*. Albuquerque: University of New Mexico Press.

Matteson, S. W., and J. O. Riley. 1981. Distribution and nesting success of Zone-tailed Hawks in west Texas. *Wilson Bulletin* 93:282–84.

Miller, L. 1952. Auditory recognition of predators. *Condor* 54:89–92.

Millsap, B. A. 1981. *Distributional status of falconiformes in west central Arizona—with notes on ecology, reproductive success and management*. Technical Note no. 355. Washington, DC: U.S. Department of the Interior, Bureau of Land Management.

Mueller, H. C. 1972. Zone-tailed Hawk and Turkey Vulture: mimicry or aerodynamics? *Condor* 74:221–22.

Murphy, J. 1978. Management considerations for some western hawks. *Transactions of the North American wildlife and natural resource conference* 43:241–51.

[NMOS] New Mexico Ornithological Society. 2008. *NMOS Field Notes* database. http://nhnm.unm.edu/partners/NMOS/ (accessed 1 May 2008).

Palmer, R. S., ed. 1988. *Handbook of North American birds*. Vol. 5. New Haven: Yale University Press.

Porter, R. D., and C. M. White. 1977. Status of some rare and lesser known hawks in western United States. In *Proceedings of the world conference on birds of prey, Vienna*, ed. R. D. Chancellor, 39–57. Cambridge, UK: International Council for Bird Preservation.

Schnell, J. H. 1994. Common Black-Hawk (*Buteogallus anthracinus*). No. 122. In *The birds of North America*, ed. A. Poole and F. Gill. Philadelphia, PA: Academy of Natural Sciences, and Washington, DC: American Ornithologists' Union.

Smith, J. P. 2004. *Spring 2004 raptor migration studies in the Sandia Mountains of central New Mexico*. Salt Lake City, UT: HawkWatch International.

Snider, P. R., ed. 1992. *New Mexico Ornithological Society Field Notes* 31(2): 1 March to 31 May.

———, ed. 1995. *New Mexico Ornithological Society Field Notes* 34(2): 1 March to 31 May.

Snyder, H. A. 1998. Zone-tailed Hawk. In *The raptors of Arizona*, ed. R. L. Glinski, 99–101. Tucson: University of Arizona Press.

Snyder, H. A., and R. L. Glinski. 1988. Zone-tailed Hawk. In *Proceedings of the southwest raptor management symposium and workshop*, ed. R. L. Glinski, B. G. Pendleton, M. B. Moss, M. N. LeFranc Jr., B. A. Millsap, and S. W. Hoffman, 105–10. Washington, DC: National Wildlife Federation.

Snyder, N. F. R., and H. A. Snyder. 1991. *Birds of prey: natural history and conservation of North American raptors.* Stillwater, MN: Voyageur Press.

Southwest Parks and Monuments Association. 1986. *A checklist of the birds of Bandelier National Monument.* Tucson, AZ: Southwest Parks and Monuments Association.

Travis, J. R. 1992. *Atlas of the breeding birds of Los Alamos County, New Mexico.* LA-12206. Los Alamos, NM: Los Alamos National Laboratory.

Williams, S.O. III. 1992. Southwest region: New Mexico. *American Birds* 46(5): 1162–65.

———. 1995. Southwest region: New Mexico. *Field Notes* 49(5): 961–63.

———. 1999. New Mexico. *North American Birds* 53(3): 312–14.

———, ed. 2000. New Mexico. *North American Birds* 54: 312–15.

———. 2003. New Mexico. *North American Birds* 57(1): 95–97.

———. 2005. New Mexico. *North American Birds* 59: 633–36.

Willis, E. 1963. Is the Zone-tailed Hawk a mimic of the Turkey Vulture? *Condor* 65:313–17.

Wilson, B. E., C. Coldren, M. Coldren, F. Chávez-Ramírez, and T. Archer. 1993. Behavior of a group of Zone-tailed Hawks. *Journal of Raptor Research* 27:127.

Wimberger, P. H. 1984. The use of green plant material in bird nests to avoid ectoparasites. *Auk* 101:615–18.

Zimmerman, D. A. 1970. Birds and bird habitat on national forest lands in the Gila River Valley, southwestern New Mexico. Unpublished report to U.S. Forest Service.

———. 1976. Comments on feeding habits and vulture-mimicry in the Zone-tailed Hawk. *Condor* 78:420–21.

18

Red-tailed Hawk

(*Buteo jamaicensis*)

JEAN-LUC E. CARTRON, RONALD P. KELLERMUELLER,
TIMOTHY REEVES, GAIL L. GARBER, AND GIANCARLO SADOTI

THE RED-TAILED HAWK (*Buteo jamaicensis*) is one of New Mexico's most conspicuous and ubiquitous hawks, occurring in nearly all habitat types and often seen soaring high overhead or perched prominently in trees, on utility poles, or on fence posts along roads. One of several hawks once colloquially referred to as the Chicken Hawk, the Red-tailed Hawk is a large raptor well adapted for soaring with a broad tail and broad, rounded wings. As is typical for birds of prey, there is reversed sexual dimorphism in the Red-tailed Hawk, the female averaging larger and heavier than the male. In females, body length ranges from 50 to 65 cm (20 to 26 in), with weight varying from 900 to 1,460 g (32 to 51 oz); in males, total length and body weight range from 45 to 56 cm (18 to 22 in) and 690 to 1,300 g (24 to 46 oz), respectively (Preston and Beane 1993).

There is considerable variation in plumage in Red-tailed Hawks, although the typical adult light-morph encountered in New Mexico is unmistakable. The topside of the tail, which is observed on a perched bird, is bright rufous with one dark subterminal band and often—but not always (see below)—additional dark bands; the head and the upperparts are brown with pale scapulars forming a "V" on the back; the pale underparts have a faint to pronounced dark belly band consisting of vertical streaking. In flight, the wings are held in a slight dihedral when soaring or gliding. Contrasting on the otherwise pale underwings are dark patagial marks (always present), dark outer tips of the primary feathers, and dark trailing edges of the flight feathers. The underside of the tail appears pale orange. Compared to the adult light-morph, the immature light-morph is slimmer with narrower wings. Its tail is also narrower, and not rufous but instead light brown above with numerous narrow, dark bands. The underparts are light-colored, but with a dark belly band typically more pronounced than in the adult. The dark patagial marks are present on the underwings. The narrow dark bands show on the undertail, which is otherwise pale-colored. Immatures and younger adults have yellow irises whereas in older adults the iris is brown.

The diagnostic traits for identifying both adult and immature light-morphs are the dark patagial marks on the underwings and the scapular "V." The patagial marks in particular are easily seen when the hawks are soaring. When present, the dark belly band is an additional important field mark in both the adult and immature light-morph hawks. Although not diagnostic, the dark outer tips of the primary feathers and the dark trailing edge of the flight feathers can also be helpful. Together with the rufous topside of the adult's tail, all of the above field marks are especially important given the

PHOTO 18.1

(*top left*) Adult light-morph Red-tailed Hawks, south of Hachita, Hidalgo Co., 7 January 2006. New Mexico's typical adult Red-tailed Hawks are unmistakable. Note the bright rufous tail and the brown upperparts with pale scapulars on the bird on the left, and the pale underparts with a faint, dark belly band on the bird on the right. PHOTOGRAPH: © ROBERT SHANTZ.

PHOTO 18.2

(*top right*) Adult light-morph in flight, Ranchos Valley, Taos Co., 11 January 2007. Note the rufous tail, brown upperparts, dark patagial mark, and dark outer tips of the primary flight feathers. PHOTOGRAPH: © GERAINT SMITH.

PHOTO 18.3

(*bottom rght*) Adult light-morph in flight, Taos Co., 5 October 2006. The dark patagial marks on the underwings and the rufous tail (pale orange on the underside) are diagnostic. Other good field marks include the dark belly band, dark outer tips of the primary flight feathers, and dark trailing edges of the flight feathers. PHOTOGRAPH: © GERAINT SMITH.

PHOTO 18.4

(*top left*) Immature light-morph, south of Animas, Hidalgo Co., 7 January 2006. Immatures tend to have more pronounced belly bands than adults. PHOTOGRAPH: © ROBERT SHANTZ.

PHOTO 18.5

(*top right*) Immature Red-tailed Hawk with very pronounced belly band north of Virden, Hidalgo Co., 1 February 2003. PHOTOGRAPH: © ROBERT SHANTZ.

PHOTO 18.6

(*bottom left*) Immature light-morph in flight, Hidalgo Co., 13 January 2008. Note the pale underparts, dark belly band, dark patagial mark on the underwings, and narrow, dark tail bands. PHOTOGRAPH: © ROBERT SHANTZ.

PHOTO 18.7

(*top*) Fuertes' Red-tailed Hawk, Steeple Rock Canyon, Hidalgo Co., 17 March 2007. In the Fuertes' race, the belly band is faint or absent. There is no rufous wash or barring on the white chest. Some white may also be present on the chin and throat. PHOTOGRAPH: © ROBERT SHANTZ.

PHOTO 18.8

(*bottom*) Adult Fuertes' Red-tailed Hawk in flight, Steeple Rock Canyon, Hidalgo Co., 3 April 2008. Note the very faint belly band, the nearly immaculate white breast, and the white on the chin and throat. PHOTOGRAPH: © ROBERT SHANTZ.

Red-tailed Hawk's extreme plumage variation. Nine or 10 subspecies (races) are generally recognized in continental North America, each tending to have its own distinguishing characteristics including color morphs, but also displaying extensive polymorphism and evidence of intergradation with adjacently distributed subspecies (see Johnsgard 1990; Preston and Beane 1993). The exact limits of the distributions of the races and of the geographic areas where intergrading occurs remain unclear. The breeding distributions of at least Western (*B. j. calurus*), Fuertes' (*fuertesi*)—and, according to Wheeler (2003), Eastern (*borealis*)—Red-tailed Hawks meet in New Mexico. During the nonbreeding season, New Mexico receives an influx of Western Red-tailed Hawks, and two additional subspecies, Krider's (*krideri*; now considered by many as merely a color morph of *B. j. borealis*) and Harlan's (*harlani*) are also occasional in the state (see below).

In the experience of the authors, most Red-tailed Hawks nesting in New Mexico are intergrades between *fuertesi* and *calurus* with dominant *fuertesi* characteristics. In the Fuertes' Red-tail, the underparts are pale, with essentially no rufous wash or barring. The whitish breast contrasts with the brown head. The chin and throat are white, brown, or a combination of the two (some white may be present on the chin, extending down to the breast through an otherwise brown throat area). The belly band is absent or faint. The back and upperwings are brown, darker than the head and

PHOTOS 18.9a and b

Erythristic (rufous-morph) adult Red-tailed Hawk trapped during the 2003 fall migration at HawkWatch International's Manzano Mountains study site. The breast is solid rufous; the underside of the wing coverts shows both rufous and dark brown.

PHOTOGRAPHS: © JOHN P. DELONG.

with conspicuously paler scapulars (forming a "V" on a perched bird). The rufous tail of the adult shows only one narrow, dark subterminal band. Fuertes' Red-tailed Hawks can occasionally be confused with light-morph Ferruginous Hawks (*Buteo regalis*), some of which have a rufous tail. However, in the latter species the tail lacks any dark band. The Fuertes' Red-tailed Hawk is known to occur only in a light color morph (Clark and Wheeler 1995), although John Hubbard (pers. comm.) and Bob Dickerman observed on 22–24 April 1993 a mixed nesting pair of light and melanistic adult Red-tailed Hawks at San Simon Cienega in Hidalgo County, well within the nesting range of *fuertesi* and outside that of *calurus*.

The Western Red-tailed Hawk ranges farther north than the Fuertes' Red-tailed Hawk—to central Alaska—and is represented primarily by a light-color morph, with dark and rufous morphs much less common. Adult light-morph Western Red-tailed Hawks are characterized by a bright rufous tail with multiple dark bands, including a dark, wide subterminal band. The dark patagial mark on the underwing and the dark belly band are both typically prominent and wide; the throat is dark, with a pronounced rufous wash on the chest; the leg feathers are not pure white but instead show brown barring. The head and back both tend to be dark brown. Adult dark-morph birds are entirely dark brown except for the topside of the tail, which is rufous with several dark bands, and the underside of the flight feathers, which is pale with dark bars and dark tips. The wing coverts are dark brown, obscuring the dark patagial marks in this color morph. Rufous-morph adults are dark brown above with rufous legs and a rufous chest, and a heavy rufous wash over the underwing coverts. Dark and rufous morphs occur with much intergradation (Wheeler and Clark 1995). Whether the Eastern Red-tailed Hawk nests in New Mexico is questionable in the authors' view (but see

PHOTO 18.10

Immature dark-morph Red-tailed Hawk near Cotton City, Hidalgo Co., 7 December 2002. PHOTOGRAPH: © ROBERT SHANTZ.

PHOTO 18.11

Adult Harlan's Red-tailed Hawk trapped during the 1996 spring migration at HawkWatch International's Sandia Mountains study site. The breast of a Harlan's Red-tailed Hawk typically shows conspicuous white streaks or a patch of solid white. On the underwings the flight feathers are dirty white with a dark trailing edge. The tail is not rufous but instead dirty white or gray with one dark subterminal band with or without additional dark bands. PHOTOGRAPH: © JOHN P. DELONG.

PHOTO 18.12

Leucistic Red-tailed Hawk along Highway 70, Otero Co., 1 December 2008. Note the plumage, mostly snow white. Unlike an albino, however, the bird in the photograph shows pigmentation of the beak and the eyes (in a complete albino, the eyes and skin appear pinkish).
PHOTOGRAPH: © DOUG BURKETT.

Wheeler 2003). Eastern Red-tailed Hawks are similar in appearance to Fuertes' Red-tailed Hawks but wingtips do not reach the tip of the tail (Wheeler and Clark 1995), and they tend to be larger in size (R. Dickerman, pers. comm.).

The Harlan's Red-tailed Hawk, which breeds in Alaska and northwestern Canada, occurs mainly in a dark morph, much more rarely in a light morph (Wheeler and Clark 1995). The dark morph is mostly black, but the chest may be streaked conspicuously with white and the underside of the flight feathers is dirty white with a dark trailing edge (adult) or white with dark bars (immature). The underside of the wing coverts is black with white speckling. Except in western intergrades the tail shows no rufous but instead is dirty white or gray with one dark subterminal band with or without additional dark bands. Finally, Krider's Red-tailed Hawk of the Great Plains occurs only in a light color morph. The adults are very pale with a partly or completely white head, white mottling on the brown back, white underparts (there is no belly band and little or no rufous wash on the chest), and a white base on the otherwise rufous tail. More information on the races and morphs of Red-tailed Hawks can be found in Clark and Wheeler (1995), Wheeler and Clark (1995), and Wheeler (2003) (and see also photos of these and other subspecies for additional information and identification tips).

Reflecting the predominance of Fuertes' Red-tailed Hawks in New Mexico, melanistic and erythristic morphs are uncommon in the state, especially during the breeding season. From 1974 through 1985, roadside counts across New Mexico from April to August found dark-morph and rufous-morph birds to comprise only 2% (or 31 birds) of 1,559 Red-tailed Hawks in which color phase was determined (Hubbard et al. 1988).

The Red-tailed Hawk's main and most familiar vocalization is its drawn-out, raspy scream, "*kree-eee-ar*." Hubbard (1974:214) described another call during courtship as a "loud, low and raspy *hrrr, hrrr, hrrr*."

Distribution

The Red-tailed Hawk has a wide distribution that spans most of North and Central America from the Pacific coast to the Atlantic coast, including the Caribbean. Breeding populations are found from across central Alaska and central Canada south through the United States, Mexico, and the mountains of Central America to Panama (AOU 1983). Northern populations are largely migratory, with the result that the species' main distribution contracts during the winter, extending from near the U.S.-Canada border south through the rest of the breeding range. However, small numbers of birds in the northernmost part of

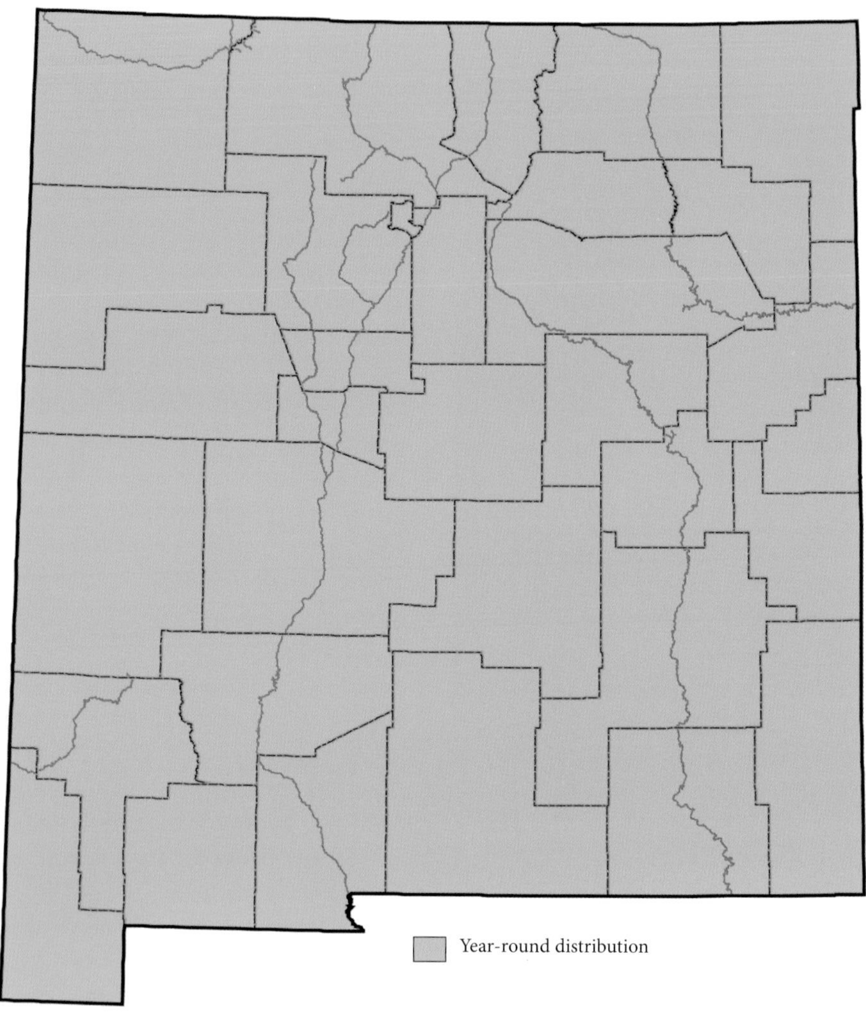

MAP 18.1

Red-tailed Hawk distribution map

the breeding distribution are sedentary (AOU 1983). The Fuertes' Red-tailed Hawk has its distribution centered on the southwestern United States, southern Texas, and northern Mexico (Ferguson-Lees and Christie 2001).

Red-tailed Hawks are found throughout New Mexico both during and outside the breeding season. Based on Bailey (1928), the upper elevational limit of the nesting distribution is about 3,050 m (10,000 ft) in New Mexico, while outside the nesting season Red-tailed Hawks may occur at all elevations.

Most of the birds breeding in the state appear to be year-round residents, and they are joined during the winter by high numbers of individuals from farther north (Hubbard et al. 1988; Preston and Beane 1993; Eakle et al. 1996). New Mexico is also part of an important migration corridor for populations breeding farther north and migrating through the state on their way south toward Mexico or Central America and northward back toward their breeding grounds (chapter 2). As a result, the Red-tailed Hawk is most abundant in New Mexico during the fall and winter. Based on vehicular surveys from 1974 to 1985 (Hubbard et al. 1988), Red-tailed Hawk numbers peak

PHOTO 18.13

Adult Red-tailed Hawk, Armendaris Ranch, Sierra Co., November 2006. PHOTOGRAPH: © PATRICK O'BRIEN.

in November–December statewide and decline notably in March; the greatest densities of birds are in the southwestern quadrant of the state (see also Status and Management).

Fall migration in New Mexico is from late August through early November; spring migration is from early March through early May (chapter 2). The median passage date for adults over the Manzanos in fall is 7 October; over the Sandias in spring it is 23 March. The median passage date for immatures is earlier in fall (26 September) and later in spring (17 April) (chapter 2).

Habitat Associations

In New Mexico, Red-tailed Hawks occur in a wide variety of open and forested environments, including Chihuahuan desertscrub, semidesert and plains grasslands, pinyon-juniper woodland, ponderosa pine (*Pinus ponderosa*) and mixed conifer forests, and deciduous woodlands of floodplain riparian areas (Eakle et al. 1996; JLEC, pers. obs.). During the nonbreeding season, large numbers of Red-tailed Hawks can be observed in flat, open areas. During the breeding season, however, we have found the species to exhibit a strong preference for rugged terrain with cliffs and canyons and for pinyon-juniper woodland especially near its upper elevational limit. Breeding pairs may also make their nests in narrow, discontinuous stringers of riparian woodlands, along irrigation canals, or in small woodlots surrounded by grasslands. Very few Red-tailed Hawk nests have been documented in the Middle Rio Grande bosque, which tends to be characterized by continuous, dense stands of cottonwood (*Populus deltoides*) trees (Cartron et al. 2008; Hawks Aloft, unpubl. data). This is in contrast with findings of an estimated eight territories (with five confirmed nests) along 26 km (16 mi) of the Gila River floodplain—mostly not in riparian cottonwood stands but alongside creeks or in copses near farmhouses—in Grant County in 2000–2001 (GS, unpubl. data). Along the Rio Grande Gorge from the Colorado state line to the John Dunn Bridge near Taos in Taos County (about 56 river km [~35 river mi]) a total of three active nests on average have been documented per year, but these nests are all on cliffs (Hawks Aloft, unpubl. data).

Despite extensive surveys, we have never found nests of Red-tailed Hawks in the Estancia Valley bottom, where breeding populations of both the Ferruginous Hawk and the Swainson's Hawk (*B. swainsoni*) occur. Instead, Red-tailed Hawks nest in pinyon-juniper woodland on nearby slopes above the valley bottom. Here Red-tailed Hawks share their nesting habitat with a variety of other raptors such as the Cooper's Hawk (*Accipiter cooperii*), Great Horned Owl (*Bubo virginianus*), and Long-eared Owl (*Asio otus*). Two occupied Red-tailed Hawk nests discovered in April 2007 in the Edgewood area, Santa Fe County, were only 0.64 km (0.39 mi) apart (Cartron and Cook, unpubl. data).

PHOTO 18.14

(*above*) Red-tailed Hawk over the Rio Grande Gorge rim, Taos Co., 16 September 2005. Typical Red-tailed Hawk habitat in New Mexico includes rugged country with cliffs and canyons. PHOTOGRAPH: © GERAINT SMITH.

PHOTO 18.15

(*left*) Typical Red-tailed Hawk nesting habitat in New Mexico: pinyon-juniper woodland at the McKinley Mine in McKinley Co. Nests are found mainly on cliffs and in mature pinyon pines (*Pinus edulis*), less often in ponderosa pines (*Pinus ponderosa*) or on power poles. An active nest was found in 2003 on a high wall. PHOTOGRAPH: © RON KELLERMUELLER/HAWKS ALOFT, INC.

PHOTO 18.16

Red-tailed Hawk perched on a utility pole, Ranchos Valley, Taos Co., 1 January 2007. Red-tailed Hawks frequently use power poles as vantage points in open, flat areas. PHOTOGRAPH: © GERAINT SMITH.

Red-tailed Hawks are common nesters in Tucson, Arizona, primarily near patches of undeveloped areas. For nest sites they use aleppo pines (*Pinus halepensis*), eucalyptus trees (*Eucalyptus camaldulensis*), and light fixtures around ballparks (Mannan et al. 2000; R. W. Mannan, pers. comm.). In contrast to Arizona, Red-tailed Hawks rarely nest in urban areas in New Mexico, although one pair once nested for many years in Albuquerque on the back of a drive-in movie screen (GLG, pers. obs.).

Life History

Pair-bonding and Courtship Displays

At the lower elevations of New Mexico, Red-tailed Hawks may be already paired and/or begin courting as early as mid January (Goodman 1983, *fide* J. Hubbard; Snider 1992, *fide* R. Fisher). It appears that some of New Mexico's birds actually remain paired and tied to a nesting territory year-round, similar to what was reported for a Red-tailed Hawk population in southeastern Wisconsin (Petersen 1979). On 30 November 1994 at San Simon Cienega, John Hubbard (pers. comm.) and Ellen Nora Cavanaugh revisited the same nest site occupied in spring 1993 by an adult of the Fuertes race and its melanistic mate. On that November day, the melanistic bird—sexed as a female based on its larger size—was noted to have circled and called.

The Red-tailed Hawk is well known for its spectacular aerial courtship displays, during which a male on a descending glide may lock talons with the female as she rolls over. Hubbard (1974:214) describes a variation of that flight courtship display, which he observed near Silver City, Grant County, on 30 April 1961. Two birds were circling overhead, the male generally above and behind the female and holding a snake. After about 15 minutes, the male began to swoop down repeatedly at the female, "trailing the snake by her as she turned over to meet him."

Nesting

Typically, Red-tailed Hawks in New Mexico begin nesting in early to mid March, but some pairs may begin as early as late February (Ligon 1961; McKnight and Snider 1967, *fide* R. Fisher) or as late as late May or early June (Bailey 1928). Most nests are on cliff ledges and in trees such as cottonwoods (*Populus* spp.), elms (*Ulmus* spp.), and ponderosa pines, as well as tall junipers (*Juniperus* spp.) and pinyon pines (*Pinus edulis*) (Bailey 1928; Ligon 1961, Hubbard 1971; Hawks Aloft data; JLEC, unpubl. data; J. Hubbard, pers. comm.). In the southern half of the state, some nests are also placed on tall yuccas (*Yucca* sp.) (Ligon 1961; photo 18.19). Although uncommon, nests on power poles and windmills have been documented (Hawks Aloft, unpubl. data; D. Griffin, pers. comm.; D. Burkett, pers. comm.;

PHOTO 18.17

(*top left*) Red-tailed Hawk nest on cliff, San Juan Co., 19 June 2008. PHOTOGRAPH: © TIM REEVES.

PHOTO 18.18

(*bottom left*) Red-tailed Hawk on nest in an elm (*Ulmus* sp.) at Maxwell National Wildlife Refuge, Colfax Co., 8 March 2005. PHOTOGRAPH: © PATTY HOBAN.

PHOTO 18.19

(*below*) Red-tailed Hawk nest in a yucca (*Yucca* sp.) in yucca grassland in Socorro Co., 17 May 2002. On that date the nest contained two young nestlings and an unhatched egg. PHOTOGRAPH: © DOUG BURKETT.

PHOTO 18.20

(*top left*) Nest on a power pole in Socorro Co., 26 May 2003. The young are approximately two weeks from fledging. PHOTOGRAPH: © DOUG BURKETT.

PHOTO 18.21

(*top right*) Red-tailed Hawk in its nest on a windmill, Otero Mesa, Otero Co., 26 April 2008. During the 2000s the windmill site was occupied on and off by both Red-tailed Hawks and Swainson's Hawks (*Buteo swainsoni*). PHOTOGRAPH: © DAVID J. GRIFFIN.

PHOTO 18.22

(*middle right*) Red-tailed Hawk taking off from its nest on a transmission double pole, San Juan Co., 19 April 1996. PHOTOGRAPH: © TIM REEVES.

PHOTO 18.23

(*bottom right*) Red-tailed Hawk in a nest in a cottonwood (*Populus* sp.), San Juan Co., March 1998. Courtship may begin as early as January in New Mexico, nesting in February. PHOTOGRAPH: © TIM REEVES.

e.g., photos 18.20 and 18.21). In San Juan County, Red-tailed Hawks now nest regularly in the crotch of timber joints on massive double poles along the transmission power line that begins at the Four Corners Power Plant at Morgan Lake (TR, unpubl. data). At the McKinley Mine (a surface mine) in McKinley County, most Red-tailed Hawk nests have been found on cliffs, in large mature pinyon pines, in ponderosa pines, or on power poles, but a pair in 2003 nested on a vertical high wall (high walls are banks of excavated areas on the uphill side) (Hawks Aloft, unpubl. data).

Although most Red-tailed Hawk nests in New Mexico are high above ground, a few exceptions have been documented. In his unpublished notes and catalog (on file at the Museum of Southwestern Biology), J. Stokley Ligon mentions finding a low nest on 10 April 1926 about 20 km (12 mi) northwest of Orange in Otero County. The nest, which contained two eggs, was only 1.8 m (5.9 ft) high in a "jointed cactus." Much more typical for the species are observations of nests 8 m (26 ft) or higher in tall trees. One of us (GS), for example, observed an active nest in the Gila River Valley about 24 m (79 ft) up in a huge (42-m-tall [138-ft-tall]) cottonwood along an irrigation ditch.

Red-tailed Hawk nests are built with sticks and lined with strips of bark and roots (Ligon 1961). As described by Bailey (1928:163), a nest in a juniper near Santa Rosa consisted of oak, pine, and juniper branches and was "heavily lined with juniper bark, making a deep soft bed for the prospective eggs and young." In New Mexico, most females have laid complete clutches, typically two eggs, less frequently three or four eggs, by the last week of March (appendix 18.1). The eggs are described by Ligon (1961) as being bluish-white and usually blotched with brown. In a clutch of more than two eggs, however, the additional egg(s)—presumably those laid last—are unmarked or nearly so. The length of the incubation period has not been determined in New Mexico in particular, but elsewhere has been estimated as being approximately 28–35 days (Bent 1937; Hardy 1939).

Based on studies conducted outside New Mexico (e.g., Petersen 1979), the eggs are incubated primarily by the adult female. After the eggs hatched, the male provides most of the food for the female and her brood, although the female does occasionally leave the nest for short periods of time to also hunt (Petersen 1979; Preston and Beane 1993). Fledging may occur as early as late May in New Mexico (Bailey 1928). Hawks Aloft monitoring data show this to be true in particular at the McKinley Mine, although at that location fledging occurs most commonly during the third and fourth weeks of June, and, rarely, as late as the first week of July. In the White Sands area in southern New Mexico, fledging is mainly during the second and third weeks of June (D. Burkett, pers. comm.).

In good years, some pairs successfully fledge three nestlings. In 2000, for example, four (29%) of 14 nests each produced three fledglings at the McKinley Mine. During 10 years of monitoring at the mine, the productivity of nesting pairs in any given year appeared linked to precipitation levels over the preceding two years (Hawks Aloft, unpubl. data).

Diet and Foraging

Despite an abundance of anecdotal reports—many of them related by Bailey (1928)—the diet of the Red-tailed Hawk has not been the focus of any comprehensive study in New Mexico. Elsewhere in the species' range, small and medium-sized mammals usually comprise the bulk of the diet, with snakes and birds also eaten regularly (Preston and Beane 1993). Hunting is conducted mainly from an elevated perch, less often while on the wing (Preston and Beane 1993).

Cully (1988, 1991) reported predation on Gunnison's prairie dogs (*Cynomys gunnisoni*), meadow voles (*Microtus pennsylvanicus*), and thirteen-lined ground squirrels (*Spermophilus tridecemlineatus*) in Moreno Valley, Colfax County. At Maxwell National Wildlife Refuge, the senior author found a nesting pair to prey extensively on Gunnison's prairie dogs, based on prey remains observed under the nest. Stomach and crop contents in two Red-tailed Hawks found dead in Torrance County and Rio Arriba County collectively included a pocket gopher, a long-tailed vole (*Microtus longicaudus*), and a prairie lizard (*Sceloporus undulatus*). Bailey (1928) reported a Red-tailed Hawk shot in Santa Clara Canyon while eating a ground squirrel, its stomach later found to also contain an Abert's squirrel (*Sciurus aberti*). Holdermann and Holdermann (1993:32) described an immature Red-tailed Hawk intercepting a

PHOTO 18.24

(*top*) Red-tailed Hawk eating on a power pole, Hidalgo Co., 30 December 2007. The Red-tailed Hawk's diet consists mostly of small and medium-sized mammals.

PHOTOGRAPH: © ROBERT SHANTZ.

PHOTO 18.25

(*bottom*) Adult Red-tailed Hawk perched on a dead Snow Goose (*Chen caerulescens*) at the Bosque del Apache National Wildlife Refuge, December 2002.

PHOTOGRAPH: © WARREN WILLIAMS.

Montezuma Quail (*Cyrtonyx montezumae*) in midair in the Dog Mountains in Hidalgo County. The quail had been flushed inadvertently by the authors of the report, and upon striking it, the Red-tailed Hawk "briefly lost some momentum as it drew the prey close to its abdomen," then disappeared out of sight.

Red-tailed Hawks prey on a wide variety of snakes (Preston and Beane 1993), including bullsnakes (*Pituophis catenifer*) and rattlesnakes (*Crotalus* spp.) in New Mexico (Bailey 1928). West of Santa Fe, Jens Jensen (Bailey 1928:164) reported a Red-tailed Hawk dropping from a cliff onto a rattlesnake on the ground and killing it only after a "fierce struggle." Red-tailed Hawks may not always win, however, as Bent (1937) reported a Red-tailed Hawk attacking a rattlesnake in Nebraska, only to be bitten and killed by the snake.

Although the Red-tailed Hawk's diet consists mainly of vertebrates, invertebrates are also taken, especially grasshoppers (Order: Orthoptera). In New Mexico, this is true in particular for young birds that recently fledged and are just starting to hunt on their own (Ligon 1961). Of two such young birds collected by the naturalist Dayton Eugene Merrill in July near Mesilla Park, Doña Ana County, one had just consumed 55 grasshoppers, the other an even larger number of these insects (Bailey 1928).

Predation and Interspecific Interactions

Because Red-tailed Hawks are so widespread in New Mexico, they likely interact regularly with most of the other raptor species in the state. In the Gila River Valley, for example, Red-tailed Hawks and Common Black-Hawks (*Buteogallus anthracinus*) likely compete for some nest sites. A nest in the Gila Bird Area was first found occupied in 2000 by a Common Black-Hawk pair, but later that same year the nest had been claimed by a Red-tailed Hawk pair (GS).

In northwestern New Mexico, the Red-tailed Hawk is only one of three species nesting on the joints of double poles, the other two species being the Great Horned Owl and the Common Raven (*Corvus corax*) (TR). At the McKinley Mine, Red-tailed Hawks again share their nesting habitat with both Great Horned Owls and Common Ravens. Great Horned Owls regularly use old nests of the other two species. In the spring of 2001, one of us (RPK) witnessed a pair of Red-tailed Hawks screaming and diving furiously at an incubating Great Horned Owl that had usurped their nest from the previous year. Quite characteristically, however, the owl paid little attention to the screaming Red-tailed Hawks. In 2002, Great Horned Owls again used the old Red-tailed Hawk nest. The resident Red-tailed Hawk pair built a new nest in a pinyon tree only about 100 m (~330 ft) away. On the San Juan Mine in 2003, a Red-tailed Hawk pair nested within 5 m (15 ft) of a Great Horned Owl active nest (Hawks Aloft, unpubl. data). The two nests were placed on ledges midway up on a small bluff about 6 m (20 ft) high, without any direct line of sight between them. The Great Horned Owl pair fledged two young successfully whereas the Red-tailed Hawk nest failed—though perhaps due to human disturbance rather than nest depredation by the owls as the area is open to the public and it receives a lot of all-terrain vehicle (ATV) traffic. The same two nests were active again in 2007, and this time both of them failed. Whether Great Horned Owls occasionally depredate Red-tailed Hawk nests or even kill adults of the latter species is unclear, but their presence in an area does not deter Red-tailed Hawks from nesting nearby.

In northwestern New Mexico, Reeves witnessed a Red-tailed Hawk defend its nest against a Golden Eagle (*Aquila chrysaetos*). The Red-tailed Hawk was returning to its nest carrying a rodent. Also flying toward the nest was the Golden Eagle. The Red-tailed Hawk dropped its prey, which tumbled some 90–150 m (300–500 ft) to splash down in a wetland near the San Juan River, and then flew at the Golden Eagle, diverting its course.

Like most raptors, the Red-tailed Hawk can be harassed or even mobbed by smaller birds. Cole Wolf (pers. comm.) and others observed an entire flock of rosy-finches—several hundred of them, mostly Black Rosy-Finches (*Leucosticte atrata*)—take off from the Crest House bird feeders on top of Sandia Peak and start flying around an approaching Red-tailed Hawk. "The rosy-finches didn't touch the hawk but they came within inches of it as they flew by. The hawk did not seem bothered by the flock of rosy-finches swirling around it, but it turned around and headed away from the Crest House. After it was 50 m away the rosy-finches stopped harassing the hawk and returned to the feeders . . ."

PHOTO 18.26

Immature Red-tailed Hawk in a tree with European Starlings (*Sturnus vulgaris*), 5 February 2007.

PHOTOGRAPH: © GERAINT SMITH.

PHOTO 18.27

Red-tailed Hawk harassed by an entire flock of rosy-finches, mostly Black Rosy-Finches (*Leucosticte astrata*) Sandia Mountains, 28 January 2007.
PHOTOGRAPH: © COLE WOLF.

The Red-tailed Hawk has few natural predators. Outside of New Mexico, Corvids have been recorded preying on eggs and nestlings (Fitch et al. 1946; Wiley 1975).

Status and Management

The Red-tailed Hawk is one of New Mexico's most common raptors year-round. Beyond the obvious (i.e., statewide, Red-tailed Hawks number in the thousands in any given season), estimating more precisely the size of New Mexico's total Red-tailed Hawk population is difficult, though not impossible. From 1985 to 2001, the mean annual count of Red-tailed Hawks migrating over the Sandias in spring was 309 birds, and over the Manzanos in fall it was nearly double that, or 616 birds (Hoffman and Smith 2003). Based on vehicular surveys from 1974 to 1985, Hubbard et al. (1988) estimated that Red-tailed Hawk numbers increased fourfold or fivefold in the state during winter. Year-round, the highest monthly numbers were in the southwestern quadrant of the state, reaching a high of 4.7 birds on average per 100 km of road in December (Hubbard et al. 1988). Using Hubbard et al.'s (1988) counts for the month of May—after nearly all spring migrants have passed through the state, but before most young fledge from nests—the density of Red-tailed Hawk pairs may be approximately twice as high in the northeastern quadrant of the state compared to the southeast and northwest. In the southwestern quadrant it may be approximately 2.67 times higher than in the southeast and northwest.

Estimates of breeding density are available from only a few areas of the state. Most of these estimates are from the northwestern quadrant of the state as defined by Hubbard et al. (1988), or all of San Juan, Rio Arriba, Taos, McKinley, Sandoval, Los Alamos, Santa Fe, Cibola, Bernalillo, and Valencia counties. During *Bird Breeding Atlas* work in Rio Arriba County in June 2007, Dale Stahlecker (pers. comm.) and collaborators found four occupied Red-tailed Hawk territories within a 25-square-kilometer (~6,180-acre) block, including two verified cliff nests and one verified tree nest. At and around the McKinley Mine, a total of 14 occupied nests were found in 2000 in an area measuring approximately 397 km^2 (98,000 acres) (Hawks Aloft unpubl. data), or about 20% the density of pairs recorded in the Rio Arriba County 25-square-km block. Lower densities still were recorded during nest searches and monitoring on the Navajo Mine and surrounding area in San Juan County (one to two breeding pairs found in a 135-km^2 [33,300-acre] area), while breeding densities in the vicinity of the San Juan Mine (one to two breeding pairs in a 39-km^2 [9,700-acre] area) were comparable to those at the McKinley Mine. Assuming that the intermediate breeding densities found at the McKinley and San Juan mines are somewhat representative of Hubbard et al.'s (1988) northwestern quadrant, the breeding population of that quadrant could be estimated at 1,786 pairs. The breeding populations of the southwestern, northeastern, and southeastern quadrants could then be estimated at 4,808 pairs, 3,056 pairs, and 1,748 pairs, respectively. New Mexico's total breeding population would then be estimated at 11,398 pairs, and the total winter population at a minimum of 91,184 birds.

Similarly to breeding densities, information on reproductive output has been reported from only a few areas of the state. In the Gila River Valley in 1989, 11 nests fledged 6 young, for a mean productivity of only 0.55 young per nest (Goodman 1989). At the McKinley Mine, mean productivity ranged from a low of 0.4 fledgling per nest ($n = 5$) in 2003 to a high of 2.14 fledglings per nest ($n = 14$) in 2000, for a mean annual productivity of 1.68 young fledged per nest over a 10-year period (Hawks Aloft, unpubl. data). For comparison, mean annual productivity per nest reported from outside New Mexico ranged from 0.91 in Michigan ($n = 22$) (Craighead and Craighead 1956) to 1.8 in Wisconsin ($n = 27$) (Orians and Kuhlman 1956). It should be noted, however, that when reported for only one or a few years, annual reproductive output estimates can be of limited value. As shown by monitoring

PHOTO 18.28

Red-tailed Hawk about to be released by Gila Wildlife Rescue, 22 September 2007. The bird in this photo had been rescued after falling into a stock tank.

PHOTOGRAPH: © DENNIS MILLER.

PHOTO 18.29

Red-tailed Hawk in Taos Co., 5 October 2006. The Red-tailed Hawk is not a species of concern either globally or in New Mexico in particular. Red-tailed Hawk populations increased through the mid 1990s in western North America.

PHOTOGRAPH: © GERAINT SMITH.

at the McKinley Mine, reproductive success can vary widely over time, probably as a function of prey abundance (Preston and Beane 1993).

At the scale of western North America, analyses of migration-counts coupled with BBS and CBC data indicated an increase in Red-tailed Hawk populations through the mid 1990s, with no further increase or decline since then (Hoffman and Smith 2003; see also White 1994; Kirk and Hyslop 1998). Until now, the species has benefited from human habitat alteration, which in many places has resulted in a mosaic of woodlands and large, open areas readily used by Red-tailed Hawks (Preston and Beane 1993). The main threats to the species, as listed by Preston and Beane (1993), consist of shooting, automobile collisions, and human disturbance around nests. This is almost certainly true in New Mexico as well, with also some mortality caused by drowning in cattle tanks—especially those steep-sided and made of metal or concrete—and by electrocutions on power poles (Dickerman 2003; D. Miller, pers. comm.). Dickerman (2003) reported the discovery of two electrocuted female Red-tailed Hawks—one was a dark-morph—with locked talons under a transformer pole in Santa Fe. One of the two birds presumably attempted to displace the other, which was perched on the pole. As they locked talons they made contact with two of the energized wires—the report served to prove that talon locking occurs in

Red-tailed Hawks not only as part of courtship displays but also during agonistic interactions. It should also remind us that although much progress has been made with the design and retrofitting of power lines in the United States, some power poles can still kill raptors. In Chihuahua, just south of the border with Mexico, Red-tailed Hawks incur high mortality on concrete power poles fitted with steel cross-arms (Cartron et al. 2000; Cartron, Rodríguez-Estrella et al. 2006; Cartron, Sierra Corona et al. 2006; Manzano-Fischer et al. 2006; see also chapter 19).

Monitoring of Red-tailed Hawks and their reproductive success remains important on New Mexico's mines and in other industrial development areas of the state. This is in large part because the Red-tailed Hawk can serve as an indicator species, its abundance and productivity indicative of ecosystem health. Mitigation of mining activities should include the establishment of buffer zones around nests, monitoring of nests and productivity at least five years beyond cessation of mining, and habitat restoration for nesting or foraging or both.

The Red-tailed Hawk is an important component of New Mexico's ecosystems . . . as well as our natural heritage. Few sounds evoke a sense of untamed nature so well as does the call of a Red-tailed Hawk echoing down the side of a canyon. Fortunately, the species appears to be doing well both globally and in our state in particular.

LITERATURE CITED

[AOU] American Ornithologists' Union. 1983. *Check-list of North American birds.* 6th ed. Washington, DC: American Ornithologists' Union.

Bailey, F. M. 1928. *Birds of New Mexico.* Santa Fe: New Mexico Department of Game and Fish.

Bent, A. C. 1937. *Life histories of North American birds of prey.* Part 1. U.S. National Museum Bulletin 167. Washington, DC: Smithsonian Institution.

Cartron, J.-L. E., G. L. Garber, C. Finley, C. Rustay, R. P. Kellermueller, M. P. Day, P. Manzano Fisher, and S. H. Stoleson. 2000. Power pole casualties among raptors and ravens in northwestern Chihuahua, Mexico. *Western Birds* 31:255–57.

Cartron, J.-L. E., D. C. Lightfoot, J. E. Mygatt, S. L. Brantley, and T. K. Lowrey. 2008. *A field guide to the plants and animals of the Middle Rio Grande bosque.* Albuquerque: University of New Mexico Press.

Cartron, J.-L. E., R. Rodríguez-Estrella, R. C. Rogers, L. B. Rivera, and B. Granados. 2006. Raptor electrocutions in northwestern Mexico: A preliminary regional assessment of the impact of concrete power poles. In *Current raptor studies in Mexico*, ed. R. Rodríguez-Estrella, 202–30. La Paz, Mexico: CIBNOR.

Cartron, J.-L. E., R. Sierra Corona, E. Ponce Guevara, R. E. Harness, P. Manzano-Fischer, R. Rodríguez-Estrella, and G. Huerta. 2006. Bird electrocutions and power poles in northwestern Mexico: an overview. *Raptors Conservation 2006* 7:4–14.

Clark, W. S., and B. K. Wheeler. 1995. *A field guide to hawks of North America.* Boston: Houghton Mifflin.

Craighead, J. J., and F. C. Craighead Jr. 1956. *Hawks, owls, and wildlife.* Harrisburg, PA: Stackpole Co.

Cully, J. F. Jr. 1988. Gunnison's prairie dog: an important autumn raptor prey species in northern New Mexico. *Proceedings of the southwest raptor management symposium and workshop. Institute for Wildlife Research.* Scientific and Technical Series no. 11. Washington, DC: National Wildlife Federation.

———. 1991. Response of raptors to reduction of a Gunnison's prairie dog population by plague. *American Midland Naturalist* 125:140–49.

Dickerman, R. W. 2003. Talon-locking in the Red-tailed Hawk. *Journal of Raptor Research* 37:176.

Eakle, W. L., E. L. Smith, S. W. Hoffman, D. W. Stahlecker, and R. B. Duncan. 1996. Results of a raptor survey in southwestern New Mexico. *Journal of Raptor Research* 30:183–88.

Ferguson-Lees, and D. A. Christie. 2001. *Raptors of the world.* New York: Houghton Mifflin.

Fitch, H. S., F. Swenson, and D. F. Tillotson. 1946. Behavior and food habits of the Red-tailed Hawk. *Condor* 48:205–57.

Goodman, R. A., ed. 1983. *New Mexico Ornithological Society Field Notes* 22(1): 1 December to 28 February.

———, ed. 1989. *New Mexico Ornithological Society Field Notes* 28(3): 1 June to 31 July.

Hardy, R. 1939. Nesting habits of the western Red-tailed Hawk. *Condor* 41:79–80.

Hoffman, S. W., and J. P. Smith. 2003. Population trends of migratory raptors in western North America, 1977–2001. *Condor* 105:397–419.

Holdermann, D. A., and C. E. Holdermann. 1993. Immature Red-tailed Hawk captures Montezuma Quail. *New Mexico Ornithological Society Bulletin* 21:31–33.

Hubbard, J. P. 1971. The summer birds of the Gila Valley, New Mexico. *Occasional Papers of the Delaware Museum of Natural History, Nemouria* 2.

———. 1974. Flight displays in two American species of *Buteo*. *Condor* 76:214–15.

Hubbard, J. P., J. W. Shipman, and S. O. Williams III. 1988. An analysis of vehicular counts of roadside raptors in New Mexico, 1974–1985. In *Proceedings of the southwest raptor management symposium and workshop*, ed. R. L. Glinski, B. G. Pendleton, M. B. Moss, M. N. LeFranc Jr., B. A. Millsap, and S. W. Hoffman, 204–9. Washington, DC: National Wildlife Federation.

Johnsgard, P. A. 1990. Goshawk. In *Hawks, eagles, and falcons of North America*, 176–82. Washington DC: Smithsonian Institution Press.

Kirk, D. A., and C. Hyslop. 1998. Population status and recent trends in Canadian raptors: a review. *Biological Conservation* 83:91–118.

Ligon, J. S. 1961. *New Mexico birds and where to find them*. Albuquerque: University of New Mexico Press.

Mannan, R. W., C. W. Boal, W. J. Burroughs, J. W. Dawson, T. S. Estabrook, and W. S. Richardson. 2000. Nest sites of five raptor species along an urban gradient. In *Raptors at risk*, ed. R. D. Chancellor and B.-U. Meyburg, 447–53. Berlin: WWGBP/Hancock House.

Manzano-Fischer, P., G. Ceballos, R. List, and J.-L. E. Cartron. 2006. Avian diversity in a priority area for conservation in North America: the Janos–Casas Grandes Prairie Dog Complex and adjacent habitats in northwestern Mexico. *Biodiversity and Conservation* 15:3801–25.

McKnight, B. C., and P. R. Snider, eds. 1967. *New Mexico Ornithological Society Field Notes* 6(1): 1 December to 31 May.

Orians, G., and F. Kuhlman. 1956. Red-tailed Hawk and Great Horned Owl populations in Wisconsin. *Condor* 58: 371–85.

Petersen, L. R. 1979. *Ecology of Great Horned Owls and Red-tailed Hawks in southeastern Wisconsin*. Technical Bulletin no. 111. Madison: Wisconsin Department of Natural Resources.

Preston, C. R., and R. D. Beane. 1993. Red-tailed Hawk (*Buteo jamaicensis*). No. 52. In *The birds of North America*, ed. A. Poole and F. Gill. Philadelphia, PA: Academy of Natural Sciences, and Washington, DC: American Ornithologists' Union.

Snider, P. R., ed. 1992. *New Mexico Ornithological Society Field Notes* 31(1): 1 December to 28 February.

Wheeler, B. K. 2003. *Raptors of western North America*. Princeton, NJ: Princeton University Press.

Wheeler, B. K., and W. S. Clark. 1995. *A photographic guide to North American raptors*. London: Academic Press.

White, C. M. 1994. Population trends and current status of selected western raptors. *Studies in Avian Biology* 15:161–72.

Wiley, J. W. 1975. The nesting and reproductive success of Red-tailed Hawks and Red-shouldered Hawks in Orange County, California, 1973. *Condor* 77:133–39.

19

Ferruginous Hawk

(*Buteo regalis*)

JEAN-LUC E. CARTRON, JAMES M. RAMAKKA, AND HART R. SCHWARZ

THE FERRUGINOUS HAWK (*Buteo regalis*) is New Mexico's largest buteo. Sibley (2000) gives an average length of 58.4 cm (23 in) for the species, a wingspan of 1.4 m (4.6 ft), and a weight of 1.59 kg (3.5 lb). Respectively, these measurements are 21%, 14%, and 46% greater than similar measurements for the Red-tailed Hawk (*Buteo jamaicensis*). The Ferruginous Hawk is also a sexually dimorphic species: females weigh approximately 53% more than males; their average wing area is 11% greater (Bechard and Schmutz 1995).

No subspecies of Ferruginous Hawks are recognized. However, the species occurs in two color morphs, light and dark. Throughout the Ferruginous Hawk's distribution, light-morph birds outnumber dark-morph birds by a ratio of at least 10 to 1 (Bechard and Schmutz 1995). In New Mexico, we have found the percentage of dark-morph birds among nesting Ferruginous Hawks to be generally higher west of the continental divide, but still reaching a maximum of only 8% in the northwestern part of the state.

Both color morphs are readily identified by the white undersides of the flight feathers and tail. The breast, throat, and flanks of light-morph Ferruginous Hawks are a nearly immaculate white, with chestnut pigmentation on the belly that is typically more pronounced in females. When viewed from above, light-morph adults

PHOTO 19.1

Adult light-morph Ferruginous Hawk at Encino, Torrance Co., 29 April 2007. Adult light-morph birds have dark brown and rufous upperparts, with rufous pigmentation especially dominant in the upperwing coverts. The underparts are mainly white. The head is light gray to brown. The irises darken as a bird becomes older, becoming chestnut brown.

PHOTOGRAPH: © DOUG BROWN.

PHOTO 19.2

(above) Adult light-morph Ferruginous Hawk in flight near Las Vegas National Wildlife Refuge, San Miguel Co., 16 November 2007. The throat, breast, and flanks are white, as are the flight feathers on the underwings. The belly and wing coverts both show some rufous pigmentation (more extensive on the wing coverts). Note the rufous legs forming a "V" on the whitish belly, distinguishing adults from immatures. PHOTOGRAPH: © MARK L. WATSON.

PHOTO 19.3

(right) Immature light-morph Ferruginous Hawk, Rio Arriba Co., 19 November 2008. The underparts are white with some dark brown pigmentation chiefly on the belly (and the wing coverts, not visible on this photo). The upperparts are mainly dark brown. Note the lack of rufous pigmentation, distinguishing this bird from an adult. Both adults and immatures have dark eye-lines. PHOTOGRAPH: © ROGER HOGAN.

PHOTOS 19.4a, b, and c

Adult dark-morph north of San Jon, Quay Co., 5 November 2007. Much of the underparts (breast, belly, and wing coverts) are a mix of dark brown and rufous, but the flight feathers are white. The tail is unbanded, gray on the upper side, white or whitish (in good light) on the underside. Note also the rufous undertail coverts. PHOTOGRAPHS: © JERRY OLDENETTEL.

have rusty colored backs and wings and a pink cast to the tail. The legs are fully feathered with rufous feathers that form a distinct dark "V" when viewed from below. Juveniles have darker gray-brown backs and lightly barred tails, and their legs are not as dark. Dark-morph adults have dark underwing coverts with a distinct white patch at the outer joint of the wing. Dark-morph juveniles appear almost black from both the front and back, whereas the adults exhibit more rust-colored feathers mixed with the blackish brown. The tails of dark-morph birds appear gray from above, but are characteristically white when seen from below. When in flight, hawks of both color morphs appear to have light patches on the upper sides of the primaries. The long pointed wings of Ferruginous Hawks almost appear gull-like when flying birds are viewed from a distance.

Distribution

The Ferruginous Hawk inhabits Canada, the United States, and Mexico. Its breeding distribution is in the Great Basin and Great Plains from southwestern Canada south to Oregon, Nevada, Arizona, New Mexico, and extreme northwestern Texas. Except in the southern part of the breeding distribution, populations are migratory, with movements to the southwest, south, and southeast. As a result, California and Mexico, neither of which is included in the breeding range, harbor wintering populations, while the species also expands its range in the fall and winter southeastward into a broader part of Texas (Bechard and Schmutz 1995). The southern end of the wintering distribution in Mexico is in Baja California, Sonora, Durango, and Coahuila, less regularly Baja California Sur, Guanajuato, and Hidalgo (Howell and Webb 1995).

PHOTO 19.5

Ferruginous Hawk north of Mosquero, Harding Co., 8 December 2008. PHOTOGRAPH: © JERRY OLDENETTEL.

The Ferruginous Hawk can be found in New Mexico year-round. However, it breeds almost exclusively in the northern two-thirds of the state. Roughly south of latitude 33° 20' N, breeding records are few and far between. In the southwestern part of the state, two egg sets were collected in 1908 at Fort Egbert [=Bayard] near Silver City, Grant County (Cartron and Polechla, unpubl. data). Ligon (1961) also reported finding a nest north of Hachita in Grant County in 1935. In south-central New Mexico, a nest with two adults and three eggs was reported in 1980 near a prairie dog (*Cynomys*) town east of Orogrande on Otero Mesa, Otero County (S. Williams, pers. comm.). In southeastern New Mexico Sandy Williams observed a nest with three young on a power pole just north of Caprock in Lea County, 9 June 1988 (Hubbard 1988). The authors are not aware of any nesting record from south of Caprock in the southeastern corner of the state. However, the ornithologist Lawry Sager (pers. comm.) observed single adults at two locales between Humble City and the Texas state line in central Lea County. The date of the observations, 4 June [2003], strongly suggests possible local breeding that year. Finally, Schwarz's observation of an occupied nest east of Estancia in Torrance County was erroneously reported in *NMOS Field Notes* (Goodman 1985) as a nest from east of Artesia, Eddy County.

According to Ligon (1961), the upper elevational range of nesting pairs in New Mexico is about 2,740 m (9,000 ft), based on nests found along the northern edge of the Plains of San Agustin and southward to Franks Mountain in the Black Range. A recent summer record (July 7) is from 7.5 km (4.7 mi) north of Cebolla in Rio Arriba, at an elevation of 2,328 m (7,638 ft) (Williams 2000). No nest was found at that particular location, but in 2007 and 2008 Ron Kellermueller (pers. comm.) recorded an active nest in a juniper (*Juniperus* sp.) at an even higher elevation of 2,518 m (8,260 ft) near San Antonio Mountain, also in Rio Arriba County.

Breeding populations have been well documented in three areas of the state north of latitude 33° 20' N: the badlands of San Juan County (Ramakka and Woyewodzic 1993), the Plains of San Agustin (Bailey 1928; Ligon 1961; Cartron et al. 2002), and the Estancia Valley (Cartron et al. 2002). The extensive short grass prairie ecosystem of northeastern New Mexico (Union, Colfax, and Harding counties)—including

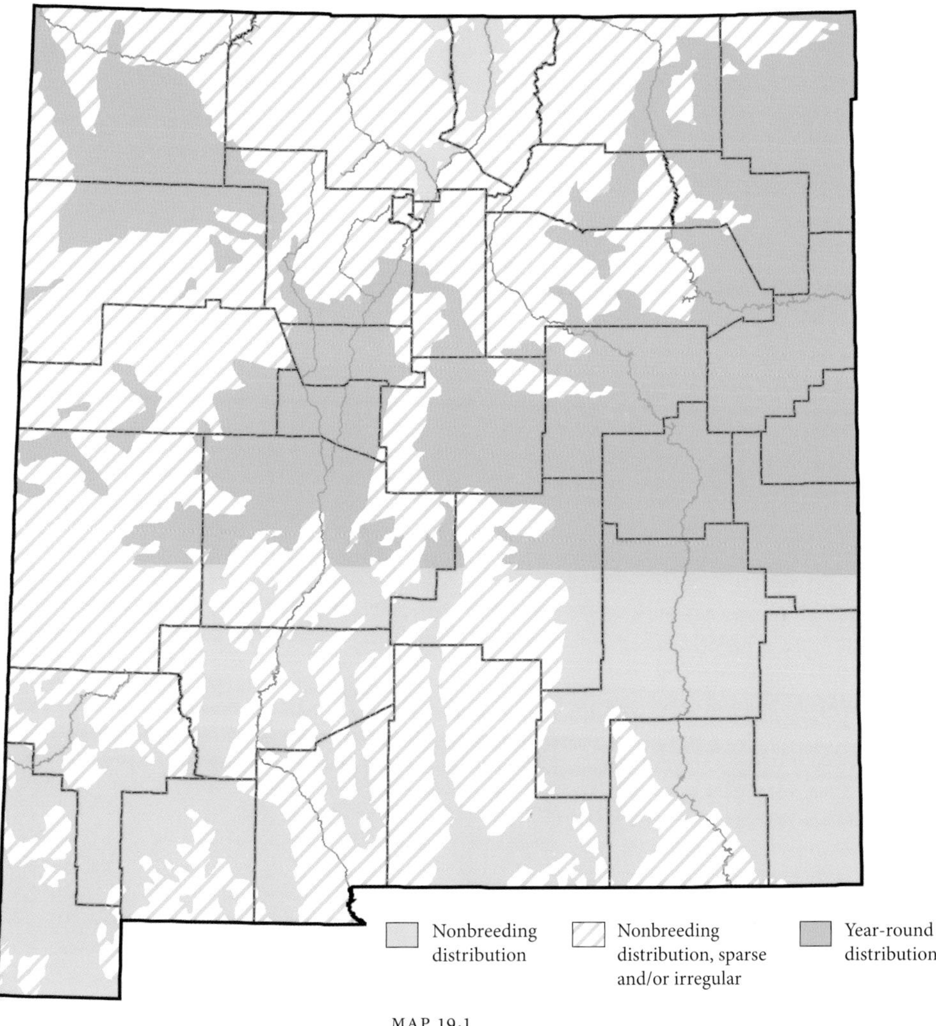

MAP 19.1

Ferruginous Hawk distribution map

the Kiowa National Grasslands—also supports what appears to be a somewhat substantial breeding population (Schwarz 2005).

The Ferruginous Hawk migrates in variable numbers across the state. It is recorded migrating in low numbers over the Sandia and Manzano mountains (Hoffman and Smith 2003; chapter 2). In Moreno Valley (Colfax County), however, Cully (1991) found the species to be abundant during fall migration, but only while Gunnison's prairie dog (*Cynomys gunnisoni*) numbers were high. Band recovery and telemetry data from New Mexico indicate that some fall migrants and/or winter residents originate from populations breeding in Alberta (Schmutz and Fyfe 1987), Montana (Harmata et al. 2001), and Wyoming (J. Watson, pers. comm.). In New Mexico, migrating Ferruginous Hawks are observed in the spring from late February through April, and fall migrants pass through in September and October (Cully 1988; Hoffman and Smith 2003).

The Ferruginous Hawk is an uncommon to locally common winter resident in New Mexico, occurring statewide but reportedly in greater numbers in the east (at least historically), northwest, and south (e.g., Hall et al. 1988; Eakle et al. 1996). The largest concentrations of wintering Ferruginous Hawks were reported from the Navajo Indian Irrigation Project area, San Juan County, during the 1980s. As many as 93 Ferruginous Hawks were sighted in that area on 31 December 1982, and 134 on 21 January 1984 (Goodman 1983, 1984)!

It remains unclear where New Mexico's Ferruginous Hawk breeding populations go for the winter. They may be nomadic rather than truly migratory, searching for prey populations in New Mexico and surrounding areas. Also poorly known is the dispersal of juveniles fledged from nests in New Mexico. One bird banded as a nestling in Socorro County on 18 June 1970 was shot on 29 October of that same year near Oakley, Kansas (Hubbard 1970). Another bird, banded as a nestling in De Baca County on 10 June 1970 was captured that same year near Odessa, Texas, on 26 October (Hubbard 1970).

Habitat Associations

The Ferruginous Hawk is a species associated with the open country, primarily grasslands and shrub-steppes, and only rarely farmland with extensive crop fields (Bechard and Schmutz 1995). In New Mexico the authors have found the species chiefly in grasslands, pinyon-juniper woodland-grassland ecotones, and

PHOTO 19.6

Example of nesting habitat in northwestern New Mexico: the Bisti/De-Na-Zin Wilderness Area. Ferruginous Hawks build their nests on pinnacles, boulders, and ledges in the badlands, and forage in surrounding open habitats. In northwestern New Mexico as elsewhere in the state, prairie dog (*Cynomys* spp.) towns are often observed within the vicinity of nests.

PHOTOGRAPH: © ROBERT SHANTZ.

PHOTO 19.7

Example of nesting habitat in northeastern New Mexico: the Kiowa National Grasslands, 28 June 2007. Ferruginous Hawks have nested in the tree shown toward the center. Note the vast expanse of grassland around the nest tree.

PHOTOGRAPH: © HART SCHWARZ.

PHOTO 19.8

Example of nesting habitat in west-central New Mexico: the Plains of San Agustin, Spring 2000. Some Ferruginous Hawk pairs nest along the ecotone between grassland and pinyon-juniper woodland.

PHOTOGRAPH: © JEAN-LUC CARTRON.

PHOTO 19.9

Example of nesting habitat on the eastern plains: nest tree in a pasture for cattle between Milnesand and Kenna, Roosevelt Co., 24 June 2003. PHOTOGRAPH: © LAWRY SAGER.

PHOTO 19.10

Ferruginous Hawk perched on a fence post in typical winter grassland habitat, northern New Mexico.

UNDATED PHOTOGRAPH: © ROGER HOGAN.

badlands, especially in the vicinity of active prairie dog towns.

Badlands are the preferred nesting habitat in northwestern New Mexico. Despite being more widespread, big sagebrush and rabbitbrush-greasewood shrublands, pinyon-juniper woodlands, and grasslands all harbor fewer nests (Ramakka and Woyewodzic 1993). Elsewhere, nest sites are found in grassland on the Kiowa National Grasslands, in the Estancia Valley, and in the Plains of San Agustin, and in the latter area also in the ecotone between pinyon-juniper woodland and grassland (Cartron et al. 2002; Schwarz 2005). A Ferruginous Hawk nest site on the Kiowa National Grasslands is often picturesquely located in a landscape as vast as the ocean in the only tree available for kilometers, sometimes on the edge of a playa or a shallow, rocky wash (Schwarz 2005).

In New Mexico during migration and winter, the Ferruginous Hawk occurs chiefly in grasslands, but may be found also in a wide variety of other vegetation types, including ponderosa pine (*Pinus ponderosa*) forest (Eakle et al. 1996).

Life History

In New Mexico, much of the ecology of Ferruginous Hawks is tied to prairie dogs. In the Estancia Valley in particular, research by Hawks Aloft, Inc. (Cook et al. 2003) showed that Ferruginous Hawks nested preferentially within 1 km (0.6 mi) of active Gunnison's prairie dog towns and that proximity to prairie dog towns was positively correlated with the number of young fledged by successful pairs. In northwestern New Mexico, the degree of spatial association between nesting Ferruginous Hawks and prairie dog towns has not been tested statistically, but several pairs monitored in that area similarly nested near active prairie dog towns. Ferruginous Hawks in New Mexico also appear to concentrate in areas with prairie dog towns during migration (Cully 1988, 1991) and in the winter (Bak et al. 2001). In Moreno Valley during migration, Ferruginous Hawk numbers seemed directly related to the status of local prairie dog towns, and declined when prairie dogs were decimated by an outburst of sylvatic plague (Cully 1991).

Nesting

In New Mexico, we have found nesting to begin as early as the first half of March, but more typically in late March through early April. The nest is built with a collection of sticks and lined with bark, grass, cow manure, rags, and strings (Ligon 1961). The Ferruginous Hawk usually builds its own nest and repairs it every year. Less often, it builds upon an old nest originally constructed by some other species, for example the Swainson's Hawk (*Buteo swainsoni*) or the Chihuahuan Raven (*Corvus cryptoleucus*) on the Kiowa National Grasslands. These two birds may, in turn, use the nest of a Ferruginous Hawk pair, but the latter likely has first dibs on any site, since it initiates nesting earlier than the other two species (HRS, pers. obs.). Nests may be used for at least up to seven consecutive years, and after years of vacancy, a historical nest site may be occupied again (JMR, unpubl. data). On the Kiowa National Grasslands, one tree nest has become gargantuan, most likely from years of buildup and repair. As documented by Schwarz (1995), it measures approximately 1.8 m (6 ft) in height!

Ferruginous Hawks use a wide variety of nesting substrates. In northwestern New Mexico's badlands they nest on pinnacles or occasionally ledges (Ramakka and Woyewodzic 1993). In that habitat type, Mikesic (2005) reported that, of over 116 nests inventoried by helicopter, 66% (77 nests) were found on sandstone and/or clay pillars, around 10% were found on each of three other substrates of ground (13 nests, 11.2%), small butte (12 nests, 10.3%), and cliff (11 nests, 9.5%), while only 3 (2.6%) were on large butte structures. Not included in this inventory were 3 nests known from steel transmission towers on the Navajo Nation. In contrast to northwestern New Mexico, most nests in the Estancia Valley are in the crowns of junipers (*Juniperus monosperma*), but elms (*Ulmus* sp.) are also often used, and at least one nest occupied in recent years was on a utility pole fitted with a nesting platform (Cartron et al. 2002). In and around the Plains of San Agustin, juniper is again the typical nesting tree, but Ferruginous Hawk nests have also been documented in pinyons (*Pinus edulis*) and ponderosa pines (Cartron et al. 2002, unpubl. data). In contrast to the Estancia Valley, the short grass prairie on the Kiowa National Grasslands is largely devoid of trees. Here the Ferruginous Hawk resorts to wizened old trees, often

PHOTO 19.11

(*top left*) Nest in what is now the Bisti/De-Na-Zin Wilderness Area, 26 January 1982. The large size of the nest suggested a long history of use for nesting. The nest was fully exposed on top of a 1.2-m-high (4-ft-high) sandstone-capped clay pillar at the top edge of a 12-m-high (40-ft-high) eroded clay hill. It consisted of a column of sticks approximately 61 cm (24 in) tall and 91 cm (36 in) in diameter. At the time the nest was located in January 1982, old feathers, down, and prey remains were still present in the nest. PHOTOGRAPH: © JIM RAMAKKA.

PHOTO 19.12

(*middle left*) Nest on a ledge in San Juan Co., 25 May 2006. PHOTOGRAPH: © RON KELLERMUELLER/HAWKS ALOFT, INC.

PHOTO 19.13

(*bottom left*) Nest tree in the Estancia Valley, spring 2000. In the Estancia Valley most Ferruginous Hawks nest in junipers (*Juniperus monosperma*) or, as shown in photo, in elms (*Ulmus* sp.). Nests in junipers are easier to find during aerial surveys than through ground searches. In contrast nests in elms are readily seen from a distance. PHOTOGRAPH: © JEAN-LUC CARTRON.

PHOTO 19.14

(*below*) Nest on a platform on a utility pole in the Estancia Valley, spring 2005. PHOTOGRAPH: © JEAN-LUC CARTRON.

PHOTO 19.15

Nest tree in the Plains of San Agustin, spring 1999. Most nest trees in and around the Plains of San Agustin are in junipers (*Juniperus monosperma*). PHOTOGRAPH: © JEAN-LUC CARTRON.

PHOTO 19.16

Ferruginous Hawks on an abandoned windmill southeast of Clayton, Union Co., along the Texas state line, 11 June 2004. PHOTOGRAPH: © LAWRY SAGER.

PHOTO 19.17

Ferruginous Hawk nest in cholla (*Cylindropuntia*) southwest of Santa Rosa, Guadalupe Co., 18 April 2000. PHOTOGRAPH: © LAWRY SAGER.

PHOTO 19.18

Nest on a transformer pole in southern Roosevelt Co., 25 July 2002. On the nest are three fledged or nearly fledged young. PHOTOGRAPH: © LAWRY SAGER.

PHOTO 19.19

Three eggs in a nest in a cholla (*Cylindropuntia*) southwest of Santa Rosa, Guadalupe Co., 18 April 2000. PHOTOGRAPH: © LAWRY SAGER.

PHOTO 19.20

Three eggs in a nest on a sandstone pinnacle in San Juan Co., 13 June 2007. Ferruginous Hawks typically lay three eggs in New Mexico. PHOTOGRAPH: © JEREMEY KNOWLTON/ HAWKS ALOFT, INC.

PHOTO 19.22

(*above*) Nest on a ledge in San Juan Co., 23 June 2005. Note the two partially feathered nestlings, which have wandered off the nest and onto the narrow ledge. A third nestling was dead on the ground below. When next visited (on 2 July), the nest had failed. The remaining two nestlings had not survived, and the adult female was also found dead on the ground below the nest. PHOTOGRAPH: © SEAMUS BRESLIN/ HAWKS ALOFT, INC.

PHOTO 19.21

(*above*) Downy nestling in a nest in a juniper (*Juniperus monosperma*) in the Plains of San Agustin, late spring 1999. PHOTOGRAPH: © JEAN-LUC CARTRON.

PHOTO 19.23

(*below*) Three partially feathered nestlings in a nest in a juniper (*Juniperus monosperma*) in the Plains of San Agustin, late spring 1999. PHOTOGRAPH: © JEAN-LUC CARTRON.

PHOTO 19.24

Nearly fledged nestling on nest in an elm (*Ulmus*) tree near Mills, Harding Co., 24 June 2004.

PHOTOGRAPH: © LAWRY SAGER.

planted long ago at homesteads now abandoned. Such trees include the exotic mulberry (*Morus* sp.) and elm, as well as the native cottonwood (*Populus* sp.) and hackberry (*Celtis* sp.) (Schwarz 2005). One known nest on the Kiowa National Grasslands is on a windmill, and just across the state line in the southwestern corner of the Oklahoma panhandle, Ferruginous Hawk pairs have taken up residence on platforms specially erected for their use (Schwarz 2005). The nest mentioned by Ligon (1961) near Hachita in southwestern New Mexico was in a tall yucca (*Yucca* sp.), as was the nest observed in 1980 on Otero Mesa (S. Williams, pers. comm.). Near Santa Rosa, Guadalupe County, in 2000 and in San Juan County in 2007, it was in chollas (*Cylindropuntia* sp.) that two nests were found (L. Sager, pers. comm.; R. Kellermueller, pers. comm.; photo 19.17). In Roosevelt County, Lawry Sager recorded a nest on a transformer pole on 25 July 2002. Historically, the Ferruginous Hawk nested on the ground, perhaps exclusively (see Bechard and Schmutz 1995), and here in New Mexico the species once built nests at the top of high sand dunes (Ligon 1961). Other than in badlands, no ground nest has been found recently in New Mexico, but one was discovered in 1994 just across the state line in the southwestern corner of the Oklahoma panhandle (HRS, unpubl. data).

Ferruginous Hawk eggs are creamy or pale greenish (Ligon 1961), with markings that are typically dark brown and tend to be more pronounced on the first and second eggs than on any third or fourth eggs. Ferruginous Hawks tend to lay more eggs than other large buteos at similar latitudes (Bechard and Schmutz 1995). Based on 25 egg sets collected in the state during the early and mid 20th century, clutch size averages 2.96 in New Mexico (appendix 19-1), or larger than the observed mean clutch size of the Swainson's Hawk (see chapter 16). Clutches consist most frequently of three eggs, less often two or four. Although uncommon (and not formally documented), clutches of five eggs are also possible (see below).

A review of estimated fledging dates for 67 nests monitored between 1981 and 1988 in northwestern New Mexico suggests that the start of the incubation period could be as early as the first week of April. This finding is confirmed by two egg sets collected in other parts of the state during the early 20th century (appendix 19-1). Incubation can also begin as late as the first week of May—based on data from northwestern New Mexico—or even later, as shown by egg sets collected in early or even mid June and by recent observations of delayed reproduction on the Kiowa National Grasslands (H. Schwarz, pers. obs.). In northwestern New Mexico, however, most (approximately 81 percent) pairs were estimated to begin incubation between 14 April and 26 April (JMR, unpubl. data), showing that reproduction in New Mexico is highly synchronous. Both the male and the female incubate the eggs.

Fledging occurs in June and early July (Ramakka and Woyewodzic 1993; JLEC, pers. obs.), and as many as four young may fledge from one nest (JMR, unpubl. data). Five fully feathered young were recorded once in New Mexico, but it is unknown if all fledged successfully (D. Mikesic, pers. comm.). One of us (JMR) found the cannibalized remains of one young in a nest, suggesting that siblicide may sometimes occur. Violent rain or hail storms likely represent another cause of nestling mortality, as documented just across the New Mexico state line in the southwestern corner of the Oklahoma panhandle (HRS, unpubl. data) or in southeastern Colorado (C. Preston, pers. comm.).

No doubt the Ferruginous Hawk can be vigorous in its defense of the nest, but this cannot be inferred from its response to human intruders. Typically, a human will be met by one of the parents flying in and circling high above while uttering a childlike, plaintive wail that can scarcely be said to intimidate, and to our ears rather sounds more like pleading for the stranger to go away. Rarely in our own experience does this hawk dive at intruders in the style of a Cooper's Hawk (*Accipiter cooperii*), a Northern Goshawk (*A. gentilis*), or a Zone-tailed Hawk (*Buteo albonotatus*).

PHOTO 19.25

(*above*) Fledgling on the ground north of Dorhman Lake, Harding Co., summer 1996.

PHOTOGRAPH: © LAWRY SAGER.

PHOTO 19.26

(*right*) Fledgling in an elm (*Ulmus*) tree in the Estancia Valley, early summer 2000.

PHOTOGRAPH: © JEAN-LUC CARTRON.

TABLE 19.1 Species documented as Ferruginous Hawk (*Buteo regalis*) prey in New Mexico.

CLASS	SPECIES	LOCATION					
		NESTING					MIGRATION
		CLAUNCH AREA	ESTANCIA VALLEY	NORTHWESTERN BADLANDS	PLAINS OF SAN AGUSTIN	QUEMADO AREA	MORENO VALLEY
INSECTA	Camel cricket (*Daihiniodes hastiferum*)				X		
	Scarab beetle (*Phanaeus* sp.)		X				
REPTILIA	Mountain short-horned lizard (*Phrynosoma hernandesi*)		X	X			
	Western diamondback rattlesnake (*Crotalus atrox*)		X				
AVES	Barn owl (*Tyto alba*)	X	X				
	Burrowing owl (*Athene cunicularia*)		X				
	Says Phoebe (*Sayornis saya*)			X			
	Horned lark (*Eremophila alpestris*)			X			
MAMMALIA	Desert cottontail (*Sylvilagus audubonii*)		X	X	X	X	
	Black-tailed jackrabbit (*Lepus californicus*)	X	X	X	X		
	Thirteen-lined ground squirrel (*Spermophilus tridecemlineatus*)		X		X		
	Rock squirrel (*Spermophilus variegatus*)		X			X	
	Spotted ground squirrel (*Spermophilus spilosoma*)		X		X		
	Gunnison's prairie dog (*Cynomys gunnisoni*)		X	X	X		X
	Botta's pocket gopher (*Thomomys bottae*)		X	X	X		
	Plains pocket gopher (*Geomys bursarius*)	X					
	Ord's kangaroo rat (*Dipodomys ordii*)	X		X			
	Deer mouse (*Peromyscus* sp.)		X		X		
	Northern grasshopper mouse (*Onychomys leucogaster*)				X		
	Woodrat (*Neotoma* sp.)		X				
	Long-tailed weasel (*Mustela frenata neomexicana*)		X				
	Antelope ground squirrel (*Citellus leucurus*)			X			

References for Claunch area (Torrance and Socorro counties), Estancia Valley (Santa Fe and Torrance counties), Plains of San Agustin (Catron and Socorro counties), and Quemado area (Catron County): Cartron et al. 2004; Northwestern badlands (San Juan County): Cartron et al. 2004; JMR, unpubl. data; Moreno Valley (Colfax County): Cully 1988.

Diet and Foraging

As documented throughout the range of the Ferruginous Hawk (Bechard and Schmutz 1995), the species' diet in New Mexico consists chiefly of small mammals (table 19.1). During the nesting season, Botta's pocket gophers (*Thomomys bottae*), desert cottontails (*Sylvilagus audubonii*), and ground squirrels (*Spermophilus* spp.) are all staple prey through much of the state. Another important prey in the northwestern badlands and in the Estancia Valley is the Gunnison's prairie dog (Cartron et al. 2004). In the Estancia Valley, pocket gophers and ground squirrels appear to be the type of prey most often taken (Cartron et al. 2004), but Gunnison's prairie dogs possibly dominate the diet of Ferruginous Hawks in terms of biomass (JLEC, unpubl. data). No published data exist for northeastern New Mexico in particular, but at the scale of the entire Kiowa/Rita Blanca National Grasslands (an area that also encompasses the southwestern corner of the Oklahoma panhandle and northwestern Texas), the diet of Ferruginous Hawks consists chiefly of black-tailed prairie dog (*Cynomys ludovicianus*), black-tailed jackrabbit (*Lepus californicus*), cottontail (*Sylvilagus*), and pocket gopher (*Geomys bursarius* and/or *Cratogeomys castanops*) (Schwarz 2005; Giovanni et al. 2007). Thirty out of 62 (48%) prey deliveries video-recorded at one nest on the Kiowa National Grasslands consisted of black-tailed prairie dog (C. Boal, pers. comm.).

PHOTO 19.27

Ferruginous Hawk, about 8 km (5 mi) south of White's City, Eddy Co., 8 November 2008.

PHOTOGRAPH: © JERRY OLDENETTEL.

Birds, reptiles, and insects are also taken in New Mexico, but seemingly in smaller numbers (Cartron et al. 2004). The Burrowing Owl (*Athene cunicularia*) and the Barn Owl (*Tyto alba*) are among the prey of Ferruginous Hawks nesting in New Mexico's grasslands, as are nonavian predators such as the long-tailed weasel (*Mustela frenata*) and the western diamondback rattlesnake (*Crotalus atrox*) (Cartron et al. 2004).

Some foraging may be conducted at dawn or dusk, as suggested by findings of crepuscular or nocturnal rodents among prey remains at Ferruginous Hawk nests (Cartron et al. 2004, but see Boal and Giovanni 2007). Ferruginous Hawks are known to perch on the ground and ambush rodents as they emerge from their burrows. They often perch on the ground in the middle of prairie dog towns, waiting for prairie dogs to exit their burrows. When the opportunity arises, they take flight and dart toward their prey, keeping low to the ground. However, Ligon (1961) describes how Ferruginous Hawks also catch prairie dogs by diving vertically toward them with tucked wings, and banking at the last second before striking. Chesser (1979) reported Ferruginous Hawks attracted by gunshots at a black-tailed prairie dog town 10 km (6 mi) east of Hayden, Union County. Prairie dogs were shot by sport hunters and often left in place, providing Ferruginous Hawks with easy food. While collecting prairie dog specimens around the prairie dog town, Chesser (1979) observed Ferruginous Hawks following his vehicle, running on the ground to secure a prairie dog he had just shot, or joining Chihuahuan Ravens (*Corvus cryptoleucus*) feeding on older carcasses. Ligon (1961) commented on Ferruginous Hawks concentrating along roads in New Mexico to feed on roadkills.

Predation and Interspecific Interactions

Raptors that share their nesting habitat with the Ferruginous Hawk in New Mexico include the Golden Eagle (*Aquila chrysaetos*), Red-tailed Hawk, Swainson's Hawk, American Kestrel (*Falco sparverius*), Prairie Falcon (*F. mexicanus*), Barn Owl, Burrowing Owl, and Great Horned Owl (*Bubo virginianus*). In northwestern New Mexico, Ferruginous Hawks and Golden Eagles likely compete for some of the same nest sites (see photo 19.28). Although Ferruginous Hawks and Red-tailed Hawks do not normally use the same nest sites in northwestern New Mexico, a pair of adult Ferruginous Hawks was observed chasing a pair of adult Red-tailed Hawks from a nesting territory in mid March (J. Kendall, pers. comm.). During migration in the Moreno Valley, Cully (1988) documented Ferruginous Hawks competing with Golden Eagles for freshly killed prairie dogs.

As mentioned above, the Ferruginous Hawk preys at least on two other raptors, the Burrowing Owl and the Barn Owl (Cartron et al. 2004). Conversely, the Ferruginous Hawk can become the target of other predators, including the bobcat (*Lynx rufus*), coyote (*Canis latrans*), and Great Horned Owl. While most tree and pinnacle nest sites are relatively safe from mammalian predators, circumstantial evidence suggests at least one nesting adult was killed by a bobcat in northwestern New Mexico (JMR, pers. obs.). In another instance, two dead nestlings found approximately 100 meters (~330 ft) from a nest inaccessible to mammalian predators exhibited distinct puncture wounds suggestive of owl talon marks (JMR, pers. obs.). Although Ehrlich et al. (1988) state that Ferruginous Hawks will chase Great Horned Owls, the two species have been documented nesting successfully, at least once, within close proximity to each other on the Navajo Nation (D. Mikesic, pers. comm.).

There is an apparent nesting association between the Western Kingbird (*Tyrannus verticalis*) and both the Ferruginous Hawk and the Swainson's Hawk in the Estancia Valley (Cartron 2005). Western Kingbirds often nest in the same trees as Ferruginous Hawks. In one case, an active Western Kingbird nest was discovered tucked under a Ferruginous Hawk nest built on a platform (Cartron 2005). In northwestern New Mexico, an occupied Say's Phoebe (*Sayornis saya*) nest was noted approximately 1 m (3.2 ft) under a Ferruginous Hawk nest (JMR, pers. obs.).

PHOTO 19.28

Bisti/De-Na-Zin Wilderness Area. Note the arrows pointing to ledge and pinnacle alternate nests within the same territory. Nest site A (ledge) was used by Golden Eagles (*Aquila Chrysaetos*) for two years in the early 1980s. It was then unoccupied for a decade before being used again, this time by a pair of Ferruginous Hawks for at least three years. A nest can be observed at Nest Site B, but it was never used for nesting during the years it was monitored. PHOTOGRAPH: © JIM RAMAKKA.

Status and Management

The Ferruginous Hawk has long suffered at the hands of humans in New Mexico. In Bailey's (1928) and Ligon's (1961) time, large numbers of Ferruginous Hawks concentrated along roads and were hit by cars or shot. This inspired Ligon (1961) to fear the species' extirpation and advocate its immediate protection. To this day, Ferruginous Hawks continue to be shot along roads, as shown by gunshot wounds in injured or dead birds (JLEC, unpubl. data; T. Gennaro, pers. comm.). Since 1980, Farmington Bureau of Land Management (BLM) biologists have received reports of at least two cases of Ferruginous Hawks being shot while perched on power lines (unpublished data, BLM files).

Other human impacts on the Ferruginous Hawk in New Mexico are indirect rather than direct. Ligon (1961) reported the widespread eradication of prairie dogs in the state, noting the resulting shift in the diet

PHOTO 19.29

Male dark-morph Ferruginous Hawk found in the Middle Rio Grande Valley with a fractured left tibia caused by a gunshot. It was brought to wildlife rehabilitator Shirley Kendall and released back into the wild on 20 May 2008.

PHOTOGRAPH: © JEAN-LUC CARTRON.

of the Ferruginous Hawk. Today there is evidence that prairie dog eradication efforts, which are still occurring in some areas, have affected the abundance and distribution of Ferruginous Hawks in New Mexico. Bailey (1928) identified the Plains of San Agustin and immediate surroundings as the area with the highest density of nesting Ferruginous Hawks. This is no longer true, as only 11 nesting pairs were found in the Plains of San Agustin during aerial and ground surveys in both 1999 and 2000 (Cartron et al. 2002). By comparison, 21 nesting pairs were counted in 1999 in the western half of the Estancia Valley in central New Mexico, or nearly four times the density of pairs found in the Plains of San Agustin (Cartron et al. 2002). Also by comparison, Ramakka and Woyewodzic (1993) reported 26 nesting pairs from the northwestern badlands (San Juan and McKinley Counties) in 1987 and 1988. Although there have been annual fluctuations, monitoring data gathered between 1988 and 2004 suggest that the number of occupied territories on both BLM and Navajo Nation lands has remained approximately the same during this period (D. Mikesic, pers. comm.; B. Wegener, pers. comm.). The lower density of nests found today in the Plains of San Agustin compared to the Estancia Valley suggests that the Ferruginous Hawk population in the general area of the Plains of San Agustin was negatively affected by particularly intensive efforts to eradicate prairie dogs (Cook et al. 2003).

Additional indirect impacts include habitat fragmentation and increasing road density in some historically important nesting areas. The Estancia Valley is primed for substantial agricultural and urban development in the next several decades and already much of the original grassland along Highway 41 has been converted to agricultural fields (Cartron et al. 2002). In 2000, a dead fledgling was found within 100 m (~330 ft) of its nest along one of the local county roads. High road density in the Estancia Valley means that many of the Ferruginous Hawk nests are now near vehicle traffic.

Human disturbance of nest sites is an increasing problem in northwestern New Mexico. As recently as the 1980s, few people visited the Bisti/De Na Zin BLM wilderness area, where several historical Ferruginous Hawk nesting territories are located. Today it is estimated that those two areas may receive a total of 12,500 visitors per year, mostly during the spring. In one case, a visitor reported to JMR that he had climbed to a nest to take photographs of the young. In another case, hikers were observed walking within 50 m (~160 ft) of an occupied nest in the Bisti/De Na Zin (B. Wegener, pers. comm.), and human disturbance has been implicated in several instances of nest failures (JMR, personal notes).

PHOTO 19.30

Example of proactive, raptor-friendly management by a utility company. Ferruginous Hawk nest being lowered onto a newly installed raptor nesting platform, late summer 2004. The power line is inactive. The installation of the nesting platform was conducted by Hawks Aloft, Inc., in partnership with Central New Mexico Electric Cooperative. Located in the Estancia Valley in Torrance Co., the nest has been active every year ever since. PHOTOGRAPH: © RON KELLERMUELLER/ HAWKS ALOFT, INC.

On the Navajo Nation, Mikesic (2005) reported the following human impacts on nesting Ferruginous Hawks: shooting, poaching of young from nests, power line strikes and electrocution, degradation of prey populations (breeding habitat quality) by drought and livestock grazing, and human disturbance and developments near nest sites (including homes and infrastructure to support them, oil and natural gas wells, power line and road development, and to a small extent, outdoor recreation [all-terrain vehicle use]).

In at least one instance, however, human presence and activities may have had a positive effect on the distribution and abundance of Ferruginous Hawks. Neither Bailey (1928) nor Ligon (1961) mentioned the Estancia Valley as an important nesting area for the species in New Mexico. This is surprising because the proximity of the Estancia Valley to Albuquerque would have made it difficult for earlier naturalists to overlook any sizeable Ferruginous Hawk nesting population in that area. It is possible that the Ferruginous Hawk increased in numbers in the Estancia Valley as a result of fire suppression and juniper and elm encroachment upon the grassland, providing the species with nesting substrates in the immediate vicinity of local prairie dog towns. Together with the apparent decline of Ferruginous Hawk numbers in the Plains of San Agustin, habitat alteration in the Estancia Valley might have resulted in the latter area's harboring the highest known density of nesting pairs anywhere in New Mexico.

Among threats to Ferruginous Hawks outside New Mexico but possibly affecting the species' population in the state is the high incidence of raptor electrocutions near Janos in northwestern Chihuahua, Mexico. The Janos area harbors one of the largest remaining prairie dog town complexes in North America, and it has a sizeable population of wintering Ferruginous Hawks (Manzano-Fischer et al. 1999). It also has power lines supplying electricity to local rural communities, and many of the power poles are built with concrete and fitted with steel cross-arms. Because such poles are conductive, a bird perched on the cross-arm is grounded and need only touch one energized wire to be electrocuted. In 2000 and 2001, a total of 19 Ferruginous Hawks were found dead under concrete poles during surveys of the area (Cartron et al. 2000, 2005; see also Cartron, Sierra Corona et al. 2006; Cartron, Rodríguez-Estrella et al. 2006; Manzano-Fischer et al. 2006). Because Janos is located less than 60 km (~37 mi) from the border with New Mexico, most of the Ferruginous Hawks wintering near Janos and exposed to the risk of electrocution likely migrate through or breed in New Mexico.

The Ferruginous Hawk is not federally listed as Endangered or Threatened, but population declines have been reported toward the northern end of the species' distribution (see Bechard and Schmutz 1995). Despite the threats mentioned above, New Mexico's Ferruginous Hawk population appears stable. Mean annual productivity levels reported recently for nesting pairs in the Estancia Valley, Plains of San Agustin, and northwestern badlands all exceeded the estimated minimal replacement level for the species (Ramakka and Woyewodzic 1993; Cartron et al. 2002).

Acknowledgments

We thank the following individuals and institutions for information on Ferruginous Hawk egg sets, skins, or skeletons from New Mexico: Rhonda Almager (San Bernardino County Museum); Joe Bopp (New York State Museum); René Corado (Western Foundation of Vertebrate Zoology); Scott Cutler (Centennial Museum, University of Texas at El Paso); Charles Dardia and Kimberly Bostwick (Cornell University Museum of Vertebrates); Krista Fahy (Santa Barbara Museum of Natural History); Moe Flannery and Douglas Long (California Academy of Sciences); Kimball Garrett (Natural History Museum of Los Angeles County); Janet Gillette (Museum of Northern Arizona); Mary Hennen (Field Museum of Natural History); Gene Hess (Delaware Museum of Natural History); Janet Hinshaw (University of Michigan Museum of Zoology); Tom Huels (University of Arizona); Mariko Kageyama (University of Colorado Museum of Natural History); John Klicka (Barrick Museum of Natural History); Josh Leal and Tony Gennaro (Natural History Museum, Eastern New Mexico University); Kathy Molina (University of California Los Angeles–Dickey Collections); Elizabeth Moore (Virginia Museum of Natural History); Amanda Person (Sam Noble Oklahoma Museum of Natural History); Nathan Rice (Academy of Natural Sciences, Philadelphia); Eric Rickart (Utah Museum of Natural History); Mark Robbins (University of Kansas Natural History Museum); Stephen Rogers (Carnegie Museum of Natural History); Jeff Stephenson (Denver Museum of Nature and Science); Stephen Sullivan (Chicago Academy of Sciences); and Tom Webber (Florida Museum of Natural History). Moe Flannery (California Academy of Sciences) and Mary Hennen (Field Museum of Natural History) both graciously photographed and measured eggs housed at their institutions. Bill Howe, Tish McDaniel, Hart Schwarz, and Sandy Williams all provided us with recent Ferruginous Hawk nesting records from New Mexico. We also thank Bob Dickerman for giving us access to the Museum of Southwestern Biology collection; and Bill Howe, Chuck Hunter, Brian Locke, and Christopher Rustay for helping us track some information on the Ferruginous Hawk in New Mexico. Clint Boal, Jim Watson, and Chuck Preston all provided helpful comments and additional information from out of state.

LITERATURE CITED

Bailey, F. M. 1928. *Birds of New Mexico*. Santa Fe: New Mexico Department of Game and Fish.

Bak, J. M., K. G. Boykin, B. C. Thompson, and D. L. Daniel. 2001. Distribution of wintering Ferruginous Hawks (*Buteo regalis*) in relation to black-tailed prairie dog (*Cynomys ludovicianus*) colonies in southern New Mexico and northern Chihuahua. *Journal of Raptor Research* 35:124–29.

Bechard, M. J., and J. K. Schmutz. 1995. Ferruginous Hawk (*Buteo regalis*). No. 172. In *The birds of North America*, ed. A. Poole and F. Gill. Philadelphia, PA: Academy of Natural Sciences, and Washington, DC: American Ornithologists' Union.

Boal, C. W., and M. D. Giovanni. 2007. Raptor predation on Ord's kangaroo rats: evidence for diurnal activity by a nocturnal rodent. *Southwestern Naturalist* 52:291–95.

Cartron, J.-L. E. 2005. Nesting associations of the Western Kingbird (*Tyrannus verticalis*) with two *Buteo* species in the Estancia Valley of central New Mexico. *New Mexico Ornithological Bulletin* 33:53–55.

Cartron, J.-L. E., R. R. Cook, G. L. Garber, and K. K. Madden. 2002. Nesting and productivity of ferruginous hawks in two areas of central and western New Mexico, 1999–2000. *Southwestern Naturalist* 47:482–85.

Cartron, J.-L. E., R. R. Cook, and P. Polechla Jr. 2004. Prey of nesting ferruginous hawks in New Mexico. *Southwestern Naturalist* 49:270–76.

Cartron, J.-L. E., G. L. Garber, C. Finley, C. Rustay, R. P. Kellermueller, M. P. Day, P. Manzano Fisher, and S. H. Stoleson. 2000. Power pole casualties among raptors and ravens in northwestern Chihuahua, Mexico. *Western Birds* 31:255–57.

Cartron, J.-L. E., R. Harness, R. Rogers, and P. Manzano. 2005. Impact of concrete power poles on raptors and ravens in northwestern Chihuahua. In *Biodiversity, ecosystems, and conservation in Northern Mexico*, ed. J.-L. E. Cartron, G. Ceballos, and R. S. Felger, 357–69. New York: Oxford University Press.

Cartron, J.-L. E., R. Rodríguez-Estrella, Robert C. Rogers, Laura B. Rivera, and B. Granados. 2006. Raptor and raven electrocutions in northwestern Mexico: a preliminary regional assessment of the impact of concrete power poles. In *Current raptor studies in Mexico*, ed. R. Rodríguez-Estrella. La Paz, Mexico: CIBNOR.

Cartron, J.-L. E., R. Sierra Corona, E. Ponce Guevara, R. E. Harness, P. Manzano-Fischer, R. Rodriguez-Estrella, and G. Huerta. 2006. Bird electrocutions and power poles in northwestern Mexico: an overview. *Raptors Conservation 2006* 7:4–14.

Chesser, R. K. 1979. Opportunistic feeding on man-killed prey by Ferruginous Hawks. *Wilson Bulletin* 91:330–31.

Cook, R. R., J.-L. E. Cartron, and P. Polechla Jr. 2003. The importance of prairie dogs to nesting ferruginous hawks in grassland ecosystems. *Wildlife Society Bulletin* 31:1073–82.

Cully, J. F. Jr. 1988. Gunnison's prairie dog: an important autumn raptor prey species in northern New Mexico. In *Proceedings of the southwest raptor management symposium and workshop*, ed. R. L. Glinski, B. G. Pendleton, M. B. Moss, M. N. LeFranc Jr., B. A. Millsap, and S. W. Hoffman. Washington, DC: National Wildlife Federation.

———. 1991. Response of raptors to reduction of a Gunnison's prairie dog population by plague. *American Midland Naturalist* 125:140–49.

Eakle, W. L., E. L. Smith, S. W. Hoffman, D. W. Stahlecker, and R. B. Duncan. 1996. Results of a raptor survey in southwestern New Mexico. *Journal of Raptor Research* 30:183–88.

Ehrlich, P. R., D. S. Dobkin, and D. Wheye. 1988. *The birders handbook*. New York: Simon and Schuster.

Giovanni, M. D., C. W. Boal, and H. A. Whitlaw. 2007. Prey use and provisioning rates of breeding Ferruginous and Swainson's Hawks on the southern Great Plains, USA. *Wilson Journal of Ornithology* 119:558–69.

Goodman, R. A., ed. 1983. *New Mexico Ornithological Society Field Notes* 22(1): 5.

———, ed. 1984. *New Mexico Ornithological Society Field Notes* 23(1): 6.

———. 1985. *New Mexico Ornithological Society Field Notes* 24(2): spring.

Hall, R. S., R. L. Glinski, D. H. Ellis, J. M. Ramakka, and D. L. Base. 1988. Ferruginous Hawk. In *Proceedings of the southwest raptor management symposium and workshop*, ed. R. L. Glinski, B. G. Pendleton, M. B. Moss, M. N. LeFranc Jr., B. A. Millsap, and S. W. Hoffman, 111–18. Scientific and Technical Series no. 11. Washington, DC: Institute for Wildlife Research, National Wildlife Federation.

Harmata, A. R., M. Restani, G. J. Montopoli, J. R. Zelenak, J. T. Ensign, and P. J. Harmata. 2001. Movements and mortality of ferruginous hawks banded in Montana. *Journal of Field Ornithology* 72:389–98.

Hoffman, S. W., and J. P. Smith. 2003. Population trends of migratory raptors in western North America, 1977–2001. *Condor* 105:397–419.

Howell, S. N. G., and S. Webb. 1995. *A guide to the birds of Mexico and northern Central America*. Oxford, UK: Oxford University Press.

Hubbard, J. P., ed. 1970. *New Mexico Ornithological Society Field Notes* 9(2): 3.

———, ed. 1988. Southwest region: New Mexico. *American Birds* 42:1326–28.

Ligon, J. S. 1961. *New Mexico birds and where to find them*. Albuquerque: University of New Mexico Press.

Manzano-Fischer, P., R. List, and G. Ceballos. 1999. Grassland birds in prairie dog towns in northwestern Chihuahua, Mexico. *Studies in Avian Biology* 19:263–71.

Manzano-Fischer, P., R. List, G. Ceballos, and J.-L. E. Cartron. 2006. Avian diversity in a priority area for conservation in North America: the Janos–Casas Grandes Prairie Dog Complex and adjacent habitats in northwestern Mexico. *Biodiversity and Conservation* 15:3801–25.

Mikesic, D. G. 2005. An inventory and protection plan for Ferruginous Hawk (*Buteo regalis*) nests on the Navajo Nation. Unpublished Report, Navajo Nation Fish. http://nnhp.nhttp://nnhp.nndfw.org/docs_reps.htmndfw.org (accessed 3 April 2009).

Ramakka, J. M., and R. T. Woyewodzic. 1993. Nesting ecology of Ferruginous Hawk in northwestern New Mexico. *Journal of Raptor Research* 27:97–101.

Schmutz, J. K., and R. W. Fyfe. 1987. Migration and mortality of Alberta ferruginous hawks. *Condor* 89:169–74.

Schwarz, H. R. 2005. Ferruginous Hawk platform monitoring summary for 2005. Unpublished report, Cibola National Forest, Albuquerque, NM.

Sibley, D. A. 2000. *The Sibley guide to the birds*. New York: Chanticleer Press.

Williams, S. O. III. 2000. *New Mexico Ornithological Society Field Notes* 39(3): Summer 2000.

20

Rough-legged Hawk

(*Buteo lagopus*)

ZACH F. JONES

THE ROUGH-LEGGED HAWK (*Buteo lagopus*) occurs in New Mexico as a sparsely distributed winter visitor. Mainly associated with grasslands, it may be observed perched on a power pole alongside a road, hovering with rapid wing beats over the ground in search of a meal, or soaring easily on long and wide wings that form a slight dihedral shape when the species is using thermal convection to climb. Although reaching the far southern boundaries of the state in some years, its annual statewide status varies considerably, from apparently absent to locally common (see below). It is a large hawk, with adult males having a mean wingspan of 407 mm (16 in) (range = 372–416 mm [14.6–16.4 in], *n* = 22) (Friedmann 1950) and a mean body mass of 822 g (1.81 lbs) (*n* = 5) (Brown and Amadon 1968). Females tend to be larger than males, with a mean wingspan of 411 mm (16.2 in) (range 395–438mm [15.6–17.2 in], *n* = 9) (Friedmann 1950) and a mean body mass of 1,080 g (2.38 lbs) (*n* = 7) (Brown and Amadon 1968).

The adult plumage varies considerably from the more common light morph to the dark melanistic morph, the latter being most common in regions of high humidity (e.g., Aleutian islands; Cade 1955), yet also recorded in New Mexico (e.g., Travis et al. 1962). Overall, leg feathers down to the top of the feet (the characteristic for which the species is named), a dark band of feathers that cover the belly, a noticeable U-shaped region of more lightly colored feathers between the breast and belly, black carpal patches on the underside of the wing, and a white band at the base of the uppertail are the most reliable field

PHOTO 20.1

Adult light-morph Rough-legged Hawk (Hawks Aloft, Inc., educational bird). Adult Rough-legged Hawks have dark irises. Dark streaking on the breast forms a bib on some birds. PHOTOGRAPH: © DOUG BROWN.

PHOTO 20.2

(*above*) Adult light-morph Rough-legged Hawk taking off north of Tres Piedras, Taos Co.–Rio Arriba Co. county line, November 2003. The legs are feathered down to the feet. The belly sports a dark band of feathers (more pronounced in females); a noticeable U-shaped region of more lightly colored feathers is present between the breast and belly; and dark streaking on the breast may form a bib. On the underwings are black carpal patches (not conspicuous on this photograph) and a dark trailing edge. Males (photo) tend to have multiple dark bands on the otherwise white tail (including a dark, wide subterminal band), whereas females typically show only one dark, wide subterminal band. PHOTOGRAPH: © BILL SCHMOKER.

PHOTO 20.3

(*left*) Juvenile Rough-legged Hawk in Utah, 21 February 2007. Both adults and juveniles show dark carpal patches on the underwings, one of the most reliable field marks for the species. Compared to adults, juveniles have light-colored irises and their underwings show a narrower, dark trailing edge. PHOTOGRAPH: © WADE REED.

characteristics identifying members of the species (Bechard and Swem 2002).

The juvenal plumage is similar to that of the adult except that the uppertail has a smaller white-colored base, and, viewed from below, the tail feathers do not possess the obvious (and abruptly defined) dark-colored subterminal band present in adults. Juveniles tend to be visibly darker on the head and neck, with less white coloration than found in adults (Cramp and Simmons 1980). The adult plumage begins replacing the juvenal plumage as young reach the end of their first winter, and both adults and juveniles molt through summer into early fall, with juveniles reaching adulthood in their second year (Bechard and Swem 2002).

Distribution

Rough-legged Hawks are panboreally distributed in summer, breeding in the tundra and taiga biomes of the Northern Hemisphere. In fall they travel southward across boreal forests to winter in open habitats at temperate latitudes of the Northern Hemisphere. In North America in particular, they winter across much of the contiguous United States except for the Southeast and coastal areas of the West (Bechard and Swem 2002). However, the limits of the winter range for this species are poorly understood. Rough-legged Hawks presumably occur as far north as Southampton Island, New York in mild winters, and in severe winters they are thought to range south into northern Sonora, Mexico

PHOTO 20.4

Rough-legged Hawk just west of Salt Lake City in Utah, 21 February 2007. The species is at the southern edge of its regular wintering distribution in New Mexico. It is more common farther north, including in Utah.

PHOTOGRAPH: © WADE REED.

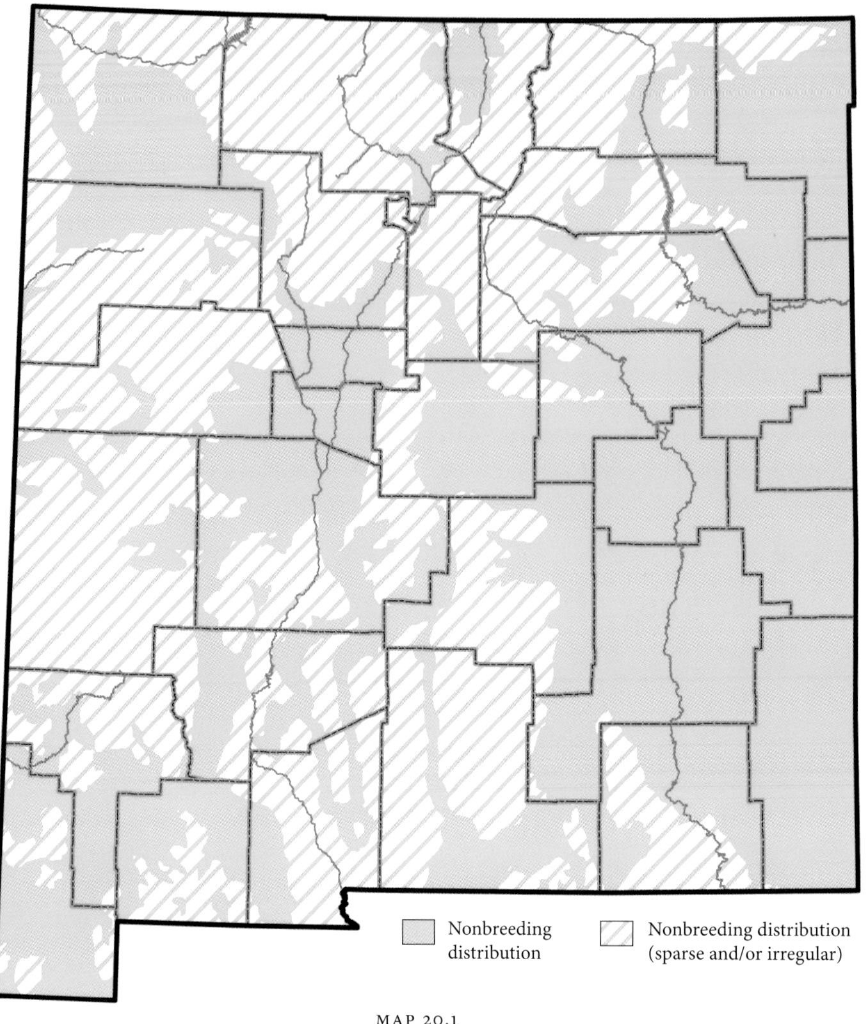

MAP 20.1

Rough-Legged Hawk distribution map

(Sutton and Parmlee 1956; Russell and Lamm 1978). Apparently climate strongly influences their winter range, and data from Christmas Bird Counts (CBCs) suggest that they avoid areas with mean minimum January temperatures below -23°C and places receiving more than 102 cm (40 in) of precipitation annually (Root 1988).

This species reaches New Mexico in most winters, but its abundance and range in the state vary considerably, likely as a function of annual climatic patterns and variation in prey availability farther north (see Bechard and Swem 2002). Christmas Bird Counts (CBCs 1951–2007 [National Audubon Society 2002]) and the *Field Notes* of the New Mexico Ornithological Society (NMOS 1962–2004) suggest a number of trends regarding winter distribution and abundance of the species in New Mexico. CBC coverage is affected annually by the numbers and skill levels of volunteers in the field, while the *Field Notes* of the NMOS focus heavily on unusual, rare, or otherwise notable sightings and are similarly influenced by the number of people in the field reporting their sightings annually. Even with

PHOTO 20.5

(*above left*) Rough-legged Hawk south of Roy, Harding Co., 8 December 2008. Although Rough-legged Hawks may be seen in all parts of New Mexico, the species is typically more common in the northeast.

PHOTOGRAPH: © JERRY OLDENETTEL.

PHOTO 20.6

(*above right*) Rough-legged Hawk at Taiban, De Baca Co., 4 November 2007.

PHOTOGRAPH: © JERRY OLDENETTEL.

PHOTO 20.7

(*right*) Rough-legged Hawk near Maxwell, Colfax Co., 9 December 2007. Local concentrations of Rough-legged Hawk have been observed in some years in New Mexico, including in Colfax County.

PHOTOGRAPH: © JERRY OLDENETTEL.

these limitations, much insightful information can be gathered from these two sets of information.

On 12 occasions, the *NMOS Field Notes* provide first fall sightings of the Rough-legged Hawk, placing the average early arrival date for the species in New Mexico around 21 October (*n* = 12, range = 9 September–24 November) (McKnight and Snider 1966; Hubbard et al. 1967; Hubbard and Baltosser 1978; Goodman 1981, 1984, 1986; Snider 1992a, 1998a, 1999a; Williams 2001a, 2002a, 2004). While it is even more difficult to know when the last bird leaves an area at the end of spring migration, 14 reports in the *NMOS Field Notes* of the last Rough-legged Hawks observed in New Mexico beyond 1 March allow an estimated average latest departure date around 7 April (*n* = 14, range = 6 March–20 May) (Hubbard 1969, 1971, 1972, 1973, 1974, 1975; Goodman 1987, 1991; Snider 1992b, 1995, 1998b, 1999b; Williams 2001b, 2002b; see also chapter 2).

Ligon (1961) viewed the Rough-legged Hawk as rare in western New Mexico. During the 1960s and early 1970s, however, both CBC results and reports contained in the *NMOS Field Notes* led to the realization that the species was more regular statewide than previously thought, with sightings being reported frequently from western and southern counties in particular (Hubbard 1975). Thereafter, reports from New Mexico, including the southern and western regions, remain steady except for the winters of 1996–97 and 1997–98, when very low numbers of sightings were reported for the species. In fact, by the early 2000s sighting reports in many of the *NMOS Field Notes* are summarized into statements such as "Once considered rare along the Mexican border, this season produced reports from Hidalgo, Luna, Doña Ana, Otero, and Eddy [counties] December–January" (Williams 2002c). Currently, the species may be best described as sparsely distributed to uncommon or even locally common, occurring primarily at lower and middle elevations especially in the northeastern section of the state (Hubbard 1978; Parmeter et al. 2002). Local concentrations have been observed sporadically, including 12 birds near Clayton, Union County, on 16 November 1972 (*fide* A. Krehbiel); 14 in Grant County, 9–13 January 1988 (*fide* R. Fisher); 10 at Angel Fire, Colfax County, 30 January 1992 (*fide* C. Rustay and S. Baldwin); and 10 at and near Maxwell National Wildlife Refuge on 22–24 November 2000 (*fide* D. Cleary).

The increase over time in the number of sightings reported in the *NMOS Field Notes* may suggest that the Rough-legged Hawk was once rare, even absent in some years, and that it is now more common. This is almost certainly not true. The species has been known to occur in New Mexico for more than 150 years, with Kennerly securing a specimen of that species near Zuni on 9 November 1853 (Bailey 1928). In fact, the early naturalist T. C. Henry did not view the Rough-legged Hawk as rare near Fort Fillmore ca. 1854 along the Rio Grande (Bailey 1928). There was, however, a hiatus of nearly 50 years before the species was recorded again, with the collection of a specimen at Tularosa on 29 November 1902 (Bailey 1928).

CBC data (which began for the Rough-legged Hawk in New Mexico in 1951) indicate that while the species has always been fairly scarce, the seven-year period beginning in 1963 (coinciding with the early publications of the *NMOS Field Notes*) records a much lower than average abundance for the species (CBCs 1963–69 mean = 0.0111, range 0.004–0.0204 observations per party hour vs. CBCs 1951–2007 mean = 0.03, range 0.0019–0.1244 observations per party hour). Perhaps this may explain why the *NMOS Field Notes* appear to document a population of wintering Rough-legged Hawks increasing through the 1960s in New Mexico. CBC counts from 1951 to 1962 (mean = 0.05, range 0.0093–0.1244 observations per party hour) recorded higher than average abundance counts, documenting that the species was at least as common in the 1950s as it has been since the 1970s.

Habitat Associations and Life History

The Rough-legged Hawk's morphology and flight pattern make it well adapted to hunting in open country, and not surprisingly the species is closely tied to areas with few trees, such as prairies, semidesert grasslands, shrub-steppes, marshlands, bogs, dunes, and open agricultural fields (Bock and Lepthien 1976). In New Mexico, vegetation types occupied by Rough-legged Hawks include semidesert and plains grasslands, Chihuahuan desertscrub, and the ecotone between grassland and pinyon-juniper woodland (Eakle et al.

PHOTO 20.8

Example of winter habitat in New Mexico: grasslands at Maxwell National Wildlife Refuge, Colfax Co.

PHOTOGRAPH: © PATTY HOBAN.

1996). In Moreno Valley in Colfax County, Cully (1988) found the species associated with hayfields and grassy meadows, but apparently not with grasslands with prairie dog towns, in contrast to Ferruginous Hawks (*Buteo regalis*).

Very little information has been recorded about the diet of the Rough-legged Hawk in New Mexico. Based on the species' habitat associations in Moreno Valley, Cully (1988:261) stated that at that location Rough-legged Hawks "were probably feeding on [meadow] voles" (*Microtus pennsylvanicus*). Without details, Ligon (1961) mentioned—or at least implied—that Rough-legged Hawks in New Mexico consumed rodent and rabbit roadkills. Throughout most of its distribution, and thus presumably in New Mexico, the species is known to hunt primarily small mammals in both summer and winter, although it also may take birds (Bechard and Swem 2002). Most hunting occurs from the wing instead of from perches. Because the small-mammal populations that the species depends upon fluctuate widely in space and time, the distribution and abundance of Rough-legged Hawks also are highly variable from place to place and among years, as previously stated. Speculation exists that this species may move nomadically in search of prey in both breeding and wintering seasons, but data on these types of movements are still lacking (Bechard and Swem 2002).

PHOTO 20.9

Rough-legged Hawk in grassland habitat in Utah, 21 February 2007.
PHOTOGRAPH: © WADE REED.

The species does not breed in New Mexico. A purported Rough-legged Hawk egg set collected 32 km (20 mi) northwest of Fairview in Socorro County on 15 April 1916 is housed at the University of New Mexico's Museum of Southwestern Biology. However, the egg set almost certainly belongs to the Ferruginous Hawk instead (J.-L. Cartron, pers. comm.). The egg set was collected by J. Stokley Ligon early in his career, at a time when his notes show him to be unsure of the identity of some of New Mexico's raptors. Ligon (1961) himself later stated that the Rough-legged Hawk was only a winter visitor to New Mexico. The collection site corresponds to that of a Ferruginous Hawk nest from which Ligon secured three eggs on 19 April 1924 (J.-L. Cartron, pers. comm.).

Status and Management

Little is known about the size of the total Rough-legged Hawk population. However, because the species is a common breeder throughout most of the northern tundra and taiga biomes—both covering vast expanses—the Rough-legged Hawk is thought to represent one of the largest populations among raptors worldwide (Bechard and Swem 2002). The species is protected by the Migratory Bird Treaty Act, but has no other special protections throughout the entirety of its breeding or wintering range. Based on a study in Colorado, Schmidt and Bock (2004) suggest that the Rough-legged Hawk can be adversely impacted by rural and exurban development, probably because such development adds structural heterogeneity to previously two-dimensional habitats (grasslands), thereby favoring Red-tailed Hawks (*Buteo jamaicensis*) over Rough-legged Hawks.

The presence of the Rough-legged Hawk in New Mexico is restricted to the winter season. While highly variable from year to year, Rough-legged Hawk numbers in the state are overall low. For this reason, New Mexico is arguably of marginal importance to the species and its overall status. While no absolute reason has ever been documented for the Rough-legged Hawk's presence or absence in New Mexico in any year, this species is most likely to occur in the myriad open spaces of the state when winters farther north are harsh and when New Mexico offers comparatively higher abundance of small mammals.

PHOTO 20.10

Rough-legged Hawk (Hawks Aloft, Inc., educational bird). The species occurs only in small numbers—and only in winter—in New Mexico, but globally the Rough-legged Hawk may have one the largest populations among raptors.

PHOTOGRAPH: © TOM KENNEDY.

LITERATURE CITED

Bailey, F. M. 1928. *Birds of New Mexico*. Santa Fe: New Mexico Department of Game and Fish.

Bechard, M. J., and T. R. Swem. 2002. Rough-legged Hawk (*Buteo lagopus*). No. 641. In *The birds of North America*, ed. A. Poole and F. Gill. Philadelphia, PA: Birds of North America, Inc.

Bock, C. E., and L. W. Lepthien. 1976. Geographical ecology of the common species of *Buteo* and *Parabuteo* wintering in North America. *Condor* 78:554–57.

Brown, L., and D. Amadon. 1968. *Eagles, hawks and falcons of the world*. Vol. 2. New York: McGraw-Hill.

Cade, T. J. 1955. Variation of the common Rough-legged Hawk in North America. *Condor* 57:313–46.

Cramp, S., and K. E. L. Simmons, eds. 1980. *The birds of the western Palearctic*. Vol. 2, *Hawks to bustards*. Oxford, UK: Oxford University Press.

Cully, J. F. 1988. Gunnison's prairie dog: an important autumn raptor prey species in northern New Mexico. In *Proceedings of the Southwest Raptor Management Symposium and Workshop*, ed. R. L. Glinski, B. G. Pendleton, M. B. Moss, M. N. LeFranc Jr., B. A. Millsap, and S. W. Hoffman. Scientific and Technical Series no. 11. Washington, DC: National Wildlife Federation.

Eakle, W. L., E. L. Smith, S. W. Hoffman, D. W. Stahlecker, and R. B. Duncan. 1996. Results of a raptor survey in southwestern New Mexico. *Journal of Raptor Research* 30:183–88.

Friedmann, H. 1950. *The birds of Middle and North America*. U.S. National Museum Bulletin 50. Washington, DC: Smithsonian Institution.

Goodman, R. A., ed. 1981. *New Mexico Ornithological Society Field Notes* 20(3): Autumn 1981.

———, ed. 1984. *New Mexico Ornithological Society Field Notes* 23(4): Autumn 1984.

———, ed. 1986. *New Mexico Ornithological Society Field Notes* 25(4): Autumn 1986.

———, ed. 1987. *New Mexico Ornithological Society Field Notes* 26(2): Spring 1987.

———, ed. 1991. *New Mexico Ornithological Society Field Notes* 30(2): Spring 1991.

Hubbard, J. P., ed. 1969. *New Mexico Ornithological Society Field Notes* 14(1): Winter 1968–69.

———, ed. 1971. *New Mexico Ornithological Society Field Notes* 10(1): Winter 1970–71.

———, ed. 1972. *New Mexico Ornithological Society Field Notes* 11(1): Winter 1971–72.

———, ed. 1973. *New Mexico Ornithological Society Field Notes* 12(1): Winter 1972–73.

———, ed. 1974. *New Mexico Ornithological Society Field Notes* 13(1): Winter 1973–74.

———, ed. 1975. *New Mexico Ornithological Society Field Notes* 14(1): Winter 1974–75.

———.1978. *Revised check-list of the birds of New Mexico*. Publ. no. 6. Albuquerque: New Mexico Ornithological Society.

Hubbard, J. P., and W. H. Baltosser, eds. 1978. *New Mexico Ornithological Society Field Notes* 17(2): Summer 1967.

Hubbard, J. P., B. C. McKnight, and P. R. Snider, eds. 1967. *New Mexico Ornithological Society Field Notes* 6(2): Summer 1967.

McKnight, B. C., and P. R. Snider, eds. 1966. *New Mexico Ornithological Society Field Notes* 5(2): Summer 1966.

National Audubon Society. 2002. The Christmas Bird Count Historical Results. http://www.audubon.org/bird/cbc (accessed 4 January 2008).

Parmeter, J., B. Neville, and D. Emkalns. 2002. *New Mexico bird finding guide*. 3rd ed. Albuquerque: New Mexico Ornithological Society.

Root, T. 1988. *Atlas of wintering North American birds: an analysis of Christmas Bird Count data*. Chicago: University of Chicago Press.

Russell, S. M., and D. W. Lamm. 1978. Notes on the distribution of birds in Sonora, Mexico. *Wilson Bulletin* 90:123–31.

Schmidt, E., and C. E. Bock. 2004. Habitat associations and population trends of two hawks in an urbanizing grassland region in Colorado. *Landscape Ecology* 20:469–78.

Snider, P. R., ed. 1992a. *New Mexico Ornithological Society Field Notes* 31(4): Autumn 1992.

———, ed. 1992b. *New Mexico Ornithological Society Field Notes* 31(2): Autumn 1992.

———, ed. 1995. *New Mexico Ornithological Society Field Notes* 34(2): Spring 1995.

———, ed. 1998a. *New Mexico Ornithological Society Field Notes* 37(4): Autumn 1998.

———, ed. 1998b. *New Mexico Ornithological Society Field Notes* 37(2): Spring 1998.

———, ed. 1999a. *New Mexico Ornithological Society Field Notes* 38(4): Autumn 1999.

———, ed. 1999b. *New Mexico Ornithological Society Field Notes* 38(2): Spring 1999.

Sutton, G. M., and D. F. Parmelee. 1956. The Rough-legged Hawk in the American Arctic. *Arctic* 9:202–7.

Travis, J. R., B. McKnight, and M. Huey, eds. 1962. *New Mexico Ornithological Field Notes* 1(1).

Williams, S. O. III., ed. 2001a. *New Mexico Ornithological Society Field Notes* 40(4): Autumn 2001.

———, ed. 2001b. *New Mexico Ornithological Society Field Notes* 40(2): Spring 2001.

———, ed. 2002a. *New Mexico Ornithological Society Field Notes* 41(4): Autumn 2002.

———, ed. 2002b. *New Mexico Ornithological Society Field Notes* 41(2): Spring 2002.

———, ed. 2002c. *New Mexico Ornithological Society Field Notes* 41(1): Winter 2001–2.

———, ed. 2004. *New Mexico Ornithological Society Field Notes* 43(4): Autumn 2004.

21

Golden Eagle

(*Aquila chrysaetos*)

DALE W. STAHLECKER, JEAN-LUC E. CARTRON, AND DAVID G. MIKESIC

PERHAPS NEW MEXICO'S most regal bird of prey, the Golden Eagle (*Aquila chrysaetos*) is conspicuously large, averaging 76 cm (30 in) in length with wingspans of 1.8 to 2.1 m (6 to 7 ft). Females weigh on average 5.2 kg (11.4 lbs), whereas males weigh on average 25% less, or about 3.8 kg (8.33 lbs) (Kochert et al. 2002). The largest Golden Eagles are those that breed at high latitudes, while in contrast New Mexico's residents are among the smallest. The species is one of the world's "booted" eagles, because the feathers of the legs extend to the toes. The adult plumage is mostly dark brown, with a golden sheen on the nape and much of the head, for which the species is named. There are also faint gray bars on the tail and variable amounts of light or even white feathers on the underside of the wings. Since an entire body molt probably requires two or more years, all adults are a mix of dark, shiny new feathers and faded and frayed older feathers. The talons are black, as is the bill at the tip. The base of the bill fades to gray and is bordered by the bright yellow cere. The feet and orbital ring are also yellow in all age-classes (Kochert et al. 2002).

Juveniles depart the nest with the diagnostic golden head. Compared to adults, however, they present an overall darker appearance since all feathers are new. They also have noticeably large areas of white feathers in the underwings at the "wrist," or the base of the primaries and secondaries. These appear like "windows," which together with the white basal third to half of the tail are a feature most visible in a soaring bird viewed from below. With each succeeding molt the amount of white in both the tail and the underwings decreases until the adult plumage is attained at about five years of age (Kochert et al. 2002).

With their large size and overall dark appearance, Golden Eagles are usually easily distinguished from all other North American raptors. Much more common in New Mexico in winter than in summer, Bald Eagles (*Haliaeetus leucocephalus*) are also large soaring birds of prey (see chapter 7). Overhead, however, they have noticeably broader wings that are held almost flat, whereas Golden Eagles soar with wings held in a dihedral, or slight V-shape. Adult Bald Eagles, with their stark white heads and tails, yellow bill, and almost black bodies, are not easily confused with Golden Eagles of any age. Immature Bald Eagles usually have extensive white on the underwings and body, unlike any Golden Eagle plumage. Juvenile Bald Eagles are all dark and could be confused with adult Golden Eagles, but in addition to differences in appearance in flight, they have bare legs, larger heads, and robust, black beaks. Although juvenile Golden Eagles have much

PHOTO 21.1

(above) Golden Eagle near Lordsburg, Hidalgo Co., March 2008. Note the golden nape, the dark brown body, and the legs feathered down to the toes. PHOTOGRAPH: © MARK L. WATSON.

PHOTO 21.2

(top right) Adult Golden Eagle on the Jornada del Muerto west of Bingham, Socorro Co., 5 December 2008. The adult's plumage is mostly dark brown, with golden crown and nape. The plumage on the back appears mottled because new feathers are darker than the old, faded ones. PHOTOGRAPH: © JERRY OLDENETTEL.

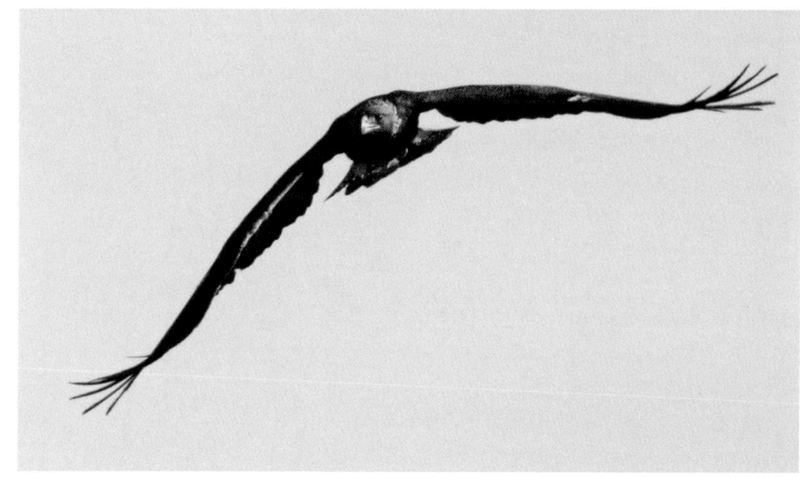

PHOTO 21.3

(above) Immature Golden Eagle in flight, south of Virden, Hidalgo Co., 1 February 2004. PHOTOGRAPH: © ROBERT SHANTZ.

PHOTO 21.4

(*top left*) Immature Golden Eagle trapped during the 1995 fall migration at HawkWatch International's study site in the Manzano Mountains. In both adults and immatures, the cere is yellow; the base of the bill is pale, its tip dark.

PHOTOGRAPH: © JOHN P. DELONG.

PHOTO 21.5

(*bottom left*) Immature Golden Eagle captured during the 1996 fall migration at HawkWatch International's study site in the Manzano Mountains. The tail of the immature Golden Eagle is blackish with a white base.

PHOTOGRAPH: © JOHN P. DELONG.

PHOTO 21.6

(*above*) Immature Golden Eagle in flight south of Virden, Hidalgo Co., 1 February 2004. The base of the tail is white. Note the contrast between the uniformly—or nearly so—dark coverts and the flight feathers, which have white patches especially at the base of the primaries and secondaries.

PHOTOGRAPH: © ROBERT SHANTZ.

white in their tails, they always have a large terminal black band and never a white head to be confused with adult Bald Eagles (Sibley 2000; Kochert et al. 2002). In summer in New Mexico, Turkey Vultures (*Cathartes aura*) are numerous in habitats shared with Golden Eagles, and are easily misidentified as eagles. They are somewhat smaller, however, and have smaller, naked heads; carry their wings, which are distinctly silver-gray on the trailing edge, in a more pronounced dihedral; and appear less graceful or stable when soaring (see chapter 3).

Golden Eagles apparently have limited need for vocal communication, so they have a relatively small vocabulary. Known calls are mostly heard from begging nestlings or from adults when delivering food. The call most often heard is a loud, sharp "*chirp*" that can be heard as far as 1.6 km (1 mi) away (Kochert et al. 2002).

There are five subspecies of Golden Eagles worldwide. Only one subspecies, *A. c. canadensis*, is found in the Western Hemisphere. Though the Golden Eagle is the only *Aquila* or "booted eagle" that occurs in the Western Hemisphere, the genus contains eight other species, most of them found in Eurasia and Africa (Ferguson-Lees and Christie 2001).

Distribution

The Golden Eagle is a northern hemispheric, circumpolar species whose distribution lies nearly entirely north of the tropics (Ferguson-Lees and Christie 2001). In the Eastern Hemisphere, the Golden Eagle is found throughout much of Eurasia, with also some small populations in northern Africa. In the Western Hemisphere, the species breeds from northern Alaska east across mainland Canada to Labrador and south through the western United States to at least Durango and San Luis Potosi in Mexico. It is mainly the breeding populations at the northernmost latitudes that migrate to warmer climes in winter. The eastern United States, which no longer has large breeding populations of Golden Eagles, is included within the species' winter distribution. Most Golden Eagle adults that hold territories south of Canada stay year-round to maintain ownership (Kochert et al. 2002).

Golden Eagles are resident through much of New Mexico, from desert grasslands to mountain meadows. Since Golden Eagles usually nest on cliffs (see below), the distribution of territories in New Mexico likely reflects the distribution of suitable nesting cliffs, which is itself quite extensive throughout most of the state. Cliff sites in New Mexico are most limited in the southeastern plains and in some scattered valleys and basins of the west and Rio Grande Valley.

Bailey (1928) summarized the distribution of 14 nests/territories documented pre-1925 in New Mexico: one in the east, six in the north-central portion of the state, seven in the west, while also stating that adults—and thus also nesting territories—occurred throughout the southwestern mountains. Ligon (1961) knew of 75 territories statewide, including some in the eastern plains, but most in the southwestern and central mountains (Black, Mogollon, and San Andres ranges). Although Kochert et al. (2002) excluded extreme southeastern New Mexico from the species' breeding distribution, nesting on cliffs of Carlsbad Caverns National Park, the southeastmost extension of the Guadalupe Mountains, has been documented regularly since at least 1965 (McKnight and Niles 1965; T. Smylie, pers. comm.; D. Roemer, pers. comm.). There are also nest records for Golden Eagles elsewhere in the adjacent Guadalupe Mountains, the Sacramento Mountains, and isolated central mountain ranges (Mollhagen et al. 1972; C. G. Schmitt, pers. comm.; DWS and DGM).

The recent status of breeding populations was best documented in northwestern New Mexico, where annual monitoring on federal, private, and tribal lands has occurred since the mid 1990s. In a 28,500 km^2 (11,000 mi^2) area encompassing much of four large counties, peak occupancy was about 110–115 territories during this period. This density of approximately one territory per 260 km^2 (100 mi^2) was about one-third that reported in a much smaller Idaho study area but equal to that reported in a Nevada study (Kochert et al. 2002). Within northwestern New Mexico, Golden Eagles locally reached densities as high as one territory per 104–129 km^2 (40–50 mi^2), and in one case in 2008, three active nests were found along a 2.9-km (1.8-mile) section of Rio Arriba cliffline (DWS). More recent work in six northeastern New Mexico counties has shown a minimum of 99 regularly occupied territories

PHOTO 21.7

Golden Eagle perched on a power pole north of Springer, Colfax Co., 6 December 2006. During the nonbreeding season, Golden Eagles are often seen on power poles. PHOTOGRAPH: © JEAN-LUC CARTRON.

in 37,555 km² (14,500 mi²) or one occupied territory per 379 km² (146 mi²). Again, densities varied depending mostly upon the availability of suitable nesting cliffs (Stahlecker 2008; DWS).

Golden Eagles migrate through and into New Mexico, generally in late autumn after most individuals of other raptor species have moved through. Between 1985 and 2001 on average 116 (+/- 30) eagles were counted over the Manzano Mountains during autumn migration and nearly three times as many, 336 (+/- 161), were counted over the Sandia Mountains in spring migration (Hoffman and Smith 2003; see chapter 2).

Migrant Golden Eagles, possibly from as far north as Alaska, spend the winter in New Mexico. Seventy-two Golden Eagles were counted on four 518-km-long (322-mi-long) surveys in southwestern New Mexico in autumn/winter, or 3.5 times higher than the count of 20 Golden Eagles during the four spring/summer surveys, when presumably only resident eagles were present (Eakle et al. 1996). Boeker and Bolen (1972) flew aerial transects within a 25,900-km² (10,000-mi²) area in east-central New Mexico in 1966 and calculated the area contained 686 eagles in January, 858 in February, but only 257 in March. Unfortunately, it seems unlikely that this 43-year-old study will be repeated, though a comparison of current wintering populations would be interesting.

Four hatch-year Golden Eagles, two males and two females, captured during their first autumn migration in the Manzano Mountains, wintered in southeastern New Mexico/west Texas, then migrated into Canada the following summer. Two returned to the same wintering area the following autumn; the other two radios expired. One male whose radio continued to operate for four years after capture spent his third and fourth winters in South Dakota and each summer in Canada. A >three-year-old female captured in the Manzanos in October 2003 spent part of the following winter near Puerto Vallarta, Mexico, then migrated 6,500 km (4,040 mi) to extreme northwestern Alaska the following summer (HawkWatch International 2006; chapter 2).

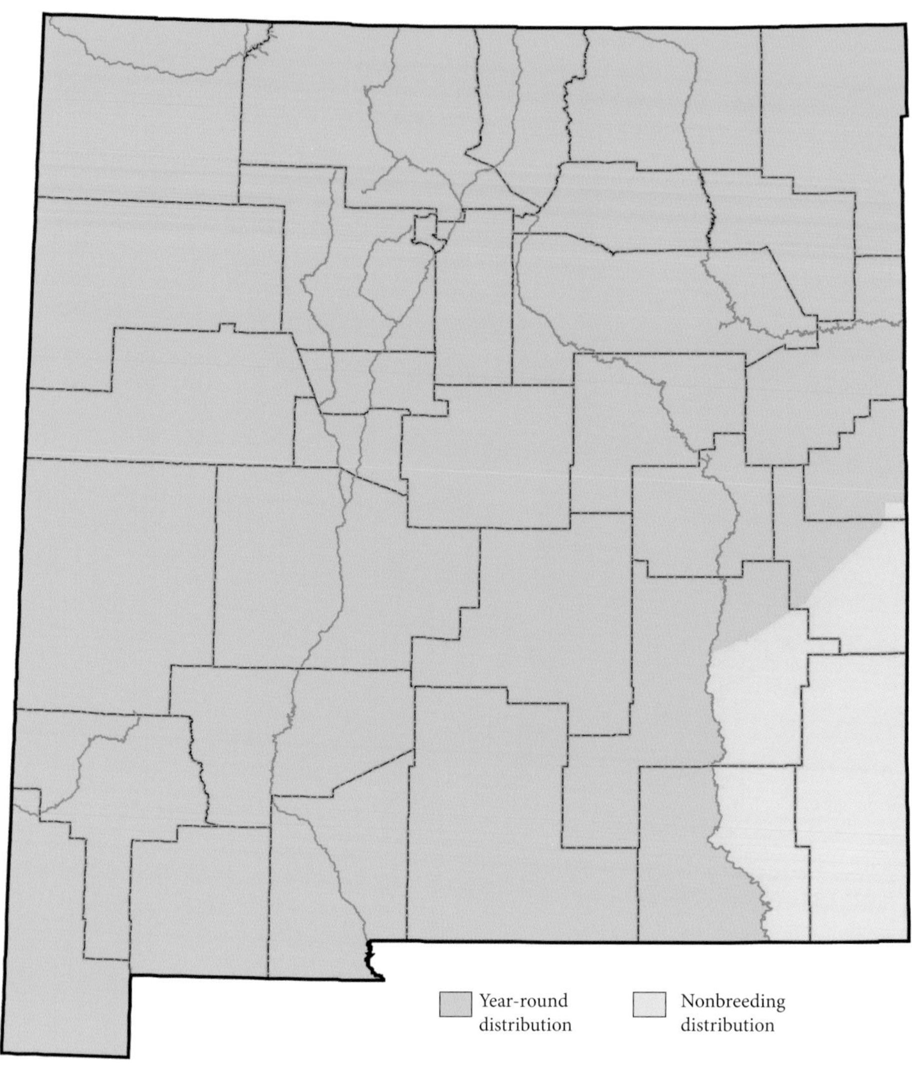

MAP 21.1

Golden Eagle distribution map

Habitat Associations

Golden Eagles forage in a variety of open habitats covering millions of hectares in New Mexico, from short grass prairie to moderately dense shrublands. Based on our experience, predominantly wooded habitat types (e.g., pinyon-juniper woodland, ponderosa pine [*Pinus ponderosa*] forest) are also often used by hunting Golden Eagles.

Throughout much of New Mexico, Golden Eagles nest along sandstone canyon walls, many of them topped with pinyon-juniper woodlands on adjoining mesas and with pinyon-juniper, sagebrush, saltbush (*Atriplex*) associations, or dry arroyo bottoms below. Nests occur also in badlands, such as those found in the northwestern part of the state. At higher elevations ponderosa pine and Douglas fir (*Pseudotsuga menziesii*) can make north-facing slopes too dense for eagle foraging, yet they can also provide nest sites. Statewide, nesting cliffs range in elevation from below 1,200 m (4,000 ft) in the southeast to 2,900 m (9,500 ft) in the northern mountains (DWS).

PHOTOS 21.8a, b, c, and d

Examples of nesting habitat in San Juan Co.

PHOTOGRAPHS: © JIM RAMAKKA.

PHOTOS 21.9a and b

Views from two Golden Eagle nests, Rio Arriba Co., 24 May 2004. Surrounding habitat consists of sagebrush flats and adjacent pinyon-juniper hills. Golden Eagles nest almost exclusively on cliffs in New Mexico. By comparison tree nesting is rare. In 2003 a nest was found on a communication tower in Taiban, De Baca Co.

PHOTOGRAPHS: © ADAM READ (9a), © MARK BLAKEMORE (9b).

Life History

Sexually mature Golden Eagles in New Mexico are either year-round residents of a territory or seeking an opening in one. Those seeking territories are known as "floaters," and their presence is evidence of a healthy population (Hunt 1998). Impatient floaters are likely constantly testing for signs of weakness in territorial eagles of like gender, regularly invading their territories. Territories are normally protected through displays, primarily undulating flight, that advertise to approaching eagles that the territory is occupied, but occasionally fights do occur (Harmata 1982).

Nesting

The egg incubation period for Golden Eagles lasts about 43 days in the wild, and young fledge between 62 and 85 days after hatching (Kochert et al. 2002). Fledged juveniles can be dependent on their parents for one to three months, and tolerated within the territory until the next breeding cycle. Thus breeding, from courtship through independence of the young, easily involves half of every year.

New Mexico's resident Golden Eagles maintain pair bonds and defend their territories year-round, but both activities intensify in midwinter. Most New Mexico territories contain one or more cliffs with ledges upon which the eagles build their stick nests. The same cliff ledges likely have been used for decades or even centuries by past generations of eagles. Nests can become quite large, in some cases involving hundreds of kilograms of sticks and softer lining. Some are on cliffs and spires that tower more than 300 m (1,000 ft) above surroundings. More often, they are found on 15- to 60-m-tall (50- to 200-ft-tall) cliffs at the top of escarpments of 90- to 180-m-tall (300- to 600-ft-tall) mesas or canyon walls. But if hunting habitat is good, Golden Eagles will maintain a territory even if the only nest site is a small cliff (<6 m or <20 ft) or tree (DWS).

Tree nesting seems uncommon in New Mexico, though it was perhaps more common historically, and elsewhere it remains common where cliffs are rare but large trees are not (Kochert et al. 2002). Ligon (1961) reported tree nesting in New Mexico in cottonwoods (*Populus* sp.) along washes, and nests belonging to three different territories were found in streamside cottonwoods in Union County as recently as 2006–8 (DWS). In the foothills of central New Mexico, nesting also occurred in lone ponderosa pines (T. Smylie, pers. comm.). One well-documented tree nest was in the Spur Lake area of Catron County, where Golden Eagles nested in "tall, ancient pines" (Ligon 1961). In his unpublished notes (on file at the University of New Mexico (UNM) Museum of Southwestern Biology), Ligon described that particular nest, visited in March 1915, as "16 feet deep, 4½ ft across, and at the top of an 80 ft high ponderosa pine." In eastern New Mexico, Ligon (1961) also noted nesting on abandoned windmills, which in our experience are now used by other raptors such as Ferruginous Hawks (*Buteo regalis*) but not by Golden Eagles. In 2003 and again in 2006, however, a Golden Eagle pair was found nesting on a 100-m-tall communication tower near Taiban, De Baca County (Williams 2003; DWS). This record, together with the lack of recent windmill records—as well as perhaps a decrease in the frequency of tree nesting—is interesting in view of Ligon's (1961) comment that human disturbance of windmill and tree nests typically resulted in nest failure.

Although each year some pairs choose not to nest, most of the others will complete nest building by mid February and begin incubation before the month is out. At one territory in northwestern New Mexico, young 60 days old were regularly present by 15 May, indicating incubation began in late January or very early February (DWS). This occurred annually from 1998 to 2005, despite the generally cold winter temperatures, well below freezing at night in that region of the state.

Golden Eagle eggs are oval-shaped and average 7.6 cm (3 in) long and 5.1 cm (2 in) wide. They are usually off-white, but speckled or blotched with reddish/brownish spots (Kochert et al. 2002); some New Mexico eggs are heavily marked, whereas others are nearly immaculate (JLEC). During the last several decades, no researcher has entered any nest during incubation. However, the UNM Museum of Southwestern Biology holds a total of 38 Golden Eagle egg sets collected mainly by J. Stokley Ligon and Willard Williams Hill from the second decade of the 20th century through the early 1960s. Thirty-five of those egg sets represent complete clutches, on the basis of which clutch size in New Mexico averages 2.06 (see appendix 21-1).

PHOTO 21.10

(*top left*) Downy eaglet in its nest, Rio Arriba Co., 25 May 2004. PHOTOGRAPH: © ADAM READ.

PHOTO 21.11

(*bottom left*) Eaglet in its nest, Rio Arriba Co., 24 May 2004. Flight feathers begin to show at about 25 days of age. PHOTOGRAPH: © SPIN SHAFFER.

PHOTOS 21.12a and b

(a, *top right*) Approximately 45-day-old eaglet, Rio Arriba Co., 25 May 2004, and (b, *bottom right*) approximately 50-day-old eaglet, Sandoval Co., 26 May 2004. At the lower elevations of northern New Mexico, most eaglets fledge by mid June. The older eaglets become noticeably more wary of humans and defensive! PHOTOGRAPHS: © SPIN SHAFFER.

Twenty-seven (77%) of the 35 clutches consist of two eggs, with only five (14%) three-egg clutches, and only three (9%) one-egg clutches. These relative proportions of three-egg, two-egg, and one-egg clutches are remarkably similar to what has been described for the western United States in general (Kochert et al. 2002).

In New Mexico, evidence of eaglets in a nest, particularly the adult female feeding unseen young, can be observed regularly in late March or early April. In fact, the adult female is on the nest almost constantly until eaglets can self-thermoregulate. She broods the eaglets at night until they are about four weeks old (range 17–42 days), and roosts on or near the nest until they are about seven weeks old (range 17–54 days) (Collopy 1984). Most eaglets from nests below 1,830 m (6,000 ft) in northern New Mexico fledge by mid June. At higher elevations fledging is generally delayed by one to three weeks (DWS).

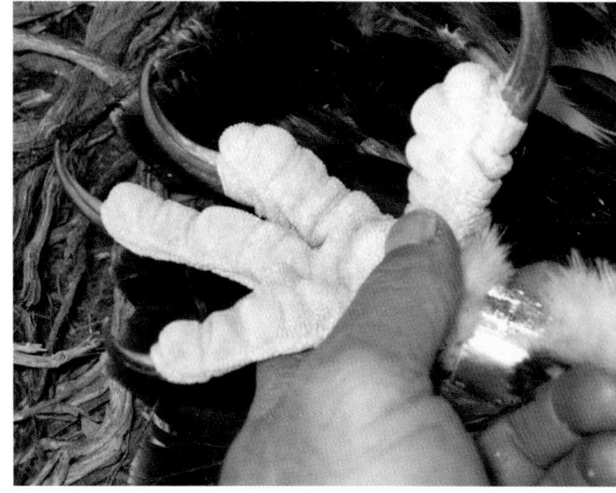

PHOTOS 21.13a and b

An unusual nest with three hatched young, Rio Arriba Co., 2006; (a, *top left*) on banding (9 May) at 35–40 days old (older two eaglets) and 30–35 days of age (younger eaglet); and (b, *bottom left*) more than a month later (19 June) when the three young are indistinguishable, and the nest ledge is heavily muted. Only about 1 nest in 100 successfully fledges three young. PHOTOGRAPHS: © DALE W. STAHLECKER.

PHOTO 21.14

(*above*) Foot of a female eaglet about 50–55 days old, Sandoval Co., 26 May 2004. A standard U.S. Fish and Wildlife Service band is being attached. Aluminum and color bands allow researchers to better track the movements of birds, study life history and demographics, and monitor the status of populations. PHOTOGRAPH: © JIM WHITE.

Only two (1.4%) of 142 New Mexico nests entered to band young between 1998 and 2008 contained three young. The average number of young banded per nest entry (all young were more than 30 days old) was 1.38 (DWS). Most Golden Eagles do not nest when prey populations are low, and since their favorite prey, rabbits of the genera *Lepus* and *Sylvilagus* (see below), are cyclic in abundance, the number of young fledged per occupied territory also fluctuates. In 10 years of monitoring in northwestern New Mexico, lows in the rate of young fledged per occupied territory were six years apart (fig. 1a). However, it is important to note that the percentage of known occupied territories did not change noticeably during those low-productivity years (fig. 1b). The mean productivity of 0.53 fledglings/occupied territory in northwestern New Mexico approximates the rate required by a stable Golden Eagle population during a long-term study in Switzerland (Haller 1996).

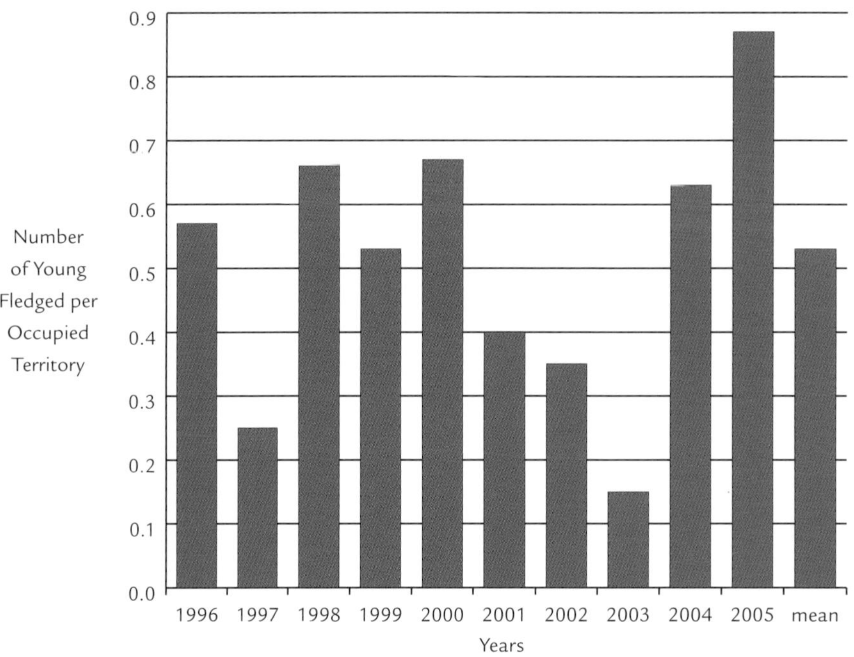

FIGURE 21.1a

Annual mean number of Golden Eagle young fledged per occupied territory in northwestern New Mexico, 1996–2005. The 1997 and 2003 breeding seasons were characterized by very low productivity.

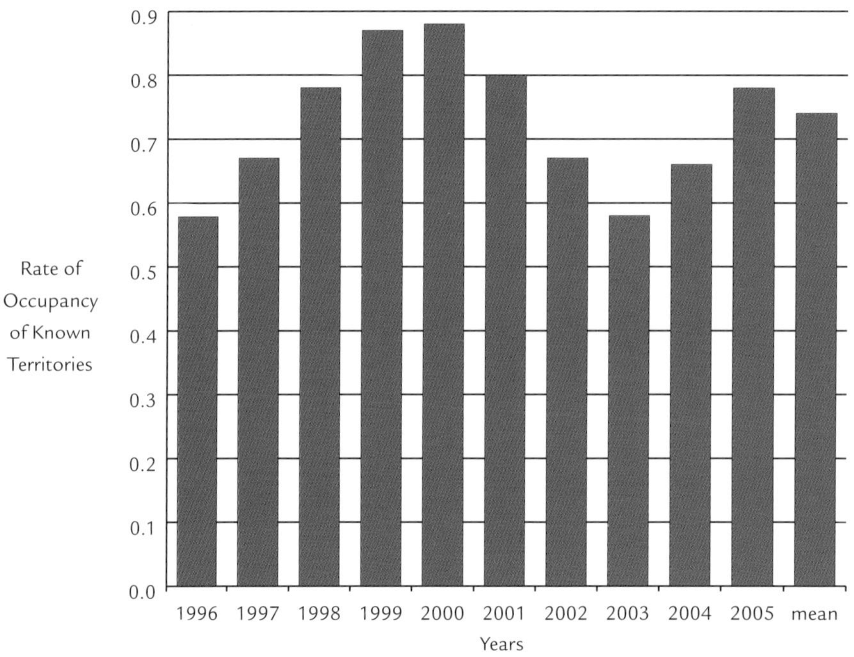

FIGURE 21.1b

Rate of Golden Eagle territory occupancy in northwestern New Mexico. Note that in 1997 and 2003, the two years of low reproductive success, most known territories were occupied as usual.

Diet and Foraging

Throughout its distribution the Golden Eagle preys mostly on mammals, particularly rabbits and ground squirrels, but it also takes a wide variety of other prey, including birds, reptiles, and even fish (Kochert et al. 2002). The Golden Eagle is a top predator capable of killing and eating both other predators and large ungulates. It is also not above eating carrion or preying upon domestic animals, particularly unprotected lambs and kids (O'Gara 1978). In New Mexico, a total of 33 prey taxa have been documented so far in the diet of Golden Eagles (table 21.1). In west-central and western New Mexico (Catron and southern Valencia counties), Mollhagen et al. (1972) found that most prey remains at nests were black-tailed jackrabbits (*Lepus californicus*, 47%) followed by Gunnison's prairie dogs (*Cynomys gunnisoni*, 40%). Among other species, they also found the remains of one ringtail (*Bassariscus astutus*) and of one gray fox (*Urocyon cinereoargenteus*). In the Capitan area (Chaves and Lincoln counties), black-tailed jackrabbits and cottontails (*Sylvilagus*) were the taxa most largely represented among prey remains at nests (Mollhagen et al. 1972). Black-tailed prairie dogs (*Cynomys ludovicianus*) were also present among remains but represented only 4.5% of the estimated total number of prey items. During 190 May and June nest entries in the Four Corners area (New Mexico, Arizona, Utah) between 1998 and 2008, 660 prey items of 24 species were documented. Most

PHOTO 21.15

A 25- to 30-day-old eaglet, an unhatched egg, a mule deer (*Odocoileus hemionus*) leg, and cottontail (*Sylvilagus* sp.) legs in a Golden Eagle nest during banding, 9 May 2007, Rio Arriba Co. Remains of a prairie dog (*Cynomys* sp.) were also in the nest. Note also the many evergreen boughs, some of which are fresh and were brought to the nest after the young hatched. PHOTOGRAPH: © CRAIG BLAKEMORE.

PHOTO 21.16

A three-week-old nestling with a cottontail rabbit (*Sylvilagus* sp.), May 2004, Rio Arriba Co. PHOTOGRAPH: © SPIN SHAFFER.

PHOTO 21.17

A partially eaten cottontail (*Sylvilagus* sp.) in a nest with a 35- to 40-day-old eaglet, Rio Arriba Co., May 2007. Between 2005 and 2008 this nest contained on average remnants of 6 cottontails or black-tailed jackrabbits (*Lepus californicus*), based on a count of similar body parts and it fledged a total of 6 young (1.5/year).

PHOTOGRAPH: © MARK BLAKEMORE.

PHOTO 21.18

Golden Eagle on a Snow Goose (*Chen caerulescens*) carcass at the Bosque del Apache National Wildlife Refuge, 15 December 2008. During the nonbreeding season, Golden Eagles can become scavengers. At carcasses they tend to be dominant even over the similarly-sized Bald Eagle (*Haliaeetus leucocephalus*). They frequently supplant smaller raptors such as the Red-tailed Hawk (*Buteo jamaicensis*) and ravens.

PHOTOGRAPH: © JAMES N. STUART.

were mammals (87%) and most of those (75%) were rabbits; birds (9%) and snakes (4%) were also observed (Stahlecker et al., in press; table 21.1).

New Mexico Golden Eagles are opportunists whose diet reflects the relative availability of prey species within their territories (DWS). At nests within proximity of Gunnison's prairie dog towns, Gunnison's prairie dogs are often present among prey remains. Within a territory encompassing a large marshy lake, a nest in 2004 contained the legs of at least a dozen American Coots (*Fulica americana*). Adults of that same territory were observed diving repeatedly on ducks flushed by a boat (10 April 1992) and on flightless moulting Canada Geese (*Branta canadensis*), particularly vulnerable as they had gone ashore to evade capture and banding by scientists (15 June 1992). In one territory in eastern New Mexico, ornate box turtles (*Terrapene ornata*) were apparently regularly preyed on (DWS).

Among ranchers Golden Eagles are known for their attacks on domestic animals. One New Mexico pair that became proficient at killing newborn calves (*Bos taurus*) had to be removed from the wild (Phillips et al. 1996). Mollhagen et al. (1972) did find sheep and goat remains at Golden Eagle nests in both the Capitan area

TABLE 21.1 Species documented in the diet of Golden Eagle (*Aquila chrysaetos*) in New Mexico.

CLASS	SPECIES
Pisces	Brown trout (*Salmo trutta*)
Reptilia	Ornate box turtle (*Terrapene ornata*)
	Gopher snake (*Pituophis melanoleucus*)
Aves	Gadwall (*Anas strepera*)
	Mallard (*Anas platyrhynchos*)
	Wild Turkey (*Meleagris gallopavo*)
	Great Blue heron (*Ardea herodias*)
	American Coot (*Fulica americana*)
	Sandhill Crane (*Grus canadensis*)
	Rock Dove (*Columba livia*)
	Great Horned Owl (*Bubo virginianus*)
	Northern Flicker (*Colaptes auratus*)
	Pinyon Jay (*Gymnorhinus cyanocephalus*)[1]
	Clark's Nutcracker (*Nucifraga columbiana*)
	Common Raven (*Corvus corax*)
Mammalia	Desert cottontail (*Sylvilagus audubonii*)
	Black-tailed jackrabbit (*Lepus californicus*)
	Chipmunk (*Tamias* sp.)
	Spotted ground squirrel (*Spermophilus spilosoma*)
	Rock squirrel (*Spermophilus variegatus*)
	Black-tailed prairie dog (*Cynomys ludovicianus*)
	Gunnison's prairie dog (*Cynomys gunnisoni*)[2]
	Mouse (*Peromyscus* sp.)
	Woodrat (*Neotoma* sp.)
	Common porcupine (*Erethizon dorsatum*)
	Gray fox (*Urocyon cinereoargenteus*)
	Ringtail (*Bassariscus astutus*)
	Badger (*Taxidea taxus*)
	Striped skunk (*Mephitis mephitis*)
	Deer (*Odocoileus* sp.)
	Domestic cow (*Bos taurus*)[3]
	Goat (*Capra hircus*)[4]
	Domestic sheep (*Ovis aries*)[4]

[1] nestlings only

[2] both during and outside the breeding season

[3] calves only

[4] both young and adults

Data are from Mollhagen et al. 1972; Cully 1988; Phillips et al. 1996; Stahlecker et al. 2009; DWS, unpubl. data; and C. Lee, pers. comm.

(4.5% of all prey) and in west-central and western New Mexico (1.2%). However, continent-wide, goats and sheep accounted for only 1.4% of 7,094 prey items in nests of multiple studies (Kochert et al. 2002).

The diet of New Mexico Golden Eagles is less well known outside the nesting season. Cully (1988) reported predation on Gunnison's prairie dogs in Colfax County's Moreno Valley in the fall. In addition, Golden Eagles are capable of preying on animals as large as pronghorn (*Antilocapra Americana*; e.g., Goodwin 1977), Whooping Cranes (*Grus americana*; Windingstad et al. 1981), and Sandhill Cranes (*Grus Canadensis*; Ellis et al. 1999). While leading an experimental flock of captive-reared Sandhill and Whooping Cranes from Grace, Idaho, to the Bosque del Apache National Wildlife Refuge in New Mexico, Ellis et al. (1999) observed several attacks by Golden Eagles, often while the cranes were crossing mountain passes. In most cases, the attack was by two or more eagles swooping down from high above on the trailing crane, which was then grabbed by one eagle from behind and held as the attacker and its prey plummeted downward. Ellis et al. (1999) fended off most of these attacks with intercepting ultralight aircraft! Although Ellis et al.'s flock was experimental and flew at lower elevations than wild cranes, attacks by Golden Eagles have been witnessed by other observers, including R. C. Drewien in the Rio Grande Valley (Ellis et al. 1999). Cook (1982) witnessed a Sandhill Crane grappling with a Golden Eagle, both birds aloft, at Clayton Lake in Union County on 11 October 1981:

> *The eagle was directly over the upside-down crane, whose upturned feet were paired with the talons of the eagle. The birds thus joined were slowly descending, wings outstretched, rotating as they came.*

While that particular attack was not successful, others most likely are. Golden Eagles feeding on Sandhill Cranes have been observed at the Bosque del Apache (C. Lee, pers. comm.).

Predation and Interspecific Interactions

Golden Eagles are often harassed by Common Ravens (*Corvus corax*), which also nest on cliffs and share the same habitat, and by smaller diurnal raptors (Kochert et al. 2002). In the Canadian Arctic, however, all other breeding raptors appeared to avoid nesting near Golden Eagles (Poole and Bromley 1988). This is likely the result of the Golden Eagle's propensity to take large prey, including nestlings of other raptorial species (Kochert et al. 2002).

Like Golden Eagles, Peregrine Falcons (*Falco peregrinus*) nest on cliffs and forage over open habitats (chapter 25). They are faster fliers and have "aerial superiority" over Golden Eagles. Despite this, Golden Eagles sometimes take large nestling, recent fledgling, or even unaware adult Peregrine Falcons (Craig and Enderson 2004). One of us (DWS) witnessed Peregrine harassment of New Mexico Golden Eagles near their aeries on two separate occasions. In late evening on 10 July 1992 an adult Golden Eagle was attacked by a male Peregrine at an apparently inactive falcon nest site. The eagle took refuge in a pothole in the cliff; the Peregrine kecked and dived at it repeatedly until light faded. The following morning the Peregrine Falcon had apparently departed to hunt at first light, and two adult eagles left the cliff before dawn. In the late afternoon of 3 July 1998 an adult female Peregrine suddenly began kecking and flew 0.8 km (0.5 mi) across a canyon to stoop repeatedly near a tall fir less than 0.4 km (0.25 mi) from the three large young in her aerie. Ravens eventually flew to the ground below the tree, revealing an adult female Golden Eagle, inaccessible to the diving Peregrine. The falcon's "anger" gradually waned, and after an hour's attention, she flew back toward her aerie. After another 15 minutes or so, the eagle sneaked away, flying low and in the opposite direction.

New Mexico Golden Eagles are typically resident year-round while Peregrines usually depart the breeding cliffs in September, not to return before March (chapter 25). Nesting Golden Eagles usually are incubating eggs by the time Peregrines return and probably cannot be dislodged easily then. In 2002 Golden Eagles nested on a cliff occupied by Peregrine Falcons in 2001; Peregrines were not seen there that year. In succeeding years, however, the eagles occupied a second cliff about 2.4 km (1.5 mi) away, and the falcons returned each year to their original cliff. The closest documented successful nesting of both species in New Mexico is 0.72 km (0.45 mi), but five sets of falcon/

eagle pairs were both successful when nesting within 1.6 km (1 mi) of each other (DWS). In general, aeries of the two species are not line-of-sight, and it appears that Peregrine Falcons ignore neighboring adult eagles as long as some invisible boundary between nest sites is not crossed, at least while Peregrines only have eggs. Through 2008 there have been no documented cases of nest-robbing of falcons by the eagles, though several falcon aeries within eagle home ranges have fledged only a single chick (DWS).

In winter Golden Eagles often interact with other scavengers at carcasses. Common Ravens are considered the smartest of birds, and an observation at Abiquiu Reservoir on 17 January 1988 showed their "respect" for Golden Eagles. During several hours of feeding at an elk (*Cervus elaphus*) hindquarter, the ravens constantly harassed several Bald Eagles while these were feeding, stealing scraps when the eagles turned to defend themselves. Later two coyotes (*Canis latrans*) approached the carcass; the Bald Eagles flew 9 m (30 ft) away, but the ravens turned their attention to the coyotes, harassing them unmercifully and continuing to steal scraps. But when an adult Golden Eagle landed on the leg quarter and fed, the coyotes left, the Bald Eagles stayed out 9–15 m (30–50 ft), and the ravens stood quietly back a respectful distance until the Golden Eagle had gorged itself and departed (DWS). Yet under different circumstances in Colorado, a coyote did not hesitate to steal prey from a Golden Eagle. On 9 February 2004, an adult male Golden Eagle captured and mortally wounded a prairie dog in about 0.3 m (1 ft) of snow (DWS). A coyote immediately jumped from a nearby arroyo and ran toward the eagle, which apparently realized it could not move effectively in the snow and flew. The coyote grabbed the prairie dog and ran back into the arroyo.

In Moreno Valley, Golden Eagles, Red-tailed Hawks (*Buteo jamaicensis*), and Ferruginous Hawks all preyed on Gunnison's prairie dogs while these rodents were abundant (Cully 1988). Once a hawk had made a kill, others attracted to the scene tried to claim the dead prairie dog for themselves, but typically failed (see also chapter 19). However, Cully (1988) reports one Golden Eagle supplanting a Ferruginous Hawk feeding on a prairie dog and states that when Golden Eagles were involved in killing a prairie dog, birds of the two hawk species typically stayed away.

Golden Eagles have no major natural predators. More than 70% of recorded North American deaths between 1965 and 1995 were caused by man, either directly or indirectly (Franson et al. 1995). The leading cause of death was blunt-force trauma from collisions with vehicles, power lines, or other structures (27%), followed by electrocution (25%), gunshot (15%), and poisoning (6%). Starvation was likely the major nonhuman cause of death.

Status and Management

Breeding Bird Survey (BBS) data for New Mexico show a decline in Golden Eagles of about 1.5% between 1966 and 2004 (Sauer et al. 2005). Population declines are suspected in the western United States (Kochert et al. 2002). However, neither spring nor autumn counts of migrants in central New Mexico showed a significant change through time (Hoffman and Smith 2003).

How many Golden Eagle territories are there in New Mexico? Even if two-thirds of the state's 315,979 km^2 (122,000 mi^2) are unsuitable for lack of nest sites or other habitat or for disturbance reasons, that still leaves about 105,325 km^2 (40,670 mi^2) of apparently suitable habitat. Assuming that the density of one territory per 259 km^2 (100 mi^2) observed in northwest New Mexico also applies elsewhere, as many as 406 Golden Eagle pairs may currently nest in the state. Using instead the density of one territory per 379 km^2 (146 mi^2) observed in northeastern New Mexico, a more conservative estimate would be approximately 275 nesting pairs statewide. The actual number of Golden Eagle pairs nesting in New Mexico is likely somewhere between these two estimates.

Golden Eagles were persecuted for centuries in the Old World because they were competitors for meat, either wild game or domestic livestock (Watson 1997). Persecution expanded into the New World through European settlers and stepped up when aerial gunning was initiated in California in 1936 (Kochert et al. 2002). Hunting clubs shot nearly 5,000 Golden Eagles in west Texas between 1941 and 1947, which may have reduced breeding populations in Texas and New Mexico (Phillips 1986). It is likely that illegal shooting continues to occur, though not so openly nor blatantly.

PHOTO 21.19

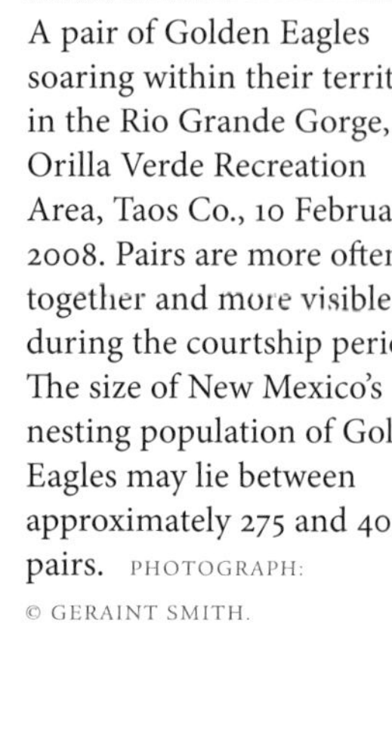

A pair of Golden Eagles soaring within their territory in the Rio Grande Gorge, Orilla Verde Recreation Area, Taos Co., 10 February 2008. Pairs are more often together and more visible during the courtship period. The size of New Mexico's nesting population of Golden Eagles may lie between approximately 275 and 400 pairs. PHOTOGRAPH: © GERAINT SMITH.

The oral histories of Native Americans indicate that they have harvested Golden Eagles for religious purposes for centuries in the southwestern United States, with no apparent impacts. The only tribe currently permitted by the U.S. Fish and Wildlife Service is the Hopis of northeastern Arizona. They remove eaglets from nests on their lands as well as from surrounding lands, most of which are part of the Navajo Nation. The Hopis then sacrifice the eagles when they are fully feathered in midsummer (Fewkes 1900). Some pueblo tribes of New Mexico claim similar traditions, and Apaches and Navajos traditionally capture or kill free-flying eagles. No New Mexico tribe is currently allowed to harvest eagles, but this could change in the future.

The importance of mortality caused indirectly by humans is often difficult to assess. This is true, for example, of mortality by lead poisoning, which is likely underreported. Yet Golden Eagles scavenge big-game carcasses at butchering sites throughout the winter. Their exposure to lead is nearly as great as that of Bald Eagles (Harmata and Restani 1995).

Electrocutions are likely declining in New Mexico, as public utilities have teamed with government and nongovernment organizations—creating, for example, the New Mexico Avian Protection Working Group, with both PNM and Hawks Aloft, Inc., acting in a prominent role—to find and retrofit problem poles. The Navajo Nation Department of Fish and Wildlife recently teamed with their largest electrical service provider, Navajo Tribal Utilities Authority, and developed the "Raptor Electrocution Prevention Regulations." These regulations, approved in September 2008,

PHOTO 21.20

Golden Eagle, north of Animas, Hidalgo Co., 10 September 2004. Once persecuted by humans, Golden Eagles are now mostly impacted indirectly by humans. Documented and suspected causes of mortality in New Mexico include electrocutions, collisions with the blades of wind turbines, and lead poisoning.

PHOTOGRAPH: © ROBERT SHANTZ.

mandate raptor-safe power poles within the expected home ranges of Golden and Bald Eagles, as well as Ferruginous Hawks, on the Navajo Nation. Of growing concern is the push to increase power generation using wind turbines, particularly in eastern New Mexico. Older, faster moving turbines have been proven to be the leading cause of mortality in some areas of California (Hunt 2001). New-generation wind turbines are so large that they can produce more energy at lower RPMs, allowing the blades to move slowly enough to be visible. However, the blades are so long that even at low RPMs the ends are moving at speeds in excess of 240 km/hr (150 mph). Golden Eagles appear to misjudge the rotating speed of these blades more than other, smaller birds of prey, and sometimes fly into their path.

The San Juan Basin lies in northwestern New Mexico, where oil and gas extraction is an important economic activity. Many canyons with Golden Eagle nesting territories now also have roads for access to well pads. Of particular concern are the compressors and air exchangers, which are a source of noise and in a number of cases lie within nesting territories (JLEC).

In 2008, Stahlecker's banding crew found a shivering eaglet in a northern New Mexico nest, pointing to possible infection by the West Nile Virus. Only one of the two young banded at that nest subsequently fledged. Though the overall fledging rate in northern New Mexico remained high in 2008, the sick eaglet plus the observed disappearance of several other young before fledging may be indicative that West Nile Virus's impact is increasing on nesting raptors in New Mexico.

Removal of the Bald Eagle from the list of Federally Endangered Species led the U.S. Fish and Wildlife Service to reevaluate its rules for "take" of both eagle species as the remaining extra protection under the Bald and Golden Eagle Protection Act (U.S. Fish and Wildlife Service 2008). Take not only includes killing of eagles, but also removal of their nests or disturbance causing them to alter their behavior and movements. Take permits will continue to be required for the range of human activities described above where direct or indirect mortality or failure to nest could result. It will be the goal of the Service to maintain or increase populations of both species, while more actively managing human activities that could impact eagles. In northwestern New Mexico, the Bureau of Land Management monitors Golden Eagle nesting pairs to detect any decrease in reproductive success. However, also needed is a spatial analysis to detect any shift in the distribution of territories over time in relation to oil and gas development. Our other recommendations include continued population monitoring and efforts to decrease and ultimately eliminate the use of lead bullets, harmful to this eagle as well as to the Bald Eagle (chapter 7).

The Golden Eagle is certainly an icon for New Mexicans past and present. Currently the population in the state appears stable and likely exceeds 275 nesting pairs. The future of this regal species is likely dependent upon how humans manage an increasingly complex world that sometimes moves too fast for this ancient, winged predator.

LITERATURE CITED

Bailey, F. M. 1928. *Birds of New Mexico*. Santa Fe: New Mexico Department of Game and Fish.

Boeker, E. L., and E. G. Bolen. 1972. Winter Golden Eagle populations in the Southwest. *Journal of Wildlife Management* 36:477–84.

Collopy, M. W. 1984. Parental care and feeding ecology of Golden Eagle nestlings. *Auk* 101:753–60.

Cook, W. 1982. Sandhill Crane grapples with Golden Eagle. *New Mexico Ornithological Society Bulletin* 10:23.

Craig, G. R., and J. H. Enderson. 2004. *Peregrine falcon biology and management in Colorado 1973–2001*. Technical Publ. no. 43. Denver: Colorado Division of Wildlife.

Cully, J. F. Jr. 1988. Gunnison's prairie dog: an important autumn raptor prey species in northern New Mexico. In *Proceedings of the southwest raptor management symposium and workshop*, ed. R. L. Glinski, B. G. Pendleton, M. B. Moss, M. N. LeFranc Jr., B. A. Millsap, and S. W. Hoffman. Scientific and Technical Series no. 11. Washington, DC: Institute for Wildlife Research, National Wildlife Federation.

Eakle, W. L., E. L. Smith, S. W. Hoffman, D. W. Stahlecker, and R. B. Duncan. 1996. Results of a raptor survey in southwestern New Mexico. *Journal of Raptor Research* 30:183–88.

Ellis, D. H., K. R. Clegg, J. C. Lewis, and E. Spaulding. 1999. Golden Eagle predation on experimental Sandhill and Whooping Cranes. *Condor* 101:664–66.

Ferguson-Lees, J., and D. A. Christie. 2001. *Raptors of the world*. New York: Houghton Mifflin.

Fewkes, J. W. 1900. Property-rights in eagles among the Hopi. *American Anthropologist* 2:690–707.

Franson, J. C., L. Sileo, and N. J. Thomas. 1995. Cause of eagle deaths. In *Our Living resources*, ed. B. T. LaRoe, G. S. Farris, C. E. Puckett, P. D. Dorna, and M. J. Mac, 68. Washington, DC: U.S. Department of the Interior.

Goodwin, G. A. 1977. Golden Eagle predation on pronghorn antelope. *Auk* 94:789–90.

Haller, H. 1996. The Golden Eagle in the Grisons: long-term studies on the population ecology of *Aquila chrysaetos* in the centre of the Alps. *Ornithol. Beob. Beiheft* 9:1–167.

Harmata, A. R. 1982. What is the function of undulating flight display in Golden Eagles? *Journal of Raptor Research* 16:103–9.

Harmata, A. R., and M. Restani. 1995. Environmental contaminants and cholinesterase in blood of vernal migrant Bald and Golden Eagles in Montana. *Intermountain Journal of Sciences* 1:1–15.

HawkWatch International. 2006. Results. http://www.hawkwatch.org/xxxx/xxc (accessed 30 March 2006).

Hoffman, S. W., and J. P. Smith. 2003. Population trends of migratory raptors in western North America, 1977–2001. *Condor* 105:397–419.

Hunt, W. G. 1998. Raptor floaters at Moffat's equilibrium. *Oikos* 82:191–97.

———. 2001. Golden Eagles in a perilous landscape: predicting the effects of mitigation for energy-related mortality. Report to California Energy Commission, PIER Program Contract no. 500-97-4033.

Kochert, M. N., K. Steenhof, C. L. McIntyre, and E. H. Craig. 2002. Golden Eagle (*Aquila chrysaetos*). No. 684. In *The birds of North America*, ed. A. Poole and F. Gill. Philadelphia, PA: Academy of Natural Sciences, and Washington, DC: American Ornithologists' Union.

Ligon, J. S. 1961. *New Mexico birds and where to find them.* Albuquerque: University of New Mexico Press.

McKnight, B. C., and D. M. Niles, eds. 1965. *New Mexico Ornithological Society Field Notes, Seasonal Report* 7:765–73.

Mollhagen, T. R., R. W. Wiley, and R. L. Packard. 1972. Prey remains in Golden Eagle nests: Texas and New Mexico. *Journal of Wildlife Management* 36:784–92.

O'Gara, B. W. 1978. Sheep predation by golden eagles in Montana. *Proceedings of the Vertebrate Pest Conference* 8:206–13.

Phillips, R. L. 1986. Current issues concerning the management of Golden Eagles in western U.S.A. In *Birds of prey*, ed. R. D. Chandellor and B.-U. Meyburg, 149–56. Bulletin no. 3. Berlin, Germany: World Working Group on Birds of Prey and Owls.

Phillips, R. L., J. L. Cummings, G. Notah, and C. Mullis. 1996. Golden Eagle predation on domestic calves. *Wildlife Society Bulletin* 24:468–70.

Poole, K. G., and R. G. Bromley. 1988. Interrelationships within a raptor guild in the central Canadian Arctic. *Canadian Journal of Zoology* 66:2275–82.

Sauer, J. R., J. E. Hines, and J. Fallon. 2005. *The North American breeding bird survey, results and analysis 1966–2004.* Version 2005.2. Laurel, MD: USGS, Patuxent Wildlife Research Center.

Sibley, D. A. 2000. *The Sibley guide to the birds.* New York: Chanticleer Press.

Stahlecker, D. W. 2008. NOT "tilting at windmills": determining a baseline resident Golden Eagle population of eastern New Mexico—2006–2008. *New Mexico Ornithological Society Bulletin* 36:34 (abstract only).

Stahlecker, D. W., D. G. Mikesic, J. N. White, S. Shaffer, J. P. DeLong, M. R. Blakemore, and C. E. Blakemore. 2009. Prey remains in nests of Four Corners Golden Eagles, 1998–2008. *Western Birds* 40:301–6.

U.S. Fish and Wildlife Service. 2008. *Draft environmental assessment: proposal to permit take under the Bald and Golden Eagle Act.* Arlington, VA: Branch of Policy and Permits, USFWS.

Watson, J. 1997. *The golden eagle.* London: T. and A. D. Poyser.

Williams, S. O. III, ed. 2003. *New Mexico Ornithological Field Notes* 42:22.

Windingstad, R. M., H. E. Stiles, and R. G. Drewien. 1981. Whooping Crane preyed upon by a Golden Eagle. *Auk* 98:393–94.

22

American Kestrel

(*Falco sparverius*)

DALE W. STAHLECKER AND JEAN-LUC E. CARTRON

THE AMERICAN KESTREL (*Falco sparverius*) is New Mexico's smallest, most abundant, most widespread, and arguably most colorful falcon. It is only 23–30 cm (9–12 in) long, with a wing span of slightly less than 60 cm (2 ft). Both sexes have warm rufous backs with black horizontal bars and boldly patterned heads. On the nape are dark large eye-like spots—known as ocelli—that are believed to deter sneak attacks by predators. Males have beautiful blue-gray wings and rufous tails with a black terminal band. On females the black barring and rufous of the back carries down the tail and onto the wings. Females are also about 10% larger than males (Sibley 2000; Smallwood and Bird 2002). Juveniles are similar to adults of the same gender, though young males have streaked breasts instead of the spotted breasts of typical adult males.

In flight, the silhouette of a kestrel is long-winged and long-tailed. Often the wings are swept back in the classic anchor shape more often associated with that of the kestrel's larger and more famous cousin, the Peregrine Falcon (*Falco peregrinus*). The underwings are pale, with darker barring evident throughout and a distal-most row of light-colored dots, especially pronounced in males. While kestrels most often hunt from a high perch, they are also well known for hovering or "kiting," wherein they face into the wind and maintain their position by flapping, vigorously if needed, while making directional adjustments with their long tails. In direct flight their wing beats are shallow, yet they are quite light and buoyant (Sibley 2000).

PHOTO 22.1

Male American Kestrel (Hawks Aloft, Inc., educational bird). Both the male and female sport two black mustaches. PHOTOGRAPH: © TOM KENNEDY.

PHOTO 22.2

(*above*) Adult male (left) and female (right) American Kestrels, dorsal view. Males have rufous backs barred with black on lower half and blue-gray upperwing coverts; the tail is rufous with a broad, black terminal band. Females (about 10% larger than males) have rufous backs, upperwing coverts, and tails, all with black barring. PHOTOGRAPH: © JEAN-LUC CARTRON.

PHOTO 22.3

(*left*) Male American Kestrel perched on a branch at the Bosque del Apache National Wildlife Refuge, 12 November 2006. The male's underparts are tinged with pale rufous and show variable amounts of black spots, often more numerous along the sides and flanks. PHOTOGRAPH: © GERAINT SMITH.

PHOTO 22.4

(*top left*) Male American Kestrel, Orilla Verde Recreation Area, Pilar, Taos Co., 16 December 2005. Note the pale underwings with darker barring and rows of light-colored dots, particularly conspicuous in males. PHOTOGRAPH: © GERAINT SMITH.

PHOTO 22.5

(*bottom left*) Female American Kestrel on the ground after diving at and catching a large grasshopper at the Bosque del Apache National Wildlife Refuge, December 2007. In contrast to the adult male, the adult female has rufous upperwing coverts and a breast and belly showing vertical rufous-brown streaking. PHOTOGRAPH: © TOM KENNEDY.

PHOTO 22.6

(*below*) An unusual-looking adult female American Kestrel at the entrance to a nest box, Eldorado, Santa Fe Co., 29 April 2009. Approximately half of the wing covert feathers are gray, giving this bird a partially male appearance. PHOTOGRAPH: © DALE W. STAHLECKER.

American Kestrels have two main calls (Smallwood and Bird 2002). Most distinctive and most used is the sharp, rapid "*killy, killy, killy*" (six notes/second), with which American Kestrels accost potential competitors or predators, including humans, when these approach their nest sites too closely. This same call is also uttered for communication between two paired birds during the breeding season. The chitter call, usually a softer, faster (20 notes/second) call, is the primary means of communication between paired birds within a relatively small distance of each other, even if one is inside a nest cavity (Smallwood and Bird 2002).

Up to 17 subspecies are recognized throughout the species' range (Smallwood and Bird 2002). One subspecies, the nominate *F. s. sparverius*, is found throughout all of Canada and the United States except Florida and adjacent areas of the Southeast.

Distribution

There are 39 species in the genus *Falco* worldwide, and 14 of them are kestrels (Ferguson-Lees and Christie 2001). The American Kestrel is the only kestrel that

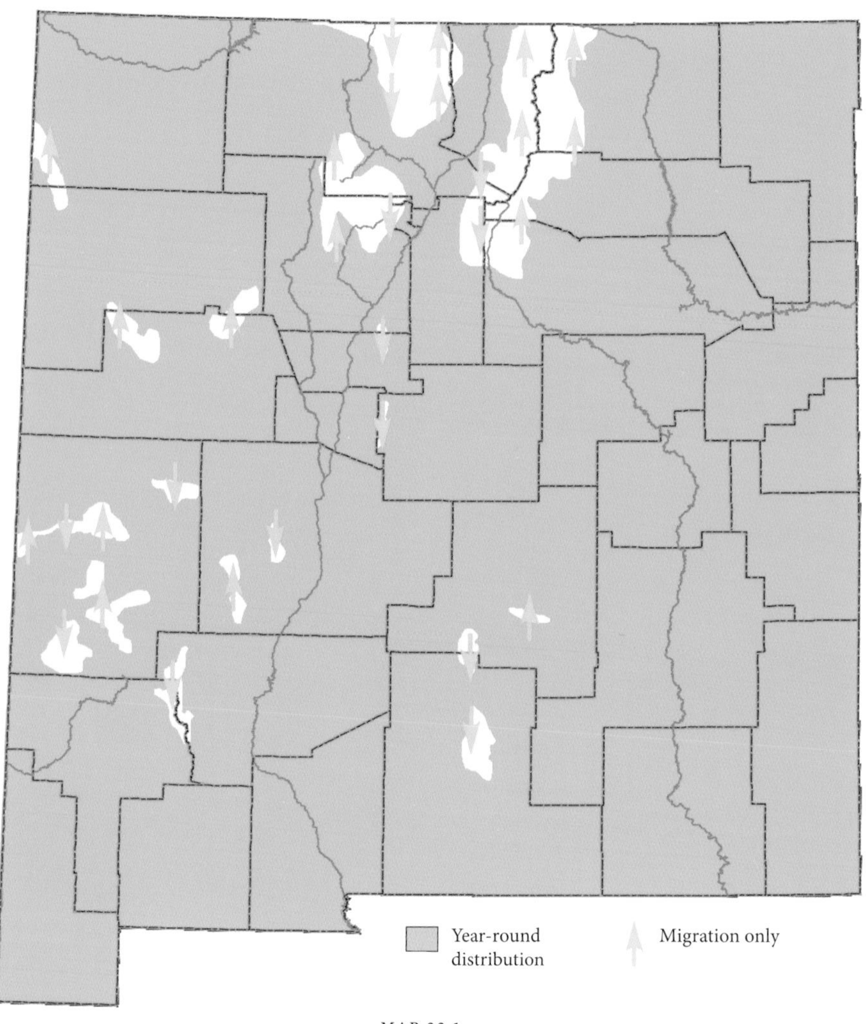

MAP 22.1

American Kestrel distribution map

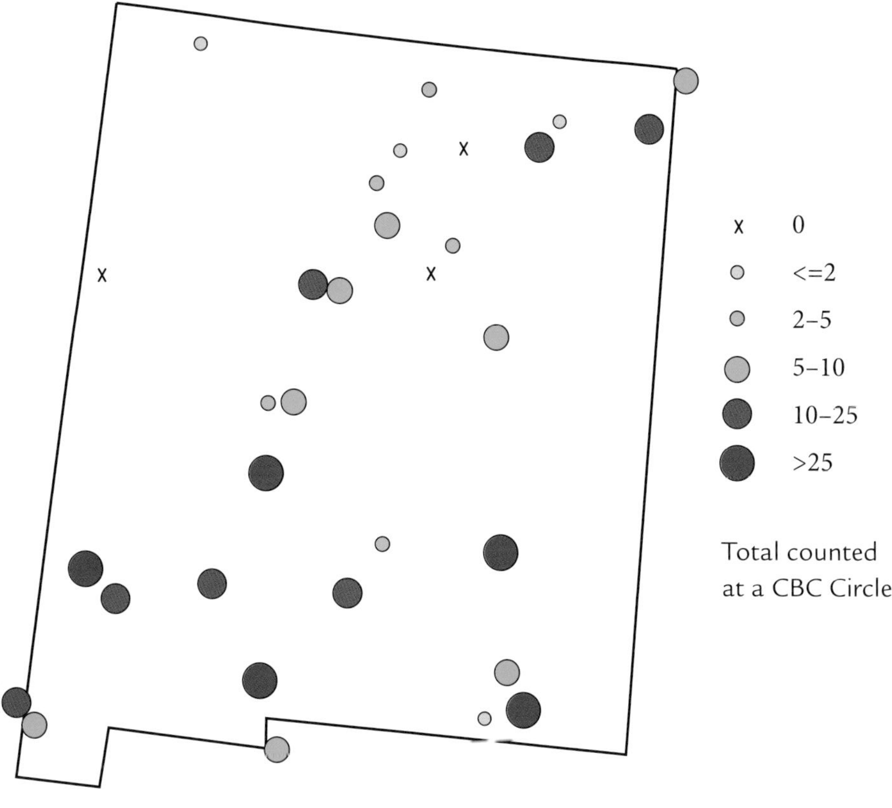

FIGURE 22.1

American Kestrels counted on New Mexico Christmas Bird Counts (CBCs) in New Mexico, 15 December 2002–5 January 2003.

occupies the Western Hemisphere, but it has a very wide geographic range, occurring as it does from Alaska's Arctic Circle to South America's tip, Tierra del Fuego (Smallwood and Bird 2002).

American Kestrels are found in New Mexico year-round. They occur in the state at all elevations from dry lowland deserts to snowcapped peaks, though the latter only during migration (Bailey 1928; Ligon 1961). During Breeding Bird Surveys (BBS) they have been documented on every route, and thus every county, in New Mexico (Sauer et al. 2008). Similarly, they are annually recorded on more than 90% of Christmas Bird Counts (CBC) in the state (National Audubon Society 2005).

At the scale of its entire distribution, the American Kestrel is known as what is called a partial migrant (chapter 2). In the northern tier of U.S. states and in Canada numbers of kestrels are lower in winter than during the breeding season (National Audubon Society 2005; Sauer et al. 2008). In New Mexico, kestrel numbers also decrease after the end of the breeding season, particularly at higher elevations and in the northern and central parts of the state (Ligon 1961; Hubbard 1978; Hink and Ohmart 1984). During an average winter, more kestrels are reported in New Mexico in the south than in the north (fig. 22.1). Whether this pattern reflects an influx of winter visitors in southern New Mexico or a greater percentage of kestrels in southern New Mexico being year-round residents, or both, is not known.

American Kestrels are regular migrants in New Mexico. On average, between 1985 and 2001, 553 (+/- 150) kestrels were counted over the Manzano

Mountains during autumn migration and 200 (+/- 95) were counted over the Sandia Mountains in spring migration. Kestrels were the fifth most common species counted in autumn migration and the sixth most common species counted in spring migration (Hoffman and Smith 2003; see chapter 2).

Habitat Associations

Range-wide, American Kestrels typically forage in mostly open areas with short ground vegetation (Smallwood 1987). How well so much of New Mexico meets that description! The state harbors hundreds of thousands of hectares of grasslands and shrublands (chapter 1), while even in predominantly wooded habitats, such as pinyon-juniper woodland or ponderosa pine forest, there are typically many openings with short vegetation.

With much of New Mexico highly suited for hunting, breeding densities of the cavity-nesting American Kestrel appear limited more by the availability of nest sites. Typical nesting habitat in New Mexico is quite diverse, ranging from the wooded edges of fields—even one lone large tree often has its own nesting pair (JLEC)—to badlands or pinyon-juniper woodland

PHOTO 22.7

(*top*) Adult female American Kestrel in flight, looking down, Separ Road, Grant Co., 30 January 2008. Kestrels can "kite," or hover, while facing into the wind, maintaining their position over the ground by flapping their wings. PHOTOGRAPH: © ROBERT SHANTZ.

PHOTO 22.8

(*middle*) Adult female American Kestrel perched on a fence post at the Bosque del Apache National Wildlife Refuge, December 2007. Typical American Kestrel habitat consists of open areas with short vegetation. During the nesting season, kestrels also need cavities for nesting. PHOTOGRAPH: © TOM KENNEDY.

PHOTO 22.9

(*bottom*) Kestrel in the wind on a wire west of Taos, Taos Co., 6 July 2007. American Kestrels are often observed perched on wires, poles, or fence posts in open areas. PHOTOGRAPH: © GERAINT SMITH.

PHOTO 22.10

(*top*) American Kestrel nesting habitat along the Rio Grande in central New Mexico. The most common diurnal raptor in the local cottonwood forest (bosque) is the Cooper's Hawk (*Accipiter cooperii*), except in burned areas, occupied instead by kestrels. PHOTOGRAPH: © JEAN-LUC CARTRON.

PHOTOS 22.11a and b

(*middle and bottom*) Two examples of typical American Kestrel nesting habitat in northwestern New Mexico: badland and open pinyon-juniper woodland, and badland and pinyon-juniper savanna. In both photographs a kestrel pair nested in a cavity along the cliff walls. Spring 2003. PHOTOGRAPHS: © RON KELLERMUELLER/HAWKS ALOFT, INC.

with rocky outcrops. Occasionally, the density of nesting pairs can be surprisingly high. In the Middle Rio Grande Valley, American Kestrels nest regularly in agricultural areas with scattered trees and in openings or along the edges of the local riparian cottonwood forest—or bosque—(Hink and Ohmart 1984; Cartron et al. 2008; JLEC). Recent wildfires in the bosque have also created multiple patches of burned areas with dead standing trees, used both during winter and for nesting by kestrels (Cartron et al. 2008). The density of nesting pairs in those burned areas remains to be quantified on a large scale though in 2004 the organization Hawks Aloft conducted intensive raptor nest surveys in the bosque around Los Lunas, Valencia County. The 18-km-long (11-mi-long) survey route crossed several post-fire bosque areas containing large numbers of snags, with adjacent fields providing suitable foraging habitat. A total of 17 documented nest sites or active nesting territories along the survey route in the bosque and adjacent areas represented a density of nearly one pair every kilometer of survey route (Hawks Aloft 2004).

In the Gila River Valley in the late 1990s and early 2000s, Scott Stoleson (pers. comm.) also found American Kestrels to be quite common if inconspicuous, nesting in large cottonwoods but hunting in adjacent fields, especially the nonfallow ones. A 5-ha (12-ac) patch of riparian wooded area with large openings—along the Gila River between Cliff and Gila—consistently had two nesting pairs and was seemingly representative of nesting densities throughout the valleys, even in narrow stringers of riparian vegetation along irrigation ditches. Nests sites were spaced about 300–400 m (~980–1,310 ft) apart on average (S. Stoleson pers. comm.).

In northern Rio Arriba County, Stahlecker et al. (1998) found six pairs of American Kestrels nesting in woodpecker holes in partially dead ponderosa pines (*Pinus ponderosa*) on the 5,260-ha (13,000-ac) Rio Chama Wildlife Area (one pair/8.62 km^2). Throughout the surrounding area, "wolf" ponderosas, unmerchantable trees that were left behind during extensive logging in the first half of the 20th century, usually have woodpecker holes, and such trees are also often occupied by kestrel pairs (DWS). During the *Los Alamos County Breeding Bird Atlas* compilation, American Kestrels were present in 76% of the 60 620-ha (2.4-mi^2) blocks, with confirmed breeding in 55% of those blocks. Kestrels nested primarily in old Northern Flicker (*Colaptes auratus*) cavities in ponderosa pines, preferring open pine stands or pine/grassland borders (Travis et al. 1992). Flood-killed cottonwoods in upper reaches of Santa Rosa Reservoir, Guadalupe County, had at least two occupying pairs of American Kestrels in 1995 (Stahlecker 1996).

American Kestrels also readily take up residence in nest boxes in low-density, semiurban areas. By placing several suitable nest boxes in and near his yard in Eldorado, a subdivision near Santa Fe in pinyon-juniper savanna with moderate human density (0.4–0.8 ha/lot with large greenbelts), one of us (DWS) has had kestrels for neighbors for more than a decade.

Life History

Nesting

Apparently most northern or higher altitude New Mexico American Kestrels are to some degree migratory, or at least they often do not maintain territories within their breeding area year-round. Indeed, adults typically return in the spring and occupy territories

PHOTO 22.12

American Kestrel nest site used in 2008 and 2009: a natural cavity approximately 2.2 m (~7 ft) aboveground along the face of an isolated rock formation in McKinley Co. PHOTOGRAPH: © RON KELLERMUELLER/HAWKS ALOFT, INC.

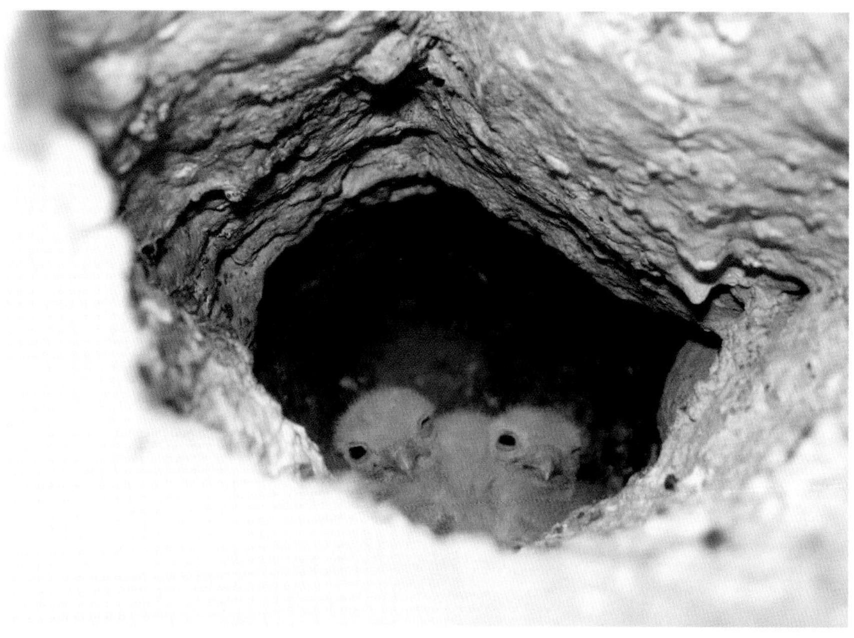

PHOTO 22.13

Two American Kestrel downy young in a nest cavity in a rocky outcrop in McKinley Co., 12 June 2009.

PHOTOGRAPH: © RON KELLERMUELLER/HAWKS ALOFT, INC.

PHOTO 22.14

The same two nestlings as in Photo 22.13, now older and partly feathered, McKinley Co., 24 June 2009.

PHOTOGRAPH: © RON KELLERMUELLER/HAWKS ALOFT, INC.

PHOTOS 22.15a and b

Occupied American Kestrel nest site in a yucca (*Yucca* sp.), Socorro Co., 26 May 2003.
The arrow in the bottom photo points at the adult female in her nest.

PHOTOGRAPH: © DOUG BURKETT.

PHOTO 22.16

Occupied nest box in Eldorado, Santa Fe Co., 28 April 2009. The adult female (peeking out of the box) is the same bird as in photo 22.6; the adult male (on top of the nest box) fathered young at both this nest box and another nest box nearby; this was apparently the first documented case of polygyny in the species in North America (DWS, unpubl. data).

PHOTOGRAPH: © DALE W. STAHLECKER.

PHOTO 22.17

Two 2-day-old kestrel chicks and four unhatched eggs in a nest box in Eldorado, Santa Fe Co., 27 June 2009. American Kestrels, evolved to nest in cavities, readily accept properly sized nest boxes.

PHOTOGRAPH: © DALE W. STAHLECKER.

PHOTO 22.18

Nestlings in a nest box, Santa Fe Co., June 1996. From 1995 through 2001, American Kestrels nesting in nest boxes in Eldorado often fledged as many as five young.

PHOTOGRAPH: © DALE W. STAHLECKER.

that contain one or more suitable nesting cavities. Courtship involves ritualized feedings, nest site examinations, and multiple copulations. Incubation for American Kestrels is usually 30 days in the wild, and young fledge 28–31 days after hatching (Smallwood and Bird 2002). As documented outside New Mexico, the relatively short incubation and nestling periods can allow a pair to produce two broods in a single breeding season, otherwise uncommon among birds of prey. Further, fledged young remain dependent upon their parents for food for as little as 12 days (Smallwood and Bird 2002); in Eldorado they drift away from the nest site within a few days after fledging. Ties between the nest site and the adults usually end quickly with the departure of the entire family group, and latitudinal/altitudinal migration occurs in early to mid autumn. However, one Eldorado male returned to the nest box snag on 25 October 2008, *killy*ing defensively.

Kestrels are secondary cavity nesters, utilizing natural or woodpecker-excavated cavities in trees throughout most of their range (Smallwood and Bird 2002). In New Mexico we have also found them nesting in cliff cavities, under the eaves of old buildings, in pilings of a bridge, and in properly sized nest boxes. More unusual was an active nest with two young in the fronds of a tall yucca in treeless desert near Gage, Luna County (Williams 1998, *fide* S. O. Williams III). Another nest, also in a yucca, was photographed in May 2003 in Socorro County by Doug Burkett (photos 22.15a, b). The yucca had been scarred by a recent fire. The resident pair of kestrels nested in a pseudo cavity formed by the dense cluster of yucca leaves (D. Burkett, pers. comm.).

While American Kestrels show reasonably strong tendencies to return to their previous year's nest site to breed, they also quickly take advantage of nest boxes placed in previously unused habitat (Smallwood and Bird 2002). Nest boxes in the Colorado high plains in an area apparently devoid of natural nest sites were immediately occupied (Stahlecker and Griese 1977). In Eldorado, nest boxes representing the two newly available territories were quickly occupied. But while they were available through at least 11 years, both territories were occupied in only 6 years, and either one or the other in the other 5 years.

In New Mexico, Bailey (1928) documented nests at elevations ranging from 1,160 m (3,800 ft) in the south, at Mesilla, Doña Ana County, to 2,130 m (7,000 ft) in Pueblo Canyon, Taos County. Ligon (1961) noted nests at Carlsbad, Eddy County, at an elevation of 945 m (3,100 ft) as well as at Chama, Rio Arriba County, at an elevation of 2,440 m (8,000 ft). Due to the elevational and latitudinal spread of nest sites in the state, the onset of breeding can be quite variable. Courtship and breeding behavior are initiated earlier in the southern part of New Mexico and at lower elevations. For example, pairs were inspecting nest cavities at 1,310 m (4,300 ft) near Deming, Luna County, on 26 January 2003 (Williams 2003, *fide* L. K. Malone). Near Santa Fe at 2,040 m (6,700 ft), a banded adult male reoccupied his previous year's territory on 16 February 1996, and an unbanded male occupied the same site on 15 January 2008, but a mid to late March return is more typical. Despite even mid March returns, the courtship period is elongated and incubation invariably starts around 1 May, based on the age of young when banded in mid June of most years. Elsewhere in the state, eggs were found at Silver City, Grant County, at 1,770 m (5,800 ft) on 21 April 1927; near Elephant Butte, Sierra County, at 1,370 m (4,500 ft) on 12 May 1913; while young were still in the nest on 15 July 1904 in Pueblo Canyon, Taos County (Bailey 1928). Also in Taos County, a nest site visited on 15 June 1958 still contained eggs (appendix 22-1).

Eggs are small, oval-shaped, and light-colored (white to light reddish-brown) with darker reddish markings, often concentrated at the blunt end. They are about 2.5 cm (1 in) wide and 3.5 cm (1 3/8 in) in length, on average. Markings on the egg suggest that the kestrel ancestral to the present-day American Kestrel nested in the open (Smallwood and Bird 2002), where camouflage might protect eggs from predators. Clutch size in New Mexico typically ranges from three to five, on the basis of nine egg sets housed at the Museum of Southwestern Biology at UNM (appendix 22-1).

Double brooding occurred in 11% ($n = 325$) of Florida pairs (Smallwood and Bird 2002). It has been documented in Colorado at 1,650 m (5,400 ft) (Stahlecker and Griese 1977), and even as far north as southern Canada (Smallwood and Bird 2002). No evidence of double brooding has been found at 2,040 m (6,700 ft) at Eldorado, where delaying clutch completion to late April may be related to availability of lizards (see below), cooler temperatures, or some unknown factor.

PHOTOS 22.19a, b, c, and d

Four fledglings (two males, two females) the morning after the pinyon (*Pinus edulis*) snag with their nest box was blown over, Eldorado, Santa Fe Co., 27 June 2007. Fortunately the box remained intact and landed with the entrance hole to one side. The four young kestrels, who spent the night in the nest box on the ground, were only slightly prematurely fledged. They were still alive and capable of flying on 1 July. PHOTOGRAPHS: © CLAUDIA WILLIAMS.

PHOTO 22.20

Fledgling American Kestrel on the ground south of Lingo, Roosevelt Co., June 2000.
PHOTOGRAPH: © LAWRY SAGER.

It is likely, however, that double brooding does occur in New Mexico, particularly at lower elevations in the southern part of the state.

Recently fledged broods were documented near Santa Fe on 22 June 1998; 1 July 2000, 2005, and 2007; and 8 July 1995 (DWS). In 2003, however, fledging at two occupied nest boxes occurred about 9 and 15 June, two to three weeks earlier than usual. Fledging seems to be most typically from the middle to the end of June in the Middle Rio Grande Valley, as well as in other lower-elevation parts of the state (Hawks Aloft 2004; R. Kellermueller, pers. comm.; JLEC).

All 18 known nesting attempts during 11 years of monitoring in the two nest box territories at Eldorado, Santa Fe County, were successful. Sixty-two banded young that could be sexed were equally split between males and females. The average number of young fledged/attempt was 4.4 (n = 16); however, observed mean productivity was higher from 1995 to 2001 (5.0 fledglings/attempt, n = 8) than from 2004 to 2008 (3.9 fledglings/attempt, n = 8), when most nesting attempts led to the production of only four or three rather than five fledglings. Premature fledging occurred occasionally. On the evening of 26 June 2007 strong winds blew over the Eldorado pinyon (*Pinus edulis*) snag and its associated nest box. After what might have been a bewildering night in the intact but grounded nest box, the four young scrambled out and up the remains of the snag (photos 22.19a–d) and were flying well by 1 July.

Diet and Foraging

Florida American Kestrels spent more than 93% of their time perched, mostly on utility lines or poles, 2% of their time pouncing on prey, about 3% of their time in other types of flight, but only 0.1% hovering (Smallwood and Bird 2002). In New Mexico, where kestrels perched on utility lines are also a common sight, the species' time budget is likely similar to that observed in Florida.

Range-wide, the American Kestrel preys upon a variety of large insects (74%), mice and voles (16%), passerine birds (9%), and reptiles (1%) (Sherrod 1978), but this diet varies regionally and seasonally (Smallwood and Bird 2002). In northern Florida, resident breeders ate large numbers of lizards (Smallwood and Bird 2002). Similarly, two-thirds of prey items ($n = 29$) brought by males to females at Eldorado were lizards, though observations were opportunistic and should be interpreted with caution. Two species of lizards, little striped whiptails (*Aspidoscelis inornatus*) and prairie lizards (*Sceloporus undulatus*), are common in the area (Degenhardt et al. 1996). Since New Mexico has a diverse and extensive reptile fauna, it would not be surprising if lizards and small snakes were more numerous in kestrel diets here than in more northerly parts of the species' range. Lizards are a part of the American Kestrel's diet in the Middle Rio Grande Valley, with invertebrates (e.g., moths and grasshoppers) also recorded (JLEC). At the Rio Grande Nature Center in Albuquerque, American Kestrels also hunt dragonflies over ponds (R. Yaksich, pers. comm.).

Kestrels have been documented taking nestling songbirds, a predatory habit called "nest-robbing" (Richards 1967). On 7 June 2008 the Eldorado male made four trips to the area where young had hatched the day before in a House Finch (*Carpodacus mexicanus*) nest (B. Squaglia, pers. comm.). Although not observed directly, nest-robbing was suspected, as all House Finch nestlings had disappeared when the nest was checked later that day.

PHOTO 22.21

Remains of a greater short-horned lizard (*Phrynosoma hernandesi*) in a rock cavity used by recently fledged American Kestrels, McKinley Co., 20 June 2008. PHOTOGRAPH: © RON KELLERMUELLER/HAWKS ALOFT, INC.

PHOTO 22.22

Male American Kestrel hovering in flight (kiting) north of Animas, Hidalgo Co., 11 April 2004. Kestrels hunt both from a perch and while kiting. PHOTOGRAPH: © ROBERT SHANTZ.

PHOTO 22.23

(*top*) Female American Kestrel eating a grasshopper near Summit Station, Hidalgo Co., 13 September 2007.

PHOTOGRAPH: © ROBERT SHANTZ.

PHOTO 22.24

(*bottom*) Adult male eating a moth at the Bosque del Apache National Wildlife Refuge, 15 November 2007.

PHOTOGRAPH: © STEPHEN INGRAHAM.

Unusual was the observation by DWS, at midday on 20 March 2001, of an adult male American Kestrel capturing a small bat in open air in a Sandoval County desert canyon. The bat also appeared to be foraging; it avoided capture about 10 times, apparently by sensing the approaching kestrel through echolocation and dodging at the last moment, before the kestrel finally snared it.

Caching of excess food for later use is well documented in American Kestrels. Thirty different kestrels cached 116 prey items in Missouri, 77% of which were later retrieved (Toland 1984). An Eldorado female cached about half of a lizard in the terminal needles of a juniper (*Juniperus monosperma*) on 30 April 2005.

Predation and Interspecific Interactions

Documented predators of the American Kestrel in New Mexico include the Swainson's Hawk (*Buteo swainsoni*) in the Middle Rio Grande Valley (Cartron et al. 2004) and the Cooper's Hawk (*Accipiter cooperii*) in both the Sandia Mountains and the Jemez Mountains/Pajarito

Plateau (Kennedy 1991; H. Schwartz, pers. comm.; see chapter 10).

Competition for nest cavities with several raptor and nonraptor species is also likely common. Within forested patches along the Middle Rio Grande, the same nest cavities may be used alternately by European Starlings (*Sturnus vulgaris*), Northern Flickers, and American Kestrels (R. Yaksich pers. comm.), suggesting competition among the three species. In 1986 a nest box in Ranchos de Albuquerque contained kestrel nestlings, grasses and twigs brought in by starlings, and two starling eggs (DWS). In the Gila River Valley, American Kestrels and Western Screech-Owls (*Megascops kennicottii*) alternated use of two nest cavities over the years (S. Stoleson, pers. comm.).

Over a decade of intermittent observations at Eldorado, American Kestrels were observed attacking Cooper's Hawks, Red-tailed Hawks (*Buteo jamaicensis*), Black-billed Magpies (*Pica hudsonia*), a rock squirrel (*Spermophilus variegatus*), coyotes (*Canis latrans*), and a neighbor's cat (*Felis domesticus*) that strayed too close to their nest or fledged young. On 4 June 2006 a Scaled Quail (*Callipepla squamata*) that chose to loudly announce his territory from the nest box tree, even though a prey exchange was occurring only 2 m (6.6 ft) away, received a "head-bop" by the kestrel male as he flew after his mate had reentered the nest box with the prey. The quail flew off immediately (B. Squaglia, pers. comm.). On 30 June 2005 a Cooper's Hawk that unknowingly chose to land in the tree containing the nest box and young was assailed with particular vigor. In 2008 when Great Horned Owls (*Bubo virginianus*) nested 75 m (240 ft) from the nest box tree, they and (eventually) their fledged young were ignored until the day the kestrels fledged, when the fledgling owl was stooped several times. However, Eldorado kestrels were not always the attackers. Flocks of 25–100 Pinyon Jays (*Gymnorhinus cyanocephalus*) would mob the adults during courtship prompting them to leave the vicinity of the nest box, which was 25 m (~80 ft) from the sunflower seed feeder visited almost daily by the jay flock. When feeders were left empty, Pinyon Jay visits decreased markedly, ending these interactions. Interestingly, the more sedentary Western Scrub Jays (*Aphelocoma californica*) were seldom seen harassing the adult kestrels.

Suburban kestrels tend to be more tolerant of humans (Smallwood and Bird 2002). Eldorado kestrels often tolerated humans outside DWS's house with no visible response. However, after DWS banded young on 27 June 1995, an adult female flew and "*killy*'d" him each time he walked outside, yet showed no response to the nonbander resident outside. Thereafter banding involved leaving home and returning in a borrowed vehicle and disguise! In both 2006 and 2007 the adults stooped repeatedly during banding, but did not recognize DWS without his disguise later the same day.

Handling of an adult female resulted in a more subtle response. The adult male banded on 8 July 1995 was likely the same banded male that returned to the territory in 1996 and 1997. However, in attempting to confirm this on 8 March 1997, the adult female was instead captured. Though the pair remained in the vicinity for a week or more, they eventually nested in a new nest box about 0.4 km (¼ mi) away. A pair of unbanded birds moved into the vacated territory and occupied the nest box. Banding may inadvertently have caused the move by the original pair . . .

All of these observations are indicative of kestrel intelligence, learning, and short-term memory, resulting in observable responses. While hand-capture of incubating females in nest boxes for marking does not seem to impact that year's reproduction (Smallwood and Natale 1998), to our knowledge no one has evaluated the possibility that the long-term memory of such handling could have an impact on return rates. Through negative mental associations, females may specifically avoid using a nest box where they were captured the year before, even if breeding was successful.

Status and Management

American Kestrels are North America's most abundant bird of prey. Dunn et al. (2005) estimated the North American population of American Kestrels in excess of 4 million—or 75% of the global population of this species—while other estimates placed population levels for kestrels breeding north of Mexico at 1.2 million (Cade 1982) and for kestrels wintering in the United States at 236,000 (Johnsgard 1990). It may come as a surprise, therefore, that the American Kestrel is a source of concern for researchers monitoring long-term raptor migration counts across North America. The American Kestrel has declined significantly in migration in many parts of its North American distribution, including in the Northeast since 1974 and the West since the early 1980s, with an acceleration of the declining trend since the late 1990s (Farmer et al. 2008). Widespread declines have also been recorded during Breeding Bird Surveys and Christmas Bird Counts (Sauer et al. 2008; National Audubon Society 2008). Proposed causes for the observed widespread declines include loss of habitat in the East; a period of drought in the West since the early 1980s; West Nile Virus; the increase in numbers of the Cooper's Hawk, a predator of the American Kestrel; or a combination of those factors (Farmer et al. 2008). In retrospect, however, West Nile Virus is probably not responsible. The decline of the American Kestrel began long before West Nile Virus arrived in North America (J. Smallwood, pers. comm.).

HawkWatch International's Sandia and Manzano migration observation sites in New Mexico stand out as among the few not registering any significant trend in American Kestrel counts (see Farmer et al. 2008). The number of American Kestrels counted during New Mexico CBCs has increased over time (1959–2004), but with more count circles and counters in recent years (National Audubon Society 2005). In analyzing CBC data, therefore, it is common to standardize data by dividing a count total by the amount of time, or party-hours, spent by all counters. While kestrels/hour for all New Mexico CBC circles has varied through the years (fig. 22.2), there was no statistically significant trend when data were regressed. BBS data statewide indicate a slight (-1.24%), though not statistically significant ($P = 0.26$), decline statewide between 1966 and 2007 (Sauer et al. 2008). Within the state, these data also indicate a range of changes, with much of the state showing positive trends, particularly from eastern New Mexico routes.

Contrary to eastern North America, there is no large-scale conversion of traditional agricultural areas reducing grassland habitat in New Mexico. The American Kestrel is readily infected by West Nile Virus (see also Golden Eagle, Peregrine Falcon, and Osprey chapters) throughout much of its North American

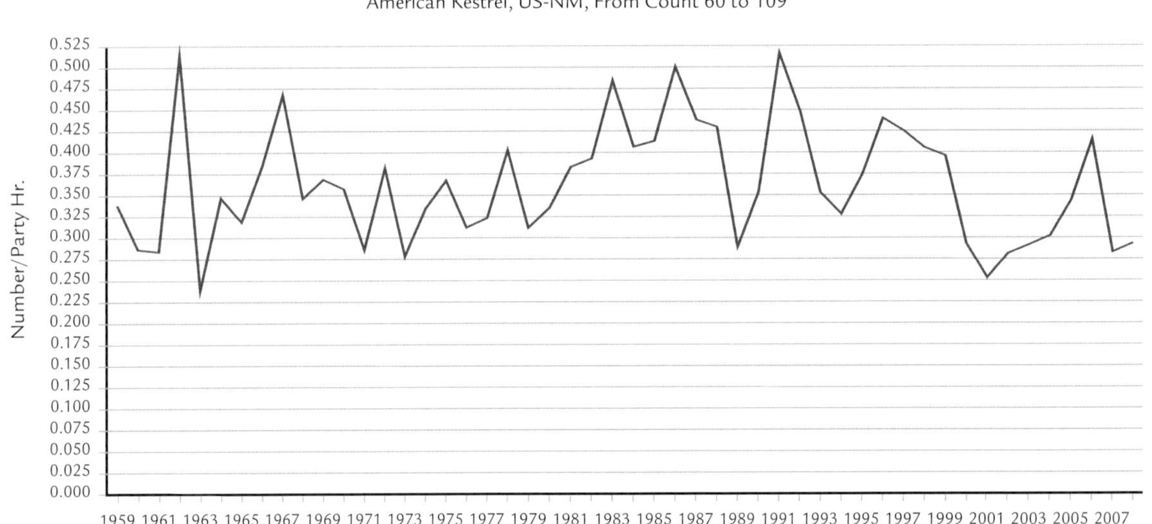

FIGURE 22.2

Number of American Kestrels per party hour on all New Mexico Christmas Bird Counts, 1959–2008.

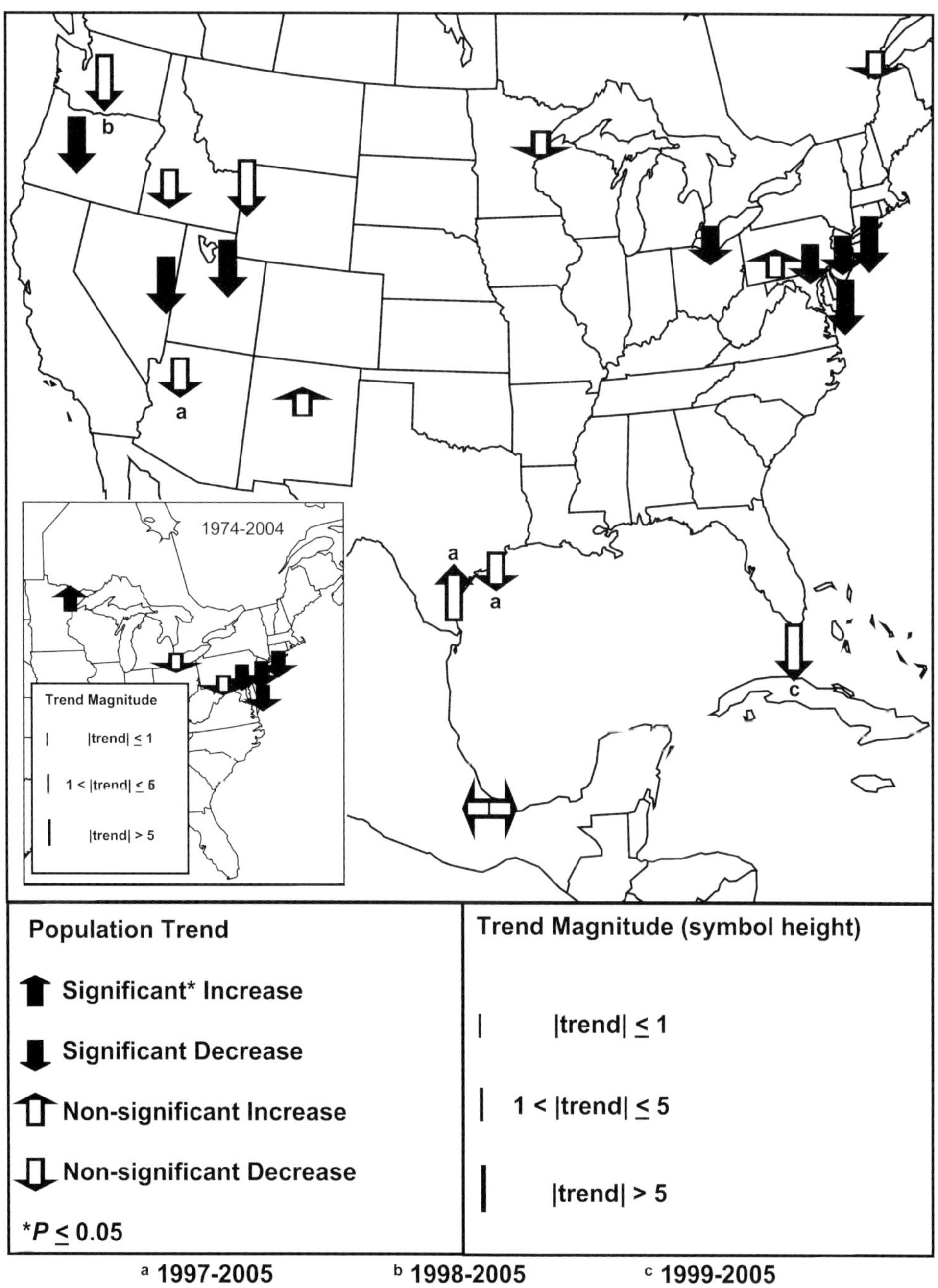

FIGURE 22.3

Population trends for American Kestrels at eight northeastern (1994–2004), eight western (1995–2005), and four Gulf of Mexico (1995–2005) raptor migration count stations in North America, and long-term trends (1974–2004) for seven northeastern counts (inset). Trends are expressed in percent change per year. A bidirectional arrow indicates that the estimated trend is 0.0% per year. Figure published originally as Figure 18 in Farmer et al. (2008) and reproduced with permission from Keith Bildstein and Christopher Farmer, Hawk Mountain Sanctuary.

PHOTO 22.25

American Kestrel (female perched and looking back) north of Animas, Hidalgo Co., 11 January 2003. The American Kestrel is a familiar sight in New Mexico but it is declining across much of North America.

PHOTOGRAPH: © ROBERT SHANTZ.

distribution including in Colorado (Nemeth et al. 2007), but impacts remain unclear (Farmer et al. 2008). Thus the outlook for the American Kestrel in New Mexico appears good for now, but continued monitoring of this species at the Manzano and Sandia migration sites should prove important. We also recommend that an intensive nesting study be initiated in one of the areas of the state with high densities of American Kestrels. No such study has been conducted to date, and yet it would yield important local natural information as well as baseline data for future monitoring.

LITERATURE CITED

Bailey, F. M. 1928. *Birds of New Mexico.* Santa Fe: New Mexico Department of Game and Fish.

Cade, T. J. 1982. *Falcons of the world.* Ithaca, NY: Cornell University Press.

Cartron, J.-L. E., D. A. Dean, S. H. Stoleson, and P. J. Polechla Jr. 2004. Nesting of the Swainson's Hawk (*Buteo swainsoni*) in riparian woodlands of New Mexico. *New Mexico Ornithological Society Bulletin* 32:91–94.

Cartron, J.-L. E., D. C. Lightfoot, J. E. Mygatt, S. L. Brantley, and T. K. Lowrey. 2008. *A field guide to the plants and animals of the Middle Rio Grande bosque*. Albuquerque: University of New Mexico Press.

Degenhardt, W. G., C. W. Painter, and A. H. Price. 1996. *Amphibians and reptiles of New Mexico*. Albuquerque: University of New Mexico Press.

Dunn, E. H., B. L. Altman, J. Bart, C. J. Beardmore, H. Berlanga, P. J. Blancher, G. S. Butcher, D. W. Demarest, R. Dettmers, W. C. Hunter, et al. 2005. High priority needs for range-wide monitoring of North American landbirds. Partners in Flight Technical Series no. 2. http://www.partnersinflight.org/pubs/ts/02-MonitoringNeeds.pdf. (last accessed February 2006).

Farmer, C. J., L. J. Goodrich, E. Ruelas, and J. Smith. 2008. Conservation status of North American raptors. In *State of North America's birds of prey*, ed. K. L. Bildstein, J. P. Smith, E. Ruelas Inzunza, and R. Veit, 303–420. Series in Ornithology no. 3. Cambridge, MA: Nuttall Ornithological Club, and Washington, DC: American Ornithologists' Union.

Ferguson-Lees, J., and D. A. Christie. 2001. *Raptors of the world*. New York: Houghton Mifflin.

Hawks Aloft, Inc. 2004. Raptor monitoring along the Rio Grande bosque: 2004 interim report. Unpublished report submitted to the Middle Rio Grande Bosque Initiative, U.S. Fish and Wildlife Service, Albuquerque, NM.

Hink, V. C., and R. D. Ohmart. 1984. Middle Rio Grande biological survey. Final report submitted to the U.S. Army Corps of Engineers, Albuquerque. Contract No. DACW47-81-C-0015.

Hoffman, S. W., and J. P. Smith. 2003. Population trends of migratory raptors in western North America, 1977–2001. *Condor* 105:397–419.

Hubbard, J. P. 1978. *Revised check-list of the birds of New Mexico*. New Mexico Ornithological Society Publ. no. 6. Albuquerque: New Mexico Ornithological Society.

Johnsgard, P. A. 1990. *Hawks, eagles, and falcons of North America: biology and natural history*. Washington, DC: Smithsonian Institute Press.

Kennedy, P. L. 1991. Reproductive strategies of Northern Goshawks and Cooper's Hawks in north-central New Mexico. Ph.D. diss. Utah State University, Logan.

Ligon, J. S. 1961. *New Mexico birds and where to find them*. Albuquerque: University of New Mexico Press.

National Audubon Society. 2005. The Christmas Bird Count historical results. http://www.audubon.org/bird/cbc (accessed 1 December 2009).

———. 2008. The Christmas Bird Counts historical results. http://www.audubon.org/bird/cbc (accessed 20 December 2008).

Nemeth, N. M., S. Beckett, E. Edwards, K. Klenk, and N. Komar. 2007. Avian mortality surveillance for West Nile Virus in Colorado. *American Journal of Tropical Medicine and Hygiene* 76:431–37.

Richards, G. L. 1967. Nest-robbing behavior of the Sparrow Hawk. *Condor* 69:88.

Sauer, J. R., J. E. Hines, and J. Fallon. 2008. *The North American breeding bird survey, results and analysis 1966–2007*. Version 5.15.2008. Laurel, MD: USGS, Patuxent Wildlife Research Center.

Sherrod, S. K. 1978. Diets of North American Falconiformes. *Journal of Raptor Research* 12:49–121.

Sibley, D. A. 2000. *The Sibley guide to the birds*. New York: Chanticleer Press.

Smallwood, J. A. 1987. Sexual segregation by habitat in American Kestrels (*Falco sparverius*) wintering in south central Florida: vegetative structure and responses to differential prey availability. *Condor* 89:842–49.

Smallwood, J. A., and D. M. Bird. 2002. American Kestrel (*Falco sparverius*). No. 602. In *The birds of North America*, ed. A. Poole and F. Gill. Philadelphia, PA: Birds of North America, Inc.

Smallwood, J. A., and C. Natale. 1998. The effect of patagial tags on breeding success in American Kestrels. *North American Bird Bander* 23:73–78.

Stahlecker, D. W. 1996. Birds of Santa Rosa Reservoir, Guadalupe County, New Mexico. *New Mexico Ornithological Society Bulletin* 24(2): 17–31.

Stahlecker, D. W., and H. J. Griese. 1977. Evidence of double brooding by American Kestrels in the Colorado high plains. *Wilson Bulletin* 89:618–19.

———. 1978. Raptor use of nest boxes and platforms on transmission towers. *Wildlife Society Bulletin* 7:59–62.

Stahlecker, D. W., J. P. DeLong, and J. V. Jewell. 1998. Breeding birds of the Rio Chama Wildlife and Fishing Area, Rio Arriba County, New Mexico. *New Mexico Ornithological Society Bulletin* 26(4): 72–99.

Toland, B. 1984. Unusual predatory and caching behavior of American Kestrels in central Missouri. *Journal of Raptor Research* 18:107–10.

Travis, J. R., D. Crowe, W. B. Lewis, D. Noveroske, A. T. Peaslee Jr., and J. Wolff. 1992. *Atlas of the breeding birds of Los Alamos County, New Mexico*. LA-12206. Los Alamos, NM: Los Alamos National Laboratory.

Williams, S. O. III, ed. 1998. *New Mexico Ornithological Field Notes* 37(2).

———, ed. 2003. *New Mexico Ornithological Field Notes* 42(1).

23

Merlin

(*Falco columbarius*)

DALE W. STAHLECKER

TO ME, ONE OF New Mexico's winter delights is a chance encounter with a Merlin (*Falco columbarius*), perhaps as a flash of pale blue as a male sprints low in pursuit of a flock of songbirds. The second smallest of our falcons, the Merlin is most often seen in New Mexico from September to April. In flight, it can be mistaken for a Rock Pigeon (*Columba livia*). The resemblance in the silhouette and flight of the two species even earned the Merlin its previous name in North America, the Pigeon Hawk. It also gave the Merlin its specific name, *columbarius*, or roughly translated, "pigeon-like." The common name is derived from the old French name for this species, *esmerillon* (Warkentin et al. 2005); it has nothing to do with the mythical sorcerer from Camelot!

Merlins are sexually dimorphic in both size and color; the smaller males are 24–27 cm (10–11 in), while the larger females are 28–30 cm (11–12 in; Clark and Wheeler 1987). Males at 160–170 g (5.75 oz) are also about 25% lighter than the 220–240 g (8 oz) females (Warkentin et al. 2005). Dorsally, juveniles and females are varying shades of brown, while adult males are varying shades of blue (Clark and Wheeler 1987; Warkentin et al. 2005).

I say "varying" because the main differences among North America's three subspecies are in the shade

PHOTO 23.1

Male Prairie Merlin with prey, Las Vegas National Wildlife Refuge, 13 December 2006. Note the light gray dorsal plumage (compare with adult female or immature Prairie Merlin) and muted facial markings. PHOTOGRAPH: © DOUG BROWN.

PHOTO 23.2

Adult female or immature Prairie Merlin in hand, caught at HawkWatch International's Manzano Mountains study site, fall 1999. Adult females and immatures of either sex have brown instead of blue or gray backs. Male and female Prairie Merlins have muted facial markings, including the lack of a distinct mustache. Note also the light areas on the outer web of the outermost primary; and the tail bands, creamy white rather than buffy. These two field marks are also indicative of the Prairie Merlin.

PHOTOGRAPH: © JOHN P. DELONG.

PHOTO 23.3

Immature female Prairie Merlin in hand, caught at HawkWatch International's Sandia Mountains study site, spring 1995. Immature Merlins resemble adult females of the corresponding subspecies.

PHOTOGRAPH: © JOHN P. DELONG.

of blue or brown of their plumages. Most common in New Mexico is the Prairie or Richardson's Merlin (*F. c. richardsonii*). Adult males are pale blue/gray on the back and wings, with muted facial markings and cinnamon streaking below. In contrast, adult females have pale tan-colored facial marks and streaking but moderate brown on the back and wings. The darkest of the North American subspecies is the Black Merlin (*F. c. suckleyi*). Adult males' backs are a deep cobalt blue, which in most light appears black, hence the name. They have heavily dark-streaked breasts and bellies. Females of this subspecies are usually a rich chocolate brown dorsally and similarly heavily streaked ventrally. The Taiga Merlin (*F. c. columbarius*) is intermediate in coloration, with its face pattern also generally standing out the most among the three subspecies.

All three subspecies have dark/black tails with terminal white bands, and generally with an additional two (Black), three (Taiga), or four (Prairie) slate blue to whitish gray (male) or buffy to white (female) bands above.

PHOTO 23.4

Immature female Taiga Merlin in hand, caught at HawkWatch International's Manzano Mountains study site, fall 1995. Note the more pronounced facial markings (including a more distinct mustache) on this bird compared to the Prairie Merlin.

PHOTOGRAPH: © JOHN P. DELONG.

PHOTO 23.5

A dark male Merlin (most likely a Black Merlin based on location, but also possibly a dark Taiga Merlin, the two being essentially indistinguishable in the field), Vancouver Island, British Columbia, Canada, July 2007. Black Merlins are found year-round along the Pacific Coast from Washington State north to Alaska. A few Black Merlins have been reported in winter as far south as California and New Mexico. New Mexico records of the Black Merlin remain disputed.

PHOTOGRAPH: © MIKE YIP (VANCOUVERISLANDBIRDS.COM).

Tail bands can become indistinct in Black Merlins. With all this plumage variation, it is at least simplifying that juveniles resemble the adult females of their corresponding race (Sibley 2000; Warkentin et al. 2005).

Distribution

The Merlin is found throughout the Northern Hemisphere. In addition to North America's Black, Taiga, and Prairie races, the species is represented by six subspecies that breed across northern Eurasia and winter south into northern Africa and southern Asia (Warkentin et al. 2005). The Prairie Merlin breeds from the northern prairies and aspen parklands of central Canada south to central Wyoming. The Black Merlin is largely a year-round resident from Alaska south to Washington along the Pacific coast. Finally, the Taiga Merlin is the race that breeds across the boreal forests of Alaska and Canada, with recent expansion into the northeastern United States (Warkentin et al. 2005; Paxton et al. 2008; Boone et al. 2008). Other than the Black Merlin and some of the Prairie Merlin's urban or southernmost breeding populations, the species is

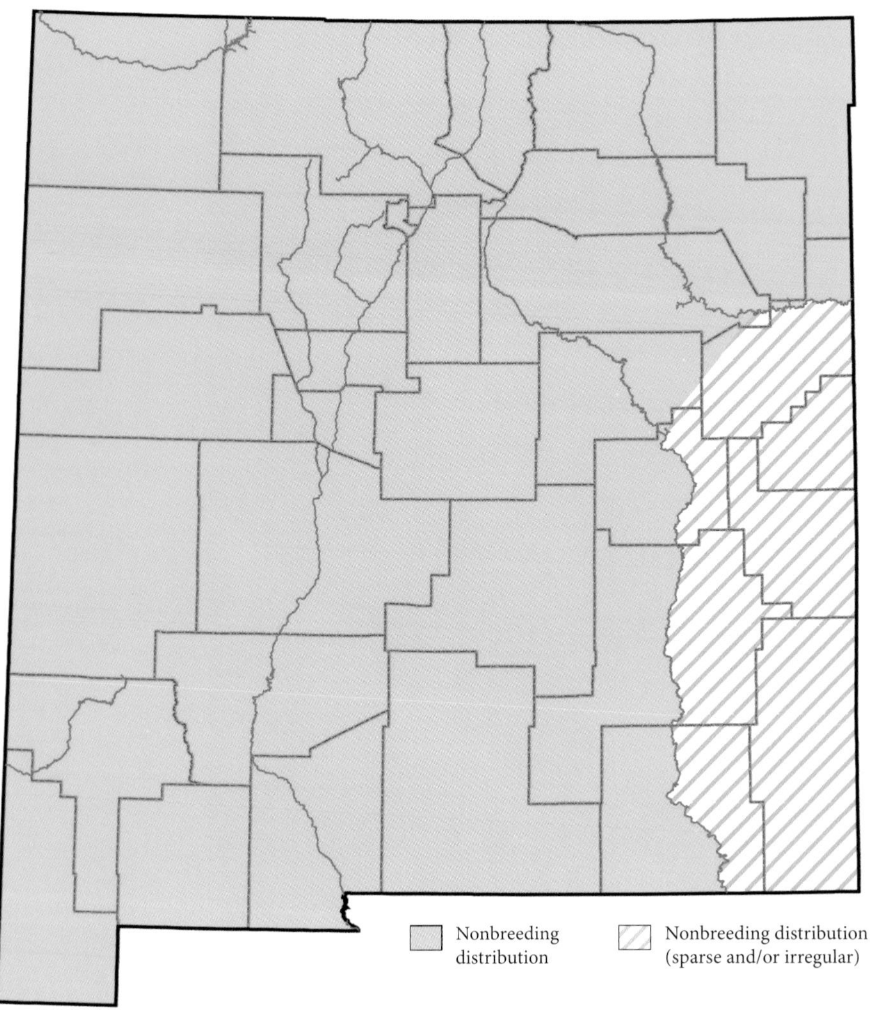

MAP 23.1

Merlin distribution map

migratory. The winter distribution includes most of the United States, primarily the West and South, and extends south through Mexico and Central America to northern South America (Warkentin et al. 2005).

Merlins are found in New Mexico during migration and throughout the winter. Apparently most Black Merlins do not migrate (Warkentin et al. 2005), but some birds of this race have been reported outside the breeding range and as far south as southern California and New Mexico in winter (Clark and Wheeler 1987). Some Prairie Merlins both breed and winter in cities of the Canadian prairies because of abundant, year-round resident prey, particularly House Sparrows (*Passer domesticus*) and Bohemian Waxwings (*Bombycilla garrulus*), but most of the members of this subspecies winter in the south-central and southwestern United States and in northern Mexico (Warkentin and Oliphant 1990; Warkentin et al. 2005). Winter band recoveries for rural nesting Prairie Merlins were concentrated in eastern Colorado (5 of 11 recoveries), but one was found in extreme southeastern New Mexico (Schmutz et al. 1991). Taiga Merlins are completely migratory; some winter in the western and southern United States with a few traveling as far south as northern Peru (Warkentin et al. 2005).

All Four Corners states have wintering Merlins. In Colorado, they are rare to uncommon migrants and winter residents. Most in that state are *F. c. richardsonii*, and "field identification of other races should be made carefully and documented thoroughly" (Andrews and Righter 1992:83). A Taiga Merlin specimen from Colorado's northeastern plains—at first incorrectly reported as *F. c. suckleyi* (Bailey 1942)—remained the only *F. c. columbarius* specimen for the state as of 1992 (Andrews and Righter 1992). Merlins also occur in Arizona, though only in low numbers (Oliphant 1998) and with no subspecific information readily available from that state. In a five-year study in north-central Utah, 10% of 171 Merlins assigned to subspecies were identified as *F. c. suckleyi*, with the rest split about equally between the other two subspecies (Haney and White 1999). *F. c. suckleyi* was "most difficult to confidently allocate to subspecies unless in hand" (Haney and White 1999:267). The same study also noted that 9 (19%) of 47 Utah specimens were *F. c. suckleyi*.

In New Mexico, Bailey (1928:192) noted that "as a

PHOTO 23.6

(*top*) Merlin at the Bosque del Apache National Wildlife Refuge, 25 November 2008. PHOTOGRAPH: © GARY K. FROEHLICH.

PHOTO 23.7

(*bottom*) Merlin at Virden, Hidalgo Co., 1 February 2003. Merlins are found in winter in all parts of New Mexico. PHOTOGRAPH: © ROBERT SHANTZ.

migrant, the Pigeon Hawk was reported as not uncommon on the upper Pecos near Willis during the fall of 1883 (Henshaw), was twice seen in Shiprock (Gilman) and once noted near Tres Piedras on 1 August 1904 (Gaut)." This last record was far out of season, and was questioned by Hubbard (1978). Bailey (1928) also reported specimens taken near La Jara Lake, Rio Arriba County, 18 September 1904; near Tularosa, Otero County, 5 November 1902; 40 km (25 mi) south of Albuquerque (presumably in present-day Valencia County), 30 October 1917 (Leopold); and from near Silver City, Grant County, 2 November 1919, 14 November 1920, 2 November 1922, and 19 October 1924.

Hubbard (1978) reviewed 175 Merlin records (specimens and sightings) from 22 of 32 (now 33) New Mexico counties between 1882 and 1977. From the winter of 1882–83 to that of 1945–46 there were several long periods of time during which no Merlins were recorded, and the species was documented in fewer than half of those winters (24 of 67). From 1946–47 to 1976–77 Merlins were reported almost every winter (23 of 27). The relative scarcity of reports in the earlier period was most likely due to a lack of observers (Hubbard 1978). Even during the latter years of this compilation, however, there were fluctuations from zero or one sighting in a winter to 15 or 16 in another winter, suggesting that the species is much more abundant in New Mexico during some winters than others.

Since 1990, with even more birdwatchers active in the state, typically 25–40 Merlins have been reported each winter just from December through February (based on a compilation of records published in *North American Birds*), and the species has now been reported in all 33 of New Mexico's counties (NMOS 2007). Between December 1994 and February 1995, 40 Merlins were seen at 30 locations, including reported sightings of all three subspecies (Williams 1995). Prairie Merlins predominate: 50% of pre-1977 specimens were *F. c. richardsonii* (Hubbard 1978), but in recent years they likely make up >67% of sightings, and in some years their proportion may reach 90%. Virtually all the rest are Taiga Merlins, but then on average once every third year a very dark, possible Black Merlin has been reported. The seven that have been reported since 1990 are unusual enough to be summarized here: one near White's City, Eddy County, on 19 September 1992 (Snider 1992, *fide* S. West); another in 1994–95 without location (Williams 1995); one at La Joya, Socorro County, on 22 December 2000 (Williams 2001); possibly the same individual at Portales, Roosevelt County, on 24 September 2005 and then at Carlsbad Caverns, Eddy County, on 2 October 2005 (Williams 2006); and one at Zuni, McKinley County, on 13 December 2007 (Williams 2008). The only multiple observations of the same dark Merlin was of an adult male I observed on four different dates between 24 January and 28 February 1996 at Eldorado, Santa Fe County (Williams 1996). Note that there is no pattern to "dark" Merlin sightings, for they were scattered throughout the state. And even if any had been photographed, in all likelihood it would still have been impossible to verify the subspecies.

As of 1977, 32 museum specimens of Merlins had been collected in New Mexico, of which 16 were identified as *F. c. richardsonii*, 4 as typical *F. c. columbarius*, 8 as *F. c. bendirei* (a former moderately dark subspecies now officially included as part of *F. c. columbarius*), and 3 considered crosses (intergrades) between two of these three forms (Hubbard 1978). The last of the 32 specimens was secured by W. L. Finley at Ghost Ranch, Rio Arriba County, on 26 December 1943 and was identified as an adult *F. c. suckleyi* (Jewett 1944). The Ghost Ranch specimen appears to be the primary piece of evidence in support of Black Merlins occasionally wintering in New Mexico. However, its identification by plumage to *F. c. suckleyi* is now disputed (Hubbard and Dickerman, *in litt.*). While I leave the resolution of this difference of opinion to peer review and (perhaps) DNA analysis, suffice it to say that separating a dark Taiga Merlin from a true Black Merlin is virtually impossible under field conditions (see also Haney and White 1999). If you are fortunate enough, as I have been, to observe such a Merlin in New Mexico, be satisfied with reporting it as a "dark" Merlin (but supply all the details possible, and take pictures!).

Merlins are found in New Mexico almost exclusively from September to April, mainly October through mid March (NMOS 2007). Hubbard (1978) was dubious of any records not bracketed between the earliest (18 September in Rio Arriba County) and latest (19 March in Grant County) collection dates

PHOTO 23.8

Female Merlin, Bosque del Apache National Wildlife Refuge, April 1993. Most New Mexico records of the species are from October through mid March. Much less frequently, Merlins have been reported in New Mexico as early as September or as late as April (as in photo), with an even smaller number of May–August reports, including one documented with videotaping on 5 August [2007].

PHOTOGRAPH: © GARY K. FROEHLICH.

of museum specimens available at that time. In fact, 85% of about 250 New Mexico Ornithological Society (NMOS 2007) records, excluding multiple sighting from migration count summaries, fall within the "specimen" limits. Further, most (11%) of the non-conforming reports originate before April 30 or after September 1, still within reason for migrating Merlins. The few remaining summer (May–August) records were, for the most part, made by reliable observers but without confirming photographs. The exception was one videotaped by Christopher Rustay on 5 August 2007 in western Socorro County (Williams 2008); it added credence to the other few summer records indicating that, indeed, a Merlin can occasionally be seen during the breeding season in New Mexico.

Just how important is New Mexico as a wintering area for Merlins? Published data from systematic counts of Merlins in New Mexico are limited, but results suggest that Merlin numbers compare favorably with those of the Colorado plains, likely the core wintering area for Prairie Merlins (Schmutz et al. 1991). In southwest New Mexico in 1988–89, Merlins were recorded only nine times in four autumn/winter surveys along a 420-km (261-mi) route through Catron, Sierra, and Luna counties, for a relative abundance of 0.55 Merlins per 100 km (0.89 Merlins per 100 mi). That abundance is low if compared to 6.84 American Kestrels (*Falco sparverius*) per 100 km (11 per 100 mi) for the same period and route (Eakle et al. 1996), but higher than the 0.33–0.38 Merlins/100 km (0.53–0.61 per

100 mi) recorded in southeastern Colorado between 1962 and 1978 (Bauer 1982). In Doña Ana County, between October and March 1983–86, no Merlins were recorded in weekly counts along a 65-km (40-mi) survey route, even though about 5,000 km (3,000 mi) were driven (Kimsey and Conley 1988). More recently, however, between 23 October and 2 December 2008, I counted 13 Prairie Merlins on 8 counts along a 86-km (53-mi) route (or 1.9/100 km) in Union County in New Mexico's northeastern highlands. Resightings of three same-aged and gendered birds, an immature male, an adult male, and an un-aged female, in three distinct locations over 17- to 24-day periods suggest that the same individuals were sedentary at least through the late autumn of 2008.

PHOTOS 23.9a and b

(a, *top*) Example of Merlin winter habitat in New Mexico: yucca (*Yucca* sp.) grassland with some creosote (*Larrea tridentata*), Doña Ana Co., December 2008. (b, *bottom*) Merlin perched on a yucca at the same location, 10 December 2008. PHOTOGRAPHS: © DOUG BURKETT.

Habitat Associations and Life History

In New Mexico Merlins are found in grasslands, whether mostly open (east), mixed with shrub stands (southeast, southwest), or interspersed with pinyons (*Pinus edulis*) and junipers (*Juniperus* sp.; west, higher elevations, and near mountains). They are also regularly encountered in suburbs or commercial areas that provide abundant prey, including Albuquerque, Santa Fe, and Las Cruces.

The only extensive study of radio-tagged Merlins was conducted on birds that remained in and near Saskatoon, Saskatchewan (Warkentin and Oliphant 1990). Winter home ranges for five adult females averaged 5 km^2 (2 mi^2), while those of three juvenile males averaged 8.5 km^2 (3.3 mi^2). While all radio-tagged Merlins wintered mostly within the city, several made excursions as far as 15 km (10 mi) to farmyards to hunt. Habitat selection by Merlins wintering elsewhere—including in New Mexico—is poorly quantified, but they generally frequent the same types of open forests, grasslands, and suburban areas as in summer, seeking vulnerable smaller (<50 g) birds (Warkentin et al. 2005). Merlins also frequent grain elevators and farm outbuildings where spillage attracts small passerine prey (Fox 1964). In Utah, most (51 of 75) Taiga Merlins were seen in urban (residential/light commercial) habitats, while most (62 of 79) Prairie Merlins, including all females,

PHOTO 23.10

Merlin southwest of Cimarron, Colfax Co., 18 December 2006. In New Mexico as elsewhere, Merlins are found in natural grasslands, in or along agricultural fields, and in urban areas, in short wherever their small-bird prey are concentrated.

PHOTOGRAPH: © ELTON M. WILLIAMS.

were seen in rural areas (fields, dairies), and the few Black Merlins showed no preference (Haney and White 1999). The authors of the study attributed these preferences to wintering Prairie and Taiga Merlins seeking habitats similar to those used during breeding.

In the short grass prairie of rural Montana and Alberta, the diet of breeding Prairie Merlins consisted predominantly of small grassland birds, particularly Horned Larks (*Eremophila alpestris*), which comprised 50% of 2,070 Alberta prey items (Hodson 1978) and 27% of 427 prey identified in Montana (Becker 1985). Chestnut-collared Longspurs (*Calcarius ornatus*: 37%) were the other important avian prey species in Alberta, while Lark Buntings (*Calamospiza melanocorys*: 18%), Vesper Sparrows (*Pooecetes gramineus*: 13%), and Mountain Bluebirds (*Sialia currucoides*: 7%) were also regularly preyed on in Montana. In New Mexico, most of these species are common to abundant migrants and/or winter residents in grasslands and along woodland edges and therefore likely prey of wintering Merlins.

Merlins often use topography to surprise prey, appearing suddenly over small hills and banks, then out-flying startled prey flushed from the ground (Warkentin et al. 2005). Saskatoon Merlins in winter were only successful on 13% (adults) and 7% (juveniles) of their hunts (Warkentin and Oliphant 1990). In New Mexico, Prairie Merlins are often found perched on ubiquitous roadside fence posts. I have also seen them waiting near corrals, stock tanks, and dairies. The dark Merlin I saw in Eldorado in 1996 likely tried to take advantage of off-guard birds at the many feeders in the subdivision.

PHOTO 23.11

Male Merlin with Bewick's Wren (*Thryomanes bewickii*), Bosque del Apache National Wildlife Refuge, November 2002. The Merlin's diet consists mostly of birds. PHOTOGRAPH: © GARY K. FROEHLICH.

PHOTO 23.12

Merlin with dragonfly on Vancouver Island, British Columbia, Canada, July 2007. The diet of Merlins also includes large insects. PHOTOGRAPH: © MIKE YIP (VANCOUVERISLANDBIRDS.COM).

Status and Management

A large amount of research and monitoring data points to a Merlin population increase and breeding range expansion in North America. Christmas Bird Counts in particular provide a series of very interesting "snapshots" of Merlin abundance during a three-week period from mid December to early January of each winter. The total Merlin counts for New Mexico have increased in recent decades; only twice since 1990 have fewer than ten been counted, with noticeable peaks of 25 in 1999–2000, 28 in 2000–2001, 29 in 2006, and 26 in 2007 (National Audubon Society 2009). Normalized to Merlins/party hour (fig. 23.1) and regressed through time, the trend is highly significant ($P < 0.01$) and positive over the last 50 years. Further, if split by the clear low point of count 74 (1973–74; fig. 23.1), the trend from 1960 to 1973 was significantly ($P < 0.05$) negative, followed by a highly significant ($P < 0.01$) positive trend from 1973 to 2007. The 1973 low approximates the banning of DDT as a pesticide (see below). The positive trend from 1960 to 2007 and particularly since 1990 mirrors similar upward trends in both urban (Sodhi et al. 1992) and rural (Houston and Hodson 1997) breeding populations of Prairie Merlins, and positive trends for both Prairie and Taiga subspecies as measured by Canadian BBS data (Kirk and Hyslop 1998; Downes 2003). They also mirror migration count trends over the Sandia and Manzano mountains. Between 1985 and 2001, an average of 23 (+/- 14) Merlins were counted over the Manzano Mountains during autumn migration and 9 (+/- 7) were counted over the Sandia Mountains in spring migration. While means were low, numbers gradually increased through the study period, resulting in statistically significant positive trends for both southward fall and northward spring migration (Hoffman and Smith 2003).

As already stated, more birdwatchers over time likely contributed—at least initially—to an increase over the years in the number of Merlin sightings in New Mexico. Nonetheless, this increase in sightings most likely also reflects positive population trends as noted previously, including in New Mexico, through CBCs and migration counts. At the same time, year-to-year fluctuations continue to be noted in the number of winter sightings in New Mexico.

Not only has the Merlin increased in numbers in North America, but it has also expanded—or perhaps recolonized—its breeding distribution, particularly southward, with some intriguing potential implications for New Mexico. Taiga Merlins expanded their breeding range several hundred kilometers into Maine, New Hampshire, and Vermont by 2005 (Warkentin et al. 2005) and into New York (Paxton et al. 2008) and north-central Pennsylvania by 2007 (Boone et al. 2008). There were no breeding records in the northeastern United States before 1993, and most new nest sites have been in suburbia (Warkentin et al. 2005). Merlins purportedly also nested in southwestern Colorado in 1887 (Morrison 1889), though confirming evidence (i.e., a collected egg set) is not available. Much more recently, an adult pair was present in June in south-central Utah in 1984, far from known breeding areas (Sailer 1987). If the species can expand (recolonize) its range in the East, why not the West, especially if some breeding occurred historically based on the purported 1887 record? While nesting in New Mexico still seems very unlikely, intermittent

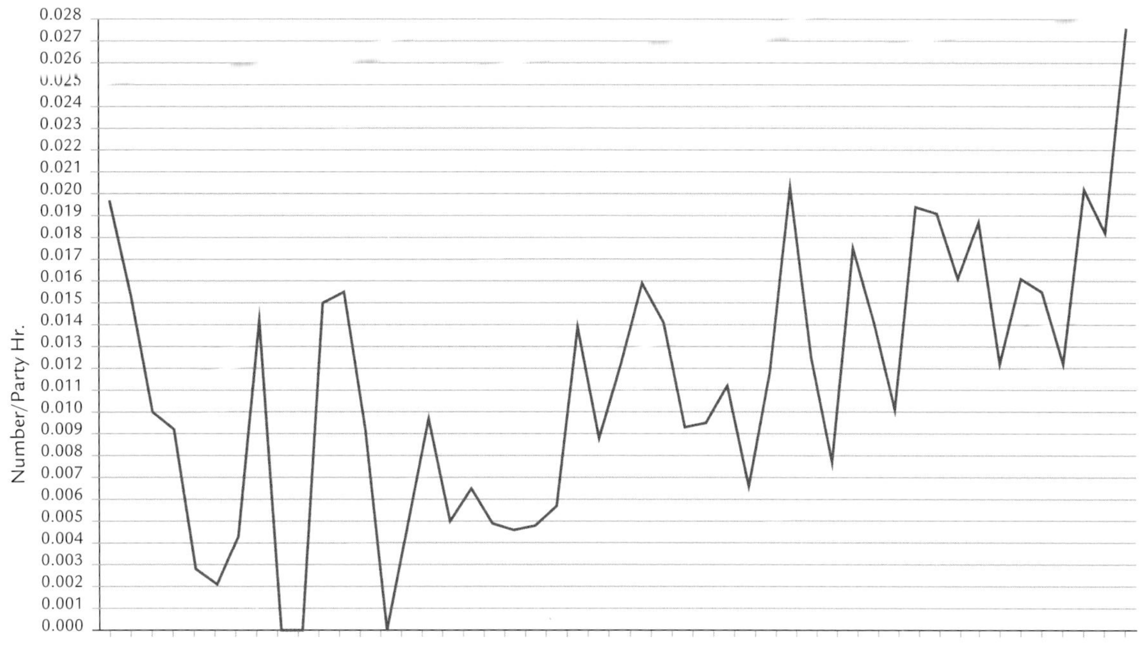

FIGURE 23.1

Number of Merlins per party hour on New Mexico Christmas Bird Counts, 1960–2008.

summer sightings leave open that remote possibility. Most intriguing were recent (2007, 2008) mid May sightings near Chama, Rio Arriba County (D. Krueper and J. Ruth, unpubl. field notes), given that the habitat is similar to that used by Merlins nesting farther north (Warkentin et al. 2005).

The Merlin's population increases coincide with the ban of DDT. Much like its cousin the Peregrine Falcon (*Falco peregrinus*), another bird-eating raptor, the Merlin suffered reproductive declines prior to the banning of DDT in the 1970s (Fox 1971). While reproduction over the past few decades has led to a population increase, DDE levels >5 mg/kg were still noted in 36% of 1980s Merlin eggs (Noble and Elliot 1990). Total pesticide use—not just organochlorines like DDT—south of the United States was statistically correlated with declines and rises in New Mexico Peregrine Falcon reproductive success through 2001, after which the data on pesticide use became unavailable (Johnson and Williams 2006). Contaminants come back into the United States and Canada in the prey that winter farther south, even if the Merlins do not. Thus the importance of southern U.S. wintering areas for both Merlins and the birds they eat cannot be overemphasized. As long as New Mexico remains a primarily rural state, with mostly grazing lands rather than tilled croplands, it should continue to provide suitable wintering habitat for Merlins and their prey.

LITERATURE CITED

Andrews, R., and R. Righter. 1992. *Colorado birds*. Denver, CO: Denver Museum of Natural History.

Bailey, A. F. M. 1942. The Black Pigeon Hawk in Colorado. *Condor* 44:37.

Bailey, F. M. 1928. *Birds of New Mexico*. Santa Fe: New Mexico Department of Game and Fish.

Bauer, E. N. 1982. Winter roadside raptor survey in El Paso County, Colorado, 1962–1979. *Raptor Research* 16:10–13.

Becker, D. M. 1985. Food habits of Richardson's Merlin in southeastern Montana. *Wilson Bulletin* 97:226–30.

Boone, A., V. W. Fazio III, and R. Wiltraut. 2008. Regional report, Eastern Highlands and Upper Ohio River Valley. *North American Birds* 61(4): 581–84.

Clark, W. S., and B. K. Wheeler. 1987. *A field guide to hawks of North America*. Boston: Houghton Mifflin Co.

Downes, C. M. 2003. Population trends in raptors from the Breeding Bird Survey. *Bird Trends* 9:9–12.

Eakle, W. L., E. L. Smith, S. W. Hoffman, D. W. Stahlecker, and R. B. Duncan. 1996. Results of a raptor survey in southwestern New Mexico. *Journal of Raptor Research* 30:183–88.

Fox, G. S. 1964. Notes on the western race of the Pigeon Hawk. *Blue Jay* 22:140–47.

———. 1971. Recent changes in the reproductive success of the Pigeon Hawk. *Journal of Wildlife Management* 35:122–28.

Haney, D. L., and C. M. White. 1999. Habitat use and subspecific status of Merlins, *Falco columbarius*, in central Utah. *Great Basin Naturalist* 59:266–71.

Hodson, K. A. 1978. Prey utilized by Merlins nesting in shortgrass prairie in southern Alberta. *Canadian Field-Naturalist* 92:76–77.

Hoffman, S. W., and J. P. Smith. 2003. Population trends of migratory raptors in western North America, 1977–2001. *Condor* 105:397–419.

Houston, C. S., and K. A. Hodson. 1997. Resurgence of breeding Merlins (*Falco columbarius richardsonii*) in Saskatchewan grasslands. *Canadian Field-Naturalist* 111:243–48.

Hubbard, J. P. 1978. The status of the Merlin in New Mexico. *New Mexico Ornithological Society Bulletin* 6(2): 12–15.

Jewett, S. G. 1944. The Black Pigeon Hawk in New Mexico. *Condor* 46:206.

Johnson, T. H., and S. O. Williams III. 2006. *The Peregrine Falcon in New Mexico—2005*. Santa Fe: New Mexico Department of Game and Fish.

Kimsey, B., and M. R. Conley. 1988. Habitat use by raptors in south central New Mexico. In *Proceedings of the southwest raptor management symposium and workshop*, ed. R. L. Glinski, B. G. Pendleton, M. B. Moss, M. N. LeFranc Jr., B. A. Millsap, and S. W. Hoffman, 197–203. Washington, DC: National Wildlife Federation.

Kirk, D. A., and C. Hyslop. 1998. Population status and recent trends in Canadian raptors: a review. *Biological Conservation* 83:91–118.

Morrison, C. F. 1888. A list of some birds of La Plata County, Colo., with annotations. *Ornithologist and Oologist* 13:70–75, 107–8, 115–16, 139–40.

National Audubon Society. 2009. The Christmas Bird Count historical results. http://www.audubon.org/bird/cbc (accessed 30 January 2009).

[NMOS] New Mexico Ornithological Society. 2007. *NMOS Field Notes* database. http://nhnm.unm.edu/partners/NMOS/ (accessed 27 January 2009).

Noble, D. G., and J. E. Elliot. 1990. Levels of contaminants in Canadian raptors, 1966 to 1988, effects and temporal trends. *Canadian Field-Naturalist* 104:222–43.

Oliphant, L. W. 1998. Merlin. In *The raptors of Arizona*, ed. R. L. Glinski, 121–23. Tucson: University of Arizona Press.

Paxton, R. O., R. R. Veit, and F. Rohrbacher. 2008. Regional report, Summer 2007: Hudson-Delaware. *North American Birds* 61(4): 569–72.

Sailer, J. 1987. Adult pair of Merlins in southern Utah in June. *Journal of Raptor Research* 21:38–39.

Schmutz, J. K., R. W. Fyfe, U. Banasch, and H. Armbruster. 1991. Routes and timing of falcons banded in Canada. *Wilson Bulletin* 103:44–58.

Sibley, D. A. 2000. *The Sibley guide to the birds*. New York: Chanticleer Press.

Snider, P. R., ed. 1992. *New Mexico Ornithological Society Field Notes* 31(4): 1 August to 30 November.

Sodhi, N. S., P. C. James, I. G. Wartentin, and L. W. Oliphant. 1992. Breeding ecology of urban Merlins (*Falco columbarius*). *Canadian Journal of Zoology* 70:1477–83.

Warkentin, I. G., and L. W. Oliphant. 1990. Habitat use and foraging behaviour of urban Merlins (*Falco columbarius*) in winter. *Journal Zoology, London* 221:539–63.

Warkentin, I. G., N. S. Sodhi, R. M. Espie, A. F. Poole, L. W. Oliphant, and P. C. James. 2005. Merlin (*Falco columbarius*). In *The birds of North America online*, ed. A. Poole. Ithaca, NY: Cornell Laboratory of Ornithology. http://bna.birds.cornell.edu/bna/species/044.

Williams, S. O. III. 1995. Regional report, winter 1994–1995: New Mexico. *National Audubon Society Field Notes* 49(2): 178–80.

———. 1996. Regional report, winter 1995–1996: New Mexico. *National Audubon Society Field Notes* 50(2): 203–6.

———. 2001. Regional report, winter 2000: New Mexico. *North American Birds* 55(2): 209–12.

———. 2006. Regional report, autumn 2005: New Mexico. *North American Birds* 60(1): 112–15.

———. 2008. Regional report, autumn 2007: New Mexico. *North American Birds* 62(1): 279–82.

24

Aplomado Falcon

(*Falco femoralis*)

KENDAL E. YOUNG AND QUINN H. YOUNG

THE APLOMADO FALCON (*Falco femoralis*) is a rare resident of New Mexico's semidesert grasslands. Where it occurs, however, this falcon stands out even from a distance with its colorful wing and tail feathers, its yellow legs, and a long, banded tail. It is a medium-sized falcon, approximately 35–45 cm (14–18 in) in length, with a wingspan of 78–102 cm (31–40 in) (Wheeler and Clark 1996). *Aplomado* is Spanish for the gray or "lead" color on the upperparts that distinguishes the species from other medium-sized falcons. Other distinctive traits of adults include a dark belly band separating the white upper breast from a cinnamon belly and a distinct, striped facial pattern that is more pronounced than in other falcons. Overall, juveniles appear similar to adults, but they have more brown in the upper back and wings with heavy dark streaks on a buff-colored breast. Females are slightly larger than males (Keddy-Hector 2000).

The Aplomado Falcon usually flies low, swiftly, and straight with rapid wing beats when hunting, much like other falcon species. However, it can be readily identified by the narrow, white trailing edge of its long, dark wings. Compared to the Peregrine Falcon (*Falco peregrinus*), Prairie Falcon (*Falco mexicanus*), or Merlin (*Falco columbarius*), its tail is also proportionally longer and narrower (Keddy-Hector 2000).

PHOTO 24.1

Adults have lead-colored upperparts, a dark belly band separating the white upper breast from a cinnamon belly, yellow eye-rings and ceres, and a distinct, striped facial pattern that is more pronounced than in other falcons. Males and females have similar plumages, but females are larger and have fine streaking on the breast (the males typically have an immaculate white breast). Here, an adult male (based on size and lack of streaking on breast), Luna Co., 1 October 2008. PHOTOGRAPH: © DAVID J. GRIFFIN.

PHOTOS 24.2a and b

Juvenile Aplomado Falcon at the Bosque del Apache National Wildlife Refuge, Socorro Co. (from the Armendaris Ranch's reintroduced population), 23 November 2007. Juveniles are similar to adults, but show wide, dark streaks on the breast. The upperparts are brown. The ceres and eye-rings are bluish. PHOTOGRAPHS: © DOUG BROWN.

PHOTO 24.3

Aplomado Falcon (reintroduced bird from the Armendaris Ranch) at the Bosque del Apache National Wildlife Refuge, 21 November 2007. The Aplomado Falcon has a long tail with six or more narrow pale bands (white in adults, buffy in juveniles). PHOTOGRAPH: © LANA HAYS.

The Aplomado Falcon is larger than the Merlin and the American Kestrel (*Falco sparverius*), the latter also sporting a double rather than single mustache and hovering more frequently and more effortlessly. None of the above falcon species has the dark belly band or the bold head pattern of the Aplomado Falcon.

Distribution

Three recognized subspecies of the Aplomado Falcon occur in the Western Hemisphere: *Falco f. pichinchae*, *F. f. femoralis*, and the Northern Aplomado Falcon, *F. f. septentrionalis*. *F. f. pichinchae* is found in western South America while *F. f. femoralis* occurs in portions of South and Central America (Keddy-Hector 1990). The Northern Aplomado Falcon is a resident of savannas and grasslands from the southwestern United States south to Nicaragua in Central America (Keddy-Hector 2000).

Documented historical occurrences indicate that the Northern Aplomado Falcon (referred to hereafter as Aplomado Falcon) occurred in Mexico along the east coast from Tamaulipas to Yucatán, and along lowlands of the west coast from Sinaloa and Nayarit to Oaxaca (Lawrence 1874; Cade et al. 1991; Keddy-Hector 2000).

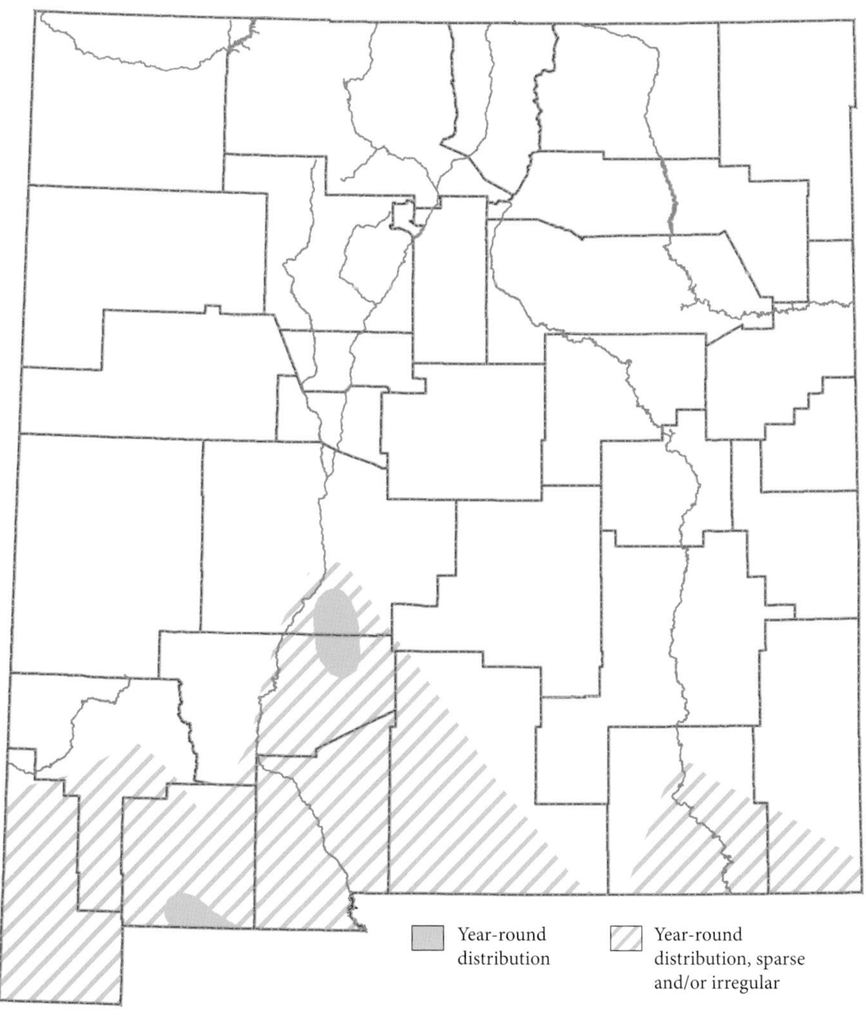

MAP 24.1

Aplomado Falcon distribution map

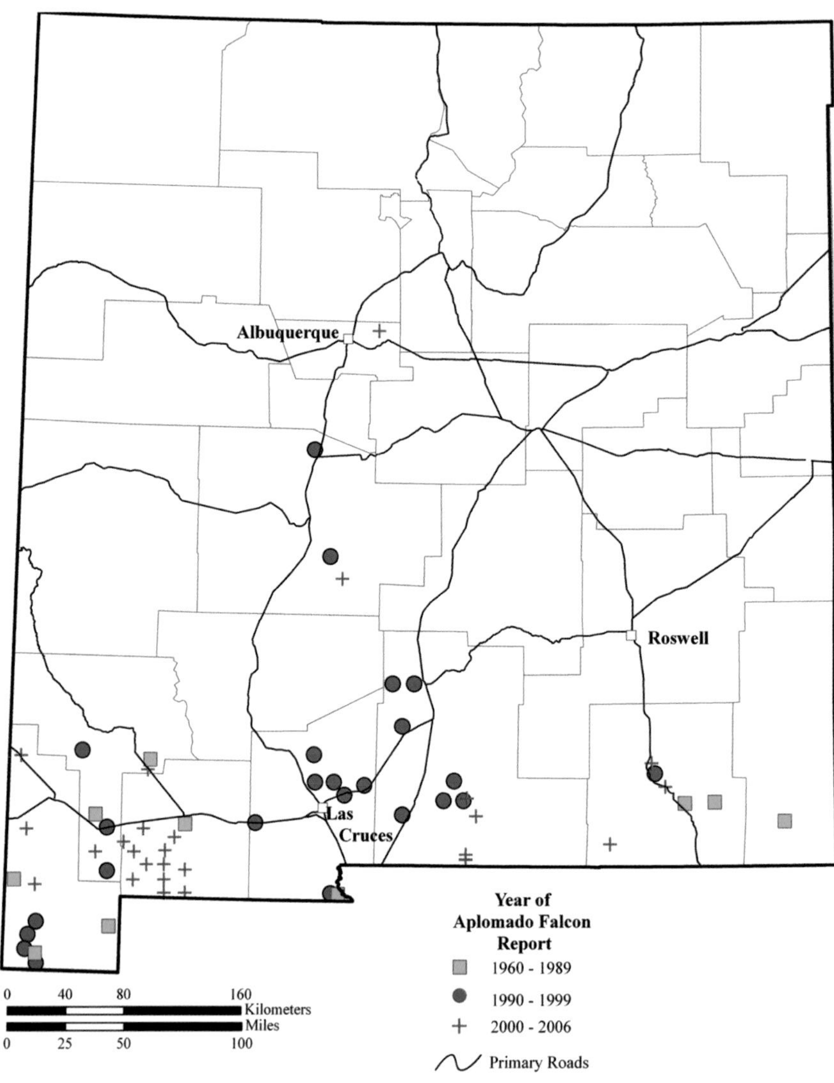

MAP 24.2

Aplomado Falcon reports in New Mexico from 1960 to 2006. Map redrawn from data provided by Meyer and Williams (2005) and unpublished data from the Bureau of Land Management.

The Aplomado Falcon's historical distribution in the interior of Mexico, however, is unclear due to a lack of documented historical occurrences in the Chihuahuan Desert of the Mexican Plateau (Young et al. 2004).

In the United States, early records indicate that Aplomado Falcons occupied savannas and grasslands in southern and western Texas, southern New Mexico, and southeastern Arizona (AOU 1998; Keddy-Hector 2000). The historical distribution of Aplomado Falcons in New Mexico included parts of Hidalgo, Grant, Luna, Doña Ana, Sierra, Socorro, Otero, Eddy, and Lea counties (Bendire 1892; Bailey 1928; Bent 1938; Ligon 1961; maps 24.1 and 24.2). Early authorities considered Aplomado Falcons fairly common throughout their

PHOTO 24.4

(*top left*) There was an increase in sightings of Aplomado Falcons in the 1990s in New Mexico. That trend has continued into the 2000s. Here, an Aplomado Falcon at Hermanas, Luna Co., November 2005.

PHOTOGRAPH: © JERRY OLDENETTEL.

PHOTOS 24.5a and b

(*top and bottom right*) An adult male in Luna Co., 1 October 2008.

PHOTOGRAPHS: © DAVID J. GRIFFIN.

range (Bent 1938). However, by the early 1900s Bailey (1928) and Ligon (1961) believed that the falcon's distribution in New Mexico had become restricted to the southwestern portion of the state. Knowledge of historical nesting locations in the northern Chihuahuan Desert, including New Mexico, is sparse. Arnold Bayne documented two nests in 1952. One nest was located in Chihuahua, Mexico, to the southwest of Antelope Wells, New Mexico (Truett 2002). The other documented nest was the last known historical falcon nest in New Mexico. This nest was located in Luna County, south of Deming (Ligon 1961).

Current resident Aplomado Falcon populations exist primarily in Mexico. Extant populations occur along the Gulf of Mexico to the Yucatán Peninsula (Howell and Webb 1995; Keddy-Hector 2000) and in the northern Chihuahuan Desert from north-central Chihuahua north to southern New Mexico (Young et al. 2004; Meyer and Williams 2005). Falcon sightings in New Mexico have become frequent since around the early 1990s (Williams 1993, 1994, 1996, 1997, 2000). There is also evidence of falcons dispersing from Chihuahua, Mexico, to New Mexico. For example, in 1999, a juvenile banded in Chihuahua was found on

Otero Mesa in southern New Mexico (approximately 280 km [174 mi] from where the bird was banded).

Meyer and Williams (2005) summarized reliable Aplomado Falcon reports in New Mexico from 1960 to 2004. Their summary documented 53 falcon reports, 10 of which occurred between 1960 and 1989. Starting in 1990, there was an increase in falcon sightings, with at least 24 reliable reports of Aplomado Falcons through 1999. This positive trend has continued into the early 2000s, with at least 19 falcon reports between 2000 and 2004, representing 24 birds, which included four pairs (Meyer and Williams 2005). In addition, there were seven reports of Aplomado Falcons in New Mexico in 2005–6 (Bureau of Land Management data on file). While most of the falcon reports between 1990 and 1999 were in Doña Ana and Otero counties, falcon reports since 2000 have been primarily in Hidalgo, Grant, Luna, and Otero counties.

PHOTO 24.6

Four juveniles hacked in 2008, together on or over a yucca at the Armendaris Ranch, 15 August 2008. The Peregrine Fund and the U.S. Fish and Wildlife Service began an Aplomado Falcon reintroduction program in New Mexico in 2006. Hacking consists of releasing birds into the wild, initially provisioning them with food while they gradually become independent. PHOTOGRAPH: © PETER STACEY.

PHOTO 24.7

Aplomado Falcon releases began in 2006 on the Armendaris Ranch in Socorro and Sierra counties. In 2007, releases also occurred on land administered by the Bureau of Land Management, White Sands Missile Range, and the New Mexico State Land Office. A total of 11 falcons were released in New Mexico in 2006, 39 in 2007. Here, a juvenile from the reintroduced population, photographed at the Bosque del Apache National Wildlife Refuge, 21 November 2007. PHOTOGRAPH: © LANA HAYS.

Perhaps the most exciting recent discovery of Aplomado Falcons in New Mexico occurred in 2000, when nesting was observed in Luna County. A falcon pair made two unsuccessful nesting attempts in 2001, but in 2002 a pair successfully fledged three young in the same territory (Meyer and Williams 2005). This was the first successful nest recorded in New Mexico since 1952.

The Peregrine Fund and the U.S. Fish and Wildlife Service began an Aplomado Falcon reintroduction program in New Mexico in 2006 to augment falcon populations in the northern Chihuahuan Desert. Falcon reintroductions were initiated on a private ranch (Armendaris Ranch) in Sierra and Socorro counties and were expanded in 2007 to include land administered by the Bureau of Land Management, White Sands Missile Range, and the New Mexico State Land Office (Angel Montoya, The Peregrine Fund, pers. comm.). A total of 11 falcons were released in New Mexico in 2006, and 39 more falcons were released in 2007. A falcon pair, originating from the 2006 release, nested and successfully fledged two young on the Armendaris Ranch in 2007 (Angel Montoya, The Peregrine Fund, pers. comm.).

Habitat Associations

Aplomado Falcons are associated with savannas and grasslands with a sparse canopy of woody vegetation (Keddy-Hector 2000). In New Mexico in particular, falcons are associated with semidesert grasslands in the Chihuahuan Desert. These grasslands are characterized by scattered yuccas, mesquite, and cactus (Bendire 1892; Ligon 1961; Meyer and Williams 2005). Because nests have been observed so infrequently in New Mexico, more detailed characterizations of nesting habitat in the state are difficult to generate. However, given the close proximity of occupied habitats in Chihuahua, Mexico, to New Mexico, and evidence of falcons dispersing from Chihuahua to New Mexico, detailed habitat associations derived from occupied sites in Chihuahua can reasonably be applied to New Mexico. In support of this, Meyer and Williams (2005) described the nest site in Luna County as similar to occupied falcon sites in northern Chihuahua. Habitat surrounding the Luna County nest site consisted of a large open tobosa *(Pleuraphis mutica)* swale adjacent to a desert grassland/shrubland mosaic.

Several studies have described Aplomado Falcon

PHOTO 24.8

Aplomado Falcon in yucca grassland on the Armendaris Ranch, 28 June 2007. In New Mexico, the primary habitat of the Aplomado Falcon consists of semidesert grasslands with scattered yuccas (*Yucca*), mesquites (*Prosopis glandulosa*), and/or cacti (*Cylinbropuntia*).

PHOTOGRAPH: © PETER STACEY.

PHOTO 24.9

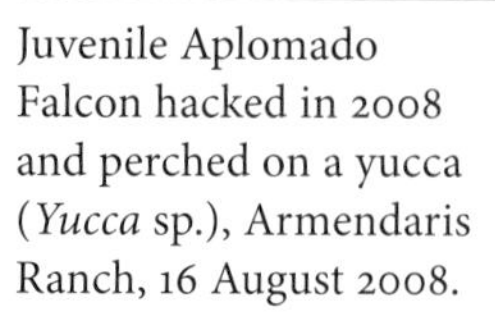
Juvenile Aplomado Falcon hacked in 2008 and perched on a yucca (*Yucca* sp.), Armendaris Ranch, 16 August 2008.
PHOTOGRAPH: © PETER STACEY.

habitat use in the northern Chihuahuan Desert of Mexico (Montoya et al. 1997; Young et al. 2002, 2004). These studies concluded that Aplomado Falcons primarily occupied grasslands comprised of grama (*Bouteloua* spp.) or tobosa that exhibit low woody plant densities of longleaf ephedra (*Ephedra trifurca*), soaptree yucca (*Yucca elata*), or honey mesquite (*Prosopis glandulosa*). Although the amount of grass cover at occupied falcon sites varies spatially and temporally, percent grass basal cover at falcon sites has been estimated to be around 40%, with an overall vegetation ground cover between 30 and 70% (Montoya et al. 1997; Young et al. 2002). Likewise, woody plant density is typically less than 300 plants per hectare at nest sites. These grassland habitat types appear to be selected disproportionately to their availability on the landscape (Young et al. 2002, 2004).

A recent study modeled the amount of suitable Aplomado Falcon habitat and predicted approximately 542,300 ha (1,340,047 ac) in New Mexico (Young et al. 2005). Key areas of suitable habitat in New Mexico include Otero Mesa, grasslands south of Silver City, the southern Animas Valley, the Stallion area on White Sands Missile Range, and the Jornada del Muerto basin to the north of Las Cruces and south of Socorro (Young et al. 2005).

Life History

Our understanding of life history requirements for Aplomado Falcons in New Mexico has been derived primarily from observations and studies of the falcon population that extends from north-central Chihuahua to southern New Mexico. Future generations of reintroduced falcons in New Mexico may provide the opportunity to learn how this species interacts with the New Mexico landscapes and human activities and disturbances.

Nesting

Aplomado Falcons are secondary nesters that rely on nests previously constructed by other raptors or ravens (Keddy-Hector 2000). These nests are typically stick nests composed of local woody plant material. Aplomado Falcon nests in the Chihuahuan Desert have been typically located in soaptree yuccas but also have occurred in Torrey yucca (*Yucca torreyi*), honey mesquite, and netleaf hackberry (*Celtis reticulata*), and on power poles (Montoya et al. 1997; Young et al. 2002, 2004). Plant structure appears to be an important element for nesting, and most nesting Aplomado Falcons have been found in large and complex soaptree yuccas (Young et al. 2002; Desmond et al. 2005). The recent falcon nests discovered in Luna County were in soaptree yuccas and in little-leaf sumac (*Rhus microphylla*) (Meyer and Williams 2005). The pair that nested on the Armendaris Ranch in 2007 used a nest on a power pole (Angel Montoya, The Peregrine Fund, pers. comm.).

Aplomado Falcons nest from February to June, and lay two or three eggs (Keddy-Hector 2000; Duarte et al. 2004). The incubation period lasts approximately 31–33 days with the female incubating most of the time; however, the male may also take part in the incubation (Keddy-Hector 2000). Nestlings fledge four to five weeks after hatching (Keddy-Hector 2000).

Diet and Foraging

The diet of Aplomado Falcons consists mainly of small birds and various insects, less predominantly small mammals, reptiles, and amphibians (Bendire 1892; Ligon 1961; Hector 1985; Montoya et al. 1997). Avian prey includes doves, cuckoos, woodpeckers, nighthawks, blackbirds, flycatchers, and thrushes. Nonavian prey includes mice, rats, bats, lizards, frogs, beetles, cicadas, grasshoppers, dragonflies, crickets, butterflies, moths, wasps, and bees (Hector 1985). Montoya et al. (1997) observed that seven bird taxa accounted for 68% (frequency of occurrence) of the falcon's avian diet in the Chihuahuan Desert in Mexico: mainly meadowlark (*Sturnella*

PHOTO 24.10

Armendaris Ranch near the 2007 power-pole nest, 29 June 2007. The power line with the nest power pole is visible in the background. From the nest the adults moved their brood to a nearby juniper (*Juniperus* sp.). Note the two birds in the center of the photograph (and see inset; apparently an adult is feeding a fledgling).

PHOTOGRAPH: © PETER STACEY.

PHOTO 24.11

Aplomado Falcons prey mainly on small birds and insects. In the reintroduction program, newly released Aplomado Falcons learn to hunt in their new natural surroundings but are also fed quails for several weeks at stations installed specially for them. Here, an Aplomado Falcon feeding on a quail at the Armendaris Ranch, Sierra Co., 1 November 2006.

PHOTOGRAPH: © GORDON FRENCH.

spp.), followed by Common Nighthawk (*Chordeiles minor*), Northern Mockingbird (*Mimus polyglottos*), Western Kingbird (*Tyrannus verticalis*), Brown-headed Cowbird (*Molothrus ater*), Scott's Oriole (*Icterus parisorum*), and Mourning Dove (*Zenaida macroura*).

Aplomado Falcons typically chase small birds and insects during flights initiated from perches. Prey is either captured in midair or forced to the ground, where pursuits may continue on foot when the prey enters thick cover (Hector 1985). Aplomado Falcons often hunt cooperatively in pairs (Hector 1986; Silvera et al. 1997) and are known to steal prey from other birds (Hector 1985; Clark et al. 1989).

Predation and Interspecific Interactions

Aplomado Falcons rely on nests constructed by corvids or by other raptors such as Swainson's Hawks (*Buteo swainsoni*), Red-tailed Hawks (*Buteo jamaicensis*), Harris's Hawks (*Parabuteo unicinctus*), and White-tailed Kites (*Elanus leucurus*). With the exception of the Harris's Hawk and the White-tailed Kite, all of these birds were common at Aplomado Falcon sites in the northern Chihuahuan Desert of Mexico (Young et al. 2004). Thus Aplomado Falcon presence may be strongly influenced by the combination of vegetation structure and local densities of nesting ravens and raptors (Desmond et al. 2005).

Is predation an important cause of Aplomado Falcon mortality? Potential predators of adults, juveniles, or eggs include Great Horned Owls (*Bubo virginianus*), Chihuahuan Ravens (*Corvus cryptoleucus*), coyotes (*Canis latrans*), and bobcats (*Lynx rufus*) (Keddy-Hector 2000; Meyer and Williams 2005). In New Mexico, however, there have been no documented cases of predation on Aplomado Falcons. Predation rates are currently unknown even outside New Mexico.

Status and Management

Specimen records and documented sightings indicate that Aplomado Falcons were common throughout their range in the United States until around the 1930s, at which point sightings began to decline (Bailey 1928; Bent 1938; Ligon 1961). The Aplomado Falcon was first considered as a candidate for federal listing under the Endangered Species Act in 1973 (U.S. Department of the Interior 1973). However, sufficient information was not available until March 1986 to support listing of the Northern Aplomado Falcon as Endangered (U.S. Fish and Wildlife Service 1986). The falcon was added to the Endangered Species List due to suspected extirpation from the United States. Factors attributed to the falcon's decline were primarily habitat degradation due to brush encroachment, and secondarily egg and specimen collecting and continued pesticide contamination (DDT) within the range of the falcon (Kiff et al. 1980; U.S. Fish and Wildlife Service 1986; Cade et al. 1991).

As of the late 1990s, the closest known free-ranging Aplomado Falcon population to New Mexico appeared to be in northern Chihuahua, Mexico (Montoya et al. 1997). In Mexico, Aplomado Falcons were listed as endangered in 1994, but in 2002, their status was changed to "Subject to Special Protection" (DOF-Semarnat 1994, 2002). Subject to Special Protection status is given to a species or population that could become threatened from factors that negatively affect their viability (DOF-Semarnat 2002). In New Mexico, the Aplomado Falcon is state-listed as Endangered (NMDGF 1996).

In 1990, the U.S. Fish and Wildlife Service released the recovery plan for the Northern Aplomado Falcon

PHOTO 24.12

Juvenile Aplomado Falcon near Cain hack tower on White Sands Missile Range, Socorro Co., 9 August 2007.

PHOTOGRAPH: © DOUG BURKETT.

PHOTO 24.13

Hack tower on the Armendaris Ranch, Sierra Co., 3 August 2006.

PHOTOGRAPH: © DOUG BURKETT.

(Keddy-Hector 1990). The main focus of the plan was to provide guidelines toward the goal of downlisting the species to Threatened status within two to four decades. The recovery plan allowed for downlisting to Threatened status after obtaining a self-sustaining population of a minimum of 60 breeding pairs in the United States. However, this number was—still is—tentative, and could be changed as new information warrants.

Aplomado Falcon recovery efforts began as early as 1977, with the development of a captive breeding and reintroduction program (Cade et al. 1991). Falcons for the captive breeding program were obtained from Chiapas, Tabasco, and Veracruz, Mexico, and sent to The Peregrine Fund's breeding facility in Idaho. Captive-bred falcons were first released by The Peregrine Fund in southern Texas in 1985 and in western Texas in 2002. There are currently breeding falcons in both these core release areas.

A controversial Aplomado Falcon reintroduction program began in 2006 in New Mexico under section 10(j) of the Endangered Species Act. At the root of the controversy was the issue of whether reintroductions in New Mexico were necessary if natural recolonization was already occurring. Evidence pointing toward natural recolonization included a nesting pair in New Mexico, an increase in falcon sightings in the southern part of the state (Williams 2000; Meyer and Williams 2005), and the presence of a falcon population that extended into southern New Mexico from Chihuahua, Mexico (Young et al. 2004).

While the U.S. Fish and Wildlife Service did not dismiss the possibility of natural recolonization, the agency believed that unaided recovery would take decades to achieve, and possibly would not occur naturally (U.S. Fish and Wildlife Service 2006). The Fish and Wildlife Service also concluded that the nesting pair in Luna County and the recent increase in falcon sightings in New Mexico did not amount to a population (U.S. Fish and Wildlife Service 2005). This was a key argument, as reintroductions under section 10(j) can only occur outside the species' current range but within its historical distribution, and reintroductions have to be "wholly separate" from any existing population (U.S. Fish and Wildlife Service 2005).

There were also concerns that the captive-bred stock (coastal birds from Veracruz, Mexico) would result in the genetic swamping of the Chihuahuan Desert population. This concern still exists, even though there are indications that the genetic variation between the two populations is low (Jenny et al. 2004).

Finally, of local concern was the removal of habitat protection that wild native Aplomado Falcons had under the Endangered Species Act. Section 10(j) effectively changed the designation of the falcon in New Mexico and Arizona from Endangered to a "Nonessential, Experimental Population" (U.S. Fish and Wildlife Service 2005). With the exception of national parks or national wildlife refuges, the Nonessential, Experimental Population designation provides Aplomado Falcons with a proposed listing status that does not require habitat protection measures on federal lands in suitable habitats.

Our current understanding of Aplomado Falcon ecology indicates that a sustainable falcon population is likely to be detrimentally affected by the alteration and loss of grassland habitat (Young et al. 2004) and a reduction or loss of a readily available and exploitable food supply. A considerable transition in vegetation communities (primarily from perennial grassland to shrubland) has occurred in the Chihuahuan Desert since the mid 1800s (Buffington and Herbel 1965; Allred 1996; Beck and Gibbens 1999). Factors that have contributed to this transition and that have negatively affected natural processes in the Chihuahuan Desert's habitats include climatic changes, diversion of surface water, livestock grazing, introduction of exotic plant and animal species, increased urban development, and soil erosion (Beck and Gibbens 1999; NMDGF 2006).

Despite an increase in falcon sightings in New Mexico, falcon productivity in northern Chihuahua is thought to be declining. Duarte et al. (2004) reported that the number of fledglings per occupied territory decreased from 1.6 in 1997 to 0.6 in 2002. The authors noted that the decrease in productivity coincided with a reduction of prey availability in falcon territories. To the extent that the falcon population in north-central Chihuahua is an important source of dispersing birds for New Mexico, declines in falcon productivity in Chihuahua could soon result in diminished reports in our state. Unfortunately, estimates of Aplomado Falcon juvenile and adult survivorship—all representing crucially needed information—are lacking in the Chihuahuan Desert. At southern Texas release sites, Perez et al. (1996) reported a minimum mortality rate of 42% among post-released falcons.

Not surprisingly, the alteration or loss of grassland habitat negatively affects species that are grassland adapted, in New Mexico as elsewhere. Grassland conservation efforts and perhaps falcon introduction efforts represent our best tools to ensure that Aplomado Falcons once again become a common raptor in New Mexico.

Allred, K. 1996. Vegetative changes in New Mexico rangelands. *New Mexico Journal of Science* 36:169–231.

[AOU] American Ornithologists' Union. 1998. *Check-list of North American birds.* 7th ed. Washington, DC: American Ornithologists' Union.

Bailey, F. M. 1928. *Birds of New Mexico.* Santa Fe: New Mexico Department of Game and Fish.

Beck, R. F., and R. P. Gibbens. 1999. The Chihuahuan Desert ecosystem. *New Mexico Journal of Science* 39:45–85.

Bendire, C. E. 1892. *Life histories of North American birds.* U.S. National Museum Bulletin 1, 551–58. Washington, DC: Smithsonian Institution.

Bent, A. C. 1938. *Life histories of North American birds of prey.* Part 2. U.S. National Museum Bulletin 170, 96–99. Washington, DC: Smithsonian Museum.

Buffington, L. C., and C. H. Herbel. 1965. Vegetational changes on a semiarid desert grassland range from 1858 to 1963. *Ecological Monographs* 35:139–64.

Cade, T. J., J. P. Jenny, and B. J. Walton. 1991. Efforts to restore the Northern Aplomado Falcon (*Falco femoralis septentrionalis*) by captive breeding and reintroduction. *The Dodo, Journal of the Jersey Wildlife Preservation Trust* 27:71–81.

Clark, W. S., P. H. Bloom, and L. W. Oliphant. 1989. Aplomado Falcon steals prey from little blue heron. *Journal of Field Ornithology* 60:380–81.

Desmond, M. J., K. E. Young, B. C. Thompson, R. Valdez, and A. Lafón Terrazas. 2005. Habitat associations and conservation of grassland birds in the Chihuahuan Desert Region: two case studies in Chihuahua, Mexico. In *Biodiversity, ecosystems, and conservation in northern Mexico*, ed. J.-L.E. Cartron, G. Ceballos, and R. S. Felger, 439–51. New York: Oxford University Press.

DOF-Semarnat. 1994. Norma Oficial Mexicana NOM-059-ECOL-1994, que determina las especies y subespecies de flora y fauna silvestre terrestres y acuáticas en peligro de extinción, amenazadas, raras y las sujetas a protección especial, y que establece especificaciones para su protección. Diario Oficial de la Federación, Primera Sección. Mexico, D.F.

———. 2002. Norma Oficial Mexicana NOM-059-ECOL-2001, protección ambiental—especies nativas de México de flora y fauna silvestres—categorías de riesgo y especificaciones para su inclusión, exclusión o cambio—lista de especies en riesgo. Diario Oficial de la Federación, Segunda Sección. Mexico, D.F.

Duarte, A. M., A. B. Montoya, W. G. Hunt, A. L. Terrazas, and R. Tafanelli. 2004. Reproduction, prey, and habitat of the Aplomado Falcon (*Falco femoralis*) in desert grasslands of Chihuahua, Mexico. *Auk* 121:1081–93.

Hector, D. P. 1985. The diet of the Aplomado Falcon (*Falco femoralis*) in eastern Mexico. *Condor* 87:336–42.

———. 1986. Cooperative hunting and its relationship to foraging success and prey size in an avian predator. *Ethology* 73:247–57.

Howell, S. N. G., and S. Webb. 1995. *A guide to the birds of Mexico and northern Central America.* New York: Oxford University Press.

Jenny, J. P., W. Heinrich, A. B. Montoya, B. Mutch, C. Standfort, and W. G. Hunt. 2004. From the field: progress in restoring the Aplomado Falcon to southern Texas. *Wildlife Society Bulletin* 32:276–85.

Keddy-Hector, D. P. 1990. *Aplomado Falcon recovery plan.* Albuquerque, NM: U.S Fish and Wildlife Service, Region 2.

———. 2000. Aplomado Falcon (*Falco femoralis*). No. 549. In *The birds of North America*, ed. A. Poole and F. Gill. Philadelphia, PA: Birds of North America, Inc.

Kiff, L. F., D. B. Peakall, and D. P Hector. 1980. Eggshell thinning and organochloride residues in bat and aplomado falcons in Mexico. *Proceedings of the international ornithological congress* 17:949–52.

Lawrence, G. N. 1874. Birds of western and northwestern Mexico, based upon collections made by Col. A. J. Grayson, Capt. J. Xantus, and Fred. Bischoff. *Memoir Boston Society of Natural History* 2:265–319.

Ligon, J. S. 1961. *New Mexico birds and where to find them.* Albuquerque: University of New Mexico Press.

Meyer, R. A., and S. O. Williams III. 2005. Recent nesting and current status of Aplomado Falcon (*Falco femoralis*) in New Mexico. *North American Birds* 59:352–56.

Montoya, A. B., P. J. Zwank, and M. Cardenas. 1997. Breeding biology of Aplomado Falcons in desert grasslands of Chihuahua, Mexico. *Journal of Field Ornithology* 68:135–43.

[NMDGF] New Mexico Department of Game and Fish. 1996. *List of threatened and endangered species.* 19 NMAC 33.1, Amendment No.1. Santa Fe: NMDGF.

———. 2006. *Comprehensive wildlife conservation strategy for New Mexico.* Santa Fe: NMDGF.

Perez, C. J., P. J. Zwank, and D. W. Smith. 1996. Survival, movements and habitat use of Aplomado Falcons released in southern Texas. *Journal of Raptor Research* 30:175–82.

Silvera, L. A., T. A. Jacoma, H. G. Rodrigues, and P. G. Crawshaw. 1997. Hunting associations between the Aplomado Falcon (*Falco femoralis*) and the maned wolf (*Chrysocyon brachyurus*) in Emas National Park, Central Brazil. *Condor* 99:201–2.

Truett, J. C. 2002. Aplomado Falcons and grazing: invoking history to plan restoration. *Southwestern Naturalist* 47:379–400.

U.S. Department of the Interior. 1973. *Threatened wildlife of the United States.* Resource Publ. 114. Washington, DC: U.S. Bureau of Sport Fisheries and Wildlife.

U.S. Fish and Wildlife Service. 1986. Final rule: listing of the Aplomado Falcon as endangered. *Federal Register* 51:6686–90.

———. 2005. Endangered and threatened wildlife and plants; establishment of a nonessential experimental population of Northern Aplomado Falcons in New Mexico and Arizona and availability of draft environmental assessment. *Federal Register* 70:6819–28.

———. 2006. *Final environmental assessment for reestablishment of the Northern Aplomado Falcon in New Mexico and Arizona.* Albuquerque: U.S. Fish and Wildlife Service, New Mexico Ecological Services Field Office.

Wheeler, B. K., and W. S. Clark. 1996. *A photographic guide to North America raptors.* San Diego, CA: Academic Press.

Williams, S.O. III. 1993. New Mexico. *American Birds* 47:130–33.

———. 1994. New Mexico. *American Birds* 48:236–38.

———. 1996. New Mexico. *National Audubon Society Field Notes* 50:980–83.

———. 1997. Recent occurrences of Aplomado Falcons in New Mexico: is natural recolonization of historic range underway? *New Mexico Ornithological Society Bulletin* 25:39.

———. 2000. New Mexico. *North American Birds* 54:86–89.

Young, K. E., B. C. Thompson, D. M. Browning, Q. H. Hodgson, J. L. Lanser, A. Láfon Terrazas, W. R. Gound, and R. Valdez. 2002. *Characterizing and predicting suitable Aplomado Falcon habitat for conservation planning in the northern Chihuahuan Desert.* Las Cruces: New Mexico Cooperative Fish and Wildlife Research Unit.

Young, K. E., B. C. Thompson, A. Lafón Terrazas, A. B. Montoya, and R. Valdez. 2004. Aplomado Falcon abundance and distribution in the northern Chihuahuan Desert of Mexico. *Journal of Raptor Research* 38:107–17.

Young, K. E., B. C. Thompson, R. Valdez, W. R. Gould, and A. Lafón Terrazas. 2005. *Assessment of predictive values from the Aplomado Falcon habitat suitability model: validation information for conservation planning in the northern Chihuahuan Desert.* Las Cruces: New Mexico Cooperative Fish and Wildlife Research Unit.

25

Peregrine Falcon

(*Falco peregrinus*)

DALE W. STAHLECKER

NO OTHER NORTH AMERICAN raptor has captured our imagination like the fast, beautiful, and spirited Peregrine Falcon (*Falco peregrinus*). The species also attracted much public attention as it experienced catastrophic population declines throughout North America and western Europe during the 1960s and 1970s. Post–World War II use of dichloro-diphenyl-trichloroethane (DDT) had induced eggshell thinning, which in turn caused massive and widespread reproductive failure (White et al. 2002). At that time the Peregrine Falcon became a rallying point for conservation biologists and the environmental movement, leading in the United States to a nationwide ban of DDT and the passing of the Endangered Species Act in 1973.

The Peregrine Falcon is a large and powerful falcon. Gender is often easily determined by size, especially in mated pairs perched or flying together. This is because within any given subspecies there is seldom any size overlap between the sexes. Ranging from 45 to 58 cm (17–23 in) in length, females average 15–20% larger than the 39- to 49-cm (15 to 19-in) males. Females are also 40–50% heavier (White et al. 2002). Other than size, however, the sexes are identical. The adult male and adult female are both deep steel blue across the back and wings, with a black cap and black cheeks that give the impression of a Roman centurion's helmet. They are light below, varying from snowy white to peach-colored, usually also with dark barring on the belly and flanks. Juveniles wear the diagnostic helmet of the species, but generally are a rich chocolate brown rather than blue, and they usually have heavy

PHOTO 25.1

Adult Peregrine Falcon in tree, Hidalgo Co., 28 January 2003. In New Mexico, the typical adult shows what looks like a blue-gray centurion's helmet. Note also the pale, unmarked breast and the pale, barred belly and flanks.

PHOTOGRAPH: © ROBERT SHANTZ.

PHOTO 25.2

Adult male American Peregrine Falcon (subspecies *anatum*) trapped during the 1996 spring migration at HawkWatch International's Sandia Mountains study site. Adults have blue-gray upperparts and yellow ceres and eye-rings. Also yellow (but not visible on this photo) are the legs. Note the blue-gray hood, in the form of a centurion's helmet. Males and females have similar plumages.

PHOTOGRAPH: © JOHN P. DELONG.

PHOTO 25.3

Close-up of a Peregrine Falcon's (subspecies *anatum*) head and breast. The underparts can vary in color from snowy white to peach (as in this bird), with no barring on the breast. Note also the yellow cere and eye-ring and the blue-gray hood in the form of a centurion's helmet.

PHOTOGRAPH: © DOUG BROWN.

PHOTO 25.4

(*top*) Immature female (subspecies *anatum*) trapped during the 1994 spring migration at HawkWatch International's Sandia Mountains study site. Immatures have heavily streaked underparts. Immatures of this subspecies have a wide, dark mustache and small, buffy cheek patches. PHOTOGRAPH: © JOHN P. DELONG.

PHOTO 25.5

(*left*) Immature male (subspecies *anatum*) trapped during the 1995 spring migration at HawkWatch International's Sandia Mountains study site. Immatures have streaked rather than barred underparts. Whereas the breast of adults is typically unmarked, in most immatures the dark streaks of the belly and flanks extend onto the breast. PHOTOGRAPH: © JOHN P. DELONG.

PHOTOS 25.6a and b

Immature male Tundra Peregrine Falcons (*Falco peregrinus tundrius*) trapped during 1993 (a, *left*) and 1994 (b, *below*) fall migration at HawkWatch International's Manzano Mountains study site. By far the most common Peregrine Falcon sub-species found in New Mexico is the American (or continental) Peregrine Falcon (*F. p. anatum*). The Tundra Peregrine Falcon does not breed in New Mexico. It is characterized by a pale head and narrower mustache. PHOTOGRAPHS: © JOHN P. DELONG.

ventral streaking. Yearling falcons retain enough juvenile feathers to be distinguishable from full adults. Occasionally subadult females occupy territories and can even successfully rear young.

In flight, Peregrine Falcons have long, pointed wings and short tails that result in a characteristic anchor silhouette (Dunn et al. 1988; Sibley 2000). Their underwings are pale, with darker barring evident throughout. Their wing beat is rapid and strong, propelling them through space quickly in pursuit of prey or competitors. Peregrines are famous for their speed in a "stoop" or steep dive from high above their prey, in which they are capable of reaching speeds of 100 m/sec (~240 mph; White et al. 2002).

Most calls of Peregrine Falcons are associated with breeding. The *cack* or *keck* is a harsh, rapid (0.15 sec), sharp, and usually loud call most often given in defense of the nesting cliff or the nest ledge (aerie). The wail is a long (0.4–2.0 sec), slightly ascending call, commonly used in communicating hunger (begging-wail) by an incubating female or large nestlings. It is also used precopulatory and for other pair communications. The *ee-chup* is a multinote call, regularly used in breeding displays, such as when the male is suggesting potential aerie locations to his mate (White et al. 2002).

There are about 20 subspecies found throughout most of the world; three occur in North America. The most widespread of the three North American subspecies, *F. p. anatum*, had its genetic distinctiveness somewhat diluted by introductions of similarly colored subspecies from Europe, South America, and Australia. Captive breeding and introduction programs were initiated during the dark days of the

1970s when Peregrine Falcons had disappeared from the eastern United States and declined precipitously through much of the West (Cade and Burnham 2003; Craig and Enderson 2004; see also below).

Distribution

Peregrine Falcons (*Falco peregrinus*) are among the most cosmopolitan of birds. They breed on every continent but Antarctica, and outside the breeding distribution migrants reach even such isolated locations as the Hawaiian Islands in the middle of the Pacific Ocean (Pratt et al. 1987). *Peregrinus* is Latin for "foreigner," emphasizing the species' ability to appear in such far-off places.

Peregrine Falcons can be encountered anywhere in New Mexico, but they are more likely to be seen along mountain ridges in migration, near cliffs and canyons during breeding, and along bodies of water—or anywhere else with concentrations of potential prey—in winter. Breeding populations inhabit most mountain ranges in the state. Because of this, however, they are seldom encountered on Breeding Bird Surveys (BBS).

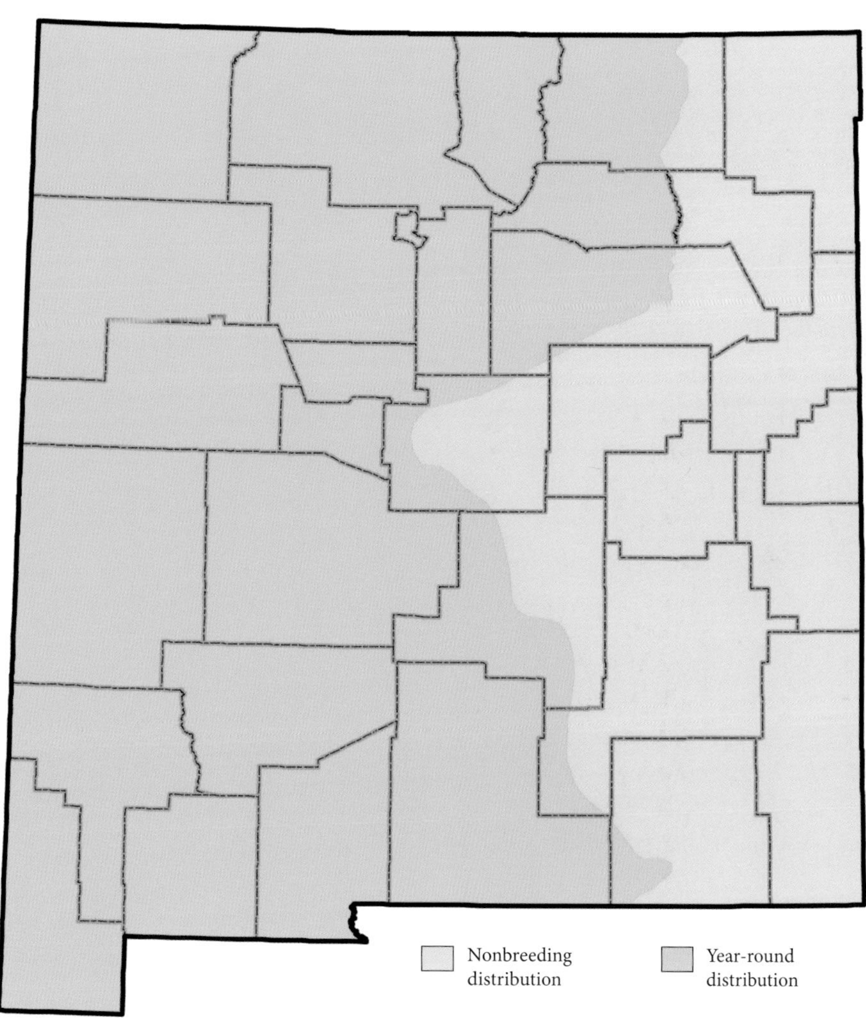

MAP 25.1

Peregrine Falcon distribution map

PHOTO 25.7

(*top left*) Peregrine Falcon in Maricopa Co., Arizona, November 2001. PHOTOGRAPH: © JIM BURNS (WWW.JIMBURNSPHOTOS.COM).

PHOTO 25.8

(*top right*) Peregrine Falcon, Hidalgo Co., 27 December 2008. PHOTOGRAPH: © ROBERT SHANTZ.

PHOTO 25.9

(*bottom left*) Peregrine Falcon at the Rio Grande Nature Center, Albuquerque, Bernalillo Co., 15 March 2007. PHOTOGRAPH: © DOUG BROWN.

PHOTO 25.10

Immature Peregrine Falcon, Bosque del Apache National Wildlife Refuge, October 2002.

PHOTOGRAPH: © GARY K. FROEHLICH.

PHOTOS 25.11a and b

Peregrine Falcon nesting habitat in San Juan Co. Peregrine Falcons nest on cliff ledges.

PHOTOGRAPHS: © JIM RAMAKKA.

Peregrine Falcons are regular spring and fall migrants in New Mexico. On average, between 1985 and 2001, 37 Peregrine Falcons were counted over the Manzano Mountains during autumn migration and 34 were counted over the Sandia Mountains in spring migration (Hoffman and Smith 2003; chapter 2). It is likely that most, though not all, of the Peregrine Falcons breeding in or observed in migration in the state have their winter grounds to the south of New Mexico. Falcons from adjacent states, for example, have been found in winter in Mexico (Enderson et al. 1991). Based on Christmas Bird Count (CBC) data, the area of New Mexico with the most consistent winter sightings of Peregrine Falcons is in and near Las Cruces, Doña Ana County. Peregrine Falcon sightings in that area were reported during CBCs for five of the seven years between 1998 and 2004, for an average of 1.6 falcons counted/year (National Audubon Society 2005).

Habitat Associations

Peregrine Falcons almost always attack their prey when it is flying in open air far from escape cover. Thus the species is more successful hunting in areas with large "gulfs" of air, such as canyons and mountains, or in large open

PHOTO 25.12

Adult Peregrine Falcon in typical New Mexico winter habitat, the Bosque del Apache National Wildlife Refuge, October 1993. In winter, the Bosque del Apache's ponds and wetlands harbor very high numbers of aquatic birds and other potential avian prey.

PHOTOGRAPH: © GARY K. FROEHLICH.

areas, like river channels and wetlands. New Mexico has considerable amounts of canyons and mountains, which, along with suitable nesting cliffs, provide abundant potential breeding habitat. Wintering habitat with concentrations of potential prey is likely limited to the major rivers, the Rio Grande and the Pecos, of the central and southern portions of the state.

Life History

Nesting

Although throughout North America most Peregrine Falcons do not winter on their breeding grounds (White et al. 2002), on 26 January 2006 I observed an adult female at a nesting cliff in north-central New Mexico. Thus it may be that a few adults remain on their breeding territories year-round in New Mexico, depending upon the severity of the winter. Although I could not verify it, the bird I observed on that occasion may have nested during the previous spring, and would again breed at that particular cliff.

My field notes indicate that in New Mexico, falcons usually return to nesting cliffs in March, occasionally earlier in the south, later in the north. Males are more likely to be the first to return to the nesting cliff; attachment to a particular cliff may be more important to an adult than who its mate is. Thus an early arrival may be chased from the cliff upon the return of the previous year's more zealous occupant. Courtship begins as soon as a pair is present and continues for several weeks, involving feedings, copulations, and the choice of the eventual nesting ledge after consideration of many options.

The eggs of Peregrine Falcons are elliptical and about 53 mm (2 in) long and 41 mm (1.5 in) wide; they average smaller in the north than in the south. The base color can range from white to light brown to reddish, and most eggs are overlaid with spots or splotches of darker pigments, usually reddish to dark brown (White et al. 2002). Mean clutch size has not been determined in New Mexico because nests have not been entered since at least 1970. However, during the last decade I have often observed broods of three so that most clutches are likely of three to four eggs, as reported in Britain (Ratcliffe 1980) and Colorado (Craig and Enderson 2004). Average clutch size is largest at mid latitudes (3.72), smaller in the Arctic (3.0) and south to Mexico (3.3; Hickey 1969; Palmer 1988). Incubation in captivity requires 33–34 days (Burnham 1983), but it can take up to 37 days in the wild with interruptions during early incubation (White et al. 2002). Typically the clutch is complete by mid April, at which time incubation begins.

In New Mexico, hatching can be backdated to mid May since based on my experience the usual peak of the fledging period is early July and fledging usually occurs about six weeks after hatching (Palmer 1988). The young remain dependent on parental care for another four to six weeks after fledging (Sherrod 1983). Most, if not all, adults and young of the year have left breeding areas by mid October.

Through the early 1960s the mean productivity of Peregrine Falcon pairs at the few aeries known from the state was in excess of three fledglings per pair, with one brood of five fledglings recorded (Johnson and Williams 2006). More recently, successful New Mexico Peregrine Falcon nests have fledged between one and four young, and mean productivity has ranged from less than one fledgling per nest to slightly above two in any given year. Reproductive output statewide in 2003–7 was estimated at 1.55 fledglings per pair based on a random sample of about 20% of known aeries, but was only 1.31 in 2007 (Johnson and Williams 2008; see also Status and Management below). In Colorado through 2001 Peregrine Falcons fledged on average 1.67 young per natural nest (Craig and Enderson 2004). Colorado also artificially augmented broods in some nests or released captive-reared young through 1995. Similar measures were deemed unnecessary in New Mexico, and in fact both populations have recovered at equal rates. Birds fledged in Colorado have occasionally been documented in breeding areas in New

PHOTO 25.13

An adult female Peregrine Falcon on a potential aerie ledge, Rio Arriba Co., 6 April 2009. Peregrine Falcon courtship involves the male bringing prey to the female and showing her potential nest sites on ledges. Courtship also includes multiple copulations, *ee-chupping*, bowing, and what seems like much deliberation! PHOTOGRAPH: © DALE W. STAHLECKER.

PHOTO 25.14

(*above*) Five-day-old Peregrine Falcons (brood of three) and an unhatched egg on an aerie in Los Alamos Co. in 1968. In that era, peregrine eggs were frequently either breaking or failing to hatch due to the effects of DDT. However, in all the years that this site was monitored (1964–71), it had only this one unhatched egg, producing three to four young per year.

PHOTOGRAPH: © DAVID A. PONTON.

PHOTO 25.15

(*right*) Nineteen-day-old Peregrine Falcon in the 1968 aerie. It already shows the large feet that are distinctive of the species.

PHOTOGRAPH: © DAVID A. PONTON.

PHOTO 25.16

(*right*) Twenty-six-day-old Peregrine Falcon at an aerie in Los Alamos Co. in 1972. At 26 days old, a young peregrine can readily stand and walk. The bird in the photograph (taken from a blind) is also showing the development of flight feathers. In the same aerie that year, two eggs failed to hatch, due to the old age of the female (in their last productive years, females produce fewer viable eggs) and/or DDT. PHOTOGRAPH: © DAVID A. PONTON.

PHOTO 25.17

(*below*) Thirty-seven-day-old Peregrine Falcon (same bird as in previous photograph), now fully feathered. It has jumped to a ledge closer to the photography blind and is standing in the sun watching every movement in the canyon. This adventitious movement out of the aerie indicates fledging is imminent. PHOTOGRAPH: © DAVID A. PONTON.

Mexico (Craig and Enderson 2004), while conversely, some Peregrine Falcons fledged from New Mexico nests likely have ended up breeding in Colorado.

Diet and Foraging

Peregrine Falcons feed almost exclusively on birds, with 429 species documented as prey in North America and an estimated 1,500–2,000 species likely taken worldwide (White et al. 2002). Documented North American prey have ranged in size from a 3.1-kg (6.83-lb) Sandhill Crane (*Grus canadensis*) down to 2.5–3.5 g (0.09–0.12 oz) hummingbirds (*Selasphorus* spp.). Peregrine Falcons are capable of capturing extremely agile flyers such as swifts and bats (Brown et al. 1992). They are also known to "pirate" or steal mammals and fish from other birds of prey (White et al. 2002). They often hunt from a high vantage point such as the nesting cliff, surprising birds that have ventured too far from cover. Such avian prey is almost always taken in flight, but Peregrine Falcons occasionally hunt nestling birds and rodents while walking on the ground (White et al. 2002).

In Colorado more than 1,000 prey remains of 83

species were collected opportunistically during banding or augmentation visits to aeries. White-throated Swifts (*Aeronautes saxatalis*: 14%), Mourning Doves (*Zenaida macroura*: 10%), and Common Nighthawks (*Chordeiles minor*: 7%) were the prey species most often documented, and only 20 species were detected among prey remains more than 12 times (Craig and Enderson 2004). As in Colorado, New Mexico Peregrine Falcons take a wide variety of birds whenever and wherever they are vulnerable. At least occasionally adult pairs hunt cooperatively: when two field companions and I flushed an adult Killdeer (*Charadrius vociferus*) from a Rio Arriba County beach in June 1992, it circled nearby, protesting loudly. Suddenly an adult female Peregrine stooped from high above, but missed. She was followed almost immediately by the adult male, who snatched the Killdeer in flight and flew quickly away with his mate now following him.

Five radio-tagged adults (two males, three females) from three Colorado Front Range aeries within a 28-km-long (18-mi) mountain range had overlapping hunting ranges, varying in size from 358 to 1,508 km^2 (138–582 mi^2). Sixty percent of all telemetry locations were within 8 km (5 mi) of the respective aeries, but 20% of all female locations were beyond 23 km (15 mi) (Enderson and Craig 1997). However, a female radio-tracked for two months in Scotland had a much smaller home range of 117 km^2 (45 mi^2) (Mearns 1985). Given the variety of nesting situations in New Mexico, home ranges could perhaps also vary from very small to very large, and in areas of nesting concentrations much overlap of hunting ranges likely occurs.

Predation and Interspecific Interactions

Peregrine Falcons regularly interact with other cliff-nesting predators, particularly Golden Eagles (*Aquila chrysaetos*) and Common Ravens (*Corvus corax*; White et al. 2002). In 1999 I watched a southwestern New Mexico pair choose to nest less than 10 m (~30 ft) from an active Common Raven nest. The ravens fledged young; the falcons did not. One unusual observation was made in Lea County on 8 August 1962, when a Peregrine was observed harassing Chihuahuan Ravens (*Corvus cryptoleucus*) at their nest, far from the normal summer haunts of the falcons (Travis et al. 1962).

In New Mexico, Golden Eagles in particular can draw reactions from nesting Peregrine Falcons if they venture too close to falcon aeries (chapter 21). Where densities of both species are high in northern New Mexico, Peregrine Falcons appear to have visual boundaries, beyond which they ignore eagles. This may change during the late nestling phase when large, active young falcons (see below) could draw the attention of hunting eagles. Other diurnal raptorial birds flying near aeries can be ignored or can draw an attack, apparently depending upon the mood of the resident falcons. I have also seen both a black bear (*Ursus americanus*) and a bobcat (*Lynx rufus*) repeatedly stooped by the resident females when they inadvertently passed above Peregrine Falcon aeries.

Very little information exists on predation in New Mexico. In Colorado between 1976 and 2001, 45% of 65 mortalities were due to predation. Golden Eagles were known to have killed 11 Peregrine Falcons, including eight large nestlings/recent fledglings. Large nestlings active on the nest ledge are, unfortunately, often visible yet inexperienced, at a time when both parents often forage away from the aerie to meet the growing demand for prey. The other three Peregrine Falcons killed by Golden Eagles were two adults and one subadult, vulnerable while perched and presumably caught off guard. Great Horned Owls (*Bubo virginianus*) killed seven Colorado nestlings and even an adult, at night in her aerie (Craig and Enderson 2004). Where Peregrine Falcons nest on the ground or on accessible cliffs, predation by mammals can occur (White et al. 2002). Possible predation of nestlings by a ringtail (*Bassariscus astutus*) was reported in New Mexico (T. Johnson, pers. comm.) and in Utah (White and Lloyd 1962).

Status and Management

Peregrine Falcons were first documented nesting in New Mexico in 1919 (Wetmore 1920). As elsewhere in North America (Enderson et al. 1995), Peregrine Falcon nest site occupancy declined during the 1960s and 1970s in New Mexico, reaching its lowest level in 1980. In 1976, at the time that a long-term study was initiated, there were 19 known Peregrine Falcon nest sites in the state, not all of them occupied (Johnson and Williams 2006). In 1980 only 30% of about 50 sites with a monitoring history were occupied, but occupancy improved through 1997 to about 75% of almost 100 such sites statewide. While more sites were discovered between 1997 and 2007, the occupancy rate statewide did not reach 85% of randomly selected historical territories until 2007 (Johnson and Williams 2008). By 2007 at least 142 sites in New Mexico had been occupied at least once in the past ~50 years (Johnson and Williams 2008). Occupancy rates have varied regionally, however. In 1980 they were: 10% of Colorado Plateau (~northwestern New Mexico) sites, 30% of Rocky Mountain sites, and 50% of Basin and Range (~central and southern New Mexico) sites (Johnson and Williams 2006). By 2007 occupancy rates ranged from 79% in Basin and Range to 91% on the Colorado Plateau (Johnson and Williams 2008).

Listed as Endangered first under the Endangered Species Conservation Act (in 1970) and later under the Endangered Species Act of 1973, the Peregrine Falcon was delisted by the U.S. Fish and Wildlife Service in 1999 (see Cade and Burnham 2003). The decision was based on comparisons of historical and recent reports of numbers of pairs and reproduction (U.S. Fish and Wildlife Service 1999), but without analysis of demography and occupancy rates (Johnson and Williams 2006). As seen above, the number of known nesting sites in New Mexico rose markedly in the last 30 years, but efforts to find them increased as well, representing a potentially important bias. In fact, unequal coverage of the state between years made it impossible to use the state or portions thereof as a study area. Johnson and Williams (2006) therefore used known Peregrine sites as a "dynamic" study area instead. They found that within New Mexico occupancy had not reached the recovery plan goal of 85% (U.S. Fish and Wildlife Service 1984), though it finally did in 2007 (Johnson and Williams 2008). Additionally, reproduction had not reached 1960s levels, albeit calculated from a much smaller sample of nests, through 31 consecutive years (1976–2006) (Johnson and Williams 2006). While the New Mexico Department of Game and Fish downlisted the Peregrine Falcon from Group 1 (Endangered) to Group 2 (Threatened) in 1994, it did not support federal delisting in 1999.

Other long-term monitoring efforts confirm or tend to confirm the increase in Peregrine Falcon numbers over the last several decades in New Mexico. Between 1985 and 2001, HawkWatch International detected a statistically significant increase in the numbers of Peregrine Falcons migrating over the Sandia and Manzano mountains (Hoffman and Smith 2003). In contrast to BBS data, Christmas Bird Count (CBC)

PHOTO 25.18

Peregrine Falcon on a dead branch, Steeple Rock Canyon north of Virden, Hidalgo Co., 10 February 2003. Although the Peregrine Falcon is found year-round in New Mexico, many of the birds nesting in New Mexico probably winter in Mexico, where contamination of food webs by pesticides continues to be a problem. PHOTOGRAPH: © ROBERT SHANTZ.

results lend themselves to calculations of Peregrine Falcon population trends—as stated earlier, the species largely escapes detection during the breeding season due to its habitat associations. More falcons have been reported on CBCs in recent years than in the more distant past, with an average of 4.7/year counted in the state between 1998 and 2004 on 25–31 counts. While only about a third as many (1.6/year) Peregrine Falcons were counted between 1959 and 1966, there were also fewer counts (7–10) being conducted at that time (National Audubon Society 2005).

Some Peregrine territories are perennially occupied, others only intermittently. Dormant territories are unoccupied for longer periods (Johnson 2001; Johnson and Williams 2008). Management of Peregrine Falcon breeding habitat on public lands in New Mexico since a 1985 Memorandum of Understanding between New Mexico Department of Game and Fish, the U.S. Fish and Wildlife Service, and the U.S. Forest Service has identified all suitable habitat and treated it as occupied unless *proven* not to be in a particular year. Thus even dormant or potential territories are protected, particularly from human disturbance, and are more attractive to pioneering individuals (Johnson and Williams 2006).

Another management decision affecting Peregrine Falcons was that New Mexico chose to follow neighboring states in allowing the take of one nestling for falconry in 2007 and two nestlings in 2008. Continued harvest is desired by falconers, but should be dependent upon continued high occupancy of nesting cliffs. And yet, while the statewide occupancy rate finally reached 85% in 2007 (Johnson and Williams 2008), lower reproduction in 2003–7 has raised concern about future regional population trends for the species (Johnson and Williams 2006). One explanation for low reproductive success might be continued insecticide contamination. Organochlorine levels were still elevated in Peregrine Falcon prey in the 1980s (Hubbard and Schmitt 1988; Kennedy et al. 1995). While Western Hemisphere DDT use decreased steadily after 1963, New Mexico Peregrine Falcon reproduction has varied through time instead of increasing as might be expected (Johnson 2001). However, Peregrine reproduction was found to be highly correlated with the previous year's index of total insecticide use, mostly organophosphates, south of the U.S. border. Regression analysis was used to calculate a decrease of 0.54 young/adult pair for every 100,000 metric tons of insecticide used in the previous year (Johnson and Williams 2006).

A more recent development is also of concern. In 2008 apparently healthy 3.5-week-old young in three northern New Mexico aeries I monitor were not present two weeks later when they should have been large, visible, and noisy. One dead young, in fact, was visible on an aerie ledge. Mysterious disappearances of half-grown Osprey (*Pandion haliaetus*) and Golden Eagle young in the same general area were additionally suggestive of a disease problem, likely West Nile Virus, though predation on Peregrine Falcon nestlings by Golden Eagles is always a possibility (Craig and Enderson 2004). In this same area, Peregrine fledgling production dropped from 2.25/pair in 2006, to 1.3 in 2007, and to 0.6 in 2008.

Craig and Enderson (2004) calculated that 1.22 young/pair would maintain the Colorado population at current levels. Johnson and Williams' (2008) annual recalculation indicated that 1.64 young/pair were needed to maintain a stable New Mexico population, but slightly fewer, 1.55 young/pair, were fledged in 2003–7. Thus they predicted a 6% decline in occupancies in 2009 (Johnson and Williams 2008). While neither estimate takes into account possible immigration (and emigration) from within the respective states, both far exceed New Mexico's statewide production in 2008 of only 0.9 young/pair (T. Johnson, pers. comm.). If pesticide contamination of prey and/or West Nile Virus have become serious threats, even higher rates of decline in occupancy seem likely in the next decade. To summarize, Peregrine Falcons are undoubtedly much more common in New Mexico than three decades ago, but if recent reproductive problems persist, the state's population of this beautiful and fearsome falcon may be in for more hard times.

> After this chapter was completed, new information was obtained on the productivity of Peregrine Falcons statewide, indicating a slight improvement in 2009.

Brown, L. H., E. K. Urban, and K. B. Newman. 1982. *The birds of Africa*. London: Academic Press.

Burnham, W. 1983. Artificial incubation of falcon eggs. *Journal of Wildlife Management* 47:158–68.

Cade, T. J., and W. Burnham. 2003. *Return of the Peregrine Falcon*. Boise, Idaho: The Peregrine Fund.

Craig, G. R., and J. H. Enderson. 2004. *Peregrine Falcon biology and management in Colorado 1973–2001*. Technical Publication 43. Denver: Colorado Division of Wildlife.

Dunn, P., D. Sibley, and C. Sutton. 1988. *Hawks in flight*. Boston: Houghton Mifflin Co.

Enderson, J. H., and G. R. Craig. 1997. Wide-ranging by nesting Peregrine Falcons (*Falco peregrinus*) determined by radio-telemetry. *Journal of Raptor Research* 31:333–38.

Enderson, J. H., C. Flatten, and J. P. Jenny. 1991. Peregrine Falcons and Merlins in Sinaloa, Mexico, in winter. *Journal of Raptor Research* 25:123–26.

Enderson, J. H., W. Heinrich, L. Kiff, and C. M. White. 1995. Population changes in North American Peregrines. *Transactions 60th North American wildlife and natural resources conference*, 24–29 March, Minneapolis, Minnesota, 142–61.

Hickey, J. J. 1969. *Peregrine Falcon populations: Their biology and decline*. Madison: University of Wisconsin Press

Hoffman, S. W., and J. P. Smith. 2003. Population trends of migratory raptors in western North America, 1977–2001. *Condor* 105:397–419.

Hubbard, J. P., and C. G. Schmitt. 1988. Organochlorine residues in avian prey of Peregrine Falcons. In *Proceedings of the southwest raptor management symposium and workshop*, ed. R. L. Glinski, B. G. Pendleton, M. B. Moss, M. N. LeFranc Jr., B. A. Millsap, and S. W. Hoffman, 176–81. Washington, DC: National Wildlife Federation.

Johnson, T. H. 2001. *The Peregrine Falcon in New Mexico—2001*. Santa Fe: New Mexico Department of Game and Fish.

Johnson, T. H., and S. O. Williams III. 2006. *The Peregrine Falcon in New Mexico—2005*. Santa Fe: New Mexico Department of Game and Fish.

———. 2008. *The Peregrine Falcon in New Mexico—2007*. Santa Fe: New Mexico Department of Game and Fish.

Kennedy, P. L., D. W. Stahlecker, and J. M. Fair. 1995. Organochlorine concentrations in potentials avian prey of breeding Peregrine Falcons in north-central New Mexico. *Southwest Naturalist* 40:94–100.

Mearns, R. 1985. The hunting of two female peregrines towards the end of a breeding season. *Journal of Raptor Research* 19:20–26.

National Audubon Society. 2005. The Christmas Bird Count historical results. http://www.audubon.org/bird/cbc (accessed 20 December 2005).

Palmer, R. S. 1988. *Handbook of North American birds*. Vol. 4, part 1, *Diurnal raptors*. New Haven: Yale University Press.

Pratt, H. D., P. L. Bruner, and D. G. Berrett. 1987. *A field guide to the birds of Hawaii and the tropical Pacific*. Princeton, NJ: Princeton University Press.

Ratcliffe, D. 1980. *The Peregrine Falcon*. Carlton, UK: T. and A. D. Poyser.

Sherrod, S. K. 1983. *Behavior of fledgling Peregrines*. Ithaca, NY: The Peregrine Fund, Inc.

Sibley, D. A. 2000. *The Sibley guide to the birds*. New York: Chanticleer Press.

Travis, J. R., B. McKnight, and M. Huey. 1962. *New Mexico Ornithological Society Field Notes* 1(1).

U.S. Fish and Wildlife Service. 1984. *American Peregrine Falcon recovery plan (Rocky Mountain/Southwest population)*. Prepared in cooperation with the American Peregrine Falcon Recovery Team. Denver: U.S. Fish and Wildlife Service.

———. 1999. Final Rule to Remove the American Peregrine Falcon from the list of endangered and threatened wildlife and to remove the similarity of appearance provision for free-flying peregrines in the conterminous United States. *Federal Register* 64(164): 46541–58.

Wetmore, F. A. 1920. Observations on the habits of birds of Lake Burford, New Mexico. *Auk* 37:221–47, 393–412.

White, C. M., and G. D. Lloyd. 1962. Predation on Peregrines by ringtails. *Auk* 79:277.

White, C. M., N. J. Clum, T. J. Cade, and W. G. Hunt. 2002. Peregrine Falcon (*Falco peregrinus*). No. 660. In *The birds of North America*, ed. A. Poole and F. Gill. Philadelphia, PA: Birds of North America, Inc.

26

Prairie Falcon

(*Falco mexicanus*)

COLE J. WOLF, JAMES M. RAMAKKA, AND BLAIR O. WOLF

RENOWNED FOR ITS hunting prowess and its aggressiveness around its nest, the Prairie Falcon (*Falco mexicanus*) is one of New Mexico's most familiar and most widespread diurnal raptors. It is a fairly large, square-headed falcon whose body length and wingspan range from 37 to 47 cm (15 to 19 in) and from 89 to 119 cm (35 to 47 in), respectively (Palmer 1988). There is little overlap in size between the sexes (Steenhof and McKinley 2006). Females of all ages are approximately one-third larger than males; adult females from the Rocky Mountains weigh on average 863 g (30 oz) (range 760–975 g [27–34 oz], n = 31), adult males from the same region only 554 g (20 oz) (500–635g [18–22 oz], n = 15) (Enderson 1964). Other than size, however, males and females are similar in appearance. The upperparts are a light gray-brown, with pale bars and edges on and across the dorsal feathers creating a nonuniform appearance. The breast, flanks, and belly are white and are spotted or streaked with dark brown. The tarsi and feet are yellow, and the talons are black. Prairie Falcons possess a distinct facial pattern with a white supercilium that is bordered by a gray-brown crown above and brownish auriculars below; the region behind the eye is whitish. A dark mustachial stripe borders a spur of white that reaches to the auriculars. The adult bill is yellowish with a dark tip and the cere is yellow (Palmer 1988). In flight, the Prairie Falcon can be easily identified from all other New Mexico raptors by its dark brown or blackish axillaries (or "armpits"). The median coverts on the underwings are also conspicuously dark, especially in females, while the undersides of the primaries are dusky with pale white striping. Apparently little geographic variation exists in plumage; rangewide the Prairie Falcon's plumage is monotypic and no color morphs are recognized (Browning 1978; Palmer 1988; Steenhof 1998).

Juvenile birds are fairly similar to adults, with only a few minor differences in plumage color. Juveniles tend to be darker brown on the upperparts and head. The facial pattern is not as well defined: the white supercilium can be indistinct, the crown is streaked with white, and the mustachial stripe is brown and poorly defined. The breast and underparts are much more heavily streaked than in adults and the underwings are darker. The bill is bluish with a dark tip, and the tarsi and feet are bluish.

Like other falcons, the Prairie Falcon is characterized by long, pointed wings. It flies with stiff, shallow wing beats, more rapid in males than in females (Steenhof 1998). When perched, Prairie Falcons are most easily confused with Peregrine Falcons (*Falco peregrinus*),

PHOTO 26.1

(*left*) Adult Prairie Falcon north of Animas, Hidalgo Co., 29 December 2005. The upperparts are grayish brown, with pale edges and bars on and across the feathers of the back. A distinct facial pattern includes a white supercilium (eyebrow) bordered by brownish auriculars below, itself separated from a dark mustachial stripe by a white patch. The adult bill is yellowish with a dark tip. The cere is yellow. PHOTOGRAPH: © ROBERT SHANTZ.

PHOTO 26.2

(*below*) Prairie Falcon trapped and banded during the 1992 spring migration at HawkWatch International's Sandia Mountains study site. The underparts are pale, with dark axillaries (or "armpits"). PHOTOGRAPH: © JOHN P. DELONG.

PHOTO 26.3

(*top*) Adult Prairie Falcon in flight, Maxwell National Wildlife Refuge, November 2005. The flight of a Prairie Falcon is characterized by shallow, stiff wing beats. Note the dark axillaries (armpits), which are diagnostic in New Mexico. PHOTOGRAPH: © BILL SCHMOKER.

PHOTO 26.4

(*bottom*) Immature Prairie Falcon north of Rodeo, Hidalgo Co., 17 December 2005. The upperparts are darker brown than in adults, the plumage on the back more uniform in appearance. PHOTOGRAPH: © ROBERT SHANTZ.

which are similar in size (Palmer 1988). However, the facial pattern easily sets the two species apart; peregrines lack the white supercilium and patch behind the eye visible in Prairie Falcons. Prairie Falcons also have shorter primary extension with the wing tips not reaching the tail tip as found in peregrines (Sibley 2000). In flight, the Prairie Falcon and the Peregrine Falcon are less likely to be confused as the Peregrine Falcon lacks the dark "armpit" of its close relative.

Prairie Falcons vocalize infrequently except around the nest during the breeding season. The most common call given is the alarm call, which is described by Brown and Amadon (1968) as "a shrill yelping *kik-kik-kik-kik*" or by Sibley (2000) as "an angry, harsh *ree-kree-kree-kree.*" The alarm call is usually given when displaying aggressive behavior near the nest (Steenhof 1998). Another call, described as a "*eechup*," "*chup*," or "*kuduchup*" is typically given by the pair when investigating nest sites, but is also uttered during food exchanges between the adult male and female and again when exhibiting aggressive behavior against nest territory intruders (Steenhof 1998).

Distribution

The Prairie Falcon is one of the most common and most widespread raptors in arid western North America. The species' breeding range extends from north-central Mexico north to southwestern Canada and includes almost all of the western United States, spilling eastward into extreme western North Dakota, South Dakota, and Nebraska (Palmer 1988; Steenhof 1998). Although seasonal movements are observed, Prairie Falcons are found year-round throughout their breeding range, with the exception of the northernmost populations in British Columbia, Alberta, and Saskatchewan, which withdraw southward during the winter (Steenhof 1998). In winter the distribution of Prairie Falcons extends eastward to Missouri and Minnesota, west to the Pacific coast, and south to include all of the Baja California peninsula and more of central Mexico (Palmer 1988).

Prairie Falcons are found in New Mexico year-round. They have been reported as breeding in nearly all areas of the state, with perhaps the stronghold of their distribution extending from the Colorado–New Mexico state line southward along the continental divide (e.g., Ligon 1961; Platt 1974; appendix 26-1). The only part of the state apparently lacking breeding Prairie Falcons is in the southeast, which lacks cliffs and other suitable nesting habitat (map 26.1). Prairie Falcons are also not known to breed at the highest elevations of the state. However, in Colorado use of tundra habitats and the occurrence of nesting at an elevation of nearly 3,700 m (12,140 ft) have been reported (Marti and Braun 1975), suggesting that the same may also be true in New Mexico.

Wintering birds range statewide except in the high-elevation forested mountains (Ligon 1961). Migrating birds range statewide and are regularly recorded at HawkWatch sites in the Sandia and Manzano mountains (chapter 2). A number of birds on migration have been tracked from the Snake River Birds of Prey Area in Idaho through New Mexico on their way to wintering sites in Texas (Steenhof et al. 2005).

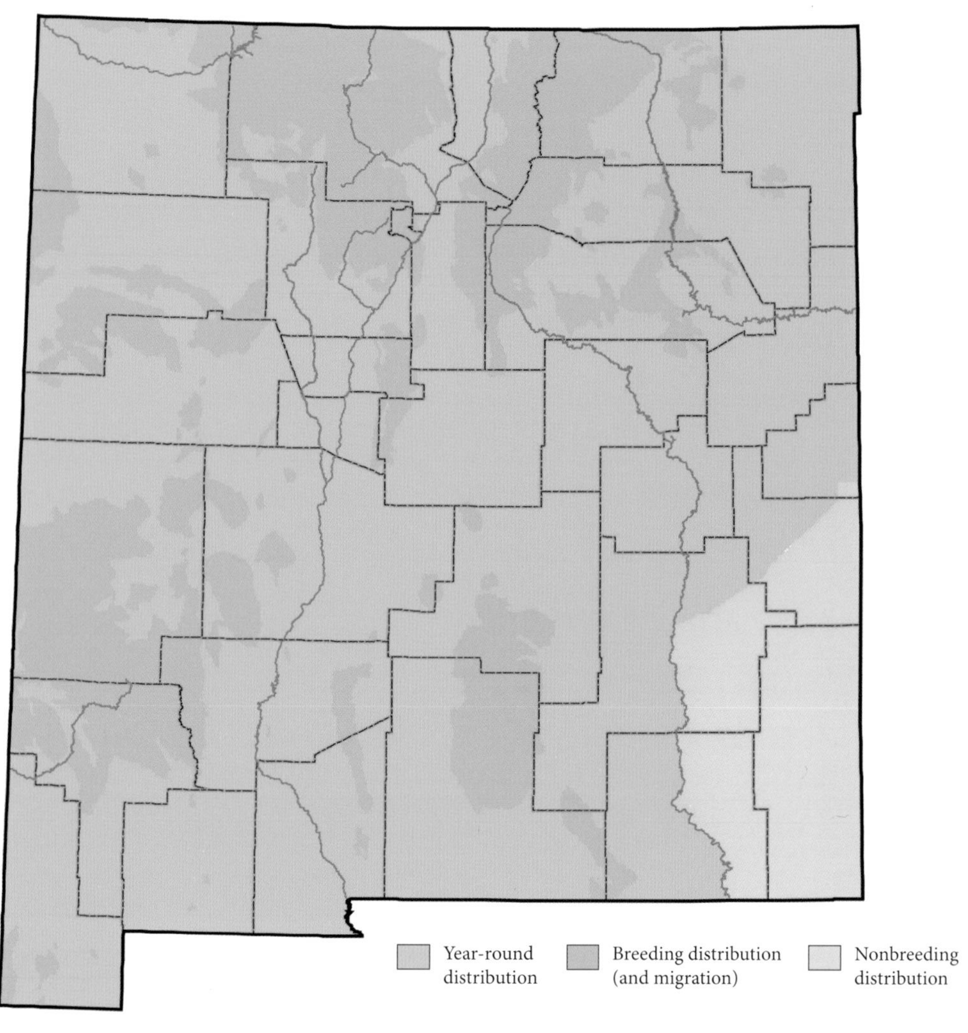

MAP 26.1

Prairie Falcon distribution map

PHOTO 26.5

Prairie Falcon in flight at the Bosque del Apache National Wildlife Refuge, 18 November 2007. Prairie Falcons are closely associated with rugged terrain only during breeding and in winter can also be found in open, flat areas. PHOTOGRAPH: © TOM KENNEDY.

PHOTO 26.6

Prairie Falcon on the crossbar of a power pole north of Rodeo, Hidalgo Co., 13 August 2006. As typical for raptors of the open country, Prairie Falcons often use power poles as vantage points. PHOTOGRAPH: © ROBERT SHANTZ.

Habitat Associations

The Prairie Falcon is a raptor adapted to dry environments (Steenhof 1998). Throughout its distribution it is found primarily in grasslands and shrublands (e.g., Enderson 1964; Millsap 1981; Peterson 1988). Outside the breeding season, some Prairie Falcons also move into agricultural areas to take advantage of concentrations of important avian prey species (Enderson 1964; White and Roseneau 1970), while during migration the species may also be encountered in or over forested areas (Schmutz et al. 1991; see chapter 2). During the nesting season, topography is an even far more restrictive habitat requirement than vegetation cover. Although Prairie Falcons nest occasionally in trees (MacLaren et al. 1984), on transmission towers (Roppe et al. 1989; Bunnell et al. 1997), and on buildings (Nelson 1974), their nests are much more typically found along cliffs. Thus, during the breeding season, topographically flat areas with otherwise suitable habitat may be unoccupied by the species.

In New Mexico, Prairie Falcons typically nest along canyon walls and mountain cliffs as well as in badlands (Ligon 1961; Platt 1974; JMR, pers. obs.). From 1981 to 1988, Ramakka (unpubl. data) found that of 15 nest sites mapped in the northwestern corner of New Mexico, 73% were on cliffs, while 27% were in clay banks.

PHOTOS 26.7a and b

Alternate nest sites along Alamo Wash in the Bisti/De-Na-Zin Wilderness, San Juan Co., ca 1983. (a, *top*) The nest ledge is in the center of the photo where white-wash is visible. In 1985, a mud flow had covered the ledge when J. Ramakka checked it in the spring. The following year, the ledge had sloughed off entirely and the site was most likely unusable. In 1987, the resident pair chose another small ledge on the bank and produced three (possibly four) young. (b, *bottom*) Clay bank nest site appropriated by Great Horned Owls (*Bubo virginianus*) in 1983 then reoccupied by the resident Prairie Falcon pair in 1985 (see discussion on p. 470).

PHOTOGRAPHS: © JIM RAMAKKA.

PHOTO 26.8

Prairie Falcon aerie and surrounding habitat north of White Rock, San Juan Co., 1986. During the breeding season, Prairie Falcons are closely associated with rugged topography, as they typically nest on cliffs.

PHOTOGRAPH: © JIM RAMAKKA.

Life History

The natural history of the Prairie Falcon has not been studied in detail in New Mexico. Based on studies conducted elsewhere, Prairie Falcons occupy home ranges that are quite large during the breeding season, covering a land area of 59 to 315 km² (23 to 122 mi²) as estimated by radio telemetry (Hamarta et al. 1978; Marzluff et al. 1997). However, core use of these home ranges is much more limited. Marzluff et al. (1997) found that 90% of prey captures occurred in a core area that made up only 38% of the home range. Limited information suggests that home ranges are smaller in winter than during breeding in east-central Colorado. In that area Beauvais et al. (1992) found an average winter home range of only 30.2 km² (11.6 mi²).

Nesting

Prairie Falcons are monogamous and typically breed at two years of age (Enderson 1964), although some females begin reproducing before the end of their first year (Webster 1944; Runde 1987). There is little information on mate fidelity between years but pairs do not stay together during the nonbreeding season. Pair bonds appear to be established or renewed each year following arrival on breeding grounds, which occurs mostly from late February through mid March in

Colorado (Enderson 1964). Courtship and nest site selection can last a month and egg-laying can begin in early March for some populations (Steenhof 1998).

A typical nest site is a scrape made on a ledge or in a crevice or cavity along a cliff or a cutbank (Steenhof 1998; JMR, unpubl. data). However, Prairie Falcons also commonly use old nests built on cliffs by Common Ravens (*Corvus corax*) or by raptors such as Golden Eagles (*Aquila chrysaetos*) (Hickman 1971; Palmer 1988). Cliff nests tend to be situated on the upper half of the cliff face, and two studies have found average nest heights ranging from 11.1 to 18.2 m (36.4 to 59.7 ft) (Enderson 1964; Runde and Anderson 1986). Nest orientation is most frequently to the south, but in low-lying deserts nest sites may face generally north to reduce afternoon heat stress (Kaiser 1986). In northwestern New Mexico, Ramakka (unpubl. data) found the mean height of nest cliffs or banks was 40 m (131 ft) (range 11–92 m [36–302 ft]) and the mean height of the nest site was 28 m (92 ft) (range 8–77 m [26–253 ft]).

Range-wide egg-laying dates vary, with birds nesting at higher latitudes and elevations usually laying their eggs later; the overall mean clutch-completion date is 18 April for 20 populations (Williams 1985). The incubation period has been estimated to be 29–31 days by Enderson (1964), but Fyfe (1972) found that it could last as long as 39 days. Range-wide the modal clutch size is five eggs (Steenhof 1998). Average clutch sizes are largest in the northernmost breeding populations and decrease southward. Of five egg sets collected in New Mexico during the early and mid 20th century, all around mid April, three consisted of five eggs, and the other two had four eggs and three eggs (appendix 26-1). Platt (1974), however, found mean clutch size in northeastern New Mexico to be only 3.2 eggs.

In northeastern New Mexico the average hatching date was 17 May, with a range of 4 May–28 June (Platt 1974). Fourteen of the 15 sites found by the Bureau of Land Management by helicopter and ground surveys in northwestern New Mexico during 1981–89 were revisited in two or more (up to nine) succeeding years. Repeat visits during a single season were only possible for a few sites. Based on observations of eggs and young during single and repeat visits, incubation was still occurring during the first week of May and as late as 16 May. Downy young were observed on nests as early as 8 May and as late as 1 June. The only territory with repeat visits later in the nesting season revealed approximate fledging dates of 11 June, 14 June, and 19 June (JMR, unpubl. data). Hatching dates after 12 June elsewhere in the species' range are most likely re-nesting attempts after a nest failure (Allen et al. 1986).

Range-wide the male and female share in the incubation of the eggs. At night, only the female incubates, but the males assist during the day. The male's contribution apparently varies greatly and can range from 18 to 76% of daytime incubation (Kaiser 1986; Holthuijzen 1989). Males also make prey deliveries to incubating females during the day and to brooding females for nestlings early in the nestling stage. Males provide 70% or more of the food required to rear the brood (Kaiser 1986; Holthuijzen 1990). Initially the female remains at the nest, but approximately four weeks after the young have hatched, she begins to hunt, and both adults feed the young (Enderson 1964). As much as 30% of the food delivered to the nest may not be eaten by the adults or the nestlings (Sitter 1983). Cannibalism has been observed, with parents feeding nestlings the remains of their dead siblings (Holthuijzen et al. 1987).

The young fledge an average of 38 days after hatching and thereafter parents continue to feed them for another 30 to 35 days (McFadzen and Marzluff 1996). The total length of the breeding cycle ranges around four to five months, from the initiation of courtship to fledgling independence. Fledgling mortality in southwestern Idaho was found to range from 28 to 34% (McFadzen and Marzluff 1996), while juvenile mortality was estimated to be 15–35% (Denton 1975; Runde 1987). A main cause of mortality among nestlings is predation by terrestrial predators, while among fledglings predation by Great Horned Owls (*Bubo virginianus*) is common (see Steenhof 1998). However, ectoparasite infections are also responsible for nest abandonment during the incubation period and for mortality among nestlings including in New Mexico (Platt 1975). Annual adult survivorship is estimated to be 65 to 81% (Denton 1975; Runde 1987).

PHOTO 26.9

Prairie Falcon after a meal along the road to Bosque del Apache National Wildlife Refuge, 22 November 2007. The diet of the Prairie Falcon has not been studied in New Mexico.

PHOTOGRAPH: © DOUG BROWN.

Diet and Foraging

Prairie Falcons consume a wide variety of prey ranging in size from grasshoppers and crickets up to large game birds such as Ring-necked Pheasants (*Phasianus colchicus*) and White-tailed Ptarmigans (*Lagopus leucura*) (Palmer 1988). In most areas, the bulk of the diet during the spring and summer seemingly consists of ground squirrels of the genus *Spermophilus*, although birds nesting at high elevations in Colorado prey mostly on American Pikas (*Ochotona princeps*) instead (Marti and Braun 1975). At least eight *Spermophilus* species have been recorded in the Prairie Falcon's diet, but only for part of the year as these mammals enter hibernation in the fall (Palmer 1988). In southwestern Idaho, small mammals made up 64% of all items brought to the nest. Townsend's Ground Squirrels (*Spermophilus townsendii*) were the favored prey and made up 37% of total prey items (Holthuijzen 1990). Prairie Falcons are also known to prey upon several species of lizards and many birds including: waterfowl (Bailey 1928), Lesser Prairie-Chickens (*Tympanuchus pallidicinctus*) (Ligon 1961), Burrowing Owls (*Athene cunicularia*) (Webster

1944), Rock Pigeons (*Columba livia*) (Webster 1944; Palmer 1988), and Brown-capped Rosy-Finches (*Leucosticte australis*) and other finch species (Webster 1944; Enderson 1964; Marti and Braun 1975; Palmer 1988). On wintering grounds, Horned Larks (*Eremophila alpestris*) and Western Meadowlarks (*Sturnella neglecta*) are the most important prey species (Steenhof 1998), although other flocking birds, such as longspurs (*Calcarius* spp.), European Starlings (*Sturnus vulgaris*), and Mourning Doves (*Zenaida macroura*) are also taken (Palmer 1988).

Prairie Falcons use several methods to attack and capture prey, often selecting an approach low to the ground and displaying great speed and maneuverability. The most common attacks are stoops or low glide attacks. In this type of attack the falcon flies directly at the prey and takes it by surprise using a low, long-angle stoop (White 1962). Prairie Falcons also use a "low coursing" attack, which is a swift, direct flight toward the prey just above the ground or vegetation. Prairie Falcons are occasionally kleptoparasitic, as they have been observed stealing prey from Northern Harriers (*Circus cyaneus*), including an American Coot (*Fulica americana*) (Parmenter 1941; Holthuijzen et al. 1987); there have also been reports of nest-robbing behavior (Holthuijzen et al. 1987).

During breeding, both males and females cache prey in grass and large clumps of vegetation, on ledges, and in small crevices. Caching rates decline as the young mature (Holthuijzen 1990).

Predation and Interspecific Interactions

Prairie Falcons frequently nest near other species, including Barn Owls (*Tyto alba*), Great Horned Owls, and Common Ravens. Cade (1982) suggested that the relationship between ravens and falcons may be symbiotic; abandoned raven nests provide nesting sites for falcons while ravens steal from falcon prey caches. During nine years of nest monitoring in New Mexico in the 1980s, Ramakka (unpubl. data) observed one instance in which a Great Horned Owl apparently displaced a pair of falcons from a previously used nest site in a clay bank. The falcons nested that same year on a ledge of another clay bank approximately 0.8 km (0.5 mi) distant. Two years later they reoccupied the original site. In the succeeding four years the pair appeared to alternate nest sites between the two banks.

Prairie Falcons are very aggressive in and near nesting territories. They are often especially aggressive toward owls and falcons, and may attack and sometimes kill owl intruders (Bond 1936; Andersen 1988). Ligon (1961:79) reports a pair of Prairie Falcons attacking a Great Horned Owl along the Upper Gila River in southwestern New Mexico. One of the falcons struck the owl, sending it "diving into heavy treetop foliage." Falcons also aggressively defend territories against intrusion by Golden Eagles and Red-tailed Hawks (*Buteo jamaicensis*) and have been recorded flying over 1.5 km (0.9 mi) to attack such birds. Turkey Vultures (*Cathartes aura*) and smaller raptors such as Northern Harriers, Sharp-shinned Hawks (*Accipiter striatus*), and American Kestrels (*Falco sparverius*) are usually tolerated. Mammalian predators such as bobcats and coyotes are usually only attacked after the young fledge (Kaiser 1986).

Prairie Falcons are not always the aggressor, and they can be attacked or robbed by other raptors. Peregrine Falcons will often attack Prairie Falcons if they enter a nesting territory, occasionally killing them (Porter and White 1973; Walton 1978). Golden Eagles have also been observed attacking Prairie Falcons (Cooper 1991). Peregrine Falcons and Red-tailed Hawks have been observed stealing prey from Prairie Falcons (Haak 1982; Seibert and Loyd 1989; Lins and Hessen 1992).

Status and Management

Cade (1982) estimated the total breeding population of Prairie Falcons in the United States to be around 5,000 to 6,000 pairs with a breeding range encompassing 4 to 4.5 million km^2. No recent population decline has been observed. In fact, Breeding Bird Survey results show a nearly significant population increase from 1966 through 2007 (Sauer et al. 2008). During the period when DDE decimated Peregrine Falcon populations, Prairie Falcons, though more sensitive to DDE than Peregrine Falcons and Merlins (*Falco columbarius*), were not significantly affected because their mammal-based diet resulted in less exposure to the chemical (Steenhof 1998). At the scale of the species' range,

PHOTO 26.10

San Juan Mine, San Juan Co. Prairie Falcons often nest in mining areas, particularly in San Juan Co. Nests are monitored every year to track any impact on nesting pairs.

PHOTOGRAPH: © MICKEY GINN.

perhaps the largest threat facing Prairie Falcons today is urbanization. Urbanization rates of only 5–7% were enough to cause most birds to avoid particular areas on the wintering grounds (Berry et al. 1998). Other anthropogenic impacts on the species include mortality by shooting and collisions with fences, motor vehicles, and power lines (see Steenhof 1998).

In New Mexico, Breeding Birds Survey results from 1966 to 2007 provide little insight into population trends because of the low numbers of birds sighted along routes. Platt (1974) listed overgrazing and associated habitat degradation through soil erosion as the primary threat to Prairie Falcons in northern New Mexico. Bednarz (1984) noted that Prairie Falcons did not breed in the Caballo Mountains, where intensive mining was occurring. However, no causal link could be established. Overall, the outlook for the Prairie Falcon in New Mexico appears to be bright.

LITERATURE CITED

Allen, G. T., R. K. Murphy, K. Steenhof, and S. W. Platt. 1986. Late fledging dates, renesting, and large clutches of Prairie Falcons. *Wilson Bulletin* 98:463–65.

Andersen, D. E. 1988. Common Barn Owl killed by a Prairie Falcon. *Southwestern Naturalist* 33:377–78.

Bailey, F. M. 1928. *Birds of New Mexico*. Santa Fe: New Mexico Department of Game and Fish.

Beauvais, G., J. H. Enderson, and A. J. Magro. 1992. Home range, habitat use and behavior of Prairie Falcons wintering in east-central Colorado. *Journal of Raptor Research* 26:13–18.

Bednarz, J. C. 1984. Effect of mining and blasting on breeding Prairie Falcon (*Falco mexicanus*) occupancy in the Caballo Mountains, New Mexico. *Raptor Research* 18:16–19.

Berry, M. E., C. E. Bock, and S. L. Haire. 1998. Abundance of diurnal raptors on open space grasslands in an urbanized landscape. *Condor* 100:601–8.

Bond, R. M. 1936. Some observations on the food of the Prairie Falcon. *Condor* 38:169–70.

Brown, L. H., and D. Amadon. 1968. *Eagles, hawks, and falcons of the world*. Vol. 2. Feltham, UK: Country Life Books.

Browning, M. R. 1978. An evaluation of the new species and subspecies proposed in Oberholser's *Bird Life of Texas*. *Proceedings of the Biological Society of Washington* 91:85–122.

Bunnell, S. T., C. M. White, D. Paul, and S. D. Bunnell. 1997. Stick nests on a building and transmission towers used for nesting by large falcons in Utah. *Great Basin Naturalist* 57:263–67.

Cade, T. 1982. *Falcons of the world*. Ithaca, NY: Comstock/Cornell University Press.

Cooper, J. M. 1991. Golden Eagle attacks Prairie Falcon. *British Columbia Birds* 1:11–12.

Denton, S. J. 1975. Status of Prairie Falcons breeding in Oregon. Master's thesis. Oregon State University, Corvallis.

Enderson, J. H. 1964. A study of the Prairie Falcon in the central Rocky Mountain region. *Auk* 81:332–52.

Fyfe, R. 1972. Breeding behavior of captive and wild Prairie and Peregrine Falcons. In *Special conference on captivity breeding of raptors. Part C, behavioral considerations to egg-laying*, ed. R. W. Nelson, 43–52. *Raptor Research* 6 (supplement).

Haak, B. A. 1982. Foraging ecology of Prairie Falcons in northern California. Master's thesis. Oregon State University, Corvallis.

Hamarta, A. R., J. E. Durr, and H. Geduldig. 1978. Home range, activity patterns and habitat use of Prairie Falcons nesting in the Mojave Desert. Contract no. YA-512-CT8-4389. Unpublished report, Colorado Wildlife Services, Fort Collins, CO, for U.S. Department of the Interior, Bureau of Land Management, Riverside, CA.

Hickman, G. L. 1971. Prairie Falcons and Red-tailed Hawks rearing young in inactive Golden Eagle nests. *Condor* 73:490.

Holthuijzen, A.M.A. 1989. *Behavior and productivity of nesting Prairie Falcons in relation to construction activities at Swan Falls Dam. Final report*. Boise: Idaho Power Co.

———. 1990. Prey delivery, caching, and retrieval rates in nesting Prairie Falcons. *Condor* 92:475–84.

Holthuijzen, A. M. A., P. A. Duley, J. C. Hagar, S. A. Smith, and K. N. Wood. 1987. Piracy, insectivory and cannibalism of Prairie Falcons (*Falco mexicanus*) nesting in southwestern Idaho. *Journal of Raptor Research* 21:32–33.

Kaiser, T. J. 1986. Behavior and energetics of Prairie Falcons (*Falco mexicanus*) breeding in the western Mojave Desert. Ph.D. diss. University of California, Los Angeles.

Ligon, J. S. 1961. *New Mexico birds and where to find them*. Albuquerque: University of New Mexico Press.

Lins, E., and J. Hessen. 1992. Prairie Falcon in Minneapolis. *Loon* 64:69.

MacLaren, P. A., D. E. Runde, and S. Anderson 1984. A Record of tree-nesting Prairie Falcons in Wyoming. *Condor* 86:487–88.

Marti, C. D., and C. E. Braun. 1975. Use of tundra habitats by Prairie Falcons in Colorado. *Condor* 77:213–14.

Marzluff, J. M., B. A. Kimsey, L. S. Schueck, M. E. McFadzen, M. S. Vekasy, and J. C. Bednarz. 1997. The influence of habitat, prey abundance, sex, and breeding success on the ranging behavior of Prairie Falcons. *Condor* 99:567–84.

McFadzen, M. E., and J. M. Marzluff. 1996. Mortality of Prairie Falcons during the fledgling-dependence period. *Condor* 98:791–800.

Millsap, B. A. 1981. *Distributional status of Falconiformes in west central Arizona—with notes on ecology, reproductive success and management*. Technical Note 355. Washington, DC: U.S. Department of the Interior, Bureau of Land Management.

Nelson, R. W. 1974. Prairie Falcons: nesting attempt on a building and the effect of weather on courtship and incubation. *Raptor Research Foundation Ethology Information Exchange* 1:10–12.

Palmer, R. S. 1988. *Handbook of North American birds*. Vol. 5. New Haven: Yale University Press.

Parmenter, H. E. 1941. Prairie Falcon parasitizing Marsh Hawk. *Condor* 43:157.

Peterson, D. L. 1988. Nesting and habitat parameters for selected raptors in the desert of northern Utah. Master's thesis. Utah State University, Logan.

Platt, S. W. 1974. Breeding status and distribution of the Prairie Falcon in northeastern New Mexico. Master's thesis. Oklahoma State University, Stillwater.

———. 1975. The Mexican chicken bug as a source of raptor mortality. *Wilson Bulletin* 87:557.

Porter, R. D., and C. M. White. 1973. Peregrine Falcon in Utah, emphasizing ecology and competition with the Prairie Falcon. *Brigham Young University Scientific Bulletin, Biological Series* 18:1–74.

Roppe, J. A., S. M. Siegel, and S. E. Wilder. 1989. Prairie Falcon nesting on transmission towers. *Condor* 91:711–12.

Runde, D. E. 1987. Population dynamics, habitat use and movement patterns of the Prairie Falcon (*Falco mexicanus*). Ph.D. diss. University of Wyoming, Laramie.

Runde, D. E., and S. H. Anderson. 1986. Characteristics of cliffs and nest sites used by breeding Prairie Falcons. *Raptor Research* 20:21–28.

Sauer, J. R., J. E. Hines, and J. Fallon. 2008. *The North American breeding bird survey, results and analysis, 1966–2007.* Version 5.15.2008. Laurel, MD: USGS, Patuxent Wildlife Research Center.

Schmutz, J. K., R. W. Fyfe, U. Banasch, and H. Armbruster. 1991. Routes and timing of migration of falcons banded in Canada. *Wilson Bulletin* 103:44–58.

Seibert, P., and J. Loyd. 1989. Notes on a Prairie Falcon. *Bulletin of the Oklahoma Ornithological Society* 22:6–7.

Sibley, D. A. 2000. *The Sibley Guide to the birds.* New York: Chanticleer Press.

Sitter, G. 1983. Feeding activity and behavior of Prairie Falcons in the Snake River Birds of Prey Natural Area in southwestern Idaho. Master's thesis. University of Idaho, Moscow.

Steenhof, K. 1998. Prairie Falcon (*Falco mexicanus*). No. 346. In *The Birds of North America*, ed. A. Poole and F. Gill. Ithaca, NY: Cornell Laboratory of Ornithology, and Philadelphia, PA: Academy of Natural Sciences.

Steenhof, K., and J. McKinley. 2006. Size dimorphism, molt status, and body mass variation of Prairie Falcons nesting in the Snake River Birds of Prey National Conservation Area. *Journal of Raptor Research* 40:71–75.

Steenhof, K., M. R. Fuller, M. N. Kochert, and K. T. Bates. 2005. Long-range movements and breeding dispersal of Prairie Falcons from southwest Idaho. *Condor* 107:481–96.

Walton, B. J. 1978. Peregrine Prairie Falcon interaction. *Raptor Research* 12:46–47.

Webster, H. Jr. 1944. Survey of the Prairie Falcon in Colorado. *Auk* 61:609–16.

White, C. M. 1962. Prairie Falcon displays Accipitrine and Circinine hunting methods. *Condor* 64:439–40.

White, C. M., and D. G. Roseneau. 1970. Observations on food, nesting and winter populations of large North American falcons. *Condor* 72:113–15.

Williams, R. N. 1985. Relationship between Prairie Falcon nesting phenology, latitude and elevation. *Raptor Research* 19:139–42.

27

Barn Owl

(*Tyto alba*)

JEAN-LUC E. CARTRON AND STEVEN W. COX

KNOWN ALSO AS the Monkey Owl or the Monkey-faced Owl, among many other names, the Barn Owl (*Tyto alba*) is all at once beautiful, ungainly, and ghost-like in appearance. With a body length of 32–40 cm (13–16 in) and a mass of approximately 470–570 g (17–20 oz) (Marti 1992), it is a medium-sized owl whose distinguishing characteristics include a cylindrical body, a large, rounded head, long wings (wingspan: 100–125 cm [39–49 in]), long legs, and a short tail. It is most readily recognizable by its white, heart-shaped face rimmed with rust or brown and the V-shaped ridge extending upward from the hooked, deceptively small-looking beak. The two dark eyes are close-set, and they too appear small compared to those of other owls. There are no ear tufts. Above, the owl's plumage is a mottled cinnamon, orange, and gray with fine white speckles and black bars, the latter only on the wings. Like the face, the rest of the owl's underparts are white, but with also some scattered dark speckles and spots. Females tend to be larger and heavier, and they have more and larger, darker ventral markings than males, some of which are pure white below (Marti 1992). The legs are covered with white feathers down to the feet. In flight, the Barn Owl lacks the dark wrist patches of the Short-eared Owl (*Asio flammeus*).

The Barn Owl has a rich vocal repertoire (Bühler and Epple 1980; Bunn et al. 1982). Unlike other owls, however, it does not hoot but instead utters eerie screams, a trait that once earned the species some of its less flattering names, including the Demon Owl,

PHOTO 27.1

Barn Owl portrait (educational bird). From up close the Barn Owl is unmistakable. Note the white, heart-shaped facial disk, the small dark eyes, and the lack of ear tufts. PHOTOGRAPH: © JEAN-LUC CARTRON.

PHOTO 27.2

(*left*) Barn Owl pair, Steeple Rock Canyon, Hidalgo Co., 28 March 2007. The female (right) has more, larger ventral markings. PHOTOGRAPH: © ROBERT SHANTZ.

PHOTO 27.3

(*below*) Barn Owl in flight, Highway 17, Chama, Rio Arriba Co., 22 January 2007. The Barn Owl lacks the dark wrist patches of the Short-eared Owl (*Asio flammeus*) PHOTOGRAPH: © ROGER HOGAN.

PHOTO 27.4

Barn Owl portrait (educational bird). Above, the plumage is mainly a mottled cinnamon, orange, and gray.

PHOTOGRAPH: © JOHN V. BROWN.

the Death Owl, or the Hobgoblin Owl. One vocalization in particular is a "*shrrreeee*," a drawn-out and often repeated hissing screech compared by many to a burst of steam being released from a valve. This screech is given by both sexes during the breeding season. According to Bühler and Epple (1980) and Bunn et al. (1982), it is the equivalent of the song of a passerine bird, serving to advertise a territory and attract a mate. Other screams have been described, some of which grade into one another. For example, the Barn Owl has a distress call and an alarm call, more drawn-out screams that vary in pitch (Marti 1992). Also part of the owl's repertoire are snores, chirrups, twitters, and squeaks, as well as nonvocal sounds, including wing-clapping and beak-clicking.

The taxonomy of the Barn Owl remains in flux. Up

to 46 races are currently recognized (del Hoyo et al. 1999). However, among them several may be elevated to species status, while the taxonomic distinctiveness of several others remains to be confirmed (del Hoyo et al. 1999). The North American Barn Owl is *T. a. pratincola* (Marti 1992; Marti et al. 2005).

Distribution

The Barn Owl is a cosmopolitan bird, occurring on all continents except Antarctica. Its distribution is centered on the tropics, however, reaching into the temperate zone only as far as the limit of the species' endurance to harsh winters (del Hoyo et al. 1999; see also Marti and Wagner 1985). Thus the Barn Owl is found throughout Australia, South America, and most of Africa, but its range excludes Scandinavia, all of Asia north of the Indian subcontinent and China, most of Canada, and Alaska (del Hoyo et al. 1999). In North America in particular, the distribution of the Barn Owl extends north only to southwestern British Columbia, North Dakota, Minnesota, Wisconsin, Michigan, extreme southern Ontario, southern Quebec, and

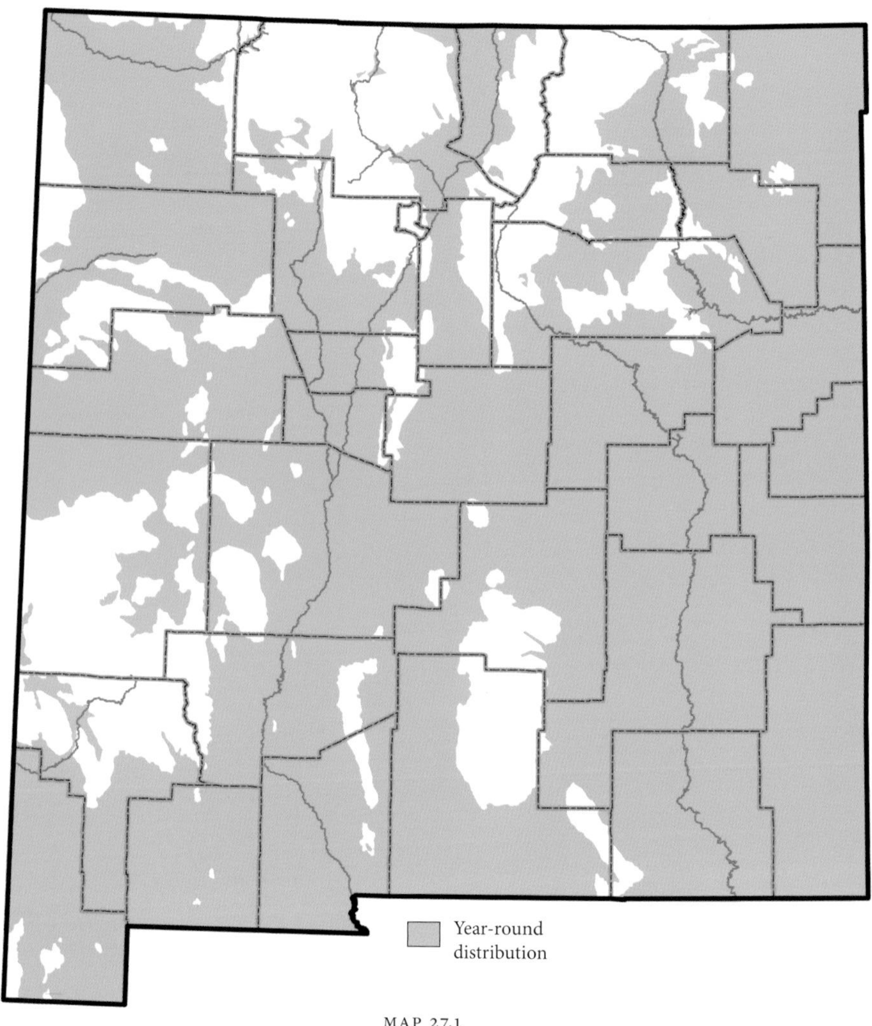

MAP 27.1

Barn Owl distribution map

Massachusetts (Marti 1992; Terres 1996). Because Barn Owl populations appear to be largely or completely sedentary, the species' winter and breeding distributions are the same.

The Barn Owl is found at low and middle elevations throughout New Mexico (Hubbard 1978; JLEC). Only excluded from its distribution are the state's mountain ranges.

Habitat Associations

In New Mexico as elsewhere, the Barn Owl is mainly a bird of the open country. However, its use of open habitats is also largely conditioned by the availability of nest or roost sites (see Marti 1992). Thus in New Mexico it seems most strongly associated with desertscrub, grassland, and agricultural fields in areas with specific topographic features (i.e., cliff walls, steep arroyo banks) or large, hollow man-made structures that have been abandoned or discarded (see below). It can also be found roosting or nesting in small groves of trees and along narrow wooded riparian corridors where these border open habitats (Cartron et al. 2008).

PHOTO 27.5

Abandoned house occupied by Barn Owls north of Grenville, Union Co., 8 September 2007. The surrounding habitat consists of grassland.

PHOTOGRAPH: © LAWRY SAGER.

PHOTOS 27.6a, b, and c

(a, *top*) Barn Owl rural habitat in Luis Lopez, Socorro Co., with nest box (doghouse-like structure in pecan tree, left rear) and roost box (doghouse-like house on end of roof of a hay shed). (b and c, *middle and bottom*) Barn Owl nest site and roost site close-ups.

PHOTOGRAPHS: © JERRY OLDENETTEL.

PHOTOS 27.7a and b

(*above*) Barn Owl nest site in abandoned house and surrounding habitat in Lea Co., 26 May 2008. The nest was on the shelf above an old closet and contained a hatchling plus three eggs. PHOTOGRAPHS: © JEAN-LUC CARTRON.

PHOTO 27.8

(*top right*) Nest site area in Estancia Valley just north of Mountainair, Torrance Co. The nest site is the cistern on the left, near an abandoned homestead with scattered elm (*Ulmus* sp.) trees. Surrounding habitat is grassland. The nest site was occupied in 2007. PHOTOGRAPH: © JEAN-LUC CARTRON.

PHOTO 27.9

(*bottom right*) Barn Owl in mine shaft north of Lordsburg, Hildago Co., September 1984. PHOTOGRAPH: © ROBERT SHANTZ.

Life History

Nesting

No doubt there is enough information on the versatile, unusual nesting habits of the Barn Owl to fill the pages of an entire book. In New Mexico the species nests in hollow trees and in recesses or small caves along the walls of arroyos, cliffs, and excavated pits (Ligon 1961; Martin 1973; Jorgensen et al. 1998; Williams 2000, Cartron et al. 2009; NM Mines and Minerals Division data on file). It also nests in a wide variety of human-made structures, including abandoned buildings, cisterns, mine shafts, nest boxes, and the underside of bridges (Ligon 1961; Havens 1998; Cartron and Stahlecker 2007; Cartron et al. 2009; SWC, unpubl. data). At the McKinley Mine, one pair of Barn Owls was reported to have nested in an old shovel, a large vehicle equipped with a boom for loading material into haul trucks (NM Mines and Minerals Division data on file). In Albuquerque in 2005, one of us (SWC) monitored a Barn Owl nest site in an old boiler. In the late 1980s or early 1990s Barn Owls nested underneath the Eastern New Mexico University football stadium halfway between Clovis and Portales (T. Gennaro, pers. comm.).

Many of the Barn Owl natural nest sites reported from New Mexico have been along arroyo walls (Martin 1973; Jorgensen et al. 1998; Williams 2003). Martin (1973) in particular found five occupied nest sites along Tijeras Arroyo, Bernalillo County, in 1971, all consisting of an elliptical chamber at the end of a horizontal tunnel 0.7–1.3 m (3–4.3 ft) long, the entrance being a ledge more than halfway up the arroyo wall. Returning on consecutive days, Martin (1973) was able to track the progress of one pair of Barn Owls as one or both of the birds excavated the tunnel and chamber themselves, scratching with their talons.

Barn Owl nest sites must be both inaccessible to predators and spacious enough to accommodate large broods. When not excavating their own nest sites along arroyo walls, Barn Owls depend on finding large natural cavities or man-made receptacles. On the eastern plains of New Mexico, isolated, abandoned buildings are often used as nest sites by Barn Owls, the nest being located in the rafters, in an attic, or even on a shelf above an old closet (Cartron et al. 2009; JLEC, unpubl. data). On Kirtland Air Force Base, Bernalillo County, Barn Owls readily take up residence in large nest boxes installed by Cox on elevated platforms and poles. The exact shape and size of the boxes varies. Most of the openings are either 15 cm (6 in) in diameter or 15 cm (6 in) square. One of the occupied boxes has an opening of 30 cm (12 in). Also illustrating the strong association with large receptacles for nesting, Cartron and Stahlecker (2007) reported two occupied nests in cisterns (receptacles holding water or other fluids) in New Mexico, one in 1998 near Liberty in extreme western Roosevelt County, the other in 2007 just north of Mountainair in Torrance County. The cistern near

PHOTOS 27.10a and b

(a, *top*) Abandoned house with Barn Owl active nest site south of Wagon Mound, Colfax Co., 25 June 2008; (b, *bottom*) six eggs in the nest in the corner of the large attic, 25 June 2008. PHOTOGRAPHS: © JEAN-LUC CARTRON.

PHOTOS 27.11a and b

Active nest site in a bucket within an abandoned cattle tank, Roosevelt Co., May 2008.

PHOTOGRAPHS: © LAWRY SAGER.

PHOTOS 27.12a and b

Active nest in old broiler in the Albuquerque area, Bernalillo Co., 5 July and 15 July 2005.

PHOTOGRAPHS: © STEVEN COX.

PHOTOS 27.13a, b, c, and d

(a, b, c) Nest boxes used as nest sites by Barn Owls on Kirtland Air Force Base, Bernalillo Co. (d) active nest in one nest box on Kirtland Air Force Base, with seven eggs, 4 April 2008. PHOTOGRAPHS: © STEVEN COX.

PHOTO 27.14

Barn Owl adult and nestlings in rafters of abandoned building east of Caprock, Lea Co., 25 May 2008.

PHOTOGRAPH: © JEAN-LUC CARTRON.

Liberty was in an abandoned farmhouse, about 3–4 m (10–13 ft) deep and bottle-shaped, and with an opening at the top less than 1 m (3 ft) across. The cistern near Mountainair was placed on a crate 1.75 m (~5.7 ft) tall, measuring 1.8 m (5.9 ft) in height and with an opening at the top only 44 cm (17 in) across (Cartron and Stahlecker 2007; JLEC, unpubl. data). Given the wingspan of Barn Owls, one must wonder just how the young manage to exit such nest sites on their first flight. It may well be that some nest sites are death traps for young and even adult Barn Owls. It is not known whether young fledged successfully at the two cisterns near Liberty and Mountainair. However, the last of two visits to the cistern near Liberty revealed not just nestlings but also the remains of one adult Barn Owl. Carl Marti (pers. comm.) has seen firsthand the difficulty with which some Barn Owls fly out from the bottom of concrete grain containers. In Utah, he once found a grain storage container resembling the cistern near Mountainair, with a narrow opening at the top. Inside that storage container were the carcasses of at least 10 Barn Owls.

PHOTO 27.15

Barn Owl nestlings on the floor of an abandoned house near Bluit in southern Roosevelt Co., 2006. Just outside the house that year was an active Great Horned Owl (*Bubo virginianus*) nest in an elm (*Ulmus* sp.) tree.

PHOTOGRAPH: © LAWRY SAGER.

Contrary to some other owls (e.g., the Long-eared Owl [*Asio otus*]), Barn Owls exhibit high nest site fidelity, returning year after year to the same place for nesting. Over time vast quantities of food pellets can accumulate on the floor of the nest site. Barn Owls do not build a nest made of sticks like most other raptors. Instead, the female makes a small scrape for the eggs in the layer of food pellets (Marti 1992).

Egg-laying in New Mexico appears to begin as early as mid March—and even earlier—for some pairs. Most pairs are with eggs in April or May (Cartron and Stahlecker 2007; SWC, unpubl. data), but Ligon (1961) reports a nest with eggs in a hollow willow in 1925 in June, and Cartron discovered a nest with six eggs on 25 June [2008] near Wagon Mound, Colfax County. However, eggs found in late spring or early summer may be indicative of re-nesting. At Bitter Lake National Wildlife Refuge, a nest box had two small nestlings and four eggs on 23 February (!) 2005 (Williams 2005a; G. Warrick, pers. comm.), indicating egg-laying had taken place as early as mid to late January (see below). That same year, two nest boxes contained a second brood of nestlings in the first week of August (Williams 2005b; G. Warrick, pers. comm.).

Clutch size typically ranges from five to seven eggs (Ligon 1961). Of nine nests monitored in Albuquerque, five held seven eggs, while the other four had five eggs, for a mean of 6.1 eggs (SWC, unpubl. data). The eggs are laid one at a time, every two to three days (Smith et al. 1974). Only the females have brood patches, and they incubate the eggs while the males bring them food (Howell 1964; Marti 1992). The incubation period lasts 29–34 days (Marti 1992).

Other than the 2005 Bitter Lake National Wildlife Refuge record, hatching in New Mexico has not been recorded any earlier than mid April, while fledging typically occurs in May or June (SWC, unpubl. data). Hatching is asynchronous in Barn Owls, with the result that nestlings are staggered in age. The eggs hatch one at a time, again every two to three days (Marti 1992). Given the large number of eggs Barn Owls lay, the oldest nestling can be one to two weeks—or more—older than the youngest one. Hatching asynchrony is presumably an adaptive strategy to maximize reproductive success in species faced with highly variable food supplies. During years of low food availability, the older, more aggressive nestlings eat most of what is brought back to the nest, their chances of survival being little affected by large brood size. The younger nestlings die of starvation. Dead Barn Owl nestlings are apparently often eaten by their siblings (Baudvin 1978; Bühler 1981; Marti 1992).

Diet and Foraging

In New Mexico the Barn Owl's diet consists mostly of small rodents, the primary prey throughout the species' distribution (Marti 1992), less often other small mammals, with crickets and birds also observed. In the foothills of the Sacramento Mountains, Jorgensen et al. (1998) found cotton rats (*Sigmodon*) to be a favored prey taxon, Barn Owls apparently discriminating among habitat types and actively seeking those where *Sigmodon* were present. Barn Owl prey remains collected along Tijeras Arroyo appeared to be predominantly pocket mice (*Perognathus*) or kangaroo rats (*Dipodomys*) (Pache 1981). On Kirtland Air Force Base, Cox noted that Ord's kangaroo rats (*Dipodomys ordii*) were seemingly very common among prey remains at occupied nest sites. Other remains consisted of southern plains woodrats (*Neotoma micropus*) and *Peromyscus* species. Species seemingly less frequently taken included the banner-tailed kangaroo rat (*Dipodomys spectabilis*), spotted ground squirrel (*Spermophilus spilosoma*), and Western Meadowlark (*Sturnella neglecta*).

Barn Owls are nocturnal. They typically hunt on the wing, patrolling back and forth over a small area, flying low to the ground and relying on their sense of hearing and low-light vision to detect prey. Not only is the Barn Owl's sense of hearing keen, it also allows accurate pinpointing of prey in terms of direction and angle. The openings and preaural flaps of the Barn Owl's right and left ears are not symmetrical, allowing sound waves to arrive and be sorted differently on the two sides of the body (Taylor 1994). Once directly above an unsuspecting prey, a Barn Owl drops on it, seizing it with its talons and killing it with a blow delivered to the back of the skull with its beak (Marti 1992). Less frequently, Barn Owls hunt from a perch.

Predation and Interspecific Interactions

In New Mexico, the Barn Owl shares its habitat with many other raptors, some of which are probable or documented predators of the owl. Among them is the Ferruginous Hawk (*Buteo regalis*), also one of the state's grassland dwellers (see chapter 19). Cartron et al. (2004) found Barn Owl remains at two Ferruginous Hawk nests in the Estancia Valley. Predation by the Great Horned Owl (*Bubo virginianus*) likely also occurs in New Mexico as it does elsewhere in North America (see Marti 1992). Barn Owls and Great Horned Owls seem to nest often in very close proximity to one another. In spring 1997, Jerry Oldenettel (pers. comm.) found a Barn Owl nest with young in a recess below the top of a bluff at the Bosque del Apache National Wildlife Refuge. The nest was facing roughly northeast, with an active Great Horned Owl nest on the other side of the bluff and facing more or less south, the two nests separated by only about 1.5 m (4.9 ft) (Snider 1997)! Near Bluit in southern Roosevelt County, Lawry Sager (pers. comm.) found Barn Owls with six or seven young in an abandoned

PHOTO 27.16

Barn Owl perched on fence, Highway 17, Chama, Rio Arriba Co., 22 January 2007.

PHOTOGRAPH: © ROGER HOGAN.

building, with an active Great Horned Owl nest in an elm (*Ulmus* sp.) tree just outside that house.

Barn Owls likely face competition with other species for at least some nest and roost sites. Barn Owl nest boxes can be used at least occasionally by American Kestrels (*Falco sparverius*). In 2005 on Kirtland Air Force Base, American Kestrels fledged three young in a nest box occupied the year before by a Barn Owl pair (SWC, unpubl. data). In 2006, 2007, and 2008, the occupants of the nest box were again Barn Owls.

Status and Management

Published accounts of the Barn Owl in New Mexico describe the species as rare or uncommon (Ligon 1961; Hubbard 1978). However, the Barn Owl can easily be overlooked as it stays out of sight during the day and it does not hoot. In fact, we now know that the species is quite common in some areas of the state, particularly where abandoned buildings, large nest boxes, or arroyos are present. On the eastern plains of New Mexico, seemingly a large number of abandoned buildings have their resident Barn Owl pair. In central and northern Lea County in 2008, Cartron et al. (2009) discovered active Barn Owl nests in two abandoned buildings, plus single birds or pairs in six additional abandoned buildings (or groups of buildings). Three additional nest sites were also found during casual inspections of abandoned buildings elsewhere in eastern New Mexico, one south of Bluit in southern Roosevelt County in 2006, one north of Grenville in Union County in 2007, and one just south of Wagon Mound in Mora County in 2008 (Cartron et al. 2009). Gordon Warrick (pers. comm.), who was the biologist at Bitter Lake National Wildlife Refuge in Chaves County for nine years, knew of at least two or three Barn Owl pairs nesting on the refuge every year, one in a large nest box in a cottonwood, the other two in nest boxes in sheds. Barn Owls were also observed in sinkholes on the refuge, suggesting additional nesting (G. Warrick, pers. comm.). Along the southern outskirts of Albuquerque, Bernalillo County, Martin (1973) found five occupied nests in 1971 along Tijeras Arroyo, while Kirtland Air Force Base (about 200 km^2 [~80 mi^2]) was home to at least seven pairs of Barn Owls in 2008 (SWC, unpubl. data). In 2000, not one but two active nests were reported from along Perico Creek southwest of Clayton, Union County (Williams 2000). There is no information on the density of Barn Owls in riparian woodlands of the state, but here again it may be higher than generally believed. While working in the Cliff–Gila Valley along the Gila River in Grant County, Scott Stoleson (pers. comm.) recorded three occupied roost or nest sites only about 2.9 km (1.8 mi) apart.

Barn Owl population declines have been reported from parts of the species' distribution, while elsewhere researchers have noted population increases or local range expansions. All of these changes have been attributed in part to habitat alterations (Marti 1992). Little can be said about any change in the status of the owl in New Mexico. Some of the grasslands inhabited by Barn Owls in New Mexico are overgrazed (e.g., Martin 1973). Overgrazing likely impacts rodent grassland populations, but may also contribute to arroyo cutting, thus perhaps increasing the local availability of nest sites.

Barn Owls have a short life span, less than two years on average (Keran 1981; Marti 1994). Among the predominant natural causes of mortality is starvation, especially during winter (Marti and Wagner 1985), while man-related mortality is due to collision with road traffic and electrocutions on power poles and perhaps also harmful exposure to pesticides (Marti 1992; Cartron et al. 2006). For Barn Owl populations to remain stable, high mortality rates must be offset by high reproductive output. In a 16-year study in Utah, Marti (1994) documented that most Barn Owls begin reproducing during their second calendar year of life and lay one clutch per year. The mean of 275 first clutches was 7.17 (and only 5.79 in 19 observed second clutches), and 71% of all nesting attempts resulted in the fledging of at least one young (Marti 1994). In the lower mainland of British Columbia, Andrusiak and Cheng (1997) found clutch size to average 6.5 and a mean productivity (mean number of young fledged per nesting pair) of 2.6. Barn Owl productivity reported in yet other North American studies ranged from 2 to 3.9 (Otteni et al. 1972; Ault 1982; Millsap and Millsap 1987). With high mortality and high productivity, many Barn Owls have low lifetime reproductive success, whereas a few can produce a large number of descendants. In Utah, Barn Owls bred only a mean of 1.3 years and produced a mean of only 5.58 fledglings over the course of their lifetime (Marti 1997). However, a female Barn Owl produced 50 fledglings during her life span. A meager 12% of female Barn Owls had descendants that bred, but one female produced 69 descendants in only three generations (Marti 1997).

Of nine nests monitored over the last three years on Kirtland Air Force Base, four fledged three and five fledged two, for a mean productivity of 2.4 per nest, within the range of productivity values mentioned above. However, more monitoring is needed to determine the status of the Barn Owl in New Mexico. Particularly valuable should be surveys in areas where nesting was historically reported and continued use of nest boxes.

LITERATURE CITED

Andrusiak, L. A., and K. M. Cheng. 1997. Breeding biology of the Barn Owl (*Tyto alba*) in the lower mainland of British Columbia. In *Biology and conservation of owls of the northern hemisphere. Second international symposium, February 5–9, 1997, Winnipeg, Manitoba, Canada*, ed. J. R. Duncan, D. H. Johnson, and T. H. Nicholls, 38–46. General Technical Report NC-190. Washington, DC: USDA Forest Service.

Ault, J. W. 1982. A quantitative estimate of Barn-Owl nesting habitat quality. M.S. thesis. Oklahoma State University, Stillwater.

Baudvin, H. 1978. Le cannibalisme chez l'Effraie Tyto alba. *Nos oiseaux* 34:223–31.

Bühler, P. 1981. Das Fütterungsverhalten der Schleiereule *Tyto alba*. *Ökologie der Vögel*. 3:183–202.

Bühler, P., and W. Epple. 1980. Die Lautäusserungen der Schleiereule (*Tyto alba*). *Journal of Ornithology* 121:36–70.

Bunn, D. S., A. B. Warburton, and R. D. S. Wilson. 1982. *The Barn Owl*. Vermillion, SD: Buteo Books.

Cartron J.-L. E., and D. W. Stahlecker. 2007. Barn owl (*Tyto alba*) use of cisterns as nest sites in New Mexico. *New Mexico Ornithological Society Bulletin* 35:91–95.

Cartron, J.-L. E., D. C. Lightfoot, J. E. Mygatt, S. L. Brantley, and T. K. Lowrey. 2008. *A field guide to the plants and animals of the Middle Rio Grande bosque*. Albuquerque: University of New Mexico Press.

Cartron, J.-L. E., P. J. Polechla Jr., and R. R. Cook. 2004. Prey of nesting ferruginous hawks in New Mexico. *Southwestern Naturalist* 49:270–76.1.

Cartron, J.-L. E., L. A. Sager III, and H. A. Walker. 2009. Notes on some breeding raptors of central and northern Lea County. *New Mexico Ornithological Society Bulletin* 37:7–14.

Cartron, J.-L. E., R. Sierra Corona, E. Ponce Guevara, R. E. Harness, P. Manzano-Fischer, R. Rodríguez-Estrella, and G. Huerta. 2006. Bird electrocutions and power poles in northwestern Mexico: an overview. *Raptors Conservation 2006* 7:4–14.

del Hoyo, J., A. Elliott, and J. Sargatal, eds. 1999. *Handbook of the birds of the world*. Vol. 5, *Barn Owls to Hummingbirds*. Barcelona, Spain: Lynx Edicions.

Havens, D. 1998. Bridge owls. *Aloft* 5(2): 8–9.

Howell, T. R. 1964. Notes on incubation and nestling temperatures and behavior of captive owls. *Wilson Bulletin* 76:28–36.

Hubbard, J. P. 1978. *Revised check-list of the birds of New Mexico*. Publ. no. 6. Albuquerque: New Mexico Ornithological Society.

Jorgensen, E. E., S. M. Sell, and S. Demarais. 1998. Barn Owl prey use in Chihuahuan Desert foothills. *Southwestern Naturalist* 43:53–56.

Keran, D. 1981. The incidence of man-caused and natural mortalities to raptors. *Raptor Research* 15:108–12.

Ligon, J. S. 1961. *New Mexico birds and where to find them*. Albuquerque: University of New Mexico Press.

Marti, C. D. 1992. Barn Owl (*Tyto alba*). No. 1. In *The birds of North America*, ed. A. Poole, P. Stettenheim, and F. Gill. Philadelphia, PA: Academy of Natural Sciences of Philadelphia.

———. 1994. Barn Owl reproduction: patterns and variation near the limit of the species' distribution. *Condor* 96:468–84.

———. 1997. Lifetime reproductive success in Barn Owls near the limit of the species' range. *Auk* 114:581–92.

Marti, C. D., and P. W. Wagner. 1985. Winter mortality in Common Barn-owls and its effect on population density and reproduction. *Condor* 87:111–15.

Marti, C. D., A. F. Poole, and L. R. Bevier. 2005. Barn Owl (*Tyto alba*). In *The birds of North America online*, ed. A. Poole. http://bna.birds.cornell.edu/BNA/account/Barn_Owl (accessed 6 March 2008).

Martin, D. J. 1973. Burrow digging by Barn Owls. *Bird-Banding* 44:59–60.

Millsap, B. A., and P. A. Millsap. 1987. Burrow nesting by Common Barn-owls in north central Colorado. *Condor* 89:668–70.

Otteni, L. C., E. G. Bolen, and C. Cottam. 1972. Predator-prey relationships and reproduction of the Barn Owl in southern Texas. *Wilson Bulletin* 84:434–48.

Pache, P. H. 1981. Prey remains in pellets from Burrowing and Barn Owls in central New Mexico. *New Mexico Ornithological Society Bulletin* 9:19–21.

Smith, D. G., C. R. Wilson, and H. H. Frost. 1974. History and ecology of a colony of Barn Owls in Utah. *Condor* 76:131–36.

Snider, P. R., ed. 1997. *New Mexico Ornithological Society Field Notes* 36(2): 25.

Taylor, I. 1994. *Barn Owls: predators-prey relationships and conservation*. Cambridge, UK: Cambridge University Press.

Terres, J. K. 1996. *The Audubon Society encyclopedia of North American birds*. New York: Alfred A. Knopf.

Williams, S. O. III, ed. 2000. *New Mexico Ornithological Society Field Notes* 39(3).

———, ed. 2003. *New Mexico Ornithological Society Field Notes* 42(2).

———. 2005a. *New Mexico Ornithological Society Field Notes* 44(1).

———. 2005b. *New Mexico Ornithological Society Field Notes* 44(4).

28

Flammulated Owl

(*Otus flammeolus*)

DAVID P. ARSENAULT

THE MOST DISTINGUISHING characteristics of the Flammulated Owl (*Otus flammeolus*) are its small size (length: 15–17 cm [5.9–6.7 in], mass: 45–63 g [~1.6–2.2 oz] [McCallum 1994]) and its dark brown eyes. Slightly larger than the Elf Owl (*Micrathene whitneyi*) and just smaller than the Whiskered Screech-Owl (*Megascops trichopsis*) (Sibley 2000), the Flammulated Owl is the second most diminutive owl in New Mexico. Only two other owls in the state have dark eyes, the Spotted Owl (*Strix occidentalis*) and the Barn Owl (*Tyto alba*), and both of these species are much larger. Also distinctive are the Flammulated Owl's partial facial disk and its ear tufts, conspicuous in camouflage posture (Grossman and Hamlet 1964).

Flammulated Owl body feathers are mostly gray and brown with a variable amount of rufous and white feathers that are most apparent around the facial disk and also along the base of each wing, in the shape of a conspicuous "V" on the back. The species name comes from the Latin *flammeus*, which means flame-colored. Many owls such as the Eastern Screech-Owl (*Megascops asio*) have a distinct rufous phase. Some authorities have concluded that no distinct rufous phase occurs in the Flammulated Owl (Voous 1988), but Marshall (1967, 1978) reported very rufous birds in Mexico and Guatemala.

The subspecific taxonomy of the Flammulated Owl remains unclear. Marshall (1967, 1978) did not recognize any subspecies, but Hekstra (1982) controversially described as many as six. Of these, however, Marshall (1997) concluded that three are clearly invalid, while the other three remain questionable. Individual plumage variation is high in Flammulated Owl populations, making recognition of potential subspecies difficult. A large-scale genetic study would help elucidate potential subspecies, as would further research in little-studied areas such as in Mexico.

The soft and ventriloquistic vocalizations of the Flammulated Owl include evenly spaced and deep territorial hoots, which become rapid and stuttered in the presence of intruding conspecifics or potential predators. Adults also utter a relatively loud, descending *meew* call, a screech-like vocalization that allies the Flammulated Owl with true screech-owls (now in the genus *Megascops*). Additionally, the adults give out low growling moans when their young are threatened, as well as quiet hoots, mews, and chittering calls during mate location, bonding, and copulation. Young birds make a raspy hiss.

PHOTO 28.1

(*above*) Adult Flammulated Owl in hand. Note the small size of the owl, its dark eyes, and the rufous and white feathers around the facial disk and along the base of each wing. Bird captured for banding (and later released) during the 2001 fall migration in the Manzano Mountains. PHOTOGRAPH: © JOHN P. DELONG.

PHOTO 28.2

(*right*) Adult Flammulated Owl perched on a branch in the Zuni Mountains, 19 June 2004. Note the dark eyes and the dense streaking and fine barring on the underparts. PHOTOGRAPH: © PETER STACEY.

Distribution

The summer distribution of Flammulated Owls extends from southern British Columbia to the mountains of central Mexico, and their winter distribution ranges from central Mexico south to northern Central America (McCallum 1994). As a neotropical migrant, the species seems adept at colonizing suitable breeding habitat even in remote areas. As evidence of this, it occurs in mountain ranges isolated by vast tracts of desert in Arizona, Nevada, and New Mexico and in forest patches as small as 40 hectares (100 acres) (Dunham et al. 1996; Arsenault et al. 2003).

In New Mexico, Flammulated Owls occur throughout montane mixed coniferous-deciduous and coniferous forests (see chapter 1 and Dick-Peddie 1993). They have been reported during the breeding season in the majority of the state's mountain ranges including the Animas, Black, Guadalupe, Jemez, Magdalena, Manzano, Mogollon, Sacramento, San Mateo, Sandia, Sangre de Cristo, Sante Fe, Tularosa, and Zuni Mountains (Ligon 1961; Balda et al. 1975; Johnson and Zwank 1990; Hurley and Gorresen 1991; McCallum 1994; McCallum et al. 1995; Arsenault et al. 2005; J. DeLong, pers. comm., B. Britton, pers. comm.; photo 28.3). During the breeding season, the species even occupies

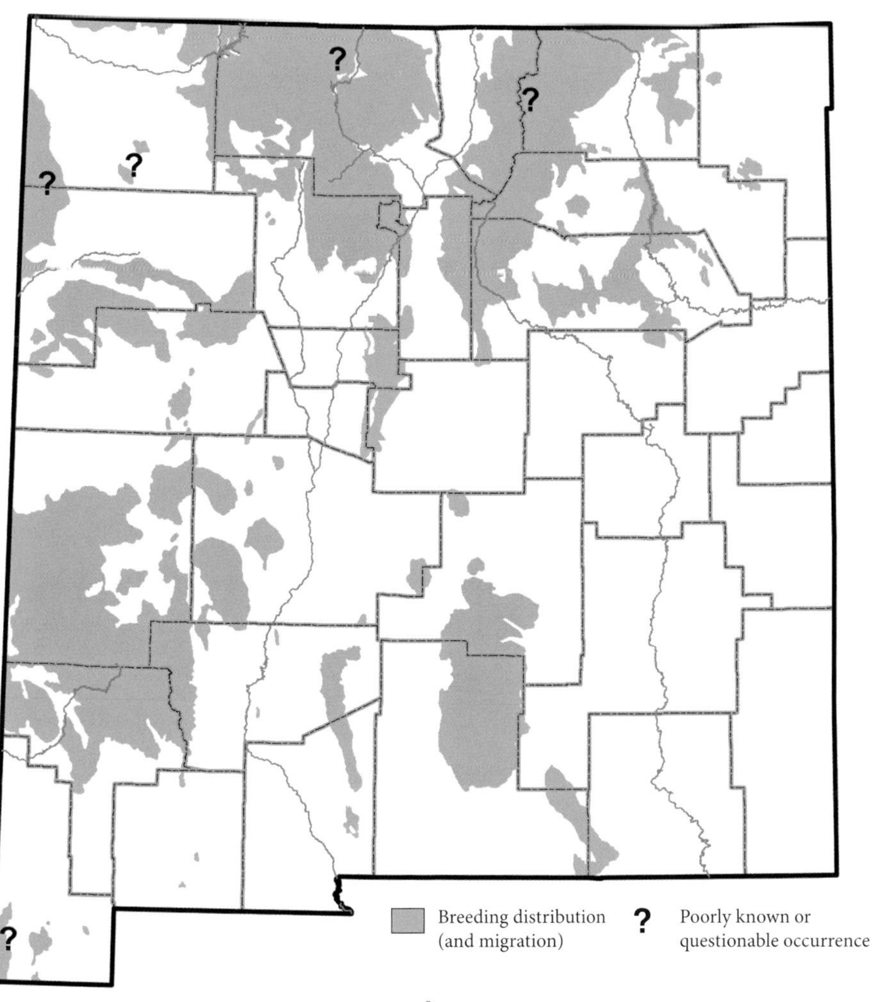

MAP 28.1

Flammulated Owl distribution map

PHOTO 28.3

Flammulated Owl, Santa Fe National Forest, Santa Fe Co., 16 June 2007.
PHOTOGRAPH: © JONATHAN BATKIN.

mountains as small and remote as Horse Mountain, located at the edge of the Plains of San Agustin near Datil, Catron County. It also likely breeds in the Peloncillo Mountains in the southwest and in the Chuska, San Pedro, Taos, and Tusas mountains in the north, although no breeding season records are known from these locations.

The Flammulated Owl is a long-distance, continental, north-south migrant (see chapter 2), which enables extensive intermixing of populations (Arsenault et al. 2005). Its migratory nature was first established based on numerous records of occurrence in nonbreeding habitat at the time of fall migration for other bird species (Collins et al. 1986) and on the rapid, sharp surge and subsequent decrease in the number of owls caught at owl-targeted migration banding stations (Balda et al. 1975; Hamilton 2002; Delong et al. 2005). The spring migration of Flammulated Owls through the southwestern United States is a rapid northward movement (Bent 1938; Balda et al. 1975). Early season records include a 15 March sighting in the Zuni Mountains, New Mexico (Rosenberg et al. 1981) and a 26 March sighting in the Santa Catalina Mountains, Arizona (Phillips 1942). Balda et al. (1975) reported that Flammulated Owl spring migrants traveled through pinyon pine (*Pinus* spp.)–juniper (*Juniperus* spp.)–Gambel oak (*Quercus gambelii*) habitat (2,040 m [6,690 ft]) on the east side of the Sandia Mountains from 16 April to 16 May. In the fall in the Manzano Mountains, DeLong (2004, 2006) and DeLong et al. (2005)

PHOTO 28.4

Flammulated Owl captured during the 2000 fall migration in the Manzano Mountains.

PHOTOGRAPH: © JOHN P. DELONG.

captured Flammulated Owls moving through ponderosa pine (*Pinus ponderosa*)–Gambel oak forest and associated montane meadows from 19 August through 18 October. Some captures reflected premigration dispersal of local hatch-year birds (i.e., recently fledged owls native of the Manzano Mountains). However, stable-isotope analyses of owl feathers also detected birds from outside the Manzano Mountains, and no local or nonlocal owls were trapped after 18 October, pointing to the end of the fall migration period by that time (DeLong et al. 2005). As documented elsewhere (e.g., Hamilton 2002), owls captured in the Manzano

Mountains in the fall had increased fat stores, no doubt for the energy to fly long distances during migration (DeLong 2006).

Flammulated Owls appear incapable of withstanding the cold temperatures found during winter in their breeding range in the United States (e.g., Banks 1964; Ligon 1968; Winter 1974). An adult male Flammulated Owl was found freshly dead in New Mexico in the Santa Fe area on 2 January 1996, following the first local snowstorm of the season and the sudden onset of cold temperatures (Williams 2007). Also in New Mexico, two adult female Flammulated Owls were found dead, again well beyond the end of the regular fall migration season, one at the village of Cochiti Lake, Sandoval County, on 28 November 2004, and the other at Bitter Lake National Wildlife Refuge, Chaves County, on 20 November 2005. The discovery of the dead owl at Cochiti Lake also followed the first snowstorm of the season (Williams 2007). While these records show that Flammulated Owls may rarely overextend their normal stay in the state, they also show the species' inability to survive cold weather conditions.

Habitat Associations

The primary nesting habitat of the Flammulated Owl is montane coniferous forest comprised of yellow pine (ponderosa pine in New Mexico) mixed with other conifers, such as Douglas fir (*Pseudotsuga menziesii*), at higher elevations, and with pinyon pine and juniper at lower elevations (McCallum 1994). Deciduous trees, including quaking aspen (*Populus tremuloides*), cottonwood (*Populus* spp.), and Gambel oak are other pine associates in Flammulated Owl habitat (Bent 1938; Marshall 1939; Johnson and Russell 1962; Winter 1974; Marcot and Hill 1980; McCallum and Gehlbach 1988; Bull et al. 1990; Dunham et al. 1996; Groves et al. 1997). Reynolds and Linkhart (1992) and Linkhart (2001) found that Flammulated Owls prefer old-growth ponderosa pine forests for nesting in Colorado. However, other studies have shown that Flammulated Owls are not limited to old-growth ponderosa pine. In Utah, Marti (1997) and Oleyar (2000) found owls in forests dominated by quaking aspen and big-toothed maple (*Acer grandidentatum*) with no yellow pine, and Powers et al. (1996) located owls in mixed deciduous forest without yellow pine in Idaho. Flammulated Owl nesting habitat must have nest cavities and a sufficient supply of insect prey, two requirements apparently met by second-growth forests.

In New Mexico in particular, the Flammulated Owl occurs in old-growth ponderosa pine forest—often

PHOTO 28.5

Typical Flammulated Owl nesting habitat in the Zuni Mountains, New Mexico, June 1996. A pair of owls occupied a nest cavity formed in the root of an old branch joint within a live Gambel oak (*Quercus gambelii*) in the center of the photograph, approximately 4 meters above the ground. After a limb drops off an oak, the joint often begins to rot. It is then expanded and shaped by primary cavity nesters, such as Acorn Woodpeckers (*Melanerpes formicivorus*) and Northern Flickers (*Colaptes auratus*). These cavities are subsequently utilized by many different secondary cavity nesters including Flammulated Owls.

PHOTOGRAPH: © PETER STACEY.

PHOTO 28.6

(*above*) The author (David Arsenault) using a fish net to capture an adult female in a nest cavity in the Zuni Mountains, June 2001. The area surrounding the cavity is typical of some of the Flammulated Owl nesting habitat in New Mexico: relatively open pine-oak woodland dominated by Gambel oak (*Quercus gambelii*) and ponderosa pine (*Pinus ponderosa*).

PHOTOGRAPH: © PETER STACEY.

PHOTO 28.7

(*right*) Flammulated Owl nesting habitat in the Zuni Mountains, New Mexico, May 1997. In the center of the photograph is a Gambel oak (*Quercus gambelii*) used by nesting Flammulated Owls.

PHOTOGRAPH: © PETER STACEY.

in association with Mexican Spotted Owls (*Strix occidentalis lucida*) (Arsenault 1999). It also nests in other vegetation types ranging widely in tree species composition, canopy closure, and tree density. Dense vegetation where Flammulated Owls have been found in New Mexico include aspen stands in Douglas fir forest with no ponderosa or other pines present and dense Gambel oak–ponderosa pine forest with high oak densities due to extensive historic logging of ponderosa pine (Arsenault 1999, 2004).

Flammulated Owls are cavity nesters. In New Mexico, they prefer tree cavities excavated by Northern Flickers (*Colaptes auratus*) but will also nest in live or dead trees in the smaller cavities made by Acorn Woodpeckers (*Melanerpes formicivorus*) or sapsuckers (*Sphyrapicus* spp.), as well as in knotholes (Arsenault 2004). Additionally, researchers have reported nest box use by Flammulated Owls in several areas within their distribution (Hasenyager et al. 1979; Cannings and Cannings 1982; Bloom 1983; Reynolds and Linkhart 1987; Marti 1997; S. Yasuda, pers. comm.). However, use of nest boxes by the owl is not as dependable as with other species, for example bluebirds (*Sialia* spp.) (Zeleny 1977). The author and Dr. Arch McCallum put up 80 nest boxes in aspen, ponderosa pine, and Douglas fir habitat in the Jemez Mountains of northern New Mexico specifically targeting use by Flammulated Owls. From 1995 to 1997, the nest boxes were used only by House Wrens (*Troglodytes aedon*), Western Bluebirds (*Sialia mexicana*),

PHOTO 28.8

Close-up of a branch joint nest in a Gambel oak (*Quercus gambelii*) in the Zuni Mountains, New Mexico, June 2000. A female Flammulated Owl is perched on the inside of the entrance to the cavity, with her eyes closed and her head blocking the opening. Because the plumage pattern of the female is very similar to that of the oak bark, this behavior makes the cavity entrance even more difficult to detect.

PHOTOGRAPH: © PETER STACEY.

PHOTO 28.9

(*above*) Female Flammulated Owl nesting in a Northern Flicker (*Colaptes auratus*) hole approximately 5 meters above the ground in a charred ponderosa pine (*Pinus ponderosa*) snag, Zuni Mountains, June 2001.

PHOTOGRAPH: © PETER STACEY.

PHOTO 28.10

(*left*) Female Flammulated Owl in a nest box in the Zuni Mountains, 29 July 2004. The nesting pair used the box the year after it was erected, in an area that had few natural cavities.

PHOTOGRAPH: © PETER STACEY.

and White-breasted Nuthatches (*Sitta carolinensis*), while Flammulated Owls nested in natural cavities as close as 5 m (16 ft) from the nest boxes! On the other hand, the author and Dr. P. Stacey put up nest boxes for Flammulated Owls in the Zuni Mountains, and 30% (3) of 10 available boxes were used in 2004, 18% (8) of 45 available boxes were used in 2005, and 8% (4) of 52 available boxes were used in 2006. The nest boxes were placed in an area that had low natural cavity and owl densities, and after only two breeding seasons the density of occupied territories doubled from 8 to 16 on 138 ha (340 ac). The other half of the study area (without boxes) had 2 ½ times the density of natural cavities (36.32/100 ha or 14.7/100 ac) and 17 occupied territories (2.9 cavities/territory), also on 138 ha (340 ac).

Life History

Nesting

Adult Flammulated Owls exhibit high fidelity to breeding sites (Linkhart 2001; appendix 28-1). In the Zuni Mountains of New Mexico, males returning from migration moved up to 360 m (1,180 ft) when occupying a new cavity (appendix 28-1). However, the majority of males reoccupied the same territory as during the previous nesting season, in most cases using the exact same nesting cavity. Some females moved farther than the males (up to 900 m [2,950 ft]), but stayed on the same territory and nested in the same cavity in 50% and 35% of all cases, respectively (appendix 28-1). Linkhart (2001) reported similar return rates and dispersal distances, but he observed higher territory fidelity (98% male,

PHOTO 28.11

Female Flammulated Owl in a nest box with two nestlings, Zuni Mountains, 4 July 2004. The female remained motionless when the nest box was opened and while the nestlings were temporarily removed from the box for banding.

PHOTOGRAPH: © PETER STACEY.

PHOTO 28.12

Two nestling Flammulated Owls, approximately one week before fledging, temporarily removed from a nest box in the Zuni Mountains, 9 July 2005. Note that the nestling on the left is significantly smaller than its nest mate. This situation is not uncommon in owl nests, and may reflect limited food availability that prevents a smaller chick from growing as fast as its sibling(s). The nestlings were reinserted inside the nest box after the photograph was taken.

PHOTOGRAPH: © PETER STACEY.

78% female) during a 19-year mark-recapture study in Colorado. In New Mexico, a juvenile male bred only 180 m (590 ft) and a female 695 m (2,280 ft) from their birth sites, but the majority of juveniles (97%) did not return to the study area (appendix 28-1). Genetic results show substantial intermixing of Flammulated Owl populations in New Mexico. Combined with mark-recapture data, results suggest that interpopulation dispersal, and ultimately gene flow, may be attributed largely to long-distance natal dispersal (Arsenault et al. 2005).

The Flammulated Owl is a monogamous species that maintains long-term pair bonds (Linkhart 2001; Arsenault et al. 2005). Reynolds and Linkhart (1990) observed one case of extra-pair copulation in Colorado, but Arsenault et al. (2002) used DNA fingerprinting to determine that extra-pair fertilization did not occur in 17 owl pairs despite sampling from groups of owls nesting in close proximity. In New Mexico, the author observed pair bonds maintained up to four years, as well as three cases of pair separation. When returning from migration, three females did not breed with their previous mate, even though that mate occupied the same territory as the year before. Instead they nested with a new mate, a male from a neighboring territory. The reasons for pair separation were unknown, though in one instance pair separation followed nest failure (due to nest depredation by a chipmunk [*Neotamias* sp.]) the year before. Males bred with up to three different females and females with up to two different males during their known life spans.

The Flammulated Owl's life history strategy revolves around high longevity and low reproductive output, including what are among the lowest and least variable clutch sizes for North American owls. Linkhart (2001) reported longevity rates of 12 years for males and 8 years for females in Colorado, and the author has recaptured eight-year-old males and seven-year-old females in New Mexico (Arsenault et al. 2005). Although breeding chronology varies annually and according to elevation and latitude, in general males in New Mexico begin singing on breeding grounds in early May and females lay one to four eggs (generally two or three) a few weeks later. The incubation and nestling stages last for about three weeks each, thus eggs hatch around the middle of June, and nestlings fledge in early July in New Mexico (McCallum et al. 1995).

Diet and Foraging

Flammulated Owls eat only insects. From 1997 to 1999, the author used miniature cameras to record prey deliveries to 15 nests in four mountain ranges of New Mexico (appendix 28-2). The prey selected by owls during the nestling stage consisted of 66% Lepidoptera adults (moths), 20% Orthopterans (crickets and grasshoppers), 10% Coleopterans (beetles), and 4% Lepidopteran larvae (caterpillars)—all three of these orders of insects are consistently dominant in the diet of the owl throughout its breeding range (McCallum 1994). Prey delivery rates were greatest during the first hour after dusk and increased steadily through the nestling stage, similar to what McCallum et al. (1995) found earlier, also in New Mexico. The males did not appear to intensify their effort to feed the young toward the end of the nestling stage. Instead, females spent progressively less time brooding and more time feeding young, thus increasing overall feeding rates until fledging (appendix 28-2). Reynolds and Linkhart (1987) made comparable observations of prey delivery in Colorado and also found that after nestlings have fledged they associate into subgroups, each attended by one parent.

A stomach content analysis revealed that the owl found dead on 2 January 1996 in the Santa Fe area had eaten a diet of earwigs (*Forficula auricularia*), ground beetles (Carabidae), and darkling beetles (Tenebrionidae), likely captured while foraging in moist litter on the ground (Williams 2007).

Predation and Interspecific Interactions

One case of nest depredation (two nestlings) by a chipmunk was documented in New Mexico in 2000, and in August 2004 a Flammulated Owl was mortally wounded by a cat in the Gallup area (J. Harden, pers. comm.). No other cases of nest depredation or predation are known for the state. Red squirrels (*Tamiasciurus*) or other rodents depredated seven nests in Colorado (Reynolds and Linkhart 1998), and a flying squirrel depredated a nest in a nest box in British Columbia (Cannings and Cannings 1982). Black bears (*Ursus americanus*) are also potential nest predators that have been documented depredating Northern Flicker and Western Bluebird nests in the Zuni Mountains in New Mexico (D. Arsenault, unpubl. data), but there are no confirmed cases for the Flammulated Owl. Predation of fledglings and adults by Great Horned Owl (*Bubo virginianus*), Mexican Spotted Owl, Cooper's Hawk (*Accipiter cooperii*), and Sharp-shinned Hawk (*Accipiter striatus*) has been documented elsewhere in the Flammulated Owl's geographic range (Zeiner et al. 1990; McCallum 1994; Reynolds and Linkhart 1998), and all of these predators coexist with Flammulated Owls in New Mexico.

Flammulated Owls often nest in close proximity to other cavity-nesting birds, even other owls such as the Northern Saw-whet Owl (*Aegolius acadicus*), sometimes in different cavities of the same tree (McCallum 1994; Arsenault 1999). Interspecific aggression can occur among cavity-nesting species, however, and Flammulated Owls have been reported to expel bluebirds from nests (McCallum 1994).

Status and Management

Flammulated Owls are very common—Marshall (1939) once suggested they were possibly the most common raptor in montane forests of the western United States. At the same time, the U.S. Forest Service lists the Flammulated Owl as a conservation sensitive species in four of its administrative regions, including Region 3, which encompasses New Mexico.

PHOTO 28.13

New Mexico wildlife rehabilitators occasionally receive into their care Flammulated Owls that have sustained injuries. In August 2004, for example, Wildlife Rescue, Inc. received a Flammulated Owl rescued in the Gallup area after being attacked by a cat. That particular owl did not survive. Others like the bird in this photograph survive but due to the severity of their injuries cannot be released into the wild. They become educational birds.

PHOTOGRAPH: © ROGER HOGAN.

PHOTO 28.14

Female owl at the entrance to a natural nest cavity in a Gambel oak (*Quercus gambelii*) in the Zuni Mountains, 25 June 2004. The author and Peter Stacey found that Flammulated Owls often used nest cavities made by Acorn Woodpeckers (*Melanerpes formicivorus*) in oaks. Because oak wood is very dense, the cavities remain intact for many years, usually until the tree itself falls down. Acorn Woodpeckers are one of the few species that can construct new nest cavities in oaks, because of the hardness of the wood. Acorn woodpeckers became extinct in the Zuni Mountains sometime in the late 1970s or early 1980s, and were absent from the area during the Flammulated Owl study. As a result, no new cavities are being made in oaks. Unfortunately, many of the larger oaks in the Zuni Mountains have recently been cut down for firewood, both on private land and under permit from the U.S. Forest Service. As a result, the number of nest sites available to both primary and secondary cavity nesters in Arsenault and Stacey's study area has been reduced. Whether or not this will eventually lead to a decline in the local populations of these species remains to be determined, but it appears likely.

PHOTOGRAPH: © PETER STACEY.

Timber management practices in the western United States certainly have the potential to influence the Flammulated Owl's viability since the species is generally found in commercially valuable forests. However, as shown in this chapter, numerous studies have found that Flammulated Owls use a wider range of forested habitat types than just yellow pine or old-growth forest, the main focus of the timber industry.

Because of the apparently high degree of interconnectedness among Flammulated Owl populations, it is important to manage this species at a broad scale, ensuring that suitable habitat is distributed throughout its range. Future studies should further delineate the Flammulated Owl's breeding distribution in part by compiling the large amount of data recorded for this species (e.g., from unpublished U.S. Forest Service survey reports and databases as well as records from the birding literature). Distribution-wide genetic studies would also be helpful to establish whether genetically distinct populations exist, analyzing any differences in the genetic makeup of the Rocky Mountain population compared to that of the Sierra Nevada, Cascade, and Great Basin populations. Further analysis of long-term mark-recapture data (e.g., Linkhart 2001; Arsenault et al. 2005) will be important in understanding the species' population dynamics.

Where owl pair densities are low, it may be possible to use nest boxes as a conservation tool to expand existing breeding populations. Nest boxes have been shown to be an effective conservation tool for other cavity-nesting raptors including the American Kestrel (*Falco sparverius*) (Hamerstrom et al. 1973) and the Boreal Owl (*Aegolius funereus*) (Hayward et al. 1992). Most important, however, research is needed to better understand the status and distribution of the Flammulated Owl in Central America in winter and in Mexico both during the breeding season and in winter. Habitat loss and pesticide use in Latin America are some of the threats that may pose the greatest conservation risk to the Flammulated Owl.

LITERATURE CITED

Arsenault, D. P. 1999. The ecology of Flammulated Owls: nest-site preferences, spatial structure, and mating system. M.Sc. thesis. University of Nevada, Reno.

———. 2004. Differentiating nest sites of primary and secondary cavity-nesting birds in New Mexico. *Journal of Field Ornithology* 75:257–65.

Arsenault, D. P., L. Neel and G. E. Wilson. 2003. Flammulated Owls in the Spring Mountains, Nevada. *Great Basin Birds* 6:45–51.

Arsenault, D. P., P. B. Stacey, and G. A. Hoelzer. 2002. No extra-pair fertilization in Flammulated Owls despite aggregated nesting. *Condor* 104:197–201.

———. 2005. Mark-recapture and DNA fingerprinting data reveal high breeding site fidelity, low natal philopatry, and low levels of genetic population differentiation in Flammulated Owls. *Auk* 122:329–37.

Balda, R. P., B. C. McKnight, and C. D. Johnson. 1975. Flammulated Owl migration in the southwestern United States. *Wilson Bulletin* 87:520–33.

Banks, R. C. 1964. An experiment on a Flammulated Owl. *Condor* 66:79.

Bent, A. C. 1938. Flammulated Screech Owl, *Otus flammeolus*. In *Life histories of North American birds of prey*. U.S. National Museum Special Bulletin 170, 291–95. Washington, DC; Smithsonian Institution.

Bloom, P. H. 1983. Notes on the distribution and biology of the Flammulated Owl in California. *Western Birds* 14:49–52.

Brawn, J. D. 1990. Interspecific competition and social behavior in Violet-green Swallows. *Auk* 107:606–8.

Bull, E. L., A. L. Wright, and M. G. Henjum. 1990. Nesting habitat of Flammulated Owls in Oregon. *Journal of Raptor Research* 24:52–55.

Cannings, R. J., and S. R. Cannings. 1982. A Flammulated Owl nest in a nest box. *Murrelet* 63:66–68.

Collins, P. W., C. Drost, and G. M. Fellers. 1986. Migratory status of Flammulated Owls in California, with recent records from the California Channel Islands. *Western Birds* 17:21–31.

DeLong, J. P. 2004. Age determination and preformative molt in hatch-year Flammulated Owls during the fall. *North American Bird Bander* 29:111–15.

———. 2006. Pre-migratory fattening and mass gain in Flammulated Owls in central New Mexico. *Wilson Journal of Ornithology* 118:187–93.

Delong, J. P., T. Meehan, and R. Smith. 2005. Investigating fall movements of Flammulated Owls (*Otus flammeolus*) in central New Mexico using stable hydrogen isotopes. *Journal of Raptor Research* 39:19–25.

Desmond, M. J. 1997. Evolutionary history of the genus Speotyto: a genetic and morphological perspective. Ph.D. diss., University of Nebraska.

Dick-Peddie, W. A. 1993. *New Mexico vegetation: past, present, and future*. Albuquerque: University of New Mexico Press.

Dunham, S., L. Butcher, D. A. Charlet, and J. M. Reed. 1996. Breeding range and conservation of Flammulated Owls (*Otus flammeolus*) in Nevada. *Journal of Raptor Research* 30:189–93.

Grossman, M. L., and J. Hamlet. 1964. *Birds of prey of the world*. New York: Bonanza Books.

Groves, C., T. Frederick, G. Frederick, E. Atkinson, M. Atkinson, J. Shepherd, and G. Servheen. 1997. Density, distribution and habitat of Flammulated Owls in Idaho. *Great Basin Naturalist* 57:116–23.

Hamerstrom, F., F. N. Hamerstrom, and J. Hart. 1973. Nest-boxes: An effective management tool for Kestrels. *Journal of Wildlife Management* 37:400–403.

Hamilton, S. L. 2002. A comparative analysis of the ecology and physiology of the Flammulated Owl (*Otus flammeolus*) and Northern Saw-whet Owl (*Aegolius acadicus*) during fall migration. M.Sc. thesis. University of Idaho, Moscow.

Hasenyager, R. N., J. C. Pederson, and A. W. Heggen. 1979. Flammulated Owl nesting in a squirrel box. *Western Birds* 10:224.

Hayward, G. D., R. K. Steinhorst, and P. H. Hayward. 1992. Monitoring Boreal Owl populations with nest-boxes: sample size and cost. *Journal of Wildlife Management* 56:777–85.

Hekstra, G. P. 1982. Description of twenty-four new subspecies of American *Otus*. *Bulletin of the Zoological Museum of the University of Amsterdam* 9:49–63.

Hurley, T., and M. Gorresen. 1991. Notes on owl species detected during the 1991 Sandia Spotted Owl survey. Unpublished report. U.S. Forest Service, Albuquerque, NM.

Johnson, E. D., and P. J. Zwank. 1990. Flammulated Owl biology on the Sacramento unit of the Lincoln National Forest. Unpublished report. U.S. Forest Service, Las Cruces, NM.

Johnson, N. K., and W. C. Russell. 1962. Distributional data on certain owls in the western Great Basin. *Condor* 64:513–14.

Ligon, J. D. 1968. Starvation of spring migrants in the Chiricahua Mountains, Arizona. *Condor* 70:387–88.

Ligon, J. S. 1961. *New Mexico birds and where to find them*. Albuquerque: University of New Mexico Press.

Linkhart, B. D. 2001. Life history characteristics and habitat quality of Flammulated Owls (*Otus flammeolus*) in Colorado. Ph.D. diss. University of Colorado, Boulder.

Linkhart, B. D., and R. T. Reynolds. 1987. Brood division and post-nesting behavior of Flammulated Owls. *Wilson Bulletin* 99:240–43.

Marcot, B. G., and R. Hill. 1980. Flammulated Owls in northwestern California. *Western Birds* 11:141–49.

Marshall, J. T. 1939. Territorial behavior of the Flammulated Screech-owl. *Condor* 41:71–78.

———. 1967. Parallel variation in North and Middle American Screech Owls. *Monographs of the Western Foundation of Vertebrate Zoology* 1:1–72.

———. 1978. Systematics of smaller Asian night birds based on voice. *Ornithological Monographs* 25:1–58.

———. 1997. Allan Phillips and the Flammulated Owl. In *The era of Allan R. Phillips: a Festschrift*, ed. R. W. Dickerman, 87–91. Albuquerque, NM: Horizon Communications.

Marti, C. D. 1997. Flammulated Owls (*Otus flammeolus*) breeding in deciduous forests. In *Biology and conservation of owls of the northern hemisphere. Second international symposium, Winnipeg, Manitoba*, ed. J. R. Duncan, D. H. Johnson, and T. H. Nicholls, 262–66. General Technical Report NC-190. Washington, DC: USDA, Forest Service.

McCallum, D. A. 1994. Flammulated Owl (*Otus flammeolus*). No. 93. In *The birds of North America*, ed. A. Poole and F. Gill. Philadelphia, PA: Academy of Natural Sciences, and Washington, DC: American Ornithologists' Union.

McCallum, D. A., and F. R. Gehlbach. 1988. Nest-site preferences of Flammulated Owls in western New Mexico. *Condor* 90:653–61.

McCallum, D. A., F. R. Gehlbach, and S. W. Webb. 1995. Life history and ecology of Flammulated Owls in a marginal New Mexico population. *Wilson Bulletin* 107:530–37.

Monroe, B. L., and C. G. Sibley. 1993. *A world checklist of birds*. New Haven: Yale University Press.

Oleyar, M. D. 2000. Flammulated Owl (*Otus flammeolus*) breeding ecology in aspen forests of northern Utah: including responses to ski area development. M.Sc. Thesis. Boise State University, Boise, ID.

Oring, L. W., and D. B. Lank. 1984. Breeding area fidelity, natal philopatry and site fidelity in the polyandrous Spotted Sandpiper. In *Shorebirds: breeding behavior and populations*, ed. J. Burger and B. L. Olla, 125–47. New York: Plenum Press.

Phillips, A. R. 1942. Notes on the migration of Elf and Flammulated Screech-Owls. *Wilson Bulletin* 54:132–37.

Powers, L. R., A. Dale, P. A. Gaede, C. Rodes, L. Nelson, J. J. Dean, and J. D. May. 1996. Nesting and food habits of the Flammulated Owl (*Otus flammeolus*) in south central Idaho. *Journal of Raptor Research* 30:15–20.

Reynolds, R. T., and B. D. Linkhart. 1987. The nesting biology of Flammulated Owls in Colorado. In *Biology and conservation of northern forest owls*, ed. R. W. Nero, R. J. Clark, R. J. Knapton, and R. H. Hamre, 239–48. General Technical Report RM-142. Washington, DC: U.S. Forest Service.

———. 1990. Extra-pair copulation and extra-range movements in Flammulated Owls. *Ornis Scandinavica* 21:74–77.

———. 1992. Flammulated Owls in ponderosa pine: evidence of preference for old growth. In *Old-growth forests in the Southwest and Rocky Mountain regions*, ed. M. R. Kaufman, W. H. Moir, and R. L. Bassett, 166–69. General Technical Report RM-213. Washington, DC: U.S. Forest Service.

———. 1998. Flammulated Owl (*Otus flammeolus*). In *The raptors of Arizona*, ed. R. L. Glinski, 140–44. Tucson: University of Arizona Press.

Rosenberg, K. V., J. P. Hubbard, and S. B. Terrill. 1981. The spring migration: Southwest region. *American Birds* 35:849–52.

Sibley, D. A. 2000. *The Sibley guide to the birds*. New York: Chanticleer Press.

Voous, K. H. 1988. *Owls of the northern hemisphere*. Cambridge, MA: MIT Press.

Williams, S. O. III. 2007. A January specimen of the Flammulated Owl from northern New Mexico. *Wilson Journal of Ornithology* 119:764–66.

Winter, J. 1974. The distribution of the Flammulated Owl in California. *Western Birds* 5:25–44.

Zeiner, D. C., W. F. Laudenslayer Jr., K. E. Mayer, and M. White, eds. 1990. *California's wildlife*. Vol. 2, *Birds*. Sacramento: California Department of Fish and Game.

Zeleny, L. 1977. Nesting box programs for bluebirds and other passerines. In *Endangered birds: management techniques for preserving threatened species*, ed. S. A. Temple, 55–60. Madison: University of Wisconsin Press.

29

Western Screech-Owl

(*Megascops kennicottii*)

FREDERICK R. GEHLBACH AND SCOTT H. STOLESON

THE WESTERN SCREECH-OWL (*Megascops kennicottii*) is a permanent resident and the largest of New Mexico's eight small owls. At an average 116 g (4.1 oz), it is 1.3 times more massive than the Whiskered Screech-Owl (*M. trichopsis*) but similar to the Eastern Screech-Owl (*M. asio*), the state's other two small owls with dark-streaked, white-flecked gray plumage resembling gray tree bark (see chapters 30 and 40). Westerns are about 10% larger north and east of the Mogollon Rim, where females are 18% more massive than males. This geographic race is *M. k. aikeni*. Southward in the state the subspecies *M. k. suttoni* is smaller and more coarsely marked, and its females are 13% larger than males. All females are slightly darker gray than males.

Western and Eastern Screech-Owls differ genetically (Proudfoot et al. 2007). Also, Westerns lack the Eastern's rufous color-morph and differ vocally, but the two occasionally mimic each other (Gehlbach 2003). Western and Eastern Screech-Owls were considered conspecific until the American Ornithologists' Union separated them in its 6th edition *Check-list of North American Birds* (AOU 1983). In the 44th supplement to the 7th edition (Banks et al. 2003), all New World relatives except the Flammulated Owl (*Otus flammeolus*) were removed from the genus *Otus* and placed in their original genus, *Megascops*.

In New Mexico generally, Western Screech-Owls measure 19–25 cm (7.5–10 in) from head to tail. Their iris (eye) is yellow, and like other screech-owls they have prominent gray facial disks, distinct ear tufts (feathers), broad rounded wings, and short rounded tail. When approached, roosting individuals enhance their cryptic appearance by sitting erect with raised ear tufts, eyes mostly closed, and wings and body feathers compressed against the body, all typical of the screech-owl "tree bark" pose. Juveniles have distinctive light and darker gray, barred breast plumage until their prebasic molt, another screech-owl characteristic (FRG).

Like many other owls in which females are larger, both sexes sing (several closely spaced notes) and call (one or more widely spaced notes), females at a higher pitch than males. Males defend nest holes and advertise for mates with a bouncing-ball song of whistles more closely spaced toward the end. Double-trill songs used in pair communication are whistles, a short pause, then more whistles. Both sexes call with single notes such as hoots at potential danger, barks if agitated, and screeches if very upset. When begging food, females and nestlings give a several-note twitter, and nest-attending males periodically give a single "*pew*" contact call that may be answered by nesting females.

Singing is more frequent during the approach to

PHOTO 29.1

(*top*) Western Screech-Owl at Kofa National Wildlife Refuge in Arizona, 3 May 2008. The Western Screech-Owl is a small owl with yellow irises, prominent gray facial disks, distinct ear tufts (feathers), broad rounded wings, and a short rounded tail. Adults have gray to black bills except for the tip, which can be cream or yellow through apparent wear. This is in contrast to the closely related Eastern Screech-Owl (*Megascops asio*), which has a yellow to olive-yellow bill. Although the Western and Eastern Screech-Owls are also—and mainly—identified by call, they can mimic each other in areas where their distributions overlap. PHOTOGRAPH: © TOM KENNEDY.

PHOTO 29.2

(*bottom*) Western Screech-Owl trapped at HawkWatch International's Manzano Mountains study site, fall 2001. Note the bill, dark except at the tip. PHOTOGRAPH: © JOHN P. DELONG.

a full moon, when neighboring pairs may chorus in response to a stranger's song or recording. Choruses were first noted in New Mexico (Bailey 1928), later in Arizona (Marshall 1964; Johnson et al. 1981), and currently in Texas, where their timing and function are being studied (S. Kennedy, pers. comm.). Chorusing may be related to aggregated nesting, which is not yet proven for this species. Western Screech-Owls are like other screech-owls in recognizing individual voices and sing defensively at intruding strangers but not at close neighbors (Lohr and Gehlbach, unpubl. data).

Distribution

In diversity and area of habitats occupied, the Western Screech-Owl is the most widespread small owl in western North America. It ranges from southeastern coastal Alaska and British Columbia, Canada, southward through all U.S. states between the Pacific coast lowlands and the Rocky Mountains, reaching the Mexico City region on the central Mexican Plateau (the most recent distribution map is in Proudfoot et al. 2007). Only extensive grassland and deserts lacking riparian habitat, lowland tropical forest, and high montane elevations are avoided.

PHOTO 29.3

(*above*) Western Screech-Owl, Abo National Monument, Torrance Co., May 1971. The Western Screech-Owl is a common owl in central New Mexico, occurring particularly in the Middle Rio Grande bosque and in wooded foothills on the eastern side of the Sandia and Manzano mountains. PHOTOGRAPH: © CALVIN SMITH.

PHOTO 29.4

(*left*) Western Screech-Owl, photographed in the backyard of a private residence in Albuquerque, Bernalillo Co., December 2006. Based on distance, the owl in this photograph may have originated in the riparian cottonwood forest (bosque) along the Rio Grande, or in a nearby city park. PHOTOGRAPH: © NANCY BACZEK.

PHOTOS 29.5a and b

Possible Eastern x Western Screech-Owl hybrid, Black Mesa State Park, Cimarron Co., Oklahoma, 24 March 2007. The bird in the photograph has the black bill of a Western Screech-Owl while it also shows the rufous feather pattern and broad breast streaks of an Eastern Screech-Owl. Its vocalizations were those of a Western Screech-Owl (but the two species can mimic each other). Note that no DNA analysis was conducted to establish that this owl is a hybrid, a first for the state of Oklahoma. Hybrids have been documented in Colorado and are not impossible in New Mexico, where the Eastern Screech-Owl has now been reported.

PHOTOGRAPHS: © STEVE METZ.

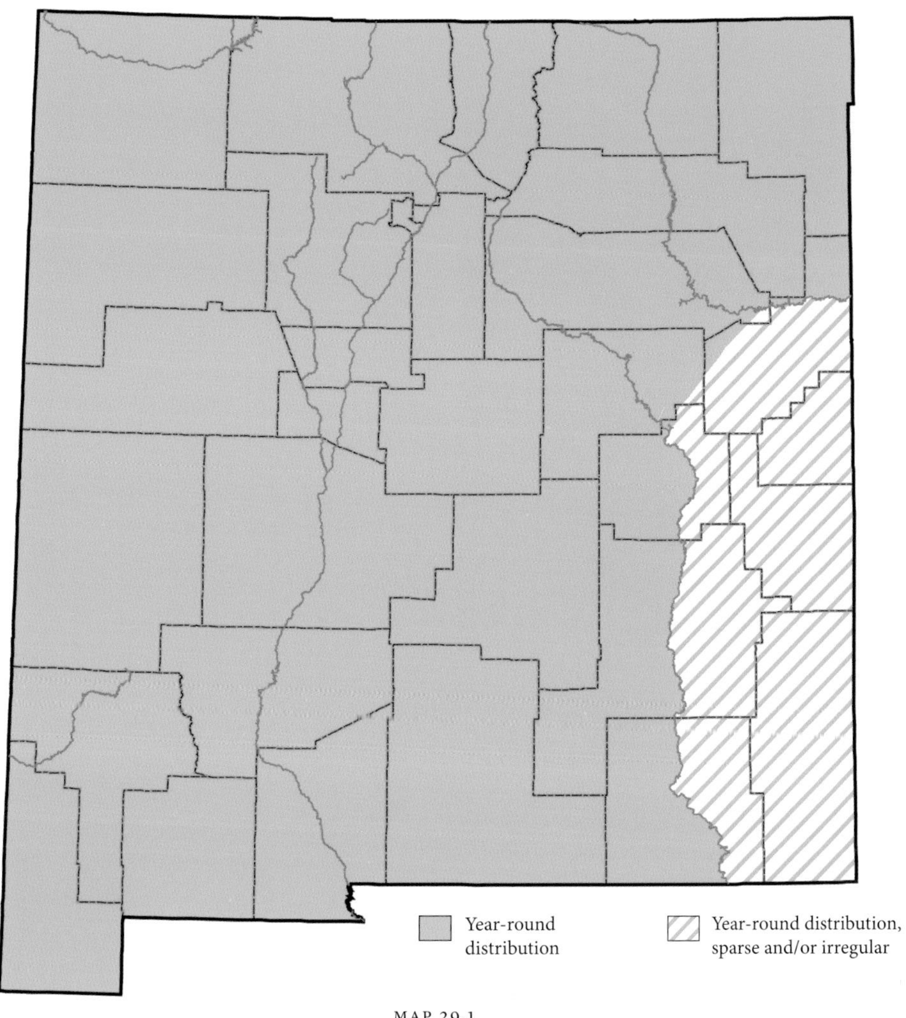

MAP 29.1

Western Screech-Owl distribution map

Western Screech-Owls nest over most of New Mexico, ranging in elevation from lowest riparian areas with tree cavities in desert and grassland mainly to about 2,360 m (7,800 ft) in the mountains, with one record at 2,700 m (8,800 ft; Bailey 1928). They are the most commonly encountered small owl overall in the state. For example, the species is common in Guadalupe Canyon, Hidalgo County; in Water Canyon in the Magdalena Mountains, Socorro County; and at Rattlesnake Springs, Eddy County. It is well known from the San Juan River Valley, San Juan County; the Lower and Middle Rio Grande Valley; (Las) Animas Creek in Sierra County; and the Gila River Valley, Grant and Hidalgo counties (e.g., Hubbard 1971; Cartron et al. 2008). Only Curry, Lea, Mora, and Guadalupe counties lack verified records, surely due to insufficient study.

This familiar owl occurs in the western panhandle of Texas and throughout Trans-Pecos Texas, where it sometimes hybridizes with the Eastern Screech-Owl. Western Screech-Owls reach an eastern limit on the Edwards Plateau of central Texas but do not hybridize with Eastern Screeches there. Biogeographically, this distribution resembles those of other east-west pairs of closely related birds such as orioles, grosbeaks, and

PHOTO 29.6

(*above*) Western Screech-Owl in a nest box at Bitter Lake National Wildlife Refuge, March 2006. The nest box (in a line of elms [*Ulmus* sp.]) was one of several installed by Bitter Lake NWR staff for American Kestrels (*Falco sparverius*). The Western Screech-Owl vacated the nest box shortly after the photo was taken. The nest box was used instead by American Kestrels. The Western Screech-Owl is not known to nest at Bitter Lake NWR, where habitat is probably unsuitable to the species.

PHOTOGRAPH: © GORDON WARRICK.

titmice, which were probably separated during the last (Wisconsin) glacial period and have reconnected in the Great Plains region during the last 10,000 years (Dixon 1989; Gehlbach 2003).

Other coexisting Western and Eastern Screech-Owls are in southeastern Colorado, where hybridization occurs rarely, and in northern Mexico, where neither species is well studied. However, in all overlap areas the two show character displacement in that their beaks ("culmen measurements") are 8% different in average length compared to only 3% east and west of overlaps (Gehlbach 2003). Such character displacement is known in other birds but not in other owls.

Habitat Associations

New Mexico's Western Screech-Owls occupy a wide range of wooded habitats from low-elevation deciduous forests along lowland creeks and rivers through juniper-oak woodlands to mixed deciduous–evergreen forest in mountain canyons and coniferous forest (Bailey 1928; Ligon 1961; Hubbard 1978). They also nest in urban parks. During FRG's studies in New Mexico's Zuni and New Mexico–Texas' Guadalupe mountains, and also in Arizona's Chiricahua and Huachuca mountains, highest nests were 1,700 m (5,600 ft) but more numerous below 1,520 m (5,000 ft; contrary information credited to FRG is misstated in Cannings and Angell 2001). Postbreeding wanderers and dispersing juveniles were occasionally observed by FRG to about 2,200 m (7,200 ft).

Optimum habitat in New Mexico is riparian woodland and forest with mixtures of deciduous or deciduous plus evergreen trees along stream edges, and on their terraces and adjacent lower slopes. Hubbard (1971) lists cottonwoods, willows, hackberries, walnuts, box elders, sycamores, and other deciduous trees along the lower Gila River, where cottonwoods and desert willows (*Chilopsis linearis*) are known nest trees (SHS). In Cherry Creek Canyon at Pinos Altos,

PHOTO 29.7

(*left*) Western Screech-Owl, Las Animas Creek west of Caballo Reservoir, Sierra Co., 19 November 2004.

PHOTOGRAPH: © ADAM D'ONOFRIO.

PHOTOS 29.8a and b

(*top left and right*) Western Screech-Owl in a private residence's yard in Corrales, Sandoval Co., 12 October 2008. PHOTOGRAPHS: © JANET RUTH.

PHOTO 29.9

(*left*) Western Screech-Owl in a saltcedar (*Tamarix* sp.) in a yard in Santa Fe Co., 9 October 2007.
PHOTOGRAPH: © WARREN BERG.

Grant County, Western Screech-Owl habitat consisted of evergreen oaks, pines, and junipers mixed with the deciduous trees (FRG). At 1,520–1,700 m (5,000–5,600 ft) in Cave Creek Canyon, Arizona, the canopy was 65% evergreen and 35% deciduous, and Arizona Sycamores (*Platanus wrightii*) had the most nests. Eastern Cottonwoods (*Populus deltoides fremontii*) and Arizona Walnuts (*Juglans major*) were also used (FRG).

Western Screech-Owls are regular residents with Elf Owls (*Micrathene whitneyi*) in the Cliff and Redrock areas and casual near Virden in New Mexico's Gila River Valley (Hubbard 1971); they are also regular residents along the San Francisco River around Glenwood (FRG). They nested near Whiskered Screech-Owls at 1,490 m (4,900 ft) in Clanton Canyon in the Peloncillo Mountains, but not higher, where Whiskered Screech-Owls were most numerous (Gehlbach 1993). A similar partial separation in higher and denser versus lower, more sparsely wooded habitat favored by Western Screech-Owls was studied in Cave Creek Canyon, Chiricahua Mountains, and Ramsey Canyon, Huachuca Mountains, in Arizona (FRG, unpubl. data) and is known in Sonora, Mexico (Russell and Monson 1998).

Western Screech-Owls belong to guilds of two to five small, insectivorous owls that nest close together as well as separately (chapters 30, 32). When nesting near Whiskered Screech, Elf, Flammulated, and Northern Pygmy (*Glaucidium gnoma*) owls above 1,520 m (5,000 ft) in Cave Creek Canyon, Western Screech-Owls were the least common species, averaging only 0.5–1.6 pairs/km^2, and were absent in branch canyons with dense forest. Below that elevation however, they averaged 4.5 pairs/km^2, second only to Elf Owls in abundance in open-canopy riparian forest dominated by cottonwoods and sycamores.

Life History

Nesting

Like other small arboreal owls, Western Screech-Owls defend tree-cavity nest sites in single-cavity territories or polyterritories, the latter defined as two or more vocally defended cavities and their immediate surroundings (see chapters 30, 32). Extra cavities are used for roosts, food storage, replacement nests, and rare polygyny (two or more females mated with the same male; Gehlbach 2008). In Cave Creek Canyon, 58% of nests were in natural-damage cavities caused by storms, gnawing rodents, and rot, while the remaining 42% were in holes drilled by Northern Flickers (*Colaptes auratus*). New Mexico nests are also known in Northern Flicker holes (Ligon 1961).

This owl begins laying eggs earlier than other nonmigratory small owls in the same habitat, possibly because its preferred low-elevation, open-canopy habitat warms up more readily. Earliest dates are 13 March–6 April in Arizona and Texas (Wise-Gervais 2005; FRG, unpubl. data; S. Kennedy, pers. comm.). Nesting in New Mexico is similarly described as beginning in mid March by Ligon (1961), who mentions a nest with "nearly grown young" at Santa Fe on 15 May, although eggs have been found as late as the first week of June (Bailey 1928).

A nest cavity 30 cm (1 ft) deep and 3.6 m (11.8 ft) high in a dead cottonwood with four incubated eggs on 22 April was noted by J. S. Ligon in unpublished notes (J.-L. Cartron). Otherwise, clutch size is poorly documented in New Mexico, but Murray (1976) provides an average of four eggs for Western and Eastern Screech-Owls, which is true in Texas (FRG). Perhaps as in Eastern Screech-Owls, 28–34 (average 30) days are required for incubation (Gehlbach 2008). Typical of all owls studied by FRG, only females incubate, brood, and feed downy nestlings, while males are the food providers, but both parents hunt and feed feathered nestlings and fledglings.

Along New Mexico's Gila River, five fledged broods had four owlets each, all observed between 4 June and 1 July (SHS). Fledging dates averaged 2 May above 1,500 m (5,000 ft) in Cave Creek Canyon, which is two weeks earlier than any other small owl and a month earlier than the migratory Elf Owl. Comparative fledging dates over several consecutive years were not available for Western Screech-Owls at that site, but other guild members fledged progressively earlier by an average 0.7 days per year during 1995–2006, coincident with increasingly warmer temperatures and earlier abundance of insects (FRG, unpubl. data). One Western Screech-Owl banded as a fledgling in Cave Creek Canyon dispersed 600 km (370 mi) to the Davis

PHOTO 29.10

Nest box used by nesting Western Screech-Owls in a juniper (*Juniperus* sp.) tree in the foothills of the Ortiz Mountains, June 2008.

PHOTOGRAPH: © LAWRY SAGER.

PHOTO 29.11

Fledgling at Kofa National Wildlife Refuge, Arizona, 13 June 2007.

PHOTOGRAPH: © TOM KENNEDY.

Mountains, Texas, where its band was found in a nest box nine years later.

Although Western Screech-Owls are not known to aggregate for nesting, they may do so occasionally, and they do join nesting clusters (chapters 30, 32). Aggregations (single species) and clusters (multispecies) contain two or more nests with a no-nest boundary between them and nearest other groups as wide as or wider than the maximum distance among nests within the group (FRG, unpubl. data). Western Screech-Owls nested in 19% of Cave Creek Canyon's 81 studied clusters, mostly with Elf Owls and Whiskered Screech-Owls, and clustered nests were more often successful than single nests, presumably because of added vigilance and deterrence of predators (FRG, unpubl. data).

Diet and Foraging

Small owls are sit-and-wait hunters, and most species eat mostly arthropods while nesting (Ross 1969; Gehlbach 2008; FRG, unpubl. data). They are typically nocturnal except the Northern Pygmy-Owl (chapter 32). The Western Screech-Owl's diet is unstudied in New Mexico, but in Cave Creek Canyon it includes 83% large insects such as June bugs (*Phyllophaga* sp.) and cicadas (*Diceroprocta* sp.) plus scorpions. Small rodents like brush mice (*Peromyscus boylei*) add 12%, and diurnal spiny lizards (*Sceloporus jarrovi, S. scalaris*) that remain active on day-warmed rocks for a short period after dark comprise the remaining 5% (Duncan et al. 2003; FRG, unpubl. data).

Along the Gila River in 1999, a routine check of a Yellow-breasted Chat (*Icteria virens*) nest revealed four recently fledged Western Screech-Owls, crops bulging, dozing in and around the nest with an adult perched higher in the tree (SHS). Presumably, adults preyed on the nestling chats. Not only birds, but also bats, amphibians, crayfish, and fish are documented in the diet of Western Screech-Owls (Cannings and Angell 2001), very similar to Eastern Screech-Owls (Gehlbach 1995), including those in Texas (Gehlbach 2008).

Predation and Interspecific Interactions

Western Screech-Owls are sympatric with many of New Mexico's other raptors, but their nocturnal habits and proclivity for tree cavities or dense foliage roosts, often adjacent to tree trunks, and their bark-mimicking plumage in the daytime may limit interactions except with larger nocturnal owls. Predation has not been documented in the state, but elsewhere Great Horned Owls (*Bubo virginianus*) may be a predator of this species. Other known predators include Spotted Owls (*Strix occidentalis*), raccoons (*Procyon lotor*), and snakes (Cannings and Angell 2001).

Although Western Screech-Owls and other owls discovered by diurnal birds during daylight often attract noisy flocks of mobbing songbirds, like most owls they tend to ignore the commotion, which, as with Eastern Screech-Owls, advertises the potential predator's location and threat but does not drive it away (Gehlbach and Leverett 1995). Westerns may compete with other cavity nesters in New Mexico as suggested by their exchanging nest cavities with American Kestrels (*Falco sparverius*) along the Gila River over several years (SHS).

Status and Management

Because of its small size, cryptic appearance, and nocturnal habits, the Western Screech-Owl has not experienced the persecution suffered by larger diurnal raptors. Nor has it been as greatly impacted by human development, including urban sprawl, given that it appears to be tolerant of general human activity and wooded habitats such as urban parks. However, it does not tolerate high-density residential urban areas (Rodríquez-Estrella and Careaga 2003). Among the most important threats in New Mexico are loss and degradation of wooded riparian habitats (Ohmart 1994; Cartron et al. 2000, 2008) and direct mortality resulting from collisions with vehicles, windows, and power lines (Cannings and Angell 2001; Harness and Wilson 2001).

Repeated physical and auditory disturbance at and around nests has included off-road vehicles, chain saws, building construction, lights, flash photography, and voice recordings used by birders to lure small owls (FRG). At birded nests, some Western Screech-Owl eggs chilled because incubating females were lured away several times by repeated recordings (females

PHOTO 29.12

Western Screech-Owl in a yard in Santa Fe Co., 27 September 2007. Western Screech-Owls are frequently seen in parks and yards, where they show themselves tolerant of human activity.

PHOTOGRAPH: © WARREN BERG.

also defend nests). Interestingly, however, a survey of 227 recreationists at random in Cave Creek Canyon showed that 90% were birders who indicated that they were less interested in seeing Western Screech-Owls than any of the four other small owls that also occurred there (FRG, unpubl. data).

The senior author has repeatedly experienced the need for education about using nonconsumable wildlife resources without damaging them. He suggests that land management agencies must be attentive to this need with area-specific exhibits, appropriate signage, and naturalist-interpretive guidance. Ecologically based nature education should begin in elementary school and continue at all grade levels. The principle of nature's irreplaceable goods and services and their local economic benefits includes nonconsumptive forms of outdoor recreation.

We do not know whether populations of Western Screech-Owls are stable, increasing, or decreasing in New Mexico, but simple management efforts such as saving snags are beneficial as shown in a four-year post-fire study at a U.S. Forest Service campground in California, where the species increased 3.5-fold after a fire (Elliot 1985). Like most small owls, Western Screech-Owls are poorly sampled by standard avian survey methods (Johnson et al. 1981). However, given the broad range of habitats occupied, their use of nest boxes, and their diverse diet, it seems likely that Western Screech-Owls will be one of New Mexico's most widespread and abundant small raptors well into the future.

Acknowledgments

FRG thanks Nancy Gehlbach for insightful and practical help in all field studies over the past 48 years and his former student Steven Kennedy for information from his ecological study of Western Screech-Owls in the Davis Mountains, Texas. SHS thanks G. Bodner, K. Brodhead, P. Chan, B. Gibbons, D. Hawksworth, R. Hunt, M. Means, G. Sadoti, B. Trussell, H. Walker, and H. Woodward for assistance with field studies along the Gila River; and the Gila National Forest, Phelps Dodge Corporation, The Nature Conservancy, and T. and D. Ogilvie for access to their lands.

[AOU] American Ornithologists' Union. 1983. *Check-list of North American birds*. 6th ed. Washington, DC: American Ornithologists' Union.

Bailey, F. M. 1928. *Birds of New Mexico*. Santa Fe: New Mexico Department of Game and Fish.

Banks, R. C., R. T. Chesser, C. Cicero, J. L. Dunn, A. W. Kratter, I. J. Lovette, P. C. Rasmussen, J. V. Remsen Jr., J. D. Rising, and D. F. Stotz. 2007. Forty-eighth supplement to the American Ornithologists' Union *Check-list of North American Birds*. Auk 124: 1109–1115.

Cannings, R. J., and T. Angell. 2001. Western Screech-Owl (*Otus kennicottii*). No. 597. In *The birds of North America*, ed. A. Poole and F. Gill. Philadelphia, PA: Birds of North America, Inc.

Cartron, J.-L. E., D. C. Lightfoot, J. E. Mygatt, S. L. Brantley, and T. K. Lowrey. 2008. *A field guide to the plants and animals of the Middle Rio Grande bosque*. Albuquerque: University of New Mexico Press.

Cartron, J.-L. E., S. H. Stoleson, P. L. L. Stoleson, and D. W. Shaw. 2000. Riparian areas. In *Livestock management in the American Southwest: ecology, society, and economics*, ed. R. Jemison and C. Raish, 281–327. Amsterdam: Elsevier.

Dixon, K. L. 1989. Contact zones of avian congeners on the southern Great Plains. *Condor* 91:15–22.

Duncan, W. W., F. R. Gehlbach, and G. A. Middendorf III. 2003. Nocturnal activity by diurnal lizards (*Sceloporus jarrovi, S. virgatus*) eaten by small owls (*Glaucidium gnoma, Otus trichopsis*). *Southwestern Naturalist* 48:205–18.

Elliot, B. 1985. Changes in distribution of owl species subsequent to habitat alteration by fire. *Western Birds* 16:25–28.

Gehlbach, F. R. 1993. *Mountain islands and desert seas: a natural history of the U.S.-Mexican borderlands*. 2nd ed. College Station: Texas A&M University Press.

———. 1995. Eastern Screech-Owl (*Otus asio*). No. 165. In *The birds of North America*, ed. A. Poole and F. Gill. Philadelphia, PA: Birds of North America, Inc.

———. 2003. Body size variation and evolutionary ecology of Eastern and Western Screech-Owls. *Southwestern Naturalist* 48:70–80.

———. 2008. *The Eastern Screech-Owl: life history, ecology, and behavior in the suburbs and countryside*. 2nd ed. College Station: Texas A&M University Press.

Gehlbach, F. R., and J. S. Leverett. 1995. Avian mobbing of Eastern Screech-Owls; predatory cues, risk to mobbers, and degree of threat. *Condor* 97:831–34.

Harness, R. E., and K. R. Wilson. 2001. Electric-utility structures associated with raptor electrocutions in rural areas. *Wildlife Society Bulletin* 29:612–23.

Hubbard, J. P. 1971. The summer birds of the Gila Valley, New Mexico. *Occasional Papers of the Delaware Museum of Natural History*, no. 2.

———. 1978. *Revised check-list of the birds of New Mexico*. Publ. no. 6. Albuquerque: New Mexico Ornithological Society.

Johnson, R. R., B. T. Brown, L. T. Haight, and J. M. Simpson. 1981. Playback recordings as a special avian census technique. *Studies in Avian Biology* 6:68–75.

Ligon, J. S. 1961. *New Mexico birds and where to find them*. Albuquerque: University of New Mexico Press.

Marshall, J. T. Jr. 1964. Typical owls: Strigidae. In *The birds of Arizona*, by A. Phillips, J. Marshall, and G. Monson, 46–55. Tucson: University of Arizona Press.

Murray, G. A. 1976. Geographic variation in the clutch sizes of seven owl species. *Auk* 93:602–13.

Ohmart, R. D. 1994. The effect of human-induced changes on the avifauna of western riparian habitats. *Studies in Avian Biology* 15:273–85.

Proudfoot, G. A., F. R. Gehlbach, and R. L. Honeycutt. 2007. Mitochondrial DNA variation and phylogeography of the Eastern and Western Screech-Owls. *Condor* 109:617–27.

Rodríguez-Estrella, R., and A. P. Careaga. 2003. The Western Screech-Owl and habitat alteration in Baja California: a gradient from urban and rural landscapes to natural habitat. *Canadian Journal of Zoology* 81:916–22.

Ross, A. 1969. Ecological aspects of the food habits of insectivorous screech-owls. *Western Foundation of Vertebrate Zoology, Proceedings* 1:301–44.

Russell, S. M., and G. Monson. 1998. *The birds of Sonora*. Tucson: University of Arizona Press.

Wise-Gervais, C. 2005. Western Screech-Owl. In *Breeding bird atlas of Arizona*, ed. T. E. Corman and C. Wise-Gervais, 210–11. Albuquerque: University of New Mexico Press.

© WAYNE LYNCH.

30

Whiskered Screech-Owl

(*Megascops trichopsis*)

FREDERICK R. GEHLBACH AND NANCY Y. GEHLBACH

AMONG THE THREE U.S. screech-owls, gray or rufous in coloration and usually less than 200 g (7 oz) in weight, the Whiskered Screech-Owl (*Megascops trichopsis*) is smallest, averaging 90 g (3 oz) (Gehlbach and Gehlbach 2000). It is one-third smaller than the Western Screech-Owl (*M. kennicottii*), and unlike that species' lemon-yellow eyes (iris) and grayish to black bill, Whiskered Screech-Owls have golden to orange-yellow eyes in natural light (not in flash photos) and a yellowish to olive bill. The gray morph (99% of U.S. individuals) also differs in having rusty or tan plumage tones, especially on the neck and throat, and fewer broader black breast streaks. Females are somewhat larger and darker in hue than males.

The species' short trill (song) advertises its nest-site territory, usually with five to seven evenly spaced,

PHOTO 30.1

Adult Whiskered Screech-Owl from Cave Creek Canyon, Chiricahua Mountains, Cochise Co., Arizona, June 2001. Like the larger Western Screech-Owl (*Megascops kennicottii*), the Whiskered Screech-Owl is characterized by yellow irises, prominent gray facial disks, and distinct ear tufts. Adults have dark vertical breast stripes whereas fledglings show pale horizontal breast bands. PHOTOGRAPH: © FRED GEHLBACH.

same-pitch, same-cadence notes followed by one or two lower notes. This song easily distinguishes it from the Western Screech-Owl. An emphatic version of the song rises and falls slightly and is given when other owls are close. A longer variation, up to 18 minutes without stopping, indicates intense territorial conflict. A second song, the telegraphic trill, resembles Morse code with alternating single and multiple notes indicating excitement about new benign stimuli. Whiskered Screech-Owls also give mild single-note whistles (mate contact), hoots (potential danger), barks (real danger), and screeches (extreme agitation). Both sexes sing, the male at a lower pitch as in other owls.

Distribution

Whiskered Screech-Owls are known in New Mexico only from the Animas, Peloncillo, and Guadalupe mountains in the "Bootheel" of Hidalgo County. Elsewhere in the United States, they are present in southeastern Arizona mountains from the Peloncillos and

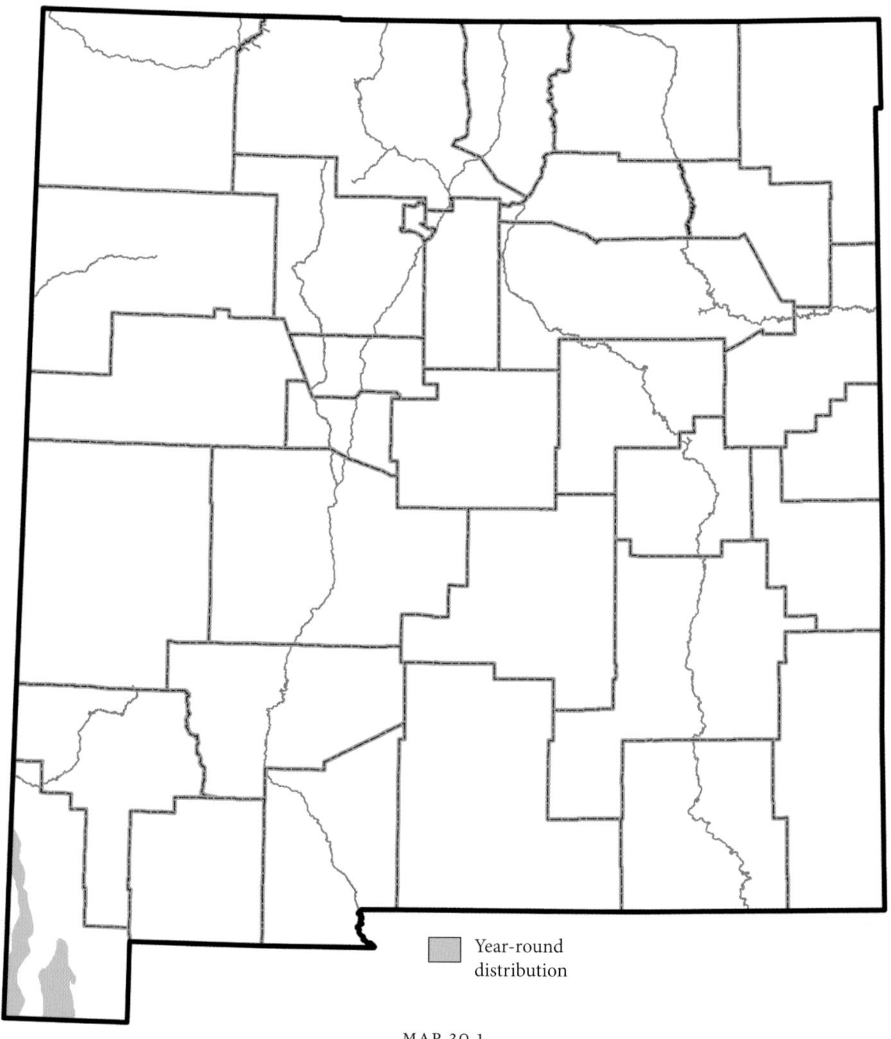

MAP 30.1

Whiskered Screech-Owl distribution map

PHOTO 30.2

(*above*) Whiskered Screech-Owl in Clanton Canyon, Hidalgo Co., 9 January 2009. Clanton Canyon is located in the Peloncillo Mountains and represents the most dependable location for finding Whiskered Screech-Owls in New Mexico. PHOTOGRAPH: © MATT BAUMANN.

PHOTO 30.3

(*left*) Clanton Canyon in the Peloncillo Mountains, New Mexico. The authors' study area (Cave Creek in the Chiricahua Mountains) is just on the other side of the New Mexico–Arizona state line and only 33 km (20 mi) distant from Clanton Canyon. PHOTOGRAPH: © MATT BAUMANN.

Chiricahuas west to the Baboquivaris and north to the Galiuros. Southward, the species nests in Mexico's Sierra Madres and high outlying ranges and in similar mountains through Central America, except Belize, to northern Nicaragua. Its nearest relatives are in Costa Rica, Panama, and South America (Proudfoot et al. 2007).

Whiskered Screech-Owls nest mainly above about 1,360 m (4,500 ft) on mountain slopes and in canyons as in Clanton and Cottonwood canyons of the Peloncillos, but youngsters disperse across lower elevations through oak-juniper woodlands such as those near the old town site of Cloverdale, New Mexico. The species was not verified in New Mexico (Marshall 1957; Hubbard 1970) until March 1971, when we found a yearling at Cloverdale and two singing territorial pairs of adults at 1,625 m (5,360 ft) among Arizona sycamores (*Platanus wrightii*) in the riparian forest of Clanton Canyon (Gehlbach 1993).

Habitat Associations

Whiskered Screech-Owls are unstudied in New Mexico but were surveyed with avian associates in southeastern Arizona and in northern Chihuahua and Sonora, Mexico, by Marshall (1957). We investigated their ecology and life history along with those of five other small owls in the Chiricahua and Huachuca mountains, Arizona in 1995–2006 (FRG, unpubl. data). Cave Creek Canyon and adjoining areas of the Chiricahuas are sources of most of the following information, supported by data from Ramsey Canyon in the Huachuca Mountains, Arizona (Gehlbach and Gehlbach 2000). Our Chiricahua study sites were only about 33 km (20 mi) northwest of Clanton Canyon across the San Simon Valley of Arizona–New Mexico.

Habitats are occupied year-round except during juvenile (natal) dispersal and temporary down-canyon movements in winter. Nesting is in the mixed riparian forest of canyon bottoms and from pinyon-oak-juniper woodland on lower slopes up to mixed coniferous forest higher in mountains. Nests are most abundant in riparian forest at about 1,500–1,950 m (4,950–6,435 ft), where 90% of them are in natural-damage cavities of Arizona sycamores on creek terraces. Oaks (*Quercus*), pines (*Pinus*), cypresses (*Cupressus*), and cottonwoods (*Populus*) provide other nest sites, some in Northern Flicker (*Colaptes auratus*) holes.

The riparian habitat is a rich mixture of evergreen and deciduous trees including junipers (*Juniperus*), pines, and cypress mixed with broadleaf evergreen oaks and madrones (*Arbutus*), and the deciduous sycamores, ashes (*Fraxinus*), chokecherries (*Prunus*), and cottonwoods. Bordering pinyon-oak-juniper woodland is less diverse and less inhabited, with border pinyons (*P. discolor*), junipers, and oaks, as is the mixed coniferous forest to at least 2,670 m (8,800 ft) with ponderosa (*P. ponderosa*) and southwestern white (*P. strobiformis*) pines and Douglas firs (*Pseudotsuga menziesii*) as leading trees.

Whiskered Screech-Owls are the most abundant members of small-owl guilds of two to five species in mid-elevation (1,520–1,980 m [5,000–6,500 ft]) riparian forest, where they average 41–61% of all nesting pairs in New Mexico–Arizona and 42% in Chihuahua and Sonora, Mexico. The guilds include various combinations of Elf (*Micrathene whitneyi*), Northern Pygmy (*Glaucidium gnoma*), Flammulated (*Otus flammeolus*), and Western Screech-Owls with Saw-whet Owls (*Aegolius acadicus*) instead of Elf Owls and Western Screech-Owls at highest elevations. Whiskered Screech-Owls comprised 11% of pairs in one guild that included similar numbers of Saw-whet and Northern Pygmy owls but many more Flammulated Owls at 2,670 m (8,800 ft) in the Chiricahua Mountains' mixed coniferous forest.

A point-count survey of all 22 species of cavity-nesting birds over 1,500–1,745 m (4,950–5,760 feet) of riparian forest showed Whiskered Screech-Owls to be the ninth most frequent species behind three small songbirds, three flycatchers, and two woodpeckers. Interyear cavity exchanges among Whiskered Screech-Owls, other small owls, Elegant Trogons (*Trogon elegans*), and Northern Flickers occurred, but only the same or another owl used an empty tree cavity the year after a Whiskered Screech-Owl nested in it. Successful Whiskered Screech-Owls nested up to six consecutive years in the same hole. This species and other small owls are typical cavity nesters in reusing only previously successful sites.

In one study plot, we found nesting Cooper's Hawks (*Accipiter cooperii*) for seven consecutive years as close

PHOTO 30.4

Adult male, Cave Creek Canyon, Arizona, June 2001. Nest densities are highest in mixed riparian forest stands along canyon bottoms, lower in pinyon-oak-juniper woodland on lower slopes and up to mixed coniferous forest higher in mountains.

as 48 m (160 ft) to the nearest Whiskered Screech-Owl nest without evidence of interaction. In another plot a Cooper's Hawk nest was only 15 m (50 ft) away one year, and the two easily seen (by us) first-day fledgling Whiskered Screech-Owls sat quietly without disturbance in a relatively open-foliage oak 6 m (20 ft) from their nest cavity. A third study plot held a Goshawk (*Accipiter gentilis*) nest about 180 m (600 ft) from the nearest nesting Whiskered Screech-Owls.

Life History

Whiskered Screech-Owls are like their guild associates in being polyterritorial; that is, they defend more than one cavity and only each cavity's immediate vicinity. Extra cavities are used for food storage, roosts, and replacement nests. They nest alone, in single-species aggregations, and most commonly in multispecies clusters that may or may not include other small owls and other cavity-nesters. For example, Whiskered Screech-Owls nested 0.5 m (1.6 ft) from Elegant Trogons that were 0.5 m (1.6 ft) from Acorn Woodpeckers (*Melanerpes formicivorus*) in the same branch. Another pair nested 3 m (10 ft) from Flammulated Owls and 4 m (15 ft) from Northern Pygmy-Owls that were 3 m (10 ft) from a Bridled Titmouse (*Baeolophus wollweberi*) nest in the same tree. Yet useable empty cavities were three times more numerous than the average owl-guild density of 16 nesting pairs per km^2 ($41/mi^2$). Such cluster nesting appears to be advantageous in detecting and deterring predators (FRG, unpubl. data).

Known predators include Ringtails (*Bassariscus astutus*) seen eating nestlings (K. Becker, pers. comm.) and Spotted Owls (*Strix occidentalis*) that ate fledglings (H. Snyder, pers. comm.). One of us (FRG) watched attempted predation on a first-day fledgling by several Mexican Jays (*Aphelocoma ultramarina*), and the

PHOTO 30.5

Whiskered Screech-Owl in tree cavity, South Fork of Cave Creek, Chiricahua Mountains, Arizona, 24 April 2004.

PHOTOGRAPH: © ROBERT SHANTZ.

PHOTO 30.6

First-day fledgling (pale horizontal breast bands) from Cave Creek Canyon, Chiricahua Mountains, Cochise Co., Arizona, June 2001.

PHOTOGRAPH: © FRED GEHLBACH.

PHOTO 30.7

(*top left*) Riparian forest habitat with many Arizona Sycamores (*Platanus wrightii*) along a side branch of Cave Creek, Chiricahua Mountains, Arizona, June 1998. PHOTOGRAPH: © FRED GEHLBACH.

PHOTO 30.8

(*top right*) The senior author on a ladder (12 m [40 ft]) in an Arizona Sycamore (*Platanus wrightii*) under a nest hole (the hole is immediately above and to his right) trying to net a nestling that just jumped out, hence the long handled net in his hand. Cave Creek Canyon, Chiricahua Mountains, Arizona, June 1998. PHOTOGRAPH: © FRED GEHLBACH.

PHOTO 30.9

(*right*) The senior author plucking a secondary feather for DNA analysis, Cave Creek Canyon, Chiricahua Mountains, Arizona, June 1998. During their study at Cave Creek Canyon, the authors and their crew of 2–4 research assistants also weighed, measured, and banded late-stage nestlings retrieved from the nest cavity one at a time. PHOTOGRAPH: © FRED GEHLBACH.

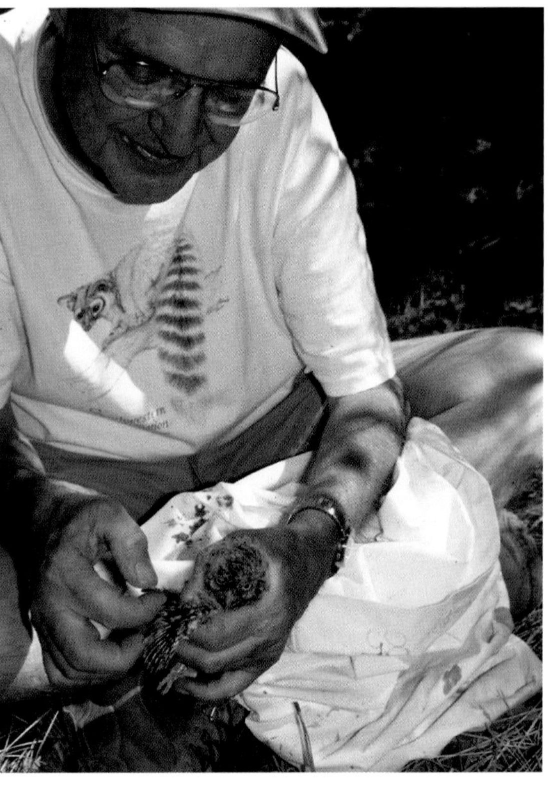

PHOTO 30.10

Adult female with three fledged young in a bare oak (*Quercus* sp.). The tree is bare due to a drought cycle phase associated with El Niño. Cave Creek Canyon, Chiricahua Mountains, Arizona, May 2001.

PHOTOGRAPH: © FRED GEHLBACH.

PHOTO 30.11

Three late-stage nestlings (often nestlings are assembled on the ground and then reinserted together in the cavity, to prevent them from jumping out). Cave Creek Canyon, Chiricahua Mountains, Arizona, May 1996.

PHOTOGRAPH: © FRED GEHLBACH.

PHOTO 30.12

Adult female feeding a fledged youngster a large moth caterpillar, Cave Creek Canyon, Chiricahua Mountains, Arizona, May 1996.

PHOTOGRAPH: © FRED GEHLBACH.

agitated adult Whiskered Screech-Owls were joined by mobbing Acorn Woodpeckers and Bridled Titmice. One owl struck a jay, and both chased others, followed by a mob that acquired more songbirds. We never saw adult Whiskered Screech-Owls strike other predators such as White-nosed Coatis (*Nasua narica*) foraging near a nest, but once a male hit our assistant as he took nestlings out of a hole for banding.

Territorial singing begins in late winter, courtship mostly in March, and two to four (mode three) eggs are laid in April. Incubation is estimated at 26 days by algebraic comparison with the Flammulated Owl using average female and egg weights. Nestlings are present 26 days and fledge 16 June on average, although fledging has been earlier by an average 0.7 days per year during 1995–2006, correlated with warmer air temperatures and the advancing mid-late June burgeoning of insect food (see also chapter 32). Whiskered Screech-Owls fledge about 10 days ahead of Northern Pygmy-Owls, two weeks ahead of Flammulated Owls, and two weeks behind Western Screech-Owls in the same nesting clusters. These differences and polyterritories probably reduce interspecific spatial competition, especially in clustered nests.

Whiskered Screech-Owl fledglings gradually move out of the nest area but remain with the adults in parental home ranges of 2–5 ha (5–12 ac) for four to seven weeks before natal dispersal. They can become reproductive adults as yearlings, three of which nested within 0.8 km (0.5 mi) of their natal sites in the same canyon. While being fed by parents, fledglings are easily distinguished by their gray cross-banded breast plumage in contrast to the black vertical breast streaks of adults (photo 30.6). Except for wing and tail feathers, fledgling body plumage (juvenal) is gradually replaced by adult-looking (Basic I) feathers before and during natal dispersal in late July–September.

While nesting, 85% of food is insects, usually about 2.5 cm (1 in) or larger in body length (Duncan et al. 2003). These are caught mostly on foliage and on the ground. Examples are cicadas (*Diceroprocta, Platypedia*), June bugs (*Phyllophaga*), giant water bugs (*Lethocerus*), and large caterpillars. Captures may be only 3–5 m (10–15 ft) from nests and roosts, some made directly from nest and roost cavity entrances. Vertebrate prey are less than 25% of average adult weight and include mice (*Peromyscus*), bats (*Myotis, Idionycteris*), spiny lizards (*Sceloporus*), and small snakes (*Leptotyphlops, Tantilla*).

Status and Management

Because of their small range in New Mexico and periodic alteration of the habitat by wildfire as in the Animas Mountains, Whiskered Screech-Owls should retain their present Threatened status until state-specific studies suggest otherwise (see Ruth et al. 2008). Even patchy tree cutting for human developments limits them and other small owls by opening the forest canopy, lowering the primary (tree) and hence secondary (insect) energy supply, and eliminating cavities. Over a 1,500- to 1,960-m (4,950- to 6,445-ft) riparian gradient in Cave Creek Canyon, human development had almost twice the negative impact on size, membership, and stability of the small-owl guild as the natural decline in size and makeup of mixed riparian forest at higher elevations (FRG, unpubl. data).

Whiskered Screech-Owls readily habituate to benign human activity around their nests, but, like other guild members, are negatively impacted by repeated physical harassment such as hitting nest trees, aiming lights on the nest, and playing recorded songs (Gehlbach and Gehlbach 2000; Wise-Gervais 2005). Deserted clutches, chilled eggs—hence fewer fledglings—and smaller, later fledglings—hence lower probability of survival—are negative effects. Disturbed adults readily abandon eggs but not nestlings. After desertion, replacement nesting occurs in another cavity but with fewer eggs and later fledglings that have less chance of survival than those from original nests (Newton 1989), possibly due to less experience before winter stress.

For effective conservation, landowners, managers, and recreationists require education about the ecological value and proper recreational use of nonconsumable wildlife like small owls (Gehlbach and Gehlbach 2000; FRG, unpubl. data). In New Mexico, Whiskered Screech-Owls benefit from being poorly known in a small area that is partly private and harder to access than parts of the larger Arizona range. They are not generally jeopardized in Arizona, because most are in remote sites. Their status is questionable in Latin America because of massive forest loss as in El Salvador, where the species is classed as Threatened (Komar 1998). We strongly advocate nature education in public and private schools and at state parks and U.S. Forest Service visitor centers.

Acknowledgments

Many individuals, agencies, and organizations have supported our biological field studies in New Mexico and Arizona. We are very grateful to all and to colleagues, friends, and research assistants, Baylor University, and the Southwestern Research Station (American Museum of Natural History) near Portal, Arizona, our frequent base for investigations of small owls and their ecological dynamics in a mix of nature and culture.

LITERATURE CITED

Duncan, W., F. R. Gehlbach, and G. A. Middendorf III. 2003. Nocturnal activity by diurnal lizards (*Sceloporus jarrovi, S. virgatus*) eaten by small owls (*Glaucidium gnoma, Otus tricopsis*). *Southwestern Naturalist* 48:218–22.

Gehlbach, F. R. 1993. *Mountain islands and desert seas: a natural history of the U.S.-Mexican borderlands*. 2nd ed. College Station: Texas A&M University Press.

Gehlbach, F. R., and N. Y. Gehlbach. 2000. Whiskered Screech-Owl (*Otus trichopsis*). No. 507. In *The birds of North America*, ed. A. Poole and F. Gill. Philadelphia, PA: Birds of North America, Inc.

Hubbard, J. P. 1970. *Check-list of the birds of New Mexico*. Publ. no. 3. Albuquerque: New Mexico Ornithological Society.

Komar, O. 1998. Avian diversity in El Salvador. *Wilson Bulletin* 110:511–33.

Marshall, J. T. Jr. 1957. Birds of pine-oak woodland in southern Arizona and adjacent Sonora. *Pacific Coast Avifauna*, no. 32.

Newton, I. 1989. Synthesis. In *Lifetime reproduction in birds*, ed. I. Newton, 441–69. London: Academic Press, Harcourt Brace Jovanovich.

Proudfoot, G. A., F. R. Gehlbach, and R. L. Honeycutt. 2007. Mitochondrial DNA variation and phylogeography of the Eastern and Western Screech-Owls. *Condor* 109:617–27.

Ruth, J. M., T. Brush, and D. J. Krueper. 2008. Preface to *Birds of the U.S.-Mexico borderlands: distribution, ecology, and conservation*, ed. J. M. Ruth, T. Brush, and D. J. Krueper. *Studies in Avian Biology* 37:1–9.

Wise-Gervais, C. 2005. Whiskered Screech-Owl. In *Arizona breeding bird atlas*, ed. T. E. Corman and C. Wise-Gervais, 212–13. Albuquerque: University of New Mexico Press.

© JEAN-LUC CARTRON.

31

Great Horned Owl

(*Bubo virginianus*)

ROBERT W. DICKERMAN, JANELLE HARDEN, AND JEAN-LUC E. CARTRON

THE GREAT HORNED OWL (*Bubo virginianus*) is the largest owl in New Mexico and is often considered the nocturnal replacement of the diurnal Red-tailed Hawk (*Buteo jamaicensis*). It can be readily recognized not only by its large and powerful appearance (body length: 46–63 cm [18–25 in]; Houston et al. 1998), but also by its large yellow eyes, white bib, and prominent ear tufts (hence the owl's common name). Above, the Great Horned Owl appears mottled gray-brown at a distance, while below it is white to buffy, with dark brown to blackish barring. There is great sexual dimorphism in the species, with females 20–28% larger than males (Craighead and Craighead 1956; Earhart and Johnson 1970). Describing the subspecies *B. v. pallescens*, Earhart and Johnson (1970) give mean body weights of 914.2 g (32 oz) (range: 724–1,257 g [25.5–44.3 oz]) and 1,142.2 g (40.3 oz) (range: 801–1,550 g [28.3–54.7 oz]) for males and females, respectively. Among the other owls found in New Mexico, only the Long-eared Owl (*Asio otus*) can be mistaken for a Great Horned Owl. However, the Long-eared Owl is much smaller and more slender. It lacks the white bib of the Great Horned Owl, has ear tufts set closer together, and shows streaking rather than barring on the underparts.

The Great Horned Owl's most familiar vocalization is its "song," a series of deep, soft hoots uttered by both males and females. The number of hoots in the series varies even within populations (Houston et al. 1998), but the song in New Mexico is most typically a "*hoo-hoo-hooo—hoo-hoo*" or variations of it. Also part of the owl's vocal repertoire are screams, screeches, barking notes, chuckles, squawks, and even loud growls (Houston et al. 1998).

It must be clearly stated that there are no color phases, morphs, or polymorphism in the Great Horned Owl. These terms were used in the past because the extent to which populations mix through migration was not yet understood. In an unfortunate taxonomic revision (in German) of the American Great Horned Owls, Weick (1999), not knowing the North American literature, illustrated several variants of six subspecies, one of which, *B. v. occidentalis*, does not exist (Dickerman 1991, 1993).

The following description of subspecies found in New Mexico is based on the careful examination of hundreds and hundreds of Great Horned Owls (Dickerman 1991, 1993, 2002, 2004; Dickerman and Johnson 2008), including 258 in the Museum of Southwestern Biology's collection—192 from New Mexico alone—with large series of nesting-season birds. The latter series sets the range of possible variation within our nesting populations, and individuals that fall outside

PHOTO 31.1

(*top left*) Adult Great Horned Owl at Eldorado, Santa Fe Co., 28 May 2007. The Great Horned Owl is the largest of New Mexico's owls. It is nearly unmistakable, with its powerful build, yellow eyes, prominent ear tufts set wider apart compared to those of the Long-eared Owl (*Asio otus*), and white bib (not always visible).

PHOTOGRAPH: © JONATHAN BATKIN.

PHOTO 31.2

(*top right*) Great Horned Owl (subspecies *pallescens*), Maxwell National Wildlife Refuge, 5 June 2003. The subspecies *pallescens* is characterized by its pale plumage: it is lightly barred below while the upperparts are pale buff with small, darker, dusky areas in the feathers. The feet are almost pure white.

PHOTOGRAPH: © PATTY HOBAN.

PHOTO 31.3

(*bottom right*) Great Horned Owl (subspecies *pinorum*) in Frijoles Canyon, Sandoval Co., 1 July 2007. *Bubo virginianus pinorum* is much darker than the subspecies *pallescens*, with heavier barring on the underparts. The subspecies was only recently described.

PHOTOGRAPH: © SALLY KING.

PHOTOS 31.4a, b, and c

Great Horned Owl inside the garden center annex of a Home Depot store, Las Cruces, Doña Anna Co., 26 December 2007. First record of the eastern subspecies (*Bubo virginianus virginianus*) for New Mexico. The owl appeared quite undisturbed despite the loud Christmas music played by the speakers. It is presumably one of the paired birds known to have nested at that location.

PHOTOGRAPHS: © CHARLOTTE STANFORD.

of that variation must represent birds from elsewhere; they need to be identified by comparison with other named populations. Fortunately, we have at hand representatives of those populations.

The American Ornithologists' Union's (AOU) checklist (1957) included New Mexico entirely within the range of *B. v. pallescens*, with no other subspecies present. However, three subspecific names must now be used for describing the state's nesting populations: *B. v. pallescens*, *B. v. pinorum*, and *B. v. virginianus*. The first two subspecies, plus intergrades of *B. v. pallescens* and *B. v. virginianus*, breed regularly in New Mexico. *Bubo virginianus pallescens* is the palest subspecies in the state (and the secondmost among all subspecies of the Great Horned Owl). It is lightly barred below, its feet are always pure white, and dorsally it is pale buff with small, darker, dusky areas in the feathers. *Bubo virginianus pallescens* towards *B. v. virginianus* shows the ochraceous underfeather coloration character of *virginianus*. *Bubo virginianus pinorum*, recently described by Dickerman and Johnson (2008), is much darker than the other two subspecies, the crown being blacker, the dorsum darker grey, and the ventrum more heavily barred; the feet are white with or without light dusky barring. While writing this chapter, the authors received a photograph of a Great Horned Owl from just outside a Home Depot store in Las Cruces, Doña Ana County, presumably one of the two birds known to have nested at that same location (photos 31.4a, b, c). The bird depicted in the photographs is a pure eastern Great Horn Owl (*B. v. virginianus*), a first record for the state.

Regularly present in New Mexico, but only during the non-nesting season, is *B. v. lagophonus*. It is similar to *pinorum*, but blacker throughout, more heavily barred on the venter and feet, and often with tan feet. More unusual in the state is *B. v. subarcticus*, the palest Great Horned Owl subspecies (Dickerman 2002), as well as an intergrade with dominant *saturatus* characteristics, *saturatus* being the darkest Great Horned Owl subspecies (Dickerman and Johnson 2008).

Distribution

The Great Horned Owl has the largest breeding distribution of any raptor in the Americas, nesting from the arctic treeline at a latitude of 68° N south to the Straits of Magellan, although rare or absent throughout most of Central and South America (Sibley and Monroe 1990; AOU 1998; Houston et al. 1998). The nonbreeding distribution of the Great Horned Owl is identical to its breeding distribution (Houston et al. 1998), although seasonal movements do occur (see below).

The Great Horned Owl is nearly ubiquitous in New Mexico, occurring throughout the lowlands and up the elevational gradient to at least ponderosa pine (*Pinus ponderosa*) forests (Hubbard 1978). *Bubo virginianus pallescens* is the nesting form of the southern and central grasslands and deserts and the riparian habitats along the larger rivers into north-central New Mexico. With the exception of the Las Cruces owl mentioned above, the birds nesting in the easternmost tier of counties are B. *v. pallescens* towards *B. v. virginianus* (with richer ochraceous undercoloring of *virginianus* genes), which although diluted reaches its terminus in eastern New Mexico, having traveled up the east-west prairie rivers. The nesting birds of the mountains, pine-oak associations and higher, belong to *B. v. pinorum* (Dickerman and Johnson 2008).

PHOTO 31.5

Great Horned Owl (subspecies *pinorum*) at Galisteo, Santa Fe Co., 31 August 2006. The elevation in Galisteo is approximately 1,850 m (6,050 ft). *Bubo virginianus pinorum* is the subspecies nesting in New Mexico's mountains. PHOTOGRAPH: © JONATHAN BATKIN.

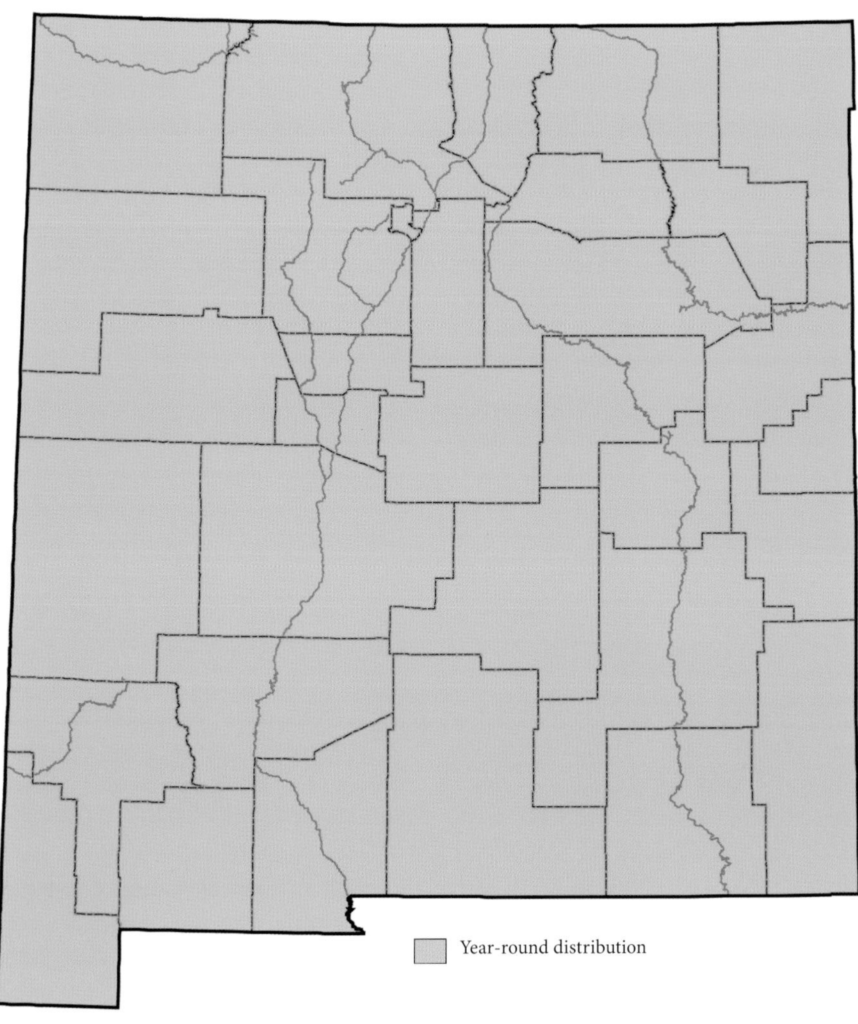

MAP 31.1

Great Horned Owl distribution map

H. C. Oberholser, in a classic revision of the species (1904), wrote, "With the exception of *occidentalis* [which is a synonym of *subarcticus*] and *wapacuthu* [old name then used for *subarcticus*, but in reality a synonym of the Snowy Owl, *Nyctea scandiaca*; Dickerman 1993], all [subspecies] seem to be nonmigratory, and thus any record safely may be considered as based upon the resident bird." Oberholser's concepts were based on his knowledge of birds in the East, where there is no morphological variation from Maine to Florida, west into central Minnesota, and south to eastern Texas! However, this mantra has persisted in the literature and was a principle of the AOU checklist (1957), which stated, "*Resident*, except where noted."

The most common migrant in the lowlands of New Mexico is *B. v. pinorum*, which nests in the Rocky Mountains south of the Snake River of southern Idaho and south to the highlands of New Mexico and Arizona. Because of the uniformity of the birds within *pinorum*'s extensive range, the distances over which migration occurs cannot be determined. We know that altitudinal movement occurs based on two lines of evidence. First, a bird banded as a nestling 29 April 1988 in Sandia Canyon on the Pajarito Plateau (ca. 2,290 m

[7,500 ft] elevation) was found dead 18 December 1988 near a bridge over the Rio Grande on the south side of Española, Rio Arriba County (Dickerman and Johnson 2008). The bird had traveled a distance of only about 25 km (16 mi) to the north and east, while also moving approximately 580 m (1,903 ft) downslope. Further evidence of altitudinal movement is provided by the morphological distinctness of fall and winter *pinorum* from the local nesting *pallescens*. On 2 February 2008, two *pinorum* were found electrocuted under a power line in Belen, Valencia County. The male had testes measuring 5 × 10 mm; however, the female's ovary, while that of an adult, had not started to enlarge. These birds either migrated down the elevational gradient as a pair, presumably originating from the slopes of the nearby Manzano Mountains, or else they mated while wintering in the Rio Grande Valley. Had they not died they would have migrated back to the mountains as a pair.

The next most frequent migrant—a long-distance migrant—is *B. v. lagophonus*, which nests from the mountains of central Idaho north in the Rocky Mountains to central-interior Alaska. At least 21 specimens have been preserved from twelve New Mexico counties as far south as Socorro County in the central part of the state and Grant County in the west. They have been found from 6 September until 7 April, but note the late date may be due to an injury that prevented migration. One young of the year, found 9 September, still retained a few juvenal feathers about the head but must have migrated a minimum of 1,900 km (1,180 mi) from central Idaho, the closest point in the nesting range of the subspecies. An immature found dead 6 September was an intermediate *lagophonus* x *pinorum* (tarsi and toes heavily barred but dorsum somewhat paler than typical *lagophonus*). Parenthetically, there are three specimens of *lagophonus* from Texas, all in the San Angelo State University collection: Brown County, Plains Cross, 10 km (6 mi) south, 2 November 1980; Irion County, no locality, 19 November 1982; and McCulloch County, Brady, 36.7 km (22.8 mi) north, 16.9 km (10.5 mi) east, 9 January 1993. There is a single specimen from Arizona reported by Rea (1983) as *saturatus*, but it is typical of *lagophonus*. It was found near Snowflake, Navajo County, 2 August 1974. When queried as to whether it was wounded in any way, Rea sent us a photograph showing a shot-hole in the front of the skull, which most likely interfered with its migration and may explain the unusual date.

Bubo virginianus subarcticus, the nesting population of the aspen parklands of sub-boreal Canada that reaches into Montana, Wyoming, and North Dakota (Dickerman 2002), is represented in New Mexico by two specimens: one from Harding County, 1.6 km (1 mi) southwest of Gallegos, 9 February 1936, and the other from Chavez County, Bitter Lake National Wildlife Refuge, 22 January 1994.

Three other specimens from New Mexico are intermediate between *subarcticus* and *lagophonus*. They have cold colors: variably black-barred venters and pale to dark dorsa with no tan undertones. These match two specimens from the intermontane regions of western Montana (Dickerman 2002). The three New Mexico specimens are from Grant County "between Silver City and Lordsburg," 20 November 1990; Edgewood, Santa Fe County, 20 October 1991; and Farmington, San Juan County, 7 November 1992.

A representative of the darkest Great Horned Owl subspecies, from the rain forests of the Pacific Northwest, was found mortally injured on a road in Los Alamos, Los Alamos County, on 3 November 1993. It is identified as *saturatus* intermediate towards *lagophonus*. Dorsally, it is ultra-typically *saturatus*, although ventrally it is nearly white, weakly barred with black; the tarsi and toes are white, and it completely lacks the rich undertones of *saturatus*.

Habitat Associations

Throughout its very broad distribution, the Great Horned Owl is catholic in its selection of habitats, shunning only humid habitats of the tropical zone (Houston et al. 1998). In New Mexico, it is found in both canyon lands and flat country and occurs in nearly all of the state's vegetation types including grasslands, shrublands, riparian woodlands, pinyon-juniper woodland, and ponderosa pine (e.g., Ligon 1961; Hubbard 1971, 1978; Cartron et al. 2008; 2009). In fact, the species' distribution is probably limited more by the availability of potential nest sites than by vegetation type.

PHOTO 31.6

A nest cliff and surrounding habitat in McKinley Co., 13 May 2008. In New Mexico, the Great Horned Owl is found both in flat areas and in rugged country. PHOTOGRAPH: © RON KELLERMUELLER/HAWKS ALOFT, INC.

on a ledge approximately 120 m (390 ft) deep into the passageway leading from the entrance to the Big Room (S. West, pers. comm.). Nesting no longer occurs inside the cave, perhaps due to the larger flow of visitors, but in Bailey's time (1928), Great Horned Owls could be heard from inside the passageway to the Big Room, and the skeleton of one owl was found as far deep as Devil's Den, more or less directly underneath the old nest (S. West, pers. comm.).

The owl everyone associates with abandoned buildings is, of course, the Barn Owl (*Tyto alba*). However, Great Horned Owls also occasionally take up residence, and nest, in buildings in New Mexico. For at least five years, a Great Horned Owl nested on a metallic table on the second floor of an abandoned building in Socorro County (D. Burkett, pers. comm.; photos 31.10a and b). Eric Greisen of the National Radio Astronomy Observatory wrote to the authors that Great Horned Owls are well-known nesters at the Very

Life History

The ecology of Great Horned Owls has not been studied in New Mexico. Most of the information below is from anecdotal observations made by the authors and by others.

Nesting

In New Mexico, most Great Horned Owl nests have been observed in trees, in yuccas (*Yucca*), along cliff walls, and in holes along cutbanks (Bailey 1928; Ligon 1961). Less frequently, nesting pairs use man-made nesting substrates such as nest platforms and transmission towers (D. Roehmer, pers. com.; T. Reeves, pers. comm.). At Bitter Lake National Wildlife Refuge, Great Horned Owls also nest occasionally in large nest boxes (G. Warrick, pers. comm.). Some nesting pairs are remarkably tolerant of human presence, as was the pair that nested successfully in the parking lot of the visitor center at the Maxwell National Wildlife Refuge, Colfax County (P. Hoban, pers. comm.; photo 31.9). Bailey (1928) reported on nesting "in high niches" far back inside the caves at Carlsbad Caverns National Park. Remains of an old nest can still be observed today

PHOTO 31.7

Great Horned Owl nest at Maxwell National Wildlife Refuge, Colfax Co., 25 April 2006. PHOTOGRAPH: © PATTY HOBAN.

PHOTOS 31.10a and b

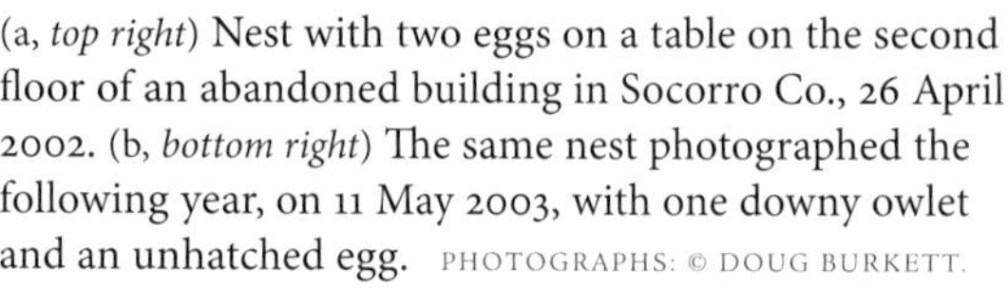

(a, *top right*) Nest with two eggs on a table on the second floor of an abandoned building in Socorro Co., 26 April 2002. (b, *bottom right*) The same nest photographed the following year, on 11 May 2003, with one downy owlet and an unhatched egg. PHOTOGRAPHS: © DOUG BURKETT.

PHOTO 31.8

(*top left*) Great Horned Owl in its nest in a yucca (*Yucca* sp.) outside the visitor center at White Sands National Monument, 25 March 2009. PHOTOGRAPH: © JEAN-LUC CARTRON.

PHOTO 31.9

(*bottom left*) Great Horned Owl nest with owlets in the parking lot of the visitor center at Maxwell National Wildlife Refuge, 11 April 2006. Great Horned Owls can be remarkably tolerant of human activities even during nesting. PHOTOGRAPH: © PATTY HOBAN.

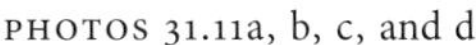

PHOTOS 31.11a, b, c, and d

Fledglings and adult in an abandoned building south of Lordsburg, Hidalgo Co., 2006. (a, *top left*) young fledgling on the floor of the building, 15 May; (b, c, and d) older fledglings and an adult, 13 June.

PHOTOGRAPHS: © ROBERT SHANTZ.

PHOTO 31.12

(*top left*) Two downy owlets in a nest in the Estancia Valley, Torrance Co., spring 2000.
PHOTOGRAPH: © JEAN-LUC CARTRON.

PHOTO 31.13

(*bottom left*) Nest with old nestling in Milnesand, Roosevelt Co., 14 April 2006. PHOTOGRAPH: © TOM KENNEDY.

PHOTO 31.14

(*top right*) Young Great Horned Owl near fledging, Maxwell National Wildlife Refuge, Colfax Co., 12 May 2003. PHOTOGRAPH: © PATTY HOBAN.

PHOTO 31.15

(*bottom right*) Fledglings, Maxwell National Wildlife Refuge, Colfax Co., 20 May 2003. PHOTOGRAPH: © PATTY HOBAN.

PHOTO 31.16

(*top left*) Young Great Horned Owl, El Vado Lake State Park, Los Ojos, Rio Arriba Co., 13 April 2007. PHOTOGRAPH: © ROGER HOGAN.

PHOTO 31.17

(*top right*) Great Horned Owl fledgling in yucca (*Yucca* sp.) approximately 30 km (20 mi) southeast of Socorro, Socorro Co., 1996. PHOTOGRAPH: © DOUG BURKETT.

PHOTO 31.18

(*bottom right*) Fledgling, east of Caprock, Lea Co., 25 May 2008. PHOTOGRAPH: © JEAN-LUC CARTRON.

PHOTO 31.19

Great Horned Owl nest in an abandoned building in Socorro Co., 21 May 2002. The nest contains one hatchling and an egg, plus food items: a decapitated glossy snake (*Arizona elegans*) and half a cottontail (*Sylvilagus* sp.). PHOTOGRAPH: © DOUG BURKETT.

Large Array (VLA) radio antenna construction barn in the Plains of San Agustin, Socorro County. "For many years, the owls spent their days, and made their nests, high in the rafters and wall-support beams of the very large, and tall, metal building. In recent years, a pair . . . preferred a certain . . . antenna that was in the barn. Unfortunately, each antenna spends only about a month or so in the barn and then is moved back into service while the next antenna comes into the barn for overhaul or significant service. In general, the nest was thus moved outside and then tilted as the antenna moved—causing the nests to be abandoned. The birds would then attempt a nest in the next antenna [that had moved into] the same location . . ."

In New Mexico, the nesting season of Great Horned Owls begins in the autumn, when territories are reestablished, and hooting begins along the Rio Grande bosque in October (RWD, pers. obs.). Egg dates range from 4 March to 17 April for 21 New Mexico clutches in the Museum of Southwestern Biology (R. Dickerman, unpubl. data; see also appendix 31-1), hatching probably occurring in April in most nests but also, occasionally, as late as mid May (photo 31.19). The 21 New Mexico egg sets include 15 two-egg clutches and 6 three-egg clutches, for a mean clutch size of 2.29 (R. Dickerman, unpubl. data). Nationwide, only eight of 155 egg sets are from outside the period from February through April (see also the discussion of mortality below); clutch size is typically two, but can range from one to four or even five (Houston et al. 1998). The eggs are incubated by the female, who also tends to the nestlings, while the male hunts and delivers prey items to the nest (see Houston et al. 1998). The length of the incubation period has not been determined in New Mexico, but elsewhere ranges from 30 to 37 days (Hoffmeister and Setzer 1947; Austing 1968; Peck and James 1983). The young move onto branches adjacent to the nest at six weeks of age and can make short flights at seven weeks (Houston et al. 1998). Along the Middle Rio Grande, fledging typically occurs around the last week of April and first week of May, and up to three young are fledged per nest (JLEC, pers. obs.).

Diet and Foraging

The Great Horned Owl is a top predator with the broadest prey base among all North American owls, indeed among all North American birds of prey (Voous 1988; Houston et al. 1998). Both opportunistic and capable of hunting at night and during the day, the Great Horned

Owl preys on an exceptionally wide variety of mammals—usually the dominant taxonomic group in the species' diet—as well as on birds, reptiles, amphibians, fish, and invertebrates. Prey as large as a Great Blue Heron (*Ardea herodias*), common raccoon (*Procyon lotor*), or common porcupine (*Erethizon dorsatum*) have been documented in the diet of the Great Horned Owl (Powell 1984; Bosakowski et al. 1989). The species is typically a perch-and-wait predator and captures its prey with its powerful talons (Marti 1974).

The Great Horned Owl's diet has not been studied in New Mexico. However, anecdotal observations conform to what has been published outside the state. Along the Middle Rio Grande, for example, prey remains collected under Great Horned Owl nests show the species to consume Ring-necked Pheasants (*Phasianus colchicus*), domestic chickens (*Gallus domesticus*), Rock Pigeons (*Columba livia*), desert cottontail rabbits (*Sylvilagus audubonii*), common muskrats (*Ondatra zibethicus*), rock squirrels (*Spermophilus variegatus*), house cats (*Felis catus*), and whiptail lizards (*Aspidoscelis* sp.) (Cartron et al. 2008; JLEC, unpubl. data; D. Dean, pers. comm.). At Carlsbad Caverns National Park, Great Horned Owls wait on cliffs near the cave entrance, only to fly through the crowd of Brazilian Free-tailed Bats (*Tadarida brasiliensis*) emerging at dusk, seizing bats in their path, then returning to the cliffs to eat their catch (S. West, pers. comm.). The nest photographed by Doug Burkett in Socorro County (photo 31.19) shows remains of a cottontail rabbit, but also a decapitated glossy snake (*Arizona elegans*).

Predation and Interspecific Interactions

Being nearly ubiquitous in New Mexico, the Great Horned Owl likely interacts frequently with—and even preys on—many of the state's other raptors. Elsewhere in the species' range, predation on an adult Osprey (*Pandion haliaetus*) and on other owls—the Barn Owl, Long-eared Owl, Short-eared Owl (*Asio flammeus*), and Northern Saw-whet Owl (*Aegolius acadicus*)—has been documented (Cold 1993; see also Houston et al. 1998). Also reported is predation on nestlings of the Northern Goshawk (*Accipiter gentilis*), Red-tailed Hawk, and Broad-winged Hawk (*B. platypterus*) (Luttich et al. 1971; Rohner and Doyle 1992). In turn, the Great Horned Owl can be the target of attacks by pairs of other raptor species defending their nesting territories (see chapter 26).

Status and Management

Over the past several decades, one of us (RWD) gained the impression that there were more female than male Great Horned Owls represented in museum collections. However, there is extensive evidence that males outnumber females among monogamous biparental birds. Breitwisch (1989:3) summarized evidence in monogamous birds of equal sex ratios in eggs, hatchlings, or nestlings, and acceptance of "the argument for the evolutionary stability of equal investment in the sexes, usually resulting in sex ratios of unity, except, perhaps, for sexually size-dimorphic species." Nonetheless, adult sex ratios are male-biased (Breitwisch 1989). For example, at the time our data set was gathered in the U.S. Museum of Natural History collections (see below), even the nondimorphic Song Sparrow (*Melospiza melodia*) showed male dominance of 75.0% in the spring and 63.3% in the fall. However, the conspicuous behavior and vocalizations of this species may have led to greater collection of male, over female, specimens.

Our present (and prior) studies of Great Horned Owls, however, indicate a potential female bias in specimen collections. Other published studies of the species yielded sex ratios ranging from essentially even (Earhart and Johnson 1970) to greatly skewed in favor of males (Craighead and Craighead 1956) or females (McGillivray 1985, 1989; Chubb 1996). The sex of birds handled by the Craigheads and by Chubb was determined indirectly, based largely on measurements and body mass; those of Earhart and Johnson and of McGillivray were museum specimens and thus gender was presumably determined by examination of the gonads.

To evaluate sex ratios in Great Horned Owls, Dickerman and Harden recorded sex and date of death for 618 birds preserved in museum collections between 1950 and 1993 (appendix 31-2). The digital data obtained were for owls from 28 states and provinces (see acknowledgments). We believe that the majority of specimens preserved in museums after 1950 were

PHOTO 31.20

Adult Great Horned Owl and nestling in the nest in winter-like weather in Eldorado, Santa Fe Co., 17 April 2009. Museum collections suggest that compared to adult males, adult females may suffer lower mortality during the nesting season and higher mortality during the rest of the year.

PHOTOGRAPH: DALE W. STAHLECKER.

from random salvage rather than from potentially male-biased collecting (RWD, pers. obs.). For example, of the 237 Great Horned Owls in the Museum of Southwestern Biology deposited since 1950, only 5% appeared to be selectively collected and another 7% were of unknown source (probably random salvage). To minimize sex-ratio bias arising from active collecting, we removed from our sample any birds that we could determine were collected as opposed to "found dead." A summary of collection data by sex and month of mortality is presented in appendix 31-2.

We divided the annual cycle of Great Horned Owls into two major seasons, "reproductive" and "dispersal/winter." The seven-month reproductive season was further split into nesting (February–April), fledging (May–June), and post-fledging (July–August). The five months of the dispersal/winter season were divided into dispersal/migration (September–November) and winter (December–January). To verify dates of nesting, we obtained data on 155 sets of eggs from seven Great Horned Owl subspecies from several museums (see acknowledgments). All but eight of the sets of eggs were from February to April, the period we defined as the nesting period.

PHOTO 31.21

Great Horned Owl fledgling on a school campus in Albuquerque, Bernalillo Co., 31 May 2009. Part of the Great Horned Owl's resilience is its ability to adapt to human-dominated environments. In New Mexico as elsewhere, however, it suffers mortality from road traffic collisions and electrocutions on power poles, and to a lesser degree also shooting.

PHOTOGRAPH: DOUG BROWN.

Male Great Horned Owl specimens outnumbered female specimens 157 to 124 during the reproductive season, a statistically significant difference between the sexes (for more detail see appendix 31-2). This dominance appeared in samples from six of the seven months in the reproductive season; only in June did female specimens outnumber males 27 to 15. In contrast, female specimens outnumbered male specimens 208 to 129 in every sample of the dispersal/migration and winter seasons, September through January (statistically significant; appendix 31-2). Overall, female specimens outnumber male specimens in museum collections, 332 to 286 respectively, another statistically significant difference. In addition, there are more total specimens (n = 337, 55%) from the five fall and winter months (September–January) than from the seven months (February–August) of the nesting season (n = 281, 45%).

We realize that different research focuses may have driven the assembling of these various data sets (Craighead and Craighead 1956; Earhart and Johnson 1970; McGillivray 1985, 1989; Chubb 1996) and that all sets are not directly comparable, but they represent the best available data. For example, the skewed sex ratio found by Craighead and Craighead (1956; males predominating 53.7%/46.3% in a series of 1,667 birds) is possibly because they were capturing adults during the nesting season. During that period, females are incubating or brooding and males would be active, feeding the female and young, and thus be easier to capture. In two studies, McGillivray (1985, 1989) measured highly disparate numbers of skeletons (presumably those available), with females dominating males 60%/40% and 58%/42%, respectively. McGillivray (1989) reported that the majority of Great Horned Owl skeletons in North American museums were acquired outside of the breeding season, when female mortality is greater.

Chubb (1996) plotted injuries of 269 Great Horned Owls, by month, from southeastern Ontario. There were 174 females and 95 males (65%/35%) in her sample. As a wildlife rehabilitator, it can be assumed that Chubb received birds with injuries that should occur randomly with respect to gender. In Chubb's data (1996), females outnumbered males every month of the year except April, but she studied a much more homogeneous population than in our data set. Chubb (1996) also reported that 44 of 46 birds injured during encounters with skunks and porcupines were female.

From a New Mexico raptor rehabilitator, we obtained cause-of-injury for 67 juvenile and adult Great Horned Owls received over the nine-year period of 1995 to 2002. Road casualties accounted for 64% of intakes, 22% were electrocuted, 6% were shot, and the remaining 8% were unknown. Chubb (1996) reported that 4% of 374 Great Horned Owls received over an 18-year period were shot.

We attribute the observed higher number of female mortalities during the fall and winter periods to the larger size of females (20–28% in data from five taxa in Earhart and Johnson 1970; 24% in Craighead and Craighead 1956). Greater size presumably creates a greater food demand, and the need for extended periods of feeding, or feeding on larger prey, results in greater temporal exposure to mortality forces. In addition, heavier females may be less agile and thus subject to higher road mortality.

The Great Horned Owl is, for a raptor, abundant, widespread, and amazingly adaptable and successful. It is, of course, state and federally protected and, therefore, shooting is no longer an important factor; electrocution on power-line poles and roadkills are far more frequent causes of mortality. Populations appear to be healthy, and no greater level of protection seems to be needed. Salvaged birds should be studied for the identification of long-distance migrants.

Acknowledgments

We would like to acknowledge the rehabilitators of New Mexico and Arizona without whom the sex-biased mortality study presented here could not have been conducted. We are also grateful to the curators of the many collections in which Great Horned Owl specimens were examined (see Dickerman 2004; Dickerman and Johnson 2008) as well as the curators who provided electronic data on death dates of skins and on egg dates of specimens in their care. We especially thank John P. Hubbard for his hours of discussion with us in relation to this project.

LITERATURE CITED

[AOU] American Ornithologists' Union. 1957. *Check-list of North American birds.* 5th ed. Washington, DC: American Ornithologists' Union.

———. 1998. *Check-list of North American birds.* 7th ed. Washington, DC: American Ornithologists' Union.

Austing, G. R. 1968. The owls and I. *Audubon* 70:72–79.

Bailey, F. M. 1928. *Birds of New Mexico.* Santa Fe: New Mexico Department of Game and Fish.

Bosakowski, T., R. Speiser, and D. G. Smith. 1989. Nesting ecology of forest-dwelling Great Horned Owls in the eastern deciduous forest biome. *Canadian Field Naturalist* 103:65–69.

Breitwisch, R. 1989. Mortality patterns, sex ratios, and parental investment in monogamous birds. *Current Ornithology* 6:1–50.

Cartron, J.-L. E., D. C. Lightfoot, J. E. Mygatt, S. L. Brantley, and T. K. Lowrey. 2008. *A field guide to the plants and animals of the Middle Rio Grande bosque.* Albuquerque: University of New Mexico Press.

Cartron, J.-L. E., L. A. Sager, Jr., and H. A. Walker. 2009. Notes on some breeding raptors of central and northern Lea County, New Mexico. *New Mexico Ornithological Society Bulletin* 37:7–14.

Chubb, K. 1996. A study of 374 Great Horned Owls. *Beaks, Brains & Bones* 1:25–40. Verona, Ontario: Avian Care and Research Foundation.

Cold, C. W. 1993. Adult male osprey killed at nest by Great Horned Owl. *Passenger Pigeon* 55:269–70.

Craighead, J. J., and F. C. Craighead Jr. 1956. Hawks, owls and wildlife. Washington, DC: Stackpole Co. and Wildlife Management Institute.

Dickerman, R. W. 1991. On the validity of *Bubo virginianus occidentalis* Stone. *Auk* 108:964–65.

———. 1993. The subspecies of the Great Horned Owls of the central Great Plains, with notes on adjacent areas. *Kansas Ornithological Society Bulletin* 44:17–21.

———. 2002. The taxonomy of the sub-arctic Great Horned Owl (*Bubo virginianus*) nesting in the United States. *American Midland Naturalist* 148:198–99.

———. 2004. Distribution of the subspecies of Great Horned Owls in Texas. *Bulletin of the Texas Ornithological Society* 37:104.

Dickerman, R. W., and Andrew B. Johnson. 2008. Notes on Great Horned Owls nesting in the Rocky Mountains, with a description of a new subspecies. *Journal of Raptor Research* 42:20–28.

Earhart, C. M., and N. K. Johnson. 1970. Size dimorphism and food habits of North American owls. *Condor* 72:251–64.

Hoffmeister, D. F., and H. W. Setzer. 1947. The postnatal development of two broods of Great Horned Owls (*Bubo virginianus*). *University of Kansas Publications of the Museum of Natural History* 1:157–73.

Houston, C. S., D. G. Smith, and C. H. Rohner. 1998. Great Horned Owl (*Bubo virginianus*). No. 372. In *The birds of North America*, ed. A. Poole and F. Gill. Philadelphia, PA: Birds of North America, Inc.

Hubbard, J. P. 1971. The summer birds of the Gila Valley, New Mexico. *Occasional Papers of the Delaware Museum of Natural History* 2:1–35.

———. 1978. *Revised check-list of the birds of New Mexico.* Publ. no. 6. Albuquerque: New Mexico Ornithological Society.

Ligon, J. S. 1961. *New Mexico birds and where to find them.* Albuquerque: University of New Mexico Press.

Luttich, S. N., L. B. Keith, and J. D. Stephenson. 1971. Population dynamics of the Red-tailed Hawk (*Buteo jamaicensis*) at Rochester, Alberta. *Auk* 88:75–87.

Marti, C. D. 1974. Feeding ecology of four sympatric owls. *Condor* 76:45–61.

McGillivray, W. B. 1985. Size, sexual dimorphism and their measurement in Great Horned Owls of Alberta. *Canadian Journal of Zoology* 63:2364–72.

———. 1989. Geographic variation in size and reverse size dimorphism of the Great Horned Owl in North America. *Condor* 91:777–86.

Oberholser, H. C. 1904. A revision of the American Great Horned Owls. *Proceedings of the U.S. National Museum* 27:177–92.

Peck, G. K., and R. D. James. 1983. *Breeding birds of Ontario: nidiology and distribution.* Vol. 1, 234–35. Toronto: Royal Ontario Museum.

Powell, B. 1984. *Labrador by choice.* St. John's, Canada: Jesperson Press.

Rea, A. M. 1983. *Once a river.* Tucson: University of Arizona Press.

Rohner, C., and F. I. Doyle. 1992. Food-stressed Great Horned Owl kills adult goshawk: exceptional observation of community process? *Journal of Raptor Research* 26:261–63.

Sibley, C. G., and B. L. Monroe Jr. 1990. *Distribution and taxonomy of birds of the world.* New Haven: Yale University Press.

Voous, K. H. 1988. *Owls of the northern hemisphere.* Cambridge, MA: MIT Press.

Weick, F. 1999. Zur taxonomie der Amerikanischen uhs (*Bubo* spp). *Okol. Vogel* [Ecol. Birds] 21:363–87.

32

Northern Pygmy-Owl

(*Glaucidium gnoma*)

FREDERICK R. GEHLBACH AND NANCY Y. GEHLBACH

OF THE TWO pygmy owls in the United States, the widespread Northern Pygmy-Owl (*Glaucidium gnoma*) differs in coloration and habitat from the much rarer Ferruginous Pygmy (*G. brasilianum*), unrecorded from New Mexico, but formerly nearby in the middle Gila River valley, Arizona (Rea 1983). Northern Pygmy-Owls have brown and white banded tails, spotted crowns, and rarely nest below 1,500 m (5,000 ft) in New Mexico and Arizona, in contrast to the Ferruginous Pygmy's all-brown tail bands, faintly streaked crown, and lower-elevation habitat. Eyespots on nape feathers distinguish both pygmy owls from other small owls in the United States. The Northern Pygmy-Owl averages 64 g (2.3 oz), but males are smaller, darker in hue, and less rusty than females.

This species gives a series of fast or slow toot-like whistles often lasting several minutes and with various cadences including single or paired notes, sometimes

PHOTO 32.1

Northern Pygmy-Owl, Santa Fe Co., 10 November 2007. The Northern Pygmy-Owl is readily identified by its small size, long banded tail, yellow eyes, and lack of ear tufts. Note also the streaked underparts.

PHOTOGRAPH: © WARREN BERG.

PHOTO 32.2

(*top left*) Northern Pygmy-Owl, Santa Fe Co., 10 November 2007. Northern Pygmy-Owls have yellow irises and a spotted crown. PHOTOGRAPH: © WARREN BERG.

both alternating in the same continuous sequence. This territorial and mate advertisement song is very different from songs of other small neighboring owls and is said to differ throughout the range, indicating different races or possibly species (König et al. 1999; Holt and Petersen 2000). The diverse vocal repertoire we recorded in southeastern Arizona reflects diverse behaviors not yet fully studied. For instance, a short fast series of toots (twitters) is the female's response to a few single toots by her mate announcing a prey delivery near the nest.

Distribution

Northern Pygmy-Owls are strictly mountain inhabitants and nonmigratory but move to lower elevations temporarily during natal dispersal and occasionally in winter. They range from southeastern Alaska and adjacent western Canada south through mountains of the western United States, Mexico, and Central America except Belize to northern Nicaragua. New Mexico's populations are widespread west of the Rio Grande. Eastward they are also known to us in the Sangre de Cristo, Sandia, Manzano, Capitan, and Sacramento

PHOTO 32.3

(*top right*) Northern Pygmy-Owl in tree, Frijoles Canyon, Bandelier National Monument, June 2006. Also distinguishing the Northern Pygmy-Owl from all other New Mexico owls is the presence of eyespots on the nape. PHOTOGRAPH: © SALLY KING.

PHOTO 32.4

(*bottom right*) Northern Pygmy-Owl, Taos Canyon, Taos Co., 7 November 2006. PHOTOGRAPH: © GERAINT SMITH.

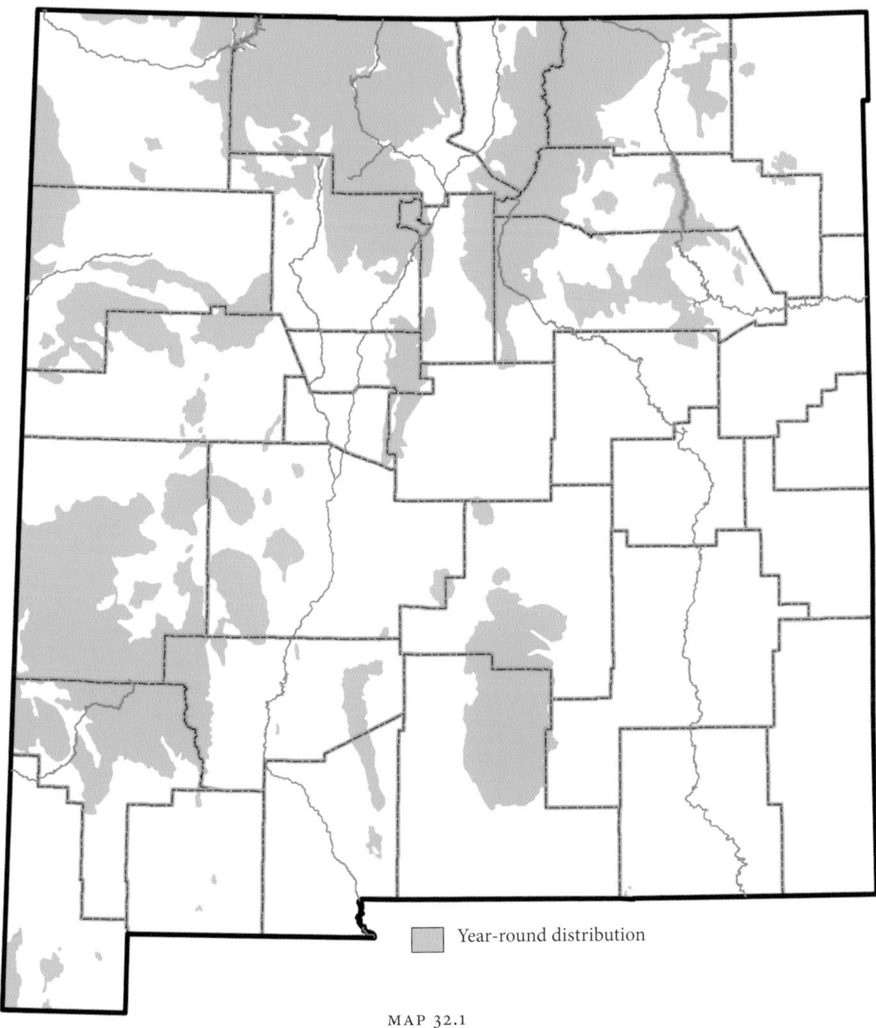

MAP 32.1

Northern Pygmy-Owl distribution map

mountains but have been strangely absent from similar forest habitat in the Guadalupe Mountains of New Mexico–Texas (Gehlbach 1993).

Nesting in New Mexico and adjacent Arizona is in mixed riparian forest and mountain-slope pinyon-oak-juniper woodland to the upslope limit of mixed coniferous forest. In the Chiricahua Mountains, Arizona, across the San Simon Valley from Northern Pygmy-Owl range in the Peloncillo Mountains, New Mexico, nests were at 1,450–2,670 m (4,800–8,800 ft). We studied this Arizona population extensively in 1995–2006, so it provides most of the following information on natural history (FRG, unpubl. data). Nesting pairs were also observed in the Zuni Mountains and on Mt. Taylor in New Mexico, at and above 2,270 m (7,500 ft), from 1951 to 2003 but not every year.

Habitat Associations

A nest in a Northern Flicker (*Colaptes auratus*) hole at 2,270 m (7,500 ft) in the Zuni Mountains was in a narrowleaf cottonwood (*Populus angustifolia*) in mixed riparian forest on Sawyer Creek below ponderosa pines (*Pinus ponderosa*) merging with Rocky Mountain pinyons (*Pinus edulis*) and Rocky Mountain junipers

PHOTO 32.5

(*above*) Northern Pygmy-Owl trapped and banded by HawkWatch International on Capilla Peak in the Manzano Mountains, September 2002. PHOTOGRAPH: © JOHN P. DELONG.

PHOTOS 32.6a, b, and c

Northern Pygmy-Owl in Frijoles Canyon in October 2005 (a, *top right*) and July 2007 (b, *middle right*), and representative surrounding habitat (c, *bottom right*). The species is found in New Mexico in mixed deciduous forest stands along canyon bottoms and from pinyon-juniper woodlands in the foothills up to mixed conifer stands on higher mountain slopes. PHOTOGRAPHS: © SALLY KING.

(*Juniperus scopulorum*) on adjacent slopes. Single pairs of Northern Pygmy-Owls were noted nesting in a 100-ha (250-ac) study area, but not annually, in contrast to Flammulated Owls (*Otus flammeolus*) (McCallum and Gehlbach 1988; McCallum et al. 1995). Long-eared Owls (*Asio otus*) and Goshawks (*Accipiter gentilis*) sometimes nested; Western Screech-Owls (*Megascops kennicottii*) and Great Horned Owls (*Bubo virginianus*) were present at times without evidence of nesting in the plot.

In our Arizona study area, nests were most common (87%) in canyon-bottom mixed riparian forest on stream terraces below 1,970 m (6,500 ft) and mostly (67%) in Arizona sycamore (*Platanus wrightii*) holes drilled by woodpeckers. Others were in cypress (*Cupressus*) and pine (*Pinus*) holes. Only Northern Flicker and Acorn Woodpecker (*Melanerpes formicivorus*) holes were used, but the Flicker holes were mostly occupied in ponderosa pines at highest elevations and the Acorn Woodpecker holes were mostly used at lowest elevations. Sites in the Chiricahua and Zuni mountains usually faced habitat edges unlike those within large tracts of continuous forest in the northwestern U.S. range (Giese and Forsman 2003).

Northern Pygmy-Owls were the third most common small owl, averaging 11–15% of all nests in three- to five-species owl guilds at about 1,500–1,970 m (5,000–6,500 ft) in the Chiricahua's mixed riparian forest. Whiskered Screech-Owls (*Megascops trichopsis*) and Elf Owls (*Micrathene whitneyi*) were more abundant, Flammulated Owls and Western Screech-Owls less abundant, at those elevations. Northern Pygmy-Owls reached maximum nesting density where riparian forest was least changed by human developments. At 2,420 m (8,000 ft) in mixed coniferous forest, Northern Pygmy-Owls averaged 11% of nesting pairs in guilds with the similarly common Whiskered and Saw-whet (*Aegolius acadicus*) owls and more abundant Flammulated Owls.

Life History

Unlike its owl associates, the Northern Pygmy-Owl is crepuscular, not nocturnal. Most activity while nesting occurs from 4 to 10 AM (52%) and 3 to 8 PM (28%) Mountain Standard Time. Northern Pygmy-Owl home ranges are 6–15 times larger than those of the other associated small owls, estimated at 36 ha (90 ac) in the Chiricahua and Zuni mountains. Like all other guild members, Northern Pygmy-Owls are polyterritorial in defending more than one tree cavity and its immediate vicinity, rather than the space encompassing all cavities (FRG, unpubl. data). Besides the nest, cavities are used for food storage, roosting, and replacement nesting.

Despite their large home ranges, Northern Pygmy-Owls may nest close to (cluster with) other cavity-nesting birds, all more or less concurrently, but they do not aggregate (form single-species groups) as do Elf, Flammulated, and Whiskered Screech owls. Examples of multispecies clusters in the Chiricahuas are Pygmy-Owls 0.4 m (1.5 ft) from Flammulated Owls in the same Arizona sycamore branch, and 2.7 m (9.0 ft) from Bridled Titmice (*Baeolophus wollweberi*) in the same tree, which was 4 m (15 ft) from a Whiskered Screech-Owl nest in another tree. Conflicts are rare, mostly vocal and with Acorn Woodpeckers that nest as close as 2 m (6.5 ft) in the same tree. After two successful years in one Acorn Woodpecker hole, Northern Pygmy-Owls were disturbed by birders and re-nested in a hole 38 m (125 ft) away, after which Acorn Woodpeckers reoccupied their hole.

Unlike northern populations that eat mostly birds and mammals, identified species in the diet we studied in the Chiricahua Mountains included only 11% birds and mammals in contrast to 14% insects and 31% lizards. Presumably this diet was because the latter two cold-blooded (ectothermic) groups are more abundant and more easily obtained in the warmer Southwestern climate (Duncan et al. 2003). Most commonly eaten were spiny (*Sceloporus virgatus*, *S. jarrovi*) and whiptail (*Aspidoscelis exsanguis*, *A. sonorae*) lizards plus katydids and cicadas. Predation by Northern Pygmy-Owls on a juvenile yellow-nosed cotton rat (*Sigmodon ochrognathus*), adult cliff chipmunk (*Eutamias dorsalis*), and adult Bewick's Wren (*Thryomanes bewickii*) was seen at 6–10 AM. The adult chipmunk, nearly equal in size to the owl, was carried in one foot by hop-flap climbing to progressively higher perches (K. Becker, pers. comm.). Food is usually delivered by males to females outside the nest.

While observing food deliveries to nests and the mobbing of Northern Pygmy-Owls by various birds,

PHOTOS 32.7a and b

Northern Pygmy-Owl in yard in Santa Fe area, Santa Fe Co., 10 November 2007 (a, *top*), and surrounding habitat (b, *bottom*). PHOTOGRAPHS: © WARREN BERG.

we learned that incidence of each avian prey species is correlated with its frequency in mobs and permanent residency in the local avifauna but primarily linked to mobbing frequency (FRG, unpubl. data) as in Eastern Screech-Owls (*Megascops asio*; Gehlbach and Leveritt 1995). There was no difference in mobbing at Pygmy-Owl models with and without typical eyespots employed alternately every three days with the same recorded territorial song at least 0.6 km (2,000 ft) from the nearest Pygmy-Owl nest (Gehlbach et al., unpubl. data; see Deppe et al. 2003).

Territorial singing usually begins in March, similar to other permanent-resident guild members; two to five (usually three to four) eggs are laid in late April–early May. The incubation period is unknown and estimated to be 24 days by algebraic comparison of female and egg size with the similar-size female Flammulated Owl. On average, three fledglings appear around 25 June, or about 10–23 days after permanent-resident Western and Whiskered screech-owls fledge and 5–7 days before the fledging of summer-resident Elf and Flammulated owls in the same study plots. The Northern Pygmy-Owl's average fledging date advanced 0.8 days per year in 1995–2006 in concert with warmer spring temperatures and earlier insect abundance, perhaps due to global warming (see also chapter 30). Unlike fledgling screech-owls that require three to seven days to follow adults, Northern Pygmy-Owl fledglings follow their parents out of the nest area in one to two days.

When Pygmy-Owls fledge the same day as a neighboring small owl, the two broods may intermingle briefly but without conflict in the 0.1–2.5 ha (0.2–6.2 ac) cluster area. The different species tend to move away from nest sites in opposite directions, which may lower chances of predation by the larger Spotted Owl (*Strix occidentalis*) and ringtail (*Bassariscus astutus*), known small-owl predators (chapter 30). Clustered nests of small owls are more often successful than solitary nests, perhaps because detection, warning, and deterrence of predators are more effective protection with more pairs and different species in proximity (FRG, unpubl. data; chapter 30).

Status and Management

Because the Northern Pygmy-Owl lives in most of New Mexico's high mountains, its geographic range in the state is broad. Because its home range is large, local pairs are widely dispersed. Thus Northern Pygmy-Owls are not endangered or threatened in New Mexico. However, they can be eliminated locally by excessive

PHOTO 32.8

Northern Pygmy-Owl glaring, Frijoles Canyon, Bandelier National Monument, June 2006. The Northern Pygmy-Owl is a perch-and-wait predator and does not hesitate to attack prey its own size or larger.

PHOTOGRAPH: © SALLY KING.

logging as happened in California (Marshall 1988), and their habitat becomes unsuitable after repeated disruptive intrusions such as by off-road vehicles and human construction. All small owls tolerate nearby nonintrusive humans, but none tolerate deforestation, excessive development, and repeated physical or auditory disturbance at the nest.

Northern Pygmy-Owls will nest close to human activity as in campgrounds if not unduly disturbed. However, this species and the Flammulated Owl are first among small owls on lists of birder desiderata (FRG, unpubl. data). Pygmy-Owls become agitated by repeated tree rubbing or tapping and audiotape or CD playing to expose the owl for listing (this caused the nest desertion and replacement nesting mentioned earlier). Like its guild associates, the Northern Pygmy-Owl readily abandons eggs but not nestlings. We have seen the negative results of attempts to expose owls but also positive changes in birder behavior with education about wildlife watching (Gehlbach and Gehlbach 2000; FRG, unpubl. data).

In Latin America, Northern Pygmy-Owls and other wildlife suffer because of logging and agrarian development (see Pliego et al. 1993). Thus we could not locate Pygmy-Owls in cut-over forest near Yécora, Sonora, Mexico, but found them in nearby mature second growth of the same type. Because size of the species' geographic range in Latin America approximates that of the U.S.-Canadian range, and habitat destruction and avifaunal impoverishment continue in Latin America, we encourage transborder environmental education that includes outdoor experiences for schoolchildren and information about the economic advantages of saving large natural areas for tourism, local enjoyment, and the support of all life.

PHOTO 32.9

Male Northern Pygmy-Owl holding a cicada nymph in an Arizona sycamore (*Platanus wrightii*), Cave Creek Canyon, Arizona, June 2000. The adult female is in a nearby nest cavity with old nestlings. The male is waiting for the female to exit the nest cavity and retrieve the prey item. PHOTOGRAPH: © DORIS HAUSLEITNER.

Acknowledgments

With gratitude we especially recognize the Cottonwood Gulch Foundation of Albuquerque, which for more than 75 years has maintained natural forested habitat in the Zuni Mountains expressly for outdoor educational purposes. We appreciate the many other organizations and persons that have supported our joint natural history investigations since 1960. In the Chiricahua Mountains, Arizona, the Southwestern Research Station of the American Museum of Natural History has fostered ecological studies of many kinds of organisms and their responses to environmental change imposed by nature and people.

LITERATURE CITED

Deppe, C., D. Holt, J. Tewksbury, L. Broberg, J. Petersen, and K. Wood. 2003. Effect of Northern Pygmy-Owl (*Glaucidium gnoma*) eyespots on avian mobbing. *Auk* 120:765–71.

Duncan, W., F. R. Gehlbach, and G. A. Middendorf III. 2003. Nocturnal activity by diurnal lizards (*Sceloporus jarrovi*, *S. virgatus*) eaten by small owls (*Glaucidium gnoma*, *Otus tricopsis*). *Southwestern Naturalist* 48:218–22.

Gehlbach, F. R. 1993. *Mountain islands and desert seas: a natural history of the U.S.-Mexican borderlands*. 2nd ed. College Station: Texas A&M University Press.

Gehlbach, F. R., and N. Y. Gehlbach. 2000. Whiskered Screech-Owl (*Otus trichopsis*). No. 507. In *The birds of North America*, ed. A. Poole and F. Gill. Philadelphia, PA: Birds of North America, Inc.

Gehlbach, F. R., and J. S. Leverett. 1995. Avian mobbing of Eastern Screech-Owls: predatory cues, risk to mobbers, and degree of threat. *Condor* 97:831–34.

Giese, A. R., and E. D. Forsman. 2003. Breeding season, habitat use, and ecology of male Northern Pygmy-Owls. *Journal of Raptor Research* 37:117–24.

Holt, D. W., and J. L. Petersen. 2000. Northern Pygmy-Owl (*Glaucidium gnoma*). No. 494. In *The birds of North America*, ed. A. Poole and F. Gill. Philadelphia, PA: Birds of North America, Inc.

König, C., F. Weick, and J-H Becking. 1999. *Owls: a guide to the owls of the world*. New Haven: Yale University Press.

Marshall, J. T. 1988. Birds lost from a giant sequoia forest during fifty years. *Condor* 90:359–72.

McCallum, D. A., and F. R. Gehlbach. 1988. Nest-site preferences of Flammulated Owls in western New Mexico. *Condor* 90:653–61.

McCallum, D. A., F. R. Gehlbach, and S. W. Webb. 1995. Life history and ecology of Flammulated Owls in a marginal New Mexico population. *Wilson Bulletin* 107:530–37.

Pliego, P. E., A. G. Navarro-Sigüenza, and A. T. Peterson. 1993. A geographic, ecological, and historical analysis of land bird diversity in Mexico. In *Biological diversity of Mexico: origins and distribution*, ed. T. P. Ramamoorthy, R. Bye, A. Lot, and J. Fa, 282–307. New York: Oxford University Press.

Rea, A. M. 1983. *Once a river: bird life and habitat changes on the middle Gila*. Tucson: University of Arizona Press.

33

Elf Owl

(*Micrathene whitneyi*)

ROBERT W. DICKERMAN,
ANDREW B. JOHNSON, AND J. DAVID LIGON

THE ELF OWL (*Micrathene whitneyi*) is the smallest owl in the world (del Hoyo et al. 1999), measuring 13–14 cm (5.5 in) in length and with an average weight of 41 g (1.4 oz; Walters 1981). Like the diminutive pygmy-owls (*Glaucidium*), the Elf Owl is characterized by a round head with no ear tufts; unlike them, however, it has a short tail. The adult male and female are alike, with overall body color speckled gray-brown dorsally, variable light streaking ventrally, some buffy spots around the face, conspicuous white eyebrows, and yellow eyes that do not reflect light at night. Juvenile Elf Owls are mottled gray above, with a pale breast and few markings on the back and the face. The adult plumage is attained at about four months of age.

Two of the recognized subspecies occur in New Mexico, *Micrathene whitneyi whitneyi* and *M. w. idonea*, the latter documented only very recently in the state (see below). *M. w. whitneyi* is a warm grayish brown, with spots of ochre about the forehead and facial disc; the ventral stripes are a diffuse and warm pale ochraceous to cinnamon. *M. w. idonea* is notably grayer than *whitneyi*, and the markings about the face and crown are substantially more buffy and less ochraceous. Some specimens collected from the Guadalupe Mountains in southeastern New Mexico are much paler and more heavily spotted with buff

PHOTO 33.1

Elf Owl (subspecies *whitneyi*) holding a scorpion, Pima Co., Arizona, 30 May 2004. Note the round head, yellow irises, lack of ear tufts, short tail, and ventral streaking.

PHOTO 33.2

Elf Owl at Mockingbird Gap approximately 70 km (45 mi) southeast of Socorro, Socorro Co., 7 July 2003. Mockingbird Gap is a wide gap between the Mockingbird Mountains and the Little Burro Mountains and separating the Tularosa Basin from the northern Jornada del Muerto. The photograph of the Elf Owl was taken near a large arroyo. Nesting at the site of the photograph has occurred for nearly a decade. The nest site lies along the extreme northeastern limit of the species' breeding distribution. PHOTOGRAPH: © DOUG BURKETT.

dorsally than any other available specimens of the two subspecies and much less richly colored than *M. w. whitneyi*. However, the taxonomic significance of this finding remains unclear. Of the two subspecies, *idonea* is smaller in size. Wing chord measurements for male Elf Owls average 106.3 mm (104–109 mm) (4.2 in [4.0–4.3 in]) for *idonea* and 110.4 mm (106–114 mm) (4.3 in [4.2–4.5 in]) for *whitneyi*.

About a dozen different calls of Elf Owls have been recognized, but only three of the most frequently heard calls of adults are briefly described here. In the spring, territorial male Elf Owls produce a high-pitched "chatter song" also sometimes described as "puppy-like yipping" and consisting of several high-pitched notes (Henry and Gelhbach 1999; Song A of Ligon 1968). Males also sing from a potential nest cavity, sometimes for minutes on end (Song B of Ligon 1968), to entice a newly mated female to enter and accept the cavity as a nest site. The song varies in volume and intensity, increasing in both as the female responds more strongly to it. When the female approaches the entrance, the male usually descends slowly to the bottom of the cavity, decreasing the volume of the song until it is barely audible to the human ear. A third vocalization is a soft "*seeu*" or "*peeu*" (Song C of Ligon 1968), given by the female. Early in the breeding cycle, this call serves to communicate the female's location to her mate. It also appears to stimulate the male to provide food to the incubating female and later to their chicks.

Distribution

The Elf Owl nests in the southwestern United States and in Mexico. In the United States, the species' nesting distribution extends from southeastern California and southern Nevada, east across southern Arizona and southern New Mexico and into western and southern Texas. The Mexican portion of the species' breeding range is poorly documented, but includes at least southern Baja California, Sonora, Chihuahua, Coahuila, Nuevo León, Tamaulipas, and Puebla (AOU 1998). The Elf Owl winters in central Mexico (Ligon 1968; AOU 1998), occasionally north to southern Texas (Lockwood and Freeman 2002). Elf Owls nesting in the United States are migratory, but farther south, at least both in southern Baja California and Puebla, the species is a year-round resident (AOU 1998).

As recently as the 1970s, the Elf Owl in New Mexico was known from only the southwestern part of the state (Hubbard 1978). Since that time, however, the species has been recorded from additional locations such as Water Canyon in the Magdalena Mountains and the eastern slopes of the Guadalupe Mountains in Eddy County. Early distributional records of the Elf Owl in the northeastern part of its range are spotty, however, and it is still unclear whether recent northerly or northeasterly records indicate a true range expansion in New Mexico—and also in Texas (see below)—as postulated by Barlow and Johnson (1967), Williams (1997), and Henry and Gehlbach (1999). Alternatively, recent records from outside the species' putative range may simply reveal previous gaps in knowledge. It was

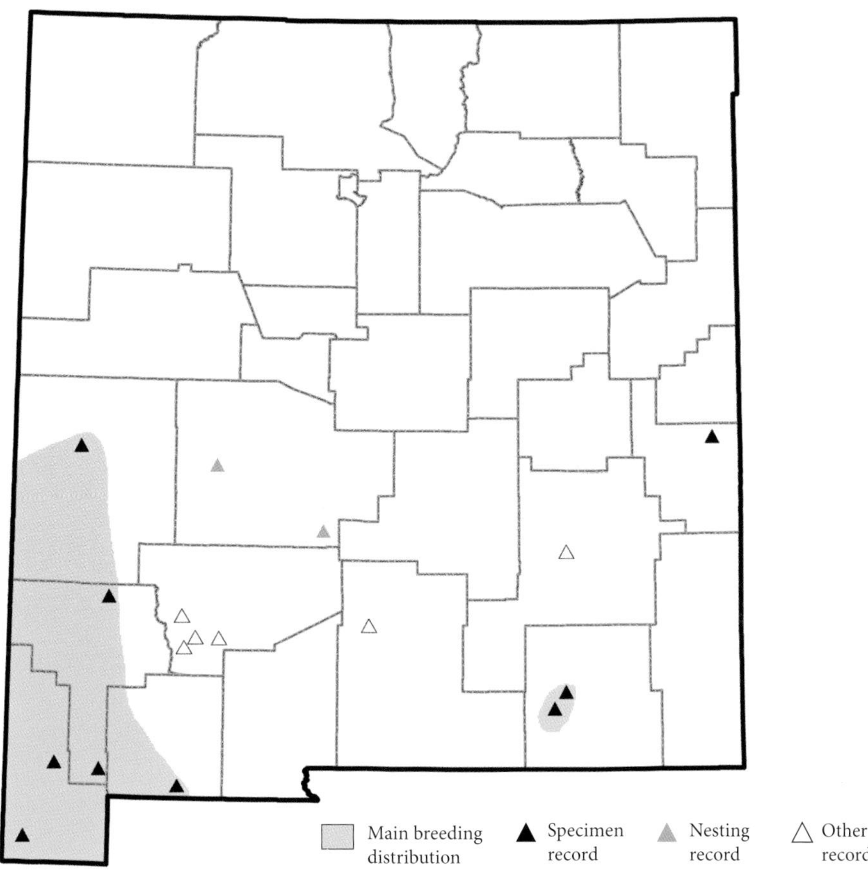

MAP 33.1

Elf Owl distribution map

only in the mid 1950s that portable, relatively inexpensive recording and playback equipment, to which Elf Owls do readily respond, became available. Marshall (1957), an expert "owl-caller," reported that Elf Owls do not respond to imitated calls, for which he blamed the then-spotty distribution record for the species. Although Marshall's statement appears to be incorrect (Henry and Gehlbach 1999), playback equipment does enhance chances of detection compared to human imitations of the call.

Of the two subspecies found in New Mexico, *whitneyi* was until recently the only one known from the state, with a well-documented range in Hidalgo, Luna, Grant, and Catron counties in the southwest. This is the owl often photographed peeping out of holes in saguaros (*Carnegia gigantia*), the tall columnar cacti found in Arizona's Sonoran Desert but not in New Mexico. A surprising record is a single specimen of *whitneyi* taken on 13 May 1980 near Portales, Roosevelt County, in eastern New Mexico, well to the east of its range in the southwestern part of the state.

The other subspecies, *M. w. idonea*, was until recently undocumented in New Mexico, known instead to occur farther south, from the Lower Rio Grande Valley in southern Texas southward into Mexico. It was first described from the Lower Rio Grande Valley based on

PHOTO 33.3

Brood of four nestlings approaching fledging and found at Water Canyon in the Magdalena Mountains, 1981.

PHOTOGRAPH: © PETER STACEY.

two specimens taken in 1889 and 1894 (Ridgway 1914; James and Hayse 1963), but was not reported from there again until 2 May 1960 (James and Hayse 1963). Although it is possible that the species was always present and simply overlooked during the seven-decade hiatus, some suggest local extirpation followed by recolonization caused by human-induced habitat alteration (Henry and Gehlbach 1999).

Elf Owls were first reported from the extreme southern end of the Guadalupe Mountains in Texas by LaVal (1969), who captured two in nets set for bats the nights of 2 and 13 June 1968. These specimens were originally identified as *M. w. whitneyi*, as was a specimen from Brewster County, Texas, on 8 May 1966 (Barlow and Johnson 1967). Based on Dickerman's examination of all three specimens, however, subspecies identification was erroneous. The three specimens belong instead to the subspecies *idonea* (RWD, pers. obs.), as also do three additional specimens collected in the Black Gap Wildlife Management Area in Texas in April 1999. Elf Owls were not found in the New Mexico portion of the Guadalupe Mountains (nor were they probably looked for) until Steve West had two respond to Elf Owl playback in Dark Canyon on the night of 16 June 1997 (Williams 1997). West saw adults with young in nearby Last Chance Canyon in June 1998 and July 1999 (Williams 1998, 1999). Competent birdwatchers heard Elf Owls on 12 May 2000 in Rocky Arroyo about 50 kilometers (~30 miles) to the northeast. All three localities (i.e., Dark Canyon, Last Chance Canyon, and Rocky Arroyo) are in drainages on the eastern slope of the mountains. In May 2004 and 2005 three Elf Owls were collected from Rocky Arroyo in Eddy County, New Mexico. One of these was identified as *M. w. idonea*, and the other two intermediate between it and *M. w. whitneyi*. During the period 21 June 2000 to 22 June 2003 six specimens were collected in Dark Canyon. The specimens collected from Rocky Arroyo and Dark Canyon suggest a zone of contact between *idonea* and *whitneyi* in the central portion of the Guadalupe Mountains.

Outlying, apparently periodic occurrences of Elf Owls are in the Magdalena Mountains (Stacey et al. 1983), and at "Mockingbird Gap" in the north end of the San Andres Mountains (Williams 1994; D. Burkett, pers. comm.; photo 33.2) and Roswell (photographed 7 April 2000) (Williams 2000). One was "described" on White Sands Missile Range in Otero County, 16 May 1993 (S. O. Williams, pers. comm.). S. O. Williams (*in litt.* 15 May 2000) wrote, "In the Rio Grande watershed there are multiple records of Elf Owls along lower Las Animas Creek, Sierra County . . . [also records] from Percha [Dam] . . . and Palomas . . . Elsewhere along the east slope of the Black Range, one on Trujillo Creek south of Hillsboro 16 May 1991, and another was along Percha Creek near Hillsboro 24 June 1994." No specimens are available from any of these localities to determine which subspecies was involved.

With regard to nesting localities within the state, probably the most interesting record comes from Water Canyon in the Magdalena Mountains, where Stacey et al. (1983) found a pair nesting in the same cavity in 1976, 1977, 1979, and 1981. In at least one year, four healthy chicks—a large brood in any part of the species' range (Ligon 1968, Henry and Gehlbach 1999)—were fledged. Other Elf Owls also responded to taped vocalizations in Water Canyon, but no additional nests were found. These records represent the northernmost known breeding of the species in New Mexico.

PHOTO 33.4

Water Canyon in the Magdalena Mountains west of Socorro is the northernmost known nesting location recorded for the Elf Owl in New Mexico. The owls at Water Canyon used an old Acorn Woodpecker (*Melanerpes formicivorus*) nest hole (indicated by the red arrow) in 1981. Elf Owls have been found to occupy dense montane (as shown in this photo) and lowland riparian habitat, in addition to lowland desert locations.

PHOTOGRAPH: © PETER STACEY.

Habitat Associations

In New Mexico Elf Owls range in elevation from 1,066 m (3,497 ft) to 1,850 m (6,069 ft), nesting from the lower Chihuahuan Desert to ponderosa pine–oak woodlands, including riparian woodlands in which cottonwoods (*Populus*) and sycamores (*Platanus*) are important elements (Ligon 1968; RWD, ABJ, pers. obs.). In suitable habitat, Elf Owls are often surprisingly common. Marshall (1957) reported 50 calling males on transects totaling 65.8 km (40.9 mi) in pine-oak woodland and riparian habitats. They were particularly abundant in the Peloncillo Mountains of Hidalgo County, New Mexico (Marshall 1957:46), where each of four dead Chihuahua pines within 0.8 km (0.5 mile) were occupied (Marshall 1957:78). "At least 30 [were heard calling] along the Gila River at the mouth of the Mogollon Creek on 6 May 1960" (J. P. Hubbard, pers. comm.).

Cavities suitable for nesting, together with sufficient vegetation to support an insect prey base, seem to be the only absolute habitat requirements of Elf Owls. Although in New Mexico this species is probably most common in riparian wooded vegetation, it

PHOTO 33.5

(*left*) Dark Canyon in the Guadalupe Mountains of southeastern New Mexico, March 2006.

PHOTO 33.6

(*below*) Nesting habitat at Mockingbird Gap, Socorro Co. The vegetation is primarily creosote (*Larrea tridentata*) and honey mesquite (*Prosopis glandulosa*). Approximately 100 m (~330 ft) to the east of the house is a nice-sized arroyo with riparian vegetation including netleaf hackberry (*Celtis reticulata*), gray oak (*Quercus grisea*), and desert willow (*Chilopsis linearis*).

PHOTOS 33.7a and b

Over the years at Mockingbird Gap, nesting Elf Owls have alternated between nest cavities in two separate power poles, one just north of the little house, and the other alongside the house on the south side. Both cavities appear to have been made by Northern Flickers (*Colaptes auratus*; based on the size of the holes) and both are also used periodically by Ladder-backed Woodpeckers (*Picoides scalaris*). PHOTOGRAPHS: © DOUG BURKETT.

is not confined to such habitat. Rather, as an obligate cavity nester, it typically cannot nest in areas devoid of large trees—or large cacti. As a general rule, cavities suitable for nesting exist only in tall, mature vegetation, and at low and middle elevations in New Mexico such vegetation is often found along drainages rather than in the surrounding uplands. However, Elf Owls also breed well away from riparian areas in pine-oak and essentially pure oak woodland (e.g., Cloverdale in the Peloncillo Mountains of Hidalgo County; JDL, pers. obs.). At Mockingbird Gap, nesting has been documented in habitat dominated by creosote (*Larrea tridentata*) and honey mesquite (*Prosopis glandulosa*). Here, Elf Owls would probably not find this vegetation type suitable for nesting were it not for the presence of wooden utility poles near a small house (see photo 33.6 and also section on diet and foraging). The pair alternates between holes excavated by woodpeckers in two of the power poles (D. Burkett, pers. comm.).

Life History

Migration

Extreme records for New Mexico are 19–20 March and 3–7 November (both dates for southwestern New Mexico), but more reliable periods are early April to mid October (from data compiled by J. P. Hubbard). Ligon (1968) mentioned five specimens in the A. R. Phillips collection from Arizona taken prior to 16 March, and Elf Owls occur in the Sonoran Desert as late as 10 October (Phillips 1942).

Some observations suggest that Elf Owls migrate in flocks. J. S. Ligon (1961:147) reported numbers of Elf Owls at the entrance of the Gila River gorge in the late summer of 1918, "sitting on the ground or on rocks beneath the thick scrub bushes of the ledges at [the] base of the canyon wall." He assumed they had concentrated there prior to fall migration. In West Texas, a flock of Elf Owls was seen flying across a highway on 19 March (Henry and Gehlbach 1999). Also reported was a large flock in trees and bushes at Kino [Sonora, Mexico] (season not stated; see Ligon 1968). Observations such as those are rare, however, and the frequency of flock migration remains unknown. Some authors have surmised that Elf Owls arrive in Arizona in large flocks only because they appear to become suddenly abundant in an area. This, however, is probably the result of mutual stimulation, whether or not the birds arrived together in a flock: on moonlit nights in early spring there is often a general outburst of calling (see Ligon 1968).

Nesting

Elf Owls will use any suitable hole to nest in, whether it is in an isolated utility pole, an *Agave* stalk, or a tall dead ponderosa pine (*Pinus ponderosa*), and whether

PHOTOS 33.8a and b

Nestling in hand, Water Canyon, Magdalena Mountains, 1981. PHOTOGRAPHS: © PETER STACEY.

it is a natural crevice or a hole excavated by a woodpecker (Henry and Gehlbach 1999). At Glenwood, Catron County; Guadalupe Canyon, Hidalgo County; and Water Canyon, Socorro County, most nest sites were woodpecker-excavated cavities in sycamore and cottonwood trees. In southwestern New Mexico, Acorn Woodpeckers (*Melanerpes formicivorus*) are probably the primary excavators of cavities used by Elf Owls. At Mockingbird Gap, however, the cavities used by nesting Elf Owls were probably excavated by Northern Flickers (*Colaptes auratus*) instead, based on the size of those holes (D. Burkett, pers. comm.).

There is only one set of eggs from New Mexico. It consists of four eggs taken from a hole in a partially live cottonwood in Rocky Arroyo, Guadalupe Mountains, Eddy County (appendix 33-1). Measurements for the eggs collected in Rocky Arroyo are similar to those for eggs from Arizona (subspecies *whitneyi*) and west Texas (subspecies *idonea*). A second set of eggs collected by H. H. Kimball on 3 May 1926, purportedly from Reserve, Catron County (J. Henshaw, pers. comm.), was apparently collected from a saguaro (based on notes on the egg set data slip) and therefore could not be from New Mexico.

As determined in southeastern Arizona (Ligon 1968), incubation of the eggs lasts an average of 24 days, and the young fledge approximately 28–33 days after hatching. At dusk, a male Elf Owl can deliver insect prey to its mate and nestlings at amazing rates, ranging from 0.75 to 1.20 feedings per minute (Ligon 1968).

Diet and Foraging

The diet of Elf Owls consists almost entirely of arthropods, with a few small vertebrate prey also documented (Henry and Gehlbach 1999). No particular group of arthropods is favored as prey. Instead Elf Owls are apparently opportunistic in their feeding habits, eating anything they can catch, based in part on time of year (Henry and Gehlbach 1999). Three food habit studies, each from a different part of the species' range, are summarized in table 33.1. These studies are from a riparian canyon in southeastern Arizona; from Chihuahuan Desert vegetation in Trans-Pecos, Texas, and from pine-oak-juniper associations in Dark Canyon, Guadalupe Mountains, New Mexico. Diet information from the last location is based on the analysis of two stomach contents (RWD and ABJ, unpubl. data). That analysis revealed the presence only of invertebrates, and overall great similarity of diet between New Mexico and the other two states. At Mockingbird Gap, Doug Burkett (pers. comm.) observed the resident nesting pair catching insects but also taking threadsnakes (*Leptotyphlops*) to its nest.

Elf Owls capture arthropods in a variety of manners. The owls may fly from a perch to take prey either from the ground (e.g., scorpions) or from the air (e.g., beetles). These birds sometimes hover close to the ground with rapidly beating wings, apparently to examine a potential food item before attacking it. Sometimes Elf Owls will land on the ground to pursue prey on foot. With all of these foraging techniques, prey is usually captured with the feet (Ligon 1968; Henry and

PHOTO 33.9

Elf Owl at Mockingbird Gap in Socorro Co. The nesting pair has been observed hunting beetles and moths attracted to the outside light of the small house near the nest site. PHOTOGRAPH: © DOUG BURKETT.

TABLE 33.1. Presence/absence of taxa in the diet of Elf Owls (*Micrathene whitneyi*) in a riparian forest in southeastern Arizona (Ligon 1968; Henry and Gehlbach 1999); a pine/oak/juniper forest in southern New Mexico (RWD and ABJ, unpubl. data); and the Chihuahuan Desert in Trans-Pecos Texas (Henry and Gehlbach 1999). Adapted from Henry and Gehlbach 1999.

	ARIZONA	NEW MEXICO	TEXAS
INSECTA			
Orthoptera			
Acrididae	Y	N	Y
Tettigoniidae	Y	N	Y
Gryllidae	Y	N	N
Mantidae	Y	N	Y
Phasmatidae	N	N	Y
Blattaria			
Blattidae	N	Y	N
Hemiptera			
Coreidae	Y	N	N
Homoptera			
Cicadidae	Y	N	Y
Cercopidae	N	Y	N
Neuroptera			
Chrysopidae	Y	N	N
Coleoptera			
Scarabaeidae	Y	Y	Y
Cerambycidae	Y	Y	Y
Tenebrionidae	N	N	Y
Curculionidae	Y	N	Y
Lepidoptera		Y*	
Nymphalidae	N	-	Y
Hesperiidae	N	-	Y
Sphingidae	Y	-	N
Noctuidae	Y	-	N
Diptera			
Tipulidae	N	Y	N
Muscoidea	N	Y	N
Hymenoptera			
Vespidae	N	N	Y
Formicidae		Y	
ISOPODA			Y
CHILOPODA	Y	N	N

	ARIZONA	NEW MEXICO	TEXAS
ARACHNIDA			
Scorpionida	Y	N	Y
Araneida	Y	Y	Y
Solpugida	Y	N	N
Uropygi	Y	N	N
REPTILIA			
Reptilia			
Phrynosomatidae	Y	N	Y
Leptotyphlopidae	Y	N	Y
MAMMALIA			
Rodentidia			
Heteromyidae	N	N	Y

* Identified to order only in the New Mexico study.

Gehlbach 1999). In late summer, after the rains have begun in southern New Mexico and Arizona, insects, particularly flying beetles, become abundant and are the most frequently captured food items. At this time the most common method of foraging is fluttering and hovering among oak foliage or other vegetation, such as agave blossoms, and capturing insects as they take flight (Ligon 1968). At Mockingbird Gap the resident pair of Elf Owls caught moths and beetles near the light at the back door of the little house near their nest site (D. Burkett, pers. comm.).

Predation and Interspecific Interactions

Predation on Elf Owls has not been documented in New Mexico. Elsewhere, known predators of adults or fledglings include the Great Horned Owl (*Bubo virginianus*) (Henry and Gehlbach 1999). Depredation of nests involving gophersnake (*Pituophis melanoleucus*) and green ratsnake (*Senticolis triaspis*) has also been documented (Boal et al. 1997; Henry and Gehlbach 1999).

Outside New Mexico, Elf Owls have been observed being mobbed by small passerine birds such as American Robins (*Turdus migratorius*), Bushtits (*Psaltriparus minimus*), Bridled Titmice (*Baeolophus wollweberi*), and Black-throated Gray Warblers (*Dendroica nigrescens*) (Ligon 1968; Henry and Gehlbach 1999). The Elf Owl competes with other birds for nesting cavities (Henry and Gehlbach 1999), and at Mockingbird Gap periodic Ladder-backed Woodpecker (*Picoides scalaris*) use of the same nesting cavities otherwise occupied by Elf Owls (D. Burkett, pers. comm.) suggests local competition between the two species.

Status and Management

The status of Elf Owls in New Mexico remains poorly known. At various times local populations are visited, and commented on, but no comprehensive statewide evaluation is available. Nest box studies have the potential to help researchers gather additional information on the species in the state. However, Elf Owl use of nest boxes may vary widely between study or management areas. In the Black Gap Wildlife Management Area in Texas, Elf Owls used 25% of all nest boxes established for the species. A study we conducted in Dark Canyon had very different results. The 30 nest boxes we used measured about 8.9 cm (3½ in) by 8.9 cm (3½ in) or 9.5 cm (3¾ in) by 10.2 cm (4 in) and were about 11.4 cm

(4½ in) or 15.2 cm (6 in) deep below the entrance hole, which measured 3.5 cm (1⅜ in) in diameter. The nest boxes were placed 2.1 m (7 ft) to 3.7 m (12 ft) above the ground in largely dead oaks (*Quercus*), juniper (*Juniperus*), or occasionally ponderosa pine on either side of the road through Dark Canyon on 31 March 2004. They were removed 2 March 2006. During the two nesting seasons, not a single nest box was used by Elf Owls! Nest boxes were used readily by Bewick's Wrens (*Thryomanes bewickii*) and also occasionally by bluebirds (*Sialia* sp.).

Our Dark Canyon experiment was a worthwhile exercise, and one that could be repeated, varying certain parameters such as box height and size and entrance hole diameter. In a riparian habitat Ligon (1968) found that natural-cavity nest sites were located on average about 10.3 m (34 ft) above ground and were almost 25.4 cm (10 in) deep, about 11.4 cm (4½ in) in internal diameter and with an entrance hole just under 5.1 cm (2 in) in diameter. The open mature forest of Dark Canyon supported many dead and dying trees with abundant woodpecker holes suitable for Elf Owls, which probably explains why no Elf Owls used our nest boxes. By contrast, the Black Gap Wildlife Management Area, Texas, is characterized by the low sparse vegetation of the Chihuahuan Desert and provided few natural nest holes (McKinney 1996).

Despite the general lack of information on Elf Owls in New Mexico, there is no reason to believe that populations are other than healthy in any region where they occur. Nonetheless, the exact status of the species in New Mexico will not be determined unless more surveys and research are conducted in the state.

Acknowledgments

We wish to thank John P. Hubbard and Sartor O. Williams III for distribution and occurrence information from their personal databases, and the Lincoln National Forest for permission to carry out the nesting-box study in Dark Canyon. Ernest Valdez conducted the food habits analysis of the Dark Canyon specimens, and Sandra Brantley correlated it with previously published data. R. B. Payne and J. Henshaw kindly provided information on Kimball's egg set, now in the University of Michigan Museum of Zoology, while Doug Burkett shared his notes on the Elf Owls nesting at Mockingbird Gap. All nest boxes used in Dark Canyon were ordered from The Bird House Depot in Tacoma, Washington. We thank the Beibelle family for their hospitality. The U.S. Fish and Wildlife Service and the New Mexico Department of Game and Fish issued collecting permits. R. Roy Johnson provided helpful comments on a draft of the chapter, and Pete Stacey, Doug Burkett, Jim Burns, and Wayne Lynch all generously contributed their photos to illustrate the chapter.

LITERATURE CITED

[AOU] American Ornithologists' Union. 1983. *Check-list of North American Birds.* 7th ed. Washington, DC: American Ornithologists' Union.

Barlow, J. C., and R. R. Johnson. 1967. Current status of the Elf Owl in the southwestern United States. *Southwestern Naturalist* 12:331–32.

Boal, C. W., B. D. Bibles, and R. W. Mannan. 1997. Nest defense and mobbing behavior of Elf Owls. *Journal of Raptor Research* 31:286–87.

del Hoyo, J., A. Elliott, and J. Sargatal. 1999. *Handbook of the birds of the world.* Vol. 5, *Barn owls to hummingbirds.* Barcelona, Spain: Lynx Ediciones.

Henry, S. G., and F. R. Gehlbach. 1999. Elf Owl (*Micrathene whitneyi*). In *The birds of North America*, ed. A. Poole and F. Gill. Philadelphia, PA: Birds of North America, Inc.

Hubbard, J. P. 1978. *Revised check-list of the birds of New Mexico.* Publ. no. 6. Albuquerque: New Mexico Ornithological Society.

James, P., and A. Hayse. 1963. Elf Owl rediscovered in Lower Rio Grande delta of Texas. *Wilson Bulletin* 75:179–82.

LaVal, R. K. 1969, Records of birds from McKittrick Canyon. *Bulletin of the Texas Ornithological Society* 3:24.

Ligon, J. D. 1968. The biology of the Elf Owl, *Micrathene whitneyi. Miscellaneous Publications of the Museum of Zoology, University of Michigan*, no. 132.

Ligon, J. S. 1961. *New Mexico birds and where to find them.* Albuquerque: University of New Mexico Press.

Lockwood, M. W., and B. Freeman. 2002. *The TOS handbook of Texas birds.* College Station: Texas A&M Press.

Marshall, J. T. Jr. 1957. Birds of the pine-oak woodlands of southern Arizona and adjacent Mexico. *Pacific Coast Avifauna* 32:1–125.

McKinney, B. R. 1996. The use of artificial nest boxes by Elf Owls in western Texas. In *Wildlife research highlights* 1:38–39. Austin: Texas Parks and Wildlife Department.

Phillips, A. R. 1942. Notes of the migrations of the Elf and Flammulated Screech Owls. *Wilson Bulletin* 54:132–37.

Ridgway, R. R. 1914. *Micropallus whitneyi idonea.* In *Birds of North and Middle America.* Bulletin of the U.S. National Museum 50, no. 6:807. Washington, DC: Smithsonian Institution.

Stacey, P. B., R. D. Arrigo, T. C. Edwards, and N. Joste. 1983. Northeastern extension of the breeding range of the Elf Owl in New Mexico. *Southwestern Naturalist* 28:99–100.

Walters, P. M. 1981. Notes on the body weight and molt of the Elf Owl (*Micrathene whitneyi*) in southeastern Arizona. *North American Bird Bander* 6:104–5.

Williams, S. O. III, ed. 1994. New Mexico. *Audubon Field Notes* 48:974.

———, ed. 1997. New Mexico. *Audubon Field Notes* 51:1032–36.

———, ed. 1998. New Mexico. *Audubon Field Notes* 52:487–90.

———, ed. 1999. New Mexico. *North American Birds* 53:418–20.

———, ed. 2000. New Mexico region. *North American Birds* 54:312–15.

34

Burrowing Owl

(*Athene cunicularia*)

MARTHA J. DESMOND

WITH ITS GROUND-DWELLING and burrowing habits, the small, long-legged Burrowing Owl (*Athene cunicularia*) is unique among all North American owls. It is the sole member of its genus in North America and interestingly, although three other *Athene* owls of similar size are found in Europe and Asia, none of them are ground- and burrow-dwelling. The Burrowing Owl is also the only North American owl that does not exhibit reversed sexual dimorphism (Earhart and Johnson 1970; McDonald et al. 2004). In fact, males are slightly larger than females in most measurements, although this size difference is not conspicuous and therefore not useful in the field (Plumpton and Lutz 1994; Desmond 1997). Total Burrowing Owl body length ranges between 19 and 25 cm (7.5–9.8 in) and mean body weight is approximately 150 g (5.3 oz) (Haug et al. 1993). It should be noted that adults lose a significant amount of weight when caring for nestlings and juveniles (MJD, pers. obs.).

Also distinguishing the Burrowing Owl from all other North American owls is its unique appearance, particularly its long legs with sparse feathering below the tibiotarsal joint. Additional characteristics of the species include a round head with bright yellow irises; the lack of ear tufts; prominent buff-colored eyebrows; a buff malar stripe (mustache); and a white chin stripe

PHOTO 34.1

Adult Burrowing Owl, Fuller Road, Hidalgo Co., 29 September 2007. The Burrowing Owl is called the "priest of the prairie dogs" by the Zuni Indians. It is a small owl that lives on the ground and in underground burrows, often those excavated by prairie dogs (*Cynomys* spp.). PHOTOGRAPH: © ROBERT SHANTZ.

PHOTO 34.2

(*left*) Adult Burrowing Owl, Bitter Lake National Wildlife Refuge, ca. June 2006. The white chin stripe visible on this bird is particularly conspicuous during courtship and territorial displays. Unlike juveniles, adults have barred chests. PHOTOGRAPH: © GORDON WARRICK.

PHOTO 34.3

(*below*) Educational bird. The head of a Burrowing Owl is characterized by yellow irises, conspicuous buff-colored eyebrows, a lack of ear tufts, and a spotted crown. PHOTOGRAPH: © DOUG BROWN.

PHOTO 34.4

(*top*) Burrowing Owl near the Sandia Casino golf course, Bernalillo Co., 11 June 2007. With its long legs sparsely feathered below the tibiotarsal joint, the Burrowing Owl is quite distinctive among New Mexico's owls. Other field marks include a short tail barred with buff and brown and a brown back spotted with white. PHOTOGRAPH: © DOUG BROWN.

PHOTO 34.5

(*bottom*) Juvenile Burrowing Owl, Kirtland Air Force Base, Bernalillo Co., 11 July 2005. Note the buffy breast that lacks barring. PHOTOGRAPH: © OCTAVIO AND KIRSTEN CRUZ.

that is most obvious when exposed during territorial displays or courtship (Columbe 1971; Haug et al. 1993; Klute et al. 2003; McDonald et al. 2004). The wings are distinctly rounded in shape with 10 brown and buff primaries that show barring (Haug et al. 1993; Klute et al. 2003; McDonald et al. 2004). The tail is short with 12 buff- and brown-colored, barred rectrices (Haug et al. 1993; Klute et al. 2003; McDonald et al. 2004). The back and head of the Burrowing Owl are dark brown with buffy white spots. The throat and undertail coverts are white. Adults have a white breast with buffy and brown barring. Burrowing Owl sexes are largely similar but the male's plumage often appears lighter in color as a result of sun-bleaching; the female appears darker because she spends a large portion of the nesting season in the burrow laying and incubating eggs and brooding the young (Columbe 1971; Haug et al. 1993). Although barring on the chest of males can be more prominent, this is not always noticeable, again as a result of sun-bleaching (Haug et al. 1993; Klute et al. 2003; McDonald et al. 2004). Juveniles differ from adults with their clear buffy to white breast and a prominent, linear buffy wing patch (Haug et al. 1993; Klute et al. 2003; McDonald et al. 2004).

The Burrowing Owl has a wide vocal repertoire. This topic was studied in detail in Albuquerque, New Mexico, during the late 1960s and early 1970s by Martin (1973a), who identified 16 Burrowing Owl vocalizations, 13 used by the adults and 3 by the young. The Burrowing Owl's primary song is a two-note call given exclusively by males during the breeding period (Martin 1973a), and playback recordings of this call can be used to elicit responses from territorial pairs during surveys (Haug and Didiuk 1993). Five additional vocalizations are associated with copulation and the last seven with nest defense and food begging (Martin 1973a). The vocalizations of juvenile owls have been described as an "*eep*" call, which is a low-intensity alarm or hunger call, a "rasp" call, which is a hunger call, and the "rattlesnake rasp," which is a distress call (Haug et al. 1993). In addition to vocalizations, Burrowing Owls also utter distinct nonvocal sounds and they can communicate through body postures and behaviors. Similar to other owls, they will give loud bill snaps and bill claps when threatened. Bill snaps often accompany vocalizations such as screams and chatters. When alarmed, Burrowing Owls also stand upright, prominently exposing their buff-colored eyebrow stripe and white chin stripe. They bob up and down, giving a single-noted cluck or chatter. This behavior earned them the name of the "Howdy Owl" from early settlers (Desmond et al. 1997).

There are 18 recognized subspecies of Burrowing Owls, 15 of them extant and 3 now extinct. Three of the 15 living subspecies are found in North America (Peters 1940; Desmond et al. 2001). The most widely distributed subspecies (*A. c. hypugaea*) is found throughout continental western North America including New Mexico (Haug et al. 1993; Desmond et al. 2001).

Distribution

The Burrowing Owl is widely distributed throughout the New World (southern Canada to Chile and the West Indies islands). It is thought to have evolved in

PHOTO 34.6

Burrowing Owl (adult) on fence in Las Cruces, Doña Ana Co. An important Burrowing Owl population is found in the mosaic of urban and rural environments of Doña Ana Co.
PHOTOGRAPH: © MARTHA DESMOND.

the central Great Plains, spreading across the North American continent and over the Isthmian Land Bridge into South America approximately 2 million years ago (Desmond et al. 2001). The western North American subspecies (*A. c. hypugaea*)—often called Western Burrowing Owl—breeds from southern Canada to central Mexico with populations extending roughly from the 100th meridian west to the Pacific coast (Johnsgard 1988; Haug et al. 1993; Desmond et al. 2001). The other two subspecies found in North America have populations that are disjunct from the western subspecies; *A. c. floridana* occurs only in the state of Florida and *A. c. rostrata* is restricted to Clarion Island off the west coast of Mexico (Desmond et al. 2001). Numerous distinct subspecies have also been recognized throughout the West Indies and on the South American continent (Desmond et al. 2001).

In North America the western subspecies is partially migratory with breeding and nonbreeding distributions that overlap in the southwestern United States and Mexico. Besides the southwestern United States and interior and coastal Mexico, the wintering grounds of the western Burrowing Owl are thought to include the Gulf Coast region east to Louisiana and to extend south into Central America south to El Salvador, more casually western Panama (Klute et al. 2003). The status of breeding and wintering populations in Central America is not well understood. Suitable habitat is not as plentiful as in areas farther north. It is likely that some year-round populations exist but this has not been documented.

The Burrowing Owl is one of New Mexico's most conspicuous and most common owls. The species is considered common throughout the state at low and middle elevations where suitable habitat exists. It has been documented in 28 of New Mexico's 33 counties (NMOS 2008) and probably occurs and breeds in all counties of the state. Populations are likely largest in the eastern third of the state, and along the U.S.-Mexico international border including the urban-agricultural matrix of Doña Ana County. The Burrowing Owl can commonly be found year-round in the extreme southern part of the state, where some individuals are resident and others are migratory. Burrowing Owls have been found to winter on their nesting territories as far north as Albuquerque, Bernalillo County, and Corrales,

PHOTO 34.7

Burrowing Owl north of Animas, Hildago Co., 22 July 2007. PHOTOGRAPH: © ROBERT SHANTZ.

Sandoval County, but only in small numbers and not in all years (NMOS 2008; J. Ruth, pers. comm.), and, in what appears to be very rare occurrences, even farther north, at Maxwell National Wildlife Refuge in Colfax County (recorded in 1977), near Las Vegas, San Miguel County (1978), and on the Navajo Indian Irrigation Project in San Juan County (21 January 1984) (Hubbard 1978; Goodman 1984, *fide* A. Nelson). Hubbard (1978) placed the northern limit of the Burrowing Owl's regular winter distribution near Deming in Luna County

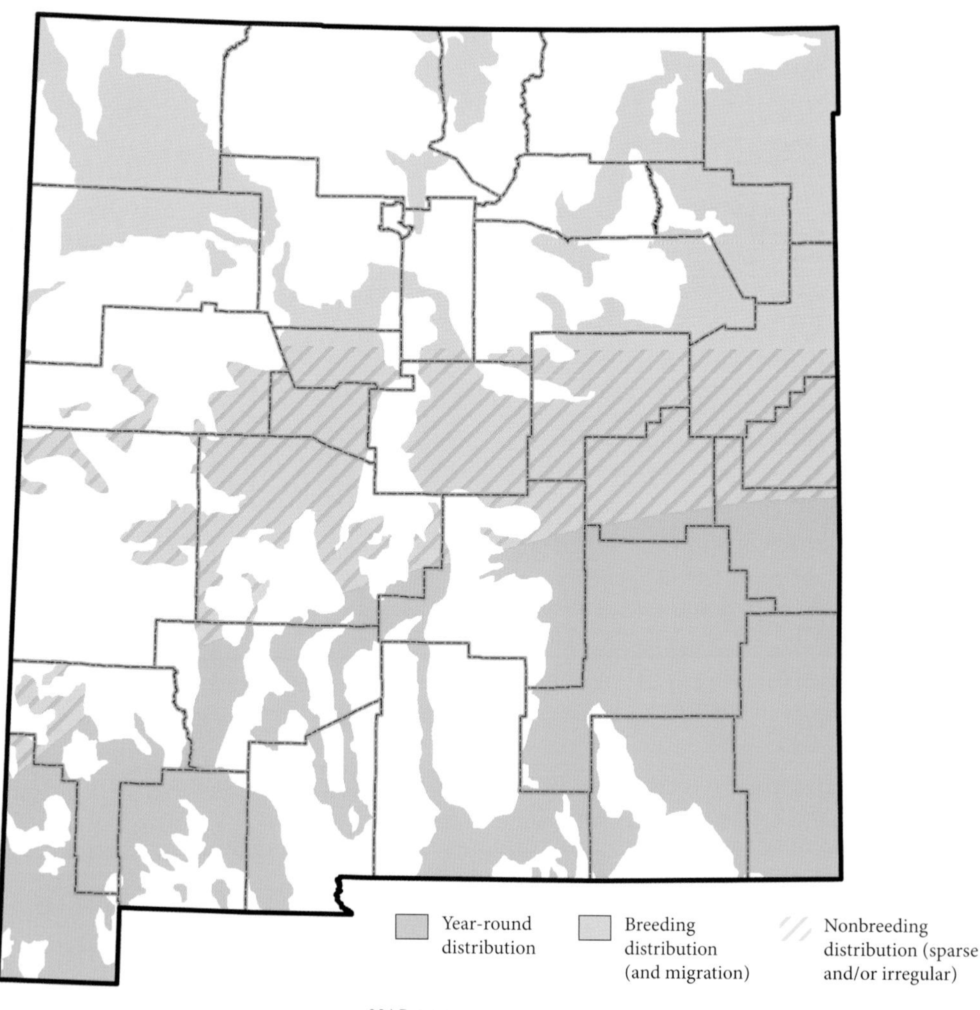

MAP 34.1

Burrowing Owl distribution map

and, eastward, in Las Cruces, Doña Ana County, in Roswell, Chaves County, and near Portales in northern Roosevelt County. Hubbard's (1978) assessment is probably not far off the mark, although a small number of Burrowing Owls appear to winter regularly farther north than Las Cruces, at the Ladder and Armendaris ranches in Sierra County (J. Truett, pers. comm.). On the eastern plains of New Mexico, Ligon (1961:147) reported Burrowing Owls occurring year-round in the Milnesand area of southern Roosevelt County. Ligon (1961:147) inferred that Burrowing Owls were year-round residents west of Portales in northern Roosevelt County, having found an owl 1.8 m (6 ft) deep underground in its burrow on 30 January 1952, and "a little later" (the exact date is not provided) as many as 40 owls aboveground when the weather was warmer.

Habitat Associations

The Burrowing Owl avoids wooded areas and instead is associated with a variety of open, sparsely vegetated habitats with suitable underground roost and nest sites (Haug et al. 1993; Klute et al. 2003; McDonald et al. 2004). Typically it occupies burrows excavated by other animals, including colonial sciurids (*Cynomys* spp. and *Spermophilus* spp.), banner-tailed kangaroo rats (*Dipodomys spectabilis*), North American badgers (*Taxidea taxus*), and desert tortoises (*Gopherus agassizii*); less often it makes its home in human-made (artificial) burrows and rock crevices or other natural cavities. (Johnsgard 1988; Haug et al. 1993; deVos 1998; Klute et al. 2003; McDonald et al. 2004).

In New Mexico the Burrowing Owl is found in desertscrub, grasslands, and savannahs, as well as in arroyos, agricultural lands, and urban and disturbed areas. Burrowing Owl densities in the state are highest in open grasslands with prairie dog (*Cynomys* spp.) colonies and in agricultural and urban areas where rock squirrel (*Spermophilus variegatus*) burrows are available (Berardelli et al., in press; MJD, unpubl. data). In open areas, including grasslands, parks, and agricultural areas, Burrowing Owls tend to nest in clusters (Desmond et al. 1995), whereas in desertscrub environments I have found Burrowing Owls usually occurring as solitary pairs. For nesting, a single burrow may not be enough; nesting pairs select areas that include not only a nest burrow but also what will serve as satellite burrows for adults and juveniles to utilize (Desmond and Savidge 1999; Berardelli et al., in press; see Nesting). In agricultural areas, the Burrowing Owl is particularly associated with irrigation canals (Botelho and Arrowood 1996; Berardelli et al., in press).

Across the eastern portion of New Mexico the Burrowing Owl is strongly tied to black-tailed prairie dog (*C. ludovicianus*) colonies. Owl densities can be locally high across Union, Colfax, Quay, Chavez, Roosevelt, Eddy, and Lea counties—strangely there appears to be no published record of Burrowing Owl occurrence in Harding County—where healthy black-tailed prairie dog populations still remain. In these counties representing most of New Mexico's eastern plains and a part of the southern Great Plains, most active prairie dog colonies are thought to harbor owl populations. With the exception of the Kiowa National Grasslands most of these colonies occur on private land.

Sidle et al. (2001) found that 93% of black-tailed prairie dog colonies on national grasslands in the southern Great Plains had Burrowing Owls present, compared to

PHOTO 34.8

Burrowing Owl grassland habitat on Kirtland Air Force Base, Bernalillo Co., 30 May 2006. Burrowing Owls are found in open, sparsely vegetated areas with suitable underground burrows. PHOTOGRAPH: © OCTAVIO AND KIRSTEN CRUZ.

PHOTO 34.9

Burrowing Owl grassland habitat north of Animas, Hidalgo Co., 26 August 2006. PHOTOGRAPH: © ROBERT SHANTZ.

PHOTO 34.10

(*top left*) Burrowing Owl in surrounding yucca grassland habitat, Oscura Breeding Bird Survey Route, Lincoln Co., May 2006. PHOTOGRAPH: © DOUG BURKETT.

PHOTO 34.11

(*bottom left*) Burrowing Owl on a shinnery oak (*Quercus havardii*) near Eunice, Lea Co., June 2006. PHOTOGRAPH: © MARK L. WATSON.

PHOTO 34.12

(*top right*) Burrowing Owl in a dry drainage pond in Las Cruces, Doña Ana Co. PHOTOGRAPH: © MARTHA DESMOND.

PHOTO 34.13

Burrowing Owl nest in a black-tailed prairie dog (*Cynomys ludovicianus*) burrow on the Kiowa National Grasslands in Union Co., June 2008. On the eastern plains of New Mexico the Burrowing Owl is associated with black-tailed prairie dog colonies.
PHOTOGRAPH: © MARTHA DESMOND.

only 59% of colonies on national grasslands of the northern Great Plains. Additionally, densities of Burrowing Owls appear to be higher in the southern Great Plains; Hanni and McLachlan (2004) reported a density of 2.55 Burrowing Owls/km^2 in native grasslands of New Mexico compared to an estimated 1.17/km^2 overall for native grassland across the entire Great Plains. In a survey of black-tailed prairie dog colonies across five counties in northeastern New Mexico, Arrowood et al. (2001) reported that Burrowing Owls occurred at 36 (73%) of 49 surveyed prairie dog colonies, for a total of 385 owls observed. The importance of black-tailed prairie dog colonies as Burrowing Owl habitat was particularly conspicuous in Sierra County on the Turner Armendaris Ranch at the time that black-tailed prairie dogs were locally reintroduced. As new black-tailed prairie dog colonies became established, Burrowing Owls immediately colonized them the following nesting season (Berardelli et al., in press).

In central New Mexico both black-tailed and Gunnison's prairie dogs (*Cynomys gunnisoni*) occur, whereas west-central and northwestern New Mexico harbors colonies of only the Gunnison's prairie dog. Here again, it is in prairie dog colonies that most Burrowing Owls have been found, with smaller numbers also detected in association with rock squirrels (e.g., Martin 1973a, 1973b; Envirological Services 2006; C. Curtin, pers. comm.). Martin (1973b) reported both Barn Owls (*Tyto alba*) and Burrowing Owls nesting along Tijeras Arroyo in Bernalillo County, the nests of the latter species found in modified rock squirrel burrows (see also chapter 27). In the grasslands of San Juan County and in the Plains of San Agustin and the Estancia Valley, Burrowing Owls most commonly occur in Gunnison's prairie dog colonies (J.-L. Cartron, pers. comm.). On Kirtland Air Force Base in the Albuquerque metropolitan area, a long-term assessment of Burrowing Owls nesting in Gunnison's prairie dog colonies showed substantial variability in the number of nesting pairs through time. Overall, however, it also revealed that fairly large Burrowing Owl populations can be found in New Mexico in association with Gunnison's prairie dogs. The number of nesting pairs ranged from 19 pairs in 2004 to 52 pairs in 1998 with an average of 31 pairs per breeding season (Envirological Services 2006).

In southern New Mexico Burrowing Owls are associated with the burrows of black-tailed prairie dogs, banner-tailed kangaroo rats, and rock squirrels, as well as with natural and artificial cavities (Berardelli et al., in press; MJD, pers. obs.). In Las Cruces and the surrounding agricultural areas of Doña Ana County, 73% of nesting owls utilized rock squirrel burrows (Berardelli et al., in press). Burrowing Owls nesting at Holloman Air Force Base in Otero County again utilized rock squirrel burrows, but also natural cavities created by soil disturbance (Arrowood et al. 2001; Envirological Services 2007a; M. Wootten, pers. comm.). Burrowing Owls have been observed nesting in banner-tailed kangaroo rat burrows on the Armendaris Ranch in Sierra County (T. Waddell, pers. comm.), in the grasslands around Hachita in Hildago County (R. Myers, pers. comm.), and on the Jornada Experimental Range in Doña Ana County (MJD, pers. obs.).

Life History

Migration and Winter

The majority of all western Burrowing Owls are migratory. However, little information exists on migratory routes and wintering areas. Burrowing Owls are known to winter in the extreme southwestern United States, including Texas and New Mexico west to southern California, along the Gulf of Mexico in coastal Louisiana and Texas, and southward into Mexico and farther south (Ligon 1961; Holroyd et al. 2001; Klute et al. 2003). Although the vast majority of Burrowing Owls nesting in New Mexico are migratory, winter sightings are regularly reported from much of the state (Ligon 1961). The most reliable wintering populations are found in southern New Mexico. From my own studies and based also on research by Envirological Services (unpubl. data; M. Wootten, pers. comm.), we know that each year a portion of the breeding populations in the Las Cruces area and at Holloman Air Force Base, Otero County, overwinters. We also know that this proportion of overwintering individuals varies substantially from one year to the next (Arrowood et al. 2001; MJD, unpubl. data). In Las Cruces in some years, very few owls are present during winter months whereas in other years up to a quarter of breeding adults from the previous summer have been observed to spend the winter (MJD, unpubl. data). At Holloman Air Force Base (Otero County), eight burrows were occupied by owls during one winter, including six that had nesting pairs the previous summer (Johnson et al. 1997).

In 2006, Envirological Services (2006, 2007b) initiated a radio telemetry study to gather more information on migration and to locate New Mexico Burrowing Owls on winter grounds. A total of 28 juveniles trapped at Kirtland Air Force Base in Bernalillo County were fitted with radio collars in the summer of 2006, 30 the next year. Envirological Services conducted aerial transects and ground searches in southern New Mexico, Texas, and Mexico, attempting to detect a radio signal from any of the owls. The amateur radio community was also solicited for help tracking radio signals. No transmitter signals were detected, but the telemetry study indicated that most juveniles began to disperse from their natal area in August (Envirological Services 2006, 2007b).

PHOTO 34.14

Burrowing Owl fledglings equipped with radio collars and color leg bands on Kirtland Air Force Base, Bernalillo Co., 12 July 2006. In 2006, Envirological Services initiated a study to identify the wintering grounds of Burrowing Owls nesting in central New Mexico. PHOTOGRAPH © OCTAVIO AND KIRSTEN CRUZ.

Nesting

Throughout most of New Mexico and on the Great Plains, Burrowing Owls arrive on breeding grounds in mid to late March. Owls establish pair bonds in April and peak nesting occurs in mid May. In the urban and agricultural areas of Doña Ana County in extreme southern New Mexico, Burrowing Owls are typically on breeding territories by late February and establish pair bonds by March; peak nesting occurs in mid to late April, nearly one month ahead of populations across the rest of the state (Berardelli et al., in press). Nesting is similarly early in northern Mexico (McNicoll 2005), and probably also in Otero County and along the New Mexico–Mexico border in Hildago and Luna counties. Immediately to the north of Doña Ana County in Sierra County, Burrowing Owls establish their nests in mid May, similar to populations across the Great Plains including northeastern New Mexico (Berardelli et al., in

PHOTO 34.15

Burrowing Owl nest in a rock squirrel (*Spermophilus variegatus*) burrow in Las Cruces, Doña Ana Co. Note the styrofoam and other trash used as nesting material.

PHOTOGRAPH: © MARTHA DESMOND.

PHOTOS 34.16a and b

Burrowing Owl nestling outside its burrow, Laguna Pueblo, Cibola Co., 5 June 2006. Even prior to fledging, young Burrowing Owls venture up to 75 m (250 ft) out of the nest burrow and may use nearby satellite burrows.

PHOTOGRAPHS: © TOM KENNEDY.

press; MJD, unpubl. data). In southeastern New Mexico there may be greater variability in the timing of nesting; however, this has not been studied. The earlier nest initiation dates in Doña Ana County in particular are likely related to the owl's partial migrant status at that location, together with a warmer climate and, in urban environments, reduced exposure to weather extremes.

Burrowing Owls typically line their nest burrows—including the entrance, tunnel, and nest chamber—with shredded cow and horse dung. They also use badger and coyote (*Canis latrans*) scat, and miscellaneous items such as grass clippings, desiccated toads, and corn cobs. In urban environments nest-lining material consists mainly of dog feces and trash. Use of excrements for nest lining is presumably designed to mask the scent of the owls and reduce the risk of predation, but may also be used to attract dung beetles and to insulate the eggs (Green and Anthony 1989).

Burrowing Owl nest sites typically have at least two satellite burrows in the immediate vicinity of the nest burrow. These satellite burrows are used by the adults for roosting and as shelters for the juveniles as they begin to venture out of the nest. In black-tailed prairie dog colonies Burrowing Owl have numerous satellite burrows; juveniles will venture up to 75 m (250 ft) from the nest prior to fledging and during that time they utilize many of those satellite burrows (Desmond and Savidge 1999).

At the scale of the species' range, mean clutch size is eight eggs with a range of three to twelve (Bent 1938; Haug et al. 1993), and in New Mexico, five egg sets collected by J. Stokley Ligon (unpubl. notes on file at the UNM Museum of Southwestern Biology) in 1926 ranged in size from six to eight eggs (but see below

PHOTO 34.17

Adult and fledglings at the entrance to their burrow along an arroyo wall in Rio Rancho, Sandoval Co., 25 May 2008. PHOTOGRAPH: JAMES N. STUART.

for evidence of larger clutches in New Mexico). The incubation period lasts approximately one month, and the female is the sole incubator of the eggs (Haug and Oliphant 1990; Haug et al. 1993; Klute et al. 2003; McDonald et al. 2004). During incubation and the nestling stage, the male provides food for the female and, after hatching, also for the nestlings. Once the nestlings are able to regulate their body temperature, however, the female also brings food to them. The juveniles typically fledge at 42 days of age (Haug and Oliphant 1990). Based on my own research, fledging typically occurs in mid to late June in southern New Mexico and early to mid July across the rest of the state. Adults and young tend to remain together as family units for at least a month following fledging. On rare occasions, whole family units in the urban areas of Las Cruces have been observed as late as December, approximately six months following fledging.

Although Burrowing Owls lay large egg clutches, they typically fledge only two to three young per nest (Desmond et al. 2000; Arrowood et al. 2001; Berardelli et al., in press). In good years, however, nesting pairs may raise on average four to five young to fledging, and I have even recorded some nests fledging as many as nine young (in Las Cruces). In New Mexico, Berardelli et al. (in press) has documented similar mean fledging rates for owl populations of urban areas and native grasslands, but with less variation in the fledging success of pairs nesting in native grasslands. In the urban and agricultural matrix of Doña Ana County, 43 nesting pairs were monitored across 19 study sites in 2000 and 37 nesting pairs across 15 sites in 2001, whereas in native grasslands of nearby Sierra County, 23 and 41 nesting pairs were located in reestablished black-tailed prairie dog colonies in 2000 and 2001, respectively (Berardelli et al., in press). The observed number of nesting pairs in prairie dog colonies increased dramatically in 2001 compared to the previous year, apparently due to the expansion and discovery of colonies by owls. Within Sierra County's native grassland environments an overall greater proportion of pairs raised at least one young to fledging, whereas in the urban and agricultural matrix of Doña Ana County pairs tended to either fail or fledge large—or very large—broods (Berardelli et al., in press). The observed higher numbers of chicks fledged per successful nest in urban areas may be related to a more plentiful and reliable food supply in these environments, while the higher incidence of nest failure is likely related to increased predation and disturbance.

In contrast to Berardelli et al. (in press), Botelho and Arrowood (1996)—also working in New Mexico—found a significantly higher number of nestlings and fledglings per nest in human-altered compared to natural nesting areas, and they speculated that natural areas experienced higher rates of predation and greater disturbance from neighboring nesting pairs due to higher pair densities. Even Berardelli et al. (in press) found that higher nest density negatively influenced productivity not only in natural grasslands where nest densities in reestablished prairie dog colonies were high, but also in urban environments characterized overall by substantially lower nest densities. A mean distance of 160 m (525 ft) was observed between nests in urban environments of Doña Ana County compared to only 54 m (177 ft) in reestablished black-tailed prairie dog colonies in Sierra County (Berardelli et al., in press). In urban landscapes, high fledging success was in general associated with fewer surrounding nests, larger nesting territories, and the occurrence of some green space, suggesting competition negatively influenced productivity, whereas land use positively influenced it. Higher productivity in areas with green space also suggested lower disturbance and/or predation in these environments.

Although owls had higher productivity where owl densities were lower, the poorest productivity was for solitary nests, suggesting there is some benefit to nesting in the vicinity of other Burrowing Owls, such as alerting each other to predators or group mobbing of predators. In grassland habitats, high productivity was associated with lower nest density, larger nesting territories, fewer satellite burrows, and lower fledgling success at the nearest nest. These results indicated productivity was highest for nests on larger territories and along colony edges (hence fewer satellite burrows) where owls had reduced competition from neighbors and easier access to foraging sites. Overall, owls may benefit from the presence of other nesting pairs as demonstrated from the urban data set. Once the density of nesting pairs exceeds a certain threshold, however, productivity becomes negatively affected, perhaps in part as a result of competition for food or more frequent antagonistic interactions. I have observed Burrowing Owls intruding upon neighboring territories at sites where densities are high; such intruders are met with alarm calls from the resident pair, with occasional fights ensuing.

Reproductive rates reported for Burrowing Owls across the state have been variable. Over a two-year period, owls nesting in the reestablished prairie dog colonies of Sierra County averaged 2.4 young fledged per nest while owls in urban and agricultural environments fledged 2.6 young per nest (Berardelli et al., in press). Arrowood et al. (2001) reported a range of 1.4 to 4.5 fledglings per nest for nests monitored in Las Cruces between 1993 and 2000. At Holloman Air Force Base, Johnson et al. (1997) reported higher rates of nest abandonment for owls nesting in highly disturbed areas. Overall, they found an average of 2.1 and 2.7 fledglings per nesting pair for 1996 and 1997, respectively; in 1996 64% of all pairs experienced nesting success while in 1997 that percentage reached 77%. At Kirtland Air Force Base in Bernalillo County, the number of fledglings per nest ranged from a low of 1.7 in 2002 to a high of 4.2 in 2001 with an average of 2.96 per breeding pair over the nine-year study period (Envirological Services 2006). Martin (1973a) reported 4.9 fledglings per pair for 15 pairs nesting in rock squirrel burrows in Tijeras Arroyo just south of Albuquerque in Bernalillo County. In 2006, 68 pairs nesting in black-tailed prairie dog colonies on the Kiowa National Grasslands in northeastern New Mexico and adjacent Rita Blanca National Grasslands in west Texas fledged on average 1.8 fledglings per nest with a 56% rate of nest success (MJD, unpubl. data).

Reuse of nest burrows over time in prairie dog colonies has been found to be low (Griebel 2000; MJD, unpubl. data). However, Martin (1973a) observed that six males and two females of nine banded pairs returned the following year to nest again along Tijeras Arroyo. The same nest burrow was selected by returning males unless it was no longer suitable. No intrapair fidelity was observed among returning owls.

Diet and Foraging

Burrowing Owls are considered generalist predators. Invertebrates have been found to make up the majority of food items in their diet but only a small percentage of the total prey biomass (Haug et al. 1993). Mammals are considered to be the most important prey items in terms of biomass, although birds, frogs, toads, lizards, snakes, and turtles have also been documented in the species' diet (Haug et al. 1993). The percentage of vertebrate prey declines as the breeding season progresses, while invertebrate prey items increase in numbers (Haug et al. 1993). Burrowing Owls are known to occasionally engage in cannibalism (Haug et al. 1993).

Pache (1981) analyzed Burrowing Owl food pellets from Tijeras Arroyo in Bernalillo County and identified mammals, birds, lizards, snakes, and invertebrates (mostly beetles) among remains. Mammals (mainly pocket mice [*Perognathus* spp.] but also deer mice [*Peromyscus*], grasshopper mice [*Onychomys*], and kangaroo rats [*Dipodomys*]) dominated the total prey biomass. In northeastern New Mexico I have also found unidentified reptile eggs at Burrowing Owl nest burrows.

Predation and Interspecific Interactions

The Burrowing Owl has many documented natural enemies. Throughout its range, its main predator is considered to be the North American badger, especially in prairie dog colonies where badgers tend to forage (Goodrich and Buskirk 1998; Desmond et al. 2000). Other mammalian predators include coyotes, foxes (*Vulpes fulva, V. velox, Urocyon cinereoargenteus*), domestic cats and dogs, and, where reintroduced, black-footed ferrets (*Mustela nigripes*). Burrowing Owls are also vulnerable to predation by several raptor species including Golden Eagles (*Aquila chrysaetos*), Ferruginous Hawks (*Buteo regalis*), Swainson's Hawks (*Buteo swainsoni*), Prairie Falcons (*Falco mexicanus*), Peregrine Falcons (*Falco peregrinus*), Great Horned Owls (*Bubo virginianus*), and Barn Owls. Burrowing Owl eggs and nestlings are vulnerable to predation by snakes, including prairie rattlesnakes (*Crotalus viridis*) in prairie dog colonies. In South Dakota an adult female was killed by a prairie rattlesnake while defending her clutch from the snake (Griebel 2000). In response to the threat of potential predators, Burrowing Owls often mob badgers, coyotes, foxes, and domestic dogs and cats. Mobbing may involve adults and juveniles from several nesting territories.

Status and Management

The Burrowing Owl is a migratory bird protected under the Migratory Bird Treaty Act. It is a national priority species of the U.S. Fish and Wildlife Service, Office of Migratory Birds (U.S. Fish and Wildlife Service 2002), while it is also listed as Endangered in Canada and Threatened in Mexico (SEDESOL 1994; Holroyd et al. 2001; Klute et al. 2003). At the scale of the species' entire distribution, primary threats to Burrowing Owl populations include or presumably include habitat loss and fragmentation as well as human-related mortality on wintering grounds and during migration (Holroyd et al. 2001; Klute et al. 2003; McDonald et al. 2004).

Across much of North America, populations of Burrowing Owls are declining, particularly in the north (Sheffield 1997; Holroyd et al. 2001; Klute et al.

PHOTO 34.18

Burrowing Owl (educational bird). Burrowing Owl populations are declining across much of North America, but population levels seem overall stable or increasing in New Mexico.

PHOTOGRAPH: © DOUG BROWN.

2003). According to long-term Breeding Bird Survey (BBS) data (1966–2007), Burrowing Owl populations are declining at an annual rate of 1.5% across the range of the bird, with an annual decline of 4.4% in the northern and central Great Plains states compared to an annual population increase of 1.5% in the southern plains. Still according to BBS data, Burrowing Owl populations across New Mexico increased in size at a rate of 3.7% per year between 1966 and 2006. Burrowing Owl numbers increased in the southeastern quadrant of the state and in the extreme northwest at a rate greater than 1.5% per year, declined in the southwest at a rate of less than 1.5% per year, and exhibited no consistent trend in the central and northeastern part of the state, with population fluctuations ranging from -1.5% to >1.5% per year (Sauer et al. 2007). However, BBS data may or may not lend themselves to accurate estimates of Burrowing Owl population trends, in part due to the patchy distribution of the species. Breeding densities also vary widely through time. At Holloman Air Force Base in Otero County, 18 and 19 pairs were documented in 1996 and 1997, respectively. By 2000, however, the breeding population had been reduced to only 2 pairs and in 2003 and 2004 no owls were documented. The population has since gradually recovered and approximately 20 nesting pairs were present in 2008 (Arrowood et al. 2001; Envirological Services 2007a; M. Wootten, pers. comm.).

Loss of colonial sciurids, most notably black-tailed prairie dog populations, is among the most significant factors that have threatened Burrowing Owl populations. Extensive populations of black-tailed prairie dogs historically occurred throughout the Great Plains. However, black-tailed prairie dog numbers have declined by an estimated 90–98% over the past 150 years, primarily due to agricultural expansion, urbanization, and prairie dog control programs (Summers and Linder 1978; Coppock et al. 1983; Miller et al. 1994). The arrival of the sylvatic plague, which is caused by the bacterium *Yersinia pestis*, has further fragmented the species' distribution and resulted in additional declines as prairie dogs are highly susceptible to the disease (Cully 1989; Cully et al. 2006).

Across the eastern half of New Mexico the incidence of sylvatic plague is currently the greatest and most unpredictable threat to the persistence of prairie dog colonies and their associated Burrowing Owl populations. The number of nesting pairs on 12 prairie dog colonies on the Kiowa National Grasslands in northeastern New Mexico and adjacent Rita Blanca National Grasslands in Texas and Oklahoma declined from 68 nests in 2006 to 35 in 2008, partly due to an outbreak of sylvatic plague (MJD, unpubl. data). Elsewhere in New Mexico, the incidence of sylvatic plague may also represent a primary threat to Burrowing Owl populations, together with prairie dog control.

Burrowing Owl populations in urban and agricultural areas face their own numerous threats, including efforts to control rock squirrel populations and disturbance due to maintenance activities along irrigation canals and drainage ponds; destruction of burrows and loss of foraging areas due to real estate development; vandalism of nests; predation by domestic dogs and cats; and road traffic mortality. Owls no longer occur in some historic nesting areas due to the loss of suitable nest burrows, particularly in Doña Ana County (Arrowood et al. 2001; MJD, pers. obs.).

There is an urgent need to better understand the winter distribution and ecology of Burrowing Owl populations (Holroyd et al. 2001; CEC 2005). Most of New Mexico's Burrowing Owls are migratory, but their winter grounds have not yet been located. Thus the specific threats they might face (e.g., mortality due to pesticides) during winter remain unknown. Efforts to maintain sustainable populations of Burrowing Owls in New Mexico and elsewhere will require a concerted and joint effort involving state and federal natural resource agencies, city and county land managers, real estate and development companies, scientists, and private landowners. Part of this effort should involve educating the public and land managers about the importance of colonial sciurids to Burrowing Owl populations and about biological diversity in general. The establishment of a New Mexico Burrowing Owl Working Group with representatives from a variety of federal and state agencies and nongovernment organizations, private consultants, and scientists is the beginning of an effort to address issues related to the conservation of the Burrowing Owl in the state. One short-term goal of the working group is to foster collaboration between all stakeholders to mitigate the loss of or damage to New Mexico's historical nesting areas.

Acknowledgments

I would like to thank P. C. Arrowood, D. Berardelli, J.-L. Cartron, C. Finley, R. Meyer, J. Ruth, J. Truett, T. VerCautern, T. Waddell, and M. Wootten for providing information on Burrowing Owl locations throughout New Mexico. Detailed unpublished reports that included substantial information on Burrowing Owl locations were graciously provided by Envirological Services and the Rocky Mountain Bird Observatory. R. Shantz, G. Warrick, D. Brown, O. Cruz and K. Cruz-McDonnell, D. Burkett, M. Watson, and T. Kennedy all contributed excellent photographs of Burrowing Owls taken in New Mexico. J.-L. Cartron provided excellent advice, support, and editorial comments. This is a New Mexico Agricultural Experiment Station publication, supported by state funds and the U.S. Hatch Act.

LITERATURE CITED

Arrowood, P., C. A. Finley, and B. C. Thompson. 2001. Analyses of Burrowing Owl populations in New Mexico. *Journal of Raptor Research* 35:362–70.

Bent, A. C. 1938. *Life histories of North American birds of prey*. Part 2. New York: Dover Publications.

Berardelli, D., M. J. Desmond, and L. Murray. In Press. Reproductive success of Burrowing Owls in urban and grassland habitats in southern New Mexico. *Wilson Journal of Ornithology*.

Botelho, E. S., and P. C. Arrowood. 1996. Nesting success of western Burrowing Owls in natural and human-altered environments. In *Raptors in human landscapes*, ed. D. Bird, D. Varland, and J. Negro, 63–68. London: Academic Press.

Columbe, H. N. 1971. Behavior and population ecology of the burrowing owl, *Speotyto cunicularia*, in the Imperial Valley of California. *Condor* 73:162–76.

[CEC] Commission for Environmental Cooperation. 2005. *North American Conservation Action Plan, Western Burrowing Owl*. Montreal: CEC.

Coppock, D. L., J. E. Ellis, J. K. Detling, and M. I. Dyer. 1983. Plant-herbivore interactions in a North American mixed-grasses prairie. *Oecologia* 56:10–15.

Cully, J. F. Jr. 1989. Plague in prairie dog ecosystems: importance for black-footed ferret management. In *The prairie dog ecosystem: managing for biological diversity*, ed. T. W. Clark, D. Hinkley, and T. Rich, 47–55. Wildlife Technical Bulletin no. 2. Billings: Montana Bureau of Land Management.

Cully, J. F., D. E. Biggins, and D. B. Seery. 2006. Conservation of prairie dogs in areas with plague. In *Conservation of the black-tailed prairie dog*, ed. J. L. Hoogland, 157–68. Washington, DC: Island Press.

Desmond, M. J. 1997. Evolutionary history of the genus *Speotyto*: a genetic and morphological perspective. Ph.D. diss. University of Nebraska, Lincoln.

Desmond, M. J., and J. A. Savidge. 1999. Satellite burrow use by burrowing owl chicks and its influence on nest fate. P. D. Vickery and J. R. Herkert, eds. Ecology and conservation of grassland birds of the western hemisphere. *Studies in Avian Biology* 19:128–30.

Desmond, M. J., T. J. Parsons, T. O. Powers, and J. A. Savidge. 2001. An initial examination of mitochondrial DNA structure in Burrowing Owl populations. *Journal of Raptor Research* 35:274–81.

Desmond, M. J., J. A. Savidge, and R. Eckstein. 1997. Prairie partners. *NEBRASKAland Magazine*, November 1, 16–25.

Desmond, M. J., J. A. Savidge, and K. Eskeridge. 2000. Correlations between Burrowing Owl and black-tailed prairie dog (*Cynomys ludovicianus*) declines: a seven year analysis. *Journal of Wildlife Management* 64:1067–75.

Desmond, M. J., J. A. Savidge, and T. F. Siebert. 1995. Spatial patterns of Burrowing Owl (*Speotyto cunicularia*) nests within black-tailed prairie dog (*Cynomys ludovicianus*) towns. *Canadian Journal of Zoology* 73:1375–79.

deVos, J. C. Jr. 1998. Burrowing Owl (Athene cunicularia). In *The raptors of Arizona*, ed. R. L. Glinski, 166–69. Tucson: University of Arizona Press.

Earhart, C. M., and N. K. Johnson. 1970. Size dimorphism and food habits of North American owls. *Condor* 72:251–64.

Envirological Services. 2006. *Annual report on population status, reproductive success, prey delivery, and site fidelity of Western Burrowing Owls (Athene cunicularia hypugaea) on Kirtland Air Force Base, 2006*. Albuquerque, NM.

Envirological Services. 2007a. *Decline of the western Burrowing Owl on Holloman Air Force Base, 2006 report*. Albuquerque, NM.

Envirological Services. 2007b. Use of radio-telemetry to determine wintering grounds of Burrowing Owls (*Athene cunicularia*). Albuquerque, NM.

Goodman, R. A., ed. 1984. *New Mexico Ornithological Society Field Notes* 23(1): 1 December 1983 to 29 February 1984.

Goodrich, J. M., and S. W. Buskirk. 1998. Spacing and ecology of North American badgers (*Taxidea taxus*) in a prairie-dog (*Cynomys leucurus*) complex. *Journal of Mammalogy* 79:171–79.

Green, G., and R. Anthony. 1989. Nesting success and habitat relationships of burrowing owls in the Columbia Basin, Oregon. *Condor* 91:347–54.

Griebel, R. L. 2000. Ecological and physiological factors affecting nesting success of burrowing owls in Buffalo Gap National Grassland. M.S. thesis. University of Nebraska, Lincoln.

Hanni, D. J., and M. McLachlan. 2004. *Section-based monitoring of breeding birds within the shortgrass prairie bird conservation region (BCR 18)*. Brighton, CO: Rocky Mountain Bird Observatory.

Haug, E. A., and A. B. Didiuk. 1993. Use of recorded calls to detect Burrowing Owls. *Journal of Field Ornithology* 64:188–94.

Haug, E. A., and L. W. Oliphant. 1990. Movements, activity patterns, and habitat use of Burrowing Owls in Saskatchewan. *Journal of Wildlife Management* 54:27–35.

Haug, E. A., B. A. Miller, and M. S. Martell. 1993. Burrowing Owl (*Speotyto cunicularia*). No. 61. In *The birds of North America*, ed. A. Poole and F. Gill. Philadelphia, PA: Academy of Natural Sciences, and Washington, DC: American Ornithologists' Union.

Holroyd, G. L., R. Rodríguez-Estrella, and S. R. Sheffield. 2001. Conservation of the Burrowing Owl in western North America: issues, challenges, and recommendations. *Journal of Raptor Research* 35:399–407.

Hubbard, J. P. 1978. *Revised check-list of the birds of New Mexico*. Publ. no. 6. Albuquerque: New Mexico Ornithological Society.

Johnsgard, P. A. 1988. *North American owls*. Washington, DC: Smithsonian Institution Press.

Johnson, K., L. DeLay, P. Mehlhoop, and K. Score. 1997. Distribution, habitat, and reproductive success of Burrowing Owls on Holloman Air Force Base. Unpublished report. Albuquerque: New Mexico Natural Heritage Program, University of New Mexico.

Klute, D. S., L. W. Ayers, M. T. Green, W. H. Howe, S. L. Jones, J. A. Shaffer, S. R. Sheffield, and T. S. Zimmerman. 2003. *Status assessment and conservation plan for the Western Burrowing Owl in the United States*. Biological Technical Publ. FWS/BTP-R6001-2003. Washington, DC: U.S. Department of the Interior, Fish and Wildlife Service.

Ligon, J. S. 1961. *New Mexico Birds and where to find them*. Albuquerque: University of New Mexico Press.

Martin, D. J. 1973a. Selected aspects of Burrowing Owl ecology and behavior. *Condor* 75:446–56.

———. 1973b. Burrow digging by Barn Owls. *Bird-Banding* 44:59–60.

McDonald, D., N. M. Korfanta, and S. J. Lantz. 2004. The Burrowing Owl (*Athene cunicularia*): a technical conservation assessment. USDA Forest Service, Rocky Mountain Region. http://www.fs.fed.us/r2/projects/scp/assessments/burrowing owl.pdf.

McNicoll, J. L. 2005. Burrowing Owl (*Athene cunicularia*) nest site selection in relation to prairie dog colony characteristics and surrounding land-use practices in Janos, Chihuahua, Mexico. Thesis. New Mexico State University, Las Cruces.

Miller, B., G. Ceballos, and R. Reading. 1994. The prairie dog and biotic diversity. *Conservation Biology* 8:677–81.

[NMOS] New Mexico Ornithological Society. 2008. *NMOS Field Notes* database. http://nhnm.unm.edu/partners/NMOS/.

Pache, P. H. 1981. Prey remains in pellets from Burrowing and Barn Owls in central New Mexico. *New Mexico Ornithological Society Bulletin* 9:19–21.

Peters, J. L. 1940. *Check-list of birds of the world*. Vol 4. Cambridge, MA: Harvard University Press.

Plumpton, D. L., and R. S. Lutz. 1994. Sexual size dimorphism, mate choice, and productivity of Burrowing Owls. *Auk* 111:724–27.

Sauer, J. R., J. E. Hines, and J. Fallon. 2007. *The North American breeding bird survey, results and analysis, 1966–2006*. Version 7.23.2007. Laurel, MD: USGS, Patuxent Wildlife Research Center.

SEDESOL 1994. Norma Oficial Mexicana NOM-059-ECOL-1994 que determina las especies y subespecies de flora y fauna silvestres terrestres y acuáticas en peligro de extinction, amenazadas, raras y las sujetas a protección especial, y que establece especificaciones para su protección. *Diario Oficial de la Federación* 431:2–60.

Sheffield, S. R. 1997. Current status, distribution, and conservation of the burrowing owl (*Speotyto cunicularia*) in midwestern and western North America. In *Biology and conservation of owls of the northern hemisphere: second international symposium*, ed. J. R. Duncan, D. H. Johnson, and T. H. Nicholls, 399–407. General Technical Report NC-190. Washington, DC: U.S. Forest Service.

Sidle, J. G., M. Ball, T. Byer, J. Chynoweth, G. Foli, R. Hodorff, G. Moravek, R. Peterson, and D. Svingen. 2001. Occurrence of burrowing owls in black-tailed prairie dog colonies on Great Plains National Grasslands. *Journal of Raptor Research* 35(4): 316–21.

Summers, C. A., and R. L. Linder. 1978. Food habits of the black-tailed prairie dog in western South Dakota. *Journal of Range Management* 31:134–36.

U.S. Fish and Wildlife Service. 2002. Birds of conservation concern 2001. Arlington, VA: U.S. Fish and Wildlife Service, Division of Migratory Bird Management. http://migratorybirds.fws.gov.

35

Spotted Owl

(*Strix occidentalis*)

PETER STACEY

THE SPOTTED OWL (*Strix occidentalis*) is a large brown and mottled owl that has black eyes and irregular brown and white spots over much of its body. The sexes are identical in plumage, but as is true of many owls, females are larger than males: in the subspecies that occurs in New Mexico, the Mexican Spotted Owl (*S. o. lucida*), adult males weighed an average 509 grams (18 oz), whereas the mean weight of adult females was 569 grams (20 oz) (Gutiérrez et al. 1995). Juveniles up to six months of age are whiter than the adults and have a downy appearance that gradually turns into the adult plumage during the first molt. First- and second-year birds can still be distinguished in the hand from older birds because they have pointed tail feathers or rectrices with a pure white terminal band, rather than the rounded tip and mottled brown and white terminal band that is found in birds older than 27 months.

PHOTO 35.1

Mexican Spotted Owl, Black Range, 20 July 1971. The Spotted Owl is a large owl with black eyes and irregular brown and white spots over much of its body. PHOTOGRAPH: © FRED GEHLBACH.

Spotted Owls have a large vocal repertoire. The vocalization most frequently heard is the "four-note" hoot or location call. This call has a very distinctive and unique time signature: "*hoo . . . hoo-hoo . . . hoo*." The owls, particularly at the beginning of the breeding season, will respond to any reasonable imitation of this call, and it is therefore widely used to locate and census the birds. Both sexes give this call, but males have a lower-pitched voice than females (even though

PHOTO 35.2

Juvenile Mexican Spotted Owl perched in a Gambel oak (*Quercus gambelii*) tree in the northern part of the Black Range, western New Mexico, June 1991. This juvenile, one of two total, was observed approximately 50 meters (~160 feet) away from its nest, and it had probably fledged within the last week. Like most juveniles at this age (note the downy appearance of the plumage), it did not attempt to fly when approached, and could be picked up by hand for banding. Both adult owls were roosting nearby.

PHOTOGRAPH: © PETER STACEY.

they are smaller) and thus the call can also be used to identify the sex of a bird at night. Juveniles frequently give the typical raptor rattle or begging call, and the adults' vocal repertoire also includes a wide variety of other calls (Forsman et al. 1984) that are heard less commonly. These include an ascending whistle and a series of sharp barks, both of which appear to be given primarily by the female during breeding. Such is the variety of Spotted Owl calls that if any strange sound is heard at night in areas where Spotted Owls are found, it is often likely to be this owl.

Although the Spotted Owl is present in many of the larger and higher-elevation mountain ranges throughout New Mexico, it is rarely encountered by most people. The owls are strictly nocturnal in their habits, including hunting. During the day, they usually roost in small and protected side canyons, often in deep shade perched 1.5 to 9 m (5 to 30 ft) above the ground, either on a horizontal side branch next to the trunk of a large conifer tree, in dense deciduous cover, or within a cave on an exposed cliff face. The spotted brown and white pattern from which the owl gets its name makes it extremely cryptic during the day, and it rarely moves or flushes even when humans approach very closely. As a result, it is possible to walk underneath a Spotted Owl that is perched motionless on a branch directly overhead and never realize that it is there. In fact, the easiest way to know that Spotted

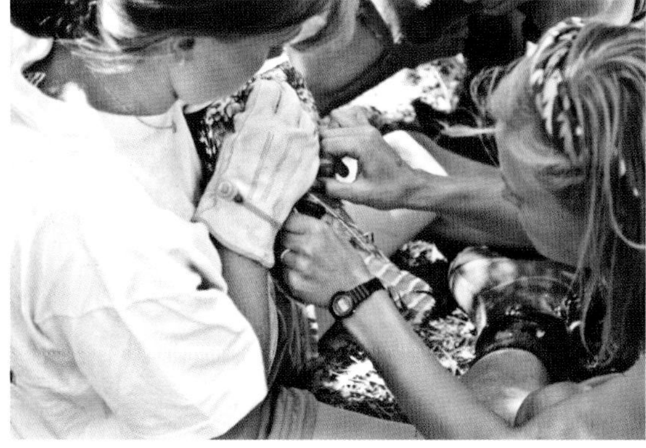

PHOTO 35.3

Some studies, including those conducted by the author, have included radio tracking of individual birds, particularly of the young during dispersal and of adults while they are foraging. A small transmitter is being attached to one of the tail feathers of the owl in this photo, taken in the Black Range, June 1993. A transmitter lasts for up to 12 months, and it falls off when the tail feather itself is molted. PHOTOGRAPH: © PETER STACEY.

PHOTO 35.4

Most studies of the Mexican Spotted Owl in New Mexico and in other parts of the Southwest have included banding in order to follow the histories of individual birds throughout their lives. Unique combinations of U.S. Fish and Wildlife Service metal and plastic bands are placed one per leg on the owl, which is then released unharmed. Banding is particularly important in this species to follow the dispersal of young birds, and because the adult owls will sometimes change territories over the winter. Studies using banding of birds have allowed researchers to develop a reliable understanding of both local and statewide changes in the sizes of Spotted Owl populations, as discussed in the text. Photo of the author and an owl about to be released after banding, San Mateo Mountains, June 1996. PHOTOGRAPH: © KERSTAN MICONE.

Owls occur in an area is to hear their territorial "four-note" hoot or call, which they frequently give during the breeding season just after dusk and just before dawn. During the fall and winter, the owls are silent, and it is almost impossible to locate them then unless their favorite roost sites are already known.

As a result of the species' cryptic appearance, secretive behavior, and frequent association with rugged canyon and mountain country, very little was known about the Spotted Owl in New Mexico prior to the 1990s. Bailey (1928) and J. Stokley Ligon (1961) summarized early distributional records, and described several nests. The extreme isolation of most Spotted Owl territories is indicated by the fact that in 1992, my research crew and I relocated one of the nest sites in a cave originally located by Ligon (1961) and found undisturbed the tree ladder that he had used to climb to the nest ledge and collect the eggs, nearly 70 years earlier!

However, as a result of the controversy over the listing under the Endangered Species Act of the Northern Spotted Owl (*S. o. caurina*) in the Pacific Northwest, considerable attention was focused during the late

PHOTO 35.5

Angie Hodgson, one of the author's students, using a portable antenna to determine the locations of owls with radio transmitters in the southern end of the San Mateo Mountains, July 1993. Once the general area where the owls were roosting was located, the author and his research crew would hike into the canyon to visually confirm the identification and check breeding status. It was possible to use the mountaintops to survey large sections of individual mountain ranges. At other times, and to study juvenile dispersal, they used a small plane with radio-tracking equipment, kindly provided by the U.S. Fish and Wildlife Service in Albuquerque, to census very wide areas, including most of the suitable owl habitat in the western part of New Mexico.

1980s and 1990s on the subspecies that occurs in the American Southwest and Mexico: the Mexican Spotted Owl. A number of research studies were initiated at that time, primarily to determine whether the Mexican Spotted Owl should also be listed as an endangered or threatened species, as well as on the potential impact that such listing would have upon the relatively small, but locally important, timber harvest industry. In New Mexico, Spotted Owl research was conducted in the Zuni, Magdalena, and San Mateo mountains and in the Black Range by the author and his collaborators (e.g., Arsenault et al. 1997; Stacey and Hodgson 1999; Stacey and Peery 2002); in the Gila Mountains by Seamans and his collaborators (e.g., Seamans et al. 1999, 2002); in the Sacramento Mountains by Ward and his collaborators (e.g., Ward 2001; Ganey et al. 2005); in the Jemez Mountains by T. Johnson (unpublished reports); and in several ranges in southern New Mexico by Zwank and his collaborators (e.g., Zwank et al. 1994). Additional studies were conducted in other southwestern states and in Mexico, including those by Young, Zwank, and their colleagues in the Sierra Madre mountains of Mexico (e.g., Young et al. 1997; Tarango et al. 1997); by Ganey and his collaborators in Arizona, particularly on the Mogollon Rim (e.g., Ganey and Balda 1994; Ganey et al. 1998, 1999); by Rinkevich and her collaborators in Zion National Park in southwestern Utah (Rinkevich and Gutiérrez 1996); and by Willey and his collaborators in Canyonlands National Park and other areas in southern Utah (e.g., Willey and van Riper 2000). A number of other studies have focused on specific issues, from behavior to population genetics (e.g., Kuntz and Stacey 1997; Delaney et al. 1999; Jenness et al. 2004; Barrowclough et al. 2006).

As a result of all this work, perhaps more is known about the Spotted Owl than any other raptor occurring in New Mexico. The newly acquired body of knowledge includes information on distribution, behavior, natural history, and diet. Because several studies have included individually marked or banded birds, there is now also good quantitative information on life history parameters and demographic trends, at least for where and when studies were conducted. However, in spite of this effort, there remain a number of unanswered questions about the Mexican Spotted Owl in New Mexico and throughout the Southwest. Perhaps the most important questions are, first, whether the declining population trends observed in two detailed demographic studies (Seamans et al. 1999, 2002; Stacey and Peery 2002; Stacey and Peery, in prep.) have continued to the present time; and, second, whether Spotted Owls have recolonized some locations where they once occurred but from which they recently disappeared (see details below). Observed population declines have occurred even in areas where there have not been any timber harvests in Spotted Owl habitat for many years. What could therefore be now causing the Mexican Spotted Owl to decline in numbers?

Perhaps their habitat is being negatively affected by some factor other than timber harvesting, but what exactly are the habitats used by the owls for hunting, as opposed to roosting and nesting? This is information extremely difficult to obtain, since it requires following the owls at night through very rough terrain without disturbing their predatory behavior. Limited observations from radio telemetry and prey remains suggest that the Spotted Owl may actually hunt primarily along the edges of forest patches, particularly those bordering meadows or riparian areas, rather than in the interior of dense forest. If true, this finding would have important implications for the effective management of the owls in New Mexico and elsewhere. Unfortunately, most of the detailed studies of the Mexican Spotted Owl that were initiated during the 1990s have now been terminated due to lack of funding. It can only be hoped that research on the owls can continue, and that the observed population declines, if they are widespread, can be reversed, since the Spotted Owl is truly one of the most extraordinary birds of New Mexico.

Distribution

The Spotted Owl occurs in forested habitats in the Pacific Northwest and California, and in the American Southwest from Utah and Colorado southward through central Mexico. Three subspecies are currently recognized: the Northern Spotted Owl, which is found in the Coastal and Cascade ranges of southwestern British Columbia, Washington, Oregon, and northern California (to near San Francisco); the California Spotted Owl (*S. o. occidentalis*), in the southern Cascade and northern Sierra Nevada mountains ranges, as well as in the

PHOTO 35.6

(*left*) Mexican Spotted Owl in the San Mateo Mountains, 25 May 2008. PHOTOGRAPH: © MARK L. WATSON.

PHOTO 35.7

(*right*) Mexican Spotted Owl pair in the Zuni Mountains, June 1988. PHOTOGRAPH: © DALE W. STAHLECKER.

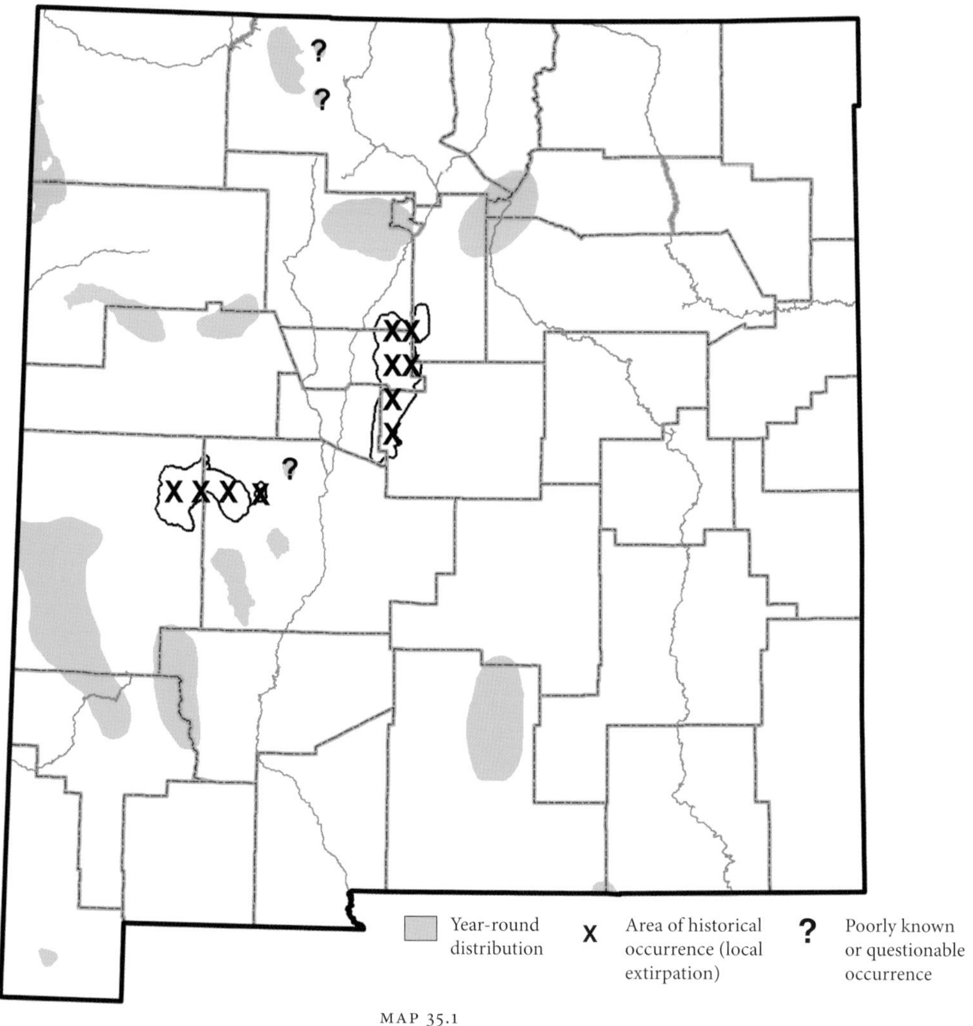

MAP 35.1

Spotted Owl distribution map

Coastal Ranges southward to northern Baja California; and the Mexican Spotted Owl, which occurs in mountains and forested canyons from the front range of the Rocky Mountains in central Colorado and southern Utah (Zion National Park and the Canyonlands region), south through New Mexico, Arizona, and extreme western Texas, and through the Sierra Madre Occidental and Sierra Madre Oriental in Mexico to Michoacán south of Mexico City (Gutiérrez et al. 1995 and references therein). The three subspecies do not overlap geographically, and most genetic studies indicate that they are currently reproductively isolated from each other, although the length of time that they have been separated remains uncertain (e.g., Haig et al. 2004). Dickerman (1997) has further suggested that the Mexican Spotted Owl itself can be divided into three subspecies, based on plumage differences of museum specimens. He proposed (a) recognizing a new subspecies, designated as the Volcano Owl (*S. o. juanaphillipsae*), from the state of Mexico, (b) restoring a previously named subspecies, *huachucae*, from the southwestern United States and northern Mexico (Sonora and Chihuahua), and (c) keeping *lucida* as only occurring in Mexico between the other two subspecies. Little is known of the genetics or the ecology of the owls in most regions of Mexico, and deciding on the usefulness of Dickerman's (1997) suggestion awaits further research.

In New Mexico, the Spotted Owl is or was once found in most of the major mountain ranges. Today it appears to be most common in the southern part of the state, including the Sacramento and Guadalupe mountains, the San Mateo Mountains and the Black Range, and the various ranges in the Gila Mountains. In the north, the largest concentration of owls is in the Jemez Mountains, with smaller numbers found locally in the Sangre de Cristo and San Juan Mountains, the Zuni Mountains, and Mt. Taylor and the nearby San Mateo Mountains. Historically, Spotted Owls also apparently occurred in a number of smaller ranges in the state, including the Sandia, Manzano, and Datil mountains, where today the owl appears to be extirpated. Similar range contractions have also occurred recently in Colorado, where the owl is no longer found at historic locations in the central Front Range near Fort Collins and Boulder (U.S. Fish and Wildlife Service 1995).

Records from the middle 1800s indicate that Mexican Spotted Owls also once lived in at least some major lowland riparian or bosque forests in the Southwest (e.g., Tucson, Arizona). It is unknown how common the owls were in floodplains of the Southwest, but there is no evidence that they continue to occur there at the present time. Lowland riparian forests prior to major European settlement probably contained many of the habitat features that are important to the Spotted Owl (see below), including deep shade, abundant rodent communities, and, at least in some areas, extensive openings or "meadows" of saltgrass (*Distichlis spicata*) and other native grasses that potentially provide good owl hunting habitat. Such habitat features have almost completely disappeared from lowland riparian forests as a result of human activities, and it is unlikely that the Spotted Owl will return to these areas in the foreseeable future.

Although some early ornithologists working in New Mexico reported the Spotted Owl to be common (e.g., Woodhouse 1851, as reported in Bailey 1928), until recently this owl was generally considered to be uncommon or rare in most parts of the state. However, the perceived rarity of the owl may have reflected more its tendency to occupy remote and often rugged mountainous areas than actual population numbers. For example, when we initiated our research on the owls in the San Mateo Mountains in Socorro County in 1989, only a single pair was known at that time from the entire mountain range, which is approximately 60 km (37 mi) long and 35 km (22 mi) wide. However, we found that between 1991 and 1998 there were actually at least 19 different territories in the San Mateos occupied by either pairs or single owls. As a result of the extensive surveys conducted after the listing of the Mexican Spotted Owl, it now appears that the Mexican Spotted Owl was relatively common for a large raptor. As mentioned already, however, there are also good data to show that during the 1990s the number of owls declined in at least some parts of the Southwest (see below). Systematic surveys for Spotted Owls mostly ended in the early 2000s, and as a result, the current numbers of owls in New Mexico and elsewhere are largely unknown. There is little doubt, however, that in some of the smaller ranges in New Mexico, like the Zuni, Magdalena, and San Juan mountains, the owls are now uncommon to rare relative to their numbers 15 years ago.

Habitat Associations

Although the Spotted Owl occurs in most of the larger mountain ranges in the central and southern parts of New Mexico, its distribution within those ranges is extremely patchy. Initially, it was assumed that the Spotted Owl in the Southwest was tied to stands of old growth or mature mixed conifer or spruce-fir forests, just as they are in the Pacific Northwest. However, once the Mexican Spotted Owl was considered for listing under the Endangered Species Act, numerous studies were undertaken to examine the actual habitat preferences of this subspecies in the Southwest, because of the obvious importance of this factor to the successful management of the owls. These studies have shown that the habitat associations of the Mexican Spotted Owl are much more complex that originally envisioned, that the birds occupy a variety of different habitat types, and that there are still a number of unanswered questions about the relative importance to the owls of different habitat features, in terms of both their reproductive success and their survival.

We now know that Mexican Spotted Owls occur

PHOTO 35.8

Mexican Spotted Owl roosting and nesting habitat in the San Mateo Mountains in Socorro Co., June 1993. The well-developed riparian vegetation in the bottom of the canyons in this region includes both deciduous and coniferous trees, which provide cool microhabitats that allow for efficient thermoregulation during the heat of the summer, and also cover that may protect the birds from predators. Owls with territories in this type of habitat often forage at night along the edges of the riparian zone and lower slopes of the canyons, where rodent prey is most abundant. PHOTOGRAPH: © PETER STACEY.

both in mountainous areas and in deep canyons that are incised within relative flat plateaus (such as near Fort Wingate in western New Mexico, or in Canyonlands National Park in southeastern Utah). Within a particular area, however, it is usually impossible to accurately predict in advance where the owls will be found on the basis of any specific vegetation or topographic feature. As a result, the way that most owls are initially located is through the use of wide-ranging "calling" surveys. The surveyor drives on roads or hikes over large areas at night and regularly gives the Spotted Owl's "four-note" location call, either with a tape recorder or by vocal imitation. Once an owl is heard, the general location of the response is noted, and the surveyor then returns to the area during the day to visually search for active nests or for roosting owls.

Calling surveys have shown that the Mexican Spotted Owl can be found within a wide variety of forest types, including mixed conifer forests, ponderosa pine–Gambel oak (*Pinus ponderosa–Quercus gambelii*) forests, Madrean pine-oak forests, encinal oak woodlands, and various types of riparian associations including those dominated by cottonwood (*Populus* spp.), by evergreen oaks (such as in the Guadeloupe Mountains), and, farther west, by Arizona cypress (*Cupressus arizonica*). The tree vegetation within the canyon and mesa systems where owls occur can be exceptionally complex, and many different tree species, including Douglas fir (*Pseudotsuga menziesii*), southwestern white pine (*P. strobiformis*), ponderosa pine, pinyon pine (*P. edulis*), juniper (*Juniperus* spp.), and various riparian species, including cottonwood

PHOTO 35.9

Overhead view of the canyon habitat shown in the previous photo, San Mateo Mountains, Socorro Co., July 1997. This canyon has not been grazed by livestock for many years, and as a result the vegetation forms a closed canopy above the canyon bottom. Mexican Spotted Owls often prefer this type of area when available, and it may be similar to the type of closed-canopy riparian forests in low-elevation areas in the Southwest where the owls were once reported to occur in the 1800s. PHOTOGRAPH: © PETER STACEY.

and aspen (*Populus tremuloides*), may all be found together within a very small area. Ironically, we have never found Spotted Owls in New Mexico occupying territories in large, continuous stands of mixed conifer forest, which is the preferred habitat of the Spotted Owl in the Pacific Northwest. This is perhaps not surprising, since the owls hunt strictly at night (see below), and in the Southwest almost all of the small mammals that occur in mixed conifer forests, including red squirrels (*Tamiasciurus hudsonicus*) and chipmunks (*Eutamias* spp.), are diurnal rather than nocturnal. In contrast, in the Pacific Northwest, the most common prey species of the Spotted Owl is the nocturnal flying squirrel (*Glaucomys sabrinus*). A second factor that may limit the occurrence of the owls in mixed conifers is that, in the Southwest, this forest type is typically extremely dense due to past timber harvests. This makes the hunting technique of the Spotted Owl, which is to typically wait on a low tree limb and then swoop down onto the ground once the prey is detected, almost impossible.

One characteristic that does appear to be common to most locations where Spotted Owls occur is a relatively steep topography, either as forested hillsides or exposed cliff faces. Areas that have the appropriate forest type(s), but which are flat, appear to be unsuitable to the birds, unless there are also some incised canyons present. For example, there is not a single Spotted Owl territory anywhere on the entire Kaibab Plateau in northern Arizona, even though it contains some of the most extensive and undisturbed mixed conifer and ponderosa pine forests in the entire Southwest. But the plateau is flat and without canyons. Spotted Owls do, however, occur in areas surrounding the Kaibab Plateau, where the same forests are present but where there are also numerous canyons or steep mountain hillsides. (U.S. Fish and Wildlife Service 1995).

The fact that the Mexican Spotted Owl actually occupies such a wide range of forest types in the Southwest suggests that a more productive way to understand the habitat associations of this subspecies may be to consider what the owls need to survive and reproduce successfully, rather than to focus upon any tree species or stand type. Using this approach, three factors seem to be particularly important for an owl to occupy and breed in a particular site: it must obtain an appropriate nest site, it must have access to suitable roosts, and it must be able to find and capture sufficient prey.

Nesting Sites

Like other owls, Spotted Owls do not build their own nests. Instead, they use either the nests of other species or natural platforms on which they can lay their eggs. As a result, the owls can only occur in areas where there already are suitable nest sites. Typically, the owls use old raven or hawk nests that are hidden in dense, interior branches of large conifer trees, usually either in or near the bottom of a canyon. These nests are often 10 m (~30 ft) aboveground or higher. They may also use natural platforms, such as provided by mistletoe, or in caves, crevices, or potholes on the sides of cliffs, particularly those that are high and isolated enough to be protected from terrestrial predators like foxes and coyotes (*Canis latrans*). Spotted Owls have also been observed to nest occasionally in natural open cavities in large oak trees (Ganey and Balda 1994). But it is unlikely that in the Southwest they are able to use one of the most important nest sites recorded for the other two subspecies along the West Coast: the top of a stub of a large Douglas fir or similar tree that has broken off 10 m (~30 ft) aboveground or higher during a windstorm or as the result of a lightning strike or other event. We have only found a single location in New Mexico that has large tree stubs of this type, in a patch of old-growth Douglas fir in the Gila Mountains. Trees that would be large enough to create this stub nesting platform simply no longer occur in most forests in the Southwest, although they undoubtedly did so in the past prior to intensive logging.

Roosting Sites

Spotted Owls typically spend the daylight hours perched motionlessly on a horizontal tree branch that is in deep shade. Roost sites are usually 3–10 m (~10–30 ft) off the ground, in a canyon or side-drainage, and either in a conifer (often one of the largest Douglas firs or ponderosa pines in the area) or deep within the spreading branches of an oak or other deciduous tree. In large canyons, they may roost on the lower parts of the slope on the side of the canyon (very rarely on the upper two-thirds of the slope), while in smaller canyons they are almost always in the bottom. When roosting in large conifers, the owls usually perch within a meter or two (~6.5 ft) of the main trunk of the tree, a habit that may make them less vulnerable to aerial attacks by predators such as Goshawks (*Accipiter gentilis*) and Great Horned Owls (*Bubo virginianus*) (see below). The owls may also occasionally roost in small caves, but trees appear to be preferred when available. The same roost sites tend to be used day after day and year after year; as a result it is often possible to locate areas of high recent owl activity by the presence of whitewash and regurgitated pellets under specific trees.

We have observed owls roosting in such a wide variety (in terms of size and species) of trees that the availability of a specific type of roost tree itself does not appear to limit the occurrence of the birds. What does seem to be important is that the site is in deep shade much of the day, and that it provides protection from the intense solar radiation typical of the Southwest. In fact, it has been suggested that Spotted Owls are particularly sensitive to thermal loading, and that their distribution may be limited by the presence of suitable microclimatic sites that enable them to remain cool during the day (e.g., Barrows 1981). In support of this possibility, Ganey (2004) found that nest sites in northern Arizona were significantly cooler than nearby randomly located areas. He also found that the expected water loss incurred by the owls from evaporation was significantly lower at sites chosen for nesting than if the nests had been placed randomly in the general surrounding areas. It is likely that the same pattern would hold true for roosting sites during the hottest time of the year. In fact the best place to search for a Spotted Owl during the summer is in the bottom of narrow canyons that are shady and humid—exactly where it is likely to be coolest.

When the owls roost on the slopes of canyons or mountains, they are typically found in areas where the canopy is densest and most structurally complex (Stacey and Hodgson 1999; Ganey et al. 2003). In western New Mexico, we have found that when they are not in the canyon bottoms, they are almost always near a natural spring, where the ground is wet and the air is moist. We have never observed the owls to roost or nest on the upper slopes or along ridgetops between canyons, even when these areas have dense forests and large trees. The importance of cool microclimates for roosting and nesting may be one explanation for the preference Spotted Owls have in the Southwest for areas with considerable topography regardless of vegetation type, and it may be one reason why they are absent in areas where there is little or no topographic relief, such as the Kaibab Plateau.

Foraging Sites

Woodrats (*Neotoma* spp.), voles (*Microtus* spp.), and deer mice (*Peromyscus* spp.) make up the bulk of the diet of Spotted Owls in most locations where it has been studied (U.S. Fish and Wildlife Service 1995; see also diet and foraging, below). In the Southwest, these prey taxa normally occur either in rocky open areas (woodrats) or in areas that have large amounts of grass, such as in meadows and along streams and other riparian systems (voles and deer mice). None of these prey species is typically found in abundance in the interior of dense mixed conifer forests (Findley et al. 1975), suggesting that the owls rarely use this type of habitat for foraging.

Spotted owls hunt almost exclusively at night, although they will also take prey during the day if it appears below their roost trees, or if it is presented to them by humans. As a result, it is very difficult to determine general areas and specific habitats where the owls conduct most of their foraging. This can only be done through radio-tracking, but as stated already, attempting to follow the owls at night through country that is often extremely rugged, and then triangulating their exact position without disturbing them, is quite challenging. We conducted radio-tracking one summer on a small sample of owls that occurred in canyons in the San Mateo Mountains and the Black Range in western New Mexico (Hodgson 1996). We found that at dusk, the birds usually moved away from their roost and nest sites in the side canyons into the major canyon system. They then perched near the side canyon entrance while giving the four-note vocalization (much like songbirds that sing in the early morning). After a variable length of time, the owls then moved up or down the main canyon, regularly stopping (presumably perching to hunt) near the boundary between the conifer forest on the sides of the canyon and riparian areas along the stream. The owls appeared to be hunting primarily along edge habitats, although this observation was not based upon a large sample size. Other studies (e.g., Ganey et al. 1999) have faced similar difficulties in obtaining detailed information on specific foraging habitats. But the potential importance of edge and riparian habitats for hunting has now been reported in radio-tracking studies in the other subspecies (e.g., the Northern Spotted Owl; Glenn et al. 2004).

To summarize, the diversity of locations in which Spotted Owls can be found in New Mexico and elsewhere in the Southwest suggests that it may be better to think of the habitat associations of this species in terms of whether a habitat provides the structural features and prey abundance that will meet the needs of the owls, rather than in terms of specific forest or stand types (e.g., mixed conifer, ponderosa pine, etc.), which is the approach that is taken by most public land managers at the present time (e.g., U.S. Fish and Wildlife Service 1995). The information outlined above suggests that the Mexican Spotted Owl can usually, if not always, occur in areas where the following conditions are present. First, there are cool and moist microsites that are suitable for roosting. Suitable locations may exist either because of dense vegetative cover, such as in riparian forests, or as the result of steep topography. This factor would explain why the owls are rarely found in forests that are open and/or flat, such as occur on mesa tops or plateaus. Second, there are trees that are large enough to provide protected nest sites, and other species already have made nests at those sites, or there are natural protected nest platforms such as in caves in cliffs (the lack of large trees would rule out most pinyon-juniper forests, as well as most areas that have recently been harvested for timber). Third, there is an abundance of *nocturnal* small and medium-size rodents that can be detected and captured by the owl using its "perch-and-pounce" hunting method, such as would be present at the edges of meadows, riparian canyon bottoms, and rocky outcropping (this would rule out dense forests at higher elevations, lowland riparian habitats that at the present time lack saltgrass meadows, large grasslands, or forest types where the trees are too small or too densely branched to provide suitable perches for hunting, as well as areas that would otherwise be suitable but which have been overgrazed by livestock and as a result have lost the grass cover needed by prey species). Viewing the owl's needs in this manner can provide useful guidance when attempting to understand the impact of different land management practices on the Mexican Spotted Owl, and perhaps facilitate the eventual recovery of the species in the Southwest (see Status and Management, below).

Life History

Migration and Dispersal

The Spotted Owl is a year-round resident in New Mexico and elsewhere (U.S. Fish and Wildlife Service 1995; Gutiérrez et al. 1995 and references therein; PS, pers. obs.). The adults tend to occupy the same general areas or territories year after year, although some birds may sometimes change territories, particularly after a nest failure. During the breeding season (March through September), the owls generally remain near the nest site if they are actively breeding, while in years during which they do not breed they tend to range farther but still are usually found within the traditional nesting area. Once the young fledge, the adults begin to expand their movements. This is particularly true of females: we found through radio-tracking that some females leave the nest area once the young are several months old, and move by themselves into an entirely different drainage. The male is then left to care for the young birds by himself until the juveniles disperse in the early fall (usually August and September).

Although the data are limited, it appears that the owls are generally solitary over the winter. Radio-tracking studies by the author have shown that some adult owls move considerable distances away from their summer nesting area during the winter months. Usually these birds move to lower elevations, including into habitats like pinyon-juniper woodlands where they would not normally breed. For example, one member of a pair that lived in Milk Ranch Canyon in the western part of the Zuni Mountains would frequently move 10 km (6 mi) downstream and spend the winter in a large cottonwood tree near Fort Wingate, in an area that was surrounded by open juniper woodland (PS, pers. obs.). The other member of the pair remained in the traditional breeding area. In some cases, the distances moved during the winter can be large. Gutiérrez et al. (1996), for example, banded an adult female owl from a territorial pair in the Tularosa Mountains in 1994. In January 1995, this bird was found dead in Chihuahuan desertscrub habitat near Deming, 187 km (116 mi) from where it was banded. Although the authors suggested that this might be an example of adult intermountain dispersal, it is also possible that it may have simply been an extreme case of seasonal movement into lower elevation habitats (see also Willey and Van Riper 2000). Given the variability in distances traveled by the owls (from one to hundreds of kilometers), it seems likely that in many cases the adults are simply showing a type of altitudinal migration, and moving into areas where there is less snow and where prey may be more abundant. There is no evidence that the Mexican Spotted Owl as a whole tends to occupy distinct summer and winter habitats, as do some other New Mexico birds.

Spotted Owls tend to breed in the same areas or territories year after year, and they rarely if ever move to a new mountain range. Even those adults that migrate to lower elevations toward the end of a breeding season return to breed in the spring from their wintering areas. The juveniles, however, have an entirely different strategy. All but one of the young that we banded in our study and that survived the winter moved to new mountain ranges in their first year, and none of them eventually established their breeding territories in the same range in which they had been fledged (Stacey and Peery 2002; Stacey and Peery, unpubl. data). Of the 57 young that we banded in four different New Mexico mountain ranges, one returned the following summer to establish a territory 18.4 km (11.4 mi) from its natal territory (see below). It then disappeared by the second summer. None of the other banded young was ever seen again after dispersal, either in their natal range or any of the other ranges included in the study. In contrast, we had 27 owls that became new breeders on territories within our study area. Twenty of these were young birds (either one or two years old as indicated by the shape of their tail feathers). Yet not one of these new breeders had been banded as a juvenile anywhere within our study area. These data indicate that most or almost all of the juveniles that are born in a particular area or mountain range disperse out of that area during their first winter. When they do establish their first breeding territories, they usually move into new populations rather than going back to where they were born.

This pattern of having most recruitment of new breeding birds come from outside the local population has been observed in other species with fragmented distributions. This is true for example in the White-tailed Ptarmigan, *Lagopus leucura*, which breeds on mountaintops in Colorado and, like the Mexican Spotted Owl, also shows altitudinal migration (Martin et al. 2001). Even in migratory species with more continuous distributions, adults usually exhibit high fidelity to a nesting territory or area, whereas juveniles are recruited into different breeding populations. A consequence of this behavior for the Mexican Spotted Owl is that the young birds that are produced in other populations are critical to the continued persistence of any local population. This means that Spotted Owl populations in the Southwest are tied together to a greater or lesser extent by the dispersal of juveniles, depending upon geographic distances between these populations. This conclusion is also supported by genetic studies, which have shown that there is a lack of genetic subdivision among local populations in many parts of the Southwest (Barrowclough et al. 2006; Stacey et al. unpubl. data).

Several studies have used radio-tracking to examine the details of juvenile dispersal in the Mexican Spotted Owls (Arsenault et al. 1997; Ganey et al. 1998; Willey and van Riper 2000). This has proven to be more difficult than originally expected, primarily because the survival of the young owls tends to be low during this period, and they also frequently fly long distances during their first winter. Most of the information on juvenile dispersal has come from locating roost sites once the young have moved to new sites in the fall. Many of the young move to lower elevations, just like some adults, but whether they occupy a broader range of habitats at that time than do adults is unknown. In 1993 and 1996, our group also followed individual juveniles through the night during the initial stages of their leaving their natal territories (Arsenault et al. 1997).

We found that dispersal was remarkably synchronous for a resident species, and that almost all young left their natal territories between late August and the end of September (nine of 12 radioed juveniles dispersed between 10 September and 29 September, including data from both years). Initially the juveniles moved within a limited area within their natal territory, foraging and roosting with one or both parents. The mean observed home range at this time was 61.8 ha (152.7 ac) ($n = 7$ owls radio-tracked for extended periods; range = 11.5 to 233 ha [28.4 to 576 ac]). As time went on, the areas used by the young first gradually increased as they explored surrounding areas, and then birds began to move longer distances as they actually left the natal territory. The mean distance moved during the first week of dispersal was 24 km (15 mi). Siblings did not disperse together, but left their natal territories independently of their nest-mates. Several of the radioed birds were followed to an adjoining mountain range, across unsuitable grassland habitats. They, and all of the other radioed owls, then either dropped their transmitters or disappeared entirely. None of the 12 owls we followed died in the area (we did not find any transmitters or dead birds) and none of them established winter territories within the study area. This indicates the juveniles (unlike most adults) tend to move long distances during their first winter. Only one owl radio-tagged as a juvenile returned to the same mountain range the following spring or summer: a female that was relocated as a one-year-old at a high-elevation, previously unused location in the San Mateo Mountains, 18.4 km (11.4 mi) distant from her original natal territory. She stayed there alone during the summer and was not present at the site the following year. We also followed three owls that were first marked as subadults. These were all females: they paired with males but did not attempt to breed. In August they began roosting large distances away from the males (up to 9 km [5.6 mi] from their mates), and eventually disappeared. None of the subadult females returned the following year, and all of the males had paired with new females.

Similar results were obtained in the other radio-tracking studies cited above: of the 61 juveniles that were followed in all of these studies, only one was followed until it paired and established a breeding territory in a different mountain range (Arsenault et al. 1997). The remaining juveniles either died, lost their transmitters, or moved to a new, unknown area to breed. This behavior is in strong contrast to what has been found in the other two Spotted Owl subspecies. In those subspecies, radio-tracking and banding studies have shown that most juveniles settle to breed within two or three territories of where they were born, although there also may be an occasional long-distance dispersal movement (LaHaye et al. 2001; Forsman et al. 2002).

These observations suggest that Mexican Spotted Owls have what may be an exceptional pattern of dispersal and seasonal migration for a year-round resident bird: juveniles disperse synchronously and early in the fall, but independently of their siblings or the adults (most of whom either do not disperse, or move only altitudinally). Again, this is very different from what is observed in most resident or nonmigratory species, where the young may disperse throughout the entire winter (e.g., Carmen 2004). Both the synchronous time of dispersal and the strong tendency for juveniles to begin breeding in different areas than where they were fledged are patterns that are much more typical of birds that migrate long distances into the neotropics each fall than of year-round resident species.

Territoriality

Adult pairs of Spotted Owls tend to remain in the same general area during the breeding season. These areas are referred to as territories, although territorial defense against intruding owls has rarely been observed. Male owls, and to a lesser extent females, give the four-note hoot or location call throughout the spring and summer, and it is likely that this alerts other owls to their presence. Individuals will fly several kilometers in response to the imitation or playback of the four-note call, and they will spend considerable time apparently trying to locate the intruding "owl." This response suggests that the owls will defend their territories if necessary. We have observed several times that a first- or second-year bird will repeatedly roost within the core area of a resident pair's territory. These young owls appear to be tolerated by the resident pair, and in one case, the extra bird, a male, became the new breeder on the territory with the old female after the original male apparently died. As occurs with many species, immature Mexican Spotted Owls may therefore sometimes form a population of "floaters" until they are able to obtain their own breeding territory and mate (e.g., Rohner 1997).

In most of the populations that have been studied to date, there are usually a number of territories that are held by single males without a mate. It is likely that this happens when the female member of the pair dies and the male is not able to attract a new mate the following year. Males of most bird species tend to stay on their territories no matter what the circumstances, whereas females will sometime move after a reproductive failure. However, the fact that at the present time it is common to have a number of single males on territories within an area suggests that there may be a shortage of females in most Mexican Spotted owl populations. This situation is typical of populations that are declining (see Status and Management).

There have been several attempts to determine the typical territory and home range size of Mexican Spotted Owls in different habitats and locations (e.g., Zwank et al. 1994; Willey 1998; Ganey et al. 2005). This information is important to the successful recovery of the Spotted Owl because it would help determine how much Spotted Owl habitat would need to be managed for the benefit of the owls themselves. The estimates have ranged from 278 to 1,028 ha (687 to 2,540 ac), depending upon the locale and dates of the study (see U.S. Fish and Wildlife Service 1995). However, the size of a territory and a home range has proven extremely difficult to determine for a number of reasons. First, radio-tracking—the only available method to examine space use when the birds are hunting (which is when they are likely to use the largest area)—must be conducted at night. Second, the longer the study continues, and the more times that a pair of birds is located, the larger the size of the area they will have been found to have used (this is generally true in telemetry studies of all birds that have large home ranges). Third, the size of the area used by Spotted Owls depends at least in part upon habitat type (Ganey et al. 2005). The owls use a great diversity of habitats, and even within a single mountain range, the amount of each potentially useable habitat type varies greatly among individual territories. This means that territory size will vary greatly depending on the types of habitat present in the territory. And finally, the birds move over much larger areas during the winter, and in some cases may move to lower elevations many kilometers away from their breeding areas. In these situations, the concept of a single year-round territory, within which all of the needs of the owls are met, or even the idea of a specific annual home range, may not be realistic.

Because of these problems, it probably does not make sense to attempt to estimate a single or typical territory or home range size of Spotted Owls in the Southwest, because the values would be so uncertain and would vary so widely as to be essentially meaningless. As an alternative to estimating the typical size of a territory or of a home range, the *Recovery Plan for the Mexican Spotted Owl* (U.S. Fish and Wildlife Service 1995) defines a core area, or Protected Activity Center, of 243 ha (640 ac) surrounding identified nests or roosting sites. The authors of the *Recovery Plan* hypothesized that most of the critical activities of the owls would occur within this area, at least during the breeding season and in most habitats. Restrictions on activities that would adversely modify owl habitat, or that might be disruptive to breeding, are then applied primarily only within this area. Whether or not 243 ha represents an appropriate estimate of the amount of protected habitat that Spotted Owls need in order to successfully survive and reproduce remains unstudied and therefore unknown.

Breeding

Mexican Spotted Owls usually breed with the same partner year after year if both sexes are still alive, although as seen already they may not spend the winter together nor even in the same general area. Breeding in New Mexico starts in late February or early March, when the members of the pair begin calling and roosting together. Eggs are laid in late March and April, and clutch size is normally two to four (Ligon 1961). Incubation lasts around 28–30 days and is performed solely by the female (e.g., Ligon 1961). The eggs most often hatch in early May, but we have found nests with new young as late as early June. The female also does all of the brooding of the young while they are still in the nest. The male provides food for both the female and the young, but he always takes it to a branch away from the nest before

PHOTO 35.10

Two juvenile Mexican Spotted Owls roosting in a Gambel oak (*Quercus gambelii*), San Mateo Mountains, New Mexico, July 1999, approximately two weeks after having left the nest. Nest mates often stay together at this time, and will roost and forage in the same area until they disperse individually in the fall. Adult males often spend more time with the fledglings than females, and radio-tracking studies by Angie Hodgson and David Arsenault of our research group have shown that some females completely abandon the young shortly after they fledge, and will move by themselves into a different part of the pair's territory.

PHOTOGRAPH: © PETER STACEY.

giving it to the female. Based on our observations and those of others, males will often store food, such as parts of rabbits, in trees near the nest, and the female will then take this food as needed. Fledging is a gradual process: the young first move to branches near the nest where they will stay for up to several weeks (they are sometimes called branchlings at this stage). The first flights often result in the juveniles' crashing to the ground after unsuccessful attempts at landing on tree limbs on the way down; the young owls then climb back up to a more protected location in the tree or on a cliff face. Learning to fly well comes slowly, and the young and one or more parents often remain in the general area of the nest for a few months . As noted above, in some cases the female will abandon the brood (and the male) after the young leave the immediate nest area, and she then moves by herself to a different part of the territory. The young then remain with the adult male until they begin to disperse in late August and September.

Spotted Owls do not breed every year. If the pair successfully produces young in one year, they often do not breed the following year. The proportion of occupied territories successfully fledging young varies greatly from year to year, and ranged in our study from 8% to 80%, largely reflecting year-to-year variation

PHOTOS 35.11a and b

Family group with fledglings, Los Alamos Co., 13 July 2005. PHOTOGRAPHS: © DAVID KELLER/LOS ALAMOS NATIONAL LABORATORY.

in the proportion of pairs that bred. Similar year to year variation in the number of pairs that breed has been reported from other areas in New Mexico (e.g., Seamans et al. 1999).

Kuntz and Stacey (1997) examined in detail the structure of the four-note territorial vocalization of owls in different mountain ranges of central New Mexico. Using sonograms and computer analysis, we found that even though there was considerable individual variation in the calls of Spotted Owls, it was not clear whether the vocalizations could be used as a reliable method to identify different individuals under field conditions. Interestingly, the calls of owls in each mountain range were different from the owls in each of the other ranges examined, and it was possible to identify the range where an owl lived just by its call alone with a high degree of reliability (Kuntz and Stacey 1997). As mentioned above, the owls do not usually breed in the same range where they were fledged, and this finding suggests that young owls are not born with an innate call, but instead learn parts of the call after they establish their first breeding territory, and that they also vocally "match" or imitate their neighbors. Such examples of vocal learning are rare in non-passerine birds.

Diet and Foraging

The Mexican Spotted Owl is primarily an ambush, or "perch-and-pounce," predator. Like many other members of the genus *Strix*, it perches on a branch, log, or rock and sits motionlessly until its prey comes into view. The owl then swoops down on the animal and if successful carries it away to a favorite site for consumption.

A number of studies have been able to examine the types of prey consumed by Spotted Owls in the Southwest using bones and hard parts contained in regurgitated pellets (e.g., Ward and Block 1995; Sureda and Morrison 1998; Ward 2001). The specific species taken varies by location, but typically small and medium-size mammals constitute the bulk of the diet. Important prey species include wood rats, voles, deer mice, pocket gophers (*Thomomys* spp.), and rabbits (*Sylvilagus* spp.). And as would be expected, Spotted Owls will consume a variety of other prey if available, including small numbers of birds, bats, insects, and reptiles. We once watched a pair of Spotted Owls in the San Mateo Mountains feeding on large beetles, and for over an hour they repeatedly swooped down and took the insects off a bare patch of ground beneath their perch tree.

Predation and Interspecific Interactions

Very little is known about the interactions between Mexican Spotted Owls and other animals. The tendency for the owls to roost in deep cover, and to

PHOTO 35.12

Older juvenile Mexican Spotted Owl, roosting in a Douglas fir (*Pseudotsuga menziesii*), located in the bottom of a narrow canyon in the Aldo Leopold Wilderness Area, Black Range, central New Mexico, August 1993. This owl had probably fledged several months previously. The wing feathers are usually well developed by this time, and the birds are efficient fliers. Both juveniles and adults that roost in conifers usually perch near the main stem of the tree. Although they are not hidden by dense vegetation, they are probably protected from most attacks by aerial predators: any Northern Goshawk (*Accipiter gentilis*) or Great Horned Owl (*Bubo virginianus*) that attempted to attack a Spotted Owl perched in this location would run the risk of serious injury if it missed its target and collided with the trunk.

PHOTOGRAPH: © PETER STACEY.

perch near the main trunk while roosting, may reflect predator avoidance as well as efforts to remain cool during the day. These roost locations not only make the owl difficult to detect, but they also pose a risk of injury to an aerial predator, since during the attack the predator might hit the tree itself if it misjudged its approach. Possible species that might attempt to prey on Spotted Owls, particularly the young, would include Goshawks and Great Horned Owls. Ganey et al. (1997) studied Spotted Owls and Great Horned Owls in ponderosa pine and ponderosa pine–Gambel oak forests in northern Arizona, and found that there was considerable overlap of their home ranges and of forest types they used, although the Spotted Owl tended to forage and roost in areas where trees were denser (i.e., greater canopy cover). In the mountains of western New Mexico, we rarely found the two species together: the Spotted Owl typically occurred either at higher elevations in the mountains or in the deep canyons, whereas Great Horned Owls tended to be found at lower elevations or in forests that had less topographic relief (including mesa tops). However, given the very highly predatory nature of Great Horned Owls, it is likely that they would take a Spotted Owl anytime that they could get one.

Recently there has been considerable concern about the possible impacts of the Barred Owl (*Strix varia*) on the Northern and California Spotted Owls (e.g., Kelly et al. 2003). A close relative of the Spotted Owl, the Barred Owl originally occurred in the eastern parts of the United States and Mexico. However, like many eastern birds, it has expanded its range across the northern states and has recently come into extensive contact with the Spotted Owl, first in the Pacific Northwest and now progressively southward. The two species use the same general types of habitats, and will hybridize with each other. Where the Barred Owl has become more common, the Spotted Owl has tended to decline, and there is concern that the larger Barred Owl may displace the Spotted Owl. There are few reports of the Barred Owl in the Southwest, although there may be some overlap between the two species in the Big Bend region of Texas (Kelly et al. 2003), and there is one recent, confirmed record from New Mexico, though not in Mexican Spotted Owl habitat (chapter 40). Although the Barred Owl is not a significant problem for the Spotted Owl in the Southwest at the present time, the ranges of many bird species are changing very rapidly as a result of human alterations of habitat and climate. It is possible that Barred Owls may begin to occur in Mexican Spotted Owl habitats at some point in the future.

Status and Management

The Mexican Spotted Owl was listed under the Endangered Species Act as a "Threatened" species in 1993. The decision to list the Spotted Owl in the Southwest was not based primarily upon known declines in population numbers, which had not been well studied at that time. Rather, there was considerable concern about potential threats to the continued survival of the birds primarily as a result of habitat alteration (U.S. Fish and Wildlife Service 1993). The major issue then was the potential impacts on Spotted Owl habitat from timber harvests, particularly the use of even-aged logging and stand management practices (i.e., clear-cutting), as well as from plans to begin logging on steep hillsides in mixed conifer forest, which at that time was believed to be the primary habitat of the owls. The listing rule also identified large or "catastrophic" fires as a major concern that could lead to the loss of Spotted Owl habitat.

Since the listing decision, there have been several detailed statistical analyses of population trends for the Mexican Spotted Owl, both in New Mexico and in other states in the Southwest. By marking the owls with individually coded aluminum and colored leg bands, it has been possible to follow specific owls throughout most or all of their lifetimes, and to determine both their survivorship and their reproductive output through time. This information, when entered into the appropriate statistical programs, has allowed us to calculate "lambda," or the extent to which individuals that are disappearing from a population (because they are either dying or moving elsewhere) are being replaced by new owls (either through the production of young, or by owls that were fledged elsewhere and have immigrated into the population). When there is a one-to-one replacement of birds (when lambda equals 1), population size remains the same through time. When lambda is less than one, more birds are dying or leaving than are being added to the population. In such a case, the population declines and over time may go extinct.

Table 35.1 shows the calculated lambda values for several Mexican Spotted Owl populations in New Mexico and Arizona, based upon our research (Stacey and Peery 2002; Stacey and Peery, unpubl. data) and on work by Seamans et al. (1999). The statistical approach used in all studies was identical. The results indicated that all of the populations under study were declining during the 1990s, with lambda values equal to less than one. And for all of the populations examined, the actual number of territorial (breeding) females declined throughout the study period. Unfortunately, the intensive banding and survey work upon which all the demographic studies were based ended when funding disappeared in the early 2000s, and it is not known whether the negative population trends have continued. However, we do know from surveys conducted by the Forest Service that the population in the Zuni Mountains has remained low and unstable (see below).

A possible exception to this pattern was reported for the Jemez Mountains in north-central New Mexico. At that location T. Johnson (2000) found that territorial occupancy rates remained relatively stable during the 1990s, and that after an initial decline at the beginning of the survey period from 1985 to 1989, approximately the same numbers of territories were held by owls in 2000 as were occupied in 1989. However, Johnson's data are based upon a different methodology than the studies discussed above. In particular, Johnson used calling surveys rather than banding of individual birds. As a result, it is not certain whether his findings are directly comparable to those of the other studies.

One trend that is apparent in the data presented in table 35.1 is that small populations of owls were declining at a faster rate than the largest ones, and the two smallest populations actually became extirpated before the end of our study. Whether extirpation of the two populations is temporary is unknown, but this phenomenon has also occurred in other areas of New Mexico. For example, prior to 1990 there were observations of small numbers of owls in the Manzano, Sandia,

TABLE 35.1. Annual rates of population change (lambda) for several Mexican Spotted Owl (*Strix occidentalis lucida*) populations in New Mexico and Arizona studied during the 1990s. Where lambda is less than "1" a population is declining, and where it is greater than "1" it is increasing. The difference between the observed value of lambda and 1 is a direct measure of the rate at which the population is declining. For example, the rate of decline in the Black Range from these data is approximately 16% per year (1.00 minus observed lambda times 100). The populations in the Magdalena and Zuni mountains both went extinct, and then were recolonized by a single pair of owls during the period of the study. Within studies, populations are listed from largest to smallest (see text).

STUDY	POPULATION	LAMBDA	ESTIMATED PERSISTENCE TIME (BASED UPON OBSERVED LAMBDA)
Stacey and Peery (2002, in preparation)	All populations combined (regional metapopulation)	0.803	23 years
	Black Range (New Mexico)	0.841	26 years
	San Mateo Mountains (New Mexico)	0.695	16 years
	Magdalena Mountains (New Mexico)	NA	Extinct-recolonized
	Zuni Mountains (New Mexico)	NA	Extinct-recolonized
Seamans et al. (1999)	Coconino National Forest (Arizona)	0.896	Not analyzed
	Tularosa Mountains (New Mexico)	0.857	Not analyzed

Datil, and possibly LaDrone mountains. The owls had disappeared from all of these small ranges by the mid 1990s and have not returned. In addition, it appears that the few Spotted Owls that occurred in the San Juan Mountains of northern New Mexico before 2000 are now mostly absent as well.

It is not entirely obvious why the Spotted Owls in the smaller ranges should decline faster, and go extinct sooner, than those in larger ranges. Like most other raptors, reproductive success and survival in Mexican Spotted Owls is probably primarily a reflection of the prey base on their territories (e.g., Ward 2001), as long as there is sufficient nesting and roosting habitat. There is no reason to believe that the rodents and rabbits that provide the bulk of the food for both adult and juvenile Spotted Owls should be more abundant on territories that happen to be located in large mountain ranges than those that happen to be in smaller ranges. Owl territories do not overlap, and even the smallest ranges contain many square kilometers of potential foraging habitat for at least some owls.

One possible explanation for this phenomenon may lie in the particular pattern of juvenile dispersal in the Mexican Spotted Owl. As described above, most or all juveniles leave their natal mountain ranges in the fall, wander over wide areas during the winter, and then settle in a new mountain range to begin breeding. Thus the number of new birds that become breeders in a range and that serve as the replacement for the older birds as they die is heavily dependent upon what happens elsewhere. In order for a local population in one mountain range to persist through time, there must be other populations in other mountain ranges that are successfully producing enough young that will disperse and then move into the target population.

In species in which the young settle in different locations than where they are born, it has been shown that one of the primary ways that the birds choose a site to establish a new breeding territory is whether or not there are other birds already present in that area. This phenomenon, which is known as "conspecific attraction" (Stamps 1988), has profound implications for species that, like the Mexican Spotted Owl, occur in the form of isolated populations and also appear to be declining throughout much of their range. If, as a result of long-distance dispersal and movements, most of the juveniles that are produced in a region are essentially mixed together into one large pool of potential recruits, and if the young are then attracted to and most likely to settle in areas that already have a number of established breeding owls, the largest populations will then be likely to obtain the most new recruits. In contrast, few if any young birds will settle in areas where there are only a few owl territories, as in the smaller ranges. Thus when there is an overall decline in the number of young produced, for whatever reason, the behaviors of wide-scale dispersal and conspecific attraction only accelerate the decline of small populations because the young will disproportionately join the larger populations. In contrast, if many young are produced, territories in the large populations will fill up, and some of the young will be required to keep moving until they settle in the smaller ranges.

A consequence of this pattern is that once a local population goes extinct, it becomes difficult for it to become reestablished if the overall population is declining (Stacey and Peery 2002). We observed an example of this phenomenon in the Zuni Mountains between 1993 and 2003. During the initial years of the study, most of the territories we were monitoring were occupied, and the population was relatively stable. The history of five of the most productive and reliably occupied territories in the Zuni Mountains is shown in figure 35.1. By 1996, the owls began to rapidly disappear from the Zunis, and by 1998, there was only a single bird on one of the five territories. After that time, owls were sometimes present in the area, including even a successful nest on one of the territories in 2002. But in complete contrast to the typical situation, territorial occupancy was very unstable: rather than the same owls holding the same territories year after year, new birds appeared in new areas and then either died or moved on (including the pair that bred successfully in 2002, but which was gone the following year in 2003). The current status of the owls throughout the Zuni Mountains is unknown, but if the pattern shown in figure 35.1 continues, it is likely that the Zuni population will become extinct in the near future, except perhaps for an occasional wandering owl that temporarily occupies a particular site.

It is not clear why Mexican Spotted Owls should be declining in most or all of the places where they occur

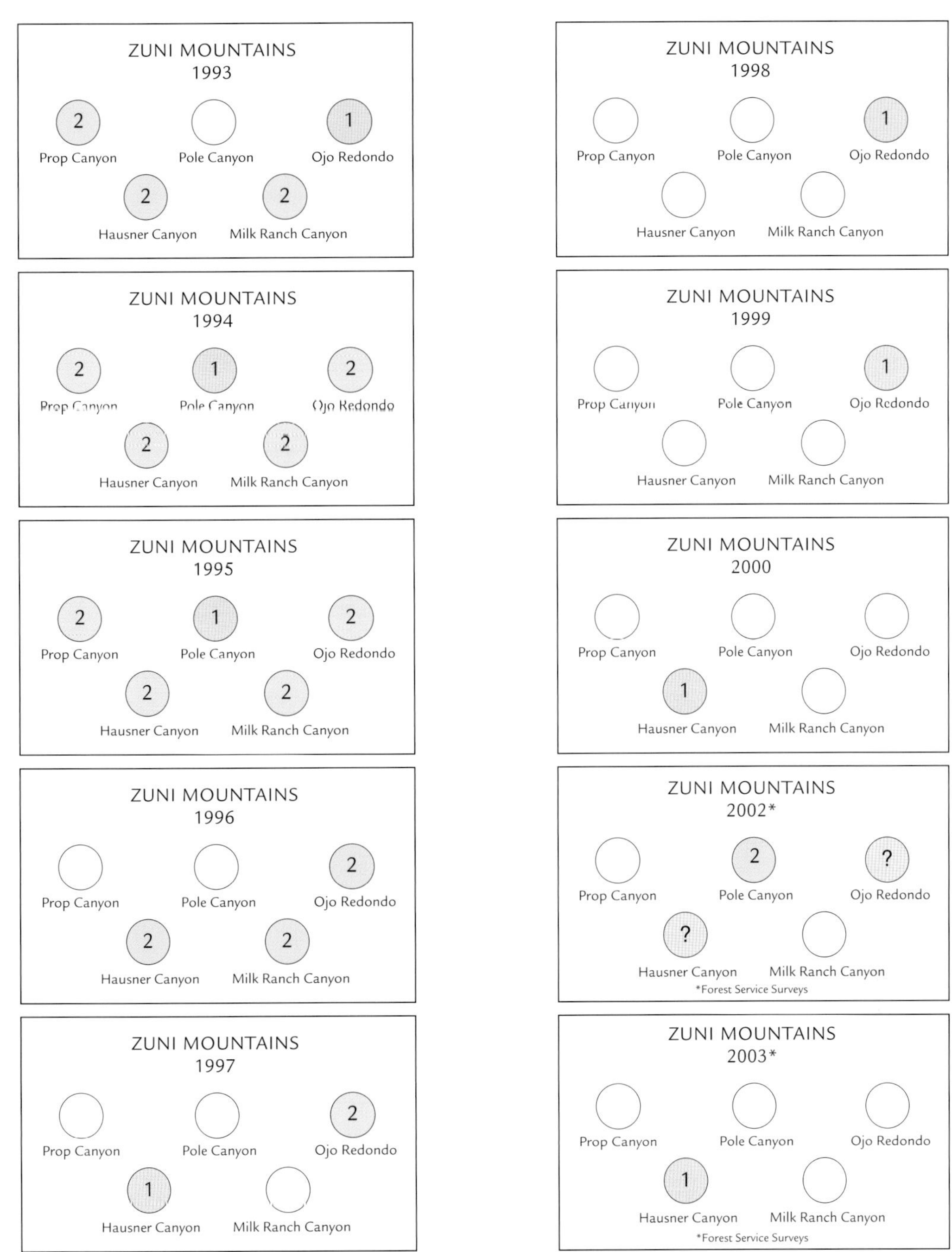

FIGURE 35.1

History of territory occupancy in the Zuni Mountains, 1993–2003.

Inside every circle is the number of owls occupying a territory. The 1993–2000 information is from the author (unpubl. data), while that from 2002–2003 is from Forest Service surveys (Cibola National Forest). No data were available for 2001. Although the territories shown were consistently occupied during the early 1990s, by 1998 there was only a single bird (yearling female) on one territory. Since that time, the population has been unstable; territories are occupied and then quickly abandoned again. In stable populations, the same territories are occupied year after year, usually by the same owls until they die.

in the American Southwest. In response to the listing of the Spotted Owl in 1993, the Forest Service and most other land-managing agencies established "Protected Activity Center," or PACs, around known nest or roost sites, as described already (see Territoriality). Within the PAC, many human activities were to be regulated, although not necessarily completely excluded. Regulated activities included harvesting of large-diameter trees, some types of thinning, timber harvests on steep slopes (>40%), construction, and some recreation uses (see U.S. Fish and Wildlife Service 1995 for details).

The available information from both the demographic studies and other research such as yearly surveys indicates that Mexican Spotted Owls have continued to decline, even after the implementation of the recommendations of the 1995 *Recovery Plan*. What is perhaps particularly telling with respect to the successful management of this species in the Southwest is that many of the population declines have occurred in areas or entire ranges where there were no timber harvests during the period of decline, or even for many decades prior to the decline. For example, Spotted Owl populations declined in the Magdalena and San Mateo Mountains and in the Black Range during our study, even though there had not been any logging in those ranges since before the 1950s. Timber harvests could not have been directly or even indirectly responsible for the decline of the owls in these areas. And while some research in Arizona has shown that the owls prefer to roost and forage in stands that have not recently been logged compared to those that have experienced timber harvests, the extent to which past logging directly affects owl survival and reproduction in those areas remains uncertain.

In addition, the second major presumed threat to the Mexican Spotted Owl, that from large or catastrophic wildfires, has now been found in numerous studies to be not as serious a problem for the owls as originally believed, and in some cases it may perhaps even be beneficial (e.g., Bond et al. 2002; Jenness et al. 2004; Stacey and Hodgson, unpubl. data). Not only do owls continue to occupy and successfully breed in areas that have been burned during large fires, but there is anecdotal evidence that they may actually be attracted to some burned areas where a ground cover of grass develops and rodent densities consequently increase (PS, pers. obs.). The finding that catastrophic fires have little or no impact on Spotted Owls is not particularly surprising, given the large size of Spotted Owl territories and the fact that even catastrophic fires typically burn in a patchy or mosaic manner in the topographically diverse types of habitats that owls utilize in the Southwest. Some roost and nest sites are almost always left even after the most severe burns, and foraging may actually be easier for the owls on the edges of the burned patches (Stacey and Hodgson, unpubl. data).

Direct impacts of humans on Mexican Spotted Owls are also unlikely to be responsible. Humans and Mexican Spotted Owls rarely interact directly with each other, primarily because the owls tend to nest in remote places in deep cover. Although it is sometimes claimed that Spotted Owls are extremely sensitive to humans, there is no direct evidence for this. In fact, when the owls are first approached, they are completely tolerant of people, and it is possible to climb within several feet of a roosting bird before it will respond and move. Although there is some evidence that hikers may have an effect on the activity patterns and time budgets of owls not used to humans (Swarthout and Steidl 2003), in other situations this does not appear to be a problem. For example, a pair of owls in southeastern Arizona has continued to roost for many years near a popular hiking trail, where they have been seen by thousands of birdwatchers. There was also concern that some military activities that are frequent in New Mexico, such as training flights by the air force, might impact the owls. While some activities appear to have a detrimental effect, such as the approach of a helicopter near a nesting site (Delaney et al. 1999), others, like low-level jet flyovers, do not (PS, pers. obs.).

Thus, despite the intensive research on the Spotted Owl over the past two decades, the reasons for the widespread and continued decline of this species in the Southwest remain unclear. One factor that may prove to be particularly important is livestock grazing. Originally, when the Mexican Spotted Owl was believed to be primarily a bird of mature mixed conifer forests, the possible impacts of cattle on owl habitats were assumed to be minor, simply because cattle rarely use mixed conifer forests in the Southwest:

they are too dense and there is little grass forage in the understory. However, as it has been recognized that the Spotted Owl may depend to a large extent upon prey that live in the grassy areas within meadows and riparian habitats, the potential impacts of livestock are now beginning to be reexamined (Stacey and Hodgson 1999; Ward 2001). Livestock grazing occurs in most of the areas where the Spotted Owl is found in the Southwest. Overgrazing by both domestic and native ungulates has the potential to significantly reduce the quality of owl habitat in a number of ways. First, it can reduce or eliminate the grass and forb cover in meadows, along the edges of riparian areas, and in other areas where the owls hunt. When vegetative cover is greatly reduced or removed entirely through grazing, the rodents and rabbits that form the bulk of the owls' diet typically disappear (e.g., Ward 2001). In addition, long-term grazing in canyon and riparian habitats will result in the loss of the deciduous tree component of the riparian forest, because livestock eat the seedlings and prevent the recruitment of new trees. Eventually the old trees die off, and because there are no new trees to replace them, the riparian habitat becomes very open, and loses the dense and structurally complex stands that the owls need for roosting and nesting (Stacey and Hodgson 1999). Interestingly, the one area of New Mexico where Spotted Owls may not be declining, the Jemez Mountains (Johnson 2000), is also one of the few areas in the state where there is only limited livestock grazing in Spotted Owl habitats (PS, pers. obs.).

The health and productivity of much of the meadow and riparian habitat in the Southwest have declined in recent years. Although the possible effects of livestock—and native ungulate—grazing were recognized in the Mexican Spotted Owl *Recovery Plan*, no specific guidelines were provided to address this concern beyond the general recommendation to maintain and restore, when necessary, riparian and meadow areas to "good" condition. As a result, to the author's knowledge, no specific management decisions and changes have been made on public lands with respect to livestock activities within Spotted Owl PACs in the Southwest. Clearly, more research on this subject is needed. However, if the large Mexican Spotted Owl population declines that were observed during the 1990s have continued to the present time—and there is no evidence that they have not—then it is essential that new approaches be instituted to the conservation and management of the habitat of this owl. Otherwise, there is considerable risk that one of the most interesting species of raptors in New Mexico will no longer be seen by future generations in the state.

Arsenault, D. P., A. Hodgson, and P. B. Stacey. 1997. Dispersal movements of juvenile Mexican Spotted Owls (*Strix occidentalis lucida*) in New Mexico. In *Biology and conservation of owls of the northern hemisphere; second international symposium*, ed. J. R. Duncan, D. H. Johnson, and T. H. Nicholls, 47–56. General Technical Report NC-190. Washington, DC: USDA Forest Service.

Bailey, F. M. 1928. *Birds of New Mexico*. Santa Fe: New Mexico Department of Game and Fish.

Barrowclough, G. F., J. G. Groth, L. A. Mertz, and R. J. Gutiérrez. (2006). Genetic structure of Mexican Spotted Owl (*Strix occidentalis lucida*) populations in a fragmented landscape. *Auk* 123:1090–1102.

Barrows, C. W. 1981. Roost selection by Spotted Owls: an adaptation to heat stress. *Condor* 83:302–9.

Bond, M. L., R. J. Gutiérrez, A. B Franklin, W. S. Lallaye, C. S. May, and M. E. Seamans. 2002. Short-term effects of wildfires on spotted owl survival, site fidelity, mate fidelity, and reproductive success. *Wildlife Society Bulletin* 30:1022–28.

Carmen, W. J. 2004. Noncooperative breeding in the California Scrub-Jay. *Studies in Avian Biology*, no. 28.

Delaney, D. K., T. G. Grubb, P. Beier, L. L. Pater, and M. H. Reiser. 1999. Effects of helicopter noise on Mexican spotted owls. *Journal of Wildlife Management* 63:60–76.

Dickerman, R. W. 1997. Geographic variation in southwestern U.S. and Mexican Spotted Owls, with the description of a new subspecies. In *The era of Allan R. Phillips: a Festschrift*, ed. R. W. Dickerman, 45–48. Albuquerque, NM: Horizon Communications.

Findley, J. S., A. H. Harris, D. E. Wilson, and C. Jones. 1975. *Mammals of New Mexico*. Albuquerque: University of New Mexico Press.

Forsman, E. D., R. G. Anthony, J. A. Reid, P. J. Loschl, S. G. Sovern, M. Taylor, B. L. Biswell, A. Ellingson, E. C. Meslow, G. S. Miller, et al. 2002. Natal and breeding dispersal of northern spotted owls. *Wildlife Monographs* 149:1–35.

Forsman, E. D., E. C. Meslow, and H. M. Wight. 1984. Distribution and biology of the spotted owl in Oregon. *Wildlife Monographs* 87:1–64.

Ganey, J. L. 2004. Thermal regimes of Mexican spotted owl nest stands. *Southwestern Naturalist* 49:478–86.

Ganey, J. L., and R. P. Balda. 1994. Habitat selection by Mexican Spotted Owls in northern Arizona. *Auk* 111:162–69.

Ganey, J. L., W. M. Block, and S. H. Ackers. 2003. Structural characteristics of forest stands within home ranges of Mexican Spotted Owls in Arizona and New Mexico. *Western Journal of Applied Forestry* 18:189–98.

Ganey, J. L., W. M. Block, J. K. Dwyer, B. E. Strohmeyer, and J. S. Jenness. 1998. Dispersal movements and survival rates of juvenile Mexican spotted owls in northern Arizona. *Wilson Bulletin* 110:206–17.

Ganey, J. L., W. M. Block, J. S. Jenness, and R. A. Wilson. 1997. Comparative habitat use of sympatric Mexican spotted and great horned owls. *Journal of Wildlife Research* 2:115–23.

———. 1999. Mexican spotted owl home range and habitat use in pine-oak forest; implications for forest management. *Forest Science* 45:127–35.

Ganey, J. L., W. M. Block, J. P. Ward, and B. E. Strohmeyer. 2005. Home range, habitat use, survival, and fecundity of Mexican spotted owls in the Sacramento Mountains, New Mexico. *Southwestern Naturalist* 50:323–33.

Glenn, E. M., M. C. Hansen, and R. G. Anthony. 2004. Spotted owl home-range and habitat use in young forests of western Oregon. *Journal of Wildlife Management* 68:33–50.

Gutiérrez, R. J., A. B. Franklin, and W. S. LaHaye. 1995. Spotted Owl. No. 652. In *The birds of North America*, ed. A. Poole and F. Gill. Philadelphia, PA: Birds of North America, Inc.

Gutiérrez, R. J., M. E. Seamans, and M. Z. Peery. 1996. Intermountain movement by Mexican spotted owls (*Strix occidentalis lucida*). *Great Basin Naturalist* 56:87–89.

Haig, S. M., T. D. Mullins, and E. D. Forsman. 2004. Subspecific relationships and genetic structure in the spotted owl. *Conservation Genetics* 5:683–705.

Hodgson, A. 1996. Dispersal and habitat use of Mexican spotted owls in New Mexico. M.S. thesis. University of Nevada, Reno.

Jenness, J. S., P. Beier, and J. L. Ganey. 2004. Associations between forest fire and Mexican spotted owls. *Forest Science* 50:765–72.

Johnson, T. H. 2000. Status of the Spotted Owl in the Jemez Mountains—2000. Unpublished report. Prepared under contract for the U.S. Geological Survey, Bandelier National Monument, New Mexico.

Kelly, E. G., E. D. Forsman, and R. G. Anthony. 2003. Are Barred Owls displacing Spotted Owls? *Condor* 105:45–53.

Kuntz, W. A., and P. B. Stacey. 1997. Preliminary investigation of vocal variation in the Mexican Spotted Owl (*Strix occidentalis lucida*): Would vocal analysis of the four-note location call be a useful field tool for individual identification? In *Biology and conservation of owls of the northern hemisphere; second international symposium*, ed. J. R. Duncan, D. H. Johnson, and T. H. Nicholls, 562–68. General Technical Report NC-190. Washington, DC: USDA Forest Service.

LaHaye, W. S., R. J. Gutiérrez, and J. R. Dunk. 2001. Natal dispersal of the Spotted Owl in southern California: dispersal profile of an insular population. *Condor* 103:691–700.

Ligon, J. S. 1961. *New Mexico birds and where to find them.* Albuquerque: University of New Mexico Press.

Martin, K., P. B. Stacey, and C.E. Braun. 2000. Recruitment, dispersal, and demographic rescue in spatially-structured White-tailed Ptarmigan populations. *Condor* 102:503–16.

Rinkevich, S. E., and R. J. Gutiérrez. 1996. Mexican Spotted Owl habitat characteristics in Zion National Park. *Journal of Raptor Research* 30:74–78.

Rohner, C. 1997. Non-territorial "floaters" in great horned owls: space use during a cyclic peak of snowshoe hares. *Animal Behaviour* 53:901–12.

Seamans, M. E., R. J. Gutiérrez, and C. A. May. 2002. Mexican spotted owl (*Strix occidentalis*) population dynamics: influence of climatic variation on survival and reproduction. *Auk* 119:321–34.

Seamans, M. E., R. J. Gutiérrez, C. A. May, and M. Z. Perry. 1999. Demography of two Mexican Spotted Owl populations. *Conservation Biology* 13:744–54.

Stacey, P. B., and A. Hodgson. 1999. Biological diversity in montane riparian ecosystems: The case of the Mexican spotted owl. In *Rio Grande ecosystems: linking land, water, and people*, ed. D. M. Finch, J. C. Whitney, J. F. Kelly, and S. R. Loftin, 204–10. Proceedings RMRS-P-7. Washington, DC: USDA Forest Service.

Stacey, P. B., and M. Z. Peery. 2002. Population trends of the Mexican Spotted Owl in west-central New Mexico. *New Mexico Ornithological Society Bulletin* 30:42.

Stamps, J. A. 1988. Conspecific attraction and aggregation in territorial species. *American Naturalist* 131:329–47.

Sureda, M., and M. L. Morrison. 1998. Habitat use by small mammals in southeastern Utah, with reference to Mexican spotted owl management. *Great Basin Naturalist* 58:76–81.

Swarthout, E.C.H., and R. J. Steidl. 2003. Experimental effects of hiking on breeding Mexican spotted owls. *Conservation Biology* 17:307–15.

Tarango, L. A., R. Valdez, P. J. Zwank, and M. Cardenas. 1997. Mexican spotted owl habitat characteristics in southwestern Chihuahua, New Mexico. *Southwestern Naturalist* 42:132–36.

U.S. Fish and Wildlife Service. 1993. Endangered and threatened wildlife and plants; final rule to list the Mexican Spotted Owl as a threatened species. *Federal Register* 58:14248–71.

———. 1995. *Recovery plan for the Mexican Spotted Owl.* Vols. 1 and 2. Albuquerque, NM.

Ward, J. P. Jr. 2001. Ecological responses by Mexican spotted owls to environmental variation in the Sacramento Mountains, New Mexico. Ph.D. diss. Colorado State University, Fort Collins.

Ward, J. P. Jr., and W. M. Block. 1995. Mexican Spotted Owl prey ecology. In Recovery plan for the Mexican Spotted Owl. Vol. 2, chap. 5, 1–48. Albuquerque, NM: U.S. Fish and Wildlife Service.

Willey, D. W. 1998. Movements and habitat utilization by Mexican spotted owls within the canyonlands of Utah. Ph.D. diss. Northern Arizona University, Flagstaff.

Willey, D. W., and C. van Riper III. 2000. First-year movements by juvenile Mexican Spotted Owls in the canyonlands of Utah. *Journal of Raptor Research* 34:1–7.

Young, K. E., P. J. Zwank, R. Valdez, J. L. Dye, and L. A. Tarango. 1997. Diet of Mexican Spotted Owls in Chihuahua and Aguascalientes, Mexico. *Journal of Raptor Research* 31:376–80.

Zwank, P. J., K. W. Kroel, D. M. Levin, G. M. Southward, and R. C. Romme. 1994. Habitat characteristics of Mexican Spotted Owls in southern New Mexico. *Journal of Field Ornithology* 65:324–34.

36

Long-eared Owl

(*Asio otus*)

CARL LUNDBLAD AND JEAN-LUC E. CARTRON

THE LONG EARED OWL (*Asio otus*) is arguably among the most charismatic of New Mexico's raptors. Although widespread, its shy habits and frequent association with thickets also make it among the most enigmatic and least known owls in the state. Physically striking, the Long-eared Owl is characterized overall by a bright yet well camouflaged plumage, a slender build, and conspicuous large ear tufts. It is a medium-sized owl with a wingspan of 90–100 cm (~35–39 in) (Cramp 1985). With a total body length of 37–40 cm (14.6–15.7 in) and a mass of 260–435 g (9.2–15.3 oz), females average larger than males, who measure 35–37 cm (13.8–14.8 in) in body length and weight 220–305 g (7.8–10.8 oz) (Mikkola 1983; Cramp 1985).

The Long-eared Owl is most similar to its closest North American relative, the Short-eared Owl (*Asio flammeus*). However, the Long-eared Owl is only approximately two-thirds the weight of its congener and has both a lighter build and shorter wings (Sibley 2000). Collectively, the *Asio* owls are most similar in size and shape to the Barn Owl (*Tyto alba*). Despite similar dimensions, however, the Long-eared Owl has a much lighter build and uniformly slender body than the top-heavy Barn Owl. The species' slim build is often accentuated, when alert, by a characteristic elongate upright posture. The other most striking aspect of the Long-eared Owl's body is its close-set, long, almost wispy ear tufts. Superficially, the Long-eared Owl is also similar to the Great Horned Owl (*Bubo virginianus*) but the latter species is much larger and has a different breast pattern and less prominent, wider-set ear tufts (Davis and Prytherch 1976; Holt and Leasure 1993).

The Long-eared Owl's plumage is relatively bright and yet surprisingly camouflaging. Overall, the species is mottled brown with variable pale and rufous highlights. The contrast is especially high on the underparts where heavy brown streaks and bars overlay an abdomen that is pale on males and tends to be more rufous on females. The Long-eared Owl's most striking plumage feature is its face pattern, which shows a solid bright rufous facial disk, again brighter in females (Bent 1938), bold black vertical eye stripes, and contrasting white "eyebrow" markings. The penetrating eyes are bright yellow with black pupils. Fledglings are mottled gray overall with bold dark "raccoon" masks. Older immatures are less heavily marked than adults, and with darker facial disks.

A conspicuous feature of the Long-eared Owl in flight is its boldly contrasting wing pattern, which shows pale rufous patches at the base of the primaries

PHOTO 36.1

Long-eared Owl, Upper Box Canyon, Hidalgo Co., 14 October 2007. Good field marks are the owl's slender body, the orange facial disk, the white "eyebrows" between and over the two yellow eyes, and the conspicuous, close-set ear tuft. PHOTOGRAPH: © ROBERT SHANTZ.

PHOTO 36.2

Long-eared Owl in the El Paso metropolitan area in Texas, near the New Mexico state line, 28 January 2007. The heavily mottled plumage does an excellent job of breaking up the Long-eared Owl's outline, which, when combined with the owl's characteristic alert posture, makes this species a master of camouflage. PHOTOGRAPH: © JEFF KAAKE.

PHOTO 36.3

Juvenile Long-eared Owl about 65 km (40 mi) east of Las Cruces, Doña Ana Co., 3 May 2007. Note the facial disk, which is darker than in the adult. PHOTOGRAPH: © DOUG BURKETT.

above, contrasting with dark carpal patches formed by the primary coverts and darker tips. The underwing pattern is similar, though overall paler, with small dark carpal patches and checkered dark tips.

In addition to its flamboyant plumage, the Long-eared Owl is made even more charismatic by its unique and varied vocal repertoire. The species' primary advertising call, given by males, consists of single low hoots spaced about three seconds apart. In response, females give a softer *sheoof* (Sibley 2000). Both sexes may produce a variety of alarm calls including nasal barking, squealing, and moaning (Vyn et al. 2006). In addition to its many vocalizations, the Long-eared Owl

PHOTOS 36.4a and b

Long-eared Owls in flight, (a, *above*) Upper Box Canyon, Hidalgo Co., 12 October 2007 (PHOTOGRAPH: © ROBERT SHANTZ); (b, *left*) El Paso metropolitan area in Texas, near the New Mexico state line, 28 January 2007 (PHOTOGRAPH: © JEFF KAAKE). A Long-eared Owl's flight is silent and consists of long glides interrupted by deep wing beats. Note that the dark wrist (or carpal) patches on the underwings (not seen on these photos) is shared by Long-eared Owl and Short-eared Owl (*Asio flammeus*), but not Barn Owl (*Tyto alba*).

produces loud cracking wing claps during flight display and possibly during nest defense (Vyn et al. 2006) as well as bill snaps when alarmed.

While six subspecies of Long-eared Owls are recognized globally, only two are described from North America, *A. o. willsonianus* in the East and *A. o. tuftsi* in the West (Cramp 1985). Both subspecies are found in New Mexico (Bailey and Niedrach 1965).

Distribution

The Long-eared Owl is among the most widely distributed owls in New Mexico and throughout the world. The species breeds throughout much of the Northern Hemisphere including northern Africa (Mikkola 1983; Cramp 1985). In North America the species breeds from southeastern Yukon east across the southern half of

Although typically rare to uncommon in any given area especially during the nesting season, the Long-eared Owl has a wide distribution in New Mexico, occurring in all parts of the state and from low to high elevations. Long-eared Owls have been found, for example: (a) at the Ladder Ranch, Sierra Co., March 2008 (PHOTOGRAPH: © MARK L. WATSON); (b) at the North Roosevelt Migrant Trap, Roosevelt Co., 4 May 2008 (PHOTOGRAPH: © JERRY OLDENETTEL); (c) in Upper Box Canyon, Hidalgo Co., 14 October 2007 (PHOTOGRAPH: © ROBERT SHANTZ); (d) near Grants, Cibola Co., 24 May 2005 (PHOTOGRAPH: © LISA SPRAY); (e) at Maxwell National Wildlife Refuge, Colfax Co., 5 April 2005 (PHOTOGRAPH: © PATTY HOBAN); and (f) at Morgan Lake, San Juan Co., April 1999 (PHOTOGRAPH: © TIM REEVES).

(a)

(b)

Canada to the Canadian maritime provinces, south into northwestern New England and Appalachia, the Great Lakes, the northern Great Plains, the Rocky Mountains, the Great Basin, and the southwestern United States (AOU 1983). The species is absent as a breeder from much of the Pacific Northwest and north to Alaska and across northern Canada as well as in the southern Great Plains, Florida, and the Caribbean. The Long-eared Owl has apparently been extirpated in areas of former occupation in California (Bloom 1994).

Many breeding areas remain occupied during the winter but some birds withdraw southward, especially from the northernmost portions of the range. During that season the species' range is extended throughout the Great Plains and southeastern United States, mostly stopping short of Florida and the Gulf Coast, and into much of Mexico (Houston 1966; AOU 1983; Enríquez-Rocha et al. 1993). Throughout its range and in all seasons, the Long-eared Owl appears sporadically and apparently opportunistically in response to prey fluctuations (Hagen 1965; Lundberg 1979; Village 1981; Korpimaki and Norrdahl 1991; S. O. Williams, pers. comm.).

In New Mexico the Long-eared Owl has been recorded breeding widely from the Colorado Plateau in the northwest south to the "Bootheel" in the extreme southwest, and east across the central part of the state and the eastern plains, spanning a wide elevation range. Despite its widespread distribution in the

(c)

(e)

(d)

(f)

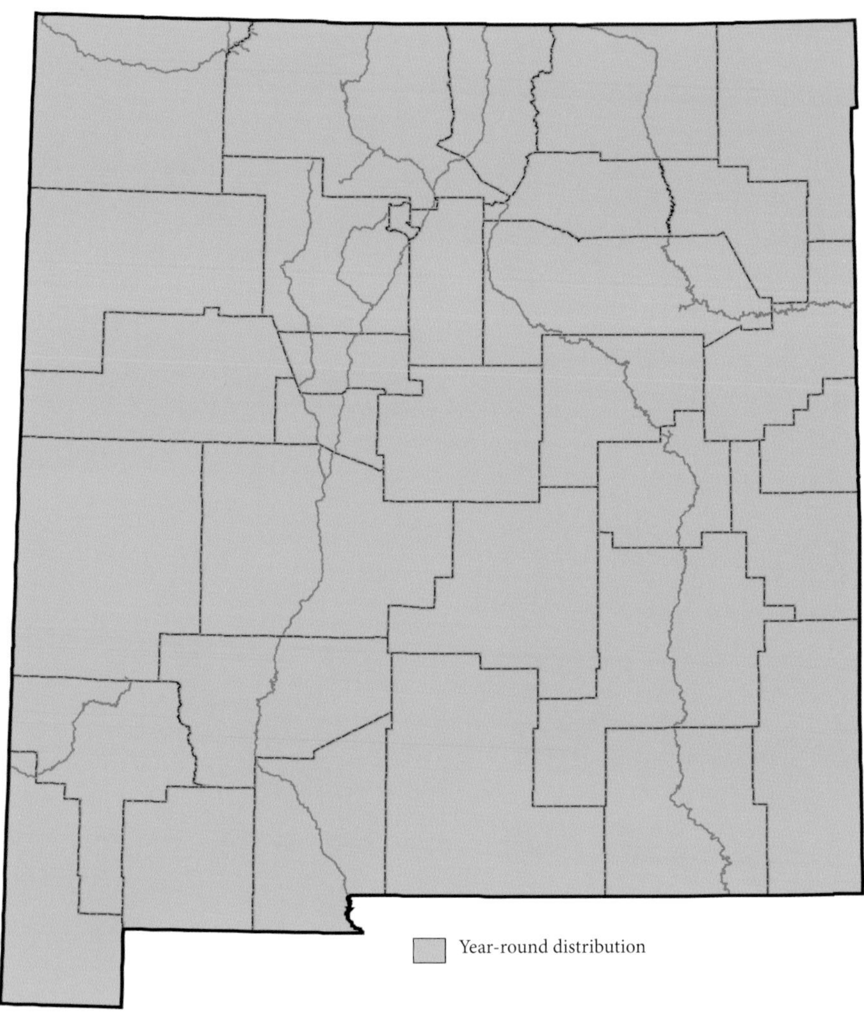

MAP 36.1

Long-eared Owl distribution map

state, the species is seldom considered common or regular at any one site (see also Status and Management; photos 36.5a–f). Among some of the earlier nesting records from New Mexico are those mentioned by Ligon (1961): a nest with young on 28 June 1916 on the rim of Mount Taylor Mesa, at an approximate elevation of 2,300 m (7,500 ft); a nest in the Tularosa Basin west of Tularosa (no date given); and a nest containing three eggs on 20 May 1955 at Abo, Torrance County. More recently, nesting has been reported from the Silver City and Fort Bayard area in Grant County; the Sandia Mountains; the Middle Rio Grande Valley, particularly near Corrales; and the Farmington area, San Juan County (Zimmerman et al. 1991; Reeves 1997; NMOS 2008). Other confirmed nesting records are from Grenville in Union County, Las Vegas and Maxwell National Wildlife Refuges, Eldorado in Santa Fe County, Cottonwood Gulch in McKinley County, Oso Ridge/Hausner Canyon in the Zuni Mountains, the Rio Puerco Valley, the Mimbres River in Luna County, the Jornada del Muerto basin, the Tularosa basin, Otero Mesa, and, on the eastern plains, Boone's Draw and Grulla and Bitter Lake National Wildlife Refuges (U.S. Fish and Wildlife Service 1994; NMOS

PHOTO 36.6

In New Mexico, typical nesting habitat includes pinyon-juniper woodland. Here, a Long-eared Owl near its nest south of Tijeras, Bernalillo Co., May 2007. PHOTOGRAPH: © DAVID DAIN.

2008; D. Arsenault, pers. comm.; P. Hoban, pers. comm.; J. Batkin, pers. comm.). Breeding along Oso Ridge (including Hausner Canyon) in the Zuni Mountains was recorded by David Arsenault (pers. comm.) in both 1999 (two fledglings and two adults roosting together in ponderosa pine [*Pinus ponderosa*] and oak trees on 30 June; two fledglings in Hausner Canyon on 1 July) and 2006 (two adults seen at night visiting a nest repeatedly on 23 June). Additionally, breeding in New Mexico has been strongly suspected or suggested near Belen in Valencia County, in the Peloncillo Mountains, and in the mountains of Rio Arriba County (NMOS 2008; CL, pers. obs.; J. Oldenettel, pers. comm.). Summering birds have been detected up to about 3,050 m (10,000 feet) (Hubbard 1988).

As in the breeding season, wintering Long-eared Owls may occur nearly statewide but are particularly noticeable in the south where communal roosts have frequently been found (Ligon 1961; NMOS 2008; CL, pers. obs.). At that time of year the species has been found from San Juan County in the far northwest east to Colfax and Mora counties in the northeast, south to Hidalgo County in the southwest, and east again across the south to Eddy County; locations associated with winter records include the eastern plains and the Rio Grande and Pecos valleys (Ligon 1961; NMOS 2008).

As mentioned already, the Long-eared Owl seems to be partially migratory (Slack et al. 1987; Russell et al. 1991; Duffy and Kerlinger 1992). In an extreme case, Houston (1966) describes a bird recovered as far as 4,000 km (2,485 mi) away from where it had been banded. The species' movements may be a combination of regular winter withdrawal, especially from the northernmost portion of its range, and nomadic wanderings in response to prey availability (Hagen 1965; Korpimaki and Norrdahl 1991). In New Mexico, some nesting may be the result of birds from farther north staying beyond the end of the winter instead of migrating back northward to breed when local conditions are favorable; however, this has not been established.

Habitat Associations

The Long-eared Owl has been found nesting in a wide variety of habitat types throughout its range including in New Mexico. The species is known to inhabit forest, forest edge, and predominately open habitats (Holt 1997). The most important habitat requirement seems to be the presence of dense woody vegetation for nesting and roosting, adjacent to relatively open areas for foraging.

In New Mexico, breeding pairs are seemingly found more regularly in mountain ranges and foothills as well as in major riparian areas. Typical breeding habitat includes pinyon-juniper (*Pinus edulis–Juniperus* sp.) and cottonwood-willow (*Populus* sp.–*Salix* sp.) woodlands (NMOS 2008; J. Batkin, pers. com.). Additional breeding records come from pine or pine-oak forests including "dog hair" stands of ponderosa pine and

PHOTO 36.7

(*top left*) Long-eared Owl in a desert willow (*Chilopsis linearis*), Upper Box Canyon, Hidalgo Co., 14 October 2007. Although Long-eared Owls hunt over open habitats, they prefer to nest and roost in dense vegetation. PHOTOGRAPH: © ROBERT SHANTZ.

PHOTO 36.8

(*bottom left*) During winter, Long-eared Owls in New Mexico are encountered mainly in the south, where they often form communal roosts. Long-eared Owl on Otero Mesa, Otero Co., 18 December 2005. PHOTOGRAPH: © JEFF KAAKE.

PHOTO 36.9

(*top right*) One of two recently fledged Long-eared Owls, from the Zuni Mountains, New Mexico, June 2000. The owls nested in a pine-oak woodland near open meadows at an elevation of approximately 2,200 m (7,200 ft). PHOTOGRAPH: © PETER STACEY.

PHOTO 36.10

(*bottom right*) Active nest with two young near fledging in a netleaf hackberry (*Celtis laevigata*) at Hackberry Well, Otero Co., 18 April 2007. In most years in New Mexico, Long-eared Owls appear to nest mainly in foothills and mountains, as well as in riparian areas. During years with high abundance of prey, they also nest in lowland desertscrub and desert grassland. PHOTOGRAPH: © JAMES E. ZABRISKIE.

from stands of Russian olive (*Elaeagnus angustifolia*) amidst short grass plains (D. Arsenault, pers. comm.; C. Rustay, pers. comm.). Nesting has also been strongly suggested in Madrean evergreen woodland of pine and oak in New Mexico's "Bootheel" (NMOS 2008; CL, pers. obs.; J. Oldenettel, pers. comm.).

In favorable years with abundant prey, Long-eared Owls in New Mexico have been known to also breed in lowland desertscrub and desert grasslands (Williams 2007; CL, pers. obs.). Nesting was detected in several such situations following the extraordinary precipitation year of 2006. During the spring and summer of 2007 at least four nests were found within lowland deserts of Doña Ana, Otero, and Grant counties with breeding suspected at two additional sites (Williams 2007; CL pers. obs., *fide* C. Britt).

In 2007, one of us (CL) discovered a nest at Hackberry Well on White Sands Missile Range in Otero County in a small stand of netleaf hackberry (*Celtis laevigata*) at an old stock tank surrounded by honey mesquite (*Prosopis glandulosa*) coppice dunes. Before summer's end two additional family groups were encountered on likely breeding territories nearby in similar habitats (CL, pers. obs.). At one site three owls were found roosting in large honey mesquites in an open run-in area within a better-developed dune system. The other area site, where six birds were detected, featured a small stand of saltcedar (*Tamarix ramosissima*) at a dirt stock tank within a matrix of fairly flat open mesquite shrubland and small patches of burrograss (*Scleropogon brevifolius*)/sacaton (*Sporobolus* sp.) grassland.

The grassland component in the vicinity of this last site was a well-developed feature of other southern New Mexico breeding sites that season. On Otero Mesa a nest was discovered in open grama grass (*Bouteloua* sp.)/tobosa (*Pleuraphis mutica*) grassland with scattered mesquites (Williams 2007, *fide* R. Meyer). In southern Grant County near Hachita a nest was discovered in yucca grassland swale (Williams 2007, *fide* R. Meyer), while a nest on the Jornada Experimental Range of Doña Ana County was in a honey mesquite shrubland-grassland matrix (*fide* C. Britt). A roost of four to five birds detected by one of us (CL) in saltcedar at a dirt tank with a matrix of grassland and desertscrub on the northern Jornada del Muerto of Socorro County during the final days of May was also suggestive of local breeding.

At the scale of the entire Long-eared Owl's distribution, winter roosts (see Life History section) are frequently found in conifer stands but also occur in saltcedar, palo verde (*Cercidium* sp.), and other vegetation types (Hillard et al. 1982; Marti et al. 1986; Barrows 1989). In New Mexico, winter roosts are most frequently found in the south and most often in saltcedar, though other roost substrates have included juniper, mesquite, little-leaf sumac (*Rhus microphylla*), and ornamental plantings of Arizona cypress (*Cupressus arizonica*) (NMOS 2008; CL, pers. obs.). In his notes and catalog (on file at Museum of Southwestern Biology, University of New Mexico), J. Stokley Ligon mentions finding a roost with several Long-eared Owls in a willow thicket at Beaverhead Range Station, 60 km (37 mi) northwest of Chloride, Sierra County on 30 October 1950, and another on 18 January 1951 in an orchard about 10 km (6 mi) southeast of Milnesand, Roosevelt County.

Life History

Nesting

Consistent with other aspects of their ecology, Long-eared Owls have been shown to be highly variable and opportunistic in their nesting habits. Individual birds have rarely been documented reoccupying the same territory in consecutive years, although nests may be reused by different individuals (Marks et al. 1994). Long-eared Owls do not build their own nests. Instead they typically lay their eggs in old nests built by other species of raptors and by corvids (Marks 1986). They also use clumps of mistletoe, old mammal nests, cliffs, and tree and cliff cavities, and even nest on the ground (Bent 1938; Craighead and Craighead 1956; Glue 1977; Marks and Yensen 1980; Bull et al. 1989; Marks and Holt 1994; Maples and Holt 1995).

Long-eared Owl nests in New Mexico have often been detected in old nests of corvids (J. Batkin, pers. comm.; C. Cook, pers. comm., *fide* D. Griffin; CL, pers. obs.) and other species, but one nest on Otero Mesa was found on a mound of flood debris at the base of a mesquite (Williams 2007, *fide* R. Meyer) while another at Bitter Lake National Wildlife Refuge was on a rocky

PHOTO 36.11

(*top left*) Nest in a juniper (*Juniperus monosperma*), McKinley Co., 28 April 2008. The female on the nest exhibits the threat display often directed at human intruders. The wings are spread to make her appear larger. PHOTOGRAPH: © RON KELLERMUELLER/HAWKS ALOFT, INC.

PHOTO 36.12

(*bottom left*) Fledglings in a Russian olive (*Elaeagnus angustifolia*) at Morgan Lake, San Juan Co., 1 May 1996. The same nest (in a Russian olive thicket) may have been used continuously from 1995 to 2007. Note the bold, dark raccoon masks of the fledglings. PHOTOGRAPH: © TIM REEVES.

PHOTO 36.13

(*bottom right*) In New Mexico, fledging appears to be from April to June. Here, one adult and two fledglings (one barely visible) at Maxwell National Wildlife Refuge, Colfax Co., 24 May 2005. PHOTOGRAPH: © PATTY HOBAN.

ledge at the edge of a sinkhole (Williams 2002). Tree and yucca species documented as substrate for Long-eared Owl nests in New Mexico include cottonwood, honey mesquite, netleaf hackberry, pinyon pine, juniper, Russian olive, soaptree yucca (*Yucca elata*), saltcedar, and ponderosa pine (Ligon 1961; Reeves 1997; Williams 2007; J. Batkin, pers. comm.; D. Arsenault, pers. comm.; C. Rustay, pers. comm.; C. Cook, pers. comm.). Nests used in two consecutive years have been documented in New Mexico. They included a nest used in 2006 and 2007 south of Tijeras, Bernalillo County, and placed about 4.5–6 m (15–20 ft) high in a pinyon tree (Williams 2006, 2007; D. Dain, pers. comm.). Also occupied at least two consecutive years (1995 and 1996) was a nest 6 m (20 ft) high in a Russian olive thicket at Morgan Lake in San Juan County (Reeves 1997; T. Reeves, pers. comm.). According to Tim Reeves (pers. comm.), the nest at Morgan Lake may well have been occupied continuously from 1995 through 2007, based on observations of owls at that location in 1997 and 1999, and reports of young in the nest during the early and mid 2000s. However, in both instances—Morgan Lake and south of Tijeras—it is not possible to state whether the nest was used by the same pair continuously, as no birds were banded.

Pair bonds among Long-eared Owls may be established on wintering grounds (Wijnandts 1984). Elaborate courtship rituals involve the male's engaging in display flight including advertising calls and wing clapping in the vicinity of the nest site (Wijnandts 1984; Vyn et al. 2006). Nesting may begin as early as late February (Kebbe 1954) and in New Mexico nestlings have been detected as early as the second week of April (CL, pers. obs.) The species typically raises only one brood per year although re-nesting after failed initial attempts has been documented (Wijnandts 1984; Marks 1986) and double clutching has been reported (Glue 1977; Marks and Perkins 1999). Incubation is performed exclusively by the female and typically lasts 26–28 days (Armstrong 1958; Cramp 1985). Clutch size ranges from 2 to 10 with an average of 4.5 eggs (Murray 1976). One of three egg sets collected in New Mexico and housed at the UNM Museum of Southwestern Biology consists of five eggs (MSB E-482, collected on 8 April 1955 in Torrance County), the other two of four eggs (MSB E-483 and MSB E-484, both collected in Torrance County, on 15 April 1956 and 29 April 1967).

The presence of three adult birds tending a nest in Utah suggested that this species might engage in polygamy or cooperative breeding (Sordahl and Tirmenstein 1980). In Idaho, communal roosting has been documented during the breeding season, and nests have been found as close as 14 m (46 ft) apart (Craig et al. 1985; Marks 1985, 1986). Marks et al. (1999), based on 12 nests in Montana, showed evidence of genetic monogamy—no extra-pair copulations—thus suggesting the occurrence of cooperative breeding. In New Mexico, the possibility that cooperative breeding occurs is suggested by CL's observation of three adults at the Hackberry Well nest in Otero County on multiple April dates in 2007.

Diet and Foraging

Throughout most of their distribution, Long-eared Owls primarily hunt small mammals, although at some locations birds are also an important part of the diet (Marti 1976; Marks et al. 1994). Reptiles, both lizards and snakes, have been documented as prey of Long-eared Owls in only a few studies (e.g., Cahn and Kemp 1930; Marks 1984). Long-eared Owls hunt at night, typically on the wing over open areas, killing small mammals with a bite to the back of the skull (Cramp 1985; Marks et al. 1994).

Nearly all diet information from New Mexico is based on Marti et al.'s (1986) analysis of 1,700 regurgitated pellets found at Long-eared Owl winter roosts in San Juan and Sandoval counties. Prey items consisted almost exclusively of mammals, primarily heteromyid rodents (*Perognathus* and *Dipodomys* spp.), with some small, unidentified passerine birds, a few lizards, one glossy snake (*Arizona elegans*), and insects belonging to the orders Orthoptera (grasshoppers and crickets) and Coleoptera (beetles). Besides pocket mice (*Perognathus* spp.) and kangaroo rats (*Dipodomys* spp.), deer mice (*Peromyscus* spp.) and western harvest mice (*Reithrodontomys megalotis*) were also found in large numbers among mammalian prey remains. Unusual mammalian prey—for the Long-eared Owl—consisted of one pallid bat (*Antrozous pallidus*), one unidentified shrew, and small numbers of Botta's pocket gophers (*Thomomys bottae*) and unidentified rabbits. The two lizards identified among remains were the short-horned lizard (*Phrynosoma douglasii*) and the sagebrush lizard (*Sceloporus graciosus*). Although most prey species identified by Marti et al. (1986) are nocturnal, some including passerine birds are not. The authors of the study mentioned the possibility that passerine birds were captured at twilight, or that Long-eared Owls seized them from their night roosts (Marti et al. 1986).

Besides Marti et al.'s (1986) study, information on the diet of Long-eared Owls in New Mexico is anecdotal and only confirms the importance of small rodents. In his unpublished notes and catalog (on file at the Museum of Southwestern Biology), the naturalist J. Stokley Ligon mentioned finding only rodents in the stomachs of a juvenile female he collected on 30 October 1950 about 65 km (40 mi) northwest of Chloride and an adult male shot near Milnesand, Roosevelt County, on 18 January 1951.

Winter Roosts

A unique habit of the Long-eared Owl throughout its entire distribution and in New Mexico specifically is

PHOTOS 36.14a and b

(*left*) Winter roost at a sewage plant in the El Paso metropolitan area along the Texas–New Mexico state line, 28 January 2007. At least three birds are visible on one of the photos. Up to 20 Long-eared Owls have been recorded at that particular winter roost. PHOTOGRAPHS: © JEFF KAAKE.

PHOTO 36.15

(*above*) Winter roost on Otero Mesa, 18 December 2005. The photographer observed five birds together, one of which is visible, taking off out of the juniper (*Juniperus monosperma*) tree. PHOTOGRAPH: © JEFF KAAKE.

its tendency to form communal wintertime roosts, typically containing anywhere from a few birds to several dozen. In rare cases, roosts of up to 100 birds have been discovered (Marks et al. 1994). Roosts of 10–30 birds have been routinely found in the Lower Rio Grande Valley from Caballo Dam to Las Cruces, Mesilla, and Sunland Park (NMOS 2008; CL, pers. obs.). Large roosts are also known from the Pecos River Valley of Chaves and Eddy counties, in the Animas Valley in the Columbus and Deming areas in Luna County, and the Ft. Bayard area in Grant County (NMOS 2008).

Predation and Interspecific Interactions

Long-eared Owls share their New Mexico habitats with many of the state's other raptors. A Long-eared Owl nest near Edgewood in Santa Fe County in 2005 was within 800 m (0.5 mi) of a nest occupied by a pair of Great Horned Owls (C. Cook, pers. comm.). The nest apparently failed. Although predation by the nearby Great Horned Owls was only one of several possible causes of the nest failure, across their range Long-eared Owls have frequently been subject to predation by raptors, corvids, and other species. Adults have reportedly been the victims, in North America, of Great Horned Owls in particular (Errington 1932; Bull et al. 1989), but also Barred Owls (*Strix varia*) (Bent 1938) and Red-tailed Hawks (*Buteo jamaicensis*) and Red-shouldered Hawks (*B. lineatus*) (Bloom 1994), among others. Nestlings are also susceptible to predation by Cooper's Hawks (*Accipiter cooperii*), Red-tailed Hawks, and Red-shouldered Hawks, and possibly Northern Goshawks (*Accipiter gentilis*) (Bull et al. 1989; Bloom 1994). Other predators may include raccoons (*Procyon lotor*) (Marks 1986) and bullsnakes (*Pituophis catenifer*) (Amstrup and McEneaney 1980). Predatory and aggressive behaviors directed by and against Long-eared Owls have not been documented in New Mexico. In California, a Long-eared Owl nest was thought to have been usurped by a Cooper's Hawk (Bloom 1994), and the species has been documented usurping American Crow (*Corvus brachyrhynchos*) nests (Sullivan 1992)

Status and Management

In New Mexico, the Long-eared Owl is rare to locally common during migration and in winter, while it is generally considered rare to uncommon during breeding (Ligon 1961; Hubbard 1978). Estimates of nesting pair densities are not available for any area of the state. Not only is the Long-eared Owl quite secretive for a bird of prey, nest fidelity is low (see Marks et al. 1994). Because few nests seem occupied for several consecutive years—presumably as a result of nomadism—monitoring of Long-eared Owl nesting pairs and their productivity is difficult. Even at the larger scale of North America, there are no estimates for Long-eared Owl numbers, which in the western United States appear to fluctuate from year to year (Marks et al. 1994).

Population declines have been substantiated or postulated in several U.S. states, including California, Minnesota, and New Jersey (Bosakowski et al. 1989; Bloom 1994; see also Marti and Marks 1989; Melvin et al. 1989; Petersen 1991). In New Mexico in particular, Christmas Bird Count data reveal a significant downward population trend between 1959 and 1988 (National Audubon Society 2008). The population decline in southern California has been linked to the loss or degradation

PHOTO 36.16

Long-eared Owl at Gila Wildlife Rescue near Silver City.

PHOTOGRAPH: © DENNIS MILLER.

of nesting habitat, including riparian areas (Marti and Marks 1989; Bloom 1994). Human-caused mortality by shooting has been documented but does not appear to be widespread or having any significant negative impact on the species. One of us (JLEC) once found a dead Long-eared Owl along Interstate 25 south of Albuquerque. Thus collision with road traffic probably represents another anthropogenic source of mortality, but here again it likely has a limited impact.

Given the population declines mentioned above, the lack of quantitative information about the ecology and status of the Long-eared Owl is all the more regrettable. In New Mexico as elsewhere, use of open-front nest boxes (used by Long-eared Owls in parts of their distribution) in historically occupied areas, banding/recapture efforts, and pooling of data by researchers would likely lead to more effective conservation of this species.

LITERATURE CITED

[AOU] American Ornithologists' Union. 1983. *Check-list of North American birds.* 6th ed. Washington DC: American Ornithologists' Union.

Amstrup, S. C., and T. P. McEneaney. 1980. Bull snake kills and attempts to eat Long-eared Owl nestlings. *Wilson Bulletin* 92:402.

Armstrong, W. H. 1958. Nesting and food habits of Long-eared Owls in Michigan. *Michigan State University Museum Biological Series* 1:61–96.

Bailey, A. M., and R. J. Niedrach. 1965. *Birds of Colorado.* Denver: Denver Museum of Natural History.

Barrows, C. W. 1989. Diets of five species of desert owls. *Western Birds* 20:1–10.

Bent, A. C. 1938. *Life histories of North American birds of prey.* Part 2. U.S. National Museum Bulletin 170. Washington, DC: Smithsonian Institution.

Bloom, P. H. 1994. The biology and current status of the Long-eared Owl in coastal southern California. *Bulletin of the Southern California Academy of Science* 93: 1–12.

Bosakowski, T., R. Kane, and D. G. Smith. 1989. Decline of the Long-eared Owl in New Jersey. *Wilson Bulletin* 101:481–85.

Bull, E. L., A. L. Wright, and M. G. Henjum. 1989. Nesting and diet of Long-eared Owls in conifer forest, Oregon. *Condor* 91:908–12.

Cahn, A. R., and J. T. Kemp. 1930. On the food of certain owls in east-central Illinois. *Auk* 47:323–28.

Craig, T. H., E. H. Craig, and L. R. Powers. 1985. Food habits of Long-eared Owls (*Asio otus*) at a communal roost site during the nesting season. *Auk* 102:193–95.

Craighead, J. J., and F. C. Craighead Jr. 1956. *Hawks, owls and wildlife.* Harrisburg, PA: Stackpole Co.

Cramp, S., ed. 1985. *The birds of the western palearctic.* Vol. 4. Oxford, UK: Oxford University Press.

Davis, A. H., and R. Prytherch. 1976. Field identification of Long-eared and Short-eared owls. *British Birds* 69:281–87.

Duffy, K., and P. Kerlinger. 1992. Autumn owl migration at Cape May Point, New Jersey. *Wilson Bulletin* 104:312–20.

Enríquez-Rocha, P., J. L. Rangel-Salazar, and D. W. Holt. 1993. Presence and distribution of Mexican owls: a review. *Journal of Raptor Research* 27:154–60.

Errington, P. L. 1932. Food habits of southern Wisconsin raptors. Part 1, Owls. *Condor* 35:176–86.

Glue, D. E. 1977. Breeding biology of Long-eared Owls. *British Birds* 70:318–31.

Hagen, Y. 1965. The food, population fluctuations, and ecology of the Long-eared Owl (*Asio otus* [L.]) in Norway. *Papers of the Norwegian State Game Research Institute,* series 2, no. 23.

Hillard, B. L., J. C. Smith, M. S. Smith, and L. R. Powers. 1982. Nocturnal activity of Long-eared Owls in southwest Idaho. *Journal of the Idaho Academy of Science* 18:29–35.

Holt, D. W. 1997. The Long-eared Owl (*Asio otus*) and forest management: a review of the literature. *Journal of Raptor Research* 31:175–86.

Holt, D. W., and S. M. Leasure. 1993. Short-eared Owl (*Asio flammeus*). No. 62. In *The birds of North America,* ed. A. Poole and F. Gill. Philadelphia, PA: Academy of Natural Sciences, and Washington, DC: American Ornithologists' Union.

Houston, C. S. 1966. Saskatchewan Long-eared Owl recovered in Mexico. *Blue Jay* 24:64.

Hubbard, J. P. 1978. *Revised check-list of the birds of New Mexico.* Publ. no. 6. Albuquerque: New Mexico Ornithological Society.

———. 1988. Southwest region: New Mexico. *American Birds* 42:1320–28.

Kebbe, C. E. 1954. Early nesting record of the Long-eared Owl. *Murrelet* 35:31.

Korpimaki, E., and K. Norrdahl. 1991. Numerical and functional responses of Kestrels, Short-eared Owls, and Long-eared Owls to vole densities. *Ecology* 72:814–26.

Ligon, J. S. 1961. *New Mexico Birds and where to find them*. Albuquerque: University of New Mexico Press.

Lundberg, A. 1979. Residency, migration and a compromise: adaptations to nest-site scarcity and food specialization in three Fennoscandian owl species. *Oecologia* 41:273–81.

Maples, M. T., and D. W. Holt. 1995. Ground nesting Long-eared Owls. *Wilson Bulletin* 107:563–65.

Marks, J. S. 1984. Feeding ecology of breeding Long-eared Owls in southwestern Idaho. *Canadian Journal of Zoology* 62:1528–33.

———. 1985. Yearling male Long-Eared Owls breed near natal nest. *Journal of Field Ornithology* 56:181–82.

———. 1986. Nest-site characteristics and reproductive success of Long-eared Owls in southwestern Idaho. *Wilson Bulletin* 98:547–60.

Marks, J. S., and E. Yensen. 1980. Nest sites and food habits of Long-eared Owls in southwestern Idaho. *Murrelet* 61:86–91.

Marks, J. S., J. L. Dickinson, and J. H. Haydock. 1999. Genetic monogamy in Long-eared Owls. *Condor* 101:854–59.

Marks, J. S., D. L. Evans, and D. W. Holt. 1994. Long-eared Owl (*Asio otus*). No. 133. In *The birds of North America*, ed. A. Poole and F. Gill. Philadelphia, PA: Academy of Natural Sciences, and Washington, DC: American Ornithologists' Union.

Marks, S. M., and A. E. H. Perkins. 1999. Double brooding in the Long-eared Owl. *Wilson Bulletin* 111:273–76.

Marti, C. D. 1976. A review of prey selection by the Long-eared Owl. *Condor* 78:331–36.

Marti, C. D., and J. S. Marks. 1989. Status of medium-sized owls in the western United States. In *Proceedings of the western raptor management symposium and workshop*, ed. B. G. Giron Pendleton, 124–33. Washington, DC: National Wildlife Federation.

Marti, C. D., J. S. Marks, T. H. Craig, and E. H. Craig. 1986. Long-eared Owl diet in northwestern New Mexico. *Southwestern Naturalist* 31:416–19.

Melvin, S. M., D. G. Smith, D. W. Holt, and G. R. Tate. 1989. Small owls. In *Proceedings of the northeast raptor management symposium and workshop*, ed. B. G. Pendleton, 88–96. Scientific and Technical Series no. 12. Washington, DC: National Wildlife Federation.

Mikkola, H. 1983. *Owls of Europe*. Vermilion, SD: Buteo Books.

Murray, G. A. 1976. Geographic variation in the clutch sizes of seven owl species. *Auk* 93:602–13.

National Audubon Society. 2008. The Christmas Bird Count Historical Results. http://www.audubon.org/bird/cbc (accessed 4 January 2008).

[NMOS] New Mexico Ornithological Society. 2008. *NMOS Field Notes* database. http://nhnm.unm.edu/partners/NMOS/ (accessed 29 September 2008).

Petersen, L. R. 1991. Mixed woodland owls. In *Proceedings of the Midwest raptor management symposium*, ed. B. G. Pendleton and D. L. Krahe, 85–95. Scientific and Technical Series no. 15. Washington, DC: National Wildlife Federation.

Reeves, T. 1997. Documented breeding and seasonal records for birds in San Juan County, New Mexico. *New Mexico Ornithological Society Bulletin* 25:23–28.

Russell, R. W., P. Dunne, C. Sutton, and P. Kerlinger. 1991. A visual study of migrating owls at Cape May Point, New Jersey. *Condor* 93:55–61.

Sibley, D. A. 2000. *The Sibley guide to the birds*. New York: Chanticleer Press.

Slack, R. S., C. B. Slack, R. N. Roberts, and D. E. Emord. 1987. Spring migration of Long-eared Owls and Northern Saw-whet Owls at Nine Mile Point, New York. *Wilson Bulletin* 99:480–85.

Sordahl, T. A., and D. A. Tirmenstein. 1980. Possible helper at a Long-eared Owl nest. *Western Birds* 11:57–59.

Sullivan, B. D. 1992. Long-eared Owls usurp newly constructed American Crown nests. *Journal of Raptor Research* 26:97–98.

U.S. Fish and Wildlife Service. 1994. Grulla National Wildlife Refuge. http://www.npwrc.usgs.gov/ resource/othrdata/chekbird/r2/grulla.htm (Version 22, May '98).

Village, A. 1981. The diet and breeding of Long-eared Owls in relation to vole numbers. *Bird Study* 28:215–24.

Vyn, G., G. F. Budney, M. Guthrie, and R. W. Grotke. 2006. *Voices of North American owls*. Ithaca, NY: Macaulay Library, Cornell Laboratory of Ornithology.

Wijnandts, H. 1984. Ecological energetic of the Long-eared Owl (*Asio otus*). *Ardea* 72:1–92.

Williams, S. O. III, ed. 2002. New Mexico. *North American Birds* 56:340–43.

———, ed. 2006. *New Mexico Ornithological Society Field Notes* 45(3): Summer 2006.

———, ed. 2007. *New Mexico Ornithological Society Field Notes* 46(2): Spring 2007.

Zimmerman, D., R. Fisher, P. Boucher, and B. Anderson. 1991. *Birds of the Gila National Forest: a checklist*. Albuquerque, NM: USDA Forest Service in cooperation with the Southwest New Mexico Audubon Society.

37

Short-eared Owl

(*Asio flammeus*)

JULIAN D. AVERY AND GREGORY S. KELLER

THE SHORT-EARED OWL (*Asio flammeus*) is a widespread medium-sized owl of North America and Eurasia that is active during both the day and night. Although males and females both measure approximately 29.6 cm (11.6 in) in wingspan, this species is sexually dimorphic in total length and body weight. Males and females average 373 mm (14.68 in) and 382 mm (15.04 in) in total length (Mikkola 1983) and 315 g (11 oz) and 378 g (13 oz) in body weight (Snyder and Wiley 1976), respectively.

Sexes are not easily distinguished in the field. Short-eared Owl females tend to be darker overall and rustier underneath, but considerable overlap exists between the sexes in plumage coloration. Short-eared Owls are named for their large, rounded head with two short feather tufts, resembling small ears, that are typically too small to be visible. The large gray or white facial disc is conspicuous, setting

PHOTO 37.1

Adult Short-eared Owl in flight, San Rafael Valley, Arizona, 12 February 2006. The underparts are pale, with a conspicuous dark comma-shaped wrist patch, brown tips to the primaries, bold streaks on the throat and breast, and dark barring on the tail. PHOTOGRAPH: © JIM BURNS (WWW.JIMBURNSPHOTOS.COM).

PHOTOS 37.2a and b

Short-eared Owl perched on a fence along Highway 17 in Chama, Rio Arriba Co., 28 January 2007. The Short-eared Owl's head is large and round, with a pale facial disk and bright yellow eyes surrounded by black rings.

off the yellow eyes rimmed with black rings and the buffy and brown-streaked throat and upper breast. The light underwing is punctuated by dark brown distal tips of the greater undercoverts, creating a comma-shaped wrist patch. Primaries have brown tips, giving the wing a distinct appearance in flight. Juvenile birds are a dark sooty brown above and light-colored below with a predominantly brownish black facial disc. Juveniles are typically darker than adults. Although up to 10 subspecies have been recognized, only the nominate subspecies, *A. f. flammeus*, occurs throughout North America, including New Mexico (Holt et al. 1999; Wiggins et al. 2006).

As adults, Short-eared Owls tend not to vocalize during winter; however, both sexes may give a "*keee-ow*" call year-round. The "*keee-ow*" call serves a variety of purposes and will be used when the adults feel threatened. During courtship, males commonly call from the ground or an elevated perch. The primary call is described as "*hoo-hoo-hoo-hoo*" given 13–16 times per bout at a rate of up to four per second (Holt 1985; Johnsgard 2002). At the nest, both males and females give a variety of sounds, including barks, screams, and whines, coupled with behavioral displays (see below).

Distribution

The Short-eared Owl is one of the most widespread bird species in the world, occurring across North America, the West Indies, South America, Hawaii, and Eurasia. In North America, the Short-eared Owl is known to breed from northern Canada and Alaska south to central California, northern Nevada, Utah, Colorado, Kansas, Missouri, Illinois, and across to Pennsylvania (AOU 1998). The species is no

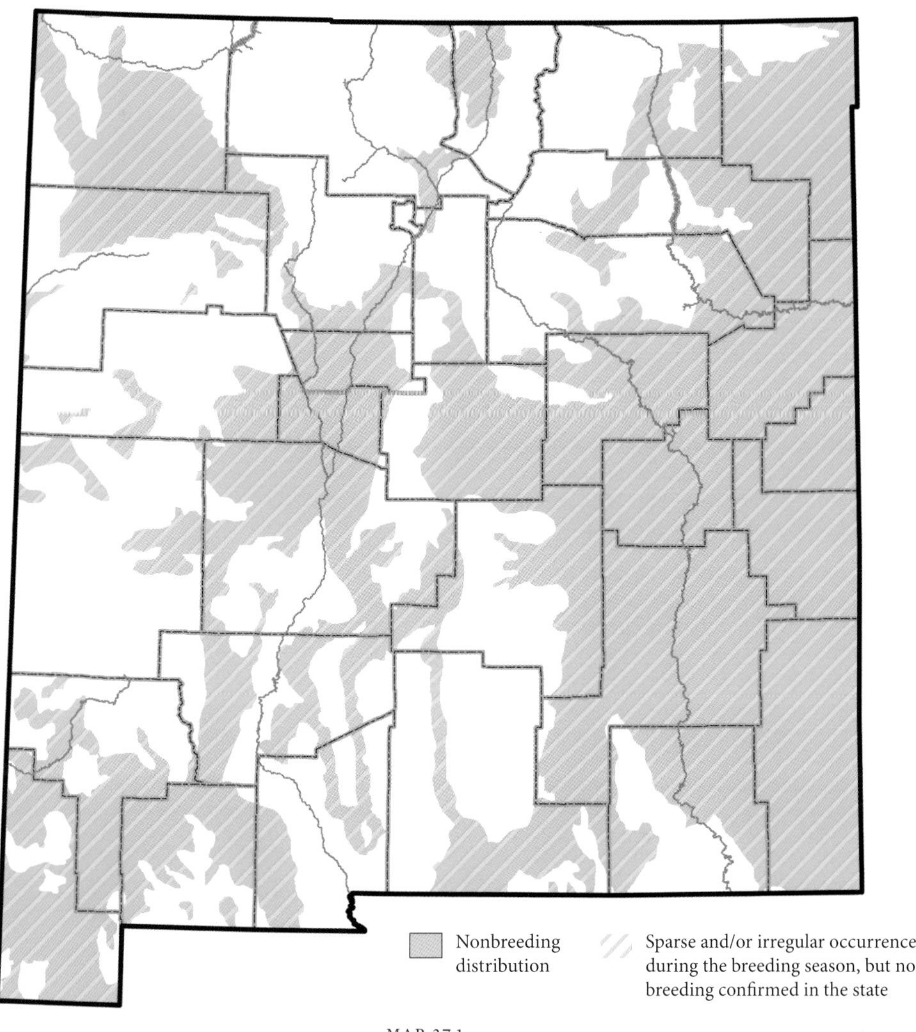

MAP 37.1

Short-eared Owl distribution map

longer present in New Jersey where it once bred in salt marshes along the coast, a pattern that is being repeated elsewhere along the northeastern coast (Wiggins et al. 2006). Although somewhat obscured by nomadism, north-south migration occurs in North American populations of the Short-eared Owl, particularly those breeding at high latitudes (Holt and Leisure 1993; see chapter 2). The North American winter range of the species extends from southern Canada to southern Baja California, Oaxaca, Puebla, and Veracruz in Mexico and the Gulf of Mexico and Florida in the United States (AOU 1998).

The Short-eared Owl has never been confirmed as a breeding bird in New Mexico. Recent Breeding Bird Atlas projects in states just north of New Mexico have shown the Short-eared Owl to be a rare breeder with very few reported nesting records. During the summer of 2005, a territorial pair of owls was observed from May to early August on the Llano Estacado west of Portales in Roosevelt County, New Mexico (Avery and Keller 2006; photo 37.4). On multiple occasions throughout that period, the two owls were seen landing in one section of a fallow field. Our attempts to locate a nest proved unsuccessful. Avery's first approach of

PHOTO 37.3

Short-eared Owl at Nanaimo Estuary, Vancouver Island, British Columbia, Canada, 21 January 2008. PHOTOGRAPH: © MIKE YIP (VANCOUVERISLANDBIRDS.COM).

the suspected nest area was met with territorial behavior, as the owls circled around him and uttered barking calls. One of the birds even dove at him. Altogether, our observations constituted New Mexico's first *probable* breeding record of Short-eared Owls (Avery and Keller 2006). Relatively recent reports of three near Clovis, Curry County, 17 June 2000 (Williams 2000); one near Bueyeros, Harding County, 16 June 2001 (Williams 2001); and one found freshly dead in a fence near Farley, Colfax County, 22 July 2005 (L. A. Sager, pers. comm.) are at least suggestive of occasional local summering. Outside of New Mexico the nearest confirmed breeding records are from southern Colorado (Kingery 1998) and the panhandle of western Oklahoma, where nesting was confirmed only in Texas County (Reinking 2004).

Outside of the breeding season, the Short-eared Owl occurs nearly statewide in New Mexico (NMOS 2008), although it is considered only a rare to locally uncommon migrant and wintering species in the state (e.g., Hubbard 1978; Parmeter et al. 2002). Localities where the species has been regularly recorded include some of New Mexico's wildlife refuges, including Bitter Lake National Wildlife Refuge and the Bosque del Apache National Wildlife Refuge (NMOS 2008). Perhaps owing to favorable environmental conditions, more Short-eared Owls than usual were reported in New Mexico during the 2004–5 winter season (October–March), with at least 14 birds at seven sites in six counties (S. O. Williams III, pers. comm.).

Habitat Associations and Life History

The Short-eared Owl is typically found in open habitats during both the breeding and wintering periods (Wiggins et al. 2006). Nesting habitat includes tundra (at high latitudes), marshes, grasslands, and sagebrush where this last habitat type borders grasslands and marshes. Grasslands suitable for nesting often consist of native prairie, swales, wet pastures, wet meadows, grain stubble fields, and hayfields (see Wiggins 2004). Suitable wintering habitat for Short-eared Owls appears to be both ungrazed and lightly grazed mid-grass prairie, dense grasslands less than 0.5 m (1.6 ft) high, and marshes (Wiggins 2004). In New Mexico, Short-eared Owls have been found during the winter months in a variety of vegetation types, including not just grasslands but also thickets of snakeweed (*Condalia*) and streamside groves of willows (*Salix* spp.) (Ligon 1961; BirdWest Archives 2004). However, Ligon (1961) characterized the primary, albeit greatly reduced, wintering habitat of the Short-eared Owl in New Mexico as "seas of grass."

The breeding habits of the Short-eared Owl are not very well known in North America. The species is a ground-nester, and nesting sites usually have enough vegetation present to obscure the nest, typically a scrape on the ground lined with grasses or down feathers (Holt and Leisure 1993). Vegetation near the nest is commonly composed of grasses (79%) and forbs (17%) averaging less than 0.5 m (1.6 ft) in height (Wiggins et al. 2006).

In eastern New Mexico, Avery and Keller (2006) noted that dominant plant species at the probable Short-eared Owl nesting site included sideoats grama (*Bouteloua curtipendula*), silverleaf nightshade (*Solanum elaeagnifolium*), sunflower (*Helianthus* spp.), Russian thistle (*Salsola* spp.), narrowleaf goosefoot (*Chenopodium desiccatum*), poverty threeawn (*Aristida divaricata*), and sand dropseed (*Sporobolus cryptandrus*).

Beginning in the late winter, male Short-eared Owls perform courtship flights known as sky dances. These involve small circles during a first phase of upward flight, followed by fanning of wings and tail, vocalizations, and shallow stoops with wing clapping (Holt and Leisure 1993; Wiggins et al. 2006). North of New Mexico in Colorado, Kansas, Wyoming, Nebraska, and South Dakota, egg-laying is in mid April through May, with occasional clutches laid in June representing possible re-nesting attempts (Bailey and Niedrach 1965; Boyle 1998; Sharpe et al. 2001; Tallman et al. 2002). Clutch size is highly variable, ranging between 3 and 11 eggs, based primarily on food resources; average clutch size is 6.9/nest in North America (Wiggins 2004). The incubation period lasts 21 to 37 days; hatching is asynchronous (Wiggins 2004). The male hunts for the brooding female and the nestlings, which disperse a short distance from the nest at the age of 14 to 17 days old. Fledging occurs later, when the young are 27 to 36 days old (Urner 1923; Clark 1975; Holt 1992).

Prey abundance is key to explaining much of the Short-eared Owl's ecology. Rather than always migrating south after breeding, some populations of this owl may be nomadic, tracking small mammals with cyclic population patterns (Holt and Leisure 1993). During the migratory and wintering periods, Short-eared Owls can form large flocks and roost communally on the ground, with roost sizes and amount of time spent in a roost influenced by prey abundance (Wiggins 2004). Also closely tied to prey abundance are nesting success and, during both breeding and wintering seasons, Short-eared owl abundance and habitat use (Wiggins 2004; Wiggins et al. 2006).

Hunting is typically conducted on the wing low over the ground. In general, small mammals make up the majority of the diet, if not its near entirety (Holt and Leisure 1993). Small mammalian prey

PHOTO 37.4

Landscape view of the probable Short-eared Owl nesting site in Roosevelt Co., eastern New Mexico, during summer 2005. PHOTOGRAPH: © JENNY RAMIREZ.

PHOTOS 37.5a and b

Short-eared Owl along Springer Lake, Colfax Co., 8 October 2005. In New Mexico, typical Short-eared Owl habitat consists of grassland, often near water, as shown in photos.

PHOTOGRAPHS: © ELTON M. WILLIAMS.

include shrews (*Sorex* and *Blarina* spp.), small rodents (*Microtus*, *Scapanus*, *Dipodomys*, and *Peromyscus* spp.), pocket gophers (*Thomomys* spp.), and rabbits (*Sylvilagus* spp., *Lepus* spp.). Voles are particularly important in the diet of some Short-eared Owl populations. Weller et al. (1955) found that prairie voles (*Microtus pennsylvanicus*) were the primary food source in Missouri, comprising 95% of regurgitated owl pellets. On the Llano Estacado, where a pair of Short-eared Owls appeared to nest in 2005 (Avery and Keller 2006), grassland habitats are dominated by western harvest mouse (*Reithrodontomys megalotis*), plains harvest mouse (*Reithrodontomys montanus*), deer mouse (*Peromyscus maniculatus*), hispid cotton rat (*Sigmodon hispidus*), and hispid pocket mouse (*Chaetodipus hispidus*), these species comprising 96% of small-mammal captures (Kuykendall 2004). Voles tend to be uncommon in this region (Davis and Schmidly 1994), but pocket gophers and rabbits would be common prey sources.

PHOTO 37.6

Short-eared Owl perched on a yucca (*Yucca* sp.) in yucca grassland, about 50 km (30 mi) southeast of Socorro, Socorro Co., 10 May 2007.

PHOTOGRAPH: © DOUG BURKETT.

Status and Management

The Short-eared Owl in North America is closely tied to diminishing habitats. In his account for the species, Ligon (1961) emphasized the substantial decrease in numbers of wintering Short-eared Owls in New Mexico during his time, linking it to the disappearance of primary habitat in the state. Rangewide in North America, the species has exhibited significant population declines over the past 38 years according to North American Breeding Bird Survey data (-4.8%/year, P = 0.01) (Sauer et al. 2007). One of the most cited explanations for this decline is the appropriation of prairie habitats for agricultural purposes. Overgrazing of grassland habitat in the Great Plains is also believed to have contributed to the decline, as it removes suitable cover for nests and impacts small-mammal populations (Wiggins 2004). The loss of old-field habitats in the northeastern United States to succession and urban development likely also played a negative role, while loss of wetland and marsh habitat to development and recreational purposes has removed critical wintering habitat (Wiggins et al. 2006).

Most recently, the population of Short-eared Owls in the Great Plains may have increased as a result of changing agricultural practices and habitat restoration through Conservation Reserve Programs (Wiggins et al. 2006). Short-eared Owls require large (>100 ha [~250 ac]) tracts of undisturbed land (Wiggins 2004), which are provided under these conservation programs. Similar management tactics emphasizing large tracts of unbroken, healthy grassland could be beneficial to Short-eared Owls in New Mexico during the nonbreeding season. A major hurdle for species that migrate is that they are often exposed to a

PHOTO 37.7

Short-eared Owl at the Bosque del Apache National Wildlife Refuge, March 1994. During the last several decades, the species has declined over much of its range, most likely due to habitat loss.

PHOTOGRAPH: © GARY K. FROEHLICH.

greater suite of threats throughout their life cycle. Large contiguous tracts of land can reduce mortality from vehicle strikes and decrease the amount of urban edge where predators can be more abundant (Heske et al. 1999; Jokimaki and Huhta 2000). Although evidence exists that suitable wintering habitat ultimately may become breeding habitat if requirements during both seasons are met (e.g., foraging, nesting, roosting) (Wilson 1995; Wiggins 2004), such a scenario in New Mexico is possible but unlikely. Nonetheless, considerable value exists in expanding monitoring efforts of wintering and migrating Short-eared Owl populations to identify presence, and hopefully nesting activity, of Short-eared Owls during the breeding season.

[AOU] American Ornithologists' Union. 1998. *Check-list of North American birds*. 7th ed. Washington, DC: American Ornithologists' Union.

Avery, J. D., and G. S. Keller. 2006. Probable breeding of the short-eared owl (*Asio flammeus*) in eastern New Mexico. *New Mexico Ornithological Bulletin* 34:14–16.

Bailey, A. M., and R. J. Niedrach. 1965. *The birds of Colorado*. Denver: Denver Museum of Natural History.

BirdWest Archives. 2004. December 2004, week 4 (#7). http://listserv.arizona.edu/archives/birdwest.html.

Boyle, S. 1998. Short-eared Owl. In *Colorado breeding bird atlas*, ed. H. E. Kingery, 226–27. Denver: Colorado Bird Atlas Partnership and Colorado Division of Wildlife.

Clark, R. J. 1975. A field study of the Short-eared Owl, *Asio flammeus* (Pontoppidan) in North America. *Wildlife Monographs* 47:1–67.

Davis, W. B., and D. J. Schmidly. 1994. *The mammals of Texas*. Austin: University of Texas Press.

Heske, E. J., S. K. Robinson, and J. D. Brawn. 1999. Predator activity and predation on songbird nests on forest-field edges in east-central Illinois. *Landscape Ecology* 14:345–54.

Holt, D. W. 1985. The Short-Eared Owl in Massachusetts. *Cape Naturalist* 14:31–35.

———. 1992. Notes on Short-eared Owl (*Asio flammeus*) nest sites, reproduction, and territory sizes in coastal Massachusetts. *Canadian Field-Naturalist* 106:352–56.

Holt, D. W., and S. M. Leasure. 1993. Short-eared Owl (*Asio flammeus*). No. 62. In *The birds of North America*, ed. A. Poole and F. Gill. Philadelphia, PA: Academy of Natural Sciences, and Washington, DC: American Ornithologists' Union.

Holt, D. W., R. Berkley, C. Deppe, P. L. Enríquez-Rocha, J. L. Petersen, J. L. Rangel-Salazar, K. P. Segars, and K. L. Wood. 1999. Short-eared Owl (Strigidae). *In Handbook of the birds of the world*. Vol. 5, *Barn-owls to hummingbirds*, ed. J. del Hoyo, A. Elliott, and J. Sargatal, 241–42. Barcelona, Spain: Lynx Edicions.

Hubbard, J. P. 1978. *Revised check-list of the birds of New Mexico*. Publ. no. 6. Albuquerque: New Mexico Ornithological Society.

Johnsgard, P. A. 2002. *North American owls*. Washington, DC: Smithsonian Institution Press.

Jokimaki, J., and E. Huhta. 2000. Artificial nest predation and abundance of birds along an urban gradient. *Condor* 102:838–47.

Kingery, H. E., ed. 1998. *Colorado breeding bird atlas*. Denver: Colorado Bird Atlas Partnership.

Kuykendall, M. T. 2004. Effects of road intensity, corridors, and landscape composition on movement of small mammals of the Llano Estacado in Texas. M.S. thesis. Eastern New Mexico University, Portales.

Ligon, J. S. 1961. *New Mexico birds and where to find them*. Albuquerque: University of New Mexico Press.

Mikkola, H. 1983. *Owls of Europe*. Vermillion, SD: Buteo Books.

[NMOS] New Mexico Ornithological Society. 2008. *NMOS Field Notes* database. http://nhnm.unm.edu/partners/NMOS/ (accessed 17 April 2008).

Parmeter, J., B. Neville, and D. Emkalns. 2002. *New Mexico bird finding guide*. Albuquerque: New Mexico Ornithological Society.

Reinking, D. L., ed. 2004. *Oklahoma breeding bird atlas*. Norman: University of Oklahoma Press.

Sauer, J. R., J. E. Hines, and J. Fallon. 2007. *The North American breeding bird survey, results and analysis, 1966–2006*. Version 7.23.2007. Laurel, MD: USGS, Patuxent Wildlife Research Center.

Sharpe, R. S., W. R. Silcock, and J. G. Jorgensen. 2001. *Birds of Nebraska*. Lincoln: University of Nebraska Press.

Snyder, N. F. R., and J. W. Wiley. 1976. Sexual size dimorphism in hawks and owls of North America. *Ornithological Monographs* 20:1–96.

Tallman, D. A., D. L. Swanson, and J. S. Palmer. 2002. *Birds of South Dakota*. Aberdeen: South Dakota Ornithologists' Union.

Urner, C. A. 1923. Notes on the Short-eared Owl. *Auk* 40:30–36.

Weller, M. W., I. C. Adams Jr., and B. J. Rose. 1955. Winter roosts of marsh hawks and short-eared owls in central Missouri. *Wilson Bulletin* 67:189–93.

Wiggins, D. 2004. Short-eared Owl (*Asio flammeus*): a technical conservation assessment. USDA Forest Service, Rocky Mountain Region. http://www.fs.fed.us/r2/projects/scp/assessments/shortearedowl.pdf.

Wiggins, D. A., D. W. Holt, and S. M. Leasure. 2006. Short-eared Owl (*Asio flammeus*). *The Birds of North America online*, ed. A. Poole. Ithaca, NY: Cornell Laboratory of Ornithology. http://bna.birds.cornell.edu/BNA/account/Short-eared_Owl/.

Williams, S. O. III. 2000. The nesting season: New Mexico. *North American Birds* 54:409.

———. 2001. The nesting season: New Mexico. *North American Birds* 55:468.

Wilson, P. W. 1995. Short-eared owls nest unsuccessfully in northeast Oklahoma. *Bulletin Oklahoma Ornithological Society* 28:24–26.

38

Boreal Owl

(*Aegolius funereus*)

DALE W. STAHLECKER

THE HAUNTING, UNDULATING song of the Boreal Owl (*Aegolius funereus*) has most likely been a part of late winter and early spring nights in the high mountains of northern New Mexico for millennia. However, while paleontological (Howard 1931) and archeological (Emslie 1981) remains document its presence in the state long before the arrival of Europeans, photographic proof that it was still an extant New Mexico species was not acquired until 1987 (Stahlecker and Rawinski 1990). This may be because the Boreal Owl is a diminutive, secretive owl found almost exclusively in spruce-fir forests. Sibley (2000) reports the average Boreal Owl as only 25 cm (10 in) long with a wingspan of 53 cm (21 in) and a mass (weight) of 135 g (5 oz). Among North American owls, Boreal Owls exhibit the most pronounced reversed sexual dimorphism (Hayward and Hayward 1993), with little overlap in weight or body length between genders. The weight of Idaho males ranged from 93 to 139 g (3.3 to 4.9 oz), while Idaho females weighed between 132 g to 215 g (4.6 to 7.6 oz) (Hayward and Hayward 1991). Eurasian males were 21–25 cm (8.3–9.8 in) long with 55–58 cm (21.6–22.8 in) wingspans while females were 25–28 cm (9.8–11 in) long and had 59–62 cm (23–24 in) wingspans (Dement'ev and Gladkov 1954).

Boreal Owls are also cryptically colored, and day-roosting adults blend in with the surrounding branches and needles. The grayish head, sprinkled liberally with black, has a large, mostly white, facial disk with black borders that gives this owl a distinctly big-headed appearance (Sibley 2000). The eyes are yellow with black pupils. Overall coloration is a dark brown; the belly and chest are streaked white and brown, with white spots more prevalent along the flanks. The back is mostly brown, with white spots fairly evenly spaced across the wings and back. Branching or recently fledged juveniles are uniformly chocolate brown overall, fading to a lighter brown on the abdomen. They have contrasting white "eyebrows" and a thin white throat. At about two months old they begin a molt of body, but not flight, feathers that is completed by September (Hayward and Hayward 1993). When this molt is close to complete, the juveniles appear very adultlike, but the black lining of the facial disk is usually incomplete, the white of the facial disk is more dusky, and chocolate brown feathers can still dominate the head and chest (Rawinski et al. 1993; Stahlecker 1997).

All small owls appear similar in size and shape in the poor light in which they are often seen. The Boreal Owl might be most easily confused with its congener, the Northern Saw-whet Owl (*Aegolius acadicus*). The latter species is about 25% smaller (40% by mass)

PHOTO 38.1

(*top*) Adult Boreal Owl, Routt Co., Colorado, August 2005. PHOTOGRAPH: © BILL SCHMOKER.

PHOTO 38.2

(*bottom*) Post-fledging Boreal Owl near Creede, Colorado, 2 September 1992. Note the mix of juvenile/subadult feathers (This bird may have fledged unusually late.) PHOTOGRAPH: © ROBIN SELL.

and exhibits lighter rufous or brown streaking on the underparts, while the white spotting on the brownish back is mostly limited to "braces" on the wings. Its facial disk is mostly buffy, as is much of the head, and with limited black feathering. Juveniles have bright buffy underparts, a full triangle of white between the eyes instead of white eyebrows, and a paler brown overall appearance (Hayward and Hayward 1993; Sibley 2000; see chapter 39). No other small owls found in New Mexico have such large facial disks, and perched *Otus* and *Megascops* owls usually show their prominent "ears."

Primary owl calls serve the same function as song in songbirds, namely to proclaim ownership of a territory and to attract a mate. Most diagnostic for the Boreal Owl is the "staccato" song, which can be short (3–6 sec) and undulating (primary song) but oft-repeated, or prolonged but nonundulating (Bondrup-Nielson 1984). The primary song is a rapid series of trilling notes that rises in pitch initially and increases in volume before ending. It is not unlike the winnowing sound made by the wings of courting Wilson's Snipes (*Gallinago delicata*). Singing bouts frequently last 20 minutes and may even continue for several hours. Singing can start in January and continue to June, but is most intensive from February to mid April. Singing males are usually very close (10–100 m [33–330 ft]) to a potential nest cavity and apparently decrease singing rates when a mate is attracted (Hayward and Hayward 1993). The typical alarm call is a sharp "*skiew-a*." It is given by owls attracted to a person using tape-playback of the staccato song during a Boreal Owl survey (Stahlecker 1997), and should not be considered territorial in nature (Hayward and Hayward 1993).

Only one subspecies of Boreal Owl is recognized in North America (*A. f. richardsonii*). Six subspecies have been identified in the Old World, where the Boreal Owl is known as Tengmalm's Owl, named after the Swedish physician and naturalist Peter Gustaf Tengmalm (1754–1803), erroneously believed to have been the first to describe the species. The three subspecies of the contiguous northern boreal forest belt of Europe and Asia intergrade in appearance, becoming paler, grayer, less rufous, more spotted with white, and larger the farther eastward they are encountered (Dement'ev and Gladkov 1954). *A. f. richardsonii* is one of the darkest subspecies (Hayward and Hayward 1993).

Distribution

The Boreal Owl is circumpolar in distribution; it is mostly sedentary, but can be nomadic during lows in prey resulting in "irruptions" of winter sightings south of, and in more varied habitats than in, its breeding range (Hayward and Hayward 1993). In North America the breeding range of Boreal Owls is continuous across the boreal forests from Alaska to eastern Canada, extending southward into Minnesota, possibly into New England and Maine (Hayward and Hayward 1993). In the West, however, the Boreal Owl's breeding range dips southward within the spruce-fir forests of the Rocky Mountains as far south as northern New

PHOTO 38.3

Boreal Owl near Duluth, Minnesota, 18 February 2005.

PHOTOGRAPH: © KENT FIALA.

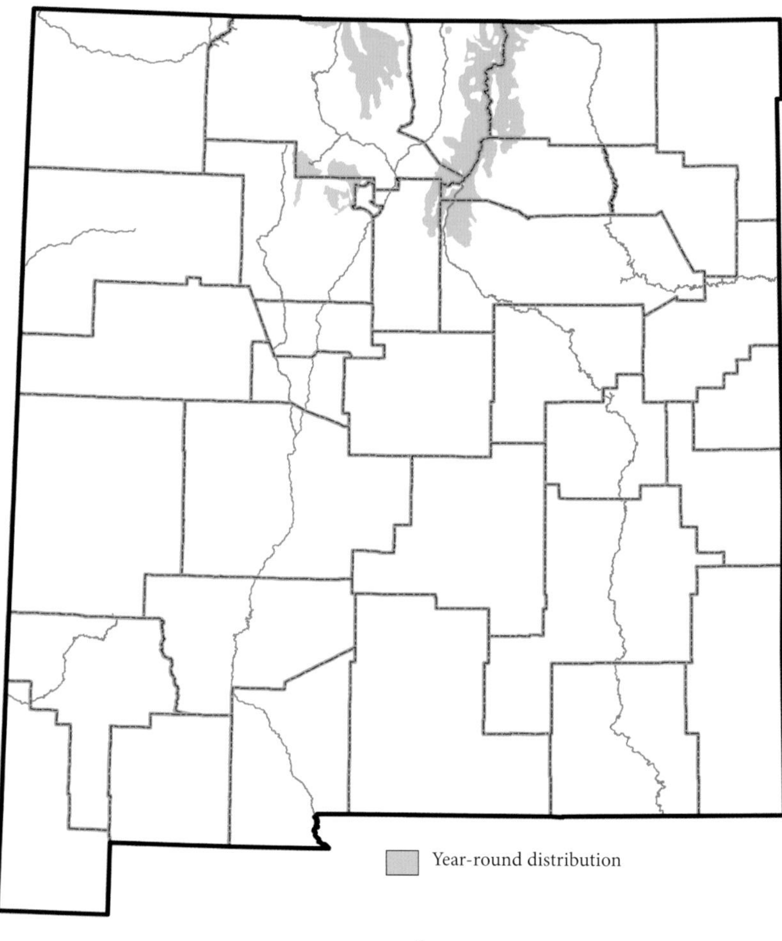

MAP 38.1

Boreal Owl distribution map

Mexico (Stahlecker and Rawinski 1990; Stahlecker and Duncan 1996).

In New Mexico, Boreal Owl bones were identified in Pleistocene (1.8 million to 11,550 years ago) remains from Shelter Cave, Doña Ana County (Howard 1931). Other Boreal Owl bones were excavated at Picuris Pueblo, Taos County, and dated to 1200–1350 AD (Emslie 1981). Thus it is likely that while the amount of spruce-fir habitat in New Mexico decreased and receded to higher elevations after the last ice age, Boreal Owls remained a hidden part of the New Mexico avifauna (Stahlecker and Duncan 1996).

It was through tape-playback surveys in the late 1980s and early 1990s that recent presence of the Boreal Owl was documented in New Mexico (Stahlecker and Rawinski 1990; Stahlecker and Duncan 1996). The first modern Boreal Owl record was one photographed just south of the Colorado Border near Cumbres Pass on 19 April 1987 (Stahlecker and Rawinski 1990; photo 38.4). By completion of surveys in 1993, at least 22 Boreal Owls had been documented on 19 separate occasions in eight montane locations, encompassing all of the Sangre de Cristo and San Juan Mountains and the northern portion of the Jemez Mountains. Additional surveys were conducted in spruce-fir habitat "islands" in the more isolated mountains of western and southern New Mexico and eastern Arizona, but no Boreal Owls were found (Stahlecker and Duncan 1996).

PHOTO 38.4

(*top*) Adult Boreal Owl looking straight down, 19 April 1987, Dixie Creek, Carson National Forest, Rio Arriba Co. The owl in this photograph represents the first modern record of the Boreal Owl in New Mexico. PHOTOGRAPH: © JOHN RAWINSKI.

PHOTO 38.5

(*bottom*) Adult Boreal Owl, Canjilon Mountain, Rio Arriba Co., 24 September 1988. PHOTOGRAPH: © TIM SMITH.

The most unusual New Mexico Boreal Owl record was of one photographed on 19 November 1989 by J. W. Garcia at 2,100 m (6,900 ft) on his property in Colfax County, 100 km (60 miles) east of suitable breeding habitat (Hubbard 1991). Mr. Garcia spent several hours admiring the bird, which perched quietly in a juniper (*Juniperus* sp.) tree. He sent a photograph to the New Mexico Department of Game and Fish for identification, or else the sighting would not have been recorded. The photo clearly showed that the owl was a young bird with an incomplete black lining of its facial disk (Stahlecker and Duncan 1996). Also unusual was the first and only New Mexico specimen (MSB #23745), a male found dead on 3 November 2000 by Hyde State Park employee L. Trujillo (Williams 2001); it was about 7.5 km (4.5 mi) southwest of the previous southernmost Boreal Owl record, a singing male heard on 13 March 1992 (Stahlecker and Duncan 1996).

Two other published records post-1993 have documented the continued presence of Boreal Owls in northern New Mexico. One was heard and briefly seen on 7 August 1998 south of Cumbres Pass (Williams 1999). Another was seen on Canjilon Mountain on 14 July 2000 (Williams 2000).

Habitat Associations and Life History

High-latitude spruce or high-elevation spruce-fir forests are always present in Boreal Owl breeding habitat (Hayward and Hayward 1993), and in New Mexico spruce-fir forests do not occur below 2,450 m (8,000 ft). All breeding habitat records for New Mexico

PHOTO 38.6

(*top left*) Occupied Boreal Owl habitat at ~3,200 m (~10,500 ft) on Canjilon Mountain, San Juan Mountains, September 1988. PHOTOGRAPH: © DALE W. STAHLECKER.

PHOTO 38.7

(*bottom left*) Adult Boreal Owl, Slumgullion Pass, Colorado, 29 September 1984. PHOTOGRAPH: © JOHN RAWINSKI.

PHOTO 38.8

(*top right*) Radio-tagged adult Boreal Owl triangulated to daytime roost, Cameron Pass, Colorado, March 1984. PHOTOGRAPH: © JOHN RAWINSKI.

PHOTO 38.9

(*bottom right*) Probable Boreal Owl habitat at 2,740 m (9,000 ft) at Laguna Larga, adjacent to Cruces Wilderness Area, San Juan Mountains, September 1988. PHOTOGRAPH: © DALE W. STAHLECKER.

PHOTO 38.10

Potential Boreal Owl habitat at ~3,050 m (~10,000 ft), Elk Mountain, southern Sangre de Cristo Mountains, July 1989. This area was heavily burned by the Viveash Fire in May/June 2000. PHOTOGRAPH: © DALE W. STAHLECKER.

PHOTOS 38.11a and b

No Boreal Owl nesting cavity has yet been located in New Mexico. Here, occupied nesting cavity and close-up of nesting bird along the Glenn Highway in Alaska, about 62°N and 147.5°W, 10 June 2006. PHOTOGRAPHS: © KENT FIALA.

Boreal Owls are actually above 2,800 m (9,190 ft), and all but one are above 3,000 m (9,840 ft) (Stahlecker and Duncan 1996). In the southern Rocky Mountains, including New Mexico, the diagnostic trees of this highest forest zone are Engelmann spruce (*Picea engelmannii*) and subalpine fir (*Abies lasiocarpa*). The greatest amount of spruce-fir forests is in the Sangre de Cristos (895 km^2 [346 mi^2]), followed by the San Juans (144 km^2 [56 mi^2]), and the Jemez Mountains (90 km^2 [35 mi^2]). Elsewhere in New Mexico, mountains either lack spruce-fir habitat or the habitat is too limited to support viable populations of Boreal Owls (Stahlecker and Duncan 1996).

In Idaho 79% of 914 prey items, and 95% of prey biomass, consisted of small mammals. Red-backed voles (*Clethrionomys gapperi*) represented 49% of all winter and 35% of all summer prey captured (Hayward et al. 1993). In northern Colorado 54% of 72 prey items were again red-backed voles and 25% were microtine (*Microtus* sp.) voles. Findley et al. (1975) documented the presence of both *Clethrionomys* and *Microtus* in mountains of New Mexico occupied by Boreal Owls.

Boreal owls nest in tree cavities throughout their range. In Idaho they foraged mostly in spruce-fir forests, but they most often nested in ponderosa pine (*Pinus ponderosa*) forest because cavity availability was greater (Hayward and Hayward 1993). The species also readily nests in nest boxes, and massive nest box programs have been successful in attracting significant numbers of nesting Boreal Owls (Mikkola 1983; Hayward et al. 1992). In Colorado, estimated egg-laying dates ranged from 17 April to 1 June (Palmer 1986). In Idaho, clutch size ranged from two to five, with a mean of 3.25 and 3.57 in two separate populations (Hayward and Hayward 1993; Hayward et al. 1993). The incubation of the eggs is conducted solely by the female (Korpimaki 1981). Incubation of the first laid egg lasts on average 29 days, while incubation of the other eggs is shorter by up to an average of 2.6 days (Korpimaki 1981). In Idaho, the first nestling to fledge left the nest 27 to 32 days after hatching (Hayward 1989). No active nest of this species has yet been found in New Mexico, but recently fledged young were documented in the San Juan Mountains on 19 August 1992 and in the Sangre de Cristo Mountains on 2 August 1993 using tape playback of the territorial staccato song (Stahlecker 1997).

In Idaho the loss of eggs or nestlings was frequently attributed to pine martens (*Martes americana*) or red squirrels (*Tamiasciurus hudsonicus*). Accipitrine hawks and larger owls are likely predators of free-flying Boreal Owls (Hayward and Hayward 1993).

Status and Management

Boreal Owls are almost certainly present year-round within their limited range in New Mexico, and have

PHOTO 38.12

Boreal Owl, Mineral Co., Colorado, July 2004. In New Mexico the Boreal Owl remains very poorly studied, but it appears for now to be secure within its small range.

PHOTOGRAPH: © BILL SCHMOKER.

been present there since the Pleistocene. They are seldom encountered because they are quiet most of the year and because as nocturnal hunters they are inactive during the day. Without tape-playback surveys, only two chance encounters would have documented their presence in the state.

The Boreal Owl was listed as Threatened (Group 2) by the New Mexico Department of Game and Fish in 1990, and remains so today. The species appears to be secure within its limited range in the state, particularly in the fairly extensive designated Wilderness Areas in the Sangre de Cristo Mountains. Because of its reclusive habits and high-altitude habitat, little work beyond distributional surveys has been done in New Mexico. The chief man-made threats would likely be the loss of habitat from either a catastrophic fire, possibly through human carelessness, or timber harvest, though regarding the later threat the Carson and Santa Fe National Forests have offered few timber sales, and none in spruce-fir habitat, in the 1990s and early 2000s.

LITERATURE CITED

Bondrup-Nielson, S. 1984. Vocalizations of the boreal owl, *Aegolius funereus richardsoni*, in North America. *Canadian Field-Naturalist* 98:191–97.

Dement'ev, G. P., and N. A. Gladkov. 1954. *Birds of the Soviet Union*. Moscow: State Publishing House.

Emslie, S. D. 1981. Birds and prehistoric agriculture: the New Mexico Pueblos. *Human Ecology* 9:305–29.

Findley, J. S., A. H. Harris, D. E. Wilson, and C. Jones. 1975. *Mammals of New Mexico*. Albuquerque: University of New Mexico Press.

Hayward, G. D. 1989. Habitat use and population biology of Boreal Owls in the northern Rocky Mountains, USA. Ph.D. diss. University of Idaho, Moscow.

Hayward, G. D., and P. H. Hayward. 1991. Body measurements of Boreal Owls in Idaho and a discriminate model to determine sex of live specimens. *Wilson Bulletin* 103:497–500.

———. 1993. Boreal Owl (*Aegolius funereus*). No. 63. In *The birds of North America*, ed. A. Poole and F. Gill. Philadelphia, PA: Academy of Natural Sciences, and Washington, DC: American Ornithologists' Union.

Hayward, G. D., P. H. Hayward, and E. O. Garton. 1993. Ecology of Boreal Owls in the Northern Rocky Mountains. *Wildlife Monograph*, no. 124.

Hayward, G. D., R. K. Steinhorst, and P. H. Hayward. 1992. Monitoring Boreal Owl populations with nest boxes: sample size and cost. *Journal of Wildlife Management* 56:776–84.

Howard, H. 1931. *Crypotoglaux funereal* in New Mexico. *Condor* 33:216.

Hubbard, J. P. 1991. Boreal Owl. *Handbook of species endangered in New Mexico*. F-321:1–2. Santa Fe: New Mexico Department of Game and Fish.

Korpimaki, E. 1981. On the ecology and biology of Tengmalm's Owl (Aegolius funereus) in southern Ostrobothnia and Suomenselka, western Finland. *Acta Universitatis Ouluensis A, Scientiae Rerum Naturalium 118, Biologica* 13:1–84.

Mikkola, H. 1983. *Owls of Europe*. Vermillion, SD: Buteo Books.

Palmer, D. A. 1986. Habitat selection, movements and activity of Boreal and Northern Saw-whet owls. Master's thesis. Colorado State University, Fort Collins.

Rawinski, J. J., R. Sell, P. Metzger, H. Kingery, and U. Kingery. 1993. Young Boreal Owls found in the San Juan Mountains, Colorado. *Colorado Field Ornithologist* 27:57–59.

Sibley, D. A. 2000. *The Sibley guide to the birds*. New York: Chanticleer Press.

Stahlecker, D. W. 1997. Using tape playback of the staccato song to document Boreal Owl reproduction. In *Biology and conservation of owls of the northern hemisphere*, ed. J. R. Duncan, D. H. Johnson, and T. H. Nicholls, 597–600. General Technical Report NC-190.7–2001. Washington, DC: USDA Forest Service.

Stahlecker, D. W., and R. B. Duncan. 1996. The boreal owl at the southern terminus of the Rocky Mountains: longtime resident or recent arrival? *Condor* 98:153–61.

Stahlecker, D. W., and J. R. Rawinski. 1990. First records of the Boreal Owl in New Mexico. *Condor* 92:517–19.

Williams, S. O. III. 1999. Changing seasons, fall 1998: New Mexico. *North American Birds* 53(1): 86–88.

———. 2000. Changing seasons, summer 2000: New Mexico. *North American Birds* 54(4): 408–11.

———. 2001. Changing seasons, fall 2000: New Mexico. *North American Birds* 55(1): 85–88.

© WAYNE LYNCH.

Northern Saw-whet Owl

(*Aegolius acadicus*)

JOHN P. DELONG

MICE BEWARE! It may be small, but the Northern Saw-whet Owl (*Aegolius acadicus*) uses its highly asymmetrical ears to pinpoint mice and voles by sound in near total darkness (Bailey 1928; Johnsgard 1988; Cannings 1993). Northern Saw-whet Owls are not often seen, but they are regular in New Mexico in woodland areas. They are easy to miss because they roost in dense vegetation, are small and active only at night, and can assume a "hiding" posture when disturbed (Cannings 1993). Females tend to be larger and heavier than males by about 25% (Cannings 1993). On average, Northern Saw-whet Owls are about 20 cm (8 in) in length, weigh about 85 g (3 oz), and have wingspans of approximately 51 cm (20 in) (Bailey 1928; Cannings 1993). Thick feathers extend down over their feet, and due also to their relatively short legs, Northern Saw-whet Owls take on a bit of a legless look.

Adult and juvenile Northern Saw-whet Owls differ in appearance, but males and females have similar

PHOTO 39.1

Northern Saw-whet Owl in hand, Manzano Mountains, fall 2001. This small, nocturnal owl is easy to miss, especially as it tends to hide in dense vegetation. Note the brown upperparts. PHOTOGRAPH: © JOHN P. DELONG.

PHOTO 39.2

Northern Saw-whet Owl, Manzano Mountains, fall 2001. The underparts are pale with brown streaking. The edges of the facial disk are brown with white streaking. The irises are yellow. PHOTOGRAPH: © JOHN P. DELONG.

plumages. Adults are brown above with brown streaks on a pale underside. They have bold yellow eyes and a prominent white facial disc that darkens to brown at the edges. Around the edges of the facial disc and across the top of the head, small white streaks frame the owl's face. Through August and sometimes September, juveniles have cinnamon-red underparts that are gradually supplanted by the adult brown streaking. The closest relative of the Northern Saw-whet Owl in New Mexico is the Boreal Owl (*Aegolius funereus*), which is distinguished from the Northern Saw-whet Owl by its larger size, black-rimmed facial disc, and white spotting, rather than streaking, along the edge of the disc and across the head. In New Mexico the distributions of these congeners overlap only in the northern mountains (Stahlecker and Duncan 1996; chapter 38).

Distribution

Northern Saw-whet Owls have a wide geographic range, occurring during the breeding season in most forested habitats across Canada, in the northeastern and mountainous western United States, and in the western mountains of Mexico (Johnsgard 1988; Cannings 1993). They have been recorded in most of the mountain ranges in New Mexico during the breeding season, including the Animas, Jemez, Magdalena, Manzano, Mogollon, Pinos Altos, Sacramento, Sandia, Sangre de Cristo, San Juan, Tularosa, and Zuni mountains (Bailey 1928; Ligon 1961; Parmeter et al. 2002; NMOS 2007). Although typically more common in the northern part of their range (Bailey 1928; Cannings 1993), in New Mexico they are more likely to be encountered in the state's southern mountains than in the northern mountains (Stahlecker and Duncan 1996). Overall, they are rare in both the summer and winter in New Mexico. However, they show a greater elevational spread in the winter, occurring in typical montane breeding habitats but also moving to lower elevations where suitable tree cover and prey are available (Bailey 1928; NMOS 2007; L. Paras, pers. comm.).

Northern Saw-whet Owls are irregular, partial migrants (see chapter 2 for more details regarding raptor migration in New Mexico). They clearly show typical southerly movements in fall and northerly movements in spring. However, not all individuals migrate. Many remain on or occupy breeding habitats in the winter or simply move downslope within the same mountain range. Further, the portion of birds migrating and/or the destination of migrants vary annually. During occasional irruptions of Northern Saw-whet Owls in northern and eastern North America there is as much as a tenfold increase in captures of migrants compared to nonirruption years (Weir et al. 1980; Whalen and Watts 2002; Stock et al. 2006). Despite regular efforts at capturing migrating owls in the Manzano Mountains, no significant pulses of Northern Saw-whet Owls have been detected (HawkWatch International unpubl. data). However, trappers in the Manzanos have broadcast primarily the calls of Flammulated Owls (*Otus flammeolus*) to lure owls into trapping stations. Although Northern Saw-whet Owls respond to these vocalizations, focusing on Flammulated Owls may have caused this study to miss irruptions of Northern Saw-whet Owls.

Irruptions are the apparent result of widespread but synchronized successful breeding. First, high prey availability is likely responsible for high reproductive success, but once there are large numbers of juvenile owls, the availability of prey per owl becomes relatively low, and many owls choose to migrate. In Quebec, for example, the number of migrating saw-whet owls was positively rather than negatively correlated with

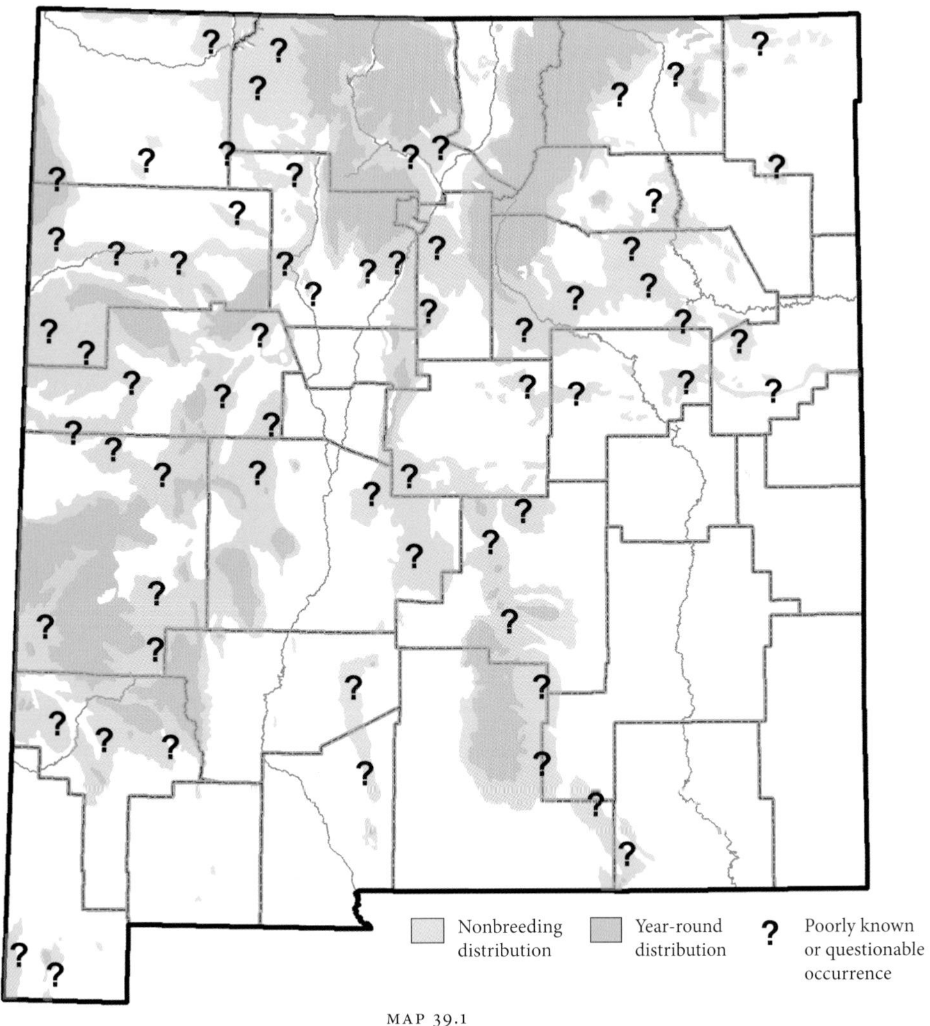

MAP 39.1

Northern Saw-whet Owl distribution map

the abundance of late summer small rodents (Côté et al. 2007). As additional evidence that irruptions are density-dependent (i.e., more likely when the density of owls is high), captured owls are in poorer physical condition (based on a measure similar to body mass index in humans) in irruption years than in nonirruption years (Whalen and Watts 2002; Stock et al. 2006). As the more experienced foragers, adults are likely to be less susceptible to food shortage than juveniles and therefore less likely to migrate. Indeed, most irruptive migrants are juveniles (Cannings 1993; Whalen and Watts 2002; Côté et al. 2007), although in Idaho most of the captured irruptive migrants were adults (Stock et al. 2006).

Fall migration generally occurs from September through November, with peak movements apparently later in October and early November (Weir et al. 1980; Duffy and Kerlinger 1992; Cannings 1993). In the West, however, Northern Saw-whet Owl migration peaks in mid October (DeLong 2003; Stock et al. 2006). As with many migratory birds during the fall, adults pass through a particular area later, on average, than juveniles. Although no significant age and sex differences in fall migration timing were found with Northern Saw-whet Owls captured in the Manzanos, adults migrated slightly later than juveniles in New Jersey and Nevada (Duffy and Kerlinger 1992; chapter 2). This pattern is magnified during irruption years (Weir et al. 1980; Stock et al. 2006). Additionally, Northern Saw-whet Owl females precede males during the fall, but males precede females during the spring (Weir et al. 1980; Carpenter 1993). Thus males spend more time on the breeding grounds than females, a pattern explained by the fact that males are primarily responsible for maintaining territorial boundaries (Cannings 1993).

Habitat Associations

Northern Saw-whet Owls are forest-dwelling owls. Nesting habitat includes all types of forests in their range (spruce-fir, pine, and deciduous forests), but

PHOTO 39.3

Roosting bird on a ponderosa pine (*Pinus ponderosa*) branch, Fillmore Canyon, Organ Mountains, Doña Ana Co., 31 December 2007.

PHOTOGRAPH: © DAVID J. GRIFFIN.

PHOTO 39.4

Adult at a nest box opening along Galisteo Creek, Santa Fe Co. about 6.5 km (4 mi) east of Cerrillos at an elevation of 1,770 m (5,810 ft), 2 March 2008. The saw-whet box was placed 3.5 m (11.5 ft) up in a dead Russian olive (*Elaeagnus angustifolia*) at the edge of the Galisteo Creek bosque.

PHOTOGRAPH: © LAWRY SAGER.

Northern Saw-whet Owls reach their highest densities in coniferous forests (Cannings 1993). In the mountains of western North America, ponderosa pine (*Pinus ponderosa*) and mixed conifer communities appear to be favored, but riparian woodlands are also used, at least in northern areas such as Idaho (Cannings 1993) but also in New Mexico, where successful nesting was recorded in 2008 along Galisteo Creek near Cerrillos, Santa Fe County, at an elevation of 1,770 m (5,800 ft) (photo 39.4). The pair recorded along Galisteo Creek occupied a nest box installed by Lawry Sager (pers. comm.) amidst dense riparian vegetation consisting of live and dead cottonwoods (*Populus deltoides* spp. *wislizenii*), saltcedar (*Tamarix* sp.), coyote willows (*Salix exigua*), and New Mexico olive (*Forestiera pubescens*). Dominant in the understory was saltcedar, which formed thickets and was used extensively by the pair (L. Sager, pers. comm.).

In New Mexico, Northern Saw-whet Owls share some of their montane coniferous forest habitat with Northern Pygmy-Owls (*Glaucidium gnoma*) and Boreal Owls, but do not range as high in elevation as these two species (Cannings 1993). Northern Saw-whet Owls show a preference for mature forest stands (Cannings 1993).

In winter, Northern Saw-whet Owls may occupy the same montane forests used in summer, but they extend their habitat use to include more southerly and lower-elevation woodlands that provide roosting cover, hunting perches, and prey (Johnsgard 1988; Cannings 1993). For example, four saw-whets were found injured between November 2007 and January 2008 in pinyon-juniper habitats in the Santa Fe–White Rock area (L. Paras, pers. comm.)

Life History

Nesting

Natural history information from New Mexico is scarce for the Northern Saw-whet Owl. Elsewhere in the species' range, breeding begins as early as early March or even the last few days of February, with nesting finished by late July (Cannings 1993). Along Galisteo Creek in 2008, the nesting pair was first observed occupying a nest box on 2 March. The first young fledged sometime between 11 May and 21 May, when it was observed outside of the nest box. Three additional young had all fledged by 24 May (L. Sager pers. comm.).

Nests are typically in old Northern Flicker (*Colaptes auratus*) cavities, but they also can be in other woodpecker cavities or, as mentioned above, in nest boxes (Bailey 1928; Johnsgard 1988; Cannings 1993; Marks and Doremus 2000). No nest is built; rather, eggs are laid on whatever former nest or woody debris is present (Cannings 1993). Northern Saw-whet Owls do not reuse nests in successive years, and if replacing lost early clutches, will even move to new nest cavities (Cannings 1993). Females lay five to six eggs, rarely four or seven (Cannings 1993).

PHOTOS 39.5a, b, c, and d

The pair that nested along Galisteo Creek in 2008 successfully raised four young. (a) nestling in nest box, approaching fledging, 11 May; (b, c, d) three fledglings in saltcedar (*Tamarix* sp.) vegetation, 24 May. PHOTOGRAPHS: © LAWRY SAGER.

PHOTO 39.6

Northern Saw-whet Owl nestlings about two to three days before fledging. The nest was found by Peter Stacey and David Arsenault on their Flammulated Owl (*Otus flammeolus*) study site in the Zuni Mountains, in pine-oak habitat at about 2,130 m (7,000 ft). The nest cavity was an old woodpecker hole in an oak (*Quercus* sp.) tree, about 6 m (20 ft) above the ground. Note that the bird on the right at that time was significantly smaller and less developed than the other two nestlings. All three nestlings were placed back in the nest cavity, where they still were the day after the photo was taken.

It appears that Northern Saw-whet Owls may be nomadic in parts of their range. They colonized the Snake River Birds of Prey National Conservation Area, Idaho, four years after nest boxes were installed at that location (Marks and Doremus 2000). Marks and Doremus suggested that wintering birds elected to stay and breed only if mice densities were high, and traveled elsewhere if prey was insufficient. Support for this idea was threefold. First, the number of breeding pairs in the area varied substantially from year to year. Second, the number of breeding pairs was highest in years of high mice availability. And finally, breeding birds in the area showed little site fidelity; few owls banded there returned to breed again. Thus it appears that Northern Saw-whet Owls are indeed nomadic in that they opportunistically nest where they find suitable nest cavities and abundant prey, and then move on when conditions are no longer favorable. Because Galisteo Creek does not represent typical Northern Saw-whet Owl nesting habitat, the 2008 nesting record might be a case of opportunistic nesting at a time of unusually abundant rodent densities.

Diet and Foraging

Throughout their distribution, Northern Saw-whet Owls tend to specialize on small mammals (Cannings 1993). A wide variety of small mammals have been recorded as Northern Saw-whet Owl prey; however, *Peromyscus* mice and *Microtus* voles make up a substantial

PHOTO 39.7

Roosting bird with partially eaten brush mouse (*Peromyscus boylei*), Fillmore Canyon, Organ Mountains, Doña Ana Co., 31 December 2007.

PHOTOGRAPH: © DAVID J. GRIFFIN.

PHOTO 39.8

Regurgitated pellets found below a roost site, Fillmore Canyon, Organ Mountains, Doña Ana Co., 6 December 2008. The pellets contained remains from at least six *Peromyscus* mice. PHOTOGRAPH: © DAVID J. GRIFFIN.

PHOTO 39.9

Fledgling with unidentified mouse along Galisteo Creek, May 2008. PHOTOGRAPH: © LAWRY SAGER.

portion of the diet (Johnsgard 1988; Cannings 1993). Of 6,507 identified prey items from throughout the owl's range, 52% were *Peromyscus* mice and 24% were *Microtus* voles (Cannings 1993). Prey selected varied substantially by habitat type and region, apparently reflecting what prey was most available (Cannings 1993). Occasionally insects, small to medium-sized birds, and medium-sized mammals also may be taken (Johnsgard 1988; Cannings 1993). Low branches, fence posts, or other perches are used to watch for ground-dwelling prey, usually near habitat edges (Cannings 1993). Very little diet and foraging information has been collected specifically in New Mexico, but recently a Northern Saw-whet Owl was documented taking brush mouse (*Peromyscus boylei*) in Fillmore Canyon, Doña Ana County (David J. Griffin, pers. comm.; see photo 39.7).

Predation and Interspecific Interactions

In New Mexico, Northern Saw-whet Owls share their forest habitats with Northern Pygmy-Owls, Western Screech-Owls (*Megascops kennicottii*), Flammulated Owls, Long-eared Owls (*Asio otus*), Great Horned Owls (*Bubo virginianus*), Spotted Owls (*Strix occidentalis*), and Boreal Owls. Elsewhere in the species' range, Long-eared Owls have predated Northern Saw-whet Owls, and aggressive attacks from Boreal Owls and Northern Pygmy-Owls have also been witnessed (Cannings 1993).

With the exception of the insect specialist Flammulated Owl, all of the above sympatric forest owls also prey on the mice and voles upon which the Northern Saw-whet Owl depends (Johnsgard 1988). Thus the Northern Saw-whet Owl may face considerable competition for food. Some of that competition may be relieved by habitat selection, with Boreal Owls tending to select higher-elevation forests in the northern part of the state, Spotted Owls often being confined to more mature mixed conifer stands, and Western Screech-Owls tending to occur in lower-elevation riparian or pinyon-juniper stands in the state. In addition, niche differentiation exists with diet, with the larger species such as Long-eared Owl, Great Horned Owl, and Spotted Owl taking a much wider size-range of prey and not specializing on the smaller mice and voles. Northern Pygmy-Owls and Boreal Owls also take a much broader variety of birds than do Northern Saw-whet Owls. Given that predator size is related to prey size selection, it is likely that Northern Saw-whet Owls compete most fiercely with other small owls. In comparison with the other small owls, Northern Saw-whet Owls take the fewest insects and are the most specialized on small mice and voles (Johnsgard 1988). This specialization probably allows some coexistence with the other more generalist owls, and may also be the foundation for some of the species' irruptive and nomadic behavior. With Northern Saw-whet Owls being more specialized, movement to take advantage of temporally and spatially variable prey would seem like an adaptive strategy that the other more flexible owls may not find worthwhile.

Status and Management

Little information on population status is available for Northern Saw-whet Owls (Johnsgard 1988). Long-term migration banding efforts are not likely to produce reliable population indices because of changing capture techniques, differing effort through time, and the variability in trapping success due to weather, irruptions, and nomadism. Further, USGS Breeding Bird Survey data are too limited for Northern Saw-whet Owls to allow analysis (Sauer et al. 2007). Given that the saw-whet is nomadic, the species may be flexible enough to occupy habitats that can support populations only temporarily. However, its dependence on tree cavities for nesting and a distinct preference for mature stands suggest that habitat loss due to forest harvest is probably a real threat to populations in many areas (Cannings 1993). In addition, it may be susceptible to collisions with automobiles, especially in winter when it may occupy lower-elevation habitats near highways. Four out of four saw-whets sent to a rehabilitation facility in Santa Fe from November 2007 to January 2008 had head traumas resulting from collisions with automobiles (L. Paras, pers. comm.). Because of these threats, and the absence of any effective monitoring of the species, some effort to gauge the status of the Northern Saw-whet Owl in New Mexico is needed.

LITERATURE CITED

Bailey, F. M. 1928. *Birds of New Mexico*. Santa Fe: New Mexico Department of Game and Fish.

Cannings, R. J. 1993. Northern Saw-whet Owl (*Aegolius acadicus*). No. 42. In *The birds of North America*, ed. A. Poole and F. Gill. Philadelphia, PA: Academy of Natural Sciences, and Washington, DC: American Ornithologists' Union.

Carpenter, T. W. 1993. Temporal differences in size of Northern Saw-whet Owls during spring migration. *Wilson Bulletin* 105:356–59.

Côté, M., J. Ibarzabal, M.-H. St.-Laurent, J. Ferron, and R. Gagnon. 2007. Age-dependent response of migrant and resident *Aegolius* owl species to small rodent population fluctuations in the eastern Canadian boreal forest. *Journal of Raptor Research* 41:16–25.

DeLong, J. P. 2003. *Flammulated Owl migration project: Manzano Mountains, New Mexico—2002 report*. Salt Lake City, UT: HawkWatch International, Inc.

Duffy, K., and P. Kerlinger. 1992. Autumn owl migration at Cape May Point, New Jersey. *Wilson Bulletin* 104:312–20.

Johnsgard, P. A. 1988. *North American owls—biology and natural history*. Washington, DC: Smithsonian Institution Press.

Ligon, J. S. 1961. *New Mexico birds and where to find them*. Albuquerque: University of New Mexico Press.

Marks, J. S., and J. H. Doremus. 2000. Are Northern Saw-whet Owls nomadic? *Journal of Raptor Research* 34:299–304.

[NMOS] New Mexico Ornithological Society. 2007. *NMOS Field Notes* database http://nhnm.unm.edu/partners/NMOS/ (accessed 5 December 2007).

Parmeter, J., B. Neville, and D. Emkalns. 2002. *New Mexico bird finding guide*. 3rd ed. Albuquerque: New Mexico Ornithological Society.

Sauer, J. R., J. E. Hines, and J. Fallon. 2007. *The North American breeding bird survey, results and analysis 1966–2006*. Version 10.13.2007. Laurel, MD: USGS, Patuxent Wildlife Research Center. http://www.pwrc.usgs.gov/ (accessed 6 December 2007).

Stahlecker, D. W., and R. B. Duncan. 1996. The Boreal Owl at the southern terminus of the Rocky Mountains: long-time resident or recent arrival? *Condor* 98:153–61.

Stock, S. L., P. J. Heglund, G. S. Kaltenecker, J. D. Carlisle, and L. Leppert. 2006. Comparative ecology of the Flammulated Owl and Northern Saw-whet Owl during fall migration. *Journal of Raptor Research* 40:120–29.

Whalen, D. M., and B. D. Watts. 2002. Annual migration density and stopover patterns of Northern Saw-whet Owls (*Aegolius acadicus*). *Auk* 119:1154–61.

Weir, R. D., F. Cooke, M. H. Edwards, and R. B. Stewart. 1980. Fall migration of Saw-whet Owls at Prince Edward Point, Ontario. *Wilson Bulletin* 92:475–88.

Casual and Accidental Raptors

JEAN-LUC E. CARTRON

BESIDES THE 37 species of vultures, hawks, falcons, and owls already presented in this volume, seven additional raptor species have been conclusively recorded in New Mexico. All may be considered vagrant or stray species occurring in New Mexico outside of their normal range. They consist of the Black Vulture (*Coragyps atratus*), Swallow-tailed Kite (*Elanoides forficatus*), Red-shouldered Hawk (*Buteo lineatus*), Short-tailed Hawk (*Buteo brachyurus*), Crested Caracara (*Caracara cheriway*), Eastern Screech-Owl (*Megascops asio*), and Barred Owl (*Strix varia*). Among those seven species, the Swallow-tailed Kite and the Red-shouldered Hawk in particular are in New Mexico far outside their normal range. Others, including the Short-tailed Hawk, have been expanding their distribution recently, and in future years may increase in both numbers and frequency of occurrence in New Mexico. Two additional raptors, the White-tailed Hawk (*Buteo albicaudatus*) and the Gyrfalcon (*Falco rusticolus*), have been reported from New Mexico. They are not described in this chapter, as both remain unverified in the state. Hubbard (1978:15) considered the White-tailed Hawk hypothetical though probable in New Mexico, with one report from Torrance County in April 1971 "especially convincing in its details and in regard to the observer's experience with the species." A small number of White-tailed Hawks have been reported from southern New Mexico since Hubbard (1978), including one bird near Cass Draw southeast of Otis, Eddy County, 22 April 2006. The observer, Steve West (pers. comm.), "had good looks [of that bird] for quite a while" and had no doubts about the species identification, having seen White-tailed Hawks in Texas as well as in numerous places in Latin America.

Black Vulture
(*Coragyps atratus*)

The Black Vulture is a large raptor with a naked gray head, an almost entirely black plumage, and white legs nearly reaching the tip of the short tail in flight. On the underside of the wings is a whitish patch in the primaries. The Black Vulture soars with a less pronounced dihedral than its relative the Turkey Vulture (*Cathartes aura*) and also alternates short glides with quick, labored wing flaps. The sexes are identical.

The Black Vulture ranges year-round throughout South and Central America. Northward it occurs throughout most of Mexico, and in the United States is found both in south-central Arizona and from southwestern and central Texas east across much of the

PHOTO 40.1

Black Vulture near Tucson, Arizona, December 2005. PHOTOGRAPH: © MARK L. WATSON.

PHOTO 40.2

Black Vulture roost in Douglas, Cochise Co., Arizona, 12 December 2008. Black Vultures have recently expanded their range eastward in Arizona, and a roost was noted during the winters of 2007–8 and 2008–9 at Douglas along the international border with Mexico. The roost is in a pine tree. The vultures (70 Black Vultures on 12 December 2008) feed across the border presumably in a garbage dump in Agua Prieta, Sonora. Black Vultures in Douglas are only about 50 km (30 mi) from the New Mexico state line. PHOTOGRAPH: © RICHARD E. WEBSTER.

South, north again along the eastern coast to New York and Pennsylvania (Buckley 1999; Corman 2005a). The Black Vulture is essentially nonmigratory although some birds near the northern limit of the species' distribution withdraw southward for the winter (Buckley 1999). The Black Vulture expanded its distribution northward beginning in the 1920s in northern Mexico and Arizona and in the 1940s along the East Coast (Greider and Wagner 1960; Coleman and Fraser 1989, 1990). The species was first reported from Arizona in 1920 (Phillips et al. 1964), with numbers and frequency of occurrence subsequently increasing steadily in that state. Breeding was first recorded in 1967 in Arizona (Monson and Phillips 1981). Although Black Vultures can often be observed soaring in large numbers in Arizona, only four active nest sites have been discovered since 1967 (Rea 1998; Corman 2005a). During preparation of the *Arizona Breeding Bird Atlas*, no possible Black Vulture nest site was recorded in the eastern third of Arizona, nearest New Mexico (Corman 2005a).

In Texas, Black Vulture numbers greatly increased beginning in the 1920s (Parmalee 1954; Buckley 1999). Although Black Vulture numbers declined after 1950

(Oberholser 1974), the species is now considered common to abundant across the eastern two-thirds of Texas. In the Trans-Pecos region—nearest New Mexico—the Black Vulture is locally uncommon, reaching the northern limit of its distribution from Val Verde County west to northern Presidio County, approximately 150 km (90 mi) south of the New Mexico state line (Lockwood and Freeman 2004). Outside the species' normal range, there have been sightings farther to the north and west, in counties adjacent to New Mexico (Oberholser 1974). Black Vultures are also found year-round near Janos in northwestern Chihuahua, Mexico, occurring both along the Rio San Pedro and in small, local towns (Manzano-Fischer et al. 2006; JLEC, pers. obs.). The 2008 Janos Christmas Bird Count recorded 150 Black Vultures (J. Parmeter, pers. comm., *fide* Michael Hilchey). In Janos, Black Vultures are within 70 km (45 mi) of the U.S. border and New Mexico.

Given the proximity of Black Vulture populations to New Mexico and the northward expansion of the species elsewhere, it is perhaps odd that only one record of occurrence exists from New Mexico. A Black Vulture was photographed 21 July 1996 on a power pole north of Rodeo, Hidalgo County, by Jon Dunn and Will Russell (Williams 1996; Parmeter et al. 2002). Clearly, however, the Black Vulture appears poised to make further appearances in New Mexico. Making this even more likely is the eastward range expansion noted in recent years in Arizona (Stevenson and Rosenberg 2008). A Black Vulture roost has been observed during the last two winters (2007–8 and 2008–9) in Douglas, Cochise County, only about 50 km (30 mi) from the New Mexico state line along the international border. During the winter 2007–8, up to 15 Black Vultures were recorded at the Douglas roost (Stevenson and Rosenberg 2008). Even more birds were observed the following winter, with Richard Webster (pers. comm.) counting as many as 70 Black Vultures on 12 December 2008. Even closer to New Mexico than Douglas, Black Vultures are also reported—though only irregularly—at San Bernardino National Wildlife Refuge, about halfway between Douglas and Guadalupe Canyon along the Mexican border (Stevenson and Rosenberg 2005; R. Webster, pers. comm.).

Swallow-tailed Kite (*Elanoides forficatus*)

The Swallow-tailed Kite is a large and conspicuous black and white kite with a long, deeply forked tail. Males and females are similar. The head of an adult is white with red eyes, and somewhat reminiscent of a dove rather than a raptor. Below, the white body and wing coverts contrast with the black flight feathers and tail. Above, the back, wings, and tail are black. Juveniles are like adults but have brown eyes and shorter tails. The species hunts and consumes food on the wing (Meyer 1995).

Two subspecies of Swallow-tailed Kites are recognized. They have disjunct breeding distributions. The northern subspecies is the nominate *forficatus*, which breeds only in the southeastern United States and is migratory, wintering in South America (Meyer 1995). The more southerly *yetapa* breeds from southern Mexico south through much of Central America and the northern two-thirds of South America. Northern populations of this second subspecies are also migratory (Meyer 1995). In the southeastern United States, breeding populations of the nominate subspecies

PHOTO 40.3

Swallow-tailed Kite, La Gamba, Costa Rica, 19 March 2009. PHOTOGRAPH: © JERRY OLDENETTEL.

occur mainly in Florida but with also some smaller, disjunct breeding areas north to South Carolina along the Atlantic Coast and west to extreme eastern Texas along the Gulf Coast (Meyer 1995).

The breeding distribution of the Swallow-tailed Kite in the United States was once much more extensive and continuous, reaching as far north as Minnesota along the Mississippi River, and including or likely including Arkansas, Oklahoma, Kentucky, Missouri, Kansas, Illinois, Iowa, and Wisconsin (Cely 1979; Meyer 1995). It also extended farther west into Texas (Strecker 1912). The dramatic range contraction observed in the United States occurred between 1880 and 1940 (Cely 1979). According to Robertson (1988), the most likely cause of the decline was human persecution, but agricultural development and logging of bottomland forests may also have been responsible (see Meyer 1995).

Post-breeding, Swallow-tailed Kites form premigration communal roosts (Meyer 1995). In the United States, departure from communal roosts begins in late July, with no bird seen after mid September. Swallow-tailed Kites typically return from migration beginning in the third week of February. Migration is thought to occur both over land, around the Gulf Coast, and over water through the Caribbean (see Meyer 1995 and references therein). Upon departure for migration in late summer and early fall, some birds may be observed far to the west and north of the current breeding distribution, including—casually—in New Mexico.

The first record from New Mexico is a specimen secured near Taos in August 1859 and sent to the U.S. National Museum (Bailey 1928; Hubbard 1978). The second record was a sighting in the Capitan Mountains in July 1903, and the third consisted of another specimen secured in the Carlsbad area around 1907 (Bailey 1928; Hubbard 1978). Parmeter et al. (2002) report two more recent late summer/early fall records from Socorro, 29 August to 7 September 1982 and 6–18 September 1993. On 6 August 2000, one Swallow-tailed Kite was photographed in Sunland Park, Doña Ana County, by James N. Paton of El Paso, Texas (Williams 2001). On 28 August 2006, Gregory Keller (pers. comm.) and students observed one Swallow-tailed Kite close together with three Mississippi Kites (*Ictinia mississippiensis*) in Portales, Roosevelt County. The four kites remained in view for less than one minute. They were flying low (less than 15 m [50 ft]) and although they circled overhead a few times, were moving in a general westerly direction (G. Keller, pers. comm.; Williams 2007b).

Red-shouldered Hawk
(*Buteo lineatus*)

The Red-shouldered Hawk is a medium-sized buteo with five recognized subspecies that exhibit noticeable differences in plumage (Crocoll 1994; Wheeler and Clark 1995). Overall, adults are readily identified by their red shoulder patches; their flight feathers, checkered black and white; rufous underwing coverts; a white crescent across the primaries, best seen on the upperwings; a pale belly barred with rufous; and a tail with alternating black and white bars. The breast is bright, solid rufous in the California Red-shouldered Hawk (*Buteo lineatus elegans*). The head and back are gray instead of brown in the race found in southern Florida (*Buteo lineatus extimus*) (Wheeler and Clark 1995). Juveniles of the California race are largely similar to the adults. In the other four subspecies, juveniles have streaked underparts, brown flight feathers, and white supercilia—when perched, juveniles can be mistaken for juvenile Northern Goshawks (*Accipiter gentilis*), but the wings reach farther toward the tip of

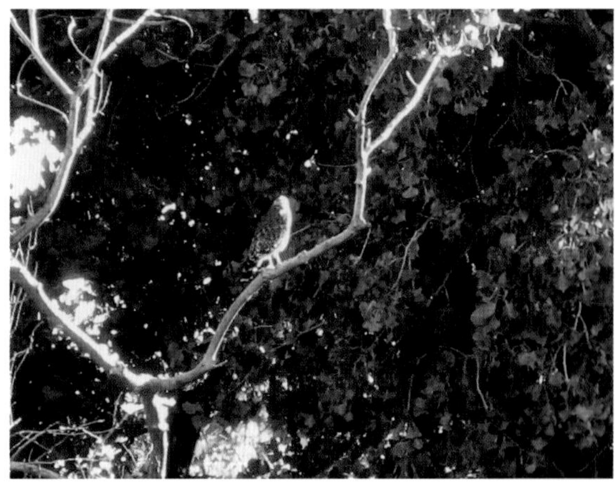

PHOTO 40.4

Red-shouldered Hawk at Rattlesnake Springs, Eddy Co., 22 September 2006. PHOTOGRAPH: © JONATHAN BATKIN.

the tail, and the pale bands on the dark tail are narrow (Clark and Wheeler 1995). Juveniles also show a pale, translucent crescent across the primaries.

The species is found mainly in woodlands throughout its distribution. It breeds in two disjunct geographic areas of North America. In the east, the species breeds from southern New Brunswick and western Ontario south between the eastern edge of the Great Plains and the Atlantic coast to Florida and the Gulf Coast, reaching eastern Texas and eastern Mexico. In the west, the Red-shouldered Hawk is represented by the California race, which breeds from southwestern Oregon and northern California south to northern Baja California (Crocoll 1994). Red-shouldered Hawks appear to be migratory only in the northern half of their breeding range, with the wintering distribution reaching farther south in Mexico. Eastern birds may stray west in migration, while birds of the California race can wander far to the east (Crocoll 1994).

Red-shouldered Hawks have been observed in New Mexico mainly in the south in spring and late summer and early fall, with also a few winter records (Parmeter et al. 2002; NMOS 2008). Accepted records include a specimen housed at the University of New Mexico's Museum of Southwestern Biology (MSB #8174). It is an immature bird of the subspecies *texanus*, found injured in Albuquerque along a canal bank on 28 November 1991 and later euthanized. Among the other accepted records is one bird found from 29 August to 6 September 1993 at Percha Dam State Park and one bird at Lovington, Lea County, 13 September 1996 (Parmeter et al. 2002; NMOS 2008). A California Red-shouldered Hawk wintering at the Bosque del Apache National Wildlife Refuge was photographed on 30 December 1999 by Jerry Oldenettel and Jon Dunn, with additional sightings including on 18 January 2000 by Jerry Oldenettel (pers. comm.) and Louis Cuellar. The most recent records include another wintering California Red-shouldered Hawk near Turn, Valencia County, first reported by Christopher Rustay on 17 February 2002 (Williams 2002a) and photographed by Jerry Oldenettel on 17 March 2002 (Williams 2002b); one bird photographed by Jonathan Batkin on 22 September 2006 at Rattlesnake Springs in Eddy County (Williams 2007a); one adult, seemingly an eastern bird, near Radium Springs, 11–15 February 2007 (Williams 2007b, photographed by James Zabriskie); and one at Lovington, Lea County, 27 November 2007 (Williams 2008, *fide* W. Wittman).

Short-tailed Hawk

(*Buteo brachyurus*)

The most recent addition to the list of raptors recorded in New Mexico is the Short-tailed Hawk, observed on 24 May 2005 and again on 28 June 2005 over the Animas Mountains in Hidalgo County (Williams et al. 2007). The Short-tailed Hawk is a small buteo that occurs in two morphs, light and dark. Light-morphs are mostly white below, with white throats contrasting with dark cheeks, an undertail that shows a dark subterminal band, and two-tone underwings—the wing coverts are white whereas the flight feathers are grayish with a dark terminal band along the trailing edge of the wings. A light-morph Short-tailed Hawk can be mistaken for a Swainson's Hawk (*Buteo swainsoni*), but the latter species lacks white ovals at the base of the outer primaries, a good field mark of the Short-tailed Hawk. Dark-morph Short-tailed Hawks have dark brown bodies. The underwings show dark brown coverts and grayish flight feathers. Dark-morphs and light-morphs share the same undertail, the dark trailing edge of the

PHOTO 40.5

Short-tailed Hawk, Animas Mountains, Hidalgo Co., 28 June 2005. First verified record of the species in New Mexico. PHOTOGRAPH: © JOHN P. DELONG.

underwings, and the white ovals at the base of the outer primaries. Juveniles are largely similar to adults. The bird sighted and photographed over the Animas Mountains was a light-morph (photo 40.5).

The Short-tailed Hawk has a wide distribution centered on the neotropics and extending as far as northern Argentina to the south and Mexico and Florida to the north (AOU 1998). In Mexico the Short-tailed Hawk appears to be spreading northward (Russell and Monson 1998), and it has even made recent, increasingly regular appearances in Arizona—where breeding was first recorded in 2001 (Rosenberg and Witzeman 1998; Rosenberg 2001)—and in Texas (Lockwood and Freeman 2004).

The migratory status of the Short-tailed Hawk remains unclear, although seasonal, short-distance movements have been documented in Florida (Ogden 1974). Breeding habitat consists primarily of dense, tall forest, but the Short-tailed Hawk forages over both wooded and nonwooded habitats (Miller and Meyer 2002).

Crested Caracara

(*Caracara cheriway*)

A year-round resident of southern Texas and southern Arizona, the Crested Caracara is only casual in the southern half of New Mexico. With its long legs and crested head, the species is unmistakable, whether perched—often on the ground—or in flight. Adults are characterized by a black cap with a short crest at the back, bright orange or red ceres and adjacent bare skin on the face, white sides along the neck, fine black barring on the white upper back and white upper breast, black wings with white panels in the primaries—much as in the Black Vulture, and a white tail tipped by a dark wide subterminal band. In the juveniles the cap, body, and wings are brown instead of black; the white of the neck is replaced with buff; the upper breast and back are streaked rather than barred.

Until recently the Crested Caracara was considered conspecific with the Southern Caracara (*Caracara plancus*), whose distribution is in South America from Tierra del Fuego north only to the Amazon River. The Crested Caracara—now also called Northern Caracara—replaces the Southern Caracara north of

PHOTO 40.6

(*top*) Crested Caracara, near Brownsville in southeastern Texas, October 2005. The Crested Caracara, also formerly known as the Mexican Eagle, is thought by some to be the bird appearing on the Mexican flag. PHOTOGRAPH: © MARK L. WATSON.

PHOTO 40.7

(*bottom*) Crested Caracara, Animas Valley, 31 December 2006. PHOTOGRAPH: © CARL LUNDBLAD.

the Amazon, its distribution extending northward through Central America and Mexico, reaching the United States mainly not only in Texas and Arizona but also in Florida (Morrison 1996). There is a narrow zone of contact between the two species along the Amazon Basin (Dove and Banks 1999).

The Crested Caracara is nonmigratory and associated with open or semiopen habitats with scattered, tall vegetation such as found in the Sonoran Desert in Arizona (Corman 2005b). It is a carrion eater, although it also preys on a wide range of small live animals, including lizards in Arizona (Morrison 1996; Corman 2005b).

The Crested Caracara was first recorded in New Mexico by Thomas Charlton Henry, at Fort Thorn [Hatch] in 1856 (Ligon 1961). The second record is a specimen secured in the vicinity of Las Cruces in 1914 (Ligon 1961). Indicative of nesting in the Middle Rio Grande Valley was an immature bird collected by James Peckumn near Belen in Valencia County in the summer of 1955, one of several birds observed together and referred to as a "family" by Ligon (1961). Peckumn's specimen was mounted and put on display in the Las Cruces office of the New Mexico Department of Game and Fish. At the request of John Hubbard (pers. comm.), Ralph Raitt (New Mexico State University) examined the specimen and confirmed its species identification and age. In the New Mexico Ornithological Society database (NMOS 2008) are 10 more recent reports from Hidalgo (1984, 2005, 2006), Grant (1990), Luna (1969, 1978), Socorro (1999, 2000), Chavez (1997), and Eddy (1985) counties, three of which are substantiated by conclusive photos. Collectively, the 10 records document Crested Caracaras in all seasons. All records but one—from the Jessie Evans Ranch in the Peloncillo Mountains on 25 November 1984—are from lowlands. A photograph of the bird sighted on the Jessie Evans Ranch by Linda Seibert was not conclusive. According to the local rancher's wife, Crested Caracaras had been sighted for two years on the ranch, and one bird had drowned in a cattle tank in 1983 (Goodman 1984, also mentioned in Hubbard 1985). Verified records include two Crested Caracaras reported in 2006: one bird photographed by Bill Howe in a field in San Acacia, Socorro County, 16 April 2006 (Williams 2006); another photographed by both Carl Lundblad and Charles R. Britt in the Animas Valley, Hidalgo County, 31 December 2006 (photo 40.7) (Williams 2007b).

Eastern Screech-Owl

(Megascops asio)

Best distinguished from its close relative the Western Screech-Owl (*Megascops kennicottii*) by its tremolo call, the Eastern Screech-Owl (*M. asio*; formerly *Otus asio*) is found primarily in the United States east of the Rocky Mountains (Gehlbach 1995). Its distribution extends throughout nearly all of the central and eastern states, east and south to the Atlantic and Gulf coasts and the southern tip of Florida including the upper Keys. From southern Texas the range of the Eastern Screech-Owl extends even farther south, into northeastern Mexico along the Atlantic Coast, reaching southeastern San Luis Potosí and southern Tamaulipas (Howell and Webb 1995). Northward, the species barely ranges into Canada, where it is found in extreme southeastern Saskatchewan, extreme southern Manitoba, southern Ontario, and extreme southern Quebec (see Gehlbach 1995). In the United States, the western edge of the species' distribution lies from west-central Montana south through north-central Wyoming, western South Dakota, southeastern Wyoming, northeastern and eastern Colorado, western Oklahoma, the Texas Panhandle, and eastern Trans-Pecos Texas (Gehlbach 1995; Seyffert 2001). Throughout its distribution, the Eastern Screech-Owl inhabits a wide variety of wooded habitats including those found in river valleys, on mountains slopes, and in urban and suburban areas (Gehlbach 1995). The species prefers trees associated with a lack of understory shrubs. It does not appear to be negatively affected by habitat fragmentation and readily habituates to the presence of humans (Gehlbach 1995).

The species *Otus asio* is mentioned in published works on the birds of New Mexico dating back more than two and a half decades (e.g., Ligon 1961), but only because the Western Screech-Owl and the Eastern Screech-Owl were formerly treated as a single species. The Western Screech-Owl's distribution essentially

PHOTO 40.8

(*top*) Akaiko, the female Eastern Screech-Owl brought to the Wildlife Center, Española, Rio Arriba, Co., from Carlsbad, Eddy Co. in July 2002. She was found with a broken wing that could not be fully repaired, thus preventing release back into the wild. Photographed 5 December 2006, Akaiko was alive and well at the time this book was completed. PHOTOGRAPH: © BRYAN J. RAMSAY.

PHOTOS 40.9a and b

(*bottom left and right*) Eastern Screech-Owl, Portales, Roosevelt Co., 20 November 2003. First verified Eastern Screech-Owl record for the state of New Mexico. The bird in the photograph exhibits an intermediate rufous plumage. In areas where they overlap, the Eastern and Western Screech-Owls can imitate each other's vocalizations, and species identification may require DNA analysis. PHOTOGRAPHS: © JULIAN D. AVERY.

begins where the Eastern Screech-Owl's distribution ends, extending westward through much of western North America (see chapter 29). The two species come into contact in river valleys along the western Great Plains, overlapping narrowly in Colorado and more broadly in Texas (Gehlbach 1995).

The first published—and only verified—Eastern Screech-Owl record from New Mexico was a single bird exhibiting an intermediate rufous plumage and found by Julian Avery on 18 November 2003 near the Eastern New Mexico University (ENMU) campus in Portales, Roosevelt County (Avery and Keller 2006). The owl, photographed and tape-recorded, appeared to remain in the same area year-round until at least March 2006, moving only 400 m (1,300 ft) from a stand of elm (*Ulmus* sp.) trees to a black locust (*Robinia pseudoacacia*) in the middle of an open lawn on the ENMU campus (Avery and Keller 2006). The owl was heard again from the same area as late as October 2006 (J. Avery, pers. comm.).

At the time that *Raptors of New Mexico* is going to press, there is no other verified Eastern Screech-Owl record from New Mexico. The Arkansas naturalist J. P. Valentik heard "at close range" an Eastern Screech-Owl on 29 December 2005 at a wooded rest stop along I-25 west of Las Vegas, San Miguel County

(Williams 2005). J. P. Valentik wrote to the author that "the call was a steadily descending whinny without the micro-phrasing (bouncing ball) of the Western['s] call. [He has] heard other Westerns since then, and find[s] that sometimes they don't have well developed micro-phrasing, and now experience[s] less certainty than [he] did at the time, though [he] still believe[s] it was an [Eastern]."

While working on this chapter, the author also learned that the Wildlife Center in Española has had in its care an Eastern Screech-Owl since July 2002. The Wildlife Center owl, a rufous-morph bird named Akaiko by the staff (photo 40.8), was sent in July 2002 from Carlsbad, Eddy County, where it was found with a broken wing (C. Bell, pers. comm.). However, the owl had apparently been brought by a truck from Texas (C. Bell, pers. comm.).

Overall, it is possible that the Eastern Screech-Owl is a more regular visitor to eastern New Mexico than is generally believed, but is overlooked because of a lack of nocturnal surveys in appropriate habitat along the edge of the state. As most Eastern Screech-Owls in New Mexico would be gray-morph birds, distinguishing an Eastern Screech-Owl from a Western Screech-Owl should be based if not on DNA analysis at least on hearing the call, although in areas of overlapping distribution the two species can hybridize and imitate each other's call (F. Gehlbach, pers. comm.). Rufous birds represent a majority of Eastern Screech-Owls in parts of the eastern United States but decline sharply in frequency toward the western edge of the species' range, comprising only 7% of the Great Plains population (Sibley 2000) and 5–8% in central Texas (Gehlbach 1995).

Barred Owl

(*Strix varia*)

The Barred Owl is a close relative of the Spotted Owl (*Strix occidentalis*). Like its very similar congener, the Barred Owl is characterized by a round head without ear tufts, a well-developed facial disk, and dark eyes. The upperparts are brownish gray with whitish to buff bars rather than the Spotted Owl's whitish spots. The best field mark is a ruff of feathers extending across the upper breast, pale with brown barring. Below the ruff

PHOTO 40.10

Barred Owl, Lake Keystone, Creek Co., Oklahoma, 19 June 2005. PHOTOGRAPH: © STEVE METZ.

PHOTO 40.11

Barred Owl at Galisteo, Santa Fe Co., 29 May 2004. First verified record of the species in New Mexico. PHOTOGRAPH: © JONATHAN BATKIN.

the underparts are whitish with long, brown streaks, not barred as found in the Spotted Owl.

The Barred Owl is a nonmigratory owl adapted to forests and woodlands and originally ranged across much of northern North America east of the Great Plains (Mazur and James 2000). During the 20th century the Barred Owl's distribution expanded northward and westward across boreal and montane forests, reaching southeastern Alaska and Yukon, then south again along the Pacific coast into California (Mazur and James 2000). The Barred Owl is also found in Mexico in the form of small, disjunct populations from Durango southward (Howell and Webb 1995). Taxonomists recognize four subspecies of Barred Owls, including *Strix varia georgica* and *S. v. helveola*, both found in Texas (Mazur and James 2000).

The Barred Owl represents another recent addition to New Mexico's avifauna. The species was not mentioned by either Ligon (1961) or Hubbard (1978) and was first reported in the state in winter 1992–93 (Snider 1993). An adult female was killed after flying into the path of an 18-wheeler reportedly just north of Albuquerque in Bernalillo or Sandoval County. Its fresh carcass was found on the grill of the truck and confiscated on 2 February in Trinidad, Colorado, and later added to the UNM Museum of Southwestern Biology bird collection (MSB #9680). This record, however, is considered unverified.

A Barred Owl was observed on several occasions from late May through early October 2004 at Galisteo, Santa Fe County (Williams 2005). It was first discovered and photographed along Galisteo Creek by Jonathan Batkin (pers. comm.), on 29 May 2004 (photo 40.11). The owl flew in front of him and swooped up to a perch high in some cottonwoods (*Populus* sp.). Until October, the Barred Owl spent most of its time at or near a pond at the bottom of the Galisteo Inn property. The pond is flanked to the north by a large stand of cottonwoods, and to the south and east by a lawn ringed with more cottonwoods (J. Batkin, pers. comm.). The Barred Owl observed at Galisteo constitutes the first and only verified record of the species in New Mexico.

LITERATURE CITED

[AOU] American Ornithologists' Union. 1998. *Check-list of North American birds*. 7th ed. Washington, DC: American Ornithologists' Union.

Avery, J. D., and G. S. Keller. 2006. First record of the Eastern Screech-Owl (*Megascops asio*) in New Mexico. *Western Birds* 37:53–54.

Buckley, N. J. 1999. Black Vulture (*Coragyps atratus*). No. 411. In *The birds of North America*, ed. A. Poole and F. Gill. Philadelphia, PA: Birds of North America, Inc.

Cely, J. E. 1979. Status of the Swallow-tailed Kite and factors affecting its distribution. In *Proceedings of the first South Carolina endangered species symposium*, ed. D. M. Forsythe and W. B. Ezell Jr., 144–50. Columbia: South Carolina Wildlife and Maritime Resources Department.

Coleman, J. S., and J. D. Fraser. 1989. Black and Turkey Vultures. In *Proceedings of the northeast raptor management symposium and workshop*, 15–21. Washington, DC: National Wildlife Federation.

Coleman, J. S., and J. D. Fraser. 1990. Black and Turkey Vultures. In *Proceedings of the southeast raptor management symposium and workshop*, 78–88. Washington, DC: National Wildlife Federation.

Corman, T. E. 2005a. Black Vulture (*Coragyps atratus*). In *Arizona breeding bird atlas*, ed. T. E. Corman and C. Wyse-Gervais, 116–17. Albuquerque: University of New Mexico Press.

———. 2005b. Crested Caracara (*Caracara cheriway*). In *Arizona breeding bird atlas*, ed. T. E. Corman and C. Wyse-Gervais, 152–53. Albuquerque: University of New Mexico Press.

Crocoll, S. T. 1994. Red-shouldered Hawk (Buteo lineatus). No. 107. In *The birds of North America*, ed. A. Poole and F. Gill. Philadelphia: Academy of Natural Sciences.

Dove, C. J., and R. C. Banks. 1999. A taxonomic study of Crested Caracaras (Falconidae). *Wilson Bulletin* 111:330–39.

Gehlbach, F. R. 1995. Eastern Screech-Owl (*Otus asio*). No. 165. In *The birds of North America*, ed. A. Poole and F. Gill. Washington, DC: American Ornithologists' Union.

Goodman, R. A., ed. 1984. *New Mexico Ornithological Society Field Notes* 23(4): 1 August–November.

Greider, M., and E. W. Wagner. 1960. Black Vulture extends breeding range northward. *Wilson Bulletin* 72:291.

Howell, S. N. G., and S. Webb. 1995. *A guide to the birds of Mexico and northern Central America*. New York: Oxford University Press.

Hubbard, J. P. 1978. *Revised check-list of the birds of New Mexico*. Publ. no. 6. Albuquerque: New Mexico Ornithological Society.

———, ed. 1985. Southwest region: New Mexico. *American Birds* 39:88–90.

Ligon, J. S. 1961. *New Mexico Birds and where to find them*. Albuquerque: University of New Mexico Press.

Lockwood, M. W., and B. Freeman. 2004. *The TOS handbook of Texas birds*. College Station: Texas A&M University Press.

Manzano-Fischer, P., G. Ceballos, R. List, and J.-L. E. Cartron. 2006. Avian diversity in a priority area for conservation in North America: the Janos–Casas Grandes prairie dog complex and adjacent habitats in northwestern Mexico. *Biodiversity and Conservation* 15:3801–25.

Mazur, K. M., and P. C. James. 2000. Barred Owl (*Strix varia*). No. 508. In *The birds of North America*, ed. A. Poole and F. Gill. Ithaca, NY: Cornell Laboratory of Ornithology, and Philadelphia: Academy of Natural Sciences.

Meyer, K. D. 1995. Swallow-tailed Kite (*Elanoides forficatus*). No. 138. In *The birds of North America*, ed. A. Poole and F. Gill. Philadelphia: Academy of Natural Sciences.

Miller, K. E., and K. D. Meyer. 2002. Short-tailed Hawk (*Buteo brachyurus*). No. 674. In *The birds of North America*, ed. A. Poole and F. Gill. Philadelphia: Birds of North America, Inc.

Monson, G., and A. R. Phillips. 1981. *Annotated checklist of the birds of Arizona*. 2nd ed. Tucson: University of Arizona Press.

Morrison, J. L. 1996. Crested Caracara (*Caracara plancus*). No. 249. In *The birds of North America*, ed. A. Poole and F. Gill. Philadelphia, PA: Academy of Natural Sciences, and Washington, DC: American Ornithologists' Union.

[NMOS] New Mexico Ornithological Society. 2008. *NMOS Field Notes* database. http://nhnm.unm.edu/partners/NMOS/ (accessed 1 May 2008).

Oberholser, H. C. 1974. *The bird life of Texas*. Austin: University of Texas Press.

Ogden, J. C. 1974. The Short-tailed Hawk in Florida. I. Migration, habitat, hunting techniques, and food habits. *Auk* 91:95–110.

Parmalee, P. W. 1954. The vultures: their movements, economic status, and control in Texas. *Auk* 71:443–53.

Parmeter, J., B. Neville, and D. Emkalns. 2002. *New Mexico bird finding guide*. 3rd ed. Albuquerque: New Mexico Ornithological Society.

Phillips, A., J. Marshall, and G. Monson. 1964. *The birds of Arizona*. Tucson: University of Arizona Press.

Rea, A. M. 1998. Black Vulture. In *The raptors of Arizona*, ed. R. L. Glinski, 24–26. Tucson: University of Arizona Press.

Robertson, W. B. Jr. 1988. American Swallow-tailed Kite. In *Handbook of North American birds*. Vol. 4, ed. R. S. Palmer, 109–31. New Haven: Yale University Press.

Rosenberg, G. H. 2001. Arizona Bird Committee report: 1996–1999 records. *Western Birds* 32:50–70.

Rosenberg, G. H., and J. L. Witzeman. 1998. Arizona Bird Committee report: 1974–1996. Part I (Nonpasserines). *Western Birds* 29:199–224.

Russell, S. M., and G. Monson. 1998. *The birds of Sonora*. Tucson: University of Arizona Press.

Seyffert, K. D. 2001. *Birds of the Texas Panhandle*. College Station: Texas A&M University Press.

Sibley, D. A. 2000. *The Sibley guide to the birds*. New York: Chanticleer Press.

Stevenson, M. M., and G. H. Rosenberg. 2005. Arizona. *North American Birds* 59:476–79.

———. 2008. Arizona. *North American Birds* 62:282–85.

Strecker, J. K. 1912. *The birds of Texas*. Vol. 15, no. 1. Waco: University of Texas.

Wheeler, B. K., and W. S. Clark. 1995. *A photographic guide to North American raptors*. London: Academic Press.

Williams, S. O. III., ed. 1996. Southwest region: New Mexico. *National Audubon Society Field Notes* 50:980–83.

———, ed. 2001. New Mexico. *North American Birds* 55:85–88.

———, ed. 2002a. New Mexico. *North American Birds* 55:207–9.

———, ed. 2002b. New Mexico. *North American Birds* 56:340–43.

———, ed. 2005. New Mexico. *North American Birds* 59:121–25.

———, ed. 2006. New Mexico. *North American Birds* 60:412–16.

———, ed. 2007a. New Mexico. *North American Birds* 61:115–18.

———, ed. 2007b. New Mexico. *North American Birds* 61:304–7.

———, ed. 2008. New Mexico. *North American Birds* 62:116–20.

Williams, S. O. III, J. P. DeLong, and W. H. Howe. 2007. Northward range expansion by the Short-tailed Hawk, with first records for New Mexico and Chihuahua. *Western Birds* 28:2–10.

Number of raptors recovered by Wildlife Rescue, Inc., by species (2004–2008)

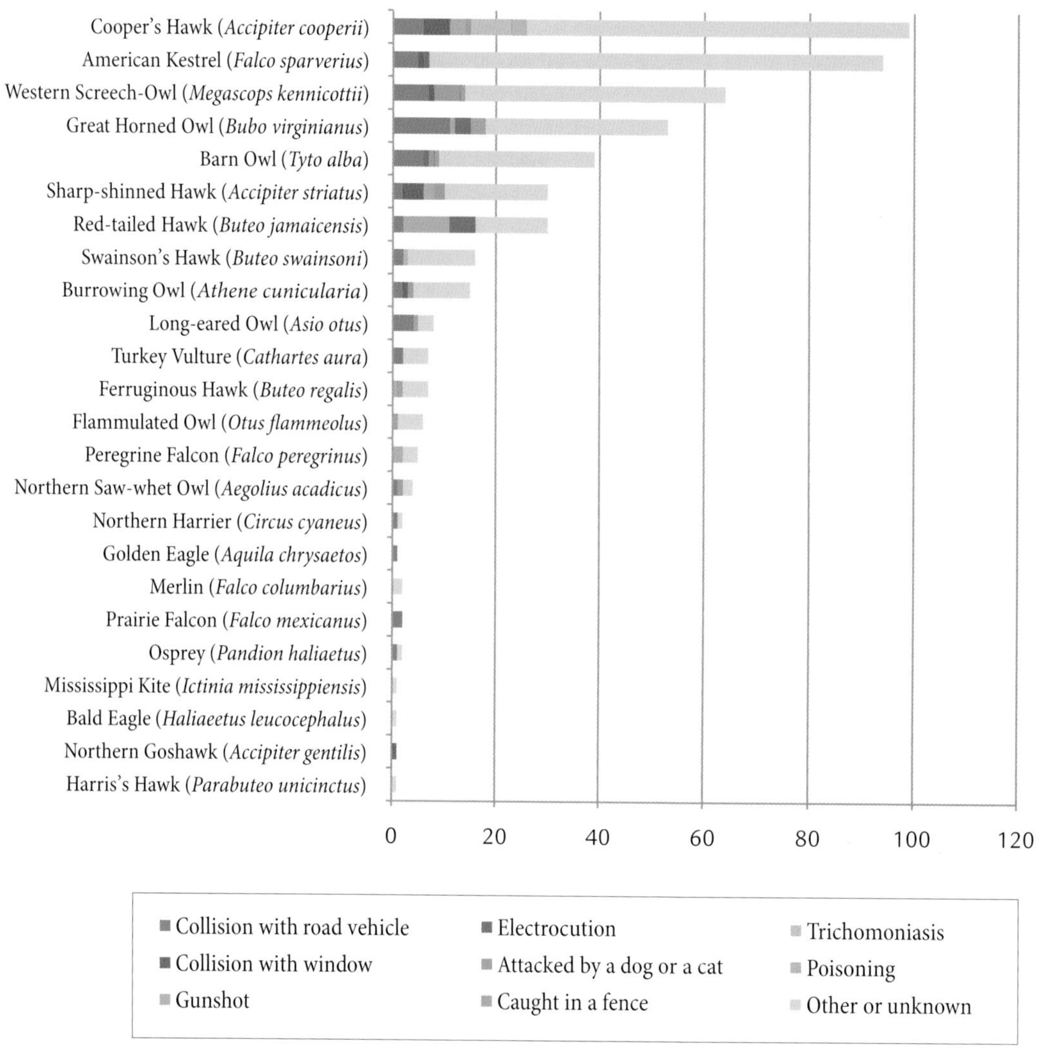

FIGURE C.1

Note that most raptor acquisitions are the result of some unknown cause. In some species, especially those nesting in the Albuquerque area, raptor acquisitions also often consist of young fallen out of the nest or fledged young not yet capable of fending for themselves. All data are from the acquisition logs of Wildlife Rescue, Inc. of Albuquerque.

Conclusion

JEAN-LUC E. CARTRON

THE CURRENT OUTLOOK seems bright for most raptors in New Mexico. As discussed in this volume, several species are expanding their range in the state, including the White-tailed Kite (*Elanus leucurus*) and the Gray Hawk (*Buteo nitidus*). In the last two decades, a breeding population of Ospreys (*Pandion haliaetus*) has colonized the shores of man-made lakes, particularly in the north. Among the species that have recently increased in numbers in New Mexico is the Merlin (*Falco columbarius*), and possibly also the Barn Owl (*Tyto alba*), while the Peregrine Falcon (*Falco peregrinus*) has staged an important comeback since the DDT era. Meanwhile, few of New Mexico's raptors are a cause for concern, at least on the basis of current population trends. Some species such as the American Kestrel (*Falco sparverius*) and the Burrowing Owl (*Athene cunicularia*) are declining throughout much of their range, but apparently not (yet) in New Mexico. The only notable exception is the Spotted Owl (*Strix occidentalis*), which has declined in the Southwest and is now seemingly extirpated in several of its historical haunts in New Mexico. One could also add the Short-eared Owl (*Asio flammeus*), declining at the scale of North America and perhaps once more common in winter in New Mexico, based on historical accounts (Ligon 1961). And, of course, there are those species not known to be declining, but with a very small distribution and therefore with a precarious conservation status in New Mexico. This is true of the Whiskered Screech-Owl (*Megascops trichopsis*) and the Boreal Owl (*Aegolius funereus*). Finally, the status of the Aplomado Falcon (*Falco femoralis*) may be seen as overall good in the short term, but it is also tied to several factors at the center of much controversy and ongoing litigation.

Threats to New Mexico's raptor populations have been described throughout this volume. The Ferruginous Hawk (*Buteo regalis*) and the Burrowing Owl, for example, are both highly dependent on prairie dogs (*Cynomys* spp.), which are still the object of eradication efforts in many parts of the state. Trichomoniasis is a source of mortality in some Cooper's Hawk (*Accipiter cooperii*) populations in New Mexico. West Nile Virus is a potential cause for concern for at least the Osprey (*Pandion haliaetus*), Golden Eagle (*Aquila chrysaetos*), and Peregrine Falcon in northern New Mexico. Habitat loss or degradation remains a threat for many species such as the Common Black-Hawk (*Buteogallus anthracinus*), Harris's Hawk (*Parabuteo unicinctus*), Spotted Owl, and Short-eared Owl.

Acquisition logs from wildlife rehabilitation facilities (e.g., Wildlife Rescue, Inc. [WRI] of Albuquerque,

Wildlife Center in Española, and Gila Wildlife Rescue in Silver City) further show that raptors in New Mexico continue to incur mortality from gunshots, electrocutions on power poles, and collisions with power lines, motor vehicles, or window panes (fig. C.1). In his time, J. S. Ligon lamented the shooting of raptors along roadways in New Mexico. Certainly this problem has abated with improved public attitudes toward raptors. Nonetheless, some people in New Mexico continue to shoot and kill not only Cooper's Hawks, Red-tailed Hawks (*Buteo jamaicensis*), or Great Horned Owls (*Bubo virginianus*), but also conservation-sensitive raptors such as the Peregrine Falcon (WRI acquisition logs). South of the border, in northwestern Chihuahua, concrete power poles are responsible for the deaths of many raptors every year (Cartron et al. 2000, 2006ab; Manzano-Fischer et al. 2006). By comparison, New Mexico's power poles are overall safe for raptors. Even so, WRI's logs record mortality by electrocution in several species, primarily the Red-tailed Hawk and the Great Horned Owl. And where raptors now live in close contact with humans, they can also be killed by dogs or cats. From 2004 through 2009, for example, five raptors were brought to WRI after being attacked by dogs: two Western Screech-Owls (*Megascops kennicottii*; one fatality), one adult Sharp-shinned Hawk (*Accipiter striatus*; fatality), one juvenile Cooper's Hawk (fatality), and one Burrowing Owl. Attacked by cats were three Western Screech-Owls (including one adult that did not survive), one fledgling Barn Owl (fatality), one adult Flammulated Owl (*Otus flammeolus*; fatality), one adult Northern Saw-whet Owl (*Aegolius acadicus*; released), and one adult Sharp-shinned Hawk (released).

Few places on Earth remain untouched by the activities of humans. Among all species—many of which are declining rapidly toward extinction—raptors are at least fortunate enough to be regarded as ecologically important, charismatic, or simply enriching of people's lives. In much of North America, birds of prey are doing far better today than they were in the early and mid 20th century, when shooting and pesticides caused many raptor populations to plummet. And yet, in New Mexico as elsewhere, only our commitment to protect natural lands will ensure that our grandchildren and great-grandchildren grow up to the familiar hoot of an owl and to the sight of the soaring hawk or the diving falcon.

Acknowledgments

Information for the conclusion was obtained from the acquisition logs of Wildlife Rescue, Inc. of New Mexico (WRI) and summarized for the author by Janelle Harden and Anne Russell. Wildlife Rescue, Inc. is supported annually by the Share with Wildlife program of the New Mexico Department of Game and Fish. In part through a research grant from Share with Wildlife, Anne Russell and Janelle Harden (WRI's RAVEN Project) have been digitizing the wildlife rehabilitation group's acquisition records; these include details for 40,000-plus injured and orphaned wildlife received by WRI since the organization's inception in the early 1980s.

LITERATURE CITED

Cartron, J.-L. E., G. L. Garber, C. Finley, C. Rustay, R. P. Kellermueller, M. P. Day, P. Manzano Fisher, and S. H. Stoleson. 2000. Power pole casualties among raptors and ravens in northwestern Chihuahua, Mexico. *Western Birds* 31:255–57.

Cartron, J.-L. E., R. Harness, R. Rogers, and P. Manzano. 2005. Impact of concrete power poles on raptors and ravens in northwestern Chihuahua. In *Biodiversity, ecosystems, and conservation in northern Mexico*, ed. J.-L. E. Cartron, G. Ceballos, and R. S. Felger, 357–69. Oxford, UK: Oxford University Press.

Cartron, J.-L. E., R. Sierra Corona, E. Ponce Guevara, R. E. Harness, P. Manzano-Fischer, R. Rodríguez-Estrella, and G. Huerta. 2006. Bird electrocutions and power poles in northwestern Mexico: an overview. *Raptors Conservation* 7:4–14.

Ligon, J. S. 1961. *New Mexico birds and where to find them.* Albuquerque: University of New Mexico Press.

Manzano-Fischer, P., G. Ceballos, R. List, and J.-L. E. Cartron. 2006. Avian diversity in a priority area for conservation in North America: the Janos–Casas Grandes Prairie Dog Complex and adjacent habitats in northwestern Mexico. *Biodiversity and Conservation* 15:3801–25.

Narca

Appendix 10-1

Cooper's Hawk Egg Sets at the Museum of Southwestern Biology, University of New Mexico

For each egg set below, the collection access number is given first, followed by the location of the nest from which the eggs were retrieved (name of the county, then name of locality), the name of the collector and collector's number, the collection date, and the number of eggs. The notations 1/1, 2/2/, 3/3, . . . indicate complete egg sets, as opposed to 1/2, 2/3, . . . resulting from the loss of one or several eggs in the original clutch.

1. MSB E-83
   Sierra Co.
   Chloride, 8 mi S
   14 May 1916
   Ligon, J. S.
   4/5
   Note: 1 egg is atypical (lightly spotted
   instead of nearly immaculate)

2. MSB E-81
   Hidalgo Co.
   Animas, 45 mi SW
   18 May 1955
   Ligon, J. S. 0653
   4/4

3. MSB E-78
   Hidalgo Co.
   Whitewater Canyon
   11 May 1956
   Hill, W. W. 0432
   4/4

4. MSB E-79
   Bernalillo Co.
   Sandia Mtns, Carlito Springs
   26 May 1956
   Hill, W. W. 0435
   4/4

5. MSB E-80
   Torrance Co.
   Abo, vicinity
   26 May 1956
   Hill, W. W. 0438
   4/4

Note that the MSB bird collection has two more Cooper's Hawk egg sets of four eggs each, collected by J. S. Ligon, but with no location or date recorded.

Appendix 18-1

New Mexico Red-tailed Hawk Egg Sets at the Museum of Southwestern Biology, University of New Mexico

For each egg set below, the collection access number is given first, followed by the location of the nest from which the eggs were retrieved (name of the county, then name of locality), the name of the collector and collector's number, the collection date, and the number of eggs. The notations 1/1, 2/2/, 3/3, . . . indicate complete egg sets, as opposed to 1/2, 2/3, . . . resulting from the loss of one or several eggs in the original clutch.

1. **MSB E-144**
   Catron Co.?
   Fairview, 28 mi NW
   16 April 1925
   Ligon, J. S. 0488
   2/2

2. **MSB E- 893**
   Otero Co.
   Orange, 12 mi NW
   10 April 1926
   Ligon, J. S. 0494A
   2/2

3. **MSB E-145**
   Catron Co.
   Datil, 20 mi NW
   20 April 1927
   Ligon, J. S. 0516
   2/2

4. **MSB E-146**
   Sierra Co.
   Hot Springs, 6 mi NW
   25 March 1928
   Ligon, J. S. 0549
   3/3

5. **MSB E-147**
   Sierra Co.
   Cuchillo, 2 mi W
   27 March 1928
   Ligon, J. S. 0550
   4/4

6. **MSB E-148**
   Eddy Co.
   Hope, 40 mi W
   1 April 1937
   Ligon, J. S. 0600
   4/4

7. **MSB E-131**
   Sandoval Co.
   Hagan, near
   15 April 1951
   Hill, W. W. 0118
   2/2

8. **MSB E-132**
   Sandoval Co.
   Hagan, near
   15 April 1951
   Hill, W. W. 0119
   2/2

9. **MSB E-138**
   San Miguel Co.
   Conchos Dam, 2 mi SW
   15 April 1955
   Hill, W. W. 0388
   2/2

10. **MSB E-139**
    Lincoln Co.
    Jct of Highway 10 and 54, 2 mi W
    14 April 1957
    Hill, W. W. 0449
    2/2

Appendix 19-1

Ferruginous Hawk Egg Sets at the Museum of Southwestern Biology, University of New Mexico, and Other North American Museums

For each egg set below, the museum acronym and collection access number are given first, followed by the location of the nest from which the eggs were retrieved (name of the county, then name of locality), the name of the collector and collector's number, the collection date, and the number of eggs. The notations 1/1, 2/2/, 3/3, . . . indicate complete egg sets, as opposed to 1/2, 2/3, . . . resulting from the loss of one or several eggs in the original clutch. Acronyms for museum institutions: CAS = California Academy of Sciences; DMNH = Delaware Museum of Natural History; FMNH = Field Museum of Natural History; MSB = Museum of Southwestern Biology; SBCM = San Bernardino County Museum; and WFVZ = Western Foundation of Vertebrate Zoology.

1. **CAS ORN-2721**
Grant Co.
Fort Egbert [Bayard]
2 May 1908
Trenholtz, C. A.
3/3

Note: According to John P. Hubbard (pers. comm.), the collector's identity remains a mystery. Other specimen records variously spell his name as Trueholtz, Turnholtz, Trueholz, . . .

2. **FMNH-6733**
Grant Co.
Fort Bayard
16 June 1908
unknown
2/2

3. **FMNH-6735**
Sierra Co.
Winston [Fairview], 20 mi NW
13 April 1914
Ligon, J. S.
3/3

4. **MSB E-157**
Socorro Co.
15 mi W Aragon
6 April 1915
Ligon, J. S.
3/3

5. **MSB E-158**
Socorro Co,
Magdalena, 40 mi SW
22 April 1915
Ligon, J. S.
4/4

6. **MSB E-159**
Socorro Co,
Magdalena, 40 mi
23 April 1915
Ligon, J. S.
4/4

7. **MSB E-166**
Socorro Co.
Fairview, 30 mi NE
29 April 1919
Ligon, J. S.
2/2

8. **MSB E-160**
Socorro Co.
Fairview, 20 mi NW
19 April 1924
Ligon, J. S.
3/3

9. **MSB E-161**
Catron Co. (?)
Magdalena, 50 mi SW
21 April 1924
Ligon, J. S.
4/4

10. **MSB E-164**
Catron Co. (?)
Magdalena, 60 mi SW
25 April 1925
Ligon, J. S.
3/3

11. **FMNH-14927**
Guadalupe Co.
Vaughn, near
2 June 1925
Pope, E. F.
3/3

12. **FMNH-14928**
Guadalupe Co.
Vaughn, near
2 June 1925
Pope, E. F.
2/2

13. **MSB E-162**
Torrance Co.
Corona, 6 mi NW
5 May 1926
Ligon, J. S.
3/3

14. **MSB E-163**
Catron Co.
Magdalena, 50 mi SW
16 April 1927
Ligon, J. S.
3/3

Appendix 19-1 (*continued*)

15. **MSB E-165**
Socorro Co. (?)
Fairview, 12 mi NE
20 April 1928
Ligon, J. S.
3/3

16. **SBCM 555**
[Catron Co.]
Magdalena, 50 mi SW
25 April 1928
Ligon, J. S.
3/3

17. **DMNH 18932**
[Harding Co.]
Mosquero
5 May 1933
More, R. L.
3/3

18. **MSB E-167**
Socorro Co.
Magdalena, 44 mi SW
6 April 1935
Ligon, J. S.
3/3

19. **DMNH 18933**
[Harding Co.]
Mosquero
25 April 1935
More, R. L.
3/3

20. **MSB E-151**
Guadalupe Co.
Vaughn, 3 mi E
6 April 1956
Hill, W. W.
2/2

21. **MSB E-152**
Sandoval Co.
La Ventana, 6 mi SW
12 April 1959
Hill, W. W.
3/3

22. **MSB E-153**
Torrance Co.
Bufford, 2 mi S
19 April 1959
Hill, W. W.
2/2

23. **MSB E-154**
De Baca Co.
Yeso, vicinity
19 April 1959
Hill, W. W.
3/3

24. **MSB E-155**
De Baca Co.
Fort Summer, 11 mi N
19 April 1959
Hill, W. W.
4/4

25. **MSB E-156**
De Baca Co.
Yeso, 2 mi W
23 April 1961
Hill, W. W.
3/3

Appendix 21-1

Golden Eagle Egg Sets at the Museum of Southwestern Biology, University of New Mexico

For each egg set below, the collection access number is given first, followed by the location of the nest from which the eggs were retrieved (name of the county, then name of locality), the name of the collector and collector's number, the collection date, and the number of eggs. The notations 1/1, 2/2/, 3/3, . . . indicate complete egg sets, as opposed to 1/2, 2/3, . . . resulting from the loss of one or several eggs in the original clutch. The notations 1/0, 2/0, 3/0, . . . indicate possibly incomplete egg sets, but there is no information on the size of the original clutch.

1. **MSB E-206**
   Socorro Co.
   Ar[a]gon, 18 mi W
   18 March 1915
   Ligon, J. S. 0246
   3/3 (1 egg nearly immaculate)

2. **MSB E-207**
   Sierra Co.
   Elephant Butte, 5 mi SE
   1 March 1916
   Ligon, J. S. 0273
   2/2

3. **MSB E-208**
   Socorro Co?
   Fairview, 18 mi NE
   25 March 1925
   Ligon, J. S. 0484
   2/2

4. **MSB E-209**
   Socorro Co.
   Monticello, 11 mi N
   3 March 1927
   Ligon, J. S. 0511
   2/2

5. **MSB E-210**
   Socorro Co.
   Monticello, 18 mi. NW
   5 March 1927
   Ligon, J. S. 0512
   1/2

6. **MSB E-211**
   Lincoln Co.
   Lincoln, 3 mi SE
   14 March 1928
   Ligon, J. S. 0546
   2/2

7. **MSB E-212**
   Socorro Co.
   Monticello, 18 mi N
   20 March 1928
   Ligon, J. S. 0547
   2/2

8. **MSB E-215**
   Socorro Co.
   Monticello, 18 mi above
   16 February 1934
   Ligon, J. S. 0589
   2 of 0

9. **MSB E-216**
   Sierra Co.
   Cuchillo, 2 mi E
   18 February 1936
   Ligon, J. S. 0596
   2 of [0]

10. **MSB E-213**
    Socorro Co.
    Monticello, 18 mi N
    29 February 1936
    Ligon, J. S. 0597
    2/2

11. **MSB E-218**
    Sierra Co.
    Cuchillo, 2 mi E
    4 April 1936
    Ligon, J. S. 0598
    2 of 0

12. **MSB E-169**
    Valencia Co.
    Los Lunas, W of
    24 February 1951
    Hill, W. W. 0115
    2/2

13. **MSB E-170**
    Valencia Co.
    El Morro Airport, 1 mi E
    4 March 1951
    Hill, W. W. 0116
    2/2

14. **MSB E-202**
    Hidalgo Co.
    Hachita, 15 mi SW
    16 February 1952
    Hill, W. W. 1951
    2/2

15. **MSB E-203**
    Grants Co.
    Hachita, 10 mi W
    17 February 1952
    Hill, W. W. 1953
    2/2

16. **MSB E-204**
    Luna Co.
    Gage, several mi SW
    17 February 1952
    Hill, W. W. 1954
    2/2

    Note: 1 egg is very heavily marked and almost brown

17. **MSB E-205**
    Sierra Co.
    Cuchillo
    23 February 1952
    Hill, W. W. 1958
    2/2

Appendix 21-1 *(continued)*

18. **MSB E-172**
Grant Co.
Hachita, 8 mi W
20 February 1953
Hill, W. W. 0209
2/2

19. **MSB E-173**
Luna Co.
Gage, 10 mi N
22 February 1953
Hill, W. W. 0210
2/2

20. **MSB E-174**
Valencia Co.
Los Lunas, 18 mi W
21 February 1954
Hill, W. W. 0350
2/2

21. **MSB E-175**
Socorro Co.
Bernardo, vicinity
7 March 1954
Hill, W. W. 0352
1/1

22. **MSB E-176**
Valencia Co.
Los Lunas, 18 mi W
5 March 1955
Hill, W. W. 0380
2/2

23. **MSB E-180**
Luna Co.
Gage, vicinity
2 March 1957
Hill, W. W. 0444
2/2

24. **MSB E-183**
McKinley Co.
Chaco Canyon,
13 mi S and W of
9 March 1957
Hill, W. W. 0447
2/2

25. **MSB E-184**
Luna Co.
Gage, 10 mi N
1 March 1958
Hill, W. W. 0461
2/2

26. **MSB E-185**
Valencia Co.
Los Lunas, about 18 mi W
6 March 1958
Hill, W. W. 0463
2/2

27. **MSB E-188**
Socorro Co.
La Joya, about 10 mi E
15 March 1958
Hill, W. W. 0466
2/2

28. **MSB E-189**
Luna Co.
Gage, 10 mi N
7 March 1959
Hill, W. W. 0485
3/3

29. **MSB E-189**
Luna Co.
Gage, 10 mi N
7 March 1959
Hill, W. W. 0485
3/3

30. **MSB E-190**
Hidalgo Co.
Hachita, about 15 mi. SW
7 March 1959
Hill, W. W. 0486
2/2

31. **MSB E-191**
Luna Co.
Nutt, 3 mi SW
8 March 1959
Hill, W. W. 0487
1/1

32. **MSB E-192**
Sierra Co.
Cuchillo
8 March 1959
Hill, W. W. 0487
2/2

33. **MSB E-194**
Luna Co.
Gage, 10 mi N
06 March 1960
Hill, W. W. 0508
3/3

34. **MSB E-195**
Sierra Co.
Truth or Consequences, S of
7 March 1960
Hill, W. W. 0509
3/3

35. **MSB E-196**
McKinley Co.
Chaco Canyon National
Monument, 6 mi S
30 March 1960
Hill, W. W. 0510
2/2

36. **MSB E-198**
Hidalgo Co.
Animas, 25 mi S
25 February 1961
Hill, W. W. 0560
1/2

37. **MSB E-197**
Grant Co.
Burro Mtns.
Hill, W. W.
[No date given]
1/1

38. **MSB E-199**
Sierra Co.
Truce or Consequences, S of
26 February 1961
Hill, W. W. 0561
2/2

Appendix 22-1

American Kestrel Egg Sets at the Museum of Southwestern Biology, University of New Mexico

For each egg set below, the collection access number is given first, followed by the location of the nest from which the eggs were retrieved (name of the county, then name of locality), the name of the collector and collector's number, the collection date, and the number of eggs. The notations 1/1, 2/2/, 3/3, . . . indicate complete egg sets, as opposed to 1/2, 2/3, . . . resulting from the loss of one or several eggs in the original clutch. The notations 1/0, 2/0, 3/0, . . . indicate possibly incomplete egg sets but there is no information on the size of the original clutch.

1. **MSB E-232**
   Santa Fe Co.
   Santa Fe
   5 May 1927
   Jensen, J. K.
   5/0

2. **MSB E-230**
   Sierra Co.
   Fairview, 2 mi S
   12 May 1929
   Ligon, J. S. 0572
   5/5

3. **MSB E-231**
   Sierra Co.
   Fairview, 3 mi N
   4 June 1929
   Ligon, J. S. 0577
   4/0

4. **MSB E-224**
   Hidalgo Co.
   Lower Animas Valley
   14 May 1955
   Hill, W. W. 0399
   3/3

5. **MSB E-225**
   Taos Co.
   Taos, 5 mi N
   15 June 1958
   Hill, W. W. 0480
   4/4

6. **MSB E-226**
   Bernalillo Co.
   Albuquerque, W
   1 May 1965
   Hill, W. W. 0614
   4/4

7. **MSB E-227**
   Bernalillo Co.
   Albuquerque, W
   1 May 1966
   Hill, W. W. 0627
   4/4

8. **MSB E-228**
   Bernalillo Co.
   Albuquerque, W
   1 May 1966
   Hill, W. W. 0628
   5/5

9. **MSB E-229**
   Torrance Co.
   Gran Quivira, 10 mi E
   29 April 1967
   Hill, W. W. 0637
   4/4

Appendix 26-1

Prairie Falcon Egg Sets Housed at the Museum of Southwestern Biology, University of New Mexico

For each egg set below, the collection access number is given first, followed by the location of the nest from which the eggs were retrieved (name of the county, then name of locality), the name of the collector and collector's number, the collection date, and the number of eggs. The notations 1/1, 2/2/, 3/3, . . . indicate complete egg sets, as opposed to 1/2, 2/3, . . . resulting from the loss of one or several eggs in the original clutch.

1. **MSB E-239**
   Socorro Co.
   Corduroy Lake, 0.75 mi below
   15 April 1915
   Ligon, J. S. 0249
   5/5

2. **MSB E-242**
   Socorro Co.
   Fairview, 20 mi NW
   19 April 1924
   Ligon, J. S. 0471
   4/4

3. **MSB E-241**
   Catron Co.
   Beaver Head P.O., 20 mi NE
   21 April 1924
   Ligon, J. S. 0473
   5/5

4. **MSB E-243**
   San Miguel Co.
   Tucumcari, 35 mi NW
   18 April 1951
   Ligon, J. S. 0636
   3/3

5. **MSB E-238**
   Torrance Co.
   Red Rock Ranch
   10 April 1965
   Hill, W. W. 0612
   5/5

Appendix 28-1

Mark-recapture results for 137 adult and juvenile Flammulated Owls banded at 47 nest sites (11 nest boxes and 36 natural cavities) on Oso Ridge in the Zuni Mountains, New Mexico, from 1996 to 2004, and recaptured from 1997 to 2005 by D. Arsenault and P. Stacey.

	NUMBER OF OWLS RECAPTURED/ NUMBER OF OWLS BANDED (RETURN RATE)	DISPERSAL DISTANCE[a] (M) MEAN ± SD (N)	TERRITORY FIDELITY[b] (N)	NEST SITE FIDELITY (N)	MATE FIDELITY (N)
Male	16/31 (52%)	232 ± 82 (5)	80% (30)	53% (30)	42% (19)
Female	14/32 (44%)	428 ± 225 (10)	50% (20)	35% (20)	41% (17)
Juvenile	2/74 (3%)	440 ± 368 (2)	NA	NA	NA
Mean ± SD	32/137 (23%)	372 ± 217 (17)	65% (50)	44% (50)	42% (36)

[a] Dispersal distance for adults was defined as the movement between years from nest sites in one territory to another, and for juveniles the movements from birth site to first known breeding site.

[b] Number of cases where an individual returned to breed on the same territory where it had bred in the previous year.

Appendix 28-2

Prey delivery rates of Flammulated Owls at 15 nests during the nestling stage recorded by D. Arsenault in New Mexico. The sex of the owl was determined at 10 of the nests, which had banded birds.

DAYS POST-HATCHING	FEEDING RATE (FIRST HOUR[a]) MEAN ± SD (N)	FEEDING RATE (SECOND HOUR) MEAN ± SD (N)	AVERAGE NUMBER OF VISITS PER HOUR (MALE) MEAN ± SD (N)	AVERAGE NUMBER OF VISITS PER HOUR (FEMALE) MEAN ± SD (N)
0-3	12.7 ± 4.8 (7)	5.8 ± 4.5 (6)	9.7 ± 5.7 (7)	0.6 ± 0.6 (7)
4-6	19.3 ± 3.8 (3)	9.5 ± 5.0 (2)	8.5 ± 0.7 (2)	5.3 ± 3.9 (2)
7-fledging	32.8 ± 10.0 (5)	-	13 ± 0.0 (1)	27 ± 0.0 (1)

[a] Number of feeding visits at nests for one hour after feeding began at dusk.

Appendix 31-1

Dates and localities for New Mexico Great Horned Owl egg sets housed in the collection of the Museum of Southwest Biology, University of New Mexico. The records are arranged chronologically by month and day of the month.

DATE COLLECTED	LOCATION COLLECTED
17 April 1926	Sierra Co., Cuchillo, 3 mi. SE
1 March 1928	Grant Co., Silver City, ca. 30 mi. W
31 April 1951	Torrance Co., Mountainair, Hwy. Overpass, 0.5 mi. N
7 March 1954	Socorro Co., Socorro, 30 mi. S
12 March 1955	Bernalillo Co., Albuquerque, 15 mi. W
11 March 1956	Torrance Co., Mountainair, Hwy. overpass
16 March 1957	San Juan Co., Chaco Canyon, 20 mi. SW
4 March 1958	San Juan Co., Chaco Canyon, vicinity
21 March 1965	Torrance Co., McIntosh, 10 mi. E

Appendix 31-2

Mortality of Great Horned Owls, 1950–1993, shown by sex and month of mortality[a]

SEASON		MALE	FEMALE	TOTALS	SEASONAL TOTALS
Nesting	FEB	22	16	38	119
	MAR	31	16	47	
	APR	21	13	34	
Fledging	MAY	15	13	28	70
	JUN	15	27	42	
Post-Fledging	JUL	23	16	39	92
	AUG	30	23	53	
TOTALS:		157	124	281	45%
Dispersal / Migration	SEP	16	27	43	186
	OCT	31	40	71	
Winter	NOV	26	46	72	
	DEC	33	55	88	151
	JAN	23	40	63	
TOTALS:		129	208	337 (n = 618)	55% (100%)

[a] Note on methodology and statistical results: 8% ($n = 56$) of the Great Horned Owl specimens of our initial data set ($n = 674$) were unsexed. For this reason, all data for these specimens had to be discarded. Statistical analyses were carried out with a Chi-square test. Male Great Horned Owl specimens outnumbered female specimens 157 to 124 ($\chi^2 = 3.8754$, df = 1, $P = 0.0490$) during the reproductive season. Female specimens outnumbered male specimens in every sample of the dispersal/migration and winter seasons, September through January, 208 to 129 ($\chi^2 = 18.4675$, df = 1, $P < 0.0000$). Overall, female specimens outnumbered male specimens in museum collections, 332 to 286 respectively ($\chi^2 = 3.4239$, df = 1, $P = 0.0643$).

Appendix 33-1

Elf Owl egg set from New Mexico, with egg measurements

EGG SET	DATE OF COLLECTION	LOCALITY	COLLECTOR	EGG MEASUREMENTS (MM)
MSB E-936	3 May 2005	Rocky Arroyo, Eddy Co.	A. B. Johnson	25.9 × 21.8
				27.3 × 22.3
				26.7 × 21.9
				26.3 × 22.3

ILLUSTRATION: © NARCA MOORE-CRAIG.

GLOSSARY

A

Active nest: nest with eggs or young.

After-hatching-year bird: bird in its second calendar year of life or older.

B

Bib: in some raptors, band across the throat and breast in the form of a bib.

Brood: young hatched from the same clutch of eggs.

C

Cere: featherless area at the base of the beak and enclosing the nostrils.

Chick: young bird.

Clutch size: number of eggs in a nest.

Congener: member of the same genus.

Conspecific: belonging to the same species.

Coverts: feathers covering the base of flight feathers on the tail and the wings.

Crest: tuft of elongated feathers rising from the crown.

Crown: top part of the head.

D

Depredation: see nest depredation.

Dihedral: shape of a bird flying with its wings above the horizontal plane formed by the rest of the body.

Dimorphic: occurring in two usually distinct appearances or sizes (see also sexual dimorphism).

Dispersal: movements by adults after breeding (post-breeding dispersal) or by young birds following fledging (juvenile dispersal).

E

Erythristic: showing red or rufous as the predominant coloration.

Extinction: dying out of a species.

Eyrie: cliff ledge nest.

F

Facial disk: in owls, circular area surrounding the eye and helping to funnel sound into the ear.

Fide: according to, on the authority of.

Flank: area of the body which is lateral to the back and posterior to the side, and extends to the base of the tail.

Fledging: time at which nestlings are able to leave the nest.

Fledgling: young bird that has fledged but is not yet capable of surviving on its own and still receives parental care.

Flight feathers: feathers on the wings and tail used during flight for lift and maneuverability; wing flight feathers are attached to the bones of the wing and include primaries and secondaries.

Food pellet: regurgitated wad of undigested food material such as bones, feathers, and fur.

Forewing: part of the wing that holds the ulna and radius, and the secondaries.

G

Gape: the line along which the mandibles of a bird close.

Guild: a group of species exploiting the same resource or with similar foraging habits within an animal community.

H

Hatch-year bird: bird in its first calendar year of life.

Hybrid: individual whose parents belong to two different species.

I

Immature: juvenile that has gone through its first molt but has not acquired the adult plumage yet.

Intergrade: In an area where two subspecies co-occur, an animal that shows characteristics of both subspecies.

Iris: colored portion of the eye.

Irruption: in fall and winter, irregular, migratory incursion of populations beyond the normal range of a species, typically caused by a crash of prey populations.

J

Jess: tether used by falconers to restrain their birds during training and in unfamiliar surroundings. The jess is attached to the leg and is traditionally made of thin straps of leather.

K

Kettle: aggregation of soaring migrant hawks.

Kiting: hovering.

Kleptoparasitic: animal that steals food from another animal.

L

Lagomorphs: order of mammals represented by rabbits, hares, and pikas.

Leucistic: lighter in coloration than normal, but not an albino. Although the plumage consists largely or mostly of white feathers, the eyes, beak, and feet show some pigmentation.

Lore: area of the face between the bill and the eye.

M

Malar stripe: see mustache.

Melanistic bird: dark-morph bird.

Migrant: a bird that migrates; spring and fall migrants are birds that neither breed nor overwinter locally but simply pass through during spring and fall migration.

Mobbing: aggregation of birds harassing or attacking an animal perceived as a threat.

Molt: periodic shedding and replacement of feathers.

Morph: color form (usually dark, rufous, and light) found within a species or a subspecies.

Mustache: stripe on each side of the chin and extending down onto the throat; also called malar stripe.

N

Nape: Back (or upper) side of neck.

Nest depredation: predation on eggs or nestlings.

Nest success: raising of at least one young to fledging.

Nocturnal: active at night.

Nomadism: act of wandering in no precise direction, typically in search of prey populations.

Nominate subspecies: subspecies that bears the name of the species, repeated. For example, the Taiga Merlin (*Falco columbarius columbarius*) is the nominate Merlin subspecies.

O

Occupied nest: nest occupied by a pair and in which eggs may or may not have been laid.

Ocelli: eye spots.

P

Patagium: the fold of skin in front of the humeral and radio-ulnar parts of a bird's wing.

Pellet: see food pellet.

Piracy: act of robbing prey from a bird carrying it.

Plumage: all of a bird's feathers.

Plume: in birds long, showy feathers displayed by males for attracting a mate.

Polymorphism: occurrence of more than one distinct plumage or form of individuals in a population.

Primaries: outermost flight feathers of birds, long and stiff, and attached to bones of the "hand."

Productivity: number of fledglings produced per pair. The mean productivity of a population corresponds to the total number of fledglings produced divided by the total number of pairs.

R

Race: subspecies.

Rectrices: flight feathers of the tail; rectrices are long and stiff feathers.

Remiges: flight feathers of the wing.

Resident: present year-round.

Rookery: breeding colony of birds.

Rump: in birds, area of the body between the uppertail coverts and the back.

S

Scapulars: in birds, short feathers covering the area of the body where the back and wing join.

Secondaries: in birds, wing flight feathers attached to the forearm; secondaries are long and stiff feathers.

Sexual dimorphism: differences in size and/or appearance between the male and the female. In most species of birds, the male is larger than the female. Many raptor species are said to exhibit reversed sexual dimorphism because the female is larger.

Siblicide: death of a nestling caused by its sibling(s).

Side: part of the body located between the belly, wing, and back; the side is anterior to the flank.

Song: series of musical notes or phrases often repeated and used during the breeding season for advertising territoriality and for courtship. Typically, only the male sings.

Subadult: in eagles, individual largely similar to adult but not yet capable of breeding.

Subspecies: within a species, subpopulations that tend to be distinguishable by morphological characteristics.

Supercilium: the overhanging margin of a bony cavity; the region of the eyebrows.

T

Tarsus: upper section of a bird's foot.

Taxonomy: in the biological sciences, discipline that concerns itself with identifying, naming, and classifying living organisms.

U

Underparts: ventral side of the body; in a bird in flight, part of the body that is visible from below.

Undertail: underside of the tail.

Undertail coverts: feathers covering the underside base of the tail.

Underwing: underside of the wing on a flying bird. On a perched bird, the underwing is tucked against the body.

Upperparts: dorsal side of the body; in a bird in flight, part of the body that is visible from above.

Uppertail: upperside of tail.

Uppertail coverts: feathers covering the upperside base of the tail.

Upperwings: upperside of the wing on a flying bird.

V

Vent: in birds, opening of the cloaca, the terminal chamber into which the intestine and urogenital ducts discharge.

Venter: underside of an animal.

Ventral: along the underside of an animal.

W

Wing coverts: in birds, wing feathers covering the bases of the flight feathers.

CONTRIBUTING AUTHORS

DAVID P. ARSENAULT
Sierra Nevada Avian Center
Quincy, CA

JULIAN D. AVERY
Department of Ecology, Evolution, and Natural Resources
Rutgers University
New Brunswick, NJ

JAMES C. BEDNARZ
Department of Biological Sciences
Arkansas State University
State University, AR

JEAN-LUC E. CARTRON
Department of Biology
University of New Mexico
Albuquerque, NM

STEVE W. COX
Rio Grande Bird Research, Inc.
Albuquerque, NM

JOHN P. DELONG
Eagle Environmental, Inc.
Santa Fe, NM

MARTHA J. DESMOND
Department of Fishery and Wildlife Sciences
New Mexico State University
Las Cruces, NM

ROBERT W. DICKERMAN
Museum of Southwestern Biology
University of New Mexico
Albuquerque NM

ROBERT H. DOSTER
Migratory Bird Program
U.S. Fish and Wildlife Service
Willows, CA

STEPHEN M. FETTIG
Los Alamos, NM

GAIL L. GARBER
Hawks Aloft, Inc.
Albuquerque, NM

FREDERICK R. GEHLBACH
Department of Biology
Baylor University
Waco, TX

NANCY Y. GEHLBACH
Department of Biology
Baylor University
Waco, TX

ANTONIO (TONY) GENNARO
Department of Biology (retired)
Eastern New Mexico University
Portales, NM

JANELLE HARDEN
Wildlife Rescue, Inc.
Albuquerque, NM

ANDREW B. JOHNSON
Museum of Southwestern Biology
University of New Mexico
Albuquerque, NM

ZACH F. JONES
Department of Biology
Eastern New Mexico University
Portales, NM

GREGORY S. KELLER
Department of Biology
Gordon College
Wenham, MA

RONALD P. KELLERMUELLER
Hawks Aloft, Inc.
Albuquerque, NM

PATRICIA L. KENNEDY
Department of Fisheries and Wildlife
Oregon State University
Union, OR

DAVID J. KRUEPER
U.S. Fish and Wildlife Service
Albuquerque, NM

J. DAVID LIGON
Department of Biology
University of New Mexico
Albuquerque, NM

TIMOTHY K. LOWREY
Museum of Southwestern Biology
University of New Mexico
Albuquerque, NM

G

H

I

J

T

U

V

W

Z